T0138653

COMPUTATIONAL INTELLIGENCE *in* MANUFACTURING HANDBOOK

The Mechanical Engineering Handbook Series

Series Editor

Frank Kreith

Consulting Engineer

Published Titles

Computational Intelligence in Manufacturing Handbook
Jun Wang and Andrew Kusiak

The CRC Handbook of Mechanical Engineering
Frank Kreith

The CRC Handbook of Thermal Engineering
Frank Kreith

The Handbook of Fluid Dynamics
Richard W. Johnson

Forthcoming Titles

Fuel Cell Technology Handbook
Gregor Hoogers

Air Pollution Control Technology Handbook
Karl B. Schnelle and Charles A. Brown

Handbook of Heating, Ventilation, and Air Conditioning
Jan F. Kreider

Handbook of Mechanical Engineering, Second Edition
Frank Kreith and Massimo Capobianchi

Hazardous and Radioactive Waste Treatment Technologies Handbook
Chang H. Oh

Handbook of Non-Destructive Testing and Evaluation Engineering
James Tulenko and David Hintenlang

Inverse Engineering Handbook
Keith A. Woodbury

MEMS Handbook
Mohamed Gad-el-Hak

COMPUTATIONAL INTELLIGENCE *in* MANUFACTURING HANDBOOK

Edited by

Jun Wang
Andrew Kusiak

CRC Press
Boca Raton London New York Washington, D.C.

Library of Congress Cataloging-in-Publication Data

Wang, Jun.
Computational intelligence in manufacturing handbook / Jun Wang and Andrew Kusiak.
p. cm. — (Mechanical engineering)
Includes bibliographical references and index.
ISBN 0-8493-0592-6 (alk. paper)
1. Production management—Data processing. 2. Computational intelligence—Industrial applications. 3. Manufacturing processes—Automation. I. Title. II. Advanced topics in mechanical engineering series

TS155.6 .W36 2000
658.5'14—dc21 00-049826
CIP

Direct all inquiries to CRC Press LLC, 2000 N.W. Corporate Blvd., Boca Raton, Florida 33431.

International Standard Book Number 0-8493-0592-6
Library of Congress Card Number 00-049826
Printed in the United States of America 1 2 3 4 5 6 7 8 9 0
Printed on acid-free paper

Preface

Computational intelligence involves science-based approaches and technologies for analyzing, designing, and developing intelligent systems. The broad usage of this term was formalized by the IEEE Neural Network Council and the IEEE World Congress on Computational Intelligence in Orlando, Florida in the summer of 1994. It represents a union of neural networks, fuzzy systems, evolutionary computation techniques, and other emerging intelligent agents and technologies.

The past two decades have witnessed the resurgence of studies in neural networks, fuzzy logic, and genetic algorithms in the areas we now call computational intelligence. Advances in theory and methodology have overcome many obstacles that previously hindered the computational intelligence research. The research has sparked considerable interest among scientists and engineers from many disciplines. As evidenced by the appealing results of numerous studies, computational intelligence has gained acceptance and popularity. In addition, computational intelligence techniques have been applied to solve numerous problems in a variety of application settings. The computational intelligence research opened many new dimensions for scientific discovery and industrial/business applications. The desirable features of computationally intelligent systems and their initial successes in applications have inspired renewed interest in practitioners from industry and service organizations. The truly interdisciplinary environment of the research and development offers rewarding opportunities for scientific breakthrough and technology innovation.

The applications of computational intelligence in manufacturing, in particular, play a leading role in the technology development of intelligent manufacturing systems. The manufacturing applications of computational intelligence span a wide spectrum including manufacturing system design, manufacturing process planning, manufacturing process monitoring control, product quality control, and equipment fault diagnosis. In the past decade, numerous publications have been devoted to manufacturing applications of neural networks, fuzzy logic, and evolutionary computation. Despite the large volume of publications, there are few comprehensive books addressing the applications of computational intelligence in manufacturing. In an effort to fill the void, this comprehensive handbook was produced to cover various topics on the manufacturing applications of computational intelligence. The aim of this handbook is to present the state of the art and highlight the recent advances on the computational intelligence applications in manufacturing. As a handbook, it contains a balanced coverage of tutorials and new results.

This handbook is intended for a wide readership ranging from professors and students in academia to practitioners and researchers in industry and business, including engineers, project managers, and R&D staff, who are affiliated with a number of major professional societies such as IEEE, ASME, SME, IIE, and their counterparts in Europe, Asia, and the rest of the world. The book is a source of new information for understanding technical details, assessing research potential, and defining future directions in the applications of computational intelligence in manufacturing.

This handbook consists of 19 chapters organized in five parts in terms of levels and areas of applications. The contributed chapters are authored by more than 30 leading experts in the fields from top institutions in Asia, Europe, North America, and Oceania.

Part I contains two chapters that present an overview of the applications of computational intelligence in manufacturing. Specifically, Chapter 1 by D. T. Pham and P. T. N. Pham offers a tutorial on computational intelligence in manufacturing to lead the reader into a broad spectrum of intelligent manufacturing applications. Chapter 2 by Wang, Tang, and Roze gives an updated survey of neural network applications in intelligent manufacturing to keep the reader informed of history and new development in the subject of study.

Part II of the handbook presents five chapters that address the issues in computational intelligence for modeling and design of manufacturing systems. In this category, Chapter 3 by Ulieru, Stefanoiu, and Norrie presents a metamorphic framework based on fuzzy logic for intelligent manufacturing. Chapter 4 by Suresh discusses the neural network applications in group technology and cellular manufacturing, which has been one of the popular topics investigated by many researchers. Chapter 5 by Kazerooni et al. discusses an application of fuzzy logic to design flexible manufacturing systems. Chapter 6 by Luong et al. discusses the use of genetic algorithms in group technology. Chapter 7 by Chang and Tsai discusses intelligent design retrieving systems using neural networks.

Part III contains three chapters and focuses on manufacturing process planning and scheduling using computational intelligence techniques. Chapter 8 by Lee, Chiu, and Fang addresses the issues on optimal process planning and sequencing of parallel machining. Chapter 9 by Zhang and Nee presents the applications of genetic algorithms and simulated annealing algorithm for process planning. Chapter 10 by Cheng and Gen presents the applications of genetic algorithms for production planning and scheduling.

Part IV of the book is composed of five chapters and is concerned with monitoring and control of manufacturing processes based on neural and fuzzy systems. Specifically, Chapter 11 by Lam and Smith presents predictive process models based on cascade neural networks with three diverse manufacturing applications. In Chapter 12, Cho discusses issues on monitoring and control of manufacturing process using neural networks. In Chapter 13, May gives a full-length discussion on computational intelligence applications in microelectronic manufacturing. In Chapter 14, Du and Xu present fuzzy logic approaches to manufacturing process monitoring and diagnosis. In Chapter 15, Li discusses the uses of fuzzy neural networks and wavelet techniques for on-line monitoring cutting tool conditions.

Part V has four chapters that address the issues on quality assurance of manufactured products and fault diagnosis of manufacturing facilities. Chapter 16 by Chen discusses an in-process surface roughness recognition system based on neural network and fuzzy logic for end milling operations. Chapter 17 by Chinnam presents intelligent quality controllers for on-line selection of parameters of manufacturing systems. Chapter 18 by Chang discusses a hybrid neural fuzzy system for statistical process control. Finally, Chapter 19 by Khoo and Zhai discusses a diagnosis approach based on rough set and genetic algorithms.

We would like to express our gratitude to all the contributors of this handbook for their efforts in preparing their chapters. In addition, we wish to thank the professionals at CRC Press LLC, which has a tradition of publishing well-known handbooks, for their encouragement and trust. Finally, we would like to thank Cindy R. Carelli, the CRC Press acquiring editor who coordinated the publication of this handbook, for her assistance and patience throughout this project.

Jun Wang
Hong Kong

Andrew Kusiak
Iowa City

Editors

Jun Wang is an Associate Professor and the Director of Computational Intelligence Lab in the Department of Automation and Computer-Aided Engineering at the Chinese University of Hong Kong. Prior to this position, he was an Associate Professor at the University of North Dakota, Grand Forks. He received his B.S. degree in electrical engineering and his M.S. degree in systems engineering from Dalian University of Technology, China and his Ph.D. degree in systems engineering from Case Western Reserve University, Cleveland, Ohio. Dr. Wang's current research interests include neural networks and their engineering applications. He has published more than 60 journal papers, 10 book chapters, 2 edited books, and numerous papers in conference proceedings. He serves as an Associate Editor of the *IEEE Transactions on Neural Networks.*

Andrew Kusiak is a Professor of Industrial Engineering at the University of Iowa, Iowa City. His interests include applications of computational intelligence in product development, manufacturing, and health-care informatics and technology. He has published research papers in journals sponsored by AAAI, ASME, IEEE, IIE, INFORMS, ESOR, IFIP, IFAC, IPE, ISPE, and SME. Dr. Kusiak speaks frequently at international meetings, conducts professional seminars, and consults for industrial corporations. He has served on the editorial boards of 16 journals, has written 15 books and edited various book series, and is the Editor-in-Chief of the *Journal of Intelligent Manufacturing.*

Contributors

K. Abhary
University of South Australia
Australia

F. T. S. Chan
University of Hong Kong
China

C. Alec Chang
University of Missouri–Columbia
U.S.A.

Shing I. Chang
Kansas State University
U.S.A.

Joseph C. Chen
Iowa State University
U.S.A.

Runwei Cheng
Ashikaga Institute of Technology
Japan

Ratna Babu Chinnam
Wayne State University
U.S.A

Nan-Chieh Chiu
North Carolina State University
U.S.A.

Hyung Suck Cho
Korea Advanced Institute
of Science and Technology
South Korea

R. Du
University of Miami
U.S.A.

Shu-Cherng Fang
North Carolina State University
U.S.A.

Mitsuo Gen
Ashikaga Institute of Technology
Japan

A. Kazerooni
University of Lavisan
Iran

M. Kazerooni
Toosi University of Technology
Iran

Li-Pheng Khoo
Nanyang Technological University
Singapore

Sarah S. Y. Lam
State University of New York
at Binghamton
U.S.A.

Yuan-Shin Lee
North Carolina State University
U.S.A.

Xiaoli Li
Harbin Institute of Technology
China

L. H. S. Luong
University of South Australia
Australia

Gary S. May
Georgia Institute of Technology
U.S.A.

A. Y. C. Nee
National University of Singapore
Singapore

Douglas Norrie
University of Calgary
Canada

D. T. Pham
University of Wales
Cardiff, U.K.

P. T. N. Pham
University of Wales
Cardiff, U.K.

Catherine Roze
IBM Global Services
U.S.A.

Alice E. Smith
Auburn University
U.S.A.

Dan Stefanoiu
University of Calgary
Canada

Nallan C. Suresh
State University of New York
at Buffalo
U.S.A.
University of Groningen
The Netherlands

Wai Sum Tang
The Chinese University
of Hong Kong
China

Chieh-Yuan Tsai
Yuan-Ze University
Taiwan

Michaela Ulieru
University of Calgary
Canada

Jun Wang
The Chinese University
of Hong Kong
China

Yangsheng Xu
The Chinese University
of Hong Kong
China

Lian-Yin Zhai
Nanyang Technological University
Singapore

Y. F. Zhang
National University of Singapore
Singapore

Table of Contents

PART I Overview

PART II Manufacturing System Modeling and Design

PART III Process Planning and Scheduling

PART IV Manufacturing Process Monitoring and Control

PART V Quality Assurance and Fault Diagnosis

I

Overview

1

Computational Intelligence for Manufacturing

D. T. Pham
University of Wales

P. T. N. Pham
University of Wales

1.1 Introduction

Computational intelligence refers to intelligence artificially realised through computation. Artificial intelligence emerged as a computer science discipline in the mid-1950s. Since then, it has produced a number of powerful tools, some of which are used in engineering to solve difficult problems normally requiring human intelligence. Five of these tools are reviewed in this chapter with examples of applications in engineering and manufacturing: knowledge-based systems, fuzzy logic, inductive learning, neural networks, and genetic algorithms. All of these tools have been in existence for more than 30 years and have found many practical applications.

1.2 Knowledge-Based Systems

Knowledge-based systems, or expert systems, are computer programs embodying knowledge about a narrow domain for solving problems related to that domain. An expert system usually comprises two main elements, a knowledge base and an inference mechanism. The knowledge base contains domain knowledge which may be expressed as any combination of "If-Then" rules, factual statements (or assertions), frames, objects, procedures, and cases. The inference mechanism is that part of an expert system which manipulates the stored knowledge to produce solutions to problems. Knowledge manipulation methods include the use of inheritance and constraints (in a frame-based or object-oriented expert system), the retrieval and adaptation of case examples (in a case-based expert system), and the application of inference rules such as *modus ponens* (If A Then B; A Therefore B) and *modus tollens* (If A Then B; Not B Therefore Not A) according to "forward chaining" or "backward chaining" control procedures and "depth-first" or "breadth-first" search strategies (in a rule-based expert system).

With *forward chaining* or data-driven inferencing, the system tries to match available facts with the If portion of the If-Then rules in the knowledge base. When matching rules are found, one of them is

0-8493-0592-6/01/$0.00+$.50

KNOWLEDGE BASE
(Initial State)

Fact :
- **F1** - A lathe is a machine tool

Rules :
- **R1** - If X is power driven Then X requires a power source
- **R2** - If X is a machine tool Then X has a tool holder
- **R3** - If X is a machine tool Then X is power driven

KNOWLEDGE BASE
(Intermediate State)

Fact :
- **F1** - A lathe is a machine tool
- **F2** - A lathe has a tool holder

Rules :
- **R1** - If X is power driven Then X requires a power source
- **R2** - If X is a machine tool Then X has a tool holder
- **R3** - If X is a machine tool Then X is power driven

KNOWLEDGE BASE
(Initial State)

Fact :
- **F1** - A lathe is a machine tool
- **F2** - A lathe has a tool holder
- **F3** - A lathe is power driven

Rules :
- **R1** - If X is power driven Then X requires a power source
- **R2** - If X is a machine tool Then X has a tool holder
- **R3** - If X is a machine tool Then X is power driven

KNOWLEDGE BASE
(Initial State)

Fact :
- **F1** - A lathe is a machine tool
- **F2** - A lathe has a tool holder
- **F3** - A lathe is power driven
- **F4** - A lathe requires a power source

Rules :
- **R1** - If X is power driven Then X requires a power source
- **R2** - If X is a machine tool Then X has a tool holder
- **R3** - If X is a machine tool Then X is power driven

FIGURE 1.1(a) An example of forward chaining.

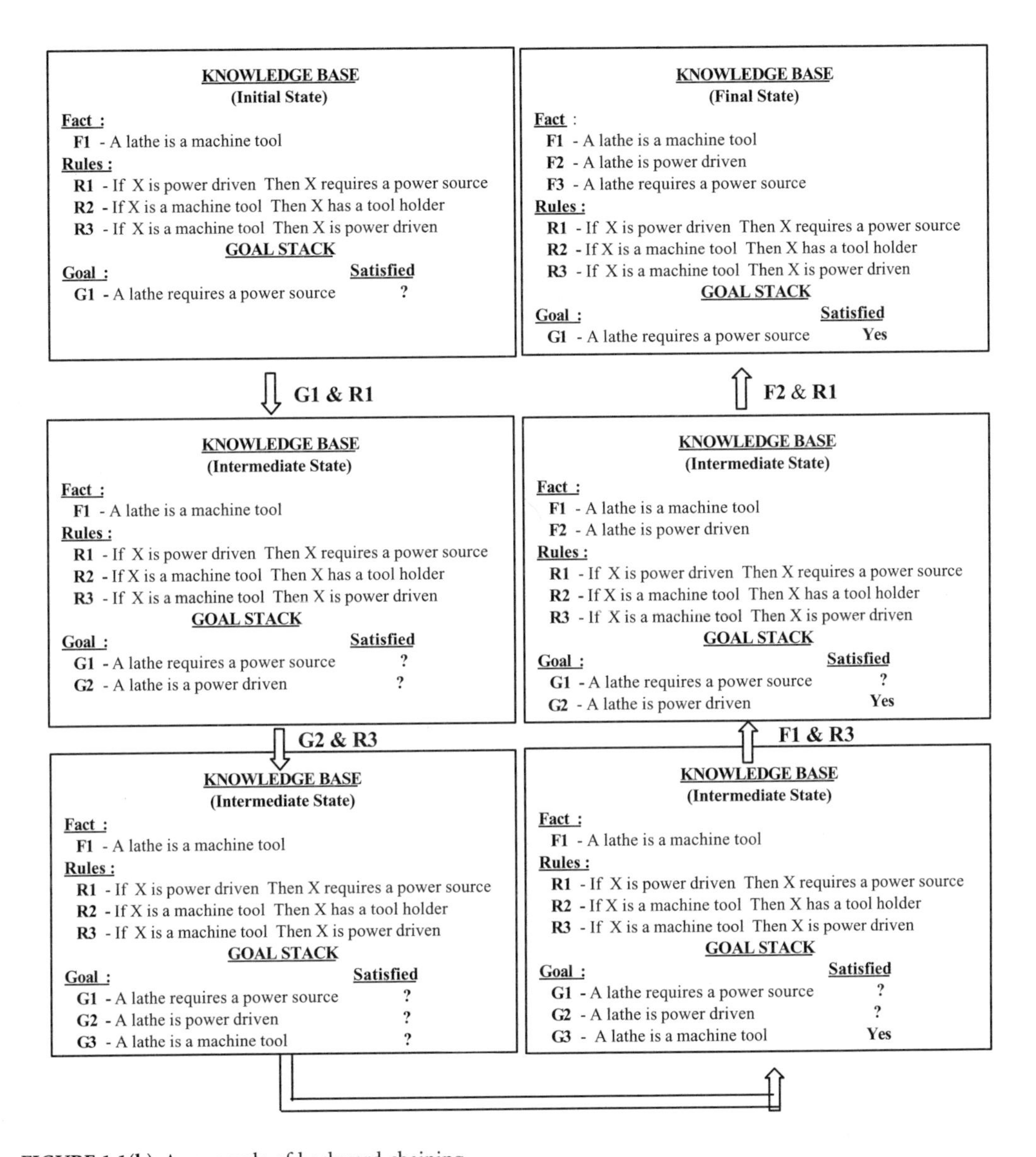

FIGURE 1.1(b) An example of backward chaining.

"fired," i.e., its Then part is made true, generating new facts and data which in turn cause other rules to "fire." Reasoning stops when no more new rules can fire. In *backward chaining* or goal-driven inferencing, a goal to be proved is specified. If the goal cannot be immediately satisfied by existing facts in the knowledge base, the system will examine the If-Then rules for rules with the goal in their Then portion. Next, the system will determine whether there are facts that can cause any of those rules to fire. If such facts are not available they are set up as subgoals. The process continues recursively until either all the required facts are found and the goal is proved or any one of the subgoals cannot be satisfied, in which case the original goal is disproved. Both control procedures are illustrated in Figure 1.1. Figure 1.1(a) shows how, given the assertion that a lathe is a machine tool and a set of rules concerning machine tools, a forward-chaining system will generate additional assertions such as "a lathe is power driven" and "a lathe has a tool holder." Figure 1.1(b) details the backward-chaining sequence producing the answer to the query "does a lathe require a power source?"

In the forward-chaining example of Figure 1.1(a), both rules R2 and R3 simultaneously qualify for firing when inferencing starts as both their If parts match the presented fact F1. Conflict resolution has to be performed by the expert system to decide which rule should fire. The conflict resolution method adopted in this example is "first come, first served": R2 fires, as it is the first qualifying rule encountered. Other conflict resolution methods include "priority," "specificity," and "recency."

The search strategies can also be illustrated using the forward-chaining example of Figure 1.1(a). Suppose that, in addition to F1, the knowledge base also initially contains the assertion "a CNC turning centre is a machine tool." Depth-first search involves firing rules R2 and R3 with X instantiated to "lathe" (as shown in Figure 1.1(a)) before firing them again with X instantiated to "CNC turning centre." Breadth-first search will activate rule R2 with X instantiated to "lathe" and again with X instantiated to "CNC turning centre," followed by rule R3 and the same sequence of instantiations. Breadth-first search finds the shortest line of inferencing between a start position and a solution if it exists. When guided by heuristics to select the correct search path, depth-first search might produce a solution more quickly, although the search might not terminate if the search space is infinite [Jackson, 1999].

For more information on the technology of expert systems, see [Pham and Pham, 1988; Durkin, 1994; Jackson, 1999].

Most expert systems are nowadays developed using programs known as "shells." These are essentially ready-made expert systems complete with inferencing and knowledge storage facilities but without the domain knowledge. Some sophisticated expert systems are constructed with the help of "development environments." The latter are more flexible than shells in that they also provide means for users to implement their own inferencing and knowledge representation methods. More details on expert systems shells and development environments can be found in [Price, 1990].

Among the five tools considered in this chapter, expert systems are probably the most mature, with many commercial shells and development tools available to facilitate their construction. Consequently, once the domain knowledge to be incorporated in an expert system has been extracted, the process of building the system is relatively simple. The ease with which expert systems can be developed has led to a large number of applications of the tool. In engineering, applications can be found for a variety of tasks including selection of materials, machine elements, tools, equipment and processes, signal interpreting, condition monitoring, fault diagnosis, machine and process control, machine design, process planning, production scheduling and system configuring. The following are recent examples of specific tasks undertaken by expert systems:

- Identifying and planning inspection schedules for critical components of an offshore structure [Peers et al., 1994]
- Training technical personnel in the design and evaluation of energy cogeneration plants [Lara Rosano et al., 1996]
- Configuring paper feeding mechanisms [Koo and Han, 1996]
- Carrying out automatic remeshing during a finite-elements analysis of forging deformation [Yano et al., 1997]
- Storing, retrieving, and adapting planar linkage designs [Bose et al., 1997]
- Designing additive formulae for engine oil products [Shi et al., 1997]

1.3 Fuzzy Logic

A disadvantage of ordinary rule-based expert systems is that they cannot handle new situations not covered explicitly in their knowledge bases (that is, situations not fitting exactly those described in the "If" parts of the rules). These rule-based systems are completely unable to produce conclusions when such situations are encountered. They are therefore regarded as shallow systems which fail in a "brittle" manner, rather than exhibit a gradual reduction in performance when faced with increasingly unfamiliar problems, as human experts would.

The use of fuzzy logic [Zadeh, 1965] that reflects the qualitative and inexact nature of human reasoning can enable expert systems to be more resilient. With fuzzy logic, the precise value of a variable is replaced by a linguistic description, the meaning of which is represented by a fuzzy set, and inferencing is carried out based on this representation. Fuzzy set theory may be considered an extension of classical set theory. While classical set theory is about "crisp" sets with sharp boundaries, fuzzy set theory is concerned with "fuzzy" sets whose boundaries are "gray."

In classical set theory, an element u_i can either belong or not belong to a set A, i.e., the degree to which element u belongs to set A is either 1 or 0. However, in fuzzy set theory, the degree of belonging of an element u to a fuzzy set $\underset{\sim}{A}$ is a real number between 0 and 1. This is denoted by $\mu_{\underset{\sim}{A}}(u_i)$, the grade of membership of u_i in A. Fuzzy set $\underset{\sim}{A}$ is a fuzzy set in U, the "universe of discourse" or "universe" which includes all objects to be discussed. $\mu_{\underset{\sim}{A}}(u_i)$ is 1 when u_i is definitely a member of A and $\mu_{\underset{\sim}{A}}(u_i)$ is 0 when u_i is definitely not a member of A. For instance, a fuzzy set defining the term "normal room temperature" might be as follows:

$$\text{normal room temperature} \quad \begin{aligned} &0.0/\text{below } 10°\text{C} + 0.3/10°\text{C} - 16°\text{C} + 0.8/16°\text{C} - 18°\text{C} + \\ &1.0/18°\text{C} - 22°\text{C} + 0.8/22°\text{C} - 24°\text{C} + 0.3/24°\text{C} - 30°\text{C} + 0.0/\text{above } 30°\text{C} \end{aligned}$$

Equation (1.1)

The values 0.0, 0.3, 0.8, and 1.0 are the grades of membership to the given fuzzy set of temperature ranges below 10°C (above 30°C), between 10°C and 16°C (24°C to 30°C), between 16°C and 18°C (22°C to 24°C), and between 18°C and 22°C. Figure 1.2(a) shows a plot of the grades of membership for "normal room temperature." For comparison, Figure 1.2(b) depicts the grades of membership for a crisp set defining room temperatures in the normal range.

Knowledge in an expert system employing fuzzy logic can be expressed as qualitative statements (or fuzzy rules) such as "If the room temperature is normal, Then set the heat input to normal," where "normal room temperature" and "normal heat input" are both fuzzy sets.

A fuzzy rule relating two fuzzy sets $\underset{\sim}{A}$ and $\underset{\sim}{B}$ is effectively the Cartesian product $\underset{\sim}{A} \times \underset{\sim}{B}$, which can be represented by a relation matrix $[\underset{\sim}{R}]$. Element R_{ij} of $[\underset{\sim}{R}]$ is the membership to $\underset{\sim}{A} \times \underset{\sim}{B}$ of pair (u_i,v_j), $u_i \in \underset{\sim}{A}$ and $v_j \in \underset{\sim}{B}$. R_{ij} is given by

$$R_{ij} = \min\left(\mu_{\underset{\sim}{A}}(u_i), \mu_{\underset{\sim}{B}}(v_j)\right) \qquad \text{Equation (1.2)}$$

For example, with "normal room temperature" defined as before and "normal heat input" described by

$$\text{normal heat input} \equiv 0.2/lkW + 0.9/2\,kW\ 0.2/3\,kW \qquad \text{Equation (1.3)}$$

$[\underset{\sim}{R}]$ can be computed as:

$$[\underset{\sim}{R}] = \begin{bmatrix} 0.0 & 0.0 & 0.0 \\ 0.2 & 0.3 & 0.2 \\ 0.2 & 0.8 & 0.2 \\ 0.2 & 0.9 & 0.2 \\ 0.2 & 0.8 & 0.2 \\ 0.2 & 0.3 & 0.2 \\ 0.0 & 0.0 & 0.0 \end{bmatrix} \qquad \text{Equation (1.4)}$$

A reasoning procedure known as the *compositional rule of inference*, which is the equivalent of the *modus ponens* rule in rule-based expert systems, enables conclusions to be drawn by generalization

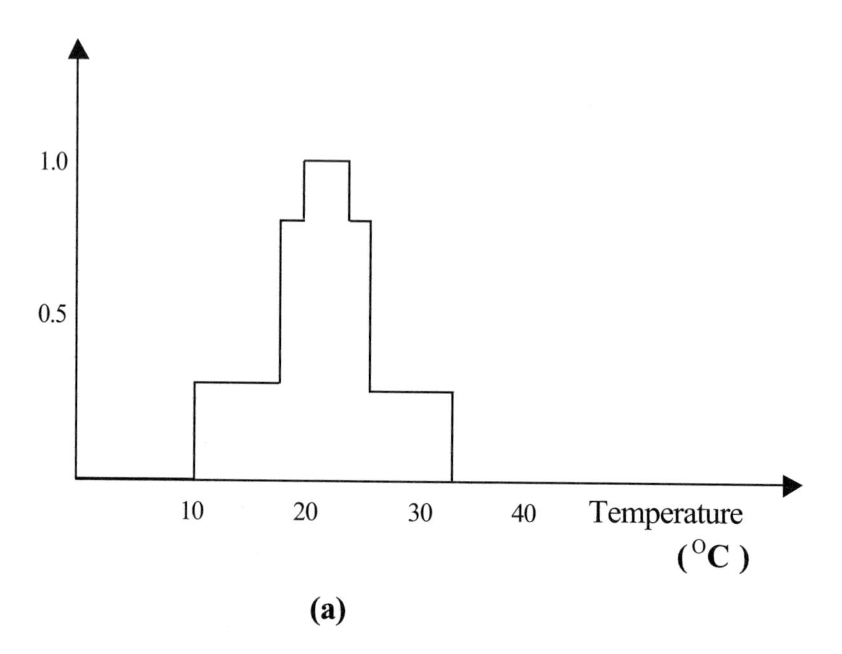

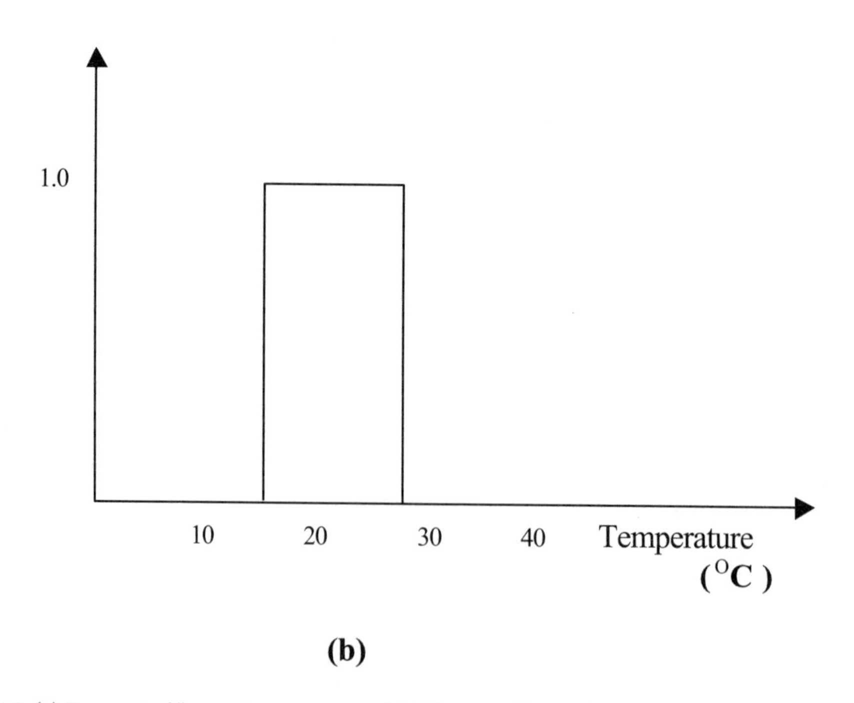

FIGURE 1.2 (a) Fuzzy set of "normal temperature." (b) Crisp set of "normal temperature."

(extrapolation or interpolation) from the qualitative information stored in the knowledge base. For instance, when the room temperature is detected to be "slightly below normal," a temperature-controlling fuzzy expert system might deduce that the heat input should be set to "slightly above normal." Note that this conclusion might not be contained in any of the fuzzy rules stored in the system. A well-known compositional rule of inference is the *max-min rule*. Let $[\underset{\sim}{R}]$ represent the fuzzy rule "If $\underset{\sim}{A}$ Then $\underset{\sim}{B}$" and $\underset{\sim}{a} \equiv \sum_i \mu_i / u_i$ a fuzzy assertion. $\underset{\sim}{A}$ and $\underset{\sim}{a}$ are fuzzy sets in the same universe of discourse. The max-min rule enables a fuzzy conclusion $\underset{\sim}{b} \equiv \sum_j \lambda_j / v_j$ to be inferred from $\underset{\sim}{a}$ and $[\underset{\sim}{R}]$ as follows:

$$\underset{\sim}{b} = \underset{\sim}{a} \text{ o } [\underset{\sim}{R}] \qquad \text{Equation (1.5)}$$

$$\lambda_j = \max_i \left[\min\left(\mu_i, R_{ij} \right) \right] \qquad \text{Equation (1.6)}$$

For example, given the fuzzy rule "If the room temperature is normal, Then set the heat input to normal" where "normal room temperature" and "normal heat input" are as defined previously, and a fuzzy temperature measurement of

$$\begin{aligned} \text{temperature} \equiv\ & 0.0/\text{below } 10°C + 0.4/10°C-16°C + 0.8/16°C-18°C \\ & + 0.8/18°C-22°C + 0.2/22°C-24°C + 0.0/24°C-30°C + 0.0/\text{above } 30°C \end{aligned} \qquad \text{Equation (1.7)}$$

the heat input will be deduced as

$$\text{heat input} = \text{temperature o } [\underset{\sim}{R}]$$

$$= 0.2/1kW + 0.8/2kW \; 0.2/3kW \qquad \text{Equation (1.8)}$$

For further information on fuzzy logic, see [Kaufmann, 1975; Zimmermann, 1991].

Fuzzy logic potentially has many applications in engineering, where the domain knowledge is usually imprecise. Notable successes have been achieved in the area of process and machine control, although other sectors have also benefited from this tool. Recent examples of engineering applications include:

- Controlling the height of the arc in a welding process [Bigand et al., 1994]
- Controlling the rolling motion of an aircraft [Ferreiro Garcia, 1994]
- Controlling a multi-fingered robot hand [Bas and Erkmen, 1995]
- Analysing the chemical composition of minerals [Da Rocha Fernandes and Cid Bastos, 1996]
- Determining the optimal formation of manufacturing cells [Szwarc et al., 1997]
- Classifying discharge pulses in electrical discharge machining [Tarng et al., 1997]

1.4 Inductive Learning

The acquisition of domain knowledge to build into the knowledge base of an expert system is generally a major task. In some cases, it has proved a bottleneck in the construction of an expert system. Automatic knowledge acquisition techniques have been developed to address this problem. Inductive learning is an automatic technique for knowledge acquisition. The inductive approach produces a structured representation of knowledge as the outcome of learning. Induction involves generalising a set of examples to yield a selected representation that can be expressed in terms of a set of rules, concepts or logical inferences, or a decision tree.

TABLE 1.1 Training Set for the Cutting Tool Problem

Example	Sensor_1	Sensor_2	Sensor_3	Tool State
1	0	–1	0	Normal
2	1	0	0	Normal
3	1	–1	1	Worn
4	1	0	1	Normal
5	0	0	1	Normal
6	1	1	1	Worn
7	1	–1	0	Normal
8	0	–1	1	Worn

An inductive learning program usually requires as input a set of examples. Each example is characterised by the values of a number of attributes and the class to which it belongs. In one approach to inductive learning, through a process of "dividing and conquering" where attributes are chosen according to some strategy (for example, to maximise the information gain) to divide the original example set into subsets, the inductive learning program builds a decision tree that correctly classifies the given example set. The tree represents the knowledge generalised from the specific examples in the set. This can subsequently be used to handle situations not explicitly covered by the example set.

In another approach known as the "covering approach," the inductive learning program attempts to find groups of attributes uniquely shared by examples in given classes and forms rules with the If part as conjunctions of those attributes and the Then part as the classes. The program removes correctly classified examples from consideration and stops when rules have been formed to classify all examples in the given set.

A new approach to inductive learning, *inductive logic programming,* is a combination of induction and logic programming. Unlike conventional inductive learning which uses propositional logic to describe examples and represent new concepts, inductive logic programming (ILP) employs the more powerful predicate logic to represent training examples and background knowledge and to express new concepts. Predicate logic permits the use of different forms of training examples and background knowledge. It enables the results of the induction process — that is, the induced concepts — to be described as general first-order clauses with variables and not just as zero-order propositional clauses made up of attribute–value pairs. There are two main types of ILP systems, the first based on the top-down generalisation/specialisation method, and the second on the principle of inverse resolution [Muggleton, 1992].

A number of inductive learning programs have been developed. Some of the well-known programs are ID3 [Quinlan, 1983], which is a divide-and-conquer program; the AQ program [Michalski, 1990], which follows the covering approach; the FOIL program [Muggleton, 1992], which is an ILP system adopting the generalisation/specialisation method; and the GOLEM program [Muggleton, 1992], which is an ILP system based on inverse resolution. Although most programs only generate crisp decision rules, algorithms have also been developed to produce fuzzy rules [Wang and Mendel, 1992].

Figure 1.3 shows the main steps in RULES-3 Plus, an induction algorithm in the covering category [Pham and Dimov, 1997] and belonging to the RULES family of rule extraction systems [Pham and Aksoy, 1994, 1995a, 1995b]. The simple problem of detecting the state of a metal cutting tool is used to explain the operation of RULES-3 Plus. Three sensors are employed to monitor the cutting process and, according to the signals obtained from them (1 or 0 for sensors 1 and 3; –1, 0, or 1 for sensor 2), the tool is inferred as being "normal" or "worn." Thus, this problem involves three attributes which are the states of sensors 1, 2, and 3, and the signals that they emit constitute the values of those attributes. The example set for the problem is given in Table 1.1.

In Step 1, example 1 is used to form the attribute–value array SETAV which will contain the following attribute–value pairs: [Sensor_1 = 0], [Sensor_2 = –1] and [Sensor_3 = 0].

In Step 2, the partial rule set PRSET and T_PRSET, the temporary version of PRSET used for storing partial rules in the process of rule construction, are initialised. This creates for each of these sets three expressions having null conditions and zero H measures. The H measure for an expression is defined as

Step 1.	Take an unclassified example and form array SETAV.
Step 2.	Initialise arrays PRSET and T_PRSET (PRSET and T_PRSET will consist of m_{PRSET} expressions with null conditions and zero H measures) and set $n_{co} = 0$.
Step 3.	IF $n_{co} < n_a$ THEN $n_{co} = n_{co} + 1$ and set m = 0; ELSE the example itself is taken as a rule and STOP.
Step 4.	DO m = m + 1; Specialise expression m in PRSET by appending to it a condition from SETAV that differs from the conditions already included in the expression; Compute the H measure for the expression; IF its H measure is higher than the H measure of any expression in T_PRSET THEN replace the expression having the lowest H measure with the newly formed expression; ELSE discard the new expression; WHILE $m \le m_{PRSET}$.
Step 5.	IF there are consistent expressions in T_PRSET THEN choose as a rule the expression that ha s the highest H measure and discard the others; ELSE copy T_PRSET into PRSET; initialise T_PRSET and go to Step 4.

n_{co} - number of conditions; n_a- number of attributes;
m_{PRSET} - number of expressions stored in PRSET (m_{PRSET} is user-provided);
T_PRSET - a temporary array of partial rules of the same dimension as PRSET.

FIGURE 1.3 Rule-formatting procedure of RULES-3 Plus.

$$H = \sqrt{\frac{E^c}{E}}\left[2 - 2\sqrt{\frac{E_i^c}{E^c}\frac{E_i}{E}} - 2\sqrt{\left(1 - \frac{E_i^c}{E^c}\right)\left(1 - \frac{E_i}{E}\right)}\right] \qquad \text{Equation (1.9)}$$

where E^c is the number of examples covered by the expression (the total number of examples correctly classified and misclassified by a given rule), E is the total number of examples, E_i^c is the number of examples covered by the expression and belonging to the target class i (the number of examples correctly classified by a given rule), and E_i is the number of examples in the training set belonging to the target class i. In Equation 1.1, the first term

$$G = \sqrt{\frac{E^c}{E}} \qquad \text{Equation (1.10)}$$

relates to the generality of the rule and the second term

$$A = 2 - 2\sqrt{\frac{E_i^c}{E^c}\frac{E_i}{E}} - 2\sqrt{\left(1 - \frac{E_i^c}{E^c}\right)\left(1 - \frac{E_i}{E}\right)} \qquad \text{Equation (1.11)}$$

indicates its accuracy.

In Steps 3 and 4, by specialising PRSET using the conditions stored in SETAV, the following expressions are formed and stored in T_PRSET:

1: [Sensor_3 = 0] ⇒ [Alarm = OFF] H = 0.2565

2: [Sensor_2 = -1] ⇒ [Alarm = OFF] H = 0.0113

3: [Sensor_1 = 0] ⇒ [Alarm = OFF] H = 0.0012

In Step 5, a rule is produced as the first expression in T_PRSET applies to only one class:

Rule 1: If [Sensor_3 = 0] Then [Alarm = OFF], H = 0.2565

Rule 1 can classify examples 2 and 7 in addition to example 1. Therefore, these examples are marked as classified and the induction proceeds.

In the second iteration, example 3 is considered. T_PRSET, formed in Step 4 after specialising the initial PRSET, now consists of the following expressions:

1: [Sensor_3 = 1] ⇒ [Alarm = ON] H = 0.0406

2: [Sensor_2 = -1] ⇒ [Alarm = ON] H = 0.0079

3: [Sensor_1 = 1] ⇒ [Alarm = ON] H = 0.0005

As none of the expressions cover only one class, T_PRSET is copied into PRSET (Step 5) and the new PRSET has to be specialised further by appending the existing expressions with conditions from SETAV. Therefore, the procedure returns to Step 3 for a new pass. The new T_PRSET formed at the end of Step 4 contains the following three expressions:

1: [Sensor_2 = -1] [Sensor_3 = 1] ⇒ [Alarm = ON] H = 0.3876

2: [Sensor_1 = 1] [Sensor_3 = 1] ⇒ [Alarm = ON] H = 0.0534

3: [Sensor_1 = 1] [Sensor_2 = -1] ⇒ [Alarm = ON] H = 0.0008

As the first expression applies to only one class, the following rule is obtained:

Rule 2: If [Sensor_2 = -1] AND [Sensor_3 = 1] Then [Alarm = ON], H = 0.3876.

Rule 2 can classify examples 3 and 8, which again are marked as classified.

In the third iteration, example 4 is used to obtain the next rule:

Rule 3: If [Sensor_2 = 0] Then [Alarm = OFF], H = 0.2565.

This rule can classify examples 4 and 5 and so they are also marked as classified.

In iteration 4, the last unclassified example 6 is employed for rule extraction, yielding

Rule 4: If [Sensor_2 = 1] Then [Alarm = ON], H = 0.2741.

There are no remaining unclassified examples in the example set and the procedure terminates at this point.

Due to its requirement for a set of examples in a rigid format (with known attributes and of known classes), inductive learning has found rather limited applications in engineering, as not many engineering problems can be described in terms of such a set of examples. Another reason for the paucity of applications is that inductive learning is generally more suitable for problems where attributes have

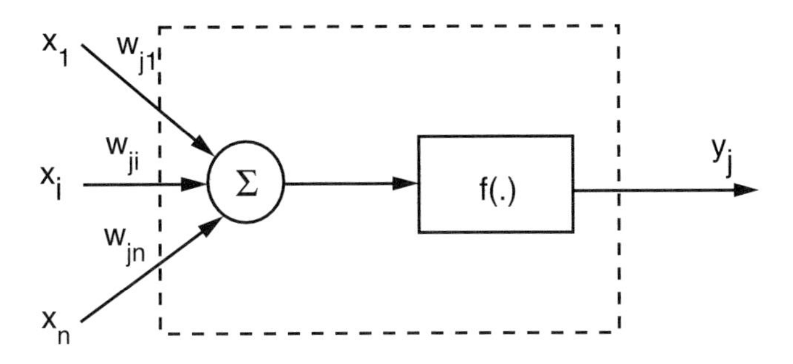

FIGURE 1.4 Model of a neuron.

discrete or symbolic values than for those with continuous-valued attributes as in many engineering problems. Some recent examples of applications of inductive learning are

- Controlling a laser cutting robot [Luzeaux, 1994]
- Controlling the functional electrical stimulation of spinally injured humans [Kostov et al., 1995]
- Classifying complex and noisy patterns [Pham and Aksoy, 1995a]
- Analysing the constructability of a beam in a reinforced-concrete frame [Skibniewski et al., 1997]

1.5 Neural Networks

Like inductive learning programs, neural networks can capture domain knowledge from examples. However, they do not archive the acquired knowledge in an explicit form such as rules or decision trees and they can readily handle both continuous and discrete data. They also have a good generalisation capability, as with fuzzy expert systems.

A neural network is a computational model of the brain. Neural network models usually assume that computation is distributed over several simple units called neurons that are interconnected and operate in parallel (hence, neural networks are also called parallel-distributed-processing systems or connectionist systems). Figure 1.4 illustrates a typical model of a neuron. Output signal y_j is a function f of the sum of weighted input signals x_i. The activation function f can be a linear, simple threshold, sigmoidal, hyberbolic tangent or radial basis function. Instead of being deterministic, f can be a probabilistic function, in which case y_j will be a binary quantity, for example, +1 or –1. The net input to such a stochastic neuron — that is, the sum of weighted input signals x_i — will then give the probability of y_j being +1 or –1.

How the interneuron connections are arranged and the nature of the connections determine the structure of a network. How the strengths of the connections are adjusted or trained to achieve a desired overall behaviour of the network is governed by its learning algorithm. Neural networks can be classified according to their structures and learning algorithms.

In terms of their structures, neural networks can be divided into two types: feedforward networks and recurrent networks. *Feedforward networks* can perform a static mapping between an input space and an output space: the output at a given instant is a function only of the input at that instant. The most popular feedforward neural network is the multi-layer perceptron (MLP): all signals flow in a single direction from the input to the output of the network. Figure 1.5 shows an MLP with three layers: an input layer, an output layer, and an intermediate or hidden layer. Neurons in the input layer only act as buffers for distributing the input signals x_i to neurons in the hidden layer. Each neuron j in the hidden layer operates according to the model of Figure 1.4. That is, its output y_j is given by

$$y_j = f\left(\Sigma w_{ji} x_i\right) \qquad \text{Equation (1.12)}$$

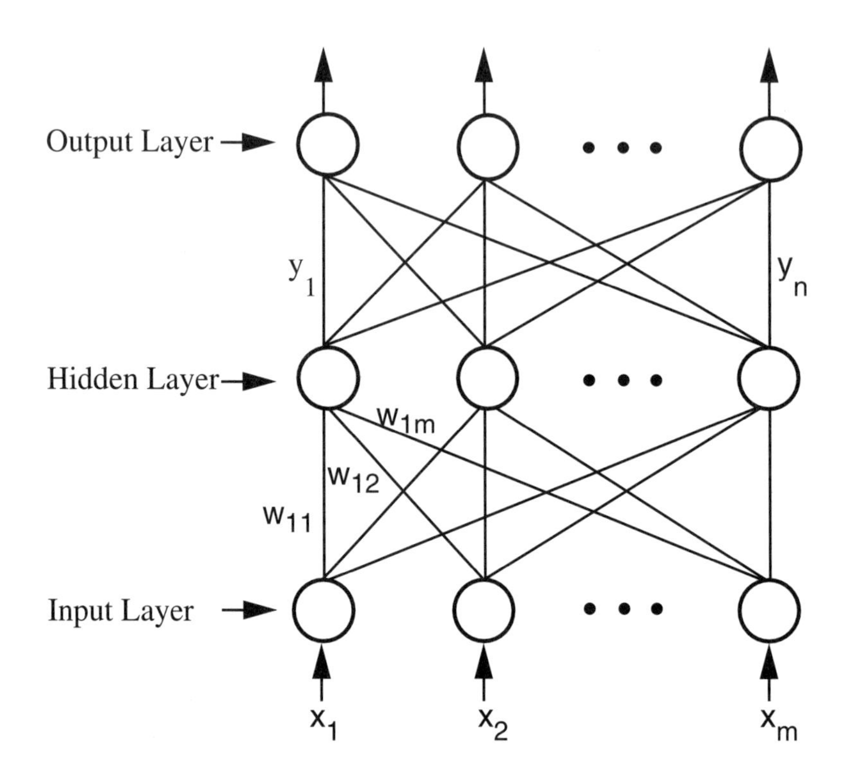

FIGURE 1.5 A multi-layer perceptron.

The outputs of neurons in the output layer are computed similarly.

Other feedforward networks [Pham and Liu, 1999] include the learning vector quantisation (LVQ) network, the cerebellar model articulation control (CMAC) network, and the group-method of data handling (GMDH) network.

Recurrent networks are networks where the outputs of some neurons are fed back to the same neurons or to neurons in layers before them. Thus signals can flow in both forward and backward directions. Recurrent networks are said to have a dynamic memory: the output of such networks at a given instant reflects the current input as well as previous inputs and outputs. Examples of recurrent networks [Pham and Liu, 1999] include the Hopfield network, the Elman network and the Jordan network. Figure 1.6 shows a well-known, simple recurrent neural network, the Grossberg and Carpenter ART-1 network. The network has two layers, an input layer and an output layer. The two layers are fully interconnected, the connections are in both the forward (or bottom-up) direction and the feedback (or top-down) direction. The vector W_i of weights of the bottom-up connections to an output neuron i forms an exemplar of the class it represents. All the W_i vectors constitute the long-term memory of the network. They are employed to select the winning neuron, the latter again being the neuron whose W_i vector is most similar to the current input pattern. The vector V_i of the weights of the top-down connections from an output neuron i is used for *vigilance* testing, that is, determining whether an input pattern is sufficiently close to a stored exemplar. The vigilance vectors V_i form the short-term memory of the network. V_i and W_i are related in that W_i is a normalised copy of V_i, viz.

$$W_i = \frac{V_i}{\varepsilon + \sum V_{ji}} \quad \text{Equation (1.13)}$$

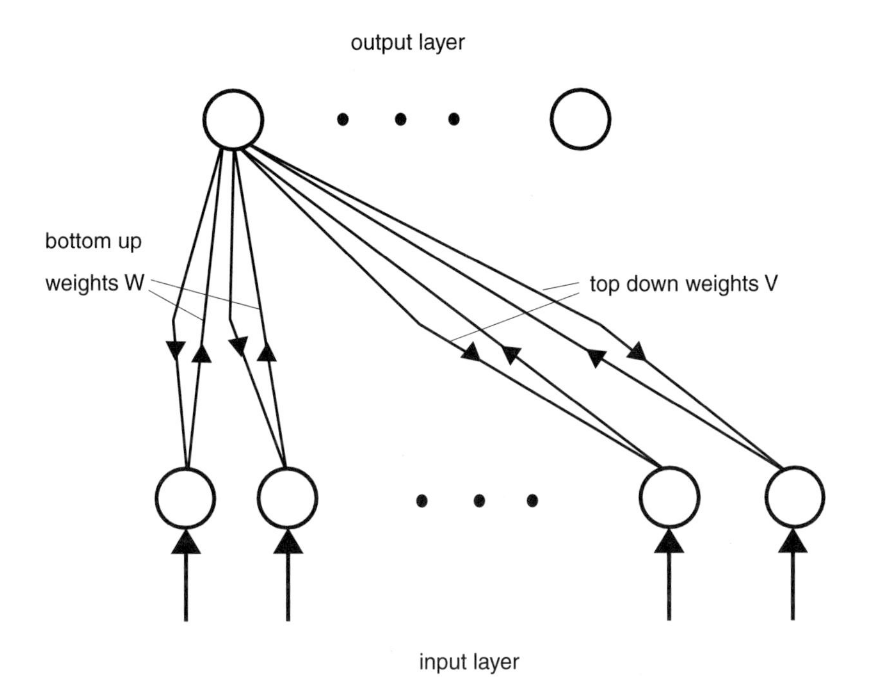

FIGURE 1.6 An ART-1 network.

where ε is a small constant and V_{ji}, the jth component of V_i (i.e., the weight of the connection from output neuron i to input neuron j).

Implicit "knowledge" is built into a neural network by training it. Neural networks are trained and categorised according to two main types of learning algorithms: supervised and unsupervised. In addition, there is a third type, reinforcement learning, which is a special case of supervised learning. In supervised training, the neural network can be trained by being presented with typical input patterns and the corresponding expected output patterns. The error between the actual and expected outputs is used to modify the strengths, or weights, of the connections between the neurons. The backpropagation (BP) algorithm, a gradient descent algorithm, is the most commonly adopted MLP training algorithm. It gives the change Δw_{ji} in the weight of a connection between neurons i and j as follows:

$$\Delta w_{ji} = \eta \delta_j x_i \qquad \text{Equation (1.14)}$$

where η is a parameter called the learning rate and δ_j is a factor depending on whether neuron j is an output neuron or a hidden neuron. For output neurons,

$$\delta_j = \left(\frac{\partial f}{\partial net_j} \right) \left(y_j^{(t)} - y_j \right) \qquad \text{Equation (1.15)}$$

and for hidden neurons,

$$\delta_j = \left(\frac{\partial f}{\partial net_j} \right) \sum_q w_{qj} \delta_q \qquad \text{Equation (1.16)}$$

In Equation 1.15, net_j is the total weighted sum of input signals to neuron j and $y_j^{(t)}$ is the target output for neuron j.

As there are no target outputs for hidden neurons, in Equation 1.16, the difference between the target and actual output of a hidden neuron j is replaced by the weighted sum of the δ_q terms already obtained for neurons q connected to the output of j. Thus, iteratively, beginning with the output layer, the δ term is computed for neurons in all layers and weight updates determined for all connections. The weight updating process can take place after the presentation of each training pattern (pattern-based training) or after the presentation of the whole set of training patterns (batch training). In either case, a training epoch is said to have been completed when all training patterns have been presented once to the MLP.

For all but the most trivial problems, several epochs are required for the MLP to be properly trained. A commonly adopted method to speed up the training is to add a "momentum" term to Equation 1.14 which effectively lets the previous weight change influence the new weight change, viz.

$$\Delta w_{ji}\left(k+1\right)=\eta\delta_j x_i+\mu\Delta w_{ji}\left(k\right) \qquad \text{Equation (1.17)}$$

where $\Delta w_{ji}(k+1)$ and $\Delta w_{ji}(k+)$ are weight changes in epochs $(k+1)$ and (k), respectively, and μ is the "momentum" coefficient.

Some neural networks are trained in an unsupervised mode where only the input patterns are provided during training and the networks learn automatically to cluster them in groups with similar features. For example, training an ART-1 network involves the following steps:

1. Initialising the exemplar and vigilance vectors W_i and V_i for all output neurons by setting all the components of each V_i to 1 and computing W_i according to Equation 1.13. An output neuron with all its vigilance weights set to 1 is known as an *uncommitted* neuron in the sense that it is not assigned to represent any pattern classes.
2. Presenting a new input pattern x.
3. Enabling all output neurons so that they can participate in the competition for activation.
4. Finding the winning output neuron among the competing neurons, i.e., the neuron for which $x.W_i$ is largest; a winning neuron can be an uncommitted neuron as is the case at the beginning of training or if there are no better output neurons.
5. Testing whether the input pattern x is sufficiently similar to the vigilance vector V_i of the winning neuron. Similarity is measured by the fraction r of bits in x that are also in V_i, viz.

$$r=\frac{x.V_i}{\sum x_i} \qquad \text{Equation (1.18)}$$

   x is deemed to be sufficiently similar to V_i if r is at least equal to vigilance threshold $\rho(0<\rho\leq 1)$.
6. Going to step 7 if $r\geq\rho$ (i.e., there is *resonance*); else disabling the winning neuron temporarily from further competition and going to step 4, repeating this procedure until there are no further enabled neurons.
7. Adjusting the vigilance vector V_i of the most recent winning neuron by logically ANDing it with x, thus deleting bits in V_i that are not also in x; computing the bottom-up exemplar vector W_i using the new V_i according to Equation 1.13; activating the winning output neuron.
8. Going to step 2.

The above training procedure ensures that if the same sequence of training patterns is repeatedly presented to the network, its long-term and short-term memories are unchanged (i.e., the network is *stable*). Also, provided there are sufficient output neurons to represent all the different classes, new patterns can always be learned, as a new pattern can be assigned to an uncommitted output neuron if it does not match previously stored exemplars well (i.e., the network is *plastic*).

In reinforcement learning, instead of requiring a teacher to give target outputs and using the differences between the target and actual outputs directly to modify the weights of a neural network, the learning algorithm employs a critic only to evaluate the appropriateness of the neural network output corresponding to a given input. According to the performance of the network on a given input vector, the critic will issue a positive or negative reinforcement signal. If the network has produced an appropriate output, the reinforcement signal will be positive (a reward). Otherwise, it will be negative (a penalty). The intention of this is to strengthen the tendency to produce appropriate outputs and to weaken the propensity for generating inappropriate outputs. Reinforcement learning is a trial-and-error operation designed to maximise the average value of the reinforcement signal for a set of training input vectors. An example of a simple reinforcement learning algorithm is a variation of the associative reward–penalty algorithm [Hassoun, 1995]. Consider a single stochastic neuron j with inputs $(x_1, x_2, x_3, \ldots, x_n)$. The reinforcement rule may be written as [Hassoun, 1995]

$$w_{ji}(k+1) = w_{ji}(k) + l\, r(k)\, [y_j(k) - E(y_j(k))]\, x_i(k) \qquad \text{Equation (1.19)}$$

w_{ji} is the weight of the connection between input i and neuron j, l is the learning coefficient, r (which is +1 or −1) is the reinforcement signal, y_j is the output of neuron j, $E(y_j)$ is the expected value of the output, and $x_i(k)$ is the ith component of the kth input vector in the training set. When learning converges, $w_{ji}(k+1) = w_{ji}(k)$ and so $E(y_j(k)) = y_j(k) = +1$ or -1. Thus, the neuron effectively becomes deterministic. Reinforcement learning is typically slower than supervised learning. It is more applicable to small neural networks used as controllers where it is difficult to determine the target network output.

For more information on neural networks, see [Hassoun, 1995; Pham and Liu, 1999].

Neural networks can be employed as mapping devices, pattern classifiers, or pattern completers (autoassociative content addressable memories and pattern associators). Like expert systems, they have found a wide spectrum of applications in almost all areas of engineering, addressing problems ranging from modelling, prediction, control, classification, and pattern recognition, to data association, clustering, signal processing, and optimisation. Some recent examples of such applications are:

- Modelling and controlling dynamic systems including robot arms [Pham and Liu, 1999]
- Predicting the tensile strength of composite laminates [Teti and Caprino, 1994]
- Controlling a flexible assembly operation [Majors and Richards, 1995]
- Recognising control chart patterns [Pham and Oztemel, 1996]
- Analysing vibration spectra [Smith et al., 1996]
- Deducing velocity vectors in uniform and rotating flows by tracking the movement of groups of particles [Jambunathan et al., 1997]

1.6 Genetic Algorithms

Conventional search techniques, such as hill-climbing, are often incapable of optimising nonlinear or multimodal functions. In such cases, a random search method is generally required. However, undirected search techniques are extremely inefficient for large domains. A genetic algorithm (GA) is a directed random search technique, invented by Holland [1975], which can find the global optimal solution in complex multidimensional search spaces. A GA is modelled on natural evolution in that the operators it employs are inspired by the natural evolution process. These operators, known as genetic operators, manipulate individuals in a population over several generations to improve their fitness gradually. Individuals in a population are likened to chromosomes and usually represented as strings of binary numbers.

The evolution of a population is described by the "schema theorem" [Holland, 1975; Goldberg, 1989]. A schema represents a set of individuals, i.e., a subset of the population, in terms of the similarity of bits at certain positions of those individuals. For example, the schema 1*0* describes the set of individuals whose first and third bits are 1 and 0, respectively. Here, the symbol "*" means any value would be acceptable. In other words, the values of bits at positions marked "*" could be either 0 or 1. A schema is characterised by two parameters: defining length and order. The defining length is the length between the first and last bits with fixed values. The order of a schema is the number of bits with specified values. According to the schema theorem, the distribution of a schema through the population from one generation to the next depends on its order, defining length and fitness.

GAs do not use much knowledge about the optimisation problem under study and do not deal directly with the parameters of the problem. They work with codes that represent the parameters. Thus, the first issue in a GA application is how to code the problem, i.e., how to represent its parameters. As already mentioned, GAs operate with a population of possible solutions. The second issue is the creation of a set of possible solutions at the start of the optimisation process as the initial population. The third issue in a GA application is how to select or devise a suitable set of genetic operators. Finally, as with other search algorithms, GAs have to know the quality of the solutions already found to improve them further. An interface between the problem environment and the GA is needed to provide this information. The design of this interface is the fourth issue.

1.6.1 Representation

The parameters to be optimised are usually represented in a string form since this type of representation is suitable for genetic operators. The method of representation has a major impact on the performance of the GA. Different representation schemes might cause different performances in terms of accuracy and computation time.

There are two common representation methods for numerical optimisation problems [Michalewicz, 1992]. The preferred method is the binary string representation method. This method is popular because the binary alphabet offers the maximum number of schemata per bit compared to other coding techniques. Various binary coding schemes can be found in the literature, for example, uniform coding, gray scale coding, etc. The second representation method uses a vector of integers or real numbers with each integer or real number representing a single parameter.

When a binary representation scheme is employed, an important step is to decide the number of bits to encode the parameters to be optimised. Each parameter should be encoded with the optimal number of bits covering all possible solutions in the solution space. When too few or too many bits are used, the performance can be adversely affected.

1.6.2 Creation of Initial Population

At the start of optimisation, a GA requires a group of initial solutions. There are two ways of forming this initial population. The first consists of using randomly produced solutions created by a random number generator, for example. This method is preferred for problems about which no *a priori* knowledge exists or for assessing the performance of an algorithm.

The second method employs *a priori* knowledge about the given optimisation problem. Using this knowledge, a set of requirements is obtained and solutions that satisfy those requirements are collected to form an initial population. In this case, the GA starts the optimisation with a set of approximately known solutions, and therefore convergence to an optimal solution can take less time than with the previous method.

1.6.3 Genetic Operators

The flowchart of a simple GA is given in Figure 1.7. There are basically four genetic operators: selection, crossover, mutation, and inversion. Some of these operators were inspired by nature. In the literature, many versions of these operators can be found. It is not necessary to employ all of these operators in a GA because each operates independently of the others. The choice or design of operators depends on

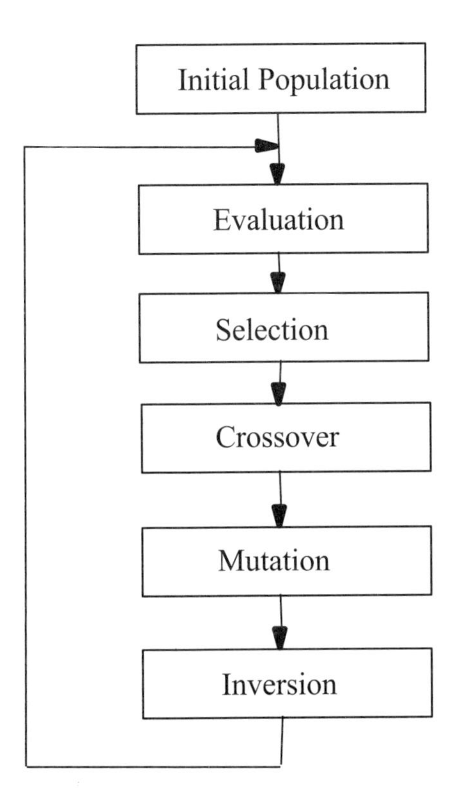

FIGURE 1.7 Flowchart of a basic genetic algorithm.

the problem and the representation scheme employed. For instance, operators designed for binary strings cannot be directly used on strings coded with integers or real numbers.

1.6.3.1 Selection

The aim of the selection procedure is to reproduce more of individuals whose fitness values are higher than those whose fitness values are low. The selection procedure has a significant influence on driving the search toward a promising area and finding good solutions in a short time. However, the diversity of the population must be maintained to avoid premature convergence and to reach the global optimal solution. In GAs there are mainly two selection procedures: proportional selection, also called stochastic selection, and ranking-based selection [Whitely, 1989].

Proportional selection is usually called "roulette wheel" selection, since its mechanism is reminiscent of the operation of a roulette wheel. Fitness values of individuals represent the widths of slots on the wheel. After a random spinning of the wheel to select an individual for the next generation, slots with large widths representing individuals with high fitness values will have a higher chance to be selected.

One way to prevent premature convergence is to control the range of trials allocated to any single individual, so that no individual produces too many offspring. The ranking system is one such alternative selection algorithm. In this algorithm, each individual generates an expected number of offspring based on the rank of its performance and not on the magnitude [Baker, 1985].

1.6.3.2 Crossover

This operation is considered the one that makes the GA different from other algorithms, such as dynamic programming. It is used to create two new individuals (children) from two existing individuals (parents) picked from the current population by the selection operation. There are several ways of doing this. Some common crossover operations are one-point crossover, two-point crossover, cycle crossover, and uniform crossover.

Parent 1	1 0 0 | 0 1 0 0 1 1 1 1 0
Parent 2	0 0 1 | 0 1 1 0 0 0 1 1 0
New string 1	1 0 0 | 0 1 1 0 0 0 1 1 0
New string 2	0 0 1 | 0 1 0 0 1 1 1 1 0

FIGURE 1.8 Crossover.

Old string	1 1 0 0 | 0 | 1 0 1 1 1 0 1
New string	1 1 0 0 | 1 | 1 0 1 1 1 0 1

FIGURE 1.9 Mutation.

Old string	1 0 | 1 1 0 0 | 1 1 1 0 1
New string	1 0 | 0 0 1 1 | 1 1 1 0 1

FIGURE 1.10 Inversion of a binary string segment.

One-point crossover is the simplest crossover operation. Two individuals are randomly selected as parents from the pool of individuals formed by the selection procedure and cut at a randomly selected point. The tails, which are the parts after the cutting point, are swapped and two new individuals (children) are produced. Note that this operation does not change the values of bits. An example of one-point crossover is shown in Figure 1.8.

1.6.3.3 Mutation

In this procedure, all individuals in the population are checked bit by bit and the bit values are randomly reversed according to a specified rate. Unlike crossover, this is a monadic operation. That is, a child string is produced from a single parent string. The mutation operator forces the algorithm to search new areas. Eventually, it helps the GA to avoid premature convergence and find the global optimal solution. An example is given in Figure 1.9.

1.6.3.4 Inversion

This operator is employed for a group of problems, such as the cell placement problem, layout problem, and travelling salesman problem. It also operates on one individual at a time. Two points are randomly selected from an individual, and the part of the string between those two points is reversed (see Figure 1.10).

1.6.3.5 Control Parameters

Important control parameters of a simple GA include the population size (number of individuals in the population), crossover rate, mutation rate, and inversion rate. Several researchers have studied the effect of these parameters on the performance a GA [Schaffer et al., 1989; Grefenstette, 1986; Fogarty, 1989]. The main conclusions are as follows. A large population size means the simultaneous handling of many solutions and increases the computation time per iteration; however, since many samples from the search space are used, the probability of convergence to a global optimal solution is higher than with a small population size.

The crossover rate determines the frequency of the crossover operation. It is useful at the start of optimisation to discover promising regions in the search space. A low crossover frequency decreases the speed of convergence to such areas. If the frequency is too high, it can lead to saturation around one solution. The mutation operation is controlled by the mutation rate. A high mutation rate introduces high diversity in the population and might cause instability. On the other hand, it is usually very difficult for a GA to find a global optimal solution with too low a mutation rate.

1.6.3.6 Fitness Evaluation Function

The fitness evaluation unit in a GA acts as an interface between the GA and the optimisation problem. The GA assesses solutions for their quality according to the information produced by this unit and not by directly using information about their structure. In engineering design problems, functional requirements are specified to the designer who has to produce a structure that performs the desired functions within predetermined constraints. The quality of a proposed solution is usually calculated depending on how well the solution performs the desired functions and satisfies the given constraints. In the case of a GA, this calculation must be automatic, and the problem is how to devise a procedure that computes the quality of solutions.

Fitness evaluation functions might be complex or simple depending on the optimisation problem at hand. Where a mathematical equation cannot be formulated for this task, a rule-based procedure can be constructed for use as a fitness function or in some cases both can be combined. Where some constraints are very important and cannot be violated, the structures or solutions that do so can be eliminated in advance by appropriately designing the representation scheme. Alternatively, they can be given low probabilities by using special penalty functions.

For further information on genetic algorithms, see [Holland, 1975; Goldberg, 1989; Davis, 1991; Pham and Karaboga, 2000].

Genetic algorithms have found applications in engineering problems involving complex combinatorial or multiparameter optimisation. Some recent examples of those applications follow:

- Configuring transmission systems [Pham and Yang, 1993]
- Generating hardware description language programs for high-level specification of the function of programmable logic devices [Seals and Whapshott, 1994]
- Designing the knowledge base of fuzzy logic controllers [Pham and Karaboga, 1994]
- Planning collision-free paths for mobile and redundant robots [Ashiru et al., 1995; Wilde and Shellwat, 1997; Nearchou and Aspragathos, 1997]
- Scheduling the operations of a job shop [Cho et al., 1996; Drake and Choudhry, 1997]

1.7 Some Applications in Engineering and Manufacture

This section briefly reviews five engineering applications of the aforementioned computational intelligence tools.

1.7.1 Expert Statistical Process Control

Statistical process control (SPC) is a technique for improving the quality of processes and products through closely monitoring data collected from those processes and products and using statistically based tools such as control charts.

XPC is an expert system for facilitating and enhancing the implementation of statistical process control [Pham and Oztemel, 1996]. A commercially available shell was employed to build XPC. The shell allows a hybrid rule-based and pseudo object-oriented method of representing the standard SPC knowledge and process-specific diagnostic knowledge embedded in XPC. The amount of knowledge involved is extensive, which justifies the adoption of a knowledge-based systems approach.

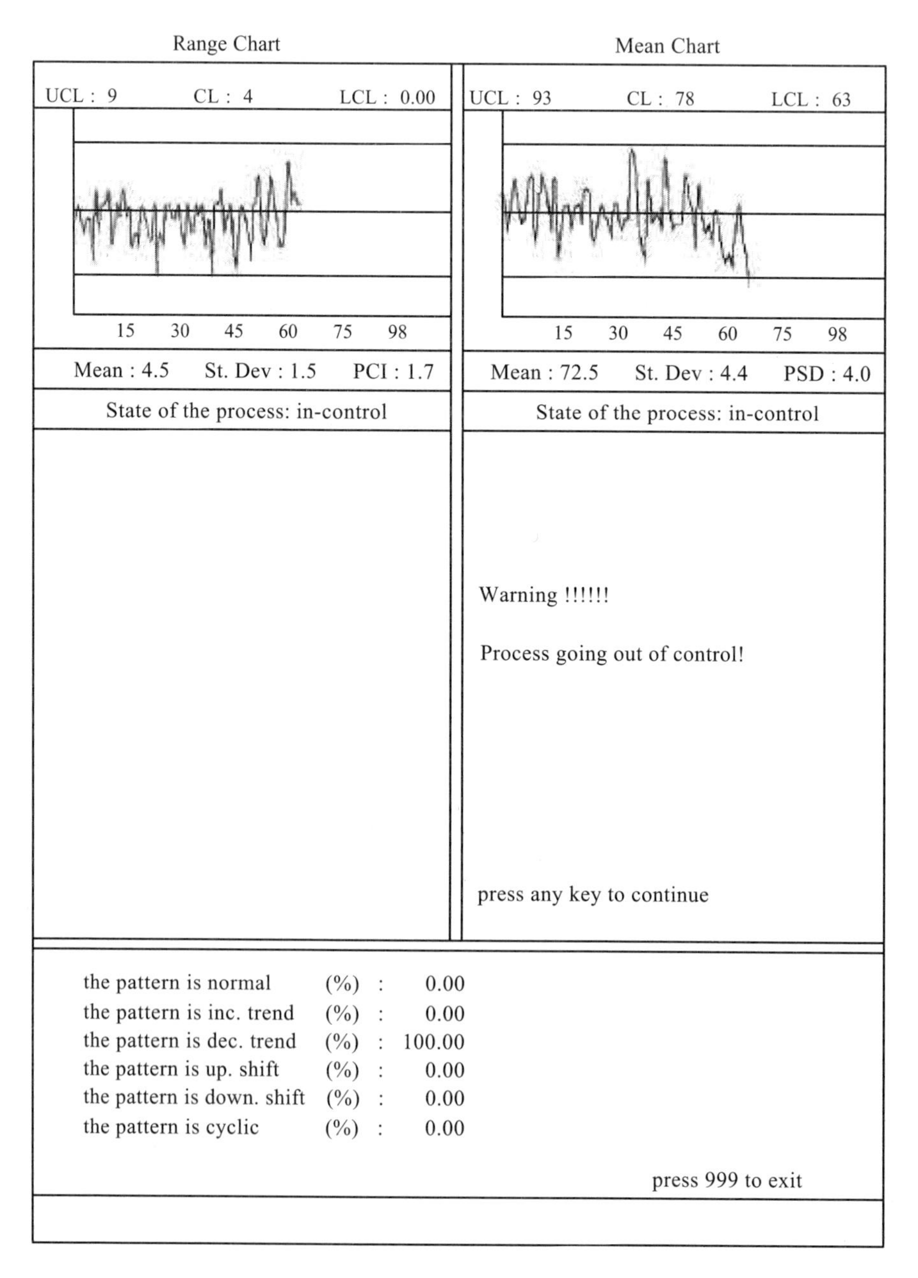

FIGURE 1.11 XPC output screen.

XPC comprises four main modules. The *construction module* is used to set up a control chart. The *capability analysis module* is for calculating process capability indices. The *on-line interpretation and diagnosis module* assesses whether the process is in control and determines the causes for possible out-of-control situations. It also provides advice on how to remedy such situations. The *modification module* updates the parameters of a control chart to maintain true control over a time-varying process. XPC has been applied to the control of temperature in an injection molding machine producing rubber seals. It has recently been enhanced by integrating a neural network module with the expert system modules to detect abnormal patterns in the control chart (see Figure 1.11).

1.7.2 Fuzzy Modelling of a Vibratory Sensor for Part Location

Figure 1.12 shows a six-degree-of-freedom vibratory sensor for determining the coordinates of the centre of mass (x_G, y_G) and orientation γ of bulky rigid parts. The sensor is designed to enable a robot to pick up parts accurately for machine feeding or assembly tasks. The sensor consists of a rigid platform (P) mounted on a flexible column (C). The platform supports one object (O) to be located at a time. O is held firmly with respect to P. The static deflections of C under the weight of O and the natural frequencies of vibration of the dynamic system comprising O, P, and C are measured and processed using a mathematical model of the system to determine x_G, y_G, and γ for O. In practice, the frequency measurements have low repeatability, which leads to inconsistent location information. The problem worsens when γ is in the region 80° to 90° relative to a reference axis of the sensor, because the mathematical model becomes ill-conditioned. In this "ill-conditioning" region, an alternative to using a mathematical model to compute γ is to adopt an experimentally derived fuzzy model. Such a fuzzy model has to be obtained for each specific object through calibration.

A possible calibration procedure involves placing the object at different positions (x_G, y_G) and orientations γ and recording the periods of vibration T of the sensor. Following calibration, fuzzy rules relating x_G, y_G and T to γ could be constructed to form a fuzzy model of the behaviour of the sensor for the given object. A simpler fuzzy model is achieved by observing that x_G and y_G only affect the reference level of T and, if x_G and y_G are employed to define that level, the trend in the relationship between T and γ is the same regardless of the position of the object. Thus, a simplified fuzzy model of the sensor consists of rules such as "If ($T-T_{ref}$) is small Then ($\gamma-\gamma_{ref}$) is small" where T_{ref} is the value of T when the object is at position (x_G, y_G) and orientation γ_{ref}. γ_{ref} could be chosen as 80°, the point at which the fuzzy model is to replace the mathematical model. T_{ref} could be either measured experimentally or computed from the mathematical model. To counteract the effects of the poor repeatability of period measurements, which are particularly noticeable in the "ill-conditioning" region, the fuzzy rules are modified so that they take into account the variance in T. An example of a modified fuzzy rule is "If ($T-T_{ref}$) is small and σ_T is small, Then ($\gamma-\gamma_{ref}$) is small." In this rule, σ_T denotes the standard deviation in the measurement of T.

Fuzzy modelling of the vibratory sensor is detailed in Pham and Hafeez (1992). Using a fuzzy model, the orientation γ can be determined to ±2° accuracy in the region 80° to 90°. The adoption of fuzzy logic in this application has produced a compact and transparent model from a large amount of noisy experimental data.

1.7.3 Induction of Feature Recognition Rules in a Geometric Reasoning System for Analysing 3D Assembly Models

Pham et al. (1999) have described a concurrent engineering approach involving generating assembly strategies for a product directly from its 3D CAD model. A feature-based CAD system is used to create assembly models of products. A geometric reasoning module extracts assembly oriented data for a product from the CAD system after creating a virtual assembly tree that identifies the components and subassemblies making up the given product (Figure 1.13(a)). The assembly information extracted by the module includes placement constraints and dimensions used to specify the relevant position of a given component or subassembly; geometric entities (edges, surfaces, etc.) used to constrain the component or subassembly; and the parents and children of each entity employed as a placement constraint. An example of the information extracted is shown in Figure 1.13(b).

Feature recognition is applied to the extracted information to identify each feature used to constrain a component or subassembly. The rule-based feature recognition process has three possible outcomes

1. The feature is recognised as belonging to a unique class.
2. The feature shares attributes with more than one class (see Figure 1.13(c)).
3. The feature does not belong to any known class.

Cases 2 and 3 require the user to decide the correct class of the feature and the rule base to be updated. The updating is implemented via a rule induction program. The program employs RULES-3 Plus, which

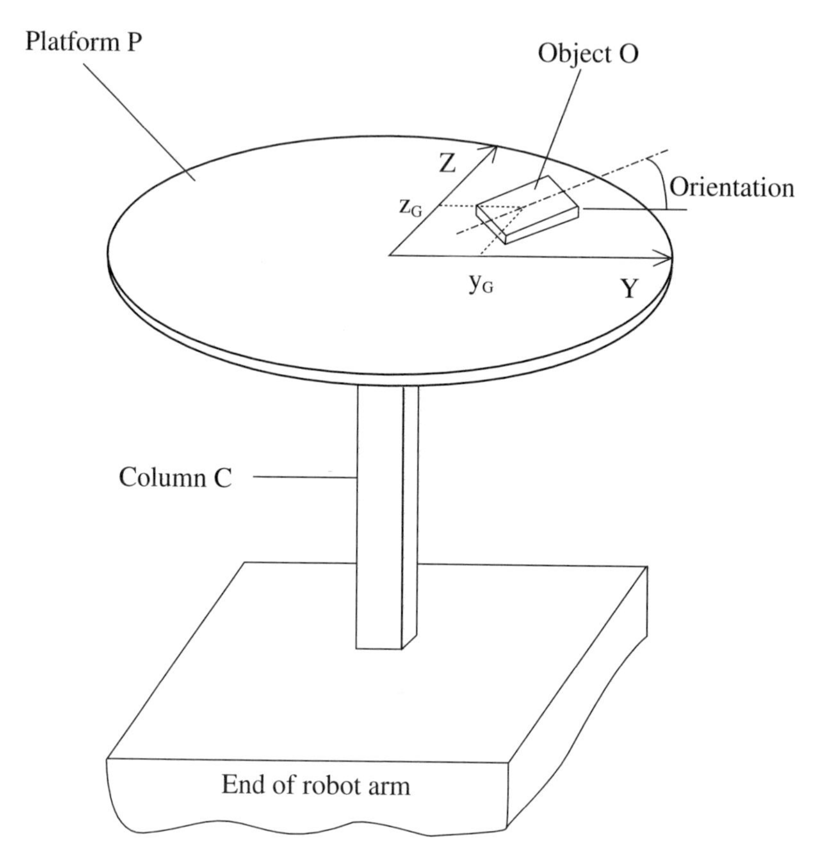

FIGURE 1.12 Schematic diagram of a vibratory sensor mounted on a robot wrist.

automatically extracts new feature recognition rules from examples provided to it in the form of characteristic vectors representing different features and their respective class labels.

Rule induction is very suitable for this application because of the complexity of the characteristic vectors and the difficulty of defining feature classes manually.

1.7.4 Neural-Network-Based Automotive Product Inspection

Figure 1.14 depicts an intelligent inspection system for engine valve stem seals [Pham and Oztemel, 1996]. The system comprises four CCD cameras connected to a computer that implements neural-network-based algorithms for detecting and classifying defects in the seal lips. Faults on the lip aperture are classified by a multilayer perceptron. The inputs to the network are a 20-component vector, where the value of each component is the number of times a particular geometric feature is found on the aperture being inspected. The outputs of the network indicate the type of defect on the seal lip aperture. A similar neural network is used to classify defects on the seal lip surface. The accuracy of defect classification in both perimeter and surface inspection is in excess of 80%. Note that this figure is not the same as that for the accuracy in detecting defective seals, that is, differentiating between good and defective seals. The latter task is also implemented using a neural network which achieves an accuracy of almost 100%. Neural networks are necessary for this application because of the difficulty of describing precisely the various types of defects and the differences between good and defective seals. The neural networks are able to learn the classification task automatically from examples.

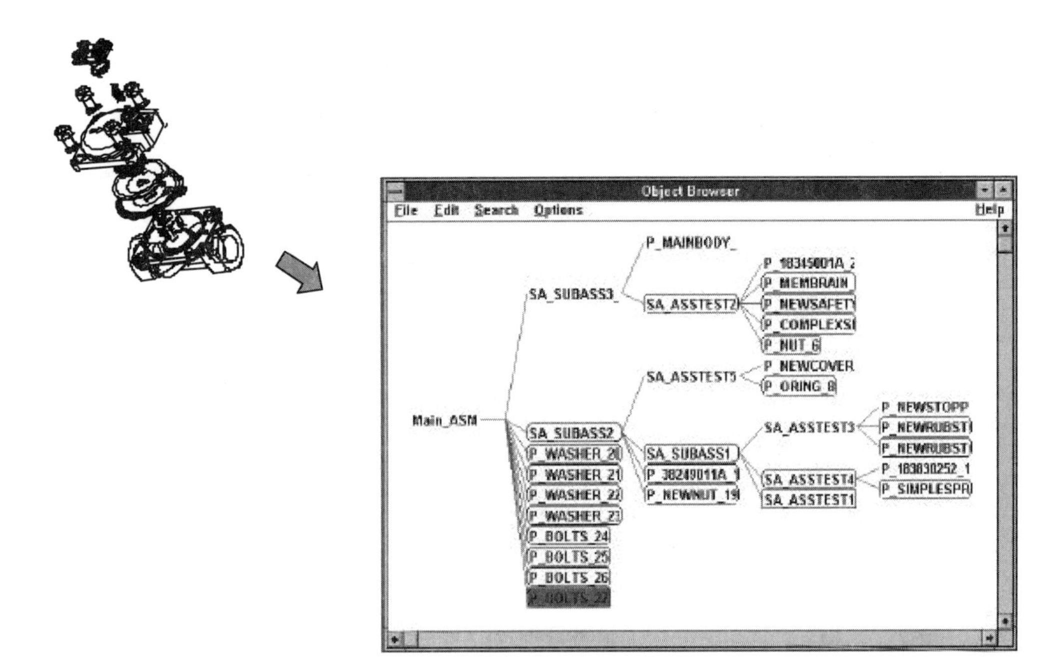

FIGURE 1.13(a) An assembly model.

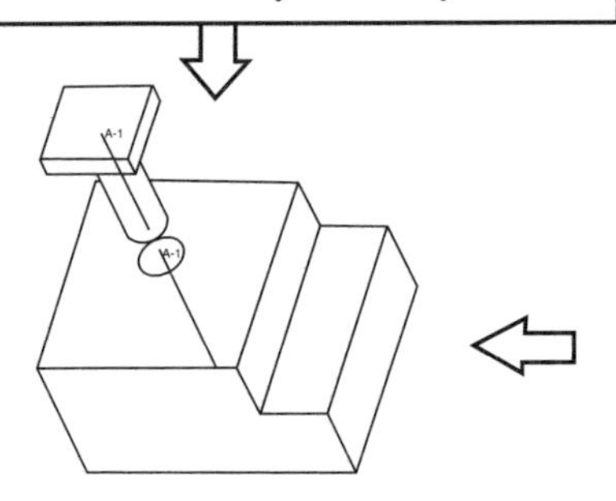

FIGURE 1.13(b) An example of assembly information.

1.7.5 GA-Based Conceptual Design

TRADES is a system using GA techniques to produce conceptual designs of transmission units [Pham and Yang, 1992]. The system has a set of basic building blocks, such as gear pairs, belt drives, and mechanical linkages, and generates conceptual designs to satisfy given specifications by assembling the building blocks into different configurations. The crossover, mutation, and inversion operators of the

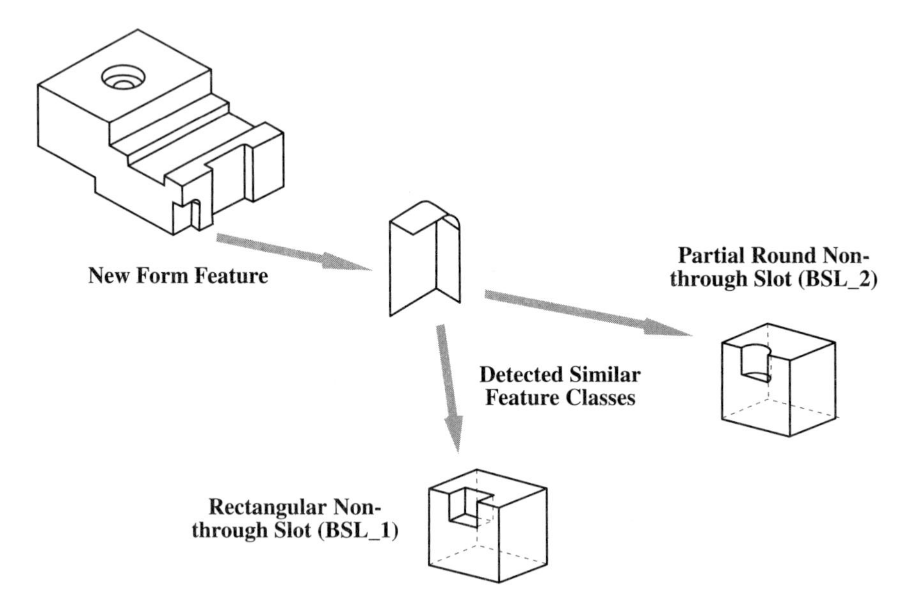

FIGURE 1.13(c) An example of feature recognition.

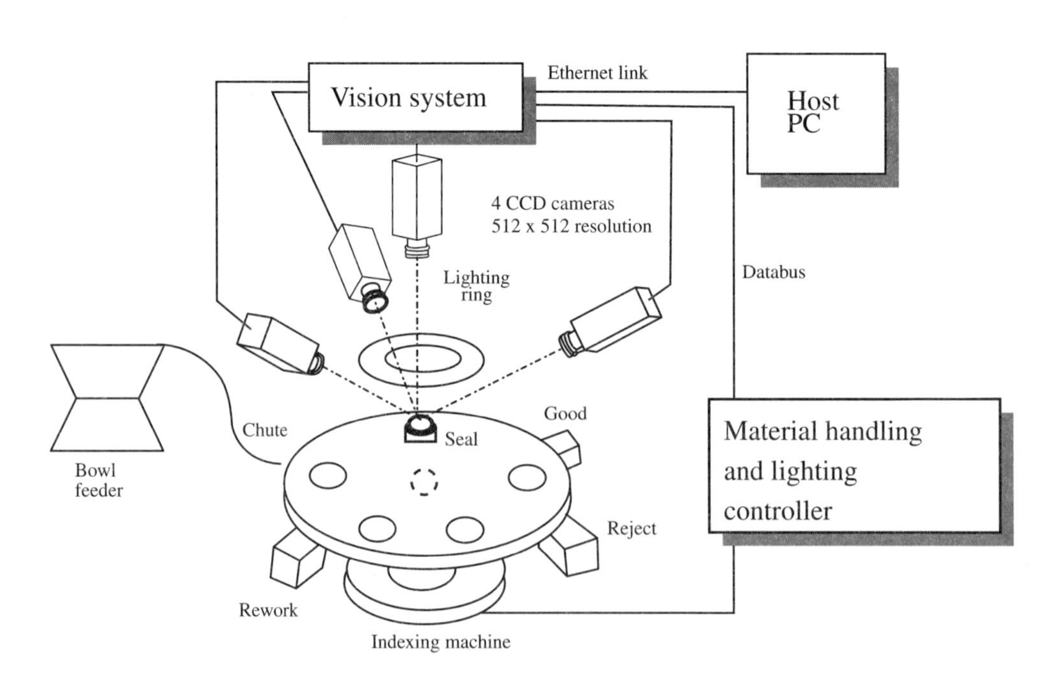

FIGURE 1.14 Valve stem seal inspection system.

GA are employed to create new configurations from an existing population of configurations. Configurations are evaluated for their compliance with the design specifications. Potential solutions should provide the required speed reduction ratio and motion transformation while not containing incompatible building blocks or exceeding specified limits on the number of building blocks to be adopted. A fitness function codifies the degree of compliance of each configuration. The maximum fitness value is assigned to configurations that satisfy all functional requirements without violating any constraints. As in a standard GA, information concerning the fitness of solutions is employed to select solutions for reproduction, thus guiding the process toward increasingly fitter designs as the population evolves. In addition to the usual GA operators, TRADES incorporates new operators to avert premature convergence to nonoptimal solutions and facilitate the generation of a variety of design concepts. Essentially, these operators reduce the chances of any one configuration or family of configurations dominating the solution population by avoiding crowding around very fit configurations and preventing multiple copies of a configuration, particularly after it has been identified as a potential solution.

TRADES is able to produce design concepts from building blocks without requiring much additional *a priori* knowledge. The manipulation of the building blocks to generate new concepts is carried out by the GA in a stochastic but guided manner. This enables good conceptual designs to be found without the need to search the design space exhaustively.

Due to the very large size of the design space and the quasi random operation of the GA, novel solutions not immediately evident to a human designer are sometimes generated by TRADES. On the other hand, impractical configurations could also arise. TRADES incorporates a number of heuristics to filter out such design proposals.

1.8 Conclusion

Over the past 40 years, computational intelligence has produced a number of powerful tools. This chapter has reviewed five of those tools, namely, knowledge-based systems, fuzzy logic, inductive learning, neural networks, and genetic algorithms. Applications of the tools in engineering and manufacture have become more widespread due to the power and affordability of present-day computers. It is anticipated that many new applications will emerge and that, for demanding tasks, greater use will be made of hybrid tools combining the strengths of two or more of the tools reviewed here [Medsker, 1995]. Other technological developments in computational intelligence that will have an impact in engineering include data mining, or the extraction of information and knowledge from large databases [Limb and Meggs, 1994], and multi-agent systems, or distributed self-organising systems employing entities that function autonomously in an unpredictable environment concurrently with other entities and processes [Wooldridge and Jennings, 1994; Rzevski, 1995; Márkus et al., 1996; Tharumarajah et al., 1996; Bento and Feijó, 1997]. The appropriate deployment of these new computational intelligence tools and of the tools presented in this chapter will contribute to the creation of more competitive engineering systems.

Acknowledgments

This review was carried out as part of the *Innovation in Manufacturing* and *Innovative Technologies for Effective Enterprises* projects supported by the European Regional Development Fund. The Fund is administered by the Welsh Assembly, the Welsh Office, and the Welsh European Programme Executive for the European Commission. Information for the review was also obtained from work on the *INFOMAN* project and the *CONFLOW* project funded by the European Commission under the ESPRIT and INCO-COPERNICUS Programmes.

References

Ashiru I., Czanecki C. and Routen T., (1995), Intelligent operators and optimal genetic-based path planning for mobile robots, *Proc. Int. Conf. Recent Advances in Mechatronics*, Istanbul, Turkey, August, 1018-1023.

Baker J. E., (1985), Adaptive selection methods for genetic algorithms, *Proc. First Int. Conf. Genetic Algorithms and Their Applications*, Pittsburgh, PA, 101-111.

Bas K. and Erkmen A. M., (1995), Fuzzy preshape and reshape control of Anthrobot-III 5-fingered robot hand, *Proc. Int. Conf. Recent Advances in Mechatronics*, Istanbul, Turkey, August, 673-677.

Bento J. and Feijó B., (1997), An agent-based paradigm for building intelligent CAD systems, *Artificial Intelligence in Engineering*, 11(3), 231-244.

Bigand A., Goureau P. and Kalemkarian J., (1994), Fuzzy control of a welding process, *Proc. IMACS Int. Symp. Signal Processing, Robotics and Neural Networks (SPRANN 94),* Villeneuve d'Ascq, France, April, 379-342.

Bose A., Gini M. and Riley D., (1997), A case-based approach to planar linkage design, *Artificial Intelligence in Engineering,* 11(2), 107-119.

Cho B. J., Hong S. C. and Okoma S., (1996), Job shop scheduling using genetic algorithm, *Proc. 3rd World Congress Expert Systems,* Seoul, Korea, February, 351-358.

Da Rocha Fernandes A. M. and Cid Bastos R., (1996), Fuzzy expert systems for qualitative analysis of minerals, *Proc. 3rd World Congress Expert Systems,* Seoul, Korea, February, 673-680.

Davis L., (1991), *Handbook of Genetic Algorithms,* Van Nostrand, New York.

Drake P. R. and Choudhry I. A., (1997), From apes to schedules, *Manufacturing Engineer,* 76 (1), 43-45.

Durkin J., (1994), *Expert Systems Design and Development,* Macmillan, New York.

Ferreiro Garcia R., (1994), FAM rule as basis for poles shifting applied to the roll control of an aircraft, *Proc. IMACS Int. Symp. Signal Processing, Robotics and Neural Networks (SPRANN 94),* 375-378.

Fogarty T. C., (1989), Varying the probability of mutation in the genetic algorithm, *Proc. Third Int. Conf. on Genetic Algorithms and Their Applications,* George Mason University, 104-109.

Goldberg D. E., (1989), *Genetic Algorithms in Search, Optimisation and Machine Learning,* Addison-Wesley, Reading, MA.

Grefenstette J. J., (1986), Optimization of control parameters for genetic algorithms, *IEEE Trans. on Systems, Man and Cybernetics,* 16(1), 122-128.

Hassoun M. H., (1995), *Fundamentals of Artificial Neural Networks,* MIT Press, Cambridge, MA.

Holland J. H., (1975), *Adaptation in Natural and Artificial Systems,* The University of Michigan Press, Ann Arbor, MI.

Jackson P., (1999), *Introduction to Expert Systems,* 3rd ed., Addison-Wesley, Harlow, Essex, U.K.

Jambunathan K., Fontama V. N., Hartle S. L. and Ashforth-Frost S., (1997), Using ART 2 networks to deduce flow velocities, *Artificial Intelligence in Engineering,* 11(2), 135-141.

Kaufmann A., (1975), *Introduction to the Theory of Fuzzy Subsets,* vol. 1, Academic Press, New York.

Koo D. Y. and Han S. H., (1996), Application of the configuration design methods to a design expert system for paper feeding mechanism, *Proc. 3rd World Congress on Expert Systems,* Seoul, Korea, February, 49-56.

Kostov A., Andrews B., Stein R. B., Popovic D. and Armstrong W. W., (1995), Machine learning in control of functional electrical stimulation systems for locomotion, *IEEE Trans. Biomedical Engineering,* 44(6), 541-551.

Lara Rosano F., Kemper Valverde N., De La Paz Alva C. and Alcántara Zavala J., (1996), Tutorial expert system for the design of energy cogeneration plants, *Proc. 3rd World Congress on Expert Systems,* Seoul, Korea, February, 300-305.

Limb P. R. and Meggs G. J., (1994), Data mining tools and techniques, *British Telecom Technology Journal,* 12(4), 32-41.

Luzeaux D., (1994), Process control and machine learning: rule-based incremental control, *IEEE Trans. Automatic Control,* 39(6), 1166-1171.

Majors M. D. and Richards R. J., (1995), A topologically-evolving neural network for robotic flexible assembly control, *Proc. Int. Conf. Recent Advances in Mechatronics,* Istanbul, Turkey, August, 894-899.

Márkus A., Kis T., Váncza J. and Monostori L., (1996), A market approach to holonic manufacturing, *CIRP Annals,* 45(1), 433-436.

Medsker L. R., (1995), *Hybrid Intelligent Systems,* Kluwer, Boston.

Michalewicz Z., (1992), *Genetic Algorithms + Data Structures = Evolution Programs,* Springer-Verlag, Berlin.

Michalski R. S., (1990), A theory and methodology of inductive learning, in *Readings in Machine Learning,* Eds. Shavlik, J. W. and Dietterich T. G., Kaufmann, San Mateo, CA, 70-95.

Muggleton S. (Ed.), (1992), *Inductive Logic Programming,* Academic Press, London.

Nearchou A. C. and Aspragathos N. A., (1997), A genetic path planning algorithm for redundant articulated robots, *Robotica,* 15(2), 213-224.

Peers S. M. C., Tang M. X. and Dharmavasan S., (1994), A knowledge-based scheduling system for offshore structure inspection, *Artificial Intelligence in Engineering IX (AIEng 9),* Eds. Rzevski G., Adey R. A. and Russell D. W., Computational Mechanics, Southampton, 181-188.

Pham D. T. and Aksoy M. S., (1994), An algorithm for automatic rule induction, *Artificial Intelligence in Engineering,* 8, 277-282.

Pham D. T. and Aksoy M. S., (1995a), RULES : A rule extraction system, *Expert Systems Applications,* 8, 59-65.

Pham D. T. and Aksoy M. S., (1995b), A new algorithm for inductive learning, *Journal of Systems Engineering,* 5, 115-122.

Pham D. T. and Dimov S. S. (1997), An efficient algorithm for automatic knowledge acquisition, *Pattern Recognition,* 30 (7), 1137-1143.

Pham D. T., Dimov S. S. and Setchi R. M. (1999), Concurrent engineering: a tool for collaborative working, *Human Systems Management,* 18, 213-224.

Pham D. T. and Hafeez K., (1992), Fuzzy qualitative model of a robot sensor for locating three-dimensional objects, *Robotica,* 10, 555-562.

Pham D. T. and Karaboga D., (1994), Some variable mutation rate strategies for genetic algorithms, *Proc. IMACS Int. Symp. Signal Processing, Robotics and Neural Networks (SPRANN 94),* 73-96.

Pham D. T. and Karaboga D., (2000), *Intelligent Optimisation Techniques: Genetic Algorithms, Tabu Search, Simulated Annealing and Neural Networks,* Springer-Verlag, London.

Pham D. T. and Liu X., (1999), *Neural Networks for Identification, Prediction and Control,* Springer-Verlag, London.

Pham D. T., Onder H. H. and Channon P. H., (1996), Ergonomic workplace layout using genetic algorithms, *J. Systems Engineering,* 6(1), 119-125.

Pham D. T. and Oztemel E., (1996), *Intelligent Quality Systems,* Springer-Verlag, London.

Pham D. T. and Pham P. T. N., (1988), Expert systems in mechanical and manufacturing engineering, Int. *J. Adv. Manufacturing Technol.,* special issue on knowledge based systems, 3(3), 3-21.

Pham D. T. and Yang Y., (1993), A genetic algorithm based preliminary design system, *Proc. IMechE, Part D: J Automobile Engineering,* 207, 127-133.

Price C. J., (1990), *Knowledge Engineering Toolkits,* Ellis Horwood, Chichester, U.K.

Quinlan J. R., (1983), Learning efficient classification procedures and their applications to chess end games, in *Machine Learning, An Artificial Intelligence Approach,* Eds. Michalski R. S., Carbonell J. G. and Mitchell T. M., Morgan Kaufmann, Palo Alto, CA, 463-482.

Rzevski G., (1995), Artificial intelligence in engineering: past, present and future, *Artificial Intelligence in Engineering X,* Eds. Rzevski G., Adey R. A. and Tasso C., Computational Mechatronics, Southampton, U.K., 3-16.

Schaffer J. D., Caruana R. A., Eshelman L. J. and Das R., (1989), A study of control parameters affecting on-line performance of genetic algorithms for function optimisation, *Proc. Third Int. Conf. on Genetic Algorithms and Their Applications,* George Mason University, 51-61.

Seals R. C. and Whapshott G. F., (1994), Design of HDL programmes for digital systems using genetic algorithms, *AI Eng 9* (ibid.), 331-338.

Shi Z. Z., Zhou H. and Wang J., (1997), Applying case-based reasoning to engine oil design, *Artificial Intelligence in Engineering,* 11(2), 167-172.

Skibniewski M., Arciszewski T. and Lueprasert K., (1997), Constructability analysis: machine learning approach, *ASCE J Computing Civil Eng.,* 12(1), 8-16.

Smith P., MacIntyre J. and Husein S., (1996), The application of neural networks in the power industry, *Proc. 3rd World Congress on Expert Systems,* Seoul, Korea, February, 321-326.

Szwarc D., Rajamani D. and Bector C. R., (1997), Cell formation considering fuzzy demand and machine capacity, *Int. J. Advanced Manufacturing Technology,* 13(2), 134-147.

Tarng Y. S., Tseng C. M. and Chung L. K., (1997), A fuzzy pulse discriminating systems for electrical discharge machining, *Int. J. Machine Tools Manufacture,* 37(4), 511-522.

Teti R. and Caprino G., (1994), Prediction of composite laminate residual strength based on a neural network approach, *AI Eng 9* (ibid), 81-88.

Tharumarajah A., Wells A. J. and Nemes L., (1996), Comparison of the bionic, fractal and holonic manufacturing system concepts, *Int. J. Computer Integrated Manfacturing,* 9(3), 217-226.

Wang L. X. and Mendel M., (1992), Generating fuzzy rules by learning from examples, *IEEE Trans. on Systems, Man and Cybernetics,* 22(6), 1414-1427.

Whitely D., (1989), The GENITOR algorithm and selection pressure: Why rank-based allocation of reproductive trials is best, *Proc. Third Int. Conf. on Genetic Algorithms and Their Applications,* George Mason University, 116-123.

Wilde P. and Shellwat H., (1997), Implementation of a genetic algorithm for routing an autonomous robot, *Robotica,* 15(2), 207-211.

Wooldridge M. J. and Jennings N. R., (1994), Agent theories, architectures and languages: a survey, *Proc. ECAI 94 Workshop on Agent Theories, Architectures and Languages,* Amsterdam, 1-32.

Yano H., Akashi T., Matsuoka N., Nakanishi K., Takata O. and Horinouchi N., (1997), An expert system to assist automatic remeshing in rigid plastic analysis, *Toyota Technical Review,* 46(2), 87-92.

Zadeh L. A., (1965), Fuzzy sets, *Information Control,* 8, 338-353.

Zimmermann H. J., (1991), *Fuzzy Set Theory and Its Applications,* 2nd ed., Kluwer, Boston.

2

Neural Network Applications in Intelligent Manufacturing: An Updated Survey

Jun Wang
The Chinese University of Hong Kong

Wai Sum Tang
The Chinese University of Hong Kong

Catherine Roze
IBM Global Services

Abstract

In recent years, artificial neural networks have been applied to solve a variety of problems in numerous areas of manufacturing at both system and process levels. The manufacturing applications of neural networks comprise the design of manufacturing systems (including part-family and machine-cell formation for cellular manufacturing systems); modeling, planning, and scheduling of manufacturing processes; monitoring and control of manufacturing processes; quality control, quality assurance, and fault diagnosis. This paper presents a survey of existing neural network applications to intelligent manufacturing. Covering the whole spectrum of neural network applications to manufacturing, this chapter provides a comprehensive review of the state of the art in recent literature.

2.1 Introduction

Neural networks are composed of many massively connected simple neurons. Resembling more or less their biological counterparts in structure, artificial neural networks are representational and computational models processing information in a parallel distributed fashion. Feedforward neural networks and recurrent neural networks are two major classes of artificial neural networks. Feedforward neural networks,

0-8493-0592-6/01/$0.00+$.50

such as the popular multilayer perceptron, are usually used as representational models trained using a learning rule based on a set of input–output sample data. A popular learning rule is the widely used backpropagation (BP) algorithm (also known as the generalized delta rule). It has been proved that the multilayer feedforward neural networks are universal approximators. It has also been demonstrated that neural networks trained with a limited number of training samples possess a good generalization capability. Large-scale systems that contain a large number of variables and complex systems where little analytical knowledge is available are good candidates for the applications of feedforward neural networks. Recurrent neural networks, such as the Hopfield networks, are usually used as computational models for solving computationally intensive problems. Typical examples of recurrent neural network applications include NP-complete combinatorial optimization problems and large-scale or real-time computation tasks. Neural networks are advantageous over traditional approaches for solving such problems because neural information processing is inherently concurrent.

In the past two decades, neural network research has expanded rapidly. On one hand, advances in theory and methodology have overcome many obstacles that hindered the neural network research a few decades ago. On the other hand, artificial neural networks have been applied to numerous areas. Neural networks offer advantages over conventional techniques for problem-solving in terms of robustness, fault tolerance, processing speed, self-learning, and self-organization. These desirable features of neural computation make neural networks attractive for solving complex problems. Neural networks can find applications for new solutions or as alternatives of existing methods in manufacturing. Application areas of neural networks include, but are not limited to, associative memory, system modeling, mathematical programming, combinatorial optimization, process and robotic control, pattern classification and recognition, and design and planning.

In recent years, the applications of artificial neural networks to intelligent manufacturing have attracted ever-increasing interest from the industrial sector as well as the research community. The success in utilizing artificial neural networks for solving various computationally difficult problems has inspired renewed research in this direction. Neural networks have been applied to a variety of areas of manufacturing from the design of manufacturing systems to the control of manufacturing processes. One top-down classification of neural network applications to intelligent manufacturing, as shown in Figure 2.1, results in four main categories without clearly cut boundaries: (1) modeling and design of manufacturing systems, including machine-cell and part-family formation for cellular manufacturing systems; (2) modeling, planning, and scheduling of manufacturing processes; (3) monitoring and control of manufacturing processes; (4) quality control, quality assurance, and fault diagnosis. The applications of neural networks to manufacturing have shown promising results and will possibly have a major impact on manufacturing in the future [1, 2].

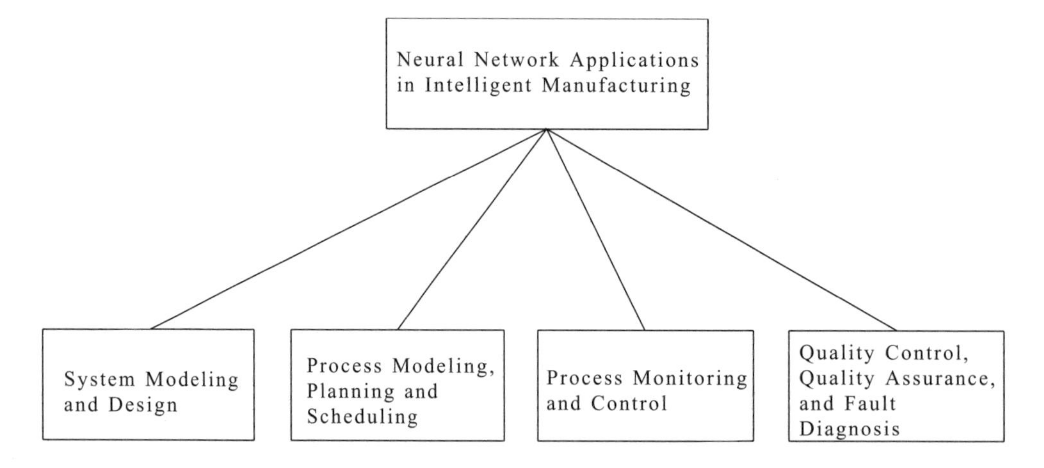

FIGURE 2.1 Hierarchy of neural network applications in intelligent manufacturing.

This chapter provides a comprehensive survey of recent neural network applications in intelligent manufacturing based on the aforementioned categorization. The aim of the chapter is to review the state of the art of the research and highlight the recent advances in research and applications of neural networks in manufacturing. Because of the vast volume of publications, this chapter considers only the works published in major archival journals and selected edited books.

2.2 Modeling and Design of Manufacturing Systems

As representational models, artificial neural networks are particularly useful for modeling systems whose underlying properties are too complex, too obscure, too costly, or too time-consuming to be modeled analytically using traditional methods. The use of neural networks for modeling and design of manufacturing systems includes manufacturing decision making, product design storage and retrieval in group technology, and formation of part families and machine cells for the design of cellular manufacturing systems.

Chryssolouris et al. [3] applied neural networks, in conjunction with simulation models, for resource allocation in job-shop manufacturing systems. Feedforward neural networks called multilayer perceptrons trained using the popular backpropagation (BP) algorithm were used to learn the inverse mapping of the simulation task: given desired performance measure levels, the neural networks output suitable values for the parameters of resources. Based on results generated by a simulator, the neural networks were demonstrated to be able to find a suitable allocation for the resources to achieve given performance levels. In a related work, Chryssolouris et al. [4] applied neural networks, also in conjunction with simulation models, to determine operational policies for hierarchical manufacturing systems under a multiple criteria decision making framework called MAnufacturing DEcision MAking (MADEMA). Multilayer perceptrons were used to generate appropriate criterion weights for an entire sequence of multiple criteria decisions on manufacturing policies. This neural network approach is more appropriate for complex applications entailing chains of decisions, such as job-shop scheduling, whereas conventional methods are preferable for single or isolated decisions. Madey et al. [5] used a neural network embeded in a general-purpose simulation system for modeling Continuous Improvement Systems (CIS) policies in manufacturing systems. A multilayer feedforward neural network trained using the BP algorithm was used to facilitate the identification of an effective CIS policy and to provide a realistic simulation framework to enhance the capabilities of simulations. The trained neural network was embedded in the simulation model code, so that the model had intrinsic advisory capability to reduce time or complexity for linking with external software. The results demonstrated not only the feasibility, but also the promising effectiveness of the combination of neural computation within simulation models for improving CIS analysis.

The crux behind group technology (GT) is to group similar parts that share common design and/or manufacturing features into part families and bring dissimilar machines together and dedicate them to the manufacture of one or more part families. GT is an important step toward the reduction of throughput time, work-in-process inventory, investment in material handling, and setup time, thus resulting in an increase of productivity vital to survive in an increasingly competitive environment and changing customer preferences. The success of GT implementation depends largely on how the part families are formed and how machines are grouped. Numerous methods exist to solve the GT problem, each with its own limitations. As alternatives, neural networks have been proposed to provide solutions to the GT problem.

Kamarthi et al. [6] used a multilayer perceptron as an associative memory for storage and retrieval of design data in group technology. Design data in the gray-level pixel representations of design drawings were stored in the neural associative memory. The simulation results reported in this paper showed that the neural network trained using the BP algorithm was able to generate the closest stored part given the geometric characteristics of new parts. The fault tolerance capability of neural networks is particularly instrumental for cases where only partial or inexact information is available. The neural network approach is useful for the standardization of product design and process planning. A weakness of the proposed approach is the lack of ability for translation, scale, and rotation invariant recognition of parts, which are essential for handling part drawings.

In Kaparthi and Suresh's work [7], a multilayer feedforward neural network trained with the BP algorithm was employed to automate the classification and coding of parts for GT applications. Given the pixel representation of a part drawing extracted from computer-aided design (CAD) systems, the neural network was able to output the Opitz codes related to the part geometric information. The work is not limited to rotational parts and may be used for nonrotational parts. Nevertheless, code generation based on features other than shapes (e.g., material type) would require the neural network to be supplemented with other algorithms/procedures.

Moon and Roy [8] introduced a neural network approach to automating part-family classification in conjunction with a feature-based solid modeling system. The part features extracted from a model or object database were used to train and test a multilayer feedforward neural network. Trained using the BP algorithm, the neural network neurons signify an appropriate part family for each part. Besides overcoming some limitations of traditional coding and classification methods, this approach offers more flexibility and faster response.

Venugopal and Narendran [9] applied the Hopfield network to design storage and retrieval for batch manufacturing systems. Binary matrix representations of parts based on geometric shapes were stored in the Hopfield network. Test cases carried out on rotational and nonrotational parts showed the high percentage of correct retrieval of stored part information using the neural network. The retrieval rapidity is another major advantage of the neural network model. Such a storage/retrieval system could benefit the design process by minimizing duplications and variety, thus increasing productivity of both designer and planner, aiding standardization, and indirectly facilitating quotations. Furthermore, this approach offers flexibility and could adjust to changes in products. Unfortunately, the limited capacity of the Hopfield network constrained the possible number of stored designs.

Chakraborty and Roy [10] applied neural networks to part-family classification based on part geometric information. The neural system consisted of two neural networks: a Kohonen's SOM network and a multilayer feedforward network trained using the BP algorithm. The former was used to cluster parts into families and provide data to train the latter to learn part-family relationships. Given data not contained in the training set, the feedforward neural network performed well with an accuracy of 100% in most of test cases.

Kiang et al. [11] used the self-organizing map (SOM) network for part-family grouping according to the operation sequence. An operation sequence based similarity coefficient matrix developed by the authors was constructed and used as the input to the SOM network, which clustered the parts into different families subsequently. The performance of the SOM network approach was compared with two other clustering techniques, the k-th nearest neighbor (KNN) and the single linkage (SLINK) clustering methods for problems varying from 19 to 200 parts. The SOM-network-based method was shown to cluster the parts more uniformly in terms of number of parts in each family, especially for large data set. The training time for the SOM network was very time-consuming, though the trained network can perform clustering in very short time.

Wu and Jen [12] presented a neural-network-based part classification system to facilitate the retrieving and reviewing similar parts from the part database. Each part was represented by its three projection views in the form of rectilinear polygons. Every polygon was encoded into a feature vector using the skeleton standard tree method, which was clustered to a six-digit polygon code by a feedforward neural network trained by the BP algorithm. By comparing the polygon codes, parts can be grouped hierarchically into three levels of similarity. For parts with all three identical polygon codes, they were grouped into a high degree similarity family. For parts shared one identical polygon code, they were grouped into a low degree similarity family. The rest of the parts were put into a medium degree similarity family. Searching from the low degree of similarity family to the high degree of similarity family would help designers to characterize a vague design.

Based on the interactive activation and competitive network model, Moon [13] developed a competitive neural network for grouping machine cells and part families. This neural network consists of three layers of neurons. Two layers correspond respectively to the machines (called machine-type pool) and parts (called part-type pool), and one hidden layer serves as a buffer between the machine-type pool and part-type pool. Similarity coefficients of machines and parts are used to form the connection weights of the neural network. One desirable feature of the competitive neural network, among others, is that it can group machine cells and part families simultaneously. In a related work, Moon [14] showed that a competitive neural network was able to identify natural groupings of part and machine into families and cells rather than forcing them. Besides routing information, design similarities such as shapes, dimensions, and tolerances can be incorporated into the same framework. Even fuzziness could be represented, by using variable connection weights. Extending the results in [13, 14], Moon and Chi [15] used the competitive neural network developed earlier for both standard and generalized part-family formation. The neural network based on Jaccard similarity coefficients is able to find near-optimal solutions with a large set of constraints. This neural network takes into account operations sequence, lot size, and multiple process plans. This approach proved to be highly flexible in satisfying various requirements and efficient for integration with other manufacturing functions. Currie [16] also used the interactive activation and competition neural network for grouping part families and machines cells. This neural network was used to define a similarity index of the pairwise comparison of parts based on various design and manufacturing characteristics. Part families were created using a bond energy algorithm to partition the matrix of part similarities. Machine cells were simply inferred from part families. The neural network simulated using a spreadsheet macro showed to be capable of forming part families.

Based on the ART-1 neural network, Kusiak and Chung [17] developed a neural network model called GT/ART for solving GT problems by block diagonalizing machine-part incidence matrices. This work showed that the GT/ART neural network is more suitable for grouping machine cells and part families than other nonlearning algorithms and other neural networks such as multilayer neural networks with the BP learning algorithm. The GT/ART model allows learning new patterns and keeping existing weights stable (plasticity vs. stability) at the same time. Kaparthi and Suresh [18] applied the ART-1 neural network for clustering part families and machine cells. A salient feature of this approach is that the entire part-machine incidence matrix is not stored in memory, since only one row is processed at a time. The speed of computation and simplicity of the model offered a reduction in computational complexity together with the ability to handle large industrial size problems. The neural network was tested using two sets of data, one set from the literature and the other artificially generated to simulate industrial size data. Further research is required to investigate and enhance the performance of this neural network in the case of imperfect data (in the presence of exceptional elements).

Liao and Chen [19] evaluated the ART-1 network for part-family and machine-cell formation. The ART-1 network was integrated with a feature-based CAD system to automate GT coding and part-family formation. The process involves a three-stage procedure, with the objective of minimizing operating and material handling costs. The first stage involved an integer programming model to determine the best part routing in order to minimize operating costs. The first stage results in a binary machine-part incidence matrix. In the second stage, the resulting incidence matrix is then input to an ART-1 network that generates machine cells. In the last stage, the STORM plant layout model, an implementation of a modified steepest descent pairwise interchange method is used to determine the optimal layout. The limitation of the approach was that the ART-1 network needs an evaluation module to determine the number of part families and machine cells.

Extending their work in [18], Kaparthi et al. [20] developed a robust clustering algorithm based on a modified ART-1 neural network. They showed that modifying the ART-1 neural network can improve the clustering performance significantly, by reversing zeros and ones in incidence matrices. Three perfectly block diagonalizable incidence matrices were used to test the modified neural network. Further research is needed to investigate the performance of this modified neural network using incidence matrices that result in exceptional elements.

Moon and Kao [21] developed a modified ART-1 neural network for the automatic creation of new part families during a part classification process. Part families were generated in a multiphase procedure interfaced with a customized coding system given part features. Such an approach to GT allows to maintain consistency throughout a GT implementation and to perform the formation and classification processes concurrently.

Dagli and Huggahalli [22] pointed out the limitations of the basic ART-1 paradigm in cell formation and proposed a modification to make the performance more stable. The ART-1 paradigm was integrated with a decision support system that performed cost/performance analysis to arrive at an optimal solution. It was shown that with the original ART-1 paradigm the classification depends largely on order of presentation of the input vectors. Also, a deficient learning policy gradually causes a reduction in the responsibility of patterns, thus leading to a certain degree of inappropriate classification and a large number of groups than necessary. These problems can be attributed to the high sensitivity of the paradigm to the heuristically chosen degree of similarity among parts. These problems can be solved by reducing the sensitivity of the network through applying the input vectors in the order of decreasing density (measured by the number of 1's in the vector) and through retaining only the vector with the greatest density as the representative patterns. The proposed modifications significantly improved the correctness of classification.

Moon [23] took into account various practical factors encountered in manufacturing companies, including sequence of operations, lot size, and the possibility of multiple process plans. A neural network trained with the BP algorithm was proposed to automate the formation of new family during the classification process. The input patterns were formed using a customized feature-based coding system. The same model could easily be adapted to take more manufacturing information into consideration.

Rao and Gu [24] combined an ART neural with an expert system for clustering machine cells in cellular manufacturing. This hybrid system helps a cell designer in deciding on the number and type of duplicate machines and resultant exceptional elements. The ART neural network has three purposes. The first purpose is to group the machines into cells given as input the desired number of cells and process plans. The second purpose is to calculate the loading on each machine given the processing time of each part. The last purpose of the neural network is to propose alternative groups considering duplicate machines. The expert system was used to reassign the exceptional elements using alternate process plans generated by the neural network based on processing time and machine utilization. The evaluation of process plans considered the cost factors of material handling, processing, and setup. Finally, the neural network was updated for future use with any changes in machine utilization or cell configuration.

Rao and Gu [25] proposed a modified version of the ART-1 algorithm to machine-cell and part-family formation. This modified algorithm ameliorates the ART-1 procedure so that the order of presentation of the input pattern no longer affects the final clustering. The strategy consists of arranging the input pattern in a decreasing order of the number of 1's, and replacing the logic AND operation used in the ART-1 algorithm, with an operation from the intersection theory. These modifications significantly improved the neural network performance: the modified ART-1 network recognizes more parts with similar processing requirements than the original ART-1 network with the same vigilance thresholds.

Chen and Cheng [26] added two algorithms in the ART-1 neural network to alleviate the bottleneck machines and parts problem in machine-part cell formation. The first one was a rearrangement algorithm, which rearranged the machine groups in descending order according to the number of 1's and their relative position in the machine-part incidence matrix. The second one was a reassignment algorithm, which reexamined the bottleneck machines and reassigned them to proper cells in order to reduce the number of exceptional elements. The extended ART-1 neural network was used to solve 40 machine-part formation problems in the literature. The results suggested that the modified ART-1 neural network could consistently produce a good quality result.

Since both original ART-1 and ART-2 neural networks have the shortcoming of proliferating categories with a very few patterns due to the monotonic nonincreasing nature of weights, Burke and Kamal [27] applied the fuzzy ART neural network to machine-part cell formation. They found that the fuzzy ART performed comparably to a number of other serial algorithms and neural network based approaches for part family and machine cell formation in the literature. In particular, for large size problem, the resulting solution of fuzzy ART approach was superior than that of ART-1 and ART-2 approaches. In an extended work, Kamal and Burke [28] developed the FACT (fuzzy art with add clustering technique) algorithm based on an enhanced fuzzy ART neural network to cluster machines and parts for cellular manufacturing. In the FACT algorithm, the vigilance and the learning rate were reduced gradually, which could overcome the proliferating cluster problem. Also, the resultant weight vector of the assigned part family were analyzed to extract the information about the machines used, which enabled FACT to cluster machines and parts simultaneously. By using the input vector that combining both the incidence matrix and other manufacturing criteria such as processing time and demand of the parts, FACT could cluster machines and parts with multiple objectives. The FACT was tested with 17 examples in the literature. The results showed that FACT outperformed other published clustering algorithms in terms of grouping efficiency.

Chang and Tsai [29] developed an ART-1 neural-network-based design retrieving system. The design being retrieved was coded to a binary matrix with the destructive solid geometry (DSG) method, which was then fed into the ART-1 network to test the similarity to those in the database. By controlling the vigilance parameter in the ART-1 network, the user can obtain a proper number of reference designs in the database instead of one. Also, the system can retrieve a similar or exact design with noisy or incomplete information. However, the system cannot process parts with protrusion features where additional operations were required in the coding stage.

Enke et al. [30] realized the modified ART-1 neural network in [22] using parallel computer for machine-part family formation. The ART-1 neural network was implemented in a distributed computer with 256 processors. Problems varying from 50×50 to 256×256 (machines $\times$ parts) were used to evaluate the performance of this approach. Compared with the serial implementation of the ART-1 neural network in one process, the distributed processor based implementation could reduce the processing time from 84.1 to 95.1%. Suresh et al. [31] applied the fuzzy ART neural network for machines and parts clustering with the consideration of operation sequences. A sequence-based incidence matrix was introduced, which included the routing sequence of each part. This incidence matrix was fed into the fuzzy ART neural network to generate the sequence-based machine-part clustering solution. The proposed approach was used to solve 20 problems with size ranging from 50×250 to 70×1400 (machines $\times$ parts) and evaluated by the measure clustering effectiveness defined by the authors. The results showed that the approach had a better performance for smaller size problems.

Lee and Fisher [32] took both design and manufacturing similarities of parts into account to part-family grouping using the fuzzy ART neural network. The design attributes, i.e., the geometrical features of the part were captured and digitalized into an array of pixels, which was then normalized to ensure scale, translation, and rotation invariant recognition of the image. The normalized pixel vectors were transformed into a five-digit characteristics vector representing the geometrical features of the part by fast Fourier transform and a dedicated spectrum analyzer. Another 8-digit vector containing the manufacturing attributes—including the processing route, processing time, demand of the part, and number of machine types—was added to the 5-digit characteristic vector to form a 13-digit attribute. By feeding the 13-digit attribute vector into a fuzzy ART network, the parts could be clustered based on both the geometric shape and manufacturing attributes. The approach was found successful in parts grouping based on both design and manufacturing attributes. However, the three input parameters in the fuzzy ART network were determined by time-consuming trial and error approach, and cannot provide optimum values when large data sets are used, since the combination of these parameters nonlinearly affected the classification results.

Malavé and Ramachandran [33] proposed a self-organizing neural network based on a modified Hebbian learning rule. In addition to proper cell formation, the neural network also identifies bottleneck machines, which is especially useful in the case of very large part-machine incidence matrices where the visual identification of bottlenecks becomes intractable. It was also possible to determine the ratio in which bottleneck machines were shared among overlapping cells. The number of groups was arbitrarily chosen, which may not result in the best cellular manufacturing system. Lee et al. [34] presented an improved self-organizing neural network based on Kohonen's unsupervised learning rule for part-family and machine-cell formation, bottleneck machine detection, and natural cluster generation. This network is able to uncover the natural groupings and produce an optimal clustering as long as homogeneous clusters exist. Besides discovering natural groupings, the proposed approach can also assign a new part not contained in the original machine-part incidence matrix to the most appropriate machine cell using the generalization ability of neural networks to maximize the cell efficiency.

Liao and Lee [35] proposed a GT coding and part family forming system composed of a feature-based CAD system and an ART-1 neural network. The geometrical and machining features of a machining part were first analyzed and identified by the user using the feature library in the feature-based CAD system, which in turn generated a binary code for the part. The assigned codes for parts were clustered into different families according to the similarity of the geometrical and machining features by the ART-1 neural network. After the part classification is completed, each part would assign a 13-digit GT code automatically, which can be used to retrieve part drawing from the database or process plan from a variant process planning system. The feasibility of the proposed system has been demonstrated by a case study. However, the system was limited to those users who knew the machining operations, since machining features of parts were required when using the feature-based CAD system.

Malakooti and Yang [36] developed a modified self-organizing neural network based on an improved competitive learning algorithm for machine-part cell formation. A momentum term was added to the weight updating equation for keeping the learning algorithm from oscillation, and a generalized Euclidean distance with adjustable coefficients were used in the learning rule. By changing the coefficients, the cluster structure can be adjusted to adopt the importance preference of machines and parts. The proposed neural network was independent of the input pattern, and hence was independent of the initial incidence matrix. On average, the neural network approach gave very good final grouping results in terms of percentage of exceptional elements, machine utilization, and grouping efficiency compared with two popular array-based clustering methods, the rank order clustering and the direct clustering analysis, to ten problems sizing from 5×7 to 16×43 (machines $\times$ parts) in the literature.

Arizono et al. [37] applied a modified stochastic neural network for machine-part grouping problem. A simplified probability function was used in the proposed neural network, which reduced the computation time compared with other stochastic neural networks. The presented neural network overcame the local minimum problem existing in deterministic neural networks. The proposed neural network was comparable to conventional methods in solving problems in the literature. However, some system parameters in the neural network were decided on trial and error basis. A general rule for determining these parameters was not found. Zolfaghari and Liang [38] presented an ortho-synapse Hopfield network (OSHN) for solving machine grouping problems. In OSHN the oblique synapses were removed to considerably reduce the number of connections between neurons, and hence shortening the computational time. Also, the objective-guided search algorithm was adopted to ease the local optima problem. The proposed neural network approach was able to automatically assign the bottleneck machines to the cells, which they had the highest belongingness without causing large cells.

Kao and Moon [39] applied a multilayer feedforward neural network trained using the BP learning algorithm for part-family formation during part classification. The proposed approach consists of four phases: seeding, mapping, training, and assigning. Learning from feature-based part patterns from a coding system with mapped binary family codes, the neural network is able to cluster parts into families, resembling how human operators perform the classification tasks. Jamal [40] also applied a multilayer feedforward neural network trained with the BP algorithm for grouping part families and machine cells for a cellular manufacturing system. The original incidence matrices and corresponding block diagonalized ones are used, respectively, as inputs and desired outputs of the feedforward neural network for training purposes. The quality of the solutions obtained by using the trained neural network is comparable to that of optimal solutions. The benefits of using neural networks were highlighted again: speed, robustness, and self-generated mathematical formulation. Nonetheless, care must be taken because the efficiency of the neural network depends on the number and type of examples with which it was trained. Chung and Kusiak [41] also used a multilayer feedforward neural network trained with the BP algorithm to group parts into families for cellular manufacturing. Given binary representations of each part shape as input, the neural network trained with standard shapes is to generate part families. The performance

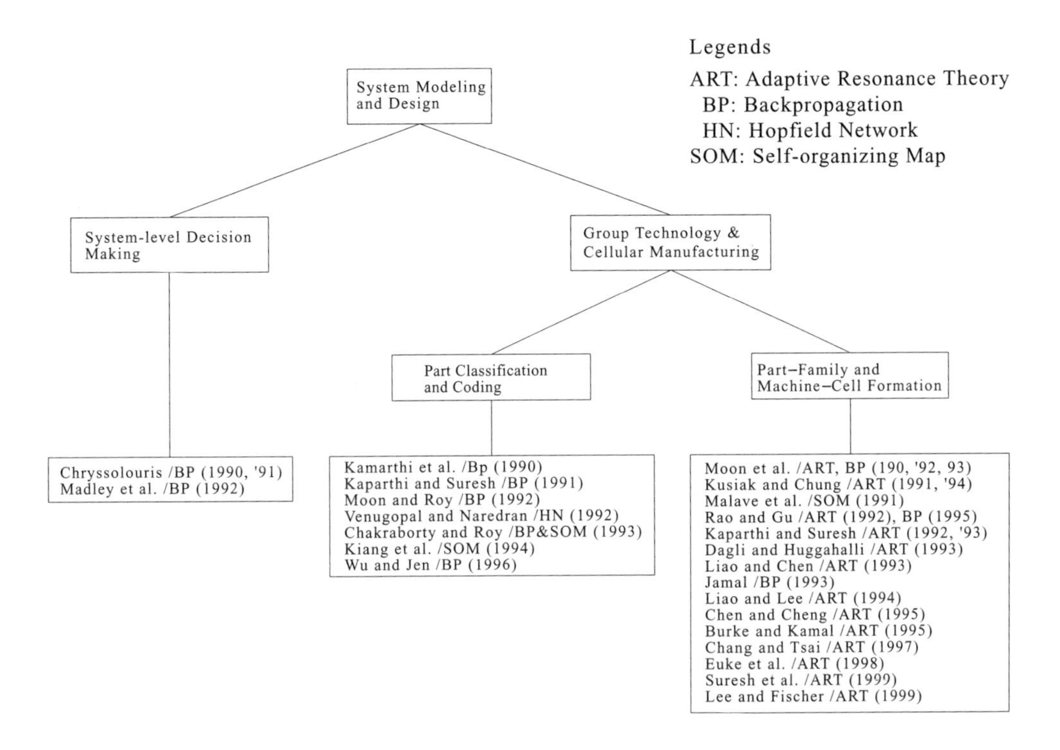

FIGURE 2.2 Hierarchy of neural network applications for manufacturing system modeling and design.

of the neural network was tested with partial and distorted shapes. The results show the effect of various design parameters on the groupings.

In summary, the applications of neural networks to modeling and design of manufacturing systems include resource allocation in job-shop manufacturing, operational policy determination for hierarchical manufacturing systems, modeling of continuous improvement systems, part classification and coding, part-family and machine-cell formation, as shown in Figure 2.2. In system-level decision making applications, simulation was used in combination with neural networks to generate data used by the neural network to implicitly model the system. In cellular manufacturing applications, neural networks used to classify parts and machines permit easy identification of part families, machine cells, and exceptional elements. Neural networks could also be used to assign new parts to an existing classification. Feedforward neural networks trained using the BP algorithm were popular for this application. Other types of neural networks included ART networks, Hopfield networks, and SOM neural networks. Weaknesses of neural networks for modeling and design of manufacturing systems result from neural networks themselves. Some parameters or constants must be determined on a trial-and-error basis. Also, neural network methods cannot always guarantee an optimal solution, and several searches must often be taken to improve the quality of the solution. Nevertheless, neural networks offer a promising alternative design method with highly computational efficiency and are able to address some of the limitations of traditional methods.

Given the ability to learn from experience and inherent parallel processing of neural networks, a neural network approach allows the implicit modeling of systems using representative data, thus eliminating the need for explicit mathematical analysis and modeling. Neural networks also have the unique ability to solve problems with incomplete or noisy data. Furthermore, neural networks are not significantly influenced by the size of the problem, because global computing is done in parallel and the local computat ion in each neuron is very simple. Neural networks are therefore appropriate for solving large industrial problems. As dedicated neurocomputing hardware emerges and improves, neural networks will become more beneficial for solving large-scale manufacturing modeling and design applications.

2.3 Modeling, Planning, and Scheduling of Manufacturing Processes

Typical tasks in process planning include material selection, process selection, process sequencing, and machining parameter selection. Planning and scheduling generally require two steps: the input–output process modeling and the selection of parameters to optimize the process with given constraints. Flexible on-demand scheduling and planning can provide a vital competitive advantage by reducing waste, improving efficiency and productivity, meeting customer due date, and reflecting the dynamic nature of increasingly competitive markets. Most planning and scheduling problems in manufacturing are NP-complete, with precedence constraints among tasks, setup costs, timing requirements, and completion deadlines. The scheduling and shop management are even more complex in flexible manufacturing systems (FMS) with on-demand production. Classical heuristic methods approach the problem by applying some priority rules based upon some easily calculated job parameters, such as due date, setup times, arrival times. Classical methods obviously cannot take into account all the variables interacting in manufacturing systems, and lack the time-dependent decision capability needed in production planning and scheduling, especially in FMS and computer-integrated manufacturing (CIM) environments, which both require an ability to deal with uncertainty and dynamic behavior. The ability of neural networks to understand temporal patterns is essential for efficient modeling, planning, and scheduling of manufacturing processes.

Andersen et al. [42] used a multilayer feedforward neural network trained with the BP algorithm to model bead geometry with recorded arc welding data. The neural network was a fairly accurate static model of the welding process and could be directly used to determine the parameters necessary to achieve a certain tool geometry. The accuracy of the neural network modeling was fully comparable with that of traditional modeling schemes. Tansel [43] developed two neural networks to model three-dimensional cutting dynamics in cylindrical turning operations. The first neural network was used to simulate the cutting-force dynamics for various operating speeds. Multilayer feedforward neural models were trained using the BP algorithm to predict the resulting cutting force given cutting speed and present (inner modulation) and previous (outer modulation) feed direction tool displacement. The neural network approach was capable of very good predictions with less than 7% errors. This approach was more advantageous than traditional methods such as time series models, which usually allow modeling of three-dimensional cutting dynamics only at one given speeds rather than over a wide range of cutting speeds and cannot represent systems nonlinearity as opposed to neural networks. In addition, the use of neural networks permits introduction of additional parameters in the model, such as the cutting speed and varying spindle speeds, that would not be easily modeled with traditional methods. A second neural network was developed to estimate the frequency response of the cutting operation. A multilayer feedforward neural network was trained using the BP algorithm with data of frequency and cutting speed to estimate inner and outer modulations at any frequency and speed in the training process. The neural network was a very accurate model of the frequency response of the cutting process realizing errors less than 5% of the defined output range. Both neural networks achieved greater accuracy for higher speeds, in contradiction to the fact that variations in cutting force are larger at higher speeds, than at lower speeds.

Dagli et al. [44] proposed an intelligent scheduling system that combined neural networks with an expert system for job scheduling applied to a newspaper printing process. The scheduling system was made of the union of two neural networks: a Hopfield network for determining the optimal job sequence and a multilayer feedforward neural network trained with the BP algorithm for job classification. The system could schedule sequence-dependent jobs given setup and processing times. The computational speed and time-dependent capability of the system make it applicable for many planning and scheduling applications including process control, cutting and packing problems, and feature-based designs. The proposed system could be modified, or integrated with additional neural networks to suit for various planning and scheduling tasks.

Arizono et al. [45] adapted a stochastic neural network for production scheduling with the objective of minimizing the total actual flow time of jobs with sequence-dependent setup times. The neural network used was a Gaussian machine. The system dynamics were designed to lead the neural network convergence to the scheduling sequence that would minimize the total actual flow-time of the system given processing and setup times. The proposed neural network was shown to converge to near-optimal (if not optimal) schedules in terms of total actual flow time. The only significant problem is that of specifying the network parameters.

Cho and Wysk [46] developed an intelligent workstation controller (IWC) within a shop floor control system. The IWC performs three main functions: real-time planning, scheduling, and execution of jobs in a shop floor. The IWC consists of a preprocessor, a feedforward neural network, and a multiprocessor simulator. The preprocessor generates input vectors for the neural network based on the workstation status, the off-line trained neural network plays the role of a decision support system in generating several part dispatching strategies, and the multi-pass simulator then selects the best strategy to maximize the system efficiency. The efficiency of this IWC was reportedly much better than that of a single-pass simulator because the choice of strategies took all the performance criteria into account.

Lo and Bavarian [47] extended the Hopfield network to job scheduling. A three-dimensional neural network called Neuro Box Network (NBN) was developed with job, machine, and time as three dimensions. The NBN is responsible for determining a sequence while minimizing the total setup costs and total time for job completion. The superiority of the NBN is that it is able to evolve in time and provide on-demand schedules each time new circumstances arise such as new job arrival or machine breakdown.

Lee and Kim [48] adopted a neural network for choosing the scaling factors to be used as a dispatching heuristic for scheduling jobs on parallel machines with sequence-dependent setup times. A multilayer feedforward neural network was trained using the BP algorithm to model the manufacturing process. Fed with various process characteristics (such as due dates, due dates range, setup times, and average number of jobs per machine), the neural network was able to determine the optimal scaling factors. The schedules generated using the predicted scaling factors were much more efficient than those generated using the scaling factors found with traditional rules. Improvements were made in at least 96% of the cases and up to 99.8% depending on the rule used to generate the schedules.

Satake et al. [49] used a stochastic neural network to find feasible production schedules in the shortest time while incorporating several manufacturing constraints. The neural network presented in this work was a Hopfield network using a Boltzmann machine mechanism to allow escapes from local minimum states. The energy function incorporated one of the constraints of the problem, while the threshold values represented the objective function and the remaining constraints. The salient feature of the Hopfield network used was that the threshold values were not predetermined but revised at each iteration. This approach circumvents the lack of guidelines for choosing the network design parameters reported elsewhere. The schedules generated by the neural system were compared with schedules generated by the branch and bound method. Results proved that the neural network solution was optimal in 67% of the cases and near optimal the rest of the time.

Wang et al. [50] proposed an FMS scheduling algorithm that determined the scheduling rules by neural network and the rule decision method used in expert system, the inductive learning. In their approach, the necessary knowledge for scheduling were obtained in two stages. In the first stage, the training examples for knowledge acquisition were generated by a simulation model that maximized the resource utilization. The generated training examples consisted of the shop floor status and dispatching rules and were classified by a neural network composed of adalines. The classified groups were used to form the decision tree by the inductive learning method to determine the scheduling rules. The approach was, however, only feasible for linearly clustered training examples.

Sabuncuoglu and Gurgun [51] applied a simplified Hopfield network to scheduling problems. The modified Hopfield network has an external processor, which was used to perform both feasibility and cost calculations. Compared with the original Hopfield network, the revised Hopfield network eliminated most of the interconnections and was more suitable to be implemented in serial computer. The relative performance of the simplified Hopfield network was evaluated against the benchmark Wilkerson and Irwin algorithm with two scheduling problems, the single machine scheduling with minimum mean tardiness, and the job shop scheduling with minimum job completion time. The results were promising that the proposed approach improved the mean tardiness in general and could find the optimal schedules in 18 out of 25 job shop scheduling problems.

Similar to the approach in [50], Li et al. [52] and Kim et al. [53] also applied neural network and the inductive learning method for FMS scheduling with multi-objectives. However, Li et al. [52] employed the ART-2 neural network to cluster the simulated training examples while Kim et al. [53] used the competitive neural network to group the unclassified training examples. Both approaches were found promising. However, systematic procedures for finding the optimal values of the parameters for ART-2 neural network and optimal number of output nodes of the competitive neural network were not developed.

Knapp and Wang [54] used two cooperative neural networks to automate the process selection and task sequencing in machining processes. After the acquisition of process planning knowledge, process sequencing was automatically prescribed using neural networks. In the first stage, a multilayer feedforward neural network trained with the BP algorithm was used to generate operation alternatives. In the second stage, a laterally inhibited MAXNET was used to make a decision among competing operation alternatives. In the last stage, the output of the MAXNET was fed back to the feedforward neural network to provide a basis for deciding the next operation in the machining sequence. Chen and Pao [55] discussed the integration of a neural network into a rule-based system applied to design and planning of mechanical assemblies. An ART-2 neural network was used to generate similar designs automatically given desired topological and geometric features of a new product. A rule-based system was then used to generate an assembly plan with the objective to minimize tool changes and assembly orientations. The rule-based system consisted of five submodules: preprocessing, liaison and detection, obstruction detection, plan formulation, and adaptation and modification. The last submodule compares existing assembly sequences with the sequence generated by the first four submodules and adapts the most similar sequences to best match the required assembly task. The proposed integrated system can increase speed and efficiency in the design and planning of mechanical assemblies.

Shu and Shin [56] formulated the tool path planning of rough-cut of pocket milling into a traveling salesman problem (TSP), in which the removal area is decomposed into a set of grid points or tool points to be visited by the tool only once, and the tool starts and ends at the same point. Then the self-organizing map was used to solve the combinatorial problem to generate the near optimal path. The simulation and real machining results showed the neural network approach can effectively and efficiently optimize the tool path regardless of the geometric complexity of pockets and the existence of many islands.

Osakada and Yang [57] applied four multilayer feedforward neural networks for process planning in cold forging. In the first module, a multilayer feedforward neural network trained using the BP algorithm was used to learn to recommend a cold forging method in order to produce a workpiece of given shape. Predictions were perfect for pieces very similar to the training set. If the neural network indicated the piece could not be produced in one stroke the next module came into action to predict the optimal number of production steps. The evaluation of the different process candidates with more than one forming step was done by using another neural network. The second neural network was trained using the BP algorithm given information on shape complexities, number of primitives, billet and dye material. The trained neural network performed perfect ranking of the different process candidates, as opposed to 68% accuracy achieved by statistical methods, as long as products were similar enough to the training patterns. The last evaluation module was to predict die fracture and surface defect of the piece in the order of priority. Two neural networks were trained using the BP algorithm with finite elements method simulations. One neural network was able to predict die fracture given important surface parameters. The other neural network was able to predict surface defect given the same surface parameters, in addition to billet and die material. The predictions of both neural networks were very reliable with accuracies of 99% for die fracture and 99% for surface defect, in contrast to 90 and 95% with statistical methods.

Eberts and Nof [58] applied a multilayer feedforward neural network trained using the BP algorithm for planning unified production in an integrated approach. The planning procedure was demonstrated through an example of advanced flexible manufacturing facility controlled by a computerized system. The neural network provided a knowledge base containing information on how to combine human and machine intelligence in order to achieve integrated and collaborative planning. The assistance of the neural network will help improve flexibility, reliability, utilization of machine, and human/machine collaboration. However, the rules to combine machines and human inputs and the effect of these rules on the neural network need to be elaborated.

Rangwala and Dornfeld [59] applied a neural network to predict optimal conditions (cutting parameters such as cutting speed, feed rate, and depth of cut) in turning operations by minimizing a performance index. A multilayer feedforward neural network was trained using the BP algorithm. The learning and optimization in the neural network were performed in either batch or incremental mode. The latter learns the process mappings and optimizes cutting parameters simultaneously and is therefore more suitable for real-time applications. Cook and Shannon [60] applied a multilayer feedforward neural network to process parameter selection for bonding treatment in a composite board manufacturing process. The neural network was trained with the BP algorithm using several process parameters to learn to model the state of control of the process. The performance of the neural network was fair, with a prediction rate of approximately 70%. The sensitivity of the performance was investigated for various network designs and learning parameters.

Sathyanarayan et al. [61] presented a neural network approach to optimize the creep feed grinding of super alloys. A multiple-objective optimization problem was formulated and transformed into a single objective one using a weighting method. Each single objective function was then easily optimized individually using the branch and bound method. A multilayer feedforward neural network was then trained using the BP algorithm to associate cutting parameters of a grinding process (feed rate, depth of cut) with its outputs (surface finish, force, and power). The neural network was able to predict the system outputs within the working conditions and overcome major limitations of conventional approaches to this task.

Matsumara et al. [62] proposed an autonomous operation planning system to optimize machining operations in a turning process. The system could accumulate machining experience and recommend process parameters of each machine tool. Machining conditions such as flank wear and surface roughness were predicted using the combination of an analytical method based on metal cutting theory and a multilayer feedforward network trained with the BP algorithm. Operations planning with adaptive prediction of tool wear and surface roughness was effective because machining processes could be evaluated simultaneously with machining time. The machining operation was optimized by minimizing the total machining cost.

Wang [63] developed a neural network approach for optimization of cutting parameters in turning operations. Considering productivity, operation costs, and cutting quality as criteria, the cutting parameter selection in turning operations was formulated as a multiple-objective optimization problem. A multilayer feedforward neural network trained using an improved learning algorithm was used to represent the manufacturer's preference structure in the form of a multiattribute value function. The trained neural network was used along with the mappings from the cutting parameter space to the criteria space to determine the optimal cutting parameters. The proposed neural network approach provides an automated paradigm for multiple-objective optimization of cutting parameters.

Roy and Liao [64] incorporated a three-layer preceptron into an automated fixture design (AFD) system for machining parameters selection. The geometry, topology, feature, and technological specification of the workpiece were given to the AFD in which the workpiece materials, hardness, carbon composition, and cutting tool materials were extracted and directed to a feedforward neural network trained by the BP algorithm to determine the cutting speed, feed rate, and depth of cut for the milling process. The estimated cutting parameters were not only for the milling process control, but also for the cutting force evaluation, which was indispensable to the stress analysis of the fixture, and hence directly help the AFD system to come up with the best fixture configuration.

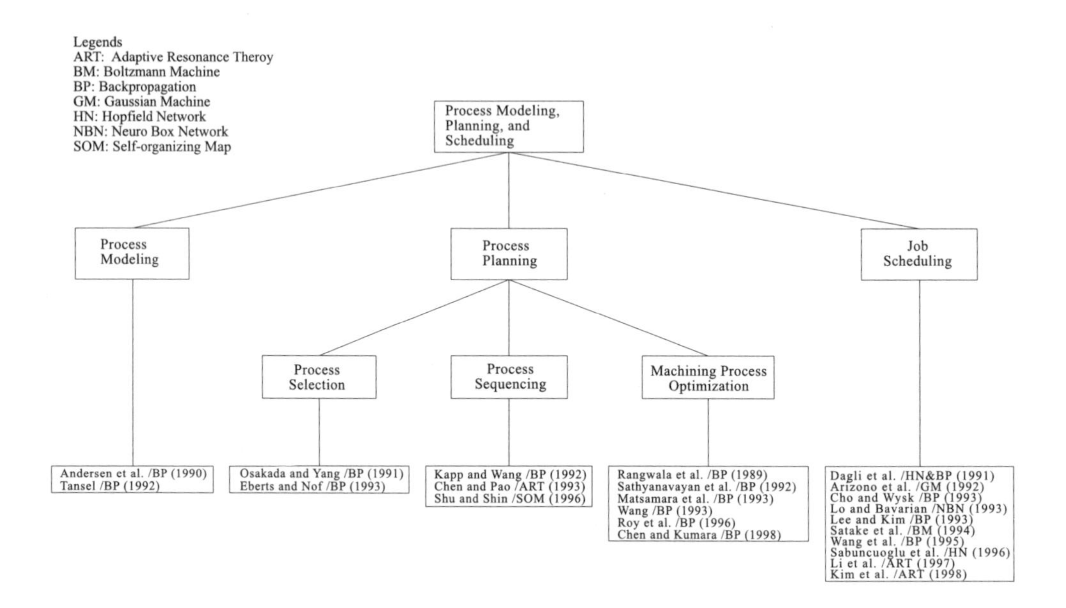

FIGURE 2.3 Hierarchy of neural network applications for process modeling, planning, and scheduling.

Chen and Kumara [65] demonstrated that fuzzy logic and neural networks are effective means for grinding process parameters selection. They built a fuzzy grinding optimizer, which can design a set of grinding process parameters to achieve desirable process conditions based on the user-defined process conditions. The fuzzy grinding optimizer was then used to generate the training sets for a multilayer feedforward neural network with the BP learning algorithm. In order to shorten the training time, they developed a procedure to decompose the neural network into a number of smaller ones, and introduced a fuzzy accelerator to adjust the learning rate, momentum coefficient, and the steepness parameter of the activation function during training. However, the theoretical analysis of the convergence of the weight due to the proposed fuzzy accelerator was not provided.

In summary, present applications of neural networks to process modeling, planning, and scheduling include process selection, process sequencing, machining process optimization, and job scheduling, as shown in Figure 2.3. The neural network models used were multilayer feedforward networks, MAXNET, Hopfield networks, ART networks, and stochastic networks. The knowledge acquisition capabilities of neural networks made them legitimate alternatives to conventional methods for most planning and scheduling applications. Some weaknesses of neural networks were due to the lack of explanation for intrinsic causal relationships existing in complex planning and scheduling applications. In order to solve such complex planning and scheduling problems, neural networks ought to be combined with knowledge-based systems such as expert systems.

2.4 Monitoring and Control of Manufacturing Processes

In driving toward automation and computer integrated manufacturing (CIM), industries are constantly seeking effective tools to monitor and control increasingly complicated manufacturing processes. The success of human operators in process monitoring and control tasks suggests that one possible approach to designing computer-based monitoring and control systems is to model the learning and decision-making abilities of human operators. An intelligent controller should possess abilities to learn from examples and use knowledge gained during a learning process to optimize the operation of machines [66]. This is analogous to the process by which a novice human machinist becomes an expert. Neural networks are promising tools for on-line monitoring of complex manufacturing processes. Their superior learning and fault tolerance capabilities enable high success rates for monitoring the machining processes.

Among the manufacturing applications of neural networks, monitoring and control can be considered in two dimensions: the monitoring and control of workpieces (e.g., surface finish, automatic setups) and machines (e.g., vibration, tool wear, thermal deflection). Neural networks are taught by examples, thus eliminating the need for explicit mathematical modeling. Neural networks can serve as black boxes that avoid an extensive study and easily lead to results.

Rangwala and Dornfeld [67] applied a multilayer feedforward neural network to recognize the occurrence of tool wear in turning operations. The neural network trained with the BP algorithm learned to perform tool-wear detection given information from the sensors on acoustic emission and cutting force. Experiments were conducted with fresh and worn data on a Tree lathe and the information was transformed between time and frequency domains using fast Fourier transformation. The superior learning and fault tolerance capabilities of the neural network contribute to the high success rates in the recognition of tool wear. However, design parameters (such as training parameters, network structure, and sensors used) affect the performance of the system.

Burke and Rangwala [68] discussed the application of neural networks for monitoring cutting tool conditions. The authors compared the performance of supervised feedforward neural networks with the BP algorithm and unsupervised ART networks for in-process monitoring of cutting tools. The raw sensor data were time representations of cutting force and acoustic emission signals. Besides excellent classification accuracy by both neural networks, the results showed that the unsupervised ART networks held greater promise in a real-world setting, since the need for data labeling is eliminated, and also the cost associated with the data acquisition for a supervised neural network was reduced. In addition, the ART networks can also remain adaptive after initial training and could easily incorporate additional patterns into the memory without having to repeat the entire training stage. Interestingly, the ART networks could distinguish between fresh and worn tools after being trained using fresh tool patterns only.

In a related work, Burke [69, 70] developed competitive learning approaches for monitoring tool-wear in a turning operation based on multiple-sensor outputs using the ART-1 network. The unsupervised system was able to process the unlabeled information with up to 95% accuracy, thus providing more efficient utilization of readily available (unlabeled) information. The success of partial labeling may lead to significant reduction in data analysis costs without substantial loss of accuracy. The speed of the system coupled with its ability to use unlabeled data rendered it a flexible on-line decision tool. Possible extensions include detection of degrees of tool wear, feature selection, and integrated neural network/expert system to incorporate higher-level capabilities.

Yao and Fang [71] applied a multilayer feedforward network to predict the development of chip breakability and surface finish at various tool wear states in a machining process. In the initial phase, chip forming patterns (i.e., chip breaking/shapes) were estimated under the condition of an unworn tool. Then the neural networks were trained with input features such as dispersion patterns, cutting parameters, and initial prediction of breakability and outputs in terms of fuzzy membership value of chip breakability and surface roughness. After off-line training using the BP algorithm, the neural network was able to successfully predict on-line machining performance such as chip breakability, chip shapes, surface finish, and tool wear. The neural network is capable of predicting chip forming patterns off line as well as updating them on line as tool wear develops. This method can be applied to any tool configuration, and/or rough machining conditions.

Tarng et al. [72] used a multilayer feedforward neural network trained with the BP algorithm to monitor tool breakage in face milling. Normalization of the cutting force signal was performed to reduce the training time required by the neural network. The output of the neural network represented the probability of having a tool breakage. The neural network was shown to be able to classify tool breakage successfully. The performance of the neural network was insensitive to variations in cutting conditions: variations in cutting speed, radial depth of cut, feed rate, and workpiece material. In other related works, Ko et al. [73, 74] used, respectively, an ART-2 neural network and a four-layer feedforward neural network trained by the BP algorithm to monitor the tool states in face milling. The cutting force signals were put into an eighth-order adaptive autoregressive function that was used to model the dynamics of the milling process. The signal patterns were classified using the ART-2 neural network [73] and the multilayer perceptron [74] to indicate the breakage of cutting tools. Both neural-network-based tool wear monitoring systems were able to successfully detect the wear of milling tools in a wide range of cutting conditions. However, the ART-2-based system had unsupervised learning capability.

Chao and Hwang [75] integrated the statistical method into the BP trained neural network for cutting tool life prediction. The variables that related to the tool life—including cutting velocity, feed rate, depth of cut, rake angle, material hardness of tool, and work material composition—were first analyzed by statistical method to identify the significant data and remove the correlation between variables. The screened data were used as the inputs of a three-layer feedforward neural network, which consequently estimated the tool life. Compared with the backward stepwise regression method, the proposed approach was shown more robust to the changes of input variables and resulted in more accurate predictions.

Jemielniak et al. [76] presented an approach for tool wear identification based on process parameters, cutting forces, and acoustic emission measures of the cutting process using a three-layer feedforward neural network with the BP learning algorithm. The multilayer perceptron initially had eight input nodes, 16 hidden nodes, and one output node that gave the crater depth to signify the tool state. A systematic pruning procedure was executed to eliminate the inputs and hidden nodes that did not affect the resulting errors. A refined neural network with five inputs, three hidden nodes, and one output resulted, which provided comparable accuracy and more uniform error distribution.

Purushothaman and Srinivasa [77] applied the BP trained multilayer feedforward neural network with an input dimension reduction technique to the tool wear monitoring. In their approach, the original six-dimensional inputs, which consisted of the cutting forces and the machining parameters, were combined to a two-dimensional input vector by using a linear mapping algorithm. The reduced dimension input vector was fed into a three-layer perceptron to evaluate the tool wear condition. Compared with the full dimension input vector approach, the proposed approach was shown to drastically reduce the number of arithmetic operations and could achieve the same accuracy of tool wear prediction.

Alguindigue et al. [78] applied a multilayer feedforward neural network for monitoring vibration of rolling elements bearings. A multilayer feedforward neural network trained with the BP algorithm learned to predict catastrophic failures to avoid forced outrages, maximize utilization of available assets, increase the life of machinery, and reduce maintenance costs. The salient asset of such a system is the possibility of automating monitoring and diagnostic processes for vibrating components, and developing diagnostic systems to complement traditional phase sensitive detection analysis.

Hou and Lin [79] used a multilayer feedforward neural network trained with the BP algorithm for monitoring manufacturing processes. Frequency domain analysis (fast Fourier transforms) was performed on periodic and aperiodic signals to detect vibrations generated by machine faults including imbalance, resonance, mechanical looseness, misalignment, oil whirl, seal rub, bearing failure, and component wear. The neural network achieved accuracy of over 95%.

Tansel et al. [80] used an ART-2 neural network in conjunction with wavelet transform to monitor drill conditions for a stepping-motor-based micro-drilling machine. Cutting force signals were sampled at two different rates to capture either two or three table-step motions (fast sample rate) or the complete drilling cycles (slow sample rate). After sampling and digitizing, cutting force signals were encoded in wavelet coefficients. The ART-2 neural network was used to classify the tool condition given as an input either 22 wavelet coefficients (direct encoding method) or six parameter representatives of the 22 wavelet coefficients (indirect encoding method). The trained neural network was able to detect severe tool damage before tool breakage occurred with both encoding methods. The direct encoding method, even though two or three times slower, was more reliable, with an accuracy greater than 98% compared with an accuracy of 95% for the indirect encoding method. Interestingly, the ART-2 network was able to classify more easily the wavelet coefficients of the data collected at the fast sampling rate, which reduces the data collection time to only a fraction of seconds and enables detection of the tool condition significantly earlier.

Lee and Kramer [81] used a neural network called the cerebellar model articulation controller (CMAC) for monitoring machine degradation and detecting faults or failures. The method integrates learning, monitoring, and recognition in order to monitor machine degradation and schedule maintenance. Machine degradation analysis and fault detection was provided by a pattern discrimination model, based on the cerebellar model articulation controller network. The controller network is in charge of the adaptive learning and the pattern discrimination model monitors the machine behavior. Machine faults are detected by comparing the conditional probability of degradation with a threshold confidence value. The innovative approach proved capable of learning fault diagnosis and performing effective maintenance, thus providing an active controller that enables preventive maintenance. The neural network played the role of a feedforward controller, which generates the conditional probabilities of machine degradation that were then compared with a threshold confidence value. The neural network learned to recognize normal machine conditions given various machine parameters such as position accuracy and straightness.

Currie and LeClair [82] applied a neural network to control product/process quality in molecular beam epitaxy processing. The neural network used was a functional-link network trained using the BP algorithm. The self-improvement, fault tolerance, and complete mapping characteristics of neural networks made the proposed system a good candidate for manufacturing process control. The trained neural network was able to predict the recipe parameters needed to achieve some desired performance. Significant misclassifications occurred due to measurement errors inherent to the complexity of the process. After enhancements, the proposed system should be able to circumvent the inaccuracies.

Balazinski et al. [83] applied a multilayer feedforward neural network trained using the BP algorithm to control a turning process. Given feed rate error and change in the error, the trained neural network was able to recommend the control actions necessary to maintain a constant cutting force (static case) in order to assure proper wear of the cutting tool. The performance of the neural network was similar to that of a fuzzy controller. The main difference between the two systems is that the neural network allowed crisp values rather than fuzzy values in input/output data. The neural network controller is more desirable than the fuzzy controller in terms of response time, steady states errors, and adaptivity. Furthermore, neural networks were more flexible (adaptive) and did not exhibit the oscillations observed with the fuzzy controller in steady states.

Lichtenwalner [84] used a neural network to control laser heating for a fiber placement composite manufacturing process. For this task, a modified version of the cerebellar model articulation controller was chosen for its unequaled speed of learning through localized weight adjustment. The neural network plays the role of a feedforward controller generating control voltage given the desired temperature and measured feed rate. The neurocontroller has superior capabilities over traditional feedforward controller, since it allows on-line learning of the control functions and accurate modeling of both linear and nonlinear control laws. The enhanced control allows fabrication of complex structures while preserving the quality of consolidation.

Ding et al. [85] applied a neural network for predicting and controlling a leadscrew grinding process. The neural network was a multilayer neural network trained with a variant of the BP algorithm called "one-by-one algorithm" that expedites the supervised learning. The neural network was used as a controller to predict and compensate for the transmission error in the grinding operation of precision leadscrews.

Chen [86] developed a neural-network-based thermal spindle error compensation system. The temperatures at 11 locations of a milling machine were monitored and fed into a multilayer feed forward neural network trained by the BP algorithm to predict the thermal deflections of the three principal spindles. The estimated thermal errors were adopted by the CNC controller, which sent out the compensated control signals to drive the milling machine. The neural network demonstrated a prediction accuracy of more than 85% in varying and new cutting conditions. In two evaluation tests, the neural-network-based system reduced the thermal spindle errors from 34 μm to 9 μm. In [87], Vanherck and Nuttin, however, used a multilayer feedforward neural network trained by the BP algorithm with momentum and adaptive learning rate for machine tools thermal deformation compensation. Unlike Chen's approach, the presented approach estimated the thermal error of each spindle by an independently multilayer perceptron. The proposed approach reduced the thermal deviations from 75 μ to 16 μ in two experimental milling tests. However, the error compensation failed in extreme high environment temperatures.

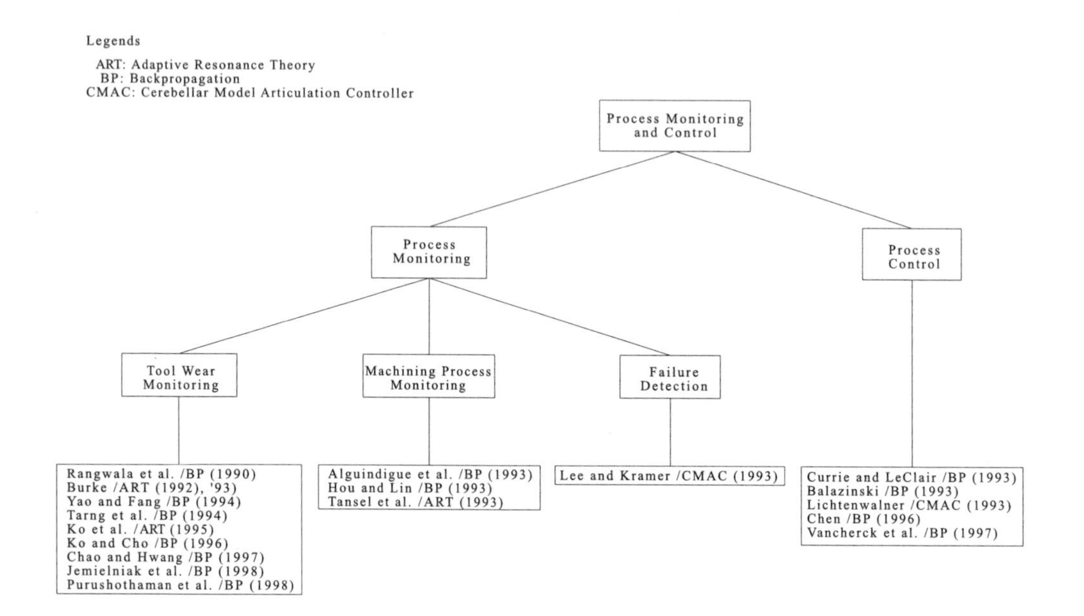

FIGURE 2.4 Hierarchy of neural network applications for process monitoring and control.

In summary, the present applications of neural networks for process monitoring and control include tool wear monitoring, machining process monitoring, process modeling, and process control. The neural network models used were multilayer feedforward networks, ART networks, and cerebellar model articulation controller, as shown in Figure 2.4. Neural networks are promising tools for on-line monitoring of complex manufacturing processes. They are appropriate in modeling cases where some information is missing, or where analytical modeling would be too complex. In addition, their superior learning and fault tolerance capabilities enable high success rates for monitoring machining processes. One important characteristic of neural networks that makes them good candidates for monitoring and control is their adaptive capability. A neural network monitor could serve as one of the most efficient tools in finding the optimum set of manufacturing parameters by predicting the effect of machining parameters to the machining process beforehand. Applications of neural networks also appear promising for real-time nonlinear mapping of distorted input data vectors. Recognition of techniques as a package of tools that could be combined in a particular application may be the key to future intelligent control. Systems analysis incorporating neural networks into real-time control systems should permit the latter to optimize the performance on line using variables that otherwise would require sophisticated models, algorithms, and complex computation. The parallel computation abilities of neural networks offer the potential for developing intelligent systems that are able to learn from examples, recognize process patterns, and initiate control actions in real-time manufacturing environment.

2.5 Quality Control, Quality Assurance, and Fault Diagnosis

Quality control and quality assurance aim at identifying defects when production is in progress or over and defective parts are being or are already manufactured. Because neural networks are especially powerful for identifying patterns and hidden relationships, they are also proposed and used for fulfilling various quality control, quality assurance, and fault diagnostics tasks.

Thomsen and Lund [88] applied a multilayer feedforward neural network trained with the BP algorithm to evaluate quality control status of composite materials based on ultrasonic test measurements. The neural network was tested on glass-epoxy laminated plates with frequently occurring flaws. Given ultrasonic power spectra of stress wave signals measured from the laminated plates, the neural network learned to classify the plate as belonging to either flaw category. The neural network performed well in classifying the different flaws. The occurring misclassifications were due to measurement configuration.

Villabos and Gruber [89] coupled a neural network with a laser scattering technique to inspect machined surface quality. A modified ART-2 neural network was used to identify surface roughness based on features extracted from the scattered angular spectrum resulting from various samples with uniform surface texture. The surface roughness determined by the neural network was compared with that determined by a profilometer measurement. The predictions of the neural network satisfied the ANSI accuracy standard with a discrepancy between 6.6 and 10.9% depending on the features used as inputs. In a related work to [89], Yan et al. [90] proposed to use a three-layer feedforward neural network with the BP learning algorithm to measure, in real time, the maximum peak-to-valley surface roughness R_{max} generated during surface finishing. The scattered angular laser light patterns reflected from the workpiece are recognized by the trained neural network to predict the R_{max}. The measurement system implemented by high-speed hardware can complete one measurement in 125 ms, which is adequate for real-time surface roughness measurement. The estimated R_{max} values have a maximum error of 10% when compared to the conventional stylus measurements.

Pugh [91] compared the performance of a multilayer feedforward neural network, trained using the BP algorithm under several conditions, with a standard bar control chart for various values of process shift. The performance of the neural network was almost equal to that of the control charts in type I (alpha) error, and was superior in type II (beta) error. Performance could be improved by careful contouring of the training data. Interestingly, if trained with the shift contour according to the Taguchi cost curve, the neural network offered a slight improvement over the traditional bar chart.

Wang and Chankong [92] developed a stochastic neural network for determining multistage and multi-attributes acceptance sampling inspection plans for quality assurance in serial production systems. A Bayesian cost model was formulated to take into account the interaction among defective attributes and between production stages. A stochastic algorithm simulated the state transition of a stochastic neural network to generate acceptance sampling plans minimizing the expected cost. This neural network generated high-quality (if not optimal) acceptance sampling plans in a reasonably short period of time. In Cook et al. [93, 94], a multilayer feedforward neural network was presented to predict the occurrence of out-of-control conditions in particle board manufacturing. Given current and past process condition parameters, the neural network was trained using the BP algorithm to predict the development of out-of-control conditions in the manufacturing process, with a success rate of up to 70%. These results were very encouraging, considering that a relatively small training set was used not representative of all possible process conditions. Payne et al. [95] used a multilayer perceptron trained with the BP algorithm to predict the quality of parts in a spray forming process. Given various process parameters, the neural network learned to predict part quality in terms of porosity and yield of future runs. The neural network predictions helped defining optimal process conditions and the correlation between input process parameters and part quality.

Wang et al. [96] applied a multilayer feedforward neural network for predicting wire bond quality in microcircuits manufacturing. The neural network trained with the BP algorithm and learned to model the relationship between process measurements (ultrasonic pulses) and bond quality. A multiple regression analysis helped identify the variables with significant influence on the wire bond quality. The performance of the system was reasonable and could be enhanced by incorporating additional variables and validating the neural network using the jackknife method. The results demonstrated the feasibility of neural networks for a high-reliability and low-cost quality assurance system for wire bonding process control.

Joseph and Hanratty [97] presented a multilayer feedforward neural network for shrinking horizon model predictive control of a batch manufacturing process. This work discusses a simulated autoclave curing process for composite manufacturing. The method was based on the model predictive control method. The models employed were derived by regressing past operational data using a feedforward neural network. The purpose of the model was to predict the outcome of a batch (a product quality) in terms of the input and processing variables. Incremental learning provided on-line adaptation to changing process conditions. The combination of the neural network, a shrinking horizon model predictive algorithms, and incremental learning strategies offered a convenient paradigm for imitating, at least in part, the role of skilled operators who learn from operational history and use the knowledge to make feedback control decisions during processing. This method is of interest in improving the batch-to-batch variation of product quality.

Smith [98] used a multilayer feedforward neural network to predict product quality from thermoplastic injection molding. The neural network trained using the BP algorithm was used to predict quality of several thermoplastic components in terms of both state and variability of the quality. The trained neural network was able to predict product quality with 100% accuracy, comparable to control charts and statistical techniques. Neural networks were advocated as more desirable than traditional quality control methods for real-world manufacturing since they allow real-time training and processing. In a related work, Smith [99] used a multilayer feedforward neural network trained using the BP algorithm to model mean $\bar{X}$ and range (R) control charts simultaneously for diagnosing and interpreting the quality status of manufacturing processes. Given statistics on product samples, the neural network was able to recognize process shifts in terms of state and variability. The performance of the neural network was sensitive to the number and type of input statistics and to the subgroup size of the raw data. For instance, the neural network performed better when trained using raw data and statistics rather than only statistics. Even with sparse and noisy data, the neural network successfully identified various shapes, with up to 99% success in the best conditions. The neural network was shown to be a good alternative to control charts and even outperformed control charts in the case of small shifts of variance and/or means and improved type II error rate.

Zhang et al. [100] applied a three-layer perceptron trained with the BP algorithm to approximate the correlation between optimal inspection sampling size and three relevant factors including machining process, hole size, and tolerance band for hole making. The neural network was shown to be capable of accurately estimating the sampling size. The deviation between the actual sample size and the estimated sample size for most tested samples was within ± 1.

Su and Tong [101] incorporated the fuzzy ART network into the quality control process for integrated circuit fabrication to reduce the false alarms. The reported wafer defects are fed into the fuzzy ART network, which generates a number of cluster of defects. Each cluster is regarded as one defect. The resulted clusters are then used to construct the c chart for quality control of wafers. The neural network-based c chart was compared with the Neyman-based c chart and the conventional c chart. The proposed approach could take account of the defect clustering phenomenon and hence reducing the false alarms.

Cook and Chiu [102] used the radial basis function (RBF) neural networks trained by the least-mean-squares algorithm for statistical process control of correlated processes. The trained RBF neural networks were used to separate the shifted and unshifted correlated papermaking and viscosity data in literature. The neural networks successfully identified data that were shifted 1.5 and 2 standard deviations from nonshifted data for both the papermaking and viscosity processes. The network for the papermaking data was able to also classify shifts of one standard deviation, while the traditional statistical process control (SPC) technique cannot achieve this because it requires a large average run length.

Guh and Tannock [103] employed a multilayer feedforward neural network trained by the BP algorithm to recognize the concurrent patterns of control chart. The trained neural network can identify the shift, trend, and cycle patterns in the control chart by taking 16 consecutive points from the control chart. The neural network was tested and the results showed it can improve the type II error perfomance while keeping the number of concurrent pattern training examples to a minimum.

Yamashina et al. [104] applied feedforward neural networks to diagnose servovalve failures. Several three-layer feedforward neural networks were trained using a learning algorithm based on the combination of the conjugate gradient and a variable metric method to expedite convergence. The neural networks learned to diagnose three types of servovalve failures given time-series vibration data with reliability of over 99%. As expected, the most reliable diagnosis was obtained for neural networks with nonlinear classification capabilities. The neural network diagnosis system was useful to circumvent the weaknesses of visual inspection, especially for multiple causes faults.

Spelt et al. [105] discussed neural networks and rule-based expert systems (ES) in a hybrid artificial intelligence system to detect and diagnose faults and/or control complex automated manufacturing processes. The hybrid system was an attempt to build a more robust intelligent system rather than using either ES or neural network alone by combining the strengths of ES and neural networks. The original hybrid system was designed for intelligent machine perception and production control. The system was tested with simulated power plant data to demonstrate its potential for manufacturing process control. A particularly useful feature of the system was its capability for self-organization through a feedback loop between the neural network and the ES. This loop allowed the modification of the knowledge contained in the neural network and/or in the ES. Further research is investigating whether the hybrid architecture would be capable of unsupervised learning without destroying or invalidating its knowledge base. The proposed system represents a significant step toward creating an intelligent, automated consultant for automated process control.

Ray [106] developed a neural-network/expert system for engine fault diagnosis in an integrated steel industry. A multilayer feedforward neural network was trained with engine fault information including maintenance history, symptoms, typical questions asked for each symptom, and causes of faults. The resulting weights of the neural network represented the knowledge base of the engine fault system. The inference was done in two steps, starting with forward chaining based on symptoms of faults and then backward chaining based on the questions asked to the user. The trained system was able to perform fairly reliable diagnosis with a 75% accuracy.

Knapp and Wang [107] used a multilayer feedforward neural network trained with the BP algorithm for machine fault diagnosis. Training data (frequency domain data of vibration signals) were collected over a period of time under artificially created machining conditions and input to the neural network. The neural network had excellent performance, correctly identifying the fault class in all test cases. Possible extensions include multiple simultaneous fault conditions, multisensor integration, and active identification of fault conditions.

Hou et al. [108] applied a multilayer feedforward neural network for quality decision making in an automated inspection system for surface mount devices on printed circuit boards (PCB). The system included a Hough transform and a multilayer neural network trained using the BP algorithm. The neural network learned to classify the quality status from image information. Hough transformation reduced the amount of data to expedite the training and recognition process, while preserving all vital information. The automated inspection system was very effective for surface-mounted assemblies and had a significantly higher detection accuracy than the traditional template-matching approach. Major defects were detected such as missing component, misaligned components, and wrong component. This automated inspection system is particularly promising, since it could lead to streamlining the entire PCB production process, from assembly to inspection.

Liu and Iyer [109] used a multilayer feedforward neural network trained with the BP algorithm to diagnose various kinds of roller bearing defects. Trained with radial acceleration features on five types of defective roller bearings as well as a normal bearing, the neural network was able to separate normal and defective bearings with a 100% accuracy, and to classify the defects into the various defect categories with an accuracy of 94%. The proposed method was demonstrated to be more reliable than traditional diagnosis techniques in identifying defective bearings.

Huang and Wang [110] used an ART-2 neural network with parametric modeling of vibration signals for machine faults monitoring and diagnosing. The parametric methods considered were the autoregressive and autoregressive and moving average models. The ART-2 neural network perfectly identified testing patterns with both models. However, the autoregressive model was shown more desirable for real-world applications in terms of computational speed and frequency resolution.

Wang et al. [111] used the multilayer feedforward neural network with the BP learning algorithm to detect the surface flaws of products. The surface images of products were skeletonized and encoded into

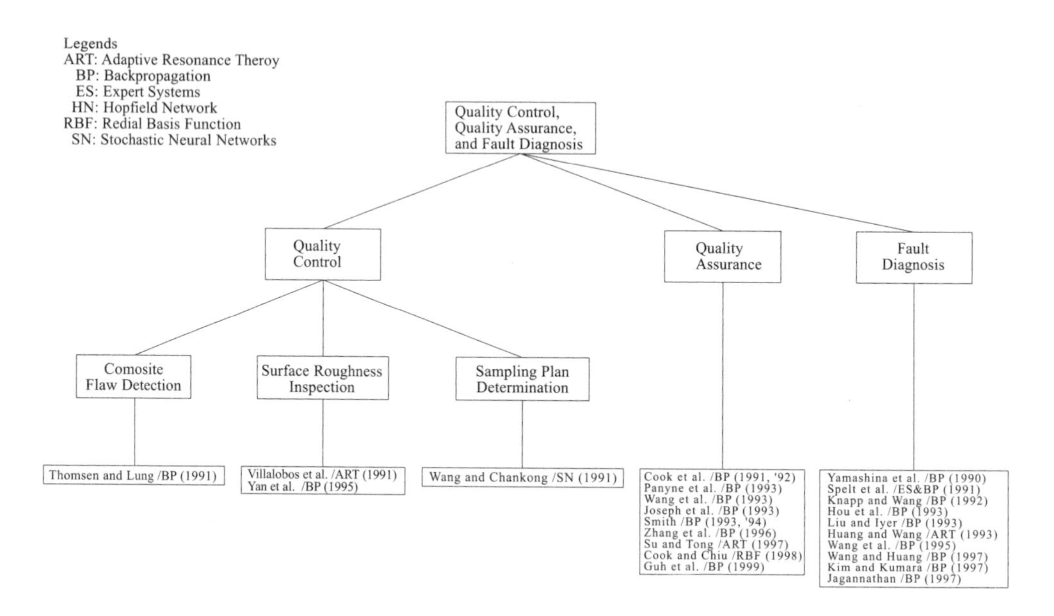

FIGURE 2.5 Hierarchy of neural network applications for quality control, quality assurance, and fault diagnosis.

a fixed number of inputs for the trained neural network to determine the surface having flaws or not. The approach was shown promising in identifying surface flaws that were not at the product boundary. In a further work, Wang and Huang [112] added to the parent inspection process an auxiliary subskeleton matching process for double confirmation of flaws, which resulted in a 97.5% correct boundary flaws identification. Moreover, the neural network connection weights were determined by the adaptive conjugate gradient learning algorithm for reducing the training time.

Kim and Kumara [113] compared the effectiveness between neural networks and traditional pattern classifiers for identification of defective boundary of casting parts. The visual image of the part boundary was captured and represented by a combination of linear and circular features using a quintuple vector. Two neural networks, multilayer perceptron trained by the BP algorithm and Hopfield network, and two traditional statistics-based methods—linear discriminant analysis and C-means algorithm—were applied to recognize whether the part boundary is defective based on the quintuple vector. The experimental results showed that the correct recognition of the multilayer perceptron and the Hopfield network ranged from 81 to 100% and 75 to 93%, respectively, while that of both the linear discriminant analysis and the C-means algorithm ranged from 57 to 75%.

Jagannathan [114] applied a multilayer feedforward neural network with the BP learning algorithm to identify and classify the defective solder joints. A modified intelligent histogram regrading technique developed by the author was used to divide the gray-level histogram of the captured image from a joint into different modes. Each mode was identified by the trained neural network to indicate the joint welding conditions of good, no solder, or excess solder. The neural-network-based inspection system was found promising in that it operated in near real-time on a 80386-based microcomputer.

In summary, the present applications of neural networks to quality control, quality assurance, and fault diagnosis include composite floor detection, surface roughness inspection, out-of-control prediction, sampling plan determination, and process and machine fault diagnosis, as shown in Figure 2.5. The neural network models used were multilayer feedforward networks, ART, and stochastic networks. Neural networks, especially when combined with expert systems, demonstrated promise as a tool for quality control, quality assurance, and fault diagnosis. The pattern recognition and parallel computation abilities of neural networks are especially beneficial for these applications.

2.6 Concluding Remarks

The factory of the future and the quality of its products will depend largely on the full integration of intelligent systems for designing, planning, monitoring, modeling, and controlling manufacturing systems and processes. Neural networks have proved able to contribute to solving many problems in manufacturing. In addition to the ability to adapt and learn in dynamic manufacturing environments, neural networks make weak assumptions regarding underlying processes. They are applicable for a wide range of real-world problems. Neural networks, however, are not a substitute for classical methods. Instead, they are viable tools that can be supplementary and used in cooperation with traditional methods, especially in instances where the expense of in-depth mathematical analysis cannot be justified. Furthermore, neural networks by no means replace the computational capabilities provided by digital computers. Instead, neural networks would provide complementary capabilities to existing computers.

A number of characteristics of some neural networks seem to limit their use in real-time, real-world manufacturing settings. Problems include lengthy training time, uncertainty of convergence, and the arbitrariness of choosing design parameters. Moreover, neural networks lack the capability for explanation of the learning outcome, and it is almost impossible to discern what has been learned from examination of the weights matrices that result from learning. Further research and development are needed before neural networks can be completely and successfully applied for real-world manufacturing. Because neural networks hardware devices are not yet commercially available for manufacturing applications, the use of neural networks is still constrained to simulations on sequential computing machines. Training a large network using a sequential machine can be time-consuming. Fortunately, training usually takes place off line, and the efficiency of training can be improved using more efficient learning algorithms. Furthermore, software tools and insert boards are currently available that permit neural network programs to run on desktop computers, making them applicable to a wide range of manufacturing applications. The advances in VLSI neural chips will eventually accelerate computation and generate solutions with minimum time, space, and energy consumption.

References

1. Wu, B., An introduction to neural networks and their applications in manufacturing, *Journal of Intelligent Manufacturing,* 3, 391, 1992.
2. Udo, G. J., Neural networks applications in manufacturing processes, *Computers and Industrial Engineering,* 23, 97, 1992.
3. Chryssolouris, G., Lee, M., Pierce, J., and Domroese, M., The use of neural networks for the design of manufacturing systems, *Manufacturing Review,* 3, 187, 1990.
4. Chryssolouris, G., Lee, M., and Domroese, M., The use of neural networks in determining operational policies for manufacturing systems, *Journal of Manufacturing Systems,* 10, 166, 1991.
5. Madey, G. R., Weinroth, J., and Shah, V., Integration of neurocomputing and system simulation for modeling continuous improvement systems in manufacturing, *Journal of Intelligent Manufacturing,* 3, 193, 1992.
6. Kamarthi, S. V., Kumara, S. T., Yu, F. T. S., and Ham, I., Neural networks and their applications in component design data retrieval, *Journal of Intelligent Manufacturing,* 1, 125, 1990.
7. Kaparthi, S., and Suresh, N. C., A neural network system for shape-based classification and coding of rotational parts, *International Journal of Production Research,* 29, 1771, 1991.
8. Moon, Y. B., and Roy, U., Learning group-technology part families from solid models by parallel distributed processing, *International Journal of Advanced Manufacturing Technology,* 7, 109, 1992.
9. Venugopal, V., and Narendran, T. T., Neural network model for design retrieval in manufacturing systems, *Computers in Industry,* 20, 11, 1992.
10. Chakraborty, K., and Roy, U., Connectionist models for part-family classifications, *Computers and Industrial Engineering,* 2, 189, 1993.

11. Kiang, M. Y., Kulkarni, U. R., and Tam, K. Y., Self-organizing map network as an interactive clustering tool: An application to group technology, *Decision Support Systems,* 15, 351, 1995.
12. Wu, M. C., and Jen, S. R., A neural network approach to the classification of 3D prismatic parts, *International Journal of Advanced Manufacturing Technology,* 11, 325, 1996.
13. Moon, Y. B., Forming part-machine families for cellular manufacturing: A neural-network approach, *International Journal of Advanced Manufacturing Technology,* 5, 278, 1990.
14. Moon, Y. B., Establishment of a neurocomputing model for part family/machine group identification, *Journal of Intelligent Manufacturing,* 3, 173, 1992.
15. Moon, Y. B., and Chi, S. C., Generalized part family formation using neural network techniques, *Journal of Manufacturing Systems,* 11, 149, 1992.
16. Currie, K. R., An intelligent grouping algorithm for cellular manufacturing, *Computers and Industrial Engineering,* 23, 109, 1992.
17. Kusiak, A., and Chung, Y., GT/ART: Using artificial neural networks to form machine cells, *Manufacturing Review,* 4, 293, 1991.
18. Kaparthi, S., and Suresh, N. C., Machine-component cell formation in group technology: A neural network approach, *International Journal of Production Research,* 30, 1353, 1992.
19. Liao, T. W., and Chen, L. J., An evaluation of ART-1 neural networks for GT part family and machine cell forming, *Journal of Manufacturing Systems,* 12, 282, 1993.
20. Kaparthi, S., Suresh, N. C., and Cerveny, R. P., An improved neural network leader algorithm for part-machine grouping in group technology, *European Journal of Operational Research,* 69, 342, 1993.
21. Moon, Y. B., and Kao, Y., Automatic generation of group technology families during the part classification process, *International Journal of Advanced Manufacturing Technology,* 8, 160, 1993.
22. Dagli, C. H., and Huggahalli, G., A neural network approach to group technology, *Neural Networks in Design and Manufacturing,* Wang, J., and Takefuji, Y., Eds., World Scientific, Singapore, 1993, 1.
23. Moon, Y. B., Neuroclustering for group technology, *Neural Networks in Design and Manufacturing,* Wang, J., and Takefuji, Y., Eds., World Scientific, Singapore, 1993, 57.
24. Rao, H. A., and Gu, P., Expert self-organizing neural network for the design of cellular manufacturing systems, *Journal of Manufacturing Systems,* 13, 346, 1994.
25. Rao, H. A., and Gu, P., A multi-constraint neural network for the pragmatic design of cellular manufacturing systems, *International Journal of Production Research,* 33, 1049, 1995.
26. Chen, S. J., and Cheng, C. S., A neural network-based cell formation algorithm in cellular manufacturing, *International Journal of Production Research,* 33, 293, 1995.
27. Burke, L., and Kamal, S., Neural networks and the part family/machine group formation problem in cellular manufacturing: A framework using fuzzy ART, *Journal of Manufacturing Systems,* 14, 148, 1995.
28. Kamal, S., and Burke, L., FACT: A new neural network-based clustering algorithm for group technology, *International Journal of Production Research,* 34, 919, 1996.
29. Chang, C. A., and Tsai, C. Y., Using ART-1 neural networks with destructive solid geometry for design retrieving systems, *Computers in Industry,* 34, 27, 1997.
30. Enke, D., Ratanapan, K., and Dagli, C., Machine-part family formation utilizing an ART-1 neural network implemented on a parallel neuro-computer, *Computers and Industrial Engineering,* 34, 189, 1998.
31. Suresh, N. C., Slomp, J., and Kaparthi, S., Sequence-dependent clustering of parts and machines: A fuzzy ART neural network approach, *International Journal of Production Research,* 37, 2793, 1999.
32. Lee, S. Y., and Fischer, G. W., Grouping parts based on geometrical shapes and manufacturing attributes using a neural network, *Journal of Intelligent Manufacturing,* 10, 199, 1999.
33. Malavé, C. O., and Ramachandran, S., Neural network-based design of cellular manufacturing systems, *Journal of Intelligent Manufacturing,* 2, 305, 1991.

34. Lee, H., Malavé, C. O., and Ramachadran, S., A self-organizing neural network approach for the design of cellular manufacturing systems, *Journal of Intelligent Manufacturing,* 3, 325, 1992.
35. Liao, T. W., and Lee, K. S., Integration of a feature-based CAD system and an ART-1 neural model for GT coding and part family forming, *Computers and Industrial Engineering,* 26, 93, 1994.
36. Malakooti, B., and Yang, Z., A variable-parameter unsupervised learning clustering neural network approach with application to machine-part group formation, *International Journal of Production Research,* 33, 2395, 1995.
37. Arizono, I., Kato, M., Yamamoto, A., and Ohta, H., A new stochastic neural network model and its application to grouping parts and tools in flexible manufacturing systems, *International Journal of Production Research,* 33, 1535, 1995.
38. Zolfaghari, S., and Liang, M., An objective-guided ortho-synapse hopfield network approach to machine grouping problems, *International Journal of Production Research,* 35, 2773, 1997.
39. Kao, Y., and Moon, Y. B., A unified group technology implementation using the backpropagation learning rule of neural networks, *Computers and Industrial Engineering,* 20, 425, 1991.
40. Jamal, A. M. M., Neural networks and cellular manufacturing: The benefits of applying a neural network to cellular manufacturing, *Industrial Management and Data Systems,* 93, 21, 1993.
41. Chung, Y., and Kusiak, A., Grouping parts with a neural network, *Journal of Manufacturing Systems,* 13, 262, 1994.
42. Andersen, K., Cook, G. E., Karsai, G., and Ramaswamy, K., Artificial neural networks applied to arc welding process modeling and control, *IEEE Transactions on Industrial Applications,* 26, 824, 1990.
43. Tansel, I. N., Modelling 3-D cutting dynamics with neural networks, *International Journal of Machine Tools and Manufacture,* 32, 829, 1992.
44. Dagli, C. H., Lammers, S., and Vellanki, M., Intelligent scheduling in manufacturing using neural networks, *Journal of Neural Networks Computing,* 2, 4, 1991.
45. Arizono, I., Yamamoto, A., and Ohta, H., Scheduling for minimizing total actual flow time by neural networks, *International Journal of Production Research,* 30, 503, 1992.
46. Cho, H., and Wysk, R. A., A robust adaptive scheduler for an intelligent workstation controller, *International Journal of Production Research,* 31, 771, 1993.
47. Lo, Z. P., and Bavarian, B., Multiple job scheduling with artificial neural networks, *Computers and Electrical Engineering,* 19, 87, 1993.
48. Lee, Y. H., and Kim, S., Neural network applications for scheduling jobs on parallel machines, *Computers and Industrial Engineering,* 25, 227, 1993.
49. Satake, T., Morikawa, K., and Nakamura, N., Neural network approach for minimizing the makespan of the general job-shop, *International Journal of Production Economics,* 33, 67, 1994.
50. Wang, L. C., Chen, H. M., and Liu, C. M., Intelligent scheduling of FMSs with inductive learning capability using neural networks, *The International Journal of Flexible Manufacturing Systems,* 7, 147, 1995.
51. Sabuncuoglu, I., and Gurgun, B., A neural network model for scheduling problems, *European Journal of Operational Research,* 93, 288, 1996.
52. Li, D. C., Wu, C., and Torng, K. Y., Using an unsupervised neural network and decision tree as knowledge acquisition tools for FMS scheduling, *International Journal of Systems Science,* 28, 977, 1997.
53. Kim, C. O., Min, H. S., and Yih, Y., Integration of inductive learning and neural networks for multi-objective FMS scheduling, *International Journal of Production Research,* 36, 2497, 1998.
54. Knapp, G. M., and Wang, H. P. B., Acquiring, storing and utilizing process planning knowledge using neural networks, *Journal of Intelligent Manufacturing,* 3, 333, 1992.
55. Chen, C. L. P., and Pao, Y. H., An integration of neural network and rule-based systems for design and planning of mechanical assemblies, *IEEE Transactions on Systems, Man, and Cybernetics,* 23, 1359, 1993.

56. Shu, S. H., and Shin, Y. S., Neural network modeling for tool path planning of rough cut in complex pocket milling, *Journal of Manufacturing Systems,* 15, 295, 1996.
57. Osakada, K., and Yang, G., Application of neural networks to an expert system for cold forging, *International Journal of Machine Tools Manufacturing,* 31, 577, 1991.
58. Eberts, R. E., and Nof, S. Y., Distributed planning of collaborative production, *International Journal of Manufacturing Technology,* 8, 258, 1993.
59. Rangwala, S. S., and Dornfeld, D. A., Learning and optimization of machining operations using computing abilities of neural networks, *IEEE Transactions on Systems, Man and Cybernetics,* 19, 299, 1989.
60. Cook, D. F., and Shannon, R. E., A sensitivity analysis of a back-propagation neural network for manufacturing process parameters, *Journal of Intelligent Manufacturing,* 2, 155, 1991.
61. Sathyanaryanan, G., Lin, I. J., and Chen, M. K., Neural networks and multiobjective optimization of creep grinding of superalloys, *International Journal of Production Research,* 30, 2421, 1992.
62. Matsumara, T., Obikawa, T., Shirakashi, T., and Usui, E., Autonomous turning operation planning with adaptive prediction of tool wear and surface roughness, *Journal of Manufacturing Systems,* 12, 253, 1993.
63. Wang, J., Multiple-objective optimization of machining operations based on neural networks, *International Journal of Advanced Manufacturing Technology,* 8, 235, 1993.
64. Roy, U., and Liao, J., A neural network model for selecting machining parameters in fixture design, *Integrated Computer-Aided Engineering,* 3, 149, 1996.
65. Chen, Y. T., and Kumara, S. R. T., Fuzzy logic and neural networks for design of process parameters: A grinding process application, *International Journal of Production Research,* 36, 395, 1998.
66. Barschdorff, D., and Monostori, L., Neural networks—Their applications and perspectives in intelligent machining, *Computers in Industry,* 17, 101, 1991.
67. Rangwala, S. S., and Dornfeld, D. A., Sensor integration using neural networks for intelligent tool condition monitoring, *Journal of Engineering for Industry,* 112, 219, 1990.
68. Burke, L. I., and Rangwala, S. S., Tool condition monitoring in metal cutting: A neural network approach, *Journal of Intelligent Manufacturing,* 2, 269, 1991.
69. Burke, L. I., Competitive learning based approaches to tool-wear identification, *IEEE Transactions on Systems, Man, and Cybernetics,* 22, 559, 1992.
70. Burke, L. I., An unsupervised neural network approach to tool wear identification, *IIE Transactions,* 25, 16, 1993.
71. Yao, Y. L., and Fang, X. D., Assessment of chip forming patterns with tool wear progression in machining via neural networks, *International Journal of Machine Tools and Manufacture,* 33, 89, 1993.
72. Tarng, Y. S., Hseih, Y. W., and Hwang, S. T., Sensing tool breakage in face milling with a neural network, *International Journal of Machine Tools and Manufacture,* 34, 341, 1994.
73. Ko, T. J., Cho, D. W., and Jung, M. Y., On-line monitoring of tool breakage in face milling using a self-organized neural network, *Journal of Manufacturing Systems,* 14, 80, 1995.
74. Ko, T. J., and Cho, D. W., Adaptive modeling of the milling process and application of a neural network for tool wear monitoring, *International Journal of Advanced Manufacturing Technology,* 12, 5, 1996.
75. Chao, P. Y., and Hwang, Y. D., An improved neural network model for the prediction of cutting tool life, *Journal of Intelligent Manufacturing,* 8, 107, 1997.
76. Jemielniak, K., Kwiatkowski, L., and Wrzosek, P., Diagnosis of tool wear based on cutting forces and acoustic emission measures as inputs to a neural network, *Journal of Intelligent Manufacturing,* 9, 447, 1998.
77. Purushothaman, S., and Srinivasa, Y. G., A procedure for training an artificial neural network with application to tool wear monitoring, *International Journal of Production Research,* 36, 635, 1998.
78. Alguindigue, I. E., Loskiewicz-Buczak, A., and Uhrig, R. E., Monitoring and diagnosis of rolling element bearing using a neural network, *IEEE Transactions on Industrial Electronics,* 40, 209, 1993.

79. Hou, T. H., and Lin, L., Manufacturing process monitoring using neural networks, *Computers and Electrical Engineering,* 19, 129, 1993.
80. Tansel, I. N., Mekdeci, C., Rodriguez, O., and Uragun, B., Monitoring drill conditions with wavelet based encoding and neural networks, *International Journal of Machine Tools and Manufacture,* 33, 559, 1993.
81. Lee, J., and Kramer, B. M., Analysis of machine degradation using a neural network based pattern discrimination model, *Journal of Manufacturing Systems,* 12, 379, 1993.
82. Currie, K. R., and LeClair, S. R., Self-improving process control for molecular beam epitaxy, *International Journal of Advanced Manufacturing Technology,* 8, 244–251, 1993.
83. Balazinski, M., Czogala, E., and Sadowski, T., Modeling of neural controllers with application to the control of a machining process, *Fuzzy Sets and Systems,* 56, 273, 1993.
84. Lichtenwalner, P. F., Neural network-based control for the fiber placement composite manufacturing process, *Journal of Materials Engineering and Performance,* 2, 687, 1993.
85. Ding, H., Yang, S., and Zhu, X., Intelligent prediction and control of a leadscrew grinding process using neural networks, *Computers in Industry,* 23, 169, 1993.
86. Chen, J. S., Neural network-based modeling and error compensation of thermally-induced spindle errors, *International Journal of Advanced Manufacturing Technology,* 12, 303, 1996.
87. Vancherck, P., and Nuttin, M., Compensation of thermal deformations in machine tools with neural network, *Computers in Industry,* 33, 119, 1997.
88. Thomsen, J. J., and Lund, K., Quality control of composite materials by neural network analysis of ultrasonic power spectra, *Materials Evaluation,* 49, 594, 1991.
89. Villabos, L., and Gruber, S., Measurement of surface roughness parameter using a neural network and laser scattering, *Industrial Metrology,* 2, 33, 1991.
90. Yan, D., Cheng, M., Popplewell, N., and Balakrishnan, S., Application of neural networks for surface roughness measurement in finish turning, *International Journal of Production Research,* 33, 3425, 1995.
91. Pugh, A. G., A comparison of neural networks to SPC charts, *Computers and Industrial Engineering,* 21, 253, 1991.
92. Wang, J., and Chankong, V., Neurally-inspired stochastic algorithm for determining multi-stage multi-attribute sampling inspection plans, *Journal of Intelligent Manufacturing,* 2, 327, 1991.
93. Cook, D. F., Massey, J. G., and Shannon, R. E., A neural network to predict particleboard manufacturing process parameters, *Forest Science,* 37, 1463, 1991.
94. Cook, D. F., and Shannon, R. E., A predictive neural network modeling system for manufacturing process parameters, *International Journal of Production Research,* 30, 1537, 1992.
95. Payne, R. D., Rebis, R. E., and Moran, A. L., Spray forming quality predictions via neural networks, *Journal of Materials Engineering and Performance,* 2, 693, 1993.
96. Wang, Q., Sun, X., Golden, B. L., DeSilets, L., Wasil, E. A., Luco, S., and Peck, A., A neural network model for the wire bonding process, *Computers and Operations Research,* 20, 879, 1993.
97. Joseph, B., and Hanratty, F. W., Predictive control of quality in a batch manufacturing process using artificial neural networks models, *Industry and Engineering Chemistry Research,* 32, 1951, 1993.
98. Smith, A. E., Predicting product quality with backpropagation: A thermoplastic injection molding case study, *International Journal of Advanced Manufacturing Technology,* 8, 252, 1993.
99. Smith, A. E., X-bar and R control chart integration using neural computing, *International Journal of Production Research,* 32, 309, 1994.
100. Zhang, Y. F., Nee, A. Y. C., Fuh, J. Y. H., Neo, K. S., and Loy, H. K., A neural network approach to determining optimal inspection sampling size for CMM, *Computer Integrated Manufacturing Systems,* 9, 161, 1996.
101. Su, C. T., and Tong, L. I., A neural network-based procedure for the process monitoring of clustered defects in integrated circuit fabrication, *Computer in Industry,* 34, 285, 1997.

102. Cook, D. F., and Chiu, C. C., Using radial basis function neural networks to recognize shift in correlated manufacturing process parameters, *IIE Transactions,* 30, 227, 1998.
103. Guh, R. S., and Tannock, J. D. T., Recognition of control chart concurrent patterns using a neural network approach, *International Journal of Production Research,* 37, 1743, 1999.
104. Yamashina, H., Kumamoto, H., Okumura, S., and Ikesak, T., Failure diagnosis of a servovalve by neural networks with new learning algorithm and structure analysis, *International Journal of Production Research,* 28, 1009, 1990.
105. Spelt, P. F., Knee, H. E., and Glover, C. W., Hybrid artificial intelligence architecture for diagnosis and decision making in manufacturing, *Journal of Intelligent Manufacturing,* 2, 261, 1991.
106. Ray, A. K., Equipment fault diagnosis: A neural network approach, *Computers in Industry,* 16, 169, 1991.
107. Knapp, G. M., and Wang, H. P. B., Machine fault classification: A neural network approach, *International Journal of Production Research,* 30, 811, 1992.
108. Hou, T. H., Lin, L., and Scott, P. D., A neural network-based automated inspection system with an application to surface mount devices, *International Journal of Production Research,* 31, 1171, 1993.
109. Liu, T. I., and Iyer, N. R., Diagnosis of roller bearing defects using neural networks, *International Journal of Advanced Manufacturing Technology,* 8, 210, 1993.
110. Huang, H. H., and Wang, H. P, Machine fault classification using an ART-2 neural network, *International Journal of Advanced Manufacturing Technology,* 8, 194, 1993.
111. Wang, C., Cannon, D., Kumara, S. R. T., and Lu G., A skeleton and neural network-based approach for identifying cosmetic surface flaws, *IEEE Transactions on Neural Networks,* 6, 1201, 1995.
112. Wang, C., and Huang, S. Z., A refined flexible inspection method for identifying surface flaws using the skeleton and neural network, *International Journal of Production Research,* 35, 2493, 1997.
113. Kim, T., and Kumara, S. R. T., Boundary defect recognition using neural networks, *International Journal of Production Research,* 35, 2397, 1997.
114. Jagannathan, S., Automatic inspection of wave soldered joints using neural networks, *Journal of Manufacturing Systems,* 16, 389, 1997.

II

Manufacturing System Modeling and Design

3

Holonic Metamorphic Architectures for Manufacturing: Identifying Holonic Structures in Multiagent Systems by Fuzzy Modeling

Michaela Ulieru
The University of Calgary

Dan Stefanoiu
The University of Calgary

Douglas Norrie
The University of Calgary

3.1 Introduction

Global competition and rapidly changing customer requirements are forcing major changes in the production styles and configuration of manufacturing organizations. Increasingly, traditional centralized and sequential manufacturing planning, scheduling, and control mechanisms are being found to be insufficiently flexible to respond to changing production styles and highly dynamic variations in product requirements. In these traditional hierarchical organizations, manufacturing resources are grouped into semipermanent, tightly coupled subgroups, with a centralized software supervisor processing information sequentially. Besides plan fragility and increased response overheads, this may result in much of the system being shut down by a single point of failure. Conventional-knowledge engineering approaches with large-scale or very-large-scale knowledge bases become inadequate in this highly distributed environment.

0-8493-0592-6/01/$0.00+$.50

The next generation of intelligent manufacturing systems is envisioned to be agile, adaptive, and fault tolerant. They need to be distributed virtual enterprises comprised of dynamically reconfigurable production resources interlinked with supply and distribution networks. Within these enterprises and their resources, both knowledge processing and material processing will be concurrent and distributed. To create this next generation of intelligent manufacturing systems and to develop the near-term transitional manufacturing systems, new and improved approaches to distributed intelligence and knowledge management are essential. Their application to manufacturing and related enterprises requires continuing exploration and evaluation.

Agent technology derived from distributed artificial intelligence has proved to be a promising tool for the design, modeling, and implementation of distributed manufacturing systems. In the past decade (Jennings et al. 1995; Shen and Norrie 1999; Shen et al. 2000), numerous researchers have shown that agent technology can be applied to manufacturing enterprise integration, supply chain management, intelligent design, manufacturing scheduling and control, material handling, and holonic manufacturing systems.

3.2 Agent-Oriented Manufacturing Systems

The requirements for twenty-first century manufacturing necessitate decentralized manufacturing facilities whose design, implementation, reconfiguration, and manufacturability allow the integration of production stages in a dynamic, collaborative network. Such facilities can be realized through agent-oriented approaches (Wooldridge and Jennings 1995) using knowledge sharing technology (Patil et al. 1992).

Different agent-based architectures have been proposed in the research literature. The autonomous agent architecture is well suited for developing distributed intelligent design and manufacturing systems in which existing engineering tools are encapsulated as agents and the system consists of a small number of agents. In the federation architecture with facilitators or mediators, a hierarchy is imposed for every specific task, which provides computational simplicity and manageability. This type of architecture is quite suitable for distributed manufacturing systems that are complex, dynamic, and composed of a large number of resource agents. These architectures, and others, have been used for agent-based design and/or manufacturing systems, some of which are reviewed in the remainder of this section.

In one of the earliest projects, Pan and Tenenbaum (1991) described a software intelligent agent (IA) framework for integrating people and computer systems in large, geographically dispersed manufacturing enterprises. This framework was based on the vision of a very large number of computerized assistants, known as intelligent agents (IAs). Human participants are encapsulated as personal assistants (PAs), a special type of IA.

ADDYMS (Architecture for Distributed Dynamic Manufacturing Scheduling) by Butler and Ohtsubo (1992) was a distributed architecture for dynamic scheduling in a manufacturing environment.

Roboam and Fox (1992) used an enterprise management network (EMN) to support the integration of activities of the manufacturing enterprise throughout the production life cycle with six levels: (1) Network Layer provides for the definition of the network structure; (2) Data Layer provides for inter-node queries; (3) Information Layer provides for invisible access to information spread throughout the EMN; (4) Organization Layer provides the primitives and elements for distributed problem solving; (5) Coordination Layer provides protocols for coordinating the activities of EMN nodes; and (6) Market Layer provides protocols for coordinating organizations in a market environment.

The SHADE project (McGuire et al. 1993) was primarily concerned with the information-sharing aspect of concurrent engineering. It provides a flexible infrastructure for anticipated knowledge-based, machine-mediated collaboration among disparate engineering tools. SHADE differs from other approaches in its emphasis on a distributed approach to engineering knowledge rather than a centralized model or knowledge base. SHADE notably avoids physically centralized knowledge, but distributes the modeling vocabulary as well, focusing knowledge representation on specific knowledge-sharing needs.

PACT (Cutkosky et al. 1993) was a landmark demonstration of both collaborative research efforts and agent-based technology. Its agent interaction relies on shared concepts and terminology for communicating knowledge across disciplines, an *interlingua* for transferring knowledge among agents, and a communication and control language that enables agents to request information and services. This technology allows agents working on different aspects of a design to interact at the knowledge level, sharing and exchanging information about the design independent of the format in which the information is encoded internally.

SHARE (Toye et al. 1993) was concerned with developing open, heterogeneous, network-oriented environments for concurrent engineering. It used a wide range of information-exchange technologies to help engineers and designers collaborate in mechanical domains.

Recently, PACT has been replaced by PACE (Palo Alto Collaborative Environment) [http://cdr.stanford.edu/PACE/] and SHARE by DSC (Design Space Colonization) [http://cdr.stanford.edu/DSC/].

First-Link (Park et al. 1994) was a system of semi-autonomous agents helping specialists to work on one aspect of the design problem. Next-Link (Petrie et al. 1994) was a continuation of the First-Link project for testing agent coordination. Process-Link (Goldmann 1996) followed on from Next-Link and provides for the integration, coordination, and project management of distributed interacting CAD tools and services in a large project.

Saad et al. (1995) proposed a production reservation approach by using a bidding mechanism based on the contract net protocol to generate the production plan and schedule. SiFA (Brown et al. 1995), developed at Worcester Polytechnic, was intended to address the issues of patterns of interaction, communication, and conflict resolution. DIDE (Shen and Barthès 1997) used autonomous cognitive agents for distributed intelligent design environments. Maturana et al. (1996) described an integrated planning-and-scheduling approach combining subtasking and virtual clustering of agents with a modified contract net protocol.

MADEFAST (Cutkosky et al. 1996) was a DARPA DSO-sponsored project to demonstrate technologies developed under the ARPA MADE (Manufacturing Automation and Design Engineering) program. MADE is a DARPA DSO long-term program for developing tools and technologies to provide cognitive support to the designer and allow an order of magnitude increase in the explored alternatives in half the time it currently takes to explore a single alternative.

In AARIA (Parunak et al. 1997a), manufacturing capabilities (e.g., people, machines, and parts) are encapsulated as autonomous agents. Each agent seamlessly interoperates with other agents in and outside of its own factory. AARIA uses a mixture of heuristic scheduling techniques: forward/backward scheduling, simulation scheduling, and intelligent scheduling. Scheduling is performed by job, by resource, and by operation. Scheduling decisions are made to minimize costs over time and production quantities.

RAPPID (Responsible Agents for Product-Process Integrated Design) (Parunak et al. 1997b) at the Industrial Technology Institute was intended to develop agent-based software tools and methods for using marketplace dynamics among members of a distributed design team to coordinate set-based design of a discrete manufactured product. AIMS (Park et al. 1993) was envisioned as integrating the U.S. industrial base and enabling it to rapidly respond, with highly customized solutions, to customer requirements of any magnitude.

3.3 The MetaMorph Project

At the University of Calgary, a number of research projects in multiagent systems have been undertaken since 1991. These include IAO (Kwok and Norrie 1993), Mediator (Gaines et al. 1995), ABCDE (Balasubramanian et al. 1996), MetaMorph I (Maturana and Norrie 1996; Maturana et al. 1998), MetaMorph II (Shen et al. 1998a), Agent-Based Intelligent Control (Brennan et al. 1997; Wang et al., 1998), and Agent-Based Manufacturing Scheduling (Shen and Norrie 1998). An overview of these projects with a summary of techniques and mechanisms developed during these projects and a discussion of key issues can be found in (Norrie and Shen 1999). The MetaMorph project is considered in some detail below. For additional details on the MetaMorph I project see (Maturana et.al. 1999).

MetaMorph incorporates planning, control and application agents that collaborate to satisfy both local and global objectives. Virtual clusters of agents are dynamically created, modified, and destroyed as needed for collaborative planning and action on tasks. Mediator agents coordinate activities both within clusters and across clusters (Maturana and Norrie, 1996.)

3.3.1 The MetaMorphic Architecture

In the first phase of the MetaMorph project (Maturana and Norrie 1996) a multiagent architecture for intelligent manufacturing was developed. The architecture has been named MetaMorphic, since a primary characteristic is reconfigurability, i.e., its ability to change structure as it dynamically adapts to emerging tasks and changing environment.

In this particular type of federation organization, intelligent agents link with mediator agents to find other agents in the environment. The mediator agents assume the role of system coordinators, promoting cooperation among intelligent agents and learning from the agents' behavior. Mediator agents provide system associations without interfering with lower-level decisions unless critical situations occur. Mediator agents are able to expand their coordination capabilities to include mediation behaviors, which may be focused upon high-level policies to break decision deadlocks. Mediation actions are performance-directed behaviors.

The generic model for mediators in MetaMorph includes the following seven meta-level activities: Enterprise, Product Specification and Design, Virtual Organizations, Planning and Scheduling, Execution, Communication and Learning, as shown in Figure 3.1. Each mediator includes some or all of these activities to a varying extent. Prototyping with this generic model and related methodology facilitates the creation of diverse types of mediators. Thus, a mediator may be specialized for organizational issues (enterprise mediator) or for shop-floor production coordination (execution mediator). Although each of these mediator types will have different manufacturing knowledge, both conform to a similar generic specification. The activity domains in Figure 3.1 are further described as follows:

- The enterprise domain globalizes knowledge of the system and represents the facility's goals through a series of objectives. Enterprise knowledge enables environment recognition and maintenance of organizational associations.
- The product specification and design domain includes encoding data for manufacturing tasks to enable mediators to recognize the tasks to be coordinated.
- The virtual organization domain is similar to the enterprise domain, but its scope is detailed knowledge of resource behavior at the shop-floor level. This activity domain dynamically establishes and recognizes dynamic relationships between dissimilar resources and agents.
- The planning and scheduling domain plays an important role in integrating technological constraints with time-dependent constraints into a concurrent information-processing model (Balasubramanian et al. 1996).
- The execution domain facilitates transactions among physical devices. During the execution of tasks, it coordinates various transactions between manufacturing devices and between the devices and other domains to complete the information requirements.
- The communication domain provides a common communication language based on the KQML protocol (Finin et al. 1993) used to wrap the message content.
- The learning domain incorporates the resource capacity planning activity, which involves repetitive reasoning and message exchange and that can be learned and automated.

Manufacturing requests associated with each domain are established under both static and dynamic conditions. The static conditions relate to the design of the products (geometrical profiles). The dynamic conditions depend upon times, system loads, system metrics, costs, customer desires, etc. A more detailed description of the generic model for mediator design can be found in (Maturana 1997).

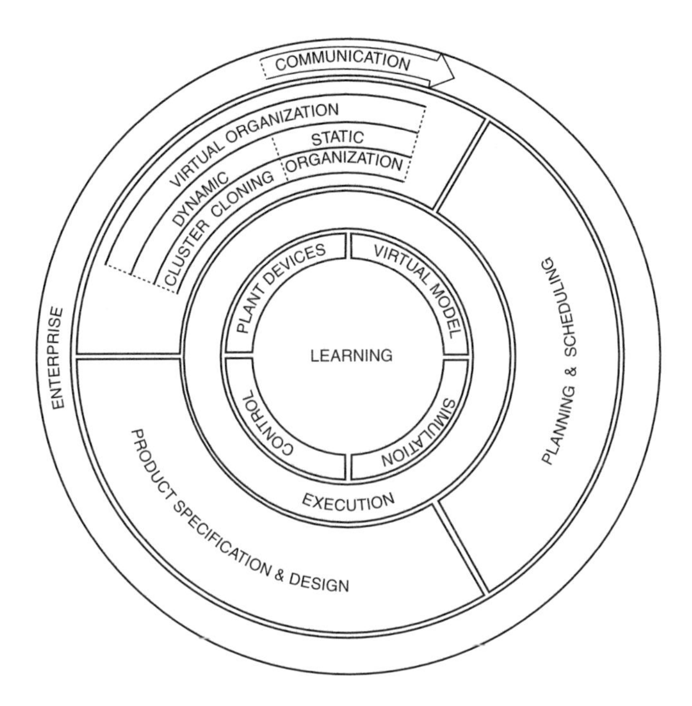

FIGURE 3.1 Generic model for mediators.

Mediators play key roles in the task decomposition and dynamic virtual clustering processes described below.

3.3.2 Agent Coalition (Clustering)

The agents may be formed into coalitions (clusters) in which dissimilar agents can work cooperatively into harmonious decision groups. Multistage negotiation and coordination protocols that can efficiently maintain the stability of these coalitions are required. Each agent has its individual representation of the external world, goals, and constraints, so diverse heterogeneous beliefs interact within a coalition through distributed cooperation models.

In MetaMorph, core reconfiguration mechanisms are based on task decomposition and dynamically formed agent groups (clusters). Mediators acting at the corresponding information level initially decompose high-level tasks. Each subtask is distributed to a subcluster with further task decomposition and clustering as necessary. As the task decomposition process is repeated, subclusters are formed and then sub-subclusters, and so on, as needed, within a dynamically interlinked structure. As the respective tasks and subtasks are solved, the related clusters and links are dissolved. However, mediators store the most relevant links, with associated task information, for future reuse. This clustering process, as described, provides scalability and aggregation properties to the system. Mediators learn dynamically from agent interactions and identify coalitions that can be used for distributed searches for the resolution of tasks.

Agents are dynamically contracted to participate in a problem-solving group (cluster). Where agents in the problem-solving group (cluster) are only able to partially complete the task's requirements, the agents will seek outside their cluster and establish conversation links with the agents in other clusters.

Mediator agents use brokering and recruiting communication mechanisms (Decker 1995) to find appropriate agents for the coordination clusters (also called collaborative subsystems or virtual clusters). The brokering mechanism consists of receiving a request message from an agent, understanding the request, finding suitable receptors for the message, and broadcasting the message to the selected group of agents. The recruiting mechanism is a superset of the brokering mechanism, since it uses the brokering mechanism to match agents. However, once appropriate agents have been found, these agents can be directly linked. The mediator agent can then step out of the scene to let the agents proceed with the communication themselves. Both mechanisms have been used in MetaMorph I. To efficiently use these mechanisms, mediator agents need to have sufficient organizational knowledge to match agent requests with needed resources. In Section 3.6, we present a mathematical solution for the grouping of agents into clusters. This can be incorporated as an algorithm within the mediator agents, to enable them to create a holonic organizational structure when forming agent coalitions.

3.3.3 Prototype Implementation

The MetaMorph architecture and coordination protocols have been used to implement a distributed concurrent design and manufacturing system in simulated form. This virtual system dynamically interconnects heterogeneous manufacturing agents in different agent-based shop floors or factories (physically separated) for concurrent manufacturability evaluation, production planning and scheduling. The system comprises the following multiagent modules: Enterprise Mediator, Design System, Shop Floors, and Execution Control & Forecasting, as shown in Figure 3.2. Each multiagent module uses common enterprise integration protocols to allow agent interoperability.

The multiagent modules are implemented within a distributed computing platform consisting of four HP Apollo 715/50 workstations, each running an HP-UX 9.0 operating system (Maturana and Norrie, 1996). The workstations communicate with each other through a local area network (LAN) and TCP/IP protocol. Graphical interfaces for each multiagent module were created in the VisualWorks 2.5 (Smalltalk) programming language, which was also used for programming the modules. The KQML protocol (Finin et al. 1993) is used as high-level agent communication language. The whole system is coordinated by high-level mediators, which provide integration mechanisms for the extended enterprise (Maturana and Norrie 1996). The Enterprise Mediator acts as the coordinator for the enterprise, and all of the manufacturing shop floors and other modules are registered with it. Registration processes are carried out through macro-level registration communications. Each multiagent-manufacturing module offers its services to the enterprise through the Enterprise Mediator. A graphical interface has been created for the Enterprise Mediator. Both human users and agents are allowed to interact with the Enterprise Mediator and registered manufacturing modules via KQML messages. Decision rules and enterprise policies can be dynamically modified by object-call protocols through input field windows by the user. Action buttons support quick access to any of the registered manufacturing modules, shown as icon-agents, as well as to the Enterprise Mediator's source code. The Enterprise Mediator offers three main services: integration, communication, and mediation. Integration permits the registration and interconnection of manufacturing components, thereby creating agent-to-agent links.

Communication is allowed in any direction among agents and between human users and agents. Mediation facilitates coordination of the registered mediators and shop floor resources. The design system module is mainly a graphical interface for retrieving design information and requesting manufacturability evaluations through the Enterprise Mediator (which also operates as shop-floor manager and message router). Designs are created in a separate intelligent design system named the Agent-Based Concurrent Design Environment (ABCDE), developed in the same research group (Balasubramanian et al. 1996).

Different shop floors can be modeled and incorporated in the system as autonomous multiagent components each containing communities of machines and tools agent. Shop-floor resources are registered in each shop floor using macro-level registration policies. Machine and tool agents are incorporated into the resource communities through micro-level registration policies. The shop-floor modules encapsulate the planning activity of the shop floor. Each shop floor interface is provided with a set of icon-agents to represent shop-floor devices. Shop-floor interfaces provide standardized communication and coordination for processing manufacturability evaluation requests. These modules communicate with the execution control and simulation module to refine promissory schedules.

The execution control and forecasting module is the container for execution agents and process-interlocking protocols. Shop floor resources are introduced as needed, thereby instantiating icon-agents

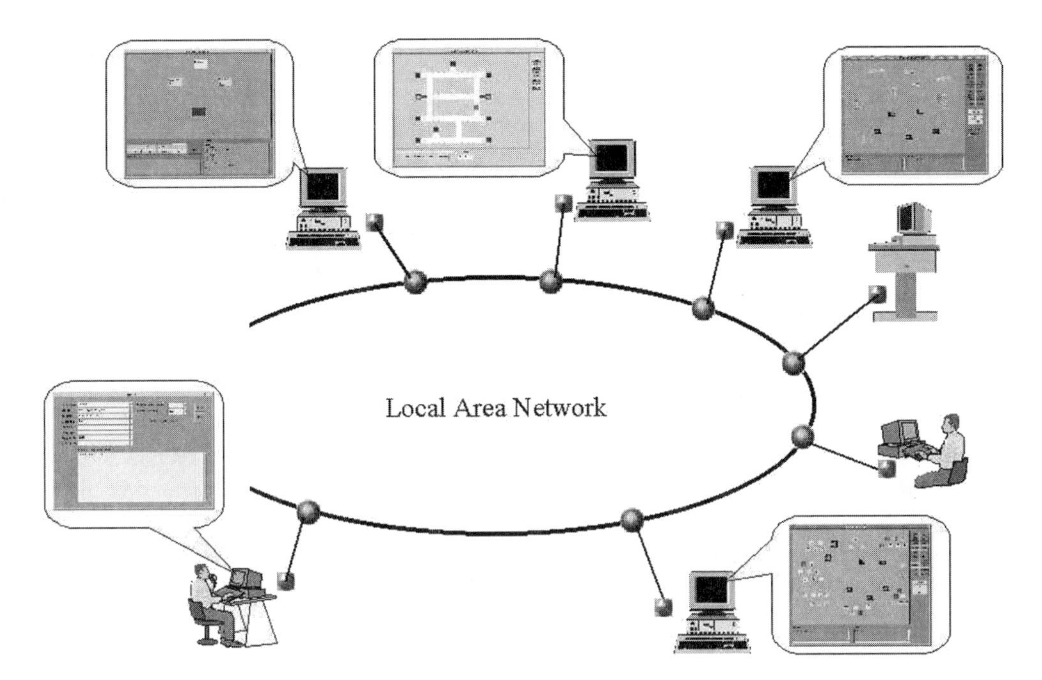

FIGURE 3.2 Prototype implementation of MetaMorph architecture.

and specifying data files for each resource. This module includes icon-agents for its graphical interface to represent machines, warehouses, collision avoidance areas, and AGV agents. Standard operation times (i.e., loading, processing, unloading, and transportation times) are already provided but can be scaled to each resource's desired characteristics. Each resource can enforce specific dispatching rules (i.e., weighted shortest processing time, earliest due date, shortest processing time, FIFO, LIFO, etc.). Parts are modeled as part agents that are implemented as background processes. A local execution mediator is embedded in the module to integrate and coordinate shop-floor resources. This local execution mediator communicates with the resource mediator to get promissory plans and to broadcast forecasting results.

The system can be run in different time modes: real-time and forecasting. In the real-time mode, the speed of the shop-floor simulation is proportional to the execution speed of the real-time system. In the forecasting mode, the simulation speed is 40 to 60 times faster than the real-time execution.

Learning mechanisms are incorporated to learn from the past as well as the future. The most significant interactions among agents are recorded during problem-solving processes, for subsequent reuse (Maturana et al. 1997).

3.3.4 MetaMorph II

The second phase of the MetaMorph project started at the beginning of 1997. Its objective is the integration of design, planning, scheduling, simulation, execution, material supply, and marketing services within a distributed intelligent open environment. The system is organized at the highest level through "subsystem" mediators (Shen et al. 1998). Each subsystem is connected (integrated) to the system through a special mediator. Each subsystem itself can be an agent-based system (e.g., agent-based manufacturing scheduling system), or any other type of system such as a functional design system or knowledge-based material management system. Agents in a subsystem may also be autonomous agents at the subsystem level. Some of these agents may also be able to communicate directly with other subsystems or the agents in other subsystems.

MetaMorph II is an extension of MetaMorph I in multiple dimensions (Shen and Norrie 1998):

a. **Integration of Design and Manufacturing:** Agent-based intelligent design systems are integrated into the MetaMorph II. Some features and mechanisms used in the DIDE project (Shen and Barthès, 1995) and ABCDE project (Balasubramanian et al. 1996) will be utilized in developing this subsystem. Each such subsystem connects within MetaMorph II with a Design Mediator that serves as the coordinator of this subsystem and its only interface to the whole system. Several design systems can be connected to MetaMorph II simultaneously. Each design system may be either an agent-based system or other type of design system.
b. **Extension to Marketing:** This is realized by several easy-to-use interfaces for marketing engineers and end customers to request product information (performance, price, manufacturing period, etc.), select a product, request modifications to a particular specification of a product, and send feedback to the enterprise.
c. **Integration of Material Supply and Management System:** A Material Mediator was developed to coordinate a special subsystem for material handling, supply, stock management, etc.
d. **Improvement of the Simulation System:** Simulation Mediators carry out production simulation and forecasting. Each Simulation Mediator corresponds to one Resource Mediator and therefore to one shop floor.
e. **Extension to Execution Control:** Execution Mediators coordinate the execution of the machines, transportation AGVs, and workers as necessary. Each shop floor is, in general, assigned with one Execution Mediator.

3.3.5 Clustering and Cloning in MetaMorph II

Clustering and cloning approaches for manufacturing scheduling were developed during the MetaMorph I project (Maturana and Norrie 1996). To reduce scheduling time through parallel computation, resources agents are cloned as needed. These clone agents are included in virtual coordination clusters where agents negotiate with each other to find the best solution for a production task. Decker et al. (1997) used a similar cloning agent approach as an information agent's response to overloaded conditions.

In MetaMorph II, both clustering and cloning have been used, with improved mechanisms (Maturana and Norrie 1996). When the Machine Mediator receives a request message from the Resource Mediator (following a request by a part agent), it creates a clone Machine Mediator, and sends "announce" messages to a group of selected machine agents according to its knowledge of their capabilities. After receiving the announce message, each machine agent creates a clone agent and participates in the negotiation cluster. During the negotiation process, the clone machine agent needs to negotiate with tool agents and worker agents. It sends a request message to the Worker Mediator and the Tool Mediator. Similarly to the Machine Mediator, the Worker Mediator and the Tool Mediator create their clone mediator agents. They send announce messages that call for bidding to worker agents and tool agents. The concerned worker agents and tool agents create clones that will then participate in the negotiation cluster

In the MetaMorph project, both clustering and cloning have proved very useful for improving manufacturing scheduling performance. When the system is scheduling in simulation mode, the resource agents are active objects with goals and associated motivations. They are, in general, located in the same computer. These clone agents are, in fact, clone objects. In the case of real on-line scheduling, the cloning mechanism can be used to "clone" resource agents from remote computers (like NC machines, manufacturing cells, and so on) to the local computer (where the resource mediators reside) so as to reduce communication time and consequently to reduce the scheduling and rescheduling time. This idea is related to mobile agent technology (Rothermel and Popescu-Zeletin 1997).

In the following, we illustrate the dynamic virtual clustering mechanism in a case study. For more details on this project see (Shen et al. 1999).

3.3.6 Case Study: Multi-Factory Production Planning

The internationally distributed manufacturing enterprise or a virtual enterprise in this case study has a headquarter (with a General Manager/CEO), a production planning center (with a Production Manager), and two factories (each with a Factory Manager), see Figure 3.3. This case study can be extended to a larger manufacturing enterprise with additional production planning centers and worldwide-distributed factories.

A Production Order A is received for 100 products B with due date D, whose description is as follows:

- One product B is composed of one part X, two parts Y, and three parts Z.
- Part *Z* has three manufacturing features (Fa, Fb, Fc), and requires three operations (Oa, Ob, Oc).

Scenario at a Glance

- CEO receives a Production Order A from a customer for 100 products B with delivery due date D.
- CEO sends the Production Order A to the Production Manager. (Actually it would not be a CEO who would handle such an order, but instead it would be staff at an order desk. The CEO appears on Figure 3.3, since this case study is to be expanded to include higher-level management activities.)
- Production Manager finds an appropriate agent for the task who arranges for Production Order A is decomposed into parts production requests.
- Production Manager sends parts production requests to suitable factories, for parts production.
- Factory Manager(s) receives a part production request, finds competent agent(s) for further (sub-) task decomposition and each part production request is decomposed into manufacturing features (with corresponding machining operations).
- Factory Manager(s) negotiates with resource agents for machining operations, awards machining operation tasks to suitable resource agents, and then sends relevant information back to Production Manager.

During this process, the *virtual clustering mechanism* is used in creating a virtual coordination group; the partial agent cloning mechanism is used to allow resource agents to be simultaneously involved in several coordination groups; and an extended contract net protocol is used for task allocation among resource agents. If the factories are not able to produce the requested parts before the due date, a new due date will be negotiated with the customer, or some subtasks will be subcontracted to other factories outside the manufacturing enterprise (e.g., through the virtual enterprise network).

3.4 Holonic Manufacturing Systems

The term "holonic" is used to characterize particular relationships that exist between holon-type agents. Autonomy and cooperativeness characterize these relationships. Holons are structured agents that act synergistically with other holon-type agents. Research in holonic systems is being carried out by the holonic manufacturing systems (HMS) research consortium, as well as by various academic and industrial researchers. The HMS consortium is industrially driven and is addressing standardization, deployment, and support of architectures and technologies for open, distributed, intelligent, autonomous and cooperating (i.e., "holonic") systems. It is one of the consortia endorsed by the Intelligent Manufacturing Systems (IMS) Steering Committee in 1995 (Parker 1997; www.ims.org). The HMS consortium includes partners from all IMS regions (Australia, Canada, Japan, EC, EFTA and the U.S.), comprising industrial companies, research institutes, and universities. Its principal goal is the advancement of the state-of-the-art in discrete, continuous and batch manufacturing through the integration of highly flexible, reusable, and modular manufacturing units.

Holon architecture and related properties — including autonomy, cooperativeness, and recursivity — have been considered by Gou et al. (1998), Mathews (1995), Brussel et al. (1998), and Bussmann (1998). Maturana and Norrie (1997) suggested an agent-based view of a holon. In the PROSA architecture

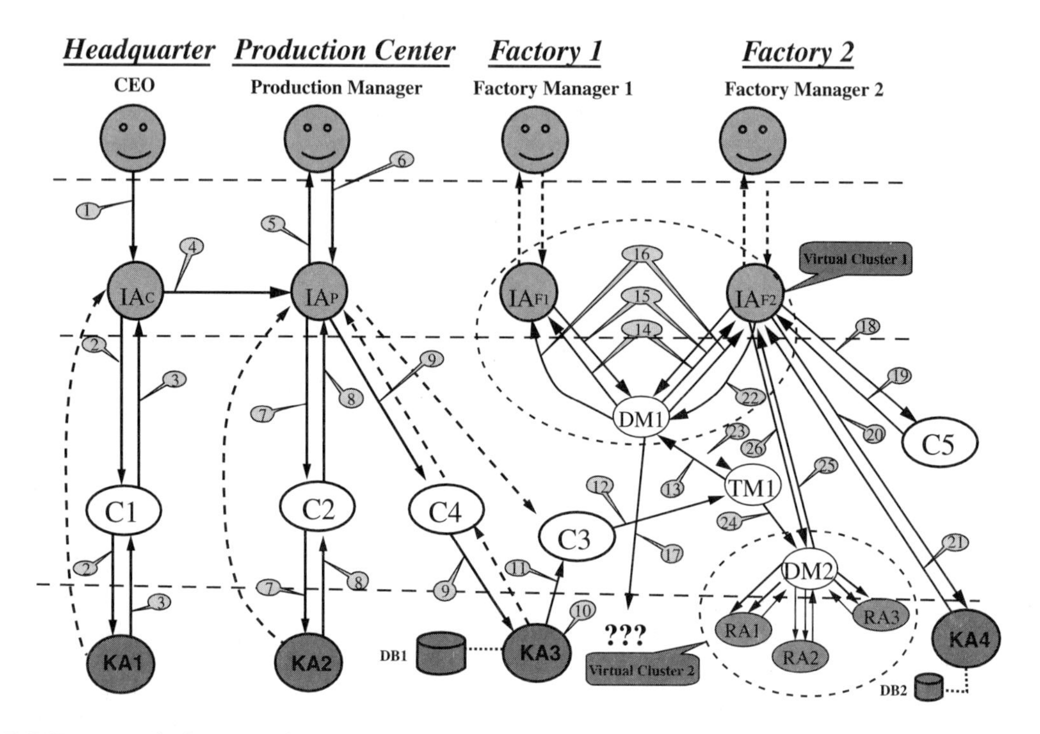

FIGURE 3.3 Multi-factory production planning scenario.

(Brussel et. al. 1998), a HMS is built from three basic holons: order holon, product holon, and resource holon. A centralized staff holon is used to assist the basic holon with expert knowledge. In the model of Gou et al. (1998), five types of holons at the factory level were suggested: product, parts, factory coordinator holons, and cell coordinator holons. The factory coordinator holon coordinates scheduling activities across cells, gathers the status of cell and product holons, and generates coordination information to guide these holons' scheduling activities for overall system performance. The cell coordinator holon gathers the status of machine-types and part holons in the cell, and coordinates scheduling activities to achieve the cell's objective.

3.4.1 Origin of the Holonic Concept

The Hungarian author and philosopher Arthur Koestler proposed the word "holon" to describe a basic unit of organization in biological and social systems (Koestler 1989). Holon is a combination of the Greek word *holos*, meaning whole, and the suffix on meaning particle or part. Koestler observed that in living organisms and in social organizations entirely self-supporting, noninteracting entities did not exist. Every identifiable unit of organization, such as a single cell in an animal or a family unit in a society, comprises more basic units (plasma and nucleus, parents and siblings) while at the same time forming a part of a larger unit of organization (a muscle tissue or a community). A holon, as Koestler devised the term, is an identifiable part of a system that has a unique identity, yet is made up of subordinate parts and in turn is part of a larger whole.

The strength of holonic organization, or holarchy, is that it enables the construction of very complex systems that are nonetheless efficient in the use of resources, highly resilient to disturbances (both internal and external), and adaptable to changes in the environment in which they exist. All these characteristics can be observed in biological and social systems.

The stability of holons and holarchies stems from holons being self-reliant units, which have a degree of independence and handle circumstances and problems on their particular level of existence without asking higher level holons for assistance. Holons can also receive instruction from and, to a certain extent, be controlled by higher-level holons. The self-reliant characteristic ensures that holons are stable, and able to survive disturbances. The subordination to higher-level holons ensures the effective operation of the larger whole.

3.4.2 Holonic Concepts in Manufacturing Systems

The task of the holonic manufacturing systems (HMS) consortium is to translate the concepts that Koestler developed for social organizations and living organisms into a set of appropriate concepts for manufacturing systems. The goal of this work is to attain in manufacturing the benefits that holonic organization provides to living organisms and societies, e.g., stability in the face of disturbances, adaptability and flexibility in the face of change, and efficient use of available resources (Christensen 1994); (Norrie and Gaines 1996).

A holonic manufacturing system should utilize the most appropriate features of hierarchical ("top down") and heterarchical ("bottom up," "cooperative") organizational structures, as the situation dictates (Dilts et al. 1991). The intent is to obtain at least some of the stability of a hierarchy while providing the dynamic flexibility of a heterarchy.

The HMS consortium has developed the following definitions to guide the translation of holonic concepts into a manufacturing setting:

Holon: An autonomous and cooperative building block of a manufacturing system for transforming, transporting, storing, and/or validating information and physical objects. The holon consists of an information processing part and often a physical processing part. A holon can be part of another holon.

Autonomy: The capability of an entity to create and control the execution of its own plans and/or strategies.

Cooperation: A process whereby a set of entities develops mutually acceptable plans and executes these plans.

Holarchy: A system of holons that can cooperate to achieve a goal or objective. The holarchy defines the basic rules for cooperation of the holons and thereby limits their autonomy.

Holonic manufacturing system (HMS): A holarchy that integrates the entire range of manufacturing activities from order booking through design, production, and marketing to realize the agile manufacturing enterprise.

Holonic attributes: The attributes of an entity that make it a holon. The minimum set is autonomy and cooperativeness.

Holonomy: The extent to which an entity exhibits holonic attributes.

From the above, it is clear that a manufacturing system having the MetaMorphic architecture is, in fact, a holonic system. In the following, we will illustrate this using MetaMorph's dynamic virtual clustering mechanism.

3.5 Holonic Self-Organization of MetaMorph via Dynamic Virtual Clustering

3.5.1 Holonic MetaMorphic Architecture

Within the HMS consortium, part of our research has focused on how to dynamically reconfigure a multiagent system, according to need, so that it develops or retains holonic structures (Zhang and Norrie 1999). For this, we have developed a mathematical framework (see Sections 3.6 and 3.7) that enables automatic holonic clustering within a generic (nonholonic) multiagent system (MAS). The method is based on uncertainty minimization via fuzzy modeling of the MAS. This method appears to have promise for reconfiguring distributed manufacturing systems as holonic structures, as well as for investigating the potential for a nonholonic manufacturing system to migrate toward a holonic one.

In this section, using metamorphic mechanisms for distributed decision-making in an agent-based manufacturing system, the concept of dynamic virtual clustering is extended to manufacturing process control at the lower levels (Zhang and Norrie 1999). Event-driven dynamic clustering of resource control services and cooperative autonomous activities are emphasized in this approach.

As mentioned in Section 3.3, virtual clustering in MetaMorph is a dynamic mechanism for organizational reconfiguration of the manufacturing system during run-time. An organization based on virtual clusters of entities can continually be reconfigured in response to changing task requirements. These tasks can include orders, production requests, as well as planning, scheduling, and control. A cluster exists for the duration of the task or subtask it was created for and is destroyed when the task is completed. Mediators play key roles in the process and manage the clusters. Instead of having preestablished and rigid layers of hierarchically organized mechanisms, a mediator-based metamorphic system can use reconfiguration mechanisms to dynamically organize its manufacturing devices. The necessary structures of control are then progressively created during the planning and execution of any production task. In this dynamically changing virtual organization, the partial control hierarchies are dynamic and transient and the number of control layers for any specific order task are task-oriented and time-dependent. It will be seen that holonic characteristics such as "clusters-within-clusters" groupings exist at different organizational levels.

3.5.2 Holon Types in MetaMorph's Holarchy

A basic HMS architecture can be based on four holon types: product holon (PH), product model holon (PMH), resource holon (RH), and mediator holon (MH). A product holon holds information about the process status of product components during manufacturing, time constraint variables, quality status, and decision knowledge relating to the order request. A product holon is a dual of a physical "component" and information "component." The physical component of the product holon develops from its initial state (raw materials or unfinished product) to an intermediate product, and then to the finished one, i.e., the end product. A product model holon holds up-to-date engineering information relating to the product life cycle (configuration, design, process plans, bills of materials, quality assurance procedures, etc.). A resource holon contains physical and information components. The physical part contains a production resource of the manufacturing system (machine, conveyor, pallet, tool, raw material, and end product, or accessories for assembling, etc.), together with controller components. The information part contains planning and scheduling components.

In the following development of a reconfigurable HMS architecture using the four basic holon types, a mediator holon serves as an intelligent logical interconnection to link and manage orders, product data, and specific manufacturing resources dynamically. The mediator holon can collaborate with other holons to search for and coordinate resource, product data, and related production tasks. A mediator holon is itself a holarchy. A mediator holon can create a dynamic mediator holon (DMH) for a new task such as a new order request or suborder task request. The dynamic mediator holon then has the responsibility for the assigned task. When the task is completed, the DMH is destroyed or terminates for reuse. DMHs identify order-related resource clusters (i.e., machine group) and manage task decomposition associated with their clusters.

3.5.3 Holonic Self-Organization

The following example will illustrate holonic clustering within this architecture. Figure 3.4 shows the initial activity sequence following the release to production of an order for 100 of a particular product. This product is composed of three identical parts (to be machined) and two identical subassemblies (each to be assembled). As shown in Figure 3.4, following the creation of the appropriate product holon, there are created the relevant part and subassembly holons. The requests for manufacturing made by these

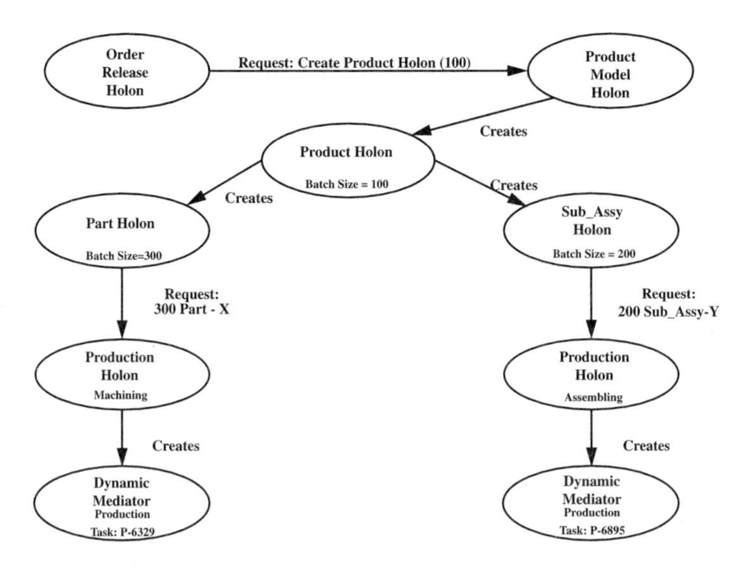

FIGURE 3.4 Holonic clustering mechanism.

latter holons to appropriate production holons (which function as high-level Production Managers for a manufacturing shop-floor plan or part dispatch) result in the creation of dynamic mediators for the machining and assembly tasks. Subsequently, each production holon coordinates inspection or assembly of the parts or subassemblies according to the production sequence prescribed by the production model holon (from its stored information). More complex situations will occur, when products having many components requiring different types of production processes are involved.

After physical and logical machine groups are derived (for example, via group-technology approaches), the necessary control structures are created and configured using control components cloned from template libraries by a DMH. The machine groups, their associated and configured controllers, then form a temporary manufacturing community, termed a virtual cluster holon (VCH), as shown in Figure 3.5. The VCH exists for the duration of the relevant job processing and is destroyed when these production processes are completed. The physical component of a VCH is composed of order-related parts, raw materials or subproducts for assembly, manufacturing machines and tools, and associated controller hardware. Within these manufacturing environments, parts develop from their initial state to an intermediate product and then to the finished one. The information component of a VCH is composed of cluster controller software-components, the associated DMH, and intermediate information on the order and the related product. Each cluster controller is further composed of multilayer control functions that execute job collaboration, control application generation and controller dynamic reconfiguration, process execution, and process monitoring, etc.

3.5.4 Holonic Clustering

The life cycle of a dynamic virtual cluster holon has four stages: resource grouping; control components creation; execution processing; and termination/destruction. The dynamic mediator holon is involved in the stages 1 and 2. The first cluster that is created is the schedule-control cluster shown in Figure 3.5. A cluster can be also considered to be a holonic grouping. The controller cluster next created is composed of three holonic parts: collaboration controller (CC), execution controller (EC), and control execution (CE) holon. One CE holon can be associated with more than one physical controller (execution platform such as real-time operation system and its hardware support devices) and functions as a distributed-node transparent-resource platform for execution of cluster control tasks at the resource level. In the prototype system under development, the CC, EC, and CE holons collaborate to control and execute the

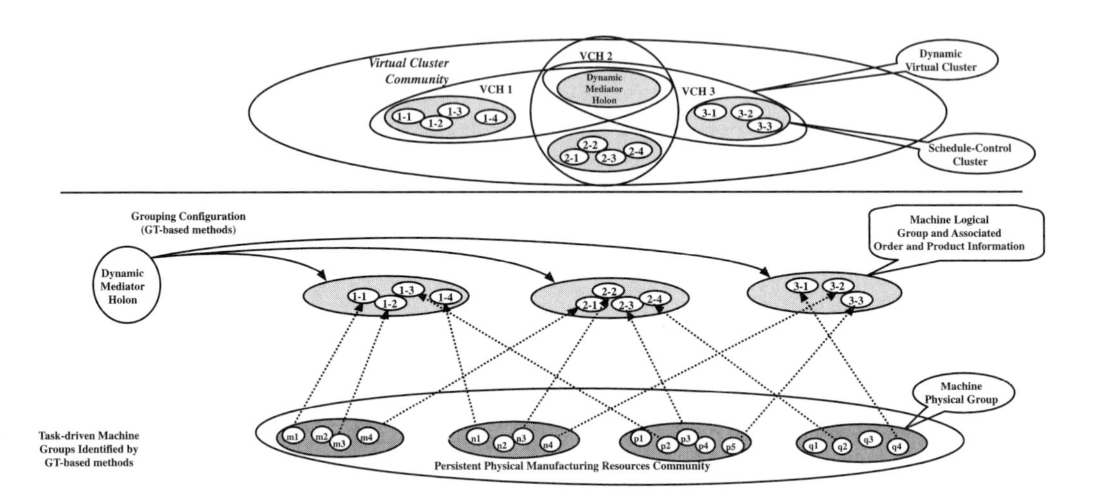

FIGURE 3.5 Virtual Cluster Holon.

distributed tasks or applications on a new type of distributed real-time operating system recently implemented (Zhang et al. 1999). The distributed tasks or applications are represented using the Function Block (FB)-1499 specification, which is a draft standard described by the IEC for distributed industrial-process measurement and control systems.

As shown in Figure 3.5, the dynamic mediator holon records and traces local dynamic information of the individual holons in its associated virtual cluster community. It is important to note that during the life cycle of the DMH, this mediator may pass instantaneous information of the partial resource holons to some new virtual cluster communities while the assigned tasks on these resource holons are being completed.

The dynamic characteristics of the event-driven holon community become more complicated as the population grows. In the next section, we present an approach for automatic grouping into holonic clusters depending on the assigned task. This approach, due to its strong mathematical foundation, should be applicable to large multiagent systems.

3.6 Automatic Grouping of Agents into Holonic Clusters

3.6.1 Rationale for Fuzzy Modeling of Multiagent Systems

In Section 3.5 we showed how resources and the associated controller components can be reconfigured dynamically into holonic structures. In the present and following sections, a novel approach to holonic clustering in a multiagent system is presented. This is applicable to systems that already have clusters as well as to those that are non-clustered.

Although there have been considerable advances in agent theory (Russell and Norwig 1995; O'Hare and Jensen 1996), a rigorous mathematical description of agent systems and their interaction is yet to be formulated. Agents can be understood as autonomous problem solvers, in general heterogeneous in nature, that interact with other agents in a given setting to progress towards solutions. Thus, capability for interaction and evolution in time are prime features of an agent. Once a meaningful framework is established for these interactions and evolution, it is natural to view the agents (in isolation and in a group) as dynamical systems. The factors that influence agent dynamics are too many and too complex to be tackled by a classical model. Also, the intrinsic stochastic nature of many of these factors introduces the dimension of uncertainty to the problem. Given the nature of the uncertainty dealt with in such a multiagent system, fuzzy set theory may be a promising approach to agent dynamics (Klir and Folger 1988; Zimmermann 1991; Subramanian and Ulieru 1999).

As already noted in Section 3.3.2, and illustrated by examples in Sections 3.3.6 and 3.5.3, agents can dynamically be contracted to a problem-solving group (cluster), through the virtual clustering mechanism. In the following, it is shown how agents can automatically be selected for such holonic clusters, using a new theoretical approach.

To model the multiagent system (MAS), we will use set theoretical concepts that extend to fuzzy set theory. Consider the set of all agents in the MAS. As already mentioned, in our metamorphic architecture, clusters and partitions or covers can change any time during the MAS evolution, according to a global strategy which aims to reach a goal.

Each group of clusters that covers the agents set is actually a *partition* of it, provided that clusters are not overlapping. Here by *cover* of a set, one understands a union of subsets at least equal to the set. Whenever an agent can belong to more than one cluster at the same time, we refer to the clusters union just as a *cover* of the agent set. Let us denote by $a \rhd b$ the relation "a and b are in the same cluster." Two types of clusters could be then defined, based on this relation: disjoint or not (i.e., overlapping), as follows:

a. If a cluster is constructed using the following axiom:
   - the agents a and b are in the same cluster if $a \rhd b$ or $b \rhd a$ or it exists c so that $a \rhd c$ and $b \rhd$ c, then the clusters are disjoint and their union is a partition of the agents set.

b. If a cluster is defined by another axiom:
   - the agents a and b are in the same cluster if $a \rhd b$ or $b \rhd$ a,
     then, when $a \rhd c$, $b \rhd c$ and no relation exists between a and b, the pairs $\{a,c\}$ and $\{b,c\}$ belong to different clusters, but c belongs to two clusters at the same time. In this case, clusters could overlap and their union is just a cover of the agents set.

Consider an MAS that evolves, transitioning from an initial state through a chain of intermediate states until it reaches its goal in a final state. A main driving force for MAS dynamics during this transition is *information exchange* among agents. While the MAS evolves through its states toward the goal, its agents associate in groups referred to as *clusters,* each cluster of agents aiming to solve a certain part of the overall task assigned to the MAS. Let us consider now the set of all agents within a MAS. Each possible group of clusters that *covers* the (agents) set is actually a *partition* of this set, provided that clusters are not overlapping. We name a plan as the succession of all states through which the MAS transitions until it reaches its goal. Each MAS state is described by a certain configuration of clusters partitioning the agent set. So, a plan is in fact a succession of such partitions describing the MAS clustering dynamics on its way toward reaching a goal. In the following discussion, we assume that clusters are not overlapping. Our findings extend to the case when one or more agents belong to different clusters simultaneously.

The succession of clusters dynamically partitioning the agent set during MAS evolution from its initial state to a final one is not known precisely. All we can do at this stage is to assign a "degree of occurrence" for each possible partition supposed to occur.

Thus, the problem we intend to solve can be stated in general terms as follows:

- Given an MAS and some vague information about the occurrence of agent clusters and partitions (or covers) during the system's evolution toward a goal, construct a fuzzy model that provides one of the least uncertain source-plans.

3.6.2 Mathematical Statement of the Problem

Denote by $\mathscr{A}_N = \{a_n\}_{n\in\overline{1,N}}$ the set of $N \geq 1$ agents acting as an MAS and by $\mathscr{P} = \{P_m\}_{m\in\overline{1,M}}$ a set of $M \geq 1$ partitions of $\mathscr{A}_N$, that seem to occur during the MAS evolution toward its goal. Notice that the number of all possible partitions covering $\mathscr{A}_N$, denoted by $\mathscr{S}_N$, increases faster with N than the number of all possible clusters (which is 2^N), as proves Theorem 1 from Appendix A. For example, if $N = 12$, then $\mathscr{S}_{12} = 4{,}213{,}597$, whereas the number of all clusters is only $2^{12} = 4{,}096$.

In our framework, one can refer to $\mathcal{P}$ as a *source-plan* in the sense that $\mathcal{P}$ can be a source of partitions for a MAS plan. The main difference between a *plan* and a *source-plan* is that, in a plan the succession of partitions is clearly specified and they can repeat in time, whereas in a source-plan the partitions order is, usually, unknown (the time coordinate is not considered) and the partitions are different from each other. The only available information about $\mathcal{P}$ is that to each of its partitions, P_m, one can assign a number $\alpha_m \in [0,1]$, assumed to represent a corresponding *degree of occurrence* during the MAS evolution.

Assume that a family $\{\mathcal{P}_k\}_{k\in\overline{1,K}}$, containing $K \geq 1$ source-plans, is constructed starting from the uncertain initial information. For each $k \in \overline{1,K}$, the source-plan $\mathcal{P}_k$ contains $M_k \in \overline{1,\mathcal{S}_N}$ partitions: $\mathcal{P}_k = \{P_{k,m}\}_{m\in\overline{1,M_k}}$. The corresponding degrees of occurrence are now members of a two-dimensional family $\{\alpha_{k,m}\}_{k\in\overline{1,K};m\in\overline{1,M_k}}$, the source plan and its constituent partitions (each $P_{k,m}$ has the degree of occurrence $\alpha_{k,m}$), that quantifies all available information about MAS.

In this framework, the aim is to construct a sound measure of uncertainty, V (from "vagueness"), fuzzy-type, real-valued, defined on the set of all source-plans of $\mathcal{A}_N$, and to optimize it in order to select the least uncertain source-plan of the family $\{\mathcal{P}_k\}_{k\in\overline{1,K}}$:

$$\mathcal{P}_{k_0} = \arg \underset{k\in\overline{1,K}}{\text{opt}}\, V(\mathcal{P}_k), \text{where } k_0 \in \overline{1,K}. \qquad \text{Equation (3.1)}$$

The cost function V will be constructed by using a *measure of fuzziness* (Klir and Folger 1988). We present hereafter the steps of this construction. The fuzzy notions used in this construction are defined in (Klir and Folger 1988; Zimmermann 1991).

3.6.3 Building an Adequate Measure of Uncertainty for MAS

3.6.3.1 Constructing Fuzzy Relations between Agents

The main goal of this first step is to construct a family of fuzzy relations, $\{\mathcal{R}_k\}_{k\in\overline{1,K}}$, between the agents of MAS ($\mathcal{A}_N$) using the numbers $\{\alpha_{k,m}\}_{k\in\overline{1,K};m\in\overline{1,M_k}}$ and the family of source-plans $\{\mathcal{P}_k\}_{k\in\overline{1,K}}$.

In order to describe how fuzzy relations between agents can be constructed, consider $k \in \overline{1,K}$ and $m \in \overline{1,M_k}$ arbitrarily fixed. In construction of the fuzzy relation $\mathcal{R}_k$, one starts from the remark that associating agents in clusters is very similar to grouping them into *equivalence classes,* given a (binary) equivalence relation between them (that is a reflexive, symmetric and transitive relation, in the crisp sets sense). It is, thus, natural to consider that every partition $P_{k,m}$ is a cover with equivalence classes of $\mathcal{A}_N$. The corresponding (unique) equivalence relation, denoted by $R_{k,m}$, can be described very succinctly: "two agents are equivalent if they belong to the same cluster of the partition $P_{k,m}$." Express by "$aR_{k,m}b$" and "$a \neg R_{k,m}b$" the facts that a and b, respectively, are not in the relation $R_{k,m}$ (where $a,b \in \mathcal{A}_N$). The relation $R_{k,m}$ can also be described by means of a $N \times N$ matrix $H_{k,m} \in \mathcal{R}^{N\times N}$— the *characteristic matrix* — whose elements are only 0 or 1, depending on whether the agents are or are not in the same cluster. (Here, $\mathcal{R}$ points to the real numbers set.) This symmetric matrix with unitary diagonal allows us to completely specify $R_{k,m}$, by enumerating only the agent pairs, which are in the same cluster (i.e., determined by the positions of the 1s inside our matrix).

Example 1

If a partition $P_{k,m}$ is defined by three clusters: $\mathcal{A}_N = \{a_1,a_4\}\cup\{a_2,a_5\}\cup\{a_3\}$, then the corresponding (5×5) matrix $(H_{k,m})$ and equivalence relation $(R_{k,m} \subseteq \mathcal{A}_N \times \mathcal{A}_N)$ are

$$H_{k,m} = \begin{bmatrix} 1 & 0 & 0 & 1 & 0 \\ 0 & 1 & 0 & 0 & 1 \\ 0 & 0 & 1 & 0 & 0 \\ 1 & 0 & 0 & 1 & 0 \\ 0 & 1 & 0 & 0 & 1 \end{bmatrix}$$

and, respectively,

$$R_{k,m} = \{(a_1, a_1), (a_2, a_2), (a_3, a_3), (a_4, a_4), (a_5, a_5), (a_1, a_4), (a_4, a_1), (a_2, a_5), (a_5, a_2)\}.$$

Denote by $X_{k,m}$ the characteristic function of $R_{k,m}$ (the matrix form of $X_{k,m}$ is exactly $H_{k,m}$):

$$\left[\begin{array}{l} \chi_{k,m} : \mathcal{A}_N \times \mathcal{A}_N \to \{0,1\} \\ \qquad (a,b) \mapsto \chi_{k,m}(a,b) = \begin{cases} 0, & aR_{k,m}b \\ 1, & a\neg R_{k,m}b \end{cases} \end{array}\right. .$$

Each equivalence relation $R_{k,m}$ can be uniquely associated to the degree of occurrence assigned to its partition: $\alpha_{k,m}$. Together, they can define a so-called *α-sharp-cut* of the fuzzy relation $\mathcal{R}_k$.

From (Klir and Folger 1988) we know that if A is a fuzzy set defined by the membership function $\mu_A : X \to [0,1]$ (where X is a crisp set), then the *grades set* of A is the following crisp set:

$$\Lambda_A \stackrel{def}{=} \{\alpha \in [0,1] | \exists x \in X : \mu_A(x) = \alpha\} \qquad \text{Equation (3.2)}$$

Moreover, the *α-cut* of A is also a crisp set, but defined as

$$A_\alpha \stackrel{def}{=} \{x \in X | \mu_A(x) \geq \alpha\}, \text{ for } \alpha \in \Lambda_A . \qquad \text{Equation (3.3)}$$

According to these notions, the *α-sharp-cut* of A can be defined here as the crisp set:

$$A_{[\alpha]} \stackrel{def}{=} \{x \in X | \mu_A(x) = \alpha\}, \text{ for } \alpha \in \Lambda_A . \qquad \text{Equation (3.4)}$$

Thus, one can consider that the α-sharp-cut of $\mathfrak{R}_k$ defined for $\alpha_{k,m}$ is exactly the crisp relation $R_{k,m}$. This can be expressed as $\mathcal{R}_{k,[\alpha k,m]} = R_{k,m}$. Next we define a fuzzy relation $\mathcal{R}_{k,m}$ with membership function $\mu_{k,m}$, expressed as the product between the characteristic function $X_{k,m}$ and the degree of occurrence $\alpha_{k,m}$, that is $\mu_{k,m} \stackrel{def}{\equiv} \alpha_{k,m} X_{k,m}$. This fuzzy set of $\mathcal{A}_N \times \mathcal{A}_N$ is uniquely associated to $\mathcal{R}_k[\alpha_{k,m}]$. More specifically,

$$\left[\begin{array}{l} \mu_{k,m} : \mathcal{A}_N \times \mathcal{A}_N \to \{0,1\} \\ \qquad (a,b) \mapsto \mu_{k,m}(a,b) = \begin{cases} \alpha_{k,m}, & aR_{k,m}b \\ 0, & a\neg R_{k,m}b \end{cases} \end{array}\right. \qquad \text{Equation (3.5)}$$

The matrix form of $\mu_{k,m}$ is exactly $\alpha_{k,m}H_{k,m}$.

If $k \in \overline{1,K}$ is kept fixed, but m varies in the range $\overline{1,M_k}$, then a family of fuzzy elementary relations is associated to $\mathcal{R}_k$. Denote by $\{\mathcal{R}_{k,m}\}_{m \in \overline{1,M_k}}$ this family. Naturally, $\mathcal{R}_k$ is then defined as the fuzzy union:

$$\mathcal{R}_k \overset{def}{=} \bigcup_{m=1}^{M_k} \mathcal{R}_{k,m} \quad . \qquad \text{Equation (3.6)}$$

Usually, the fuzzy union in Equation 3.6 is computed by means of max operator (although some other definitions of fuzzy union could be considered as well (Klir and Folger 1988). This involves the membership function of $\mathcal{R}_k$ being expressed as follows (using the max operator):

$$\mu_k(a,b) \overset{def}{=} \max_{m \in \overline{1,M_k}} \{\mu_{k,m}(a,b)\}, \ \forall a,b \in \mathcal{A}_N . \qquad \text{Equation (3.7)}$$

Consequently, the matrix form of μ_k is obtained (according to Equation 3.7) by applying the max operator on the matrices $\alpha_{k,m}H_{k,m}$, for $m \in \overline{1,M_k}$:

$$\mathcal{M}_k \overset{def}{=} \max_{m \in \overline{1,M_k}} \bullet \{\alpha_{k,m}H_{k,m}\} \in \mathcal{R}^{N \times N} , \qquad \text{Equation (3.8)}$$

where " $\max\bullet$ " means that the operator acts on matrix elements and not globally, on matrices.

Actually,

$$\mathcal{R}_k = \left|\mu_k(a_i,a_j)\right|_{i,j \in \overline{1,N}} \qquad \text{Equation (3.9)}$$

and it is often referred to as the *membership matrix* of the fuzzy relation $\mathcal{R}_k$.

Equation 3.6 is very similar to the *resolution form* of $\mathcal{R}_k$, as defined in (Klir and Folger 1988). Indeed, if we consider that the numbers $\{\alpha_{k,m}\}_{m \in \overline{1,M_k}}$ are arranged in increasing order and that they are all grades of $\mathcal{R}_k$ (which is not always verified, as shown in Example 2 below), then all the α–cuts of $\mathcal{R}_k$ are

$$\mathcal{R}_{k,\alpha_{k,m}} = \bigcup_{i=m}^{M_k} R_{k,m} , \qquad \text{Equation (3.10)}$$

where, here, the union is classical, between crisp sets. Consequently, the fuzzy sets from the resolution form of $\mathcal{R}_k$ (i.e., $\alpha_{k,m}\mathcal{R}_{k,\alpha_{k,m}}$, for $m \in \overline{1,M_k}$) are defined by the membership functions below (denoted simply $\mu_{k,m}$, for $m \in \overline{1,M_k}$ and very similar to those expressed in Equation 3.5):

$$\left[\begin{array}{l} \bar{\mu}_{k,m}: \mathcal{A}_N \times \mathcal{A}_N \to \{0,1\} \\ \qquad (a,b) \mapsto \bar{\mu}_{k,m}(a,b) = \begin{cases} \alpha_{k,m}, & \exists i \in \overline{m,M_k}: aR_{k,i}b \\ 0, & a \neg R_{k,i} b, \forall i \in \overline{m,M_k} \end{cases} \end{array}\right. .$$

This property is due to the fact that the characteristic function of $\mathcal{R}_{k,\alpha_{k,m}}$ is

$$\bar{\chi}_{k,m}(a,b) = \max_{i \in \overline{m,M_k}} \chi_{k,i}(a,b), \quad \forall a,b \in \mathcal{A}_N$$

and $\alpha_{k,1} \leq \alpha_{k,2} \leq \ldots \alpha_{k,M_k}$. (As stated in (Klir and Folger, 1988), $\overline{\mu}_{k,m}$ are defined by $\overline{\mu}_{k,m} \equiv \alpha_{k,m}\overline{\chi}_{k,m}$, $\forall m \in \overline{1, M_k}$.)

The resolution form is then:

$$\mathcal{R}_k = \bigcup_{m=1}^{M_k} \alpha_{k,m} \mathcal{R}_{k,\alpha_{k,m}}, \qquad \text{Equation (3.11)}$$

preserving the same fuzzy union as in Equation 3.6 (max-union, in fact). Moreover, each α-cut of $\mathcal{R}_k$ is, actually, a (crisp) union of its α-sharp-cuts:

$$\mathcal{R}_{k,\alpha_{k,m}} = \bigcup_{i=m}^{M_k} \mathcal{R}_{k,[\alpha_{k,m}]} = \bigcup_{i=m}^{M_k} R_{k,i}, \quad \forall m \in \overline{1, M_k} \qquad \text{Equation (3.12)}$$

If $\alpha_{k,m}$ disappear from membership grades of $\mathcal{R}_k$, then the corresponding α-cut are identical with other α-cut (for a superior grade) and cannot be revealed. This vanishing effect of $\alpha_{k,m}$ is due to the fact that the corresponding equivalence relation $R_{k,m}$ is included in the union of next equivalence relations: $\bigcup_{i=m}^{Mk} R_{k,i}$ (remember that $\alpha_{k,1} \leq \alpha_{k,2} \leq \ldots \alpha_{k,m_k}$).

The following example shows how a fuzzy relation between agents can be constructed, starting from a source-plan and the associated degrees of occurrence.

Example 2

Consider $\mathcal{A}_N = \{a_1, a_2, a_3, a_4, a_5\}$ and the following set of partitions with corresponding degrees of occurrence:

$$P_1 = \{\{a_1, a_2\}, \{a_3\}, \{a_4, a_5\}\}, \quad \alpha_1 = 0.15$$
$$P_2 = \{\{a_1\}, \{a_2, a_3\}, \{a_4, a_5\}\}, \quad \alpha_2 = 0.25$$
$$P_3 = \{\{a_1, a_2, a_3, a_4\}, \{a_5\}\}, \quad \alpha_3 = 0.57$$
$$P_4 = \{\{a_1, a_4\}, \{a_2, a_5\}, \{a_3\}\}, \quad \alpha_4 = 0.7$$

Then the four corresponding 5×5 matrices $H_{\overline{1,4}}$ describing the associated equivalence relations are

$$\overset{H_1}{\begin{bmatrix} 1 & 1 & 0 & 0 & 0 \\ 1 & 1 & 0 & 0 & 0 \\ 0 & 0 & 1 & 0 & 0 \\ 0 & 0 & 0 & 1 & 1 \\ 0 & 0 & 0 & 1 & 1 \end{bmatrix}} \overset{H_2}{\begin{bmatrix} 1 & 0 & 0 & 0 & 0 \\ 0 & 1 & 1 & 0 & 0 \\ 0 & 1 & 1 & 0 & 0 \\ 0 & 0 & 0 & 1 & 1 \\ 0 & 0 & 0 & 1 & 1 \end{bmatrix}} \overset{H_3}{\begin{bmatrix} 1 & 1 & 1 & 1 & 0 \\ 1 & 1 & 1 & 1 & 0 \\ 1 & 1 & 1 & 1 & 0 \\ 1 & 1 & 1 & 1 & 0 \\ 0 & 0 & 0 & 0 & 1 \end{bmatrix}} \overset{H_4}{\begin{bmatrix} 1 & 0 & 0 & 1 & 0 \\ 0 & 1 & 0 & 0 & 1 \\ 0 & 0 & 1 & 0 & 0 \\ 1 & 0 & 0 & 1 & 0 \\ 0 & 1 & 0 & 0 & 1 \end{bmatrix}}$$

Actually, $H_{\overline{1,4}}$ are the matrix forms of characteristic functions $\chi_{\overline{1,4}}$. The matrix form of the membership function defining the fuzzy relation $\mathcal{R}$ is then

$$\mathcal{M} \overset{def}{=} \underset{m\in 1,4}{\max}\bullet\left\{\alpha_m H_m\right\} = \begin{bmatrix} 0.70 & 0.57 & 0.57 & 0.70 & 0 \\ 0.57 & 0.70 & 0.57 & 0.57 & 0.70 \\ 0.57 & 0.57 & 0.70 & 0.57 & 0 \\ 0.70 & 0.57 & 0.57 & 0.70 & 0.25 \\ 0 & 0.70 & 0 & 0.25 & 0.70 \end{bmatrix}.$$

Thus, for example, the agents a_4 and a_5 share the same cluster with the degree of occurrence 0.25, whereas a_2 and a_5 share the same cluster with the degree of 0.7. We have chosen a set of partitions and corresponding degrees of occurrence such that the degree 0.15 vanishes in $\mathcal{M}$. It is easy to remark that the equivalence relation R_1 is included in the union $R_2 \cup R_3$ and this forces α_1 to vanish in $\mathcal{M}$. It is suitable to set the degrees of occurrence so that all of them appear in $\mathcal{M}$; otherwise some partitions can be removed from the source-plan (those for which the degrees of occurrence do not appear in $\mathcal{M}$). Here, the partition P_1 vanishes completely, if we look only at $\mathcal{M}$. If for example, 0.57 is replaced by 0.07, then all degrees of occurrence will appear in $\mathcal{M}$, because the increasing order of αs is now $\alpha_3 \leq \alpha_1 \leq \alpha_2 \leq \alpha_4$ and no equivalence relation is included in the union of "next" equivalence relations (according to the order of αs).

Obviously, since all matrices $\alpha_{k,m}H_{k,m}$ are symmetric $\mathcal{M}_k$, from Equation 3.8 is symmetric as well, which means that $\mathcal{R}_k$ is a fuzzy symmetric relation. The fuzzy reflexivity is easy to ensure, by adding to each source-plan the trivial partition containing only singleton clusters, with the maximum degree of occurrence, 1. Naturally, this could be considered the partition associated to the initial state, when no clusters are yet observed. Thus, $\mathcal{R}_k$ is at least a proximity relation (i.e., fuzzy reflexive and symmetric) between agents.

The manner in which the degrees of occurrence are assigned to partitions greatly affects the quality of the fuzzy relation. Although all its α-sharp-cuts are equivalence relations, it is not necessary that the resulting fuzzy relation be a *similarity* relation (i.e., fuzzy reflexive, symmetric, and transitive). But it is at least a proximity relation, as explained above.

The fuzzy transitivity expressed (for example) as follows (for each $k \in \overline{1,K}$),

$$\mu_k\left(a_p,a_q\right) \geq \underset{m\in 1,N}{\max}\min\left\{\mu_k\left(a_p,a_n\right),\mu_k\left(a_n,a_q\right)\right\} = \forall p,q \in \overline{1,N}, \qquad \text{Equation (3.13)}$$

is the most difficult to ensure. This is the *max–min (fuzzy) transitivity.* (Notice that other forms of fuzzy transitivity properties could be defined (Klir and Folger 1988). A matrix form of the Equation 3.13 can be straightforwardly derived (due to Equation 3.9)):

$$\mathcal{M}_k \geq \bullet\left(\mathcal{M}_k \circ \mathcal{M}_k\right). \qquad \text{Equation (3.14)}$$

Here, "$\circ$" points to *fuzzy multiplication* (*product*) between matrices with compatible dimensions, involved by the composition of the corresponding fuzzy relations (see Klir and Folger 1988 for details). This multiplication is expressed starting from classical matrix multiplication, where max operator is used instead of summation and min operator is used instead of product. The equivalent Equations 3.13 and (especially) 3.14 suggest an interesting procedure to construct similarity relations starting from proximity relations, using the notion of transitive closure (Klir and Folger 1988). A transitive closure of a fuzzy relation $\mathcal{R}$ is, by definition, the minimal transitive fuzzy relation that includes $\mathcal{R}$. (Here, "minimal" is considered with respect to inclusion on fuzzy sets.)

It is interesting that the composition of fuzzy relations preserves both reflexivity and symmetry, if the relations are not necessarily identical, and it preserves even the transitivity, if relations are identical. This is proven by the Theorem 2 in Appendix B.

It is very important if we preserve the proximity property of relation $\mathcal{R}_k$ by composition with itself, because, thus, the following simple procedure allows us to transform $\mathcal{R}_k$ into a similarity relation:

Step 1. Compute the following fuzzy relation: $\mathcal{Q}_k = \mathcal{R}_k \cup (\mathcal{R}_k \circ \mathcal{R}_k)$.

Step 2. If $\mathcal{Q}_k \neq \mathcal{R}_k$, then replace $\mathcal{R}_k$ by $\mathcal{Q}_k$, i.e., $\mathcal{R}_k \leftarrow \mathcal{Q}_k$, and go to Step 1. Otherwise, $\mathcal{Q}_k = \mathcal{R}_k$ is the transitive closure of the initial $\mathcal{R}_k$.

The first step consists of two operations: one fuzzy matrix multiplication and one fuzzy union (expressed by the "max•" operator, as in Equation 3.8, in matrix notation). The second step is actually a simple and efficient test of fuzzy transitivity, for any fuzzy relation, avoiding the inequality Equation 3.13 or 3.14. To clarify this we give the following example:

Consider the fuzzy relation $\mathcal{R}$, constructed at the previous example. A very simple test using one single pass of the steps in procedure before shows that the new relation $\mathcal{Q}$ is different of $\mathcal{R}$, so that $\mathcal{R}$ is not transitive. Indeed, the matrix membership of $\mathcal{R} \circ \mathcal{R}$ is $\mathcal{M} \circ \mathcal{M}$, whereas the matrix membership of $\mathcal{Q}$ is $\mathcal{M} \cup (\mathcal{M} \circ \mathcal{M})$. Both matrices are depicted below and, obviously, $\mathcal{M} \neq \mathcal{M} \cup (\mathcal{M} \circ \mathcal{M})$. But if a second pass is initiated, the matrix $\mathcal{Q}$ is unchanged. Thus, $\mathcal{Q}$ is the transitive closure of $\mathcal{R}$.

$$\mathcal{M} \circ \mathcal{M} = \begin{bmatrix} 0.70 & 0.57 & 0.57 & 0.70 & 0.57 \\ 0.57 & 0.70 & 0.57 & 0.57 & 0.70 \\ 0.57 & 0.57 & 0.70 & 0.57 & 0.57 \\ 0.70 & 0.57 & 0.57 & 0.70 & 0.57 \\ 0.57 & 0.70 & 0.57 & 0.25 & 0.70 \end{bmatrix} = \mathcal{M} \cup (\mathcal{M} \circ \mathcal{M}) \neq \mathcal{M}$$

Observe that $\mathcal{Q}$ is coarser than $\mathcal{R}$, because the membership grade 0.25 is also disappeared. On one hand, this is probably the price for transitivity: the loss of refinement. On the other hand, the transitive closure may eliminate those degrees of occurrence that are parasites, due to subjective observations.

The argument presented in the above paragraph can be used identically to construct proximity relations starting from covers of $\mathcal{A}_N$ (with overlapping clusters), but, this time, the crisp relations are only compatibility (tolerance) type (i.e., only reflexive and symmetric). However, the procedure before could be invoked to transform the resulting proximity relations into similarity ones, if desired.

In conclusion, at this step, a family of fuzzy relations (at least of proximity type) was defined for further constructions, $\{\mathcal{R}_k\}_{k \in \overline{1,K}}$. Obviously, a one-to-one map between $\{\mathcal{P}_k\}_{k \in \overline{1,K}}$ and $\{\mathcal{R}_k\}_{k \in \overline{1,K}}$, say T, was thus constructed:

$$T(\mathcal{P}_k) = \mathcal{R}_k, \ \forall k \in \overline{1,K}. \qquad \text{Equation (3.15)}$$

3.6.3.2 Building an Appropriate Measure of Fuzziness

3.6.3.2.1 On Measures of Fuzziness

The next step aims to construct a measure of fuzziness over the fuzzy relations on the Cartesian product $\mathcal{A}_N \times \mathcal{A}_N$. This measure will be used to select the "minimally fuzzy" relation within the set constructed above.

According to Klir and Folger 1988, in general, if X is a crisp set and $\mathcal{F}(X)$ is the set of its fuzzy parts, then a *measure of fuzziness* is a map $f: \mathcal{F}(X) \rightarrow \mathcal{R}_+$ that verifies the following properties:

fa) $f(A) = 0 \Leftrightarrow A \in \mathcal{F}(X) \rightarrow$ is a crisp set.

fb) Suppose that a "sharpness relation" between fuzzy sets is defined and denote it by "$\prec$" ($A \prec B$ meaning "A is sharper than B," where $A, B \in \mathcal{F}(X)$). Then with $A, B \in \mathcal{F}(X)$ with $A \prec B$, f must verify the inequality $f(A) \leq f(B)$.

c) Suppose that, according to the "sharpness relation" defined before, there is at least a fuzzy set that is *maximally fuzzy* , i.e., $A^{MAX} \in \mathcal{F}(X)$ for which $A \prec A^{MAX}$, $\forall A \in \mathcal{F}(X)$.Then $A \in \mathcal{F}(X)$ is maximally fuzzy if and only if $f(B) \leq f(A)$, $\forall B \in \mathcal{F}(X)$.

Accordingly, we can define $A \in Y \subseteq \mathcal{F}(X)$ as *minimally fuzzy* in Y if, given f, the following property is verified: $f(A) \leq f(B)$, $\forall B \in Y$. Minimally fuzzy sets are the closest to the crisp state, according to f a), that is they have the least fuzzy intrinsic structure. All the crisp sets of Y (if exist) are minimally fuzzy and none of its fuzzy sets are minimally fuzzy. However, it is not mandatory that Y have a minimally fuzzy set, and several related definitions about "infimumly fuzzy" sets could be stated. But if Y is a finite set — and this is the case in our framework — then always at least one minimally fuzzy set can be pointed out.

3.6.3.2.2 The Role of Sharpness in our Construction

It is not necessary either that a maximally fuzzy set exists for the entire $\mathcal{F}(X)$, because the sharpness relation is only a partial ordering defined on $\mathcal{F}(X)$. In this case, when constructing the measure of fuzziness, we can skip the requirement f c). Since the classical ordering of numbers is a total ordering relation on $\mathcal{R}$, there are only two possibilities that we may have:

a. The set $\{f(A) \mid A \in \mathcal{F}(A)\} \subseteq \mathcal{R}_+$ is not bounded and, thus, maximally fuzzy sets do not exist.
b. It exists $A \in \mathcal{F}(X)$ so that $f(B) \leq (A)$, $\forall B \in \mathcal{F}(X)$ and, in this case, A can be considered as maximally fuzzy.

But, even so, minimally fuzzy sets (as defined before) exist in finite subsets of $\mathcal{F}(X)$. However, it is important to define the sharpness relation so that maximally fuzzy sets exist, because the measure of fuzziness is, thus, bounded on $\mathcal{F}(X)$. The existence of the maximally fuzzy sets is determined not only by the sharpness relation itself, but also by the set $\mathcal{F}(X)$.

One of most usual (classical) sharpness relations between fuzzy sets is the following:

$$A \prec B \text{ in } \mathcal{F}(X) \text{ if: } \begin{cases} \mu_A(x) \leq \mu_B(x), & \text{for } x \in X \text{ so that } \mu_B(x) \leq \frac{1}{2} \\ \mu_A(x) \geq \mu_B(x), & \text{for } x \in X \text{ so that } \mu_B(x) \geq \frac{1}{2}. \end{cases} \qquad \text{Equation (3.16)}$$

Figure 3.6 shows an example of classical sharpness order between two sets. Obviously, the maximally fuzzy set is defined by the constant membership function equal with 1/2. This sharpness relation is not a total ordering relation, because there are fuzzy sets, which are non-comparable. Looking again at Figure 3.6, the sets A and B become non-comparable if, for example, μ_B has the maximum value equal with $\frac{1}{2}$, at $x = \frac{1}{2}$.

This sharpness relation is not the most unique that we can define, but it helps us to select an interesting measure of fuzziness. One very important class consists of measures that evaluate "the fuzziness" of a fuzzy set by taking into consideration both the set and its (fuzzy) complement.

3.6.3.2.3 Shannon Entropy as an Adequate Measure of Fuzziness

From this large class, we have selected the *Shannon measure,* based on Shannon's function:

$$\left[\begin{array}{l} S:[0,1] \to \mathcal{R} \\ x \mapsto S(x) \overset{def}{=} -x\log_2 x - (1-x)\log_2(1-x) \end{array} \right.$$

It is easy to observe that S has one maximum, (0.5;1), and two minima, (0;0) and (1;0). It looks very similar to μ_A of Figure 3.6, only the aperture is bigger because the marginal derivatives are infinite.

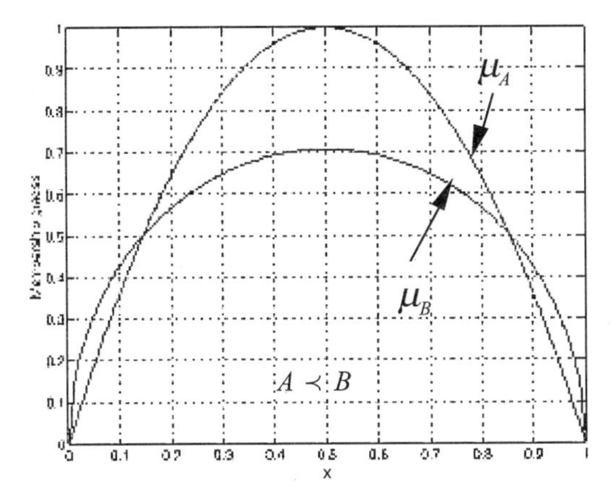

FIGURE 3.6 An example of classical sharpness relation.

If the argument of S is replaced by a (vectorial) set of variables, then its definition extends as follows:

$$\left[\begin{array}{l} S:[0,1]^M \to \mathscr{R} \\ \quad x \mapsto S(x) \stackrel{def}{=} -\sum_{m=1}^{M} x_m \log_2 x_m - \sum_{m=1}^{M} (1-x_m)\log_2(1-x_m) \end{array}\right. \qquad \text{Equation (3.17)}$$

where $x = (x_1, x_2, \ldots, x_M)$. This function also has one single maximum (equal by M for x in the middle of the hyper-cube$[0,1]^M$) and 2^M null minima (one for each apex of the hyper-cube). For example, if $M = 2$, the shape of the Shannon function looks like that in Figure 3.7(a).

The gradient of the function 3.17 is also useful for some interpretations. Its expression can be derived by simple manipulations:

$$\nabla S(x) \stackrel{def}{=} \left(\frac{\partial S}{\partial x_1}\frac{\partial S}{\partial x_2}\cdots\frac{\partial S}{\partial x_M}\right)(x) = \left(\log_2\frac{1-x_1}{x_1}\log_2\frac{1-x_2}{x_2}\cdots\log_2\frac{1-x_M}{x_M}\right), \qquad \text{Equation (3.18)}$$

for $x_m \in (0,1)$, $\forall m \in \overline{1,M}$. If there is $x_{m_0} \in \{0,1\}$, but not all x_m are 0 or 1, then the m_0-th component of $\nabla S(x)$ is *a priori* null. If all x_m are 0 or 1, then all components of $\nabla S(x)$ are infinity.

As in Subramanian and Ulieru [1999], we refer to this gradient as *force*, for reasons that we will explain later. For $M = 2$, this force has the shape in Figure 3.7(b). It is infinite for each apex of the hyper-cube and null in the middle of hyper-cube.

When the argument of Shannon function 3.17 is a discrete probability distribution, it is referred to as Shannon entropy.* This denomination extends even if the argument is a discrete membership function defining a fuzzy set, in which case it is denoted by S_μ and named *Shannon fuzzy entropy* [Zimmermann, 1991]:

$$\left[\begin{array}{l} S_\mu:\mathscr{F}(X) \to \Re_+ \\ \quad x \mapsto S_\mu(A) \stackrel{def}{=} -\sum_{x\in X} \mu_A(x)\log_2\mu_A(x) - \sum_{x\in X} [1-\mu_A(x)]\log_2[1-\mu_A(x)] \end{array}\right. \qquad \text{Equation (3.19)}$$

*In the continuous case, the sum in Equation 3.17 is replaced by the integral.

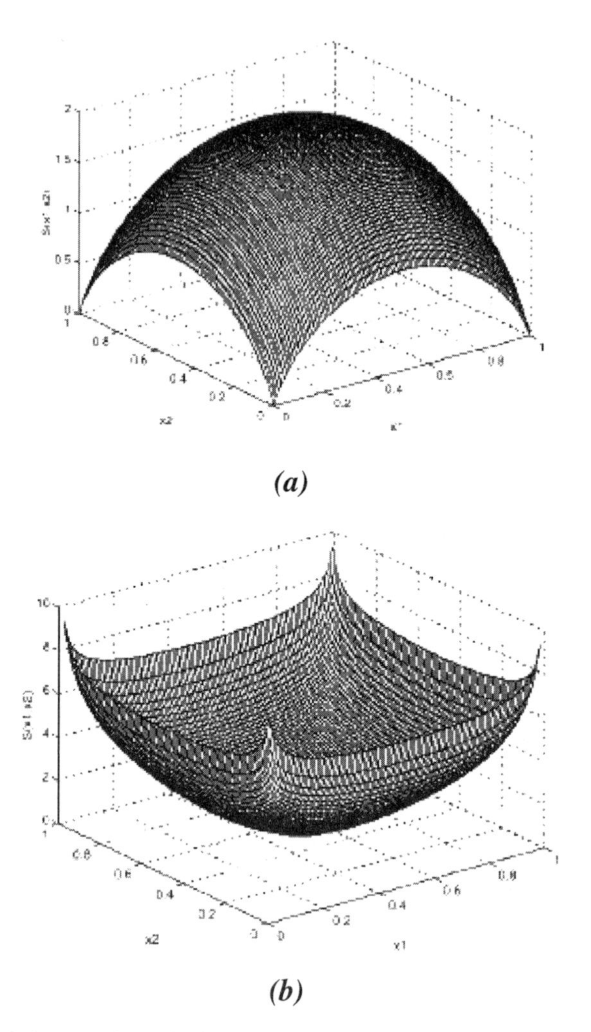

FIGURE 3.7 (a) Shannon's function ($M = 2$). (b) Amplitude of the gradient for Shannon's function ($M = 2$).

It is easy to prove that S_μ is a measure of fuzziness, according to requirements f a), f b), and f c), for the sharpness relation given by Equation 3.16. If the fuzzy complement of a set $A \in \mathcal{F}(X)$ is classically defined (i.e., by $\overline{A} \in \mathcal{F}(X)$ with the membership function $\mu_{\overline{A}} \overset{def}{\equiv} 1 - \mu_A$), then $S_\mu(A)$ can be viewed as a total entropy of the set and its complement. As a matter of fact, $S_\mu(A) = S_\mu(\overline{A})$. There are some other measures of fuzziness, e.g., based on sharpness relations that evaluate the distance between the fuzzy set and a well-defined complement (or even "the closest" crisp set; see Klir and Folger 1988).

In our framework, X is identified with the Cartesian product $A_N \times A_N$ and the fuzzy sets are fuzzy relations between agents. Thus, according to Equation 3.19, for each $k \in \overline{1,K}$, the Shannon (fuzzy) entropy of the relation $\mathcal{R}_k$ can be evaluated as follows:

$$S_\mu(\mathcal{R}_k) = -\sum_{i=1}^{N}\sum_{j=1}^{N}\mu_k(a_i,a_j)\log_2\mu_k(a_i,a_j) - \sum_{i=1}^{N}\sum_{j=1}^{N}\left[1-\mu_k(a_i,a_j)\right]\log_2\left[1-\mu_k(a_i,a_j)\right]. \quad \text{Equation (3.20)}$$

Two main reasons motivate this choice (and they reside in the fact that Shannon fuzzy entropy provides very interesting interpretations, closely related to the concept of "uncertainty"). First, S helps us to make a direct connection between "how fuzzy" is a set and "how much uncertainty" it contains (using the property f (b)). Thus, since S computes the quantity of information of an informational entity (like, e.g., a set of events or a fuzzy set) as the estimated uncertainty that the entity contains, the minimally fuzzy sets contain also the minimally uncertain information.* Secondly, the "total ignorance" (or uncertain) information is pointed out by the unique maximum of S (see Figures 3.7), whereas multiple minimum points (actually, the apexes of the hyper-cube) belong to the "perfect knowledge zone" (as less uncertain information as possible). Between "total ignorance" (which, interestingly, is unique) and "perfect knowledge zone" (which is always multiple) there are many intermediate points associated with different knowledge/uncertainty levels about the entity. The hyper-surface S_μ can be considered as a potential knowledge surface, where the minimum potential stands for the knowledge area, whereas the (unique) maximum potential is the ignorance point. Moreover, according to this interpretation, a *force driving toward knowledge* can be determined, by computing the gradient of the potential knowledge surface. Obviously, using Equations 3.18 and 3.20 the expression of the amplitude (the Euclidean norm) of this force can be straightforwardly derived (for each $k \in \overline{1,K}$):

$$\left\|\nabla S_\mu\left(\mathcal{R}_k\right)\right\| = \sqrt{\sum_{i=1}^{N}\sum_{j=1}^{N}\left[\log_2 \frac{1-\mu_k\left(a_i,a_j\right)}{\mu_k\left(a_i,a_j\right)}\right]^2}. \qquad \text{Equation (3.21)}$$

The amplitude increases very fast in the vicinity of any "perfect knowledge" point.

3.6.3.3 Constructing and Optimizing the Uncertainty Measure

Despite that a unique maximum of Shannon fuzzy entropy exists, we are searching for one of its minima. More specifically, at this step one can construct the required measure of uncertainty, V, by composing S_μ (definition 3.20) with T (Equation 3.15), that is: $V = S_\mu \circ T$. Notice that V is not a measure of fuzziness, because its definition domain is the set of source-plans (crisp sets) and not the set of fuzzy relations between agents (fuzzy sets). But, since T is a bijection, the optimization problem (1) (stated in the previous section) becomes equivalent with

$$\mathcal{P}_{k_0} = T^{-1}\left(\underset{k\in\overline{1,K}}{\arg\min}\, S_\mu\left(\mathcal{P}_k\right)\right), \text{ where } k_0 \in \overline{1,K}. \qquad \text{Equation (3.22)}$$

Normally, this new problem does not require a special optimization algorithm, since K is a finite number and the maximum point is not the focus, but rather a minimum point. Problems could appear only if K is very large. This may happen because, as mentioned before, the number of possible partitions covering A_N increases rapidly with N. In this case, a *genetic algorithm* can be used to find the optimum. According to the previous interpretations, P_{k_0} is the least fuzzy (minimally fuzzy), i.e., the least uncertain source-plan from the family and the most attracted by the knowledge zone. Its corresponding optimum fuzzy relation R_{k_0}, might be useful in the construction of a *least uncertain plan* of MAS.

3.6.3.4 Identifying Holonic Structures by Constructing the Least Uncertain Source-Plan

Once one pair ($\mathcal{P}_{k_0}$, $\mathcal{R}_{k_0}$) has been selected by solving the problem 3.22 (multiple choices could be possible), two options are available:

1. Stop the procedure, by listing all the partitions/covers of P_{k_0}:

$$\mathcal{P}_{k_0} = \left\{P_{k_0,1}, P_{k_0,2}, \ldots, P_{k_0,Mk_0}\right\}.$$

---

*Notice, however, that only the vagueness facet of the uncertainty is measured here. Ambiguity requires more sophisticated measures.

Although these partitions/covers are most likely to appear during the MAS evolution, the order in which they succeed remains unknown and their consistency is often very poor, corrupted by parasite partitions/covers due to subjective observations. This is the main drawback of the option.

2. Construct a source-plan starting not from $\mathcal{P}_{k_0}$, but from $\mathcal{R}_{k_0}$. This option will be detailed below. Its main drawback is that the plan could contain also other partitions/covers, not included in $\mathcal{P}_{k_c}$. Moreover the plan may not contain any of the partitions/covers of $\mathcal{P}_{k_0}$. Also, the succession order of partitions/covers is still unknown, in general.

There is a reason for the second option. Usually, the initial available information about MAS is so vague that it is impossible to construct even consistent source-plans. This is the case, for example, when all we can set are the degrees of occurrence corresponding to clusters created only by couples of agents. (Case Study 1 in Section 3.7.1 shows this situation). However, it is suitable to point out at least a source-plan, in order to solve the problem.

The main idea in constructing source-plans other than the initial ones is to evaluate the α-cuts of $\mathcal{R}_{k_0}$ (recall Equation 3.3), and to arrange them in decreasing order of membership grades $\alpha \in \Lambda_{\mathcal{R}k_0}$. This ordering is unique. Since the time dimension of MAS evolution was not taken into consideration when constructing the model, no time ordering criterion is yet available. Thus, basically, plans are not constructible with this model. However, it is possible for a plan to be coincident with the source-plan generated in this manner (especially when the relation is a similarity one).

Two categories of source-plans can be generated using the α-cuts of $\mathcal{R}_{k_0}$:

1. If within the source-plan there are clusters that overlap, then, obviously, the α-cuts are *compatibility/tolerance* (crisp) relations (that is, reflexive and symmetric, but not necessarily transitive). In this case, we refer to covers as *tolerance covers* and to the source-plan as the *compatibility source-plan.*
2. If the source-plan contains only partitions (covers with disjoint clusters), then it is easy to prove that the α-cuts are *equivalence* (crisp) relations (that is, reflexive, symmetric, and transitive). (Actually, the crisp transitivity is not obvious, but it follows straightforwardly from fuzzy (max–min) transitivity, after some elementary manipulations.) In this case, we refer to the source-plan as the *equivalence source-plan* or the *holonic source-plan* .

If $\mathcal{R}_{k_0}$ is a proximity relation, two source-plans can be generated: one from the compatibility category (using $\mathcal{R}_{k_0}$) and another from the equivalence (holonic) category (using the transitive closure of $\mathcal{R}_{k_0}$). The fact that clusters could be overlapped reveals the capacity of some agents to play multiple roles simultaneously by being involved in several actions at the same time, and this is not necessarily a drawback.

If $\mathcal{R}_{k_0}$ is a similarity relation, then the succession of partitions reveals an interesting behavior of the MAS, where clusters are associated in order to form new clusters, as if MAS would be *holonic type*. (Obviously, the same behavior is proven by the transitive cover of a proximity relation, which is a similarity relation.) In this case, the source-plan may indicate exactly the evolution of MAS and it could be identified by a *holonic plan* (with clusters within clusters). This important result gives the mathematical framework for identifying holonic (potential) structure within a multiagent system and can be successfully used as well to isolate holonic systems out of less structured multiagent systems in manufacturing organizations, as the following simulation studies prove.

3.7 MAS Self-Organization as a Holonic System: Simulation Results

Two examples are presented here by using a simulator designed in MATLAB. Both are founded on a MAS with $N = 7$ agents.

The MAS is considered initially as a totally unknown black box, a system about which we know nothing. We can stimulate it by setting some goals and letting it reach them. When the goals have been input, the observed output is the succession of agent clusters that occur each time the MAS transitions from the initial state to the final one. In order to get the best response from the MAS (in terms of having it go through as many states as possible), we shall stimulate it with a set of goals equivalent to the "white noise" in systems identification (i.e., a class of goals, as large as possible, that might be set for the system). By inputting these goals one by one (or in a mixed manner — if this is possible), we can count the number of occurrences for each cluster. By normalizing the numbers (i.e., dividing them by their sum),one obtains the occurrence degrees of each cluster, i.e., the numbers $\{\alpha_{k,m}\}_{k\in\overline{1,K};m\in\overline{1,Mk}}$ that are, in fact, occurrence frequencies of clusters. For these simulation experiments, quasi-random (Gauss type) numbers were generated as occurrence degrees. After ~200 simulations, the final holonic clustering patterns emerged. Moreover, some information about the nature of the agents (i.e., whether they are knowledge agents, or yellow-page agents, or interface agents, etc.) became revealed without any *a priori* knowledge about the agent's type, from their cluster involvements.

In these experiments, we considered the most general case and the most parsimonious in our knowledge about it. Therefore, we chose to determine the occurrence degrees by using a system identification technique of stimulation, as being the most general one applicable. If more is known about the agents, different clustering criteria will emerge and the agents will group according to these. For example, if we have information on the agents' knowledge bases, internal goals, and the global goal of the system, then on their way to reaching the global goal of the system those agents with similar knowledge about some specific topic will cluster together, agents with similar partial goals will also cluster together, and the degrees of cluster occurrence will be determined and assigned accordingly.

3.7.1 Case Study 1

The first example starts from very parsimonious information about clusters created by couples of agents. Every degree of occurrence is associated with only a pair of agents. Although this information is vague enough, compatibility and holonic plans are still constructed.

- Since $N = 7$, there are maximum $C_7^2 = 21$ possible couples between agents. Each couple of agents has a certain degree of occurrence but, in general, it is not necessary that all 21 couples appear during the MAS evolution. A null degree of occurrence will be assigned to every impossible couple. However, in the example, all 21 couples were graded by non-null values, in order to avoid triviality.
- Starting from this vague information, one constructs first the corresponding fuzzy relation $\mathcal{R}$ between the seven agents. None of the occurrence degrees has disappeared in $\mathcal{R}$. Then, since this relation is only proximity one, its transitive cover $\mathcal{Q}$ is generated, according to two-step the procedure proposed in the previous section. Actually, $\mathcal{Q}$ is a similarity relation between agents.
- Finally, two types of source-plans are generated: one emerging from $\mathcal{R}$ and including (tolerance) covers (with overlapped clusters) and another — from $\mathcal{Q}$, including partitions (with disjoint clusters).
- The MAS evolution starts from the highest degree of occurrence (which is of course 1) and completes when the smallest degree is reached. Only the first six tolerance covers are presented starting from $\mathcal{R}$, but all seven partitions constructed from $\mathcal{Q}$ are depicted. The possible holonic plan generated by $\mathcal{Q}$ is shown also in Figure 3.8.
- Shannon fuzzy entropy of the source-plan: 36.6784 (max = 49) ($N = 7$ agents)
  Amplitude of force toward knowledge: 12.2942 (max = ∞).

- The membership matrix of the corresponding proximity relation $\mathcal{R}$:

$$\mathcal{M}_{\mathcal{R}}: \begin{array}{ccccccc} 1.0000 & 0.5402 & 0.6343 & 0.1877 & 0.3424 & 0.2001 & 0.4863 \\ 0.5402 & 1.0000 & 0.3561 & 0.4797 & 0.7651 & 0.2092 & 0.2794 \\ 0.6343 & 0.3561 & 1.0000 & 0.4780 & 0.1389 & 0.6858 & 0.6414 \\ 0.1877 & 0.4797 & 0.4780 & 1.0000 & 0.3191 & 0.4888 & 0.3347 \\ 0.3424 & 0.7651 & 0.1389 & 0.3191 & 1.0000 & 0.2784 & 0.4291 \\ 0.2001 & 0.2092 & 0.6858 & 0.4888 & 0.2784 & 1.0000 & 0.1666 \\ 0.4862 & 0.2794 & 0.6414 & 0.3347 & 0.4291 & 0.1666 & 1.0000 \end{array}$$

- The membership matrix of the corresponding proximity relation $\mathcal{Q}$:

$$\mathcal{M}_{\mathcal{Q}}: \begin{array}{ccccccc} 1.0000 & 0.5402 & 0.6343 & 0.4888 & 0.5402 & 0.6343 & 0.6343 \\ 0.5402 & 1.0000 & 0.5402 & 0.4888 & 0.7651 & 0.5402 & 0.5402 \\ 0.6343 & 0.5402 & 1.0000 & 0.4888 & 0.5402 & 0.6858 & 0.6414 \\ 0.4888 & 0.4888 & 0.4888 & 1.0000 & 0.4888 & 0.4888 & 0.4888 \\ 0.5402 & 0.7651 & 0.5402 & 0.4888 & 1.0000 & 0.5402 & 0.5402 \\ 0.6343 & 0.5402 & 0.6858 & 0.4888 & 0.5402 & 1.0000 & 0.6414 \\ 0.6343 & 0.5402 & 0.6414 & 0.4888 & 0.5402 & 0.6414 & 1.0000 \end{array}$$

- First five tolerance covers generated by the proximity relation $\mathcal{R}$:

# Characteristic matrix (α =1.0000)	Clusters:
1 0 0 0 0 0 0	{a1}
0 1 0 0 0 0 0	{a2}
0 0 1 0 0 0 0	{a3}
0 0 0 1 0 0 0	{a4}
0 0 0 0 1 0 0	{a5}
0 0 0 0 0 1 0	{a6}
0 0 0 0 0 0 1	{a7}

# Characteristic matrix (α=0.7651)	Clusters:
1 0 0 0 0 0 0	{a1}
0 1 0 0 1 0 0	{a2,a5}
0 0 1 0 0 0 0	{a3}
0 0 0 1 0 0 0	{a4}
0 1 0 0 1 0 0	
0 0 0 0 0 1 0	{a6}
0 0 0 0 0 0 1	{a7}

# Characteristic matrix (α=0.6858)	Clusters:
1 0 0 0 0 0 0	{a1}
0 1 0 0 1 0 0	{a2,a5}
0 0 1 0 0 1 0	{a3,a6}
0 0 0 1 0 0 0	{a4}
0 1 0 0 1 0 0	
0 0 1 0 0 1 0	
0 0 0 0 0 0 1	{a7}

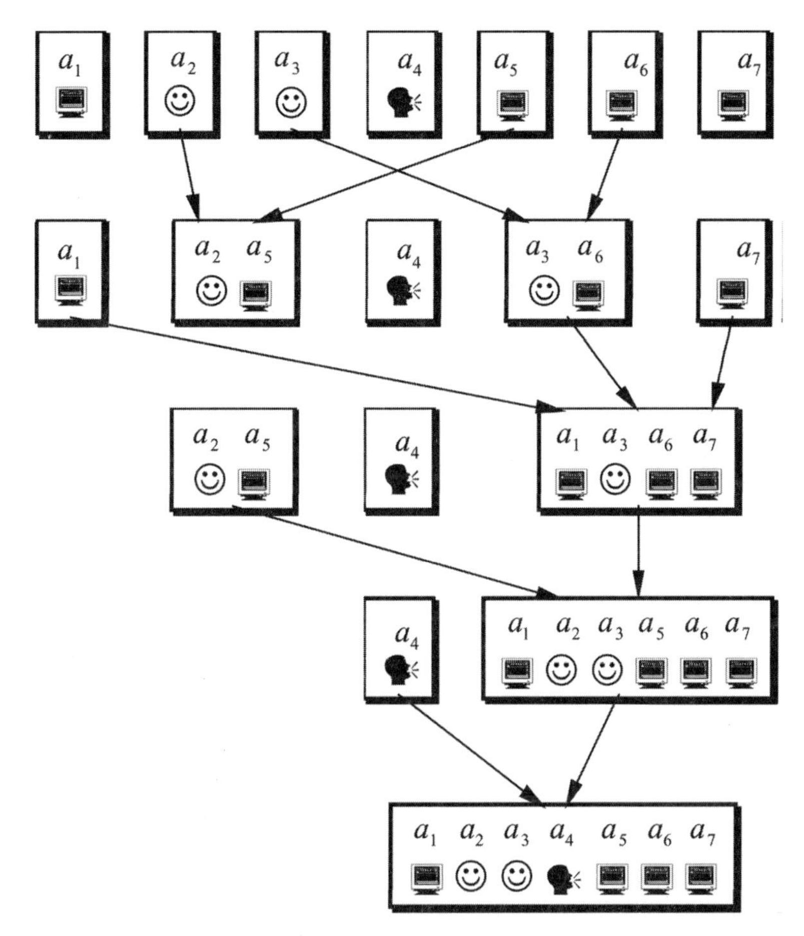

FIGURE 3.8 An illustration of the holonic plan revealed by Case Study 1.

# Characteristic matrix	
(α=0.6414)	Clusters:
1 0 0 0 0 0 0	{a1}
0 1 0 0 1 0 0	{a2,a5}
0 0 1 0 0 1 1	{a3,a6,a7}
0 0 0 1 0 0 0	{a4}
0 1 0 0 1 0 0	
0 0 1 0 0 1 0	{a3,a6}
0 0 1 0 0 0 1	{a3,a7}

# Characteristic matrix	
(α=0.6343)	Clusters:
1 0 1 0 0 0 0	{a1,a3}
0 1 0 0 1 0 0	{a2,a5}
1 0 1 0 0 1 1	{a1,a3,a6,a7}
0 0 0 1 0 0 0	{a4}
0 1 0 0 1 0 0	
0 0 1 0 0 1 0	{a3,a6}
0 0 1 0 0 0 1	{a3,a7}

- Partitions generated by the similarity relation $\mathcal{Q}$:

Characteristic matrix

($\alpha = 1.0000$)	Clusters:
1 0 0 0 0 0 0	{a1}
0 1 0 0 0 0 0	{a2}
0 0 1 0 0 0 0	{a3}
0 0 0 1 0 0 0	{a4}
0 0 0 0 1 0 0	{a5}
0 0 0 0 0 1 0	{a6}
0 0 0 0 0 0 1	{a7}

Characteristic matrix

($\alpha = 0.7651$)	Clusters:
1 0 0 0 0 0 0	{a1}
0 1 0 0 1 0 0	{a2,a5}
0 0 1 0 0 0 0	{a3}
0 0 0 1 0 0 0	{a4}
0 1 0 0 1 0 0	
0 0 0 0 0 1 0	{a6}
0 0 0 0 0 0 1	{a7}

Characteristic matrix

($\alpha = 0.6858$)	Clusters:
1 0 0 0 0 0 0	{a1}
0 1 0 0 1 0 0	{a2,a5}
0 0 1 0 0 1 0	{a3,a6}
0 0 0 1 0 0 0	{a4}
0 1 0 0 1 0 0	
0 0 1 0 0 1 0	
0 0 0 0 0 0 1	{a7}

Characteristic matrix

($\alpha = 0.6414$)	Clusters:
1 0 0 0 0 0 0	{a1}
0 1 0 0 1 0 0	{a2,a5}
0 0 1 0 0 1 1	{a3,a6,a7}
0 0 0 1 0 0 0	{a4}
0 1 0 0 1 0 0	
0 0 1 0 0 1 1	
0 0 1 0 0 1 1	

Characteristic matrix

($\alpha = 0.6343$)	Clusters:
1 0 1 0 0 1 1	{a1,a3,a6,a7}
0 1 0 0 1 0 0	{a2,a5}
1 0 1 0 0 1 1	
0 0 0 1 0 0 0	{a4}
0 1 0 0 1 0 0	
1 0 1 0 0 1 1	
1 0 1 0 0 1 1	

# Characteristic matrix ($\alpha = 0.5402$)	Clusters:
1 1 1 0 1 1 1	{a1,a2,a3,a5,a6,a7}
1 1 1 0 1 1 1	
1 1 1 0 1 1 1	
0 0 0 1 0 0 0	{a4}
1 1 1 0 1 1 1	
1 1 1 0 1 1 1	
1 1 1 0 1 1 1	

# Characteristic matrix ($\alpha = 0.4888$)	Clusters:
1 1 1 1 1 1 1	{a1,a2,a3,a4,a5,a6,a7}
1 1 1 1 1 1 1	
1 1 1 1 1 1 1	
1 1 1 1 1 1 1	
1 1 1 1 1 1 1	
1 1 1 1 1 1 1	
1 1 1 1 1 1 1	

- **Holonic structure:** The holonic behavior can be observed for the equivalence source-plan: clusters associate together in order to form larger clusters and, finally, the whole agents set is grouped in one single cluster. In Figure 3.8, one starts from a MAS with one manager (a_4), two executive agents (a_2 and a_3) and four resource agents (a_1, a_5, a_6, and a_7). Recall that all we know about the clustering capacity of these agents is the degree of occurrence of their couples (see the matrix $\mathcal{M}_{\mathcal{R}}$ before). For example, the manager is tempted to work in association with the executive a_2 (0.4797) rather than with a_3 (0.4780), but also is oriented to solve problems by himself using the resource a_6 (0.4888). In our example, first, the manager states a goal. Immediately, the executives a_2 and a_3 reach for resources a_5 and a_6, respectively (that they prefer mostly, see the matrix $\mathcal{M}_{\mathcal{R}}$ before). The executive a_3 realizes that he needs more resources and he starts to use both a_1 and a_7 (therefore, maybe, the manager prefers a_2). The next step shows that the two executives associate together (including their resources) in order to reach the goal. The manager associates with them only in the final phase, when the goal is reached.

3.7.2 Case Study 2

The second example operates with vaguer information than the previous one (in the sense of Shannon fuzzy entropy, which is larger), although, apparently, it reveals more about the agents clustering capacity (clusters with two agents can appear simultaneously and even clusters with three agents have non-null occurrence degrees). The construction steps are similar and the holonic behavior is presented also in Figure 3.9.

- Shannon fuzzy entropy of the source-plan: 39.5714 (max = 49) ($N = 7$ agents)
  Amplitude of force towards knowledge: 7.7790 (max = ∞).

- The membership matrix of the corresponding proximity relation $\mathscr{R}$:

$$
\mathcal{M}_{\mathscr{R}}:\begin{matrix}
1.0000 & 0.3420 & 0.6988 & 0.5666 & 0.5179 & 0.2372 & 0.6694 \\
0.3420 & 1.0000 & 0.3568 & 0.4340 & 0.4802 & 0.7034 & 0.6032 \\
0.6988 & 0.3568 & 1.0000 & 0.3543 & 0.6901 & 0.5926 & 0.4601 \\
0.5666 & 0.4340 & 0.3543 & 1.0000 & 0.4028 & 0.3538 & 0.6791 \\
0.5179 & 0.4802 & 0.6901 & 0.4028 & 1.0000 & 0.6358 & 0.3887 \\
0.2372 & 0.7034 & 0.5926 & 0.3538 & 0.6358 & 1.0000 & 0.4275 \\
0.6694 & 0.6032 & 0.4601 & 0.6791 & 0.3887 & 0.4275 & 1.0000
\end{matrix}
$$

- The membership matrix of the covering similarity relation $\mathscr{Q}$:

$$
\mathcal{M}_{\mathscr{Q}}:\begin{matrix}
1.0000 & 0.6358 & 0.6988 & 0.6694 & 0.6901 & 0.6358 & 0.6694 \\
0.6358 & 1.0000 & 0.6358 & 0.6358 & 0.6358 & 0.7034 & 0.6358 \\
0.6988 & 0.6358 & 1.0000 & 0.6694 & 0.6901 & 0.6358 & 0.6694 \\
0.6694 & 0.6358 & 0.6694 & 1.0000 & 0.6694 & 0.6358 & 0.6791 \\
0.6901 & 0.6358 & 0.6901 & 0.6694 & 1.0000 & 0.6358 & 0.6694 \\
0.6358 & 0.7034 & 0.6358 & 0.6358 & 0.6358 & 1.0000 & 0.6358 \\
0.6694 & 0.6358 & 0.6694 & 0.6791 & 0.6694 & 0.6358 & 1.0000
\end{matrix}
$$

- First seven tolerance covers generated by the proximity relation $\mathscr{R}$:

# Characteristic matrix (α= 1.0000)	Clusters:
1 0 0 0 0 0 0	{a1}
0 1 0 0 0 0 0	{a2}
0 0 1 0 0 0 0	{a3}
0 0 0 1 0 0 0	{a4}
0 0 0 0 1 0 0	{a5}
0 0 0 0 0 1 0	{a6}
0 0 0 0 0 0 1	{a7}

# Characteristic matrix (α= 0.7034)	Clusters:
1 0 0 0 0 0 0	{a1}
0 1 0 0 0 1 0	{a2,a6}
0 0 1 0 0 0 0	{a3}
0 0 0 1 0 0 0	{a4}
0 0 0 0 1 0 0	{a5}
0 1 0 0 0 1 0	
0 0 0 0 0 0 1	{a7}

# Characteristic matrix (α= 0.6988)	Clusters:
1 0 1 0 0 0 0	{a1,a3}
0 1 0 0 0 1 0	{a2,a6}
1 0 1 0 0 0 0	
0 0 0 1 0 0 0	{a4}
0 0 0 0 1 0 0	{a5}
0 1 0 0 0 1 0	
0 0 0 0 0 0 1	{a7}

Characteristic matrix

($\alpha = 0.6901$)	Clusters:
1 0 1 0 0 0 0	{a1,a3}
0 1 0 0 0 1 0	{a2,a6}
1 0 1 0 1 0 0	{a1,a3,a5}
0 0 0 1 0 0 0	{a4}
0 0 1 0 1 0 0	{a3,a5}
0 1 0 0 0 1 0	
0 0 0 0 0 0 1	{a7}

Characteristic matrix

($\alpha = 0.6791$)	Clusters:
1 0 1 0 0 0 0	{a1,a3}
0 1 0 0 0 1 0	{a2,a6}
1 0 1 0 1 0 0	{a1,a3,a5}
0 0 0 1 0 0 1	{a4,a7}
0 0 1 0 1 0 0	{a3,a5}
0 1 0 0 0 1 0	
0 0 0 1 0 0 1	

Characteristic matrix

($\alpha = 0.6694$)	Clusters:
1 0 1 0 0 0 1	{a1,a3,a7}
0 1 0 0 0 1 0	{a2,a6}
1 0 1 0 1 0 0	{a1,a3,a5}
0 0 0 1 0 0 1	{a4,a7}
0 0 1 0 1 0 0	{a3,a5}
0 1 0 0 0 1 0	
1 0 0 1 0 0 1	{a1,a4,a7}

Characteristic matrix

($\alpha = 0.6358$)	Clusters:
1 0 1 0 0 0 1	{a1,a3,a7}
0 1 0 0 0 1 0	{a2,a6}
1 0 1 0 1 0 0	{a1,a3,a5}
0 0 0 1 0 0 1	{a4,a7}
0 0 1 0 1 1 0	{a3,a5,a6}
0 1 0 0 1 1 0	{a2,a5,a6}
1 0 0 1 0 0 1	{a1,a4,a7}

- Partitions generated by the similarity relation $\mathcal{Q}$:

Characteristic matrix

($\alpha = 1.0000$)Clusters:	
1 0 0 0 0 0 0	{a1}
0 1 0 0 0 0 0	{a2}
0 0 1 0 0 0 0	{a3}
0 0 0 1 0 0 0	{a4}
0 0 0 0 1 0 0	{a5}
0 0 0 0 0 1 0	{a6}
0 0 0 0 0 0 1	{a7}

Characteristic matrix

(α= 0.7034)	Clusters:
1 0 0 0 0 0 0	{a1}
0 1 0 0 0 1 0	{a2,a6}
0 0 1 0 0 0 0	{a3}
0 0 0 1 0 0 0	{a4}
0 0 0 0 1 0 0	{a5}
0 1 0 0 0 1 0	
0 0 0 0 0 0 1	{a7}

Characteristic matrix

(α= 0.6988)	Clusters:
1 0 1 0 0 0 0	{a1,a3}
0 1 0 0 0 1 0	{a2,a6}
1 0 1 0 0 0 0	
0 0 0 1 0 0 0	{a4}
0 0 0 0 1 0 0	{a5}
0 1 0 0 0 1 0	
0 0 0 0 0 0 1	{a7}

Characteristic matrix

(α=0.6901)	Clusters:
1 0 1 0 1 0 0	{a1,a3,a5}
0 1 0 0 0 1 0	{a2,a6}
1 0 1 0 1 0 0	
0 0 0 1 0 0 0	{a4}
1 0 1 0 1 0 0	
0 1 0 0 0 1 0	
0 0 0 0 0 0 1	{a7}

Characteristic matrix

(α=0.6791)	Clusters:
1 0 1 0 1 0 0	{a1,a3,a5}
0 1 0 0 0 1 0	{a2,a6}
1 0 1 0 1 0 0	
0 0 0 1 0 0 1	{a4,a7}
1 0 1 0 1 0 0	
0 1 0 0 0 1 0	
0 0 0 1 0 0 1	

Characteristic matrix

(α= 0.6694)	Clusters:
1 0 1 1 1 0 1	{a1,a3,a4,a5,a7}
0 1 0 0 0 1 0	{a2,a6}
1 0 1 1 1 0 1	
1 0 1 1 1 0 1	
1 0 1 1 1 0 1	
0 1 0 0 0 1 0	
1 0 1 1 1 0 1	

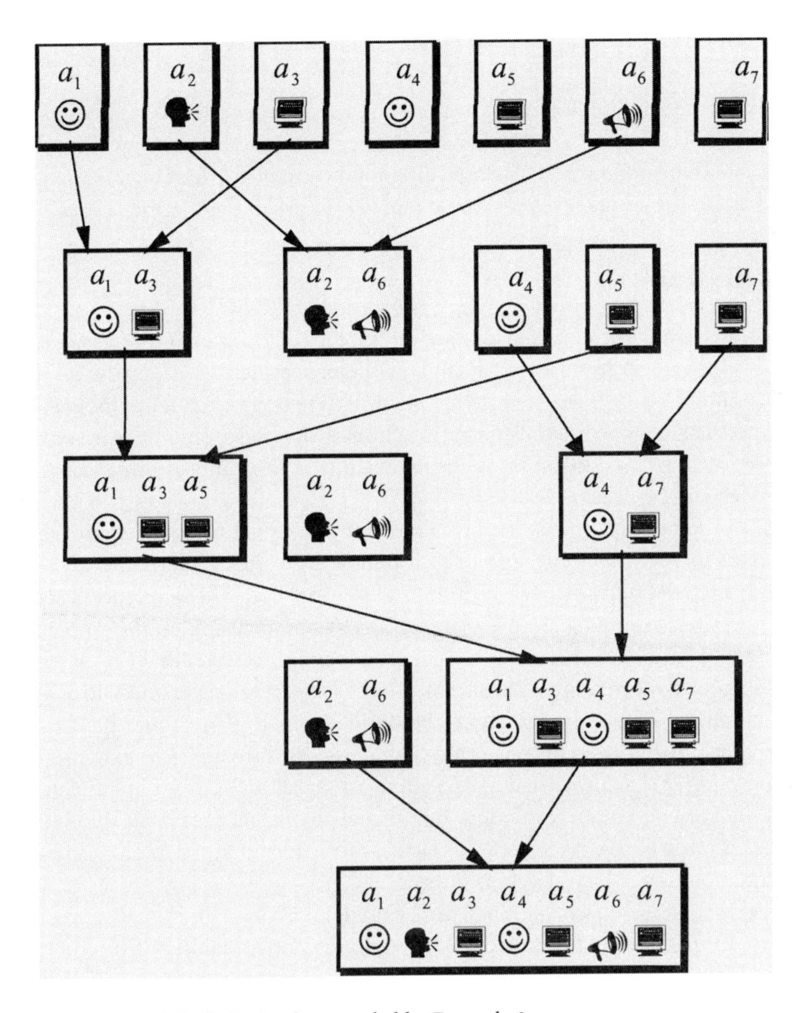

FIGURE 3.9 An illustration of the holonic plan revealed by Example 2.

\# Characteristic matrix
(α= 0.6358)
1 1 1 1 1 1 1
1 1 1 1 1 1 1
1 1 1 1 1 1 1
1 1 1 1 1 1 1
1 1 1 1 1 1 1
1 1 1 1 1 1 1
1 1 1 1 1 1 1

Clusters:
{a1,a2,a3,a4,a5,a6,a7}

- **Holonic Structure:** In Figure 3.9, the MAS includes one manager (a_2), two executive agents (a_1 and a_4), one transmitter agent (a_6), and three resource agents (a_3, a_5, and a_7). Once the manager has established the goal, the executive a_1 starts its job together with the resource a_3. But the other executive is not yet informed about the goal, because he only can receive information from the manager through a transmitter. Thus, the manager a_2 associates to the transmitter a_6 to inform his second executive about the goal. Meantime, the executive a_1 realizes that he needs more resources and he also starts to use a_5. Once the information reaches the receiver a_4, this one starts to work together with the resource a_7. Next, the two executives associate together (including their resources) in order to reach the goal. Finally, the manager and the transmitter associate to the cluster above, when the goal is reached.

In both examples, the amplitude of the force driving toward knowledge is also computed (according to (21)) and depicted. The second amplitude represents only 63% of the first one. Thus, we can say that the second available information is 37% closer to ignorance than the first one, although we would be tempted to say that, in the second case, we know more about MAS than in the first case.

3.8 Conclusions

This chapter described a holonic metamorphic architecture for distributed reconfigurable manufacturing systems, together with a novel approach for identifying holonic structures in generic multiagent systems, using uncertainty minimization. Fuzzy sets theory has proven to be a very suitable tool in the construction of appropriate models able to extract adequate structural properties in multiagent systems. Depending on the measure of fuzziness used to minimize the uncertainty in the initial information about the system, several fuzzy models can be used.

To the best of our knowledge this is the first time that a fuzzy set theoretic approach has been used successfully to extract structural information about multiagent systems. Currently we are working on a comparative study between different measures of fuzziness that could be developed starting from a real application. We are also developing a more complex analysis by focusing on the ambiguity facet of the uncertainty. So far, we considered here only the aspect of vagueness that deals with inconsistency in the information. Ambiguous information is consistent, but the facts revealed are mixed in an unclear manner. For an MAS this can be useful when it cannot be decided which plan should be chosen as the most efficient due to a lack of information regarding plan costs. In this case, one may use quasi-empirical information on the *degree of evidence* that a plan belongs to a well-defined group of plans having similar estimated costs. At present, we are extending our model to embrace, as well, the ambiguity facet of uncertainty. To this we will add time as a supplementary coordinate, to more accurately model the dynamics of MAS evolution, which develops in time.

It may be noted that the approach described can be used not only for an initially non-clustered system, but also for a system that already has clusters, to investigate system evolution from that stage onwards.

References

Balasubramanian, S., Maturana, F. and Norrie, D.H. (1996). Multiagent Planning and Coordination for Distributed Concurrent Engineering, *International Journal of Cooperative Information Systems,* Vol. 5, Nos. 2-3, pp. 153-179.

Brennan, R., Balasubramanian, S. and Norrie, D.H. (1997). Dynamic Control Architecture for Advanced Manufacturing Systems, *Proceedings of International Conference on Intelligent Systems for Advanced Manufacturing,* Pittsburgh, PA, pp. 213-223.

Brown, D., Dunskus, B., Grecu, D. and Berker, I. (1995). SINE: Support for Single Function Agents, *Proceedings of AIENG'95, Applications of AI in Engineering,* Udine, Italy.

Brussel, H.V., Wyns, J., Valckenaers, P., Bongaerts, L. and Peeters, P. (1998). Reference Architecture for Holonic Manufacturing Systems: PROSA, *Computer in Industrial,* Vol. 37, pp. 255-274.

Bussmann, S. (1998). An Agent-Oriented Architecture for Holonic Manufacturing Control, *Proceedings of the 1st Int. Workshop on Intelligent Manufacturing Systems,* EPFL, Lausanne, Switzerland.

Butler, J. and Ohtsubo, H. (1992). ADDYMS: Architecture for Distributed Dynamic Manufacturing Scheduling, *Artificial Intelligence Applications in Manufacturing,* (Eds. Famili, A., Nau, D.S. and Kim, S.H.), The AAAI Press, pp. 199-214.

Christensen, J.H. (1994). Holonic Manufacturing Systems: Initial Architecture and Standard Directions, *First European Conference on HMS,* Hanover, Germany.

Cutkosky, M.R., Engelmore, R.S., Fikes, R.E., Genesereth, M.R., Gruber, T.R., Mark, W.S., Tenenbaum, J.M. and Weber, J.C. (1993). PACT: An Experiment in Integrating Concurrent Engineering Systems, *IEEE Computer,* Vol. 26, No. 1, pp. 28-37.

Cutkosky, M.R., Tenenbaum, J.M. and Glicksman, J. (1996). Madefast: Collaborative Engineering over the Internet, *Communication of the ACM,* Vol. 39, No. 9, pp. 78-87.

Decker, K. (1995). Environment Centered Analysis and Design of Coordination Mechanisms, Ph.D. thesis, Dept. of Computer Science, University of Massachusetts, Amherst.

Decker, K., Sycara, K. and Williamson, M. (1997). Cloning for Intelligent Adaptive Information Agents, *Multi-Agent Systems: Methodologies and Applications,* Lecture Notes in Artificial Intelligence 1286, (eds. Zhang, C. and Lukose, D.), Springer-Verlag, pp. 63-75.

Dilts, D.M., Boyd, N.P. and Whorms, H.H. (1991). The Evolution of Control Architectures for Automated Manufacturing Systems, *Journal of Manufacturing Systems,* Vol. 10, No. 1.

Finin, T., Fritzon, R., McKay, D. and McEntire, R. (1993). KQML — A Language and Protocol for Knowledge and Information Exchange, Tech. Report, University of Maryland, Baltimore.

Gaines, B.R., Norrie, D.H. and Lapsley, A.Z. (1995). Mediator: An Intelligent Information System Supporting the Virtual Manufacturing Enterprise, *Proceedings of 1995 IEEE International Conference on Systems, Man and Cybernetics,* New York, IEEE, pp. 964-969.

Goldmann, S. (1996). Procura: A Project Management Model of Concurrent Planning and Design, *Proceedings of WET ICE'96,* Stanford, CA.

Gou, L., Luh, P. B. and Kyoya, Y. (1998). Holonic Manufacturing Scheduling: Architecture, Cooperation Mechanism, and Implementation, *Computer in Industrial,* Vol. 37, pp. 213-231.

Jennings, N.R., Corera, J.M. and Laresgoiti, I. (1995). Developing Industrial Multiagent Systems, *Proceedings of the First International Conference on Multi-Agent Systems,* San Francisco, CA, AAAI Press/MIT Press, pp. 423-430.

Klir, G.J. and Folger, T.A. (1988). *Fuzzy Sets, Uncertainty, and Information,* Prentice-Hall, Englewood Cliffs, NJ.

Koestler, A. (1967). *The Ghost in the Machine,* Arkana Books, London.

Kwok, A. and Norrie, D.H. (1993). Intelligent Agent Systems for Manufacturing Applications, *Journal of Intelligent Manufacturing,* Vol. 4, pp. 285-293.

Mathews, J. (1995). Organization Foundations of Intelligent Manufacturing Systems — The Holonic Viewpoint, *Computer Integrated Manufacturing Systems,* Vol. 8, No. 4, pp. 237-243.

Maturana, F. (1997). MetaMorph: An Adaptive Multi-Agent Architecture for Advanced Manufacturing Systems, Ph.D. thesis, University of Calgary.

Maturana, F. and Norrie, D.H. (1996). Multi-Agent Mediator Architecture for Distributed Manufacturing, *Journal of Intelligent Manufacturing,* Vol. 7, pp. 257-270.

Maturana, F. and Norrie, D.H. (1996). Multi-Agent Coordination Using Dynamic Virtual Clustering in a Distributed Manufacturing System, *Proceedings of Fifth Industrial Engineering Research Conference (IERC5),* Institute of Industrial Engineering, Minneapolis, May 18-20, pp. 473-478.

Maturana, F.P. and Norrie, D.H. (1997). Distributed Decision-Making Using the Contract Net within a Mediator Architecture, *Decision Support Systems,* Vol. 20, pp. 53-64.

Maturana, F., Balasubramanian, S. and Norrie, D.H. (1997). Learning Coordination Patterns from Emergent Behavior in a Multi-Agent Manufacturing System, *Intelligent Systems and Semiotics '97: A Learning Perspective,* Gaithersburg, Maryland, September 23-25.

Maturana F., Shen, W. and Norrie, D.H. (1998). MetaMorph: An Adaptive Agent-Based Architecture for Intelligent Manufacturing, *International Journal of Production Research.*

Maturana, F., Shen, W. and Norrie, D.H. (1999). MetaMorph: An Adaptive Agent-Based Architecture for Intelligent Manufacturing, *International Journal of Production Research,* Vol. 37, No. 10, pp. 2159-2174.

McGuire, J., Huokka, D., Weber, J., Tenenbaum, J., Gruber, T. and Olsen, G. (1993). SHADE: Technology for Knowledge-Based Collaborative Engineering, *Journal of Concurrent Engineering: Applications and Research,* Vol. 1, No. 3.

Norrie, D.H. and Gaines, B. (1996). Distributed Agents Systems for Intelligent Manufacturing, *Canadian Artificial Intelligence*, No. 40, pp. 31-33.

Norrie, D.H. and Shen, W. (1999). Applications of Agent Technology for Agent-Based Intelligent Manufacturing Systems, *Proceedings of IMS'99*, Leuven, Belgium.

O'Hare, G.M.P. and Jennings, N.R. (1996). *Foundations on Distributed Artificial Intelligence*, John Wiley & Sons Interscience, New York.

Pan, J.Y.C. and Tenenbaum, M.J. (1991). An Intelligent Agent Framework for Enterprise Integration, *IEEE Transactions on Systems, Man, and Cybernetics*, Vol. 21, No. 6, pp. 1391-1408.

Park, H., Tenenbaum, J. and Dove, R. (1993). Agile Infrastructure for Manufacturing Systems (AIMS): A Vision for Transforming the US Manufacturing Base, *Defense Manufacturing Conference*.

Parker, M. (1997). The IMS Initiative, *Manufacturing Engineer*, February.

Parunak, H.V.D., Baker, A.D. and Clark, S.J. (1997a). The AARIA Agent Architecture: An Example of Requirements-Driven Agent-Based System Design, *Proceedings of the First International Conference on Autonomous Agents*, Marina del Rey, CA.

Parunak, H.V.D., Ward, A., Fleischer, M. and Sauter, J. (1997b). A Marketplace of Design Agents for Distributed Concurrent Set-Based Design, *Proceedings of the Fourth ISPE International Conference on Concurrent Engineering: Research and Applications*, Troy, MI.

Patil, R., Fikes, R., Patel-Schneider, P., Mckay, D., Finin, T., Gruber, T. and Neches, R. (1992). The DARPA Knowledge Sharing Effort: Progress Report, *Principles of Knowledge Representation and Reasoning: Proceedings of the Third International Conference* (Eds. Rich, C., Nebel, B. and Swartout, W.), Morgan Kaufmann, Cambridge, MA.

Petrie, C., Cutkosky, M., Webster, T., Conru, A. and Park, H. (1994). Next-Link: An Experiment in Coordination of Distributed Agents, Position Paper for the AID-94 Workshop on Conflict Resolution, Lausanne.

Roboam, M. and Fox, M.S. (1992). Enterprise Management Network Architecture, *Artificial Intelligence Applications in Manufacturing*, (Eds. Famili, A., Nau, D.S. and Kim, S.H.), The AAAI Press, pp. 401-432.

Rothermel, K. and Popescu-Zeletin, R. (Eds.) (1997). Mobile Agents, *Lecture Notes in Computer Science 1219*, Springer-Verlag, New York.

Russell, S. and Norvig, P. (1995). Artificial Intelligence — A Modern Approach, Prentice-Hall, Upper Saddle River, NJ.

Saad, A., Biswas, G., Kawamura, K., Johnson, M.E. and Salama, A. (1995). Evaluation of Contract Net-Based Heterarchical Scheduling for Flexible Manufacturing Systems, *Proceedings of the 1995 International Joint Conference on Artificial Intelligence (IJCAI'95)*, Workshop on Intelligent Manufacturing, Montreal, Canada, pp. 310-321.

Shen, W. and Barthès, J.P. (1995). DIDE: A Multi-Agent Environment for Engineering Design, *Proceedings of the First International Conference on Multi-Agent Systems*, San Francisco, USA, The AAAI Press/MIT Press, pp. 344-351.

Shen, W. and Barthès, J.P. (1997). An Experimental Environment for Exchanging Engineering Design Knowledge by Cognitive Agents, *Knowledge Intensive CAD-2*, (Eds. Mantyla, M., Finger, S. and Tomiyama, T.), Chapman & Hall, London, pp. 19-38.

Shen, W., Xue, D. and Norrie, D.H. (1998). An Agent-Based Manufacturing Enterprise Infrastructure for Distributed Integrated Intelligent Manufacturing Systems, *Proceedings of PAAM'98*, London, pp. 533-548.

Shen, W. and Norrie, D.H. (1996). A Hybrid Agent-Oriented Infrastructure for Modeling Manufacturing Enterprises, *Proceedings of KAW'98*, Banff, Canada.

Shen, W. and Norrie, D.H. (1998). An Agent-Based Approach for Dynamic Manufacturing Scheduling, *Working Notes of the Agent-Based Manufacturing Workshop*, Minneapolis, MN, pp. 117-128.

Shen, W. and Norrie, D.H. (1999). Agent-Based Systems for Intelligent Manufacturing: A State-of-the-Art Survey, *Knowledge and Information Systems: An International Journal*, Vol. 1, No. 2, pp. 129-156.

Shen, W., Norrie, D.H. and Kremer, R. (1999). Developing Intelligent Manufacturing Systems Using Collaborative Agents, *Proceedings of IMS'99*, Leuven, Belgium.

Shen W., Barthès, J-P. and Norrie, D.H. (2000). *Multi-Agent Systems for Concurrent Design and Manufacturing*, Taylor & Francis, London.

Subramanian, R. and Ulieru, M. (1999). An Approach to the Modeling of Multi-Agent Systems as Fuzzy Dynamical Systems, *International Conference on Systems Research, Informatics and Cybernetics*, Baden-Baden, Germany, August 2-7.

Toye, G., Cutkosky, M., Leifer, L., Tenenbaum, J. and Glicksman, J. (1993). SHARE: A Methodology and Environment for Collaborative Product Development, *Proceedings of 2nd Workshop on Enabling Technologies: Infrastructure for Collaborative Enterprises*, IEEE Computer Press.

Wooldridge, M. and Jennings, N.R. (1995). Intelligent Agents: Theory and Practice, *The Knowledge Engineering Review*, Vol. 10, No. 2, pp.115-152.

Zimmermann, H.J. (1991). *Fuzzy Set Theory and Its Applications*, Kluwer Academic Publishers, Boston.

Zhang, X. and Norrie, D.H. (1999). Holonic Control at the Production and Controller Levels, *IMS'99*, Leuven, Belgium, Sept. 22-24.

Zhang, X., Balasubramanian, S. and Norrie, D.H. (1999). An Intelligent Controller Implementation for Holonic Systems: DCOS-1 Architecture, Submitted to 14th IEEE International Symposium on Intelligent Control, Cambridge, MA, September 15-17.

Appendix A The Number of All Possible Source-Plans

Theorem 1

Let P_N^n be the number of distinctive partitions with n clusters over N members of the agents set $\mathcal{A}_N$ and $\mathcal{S}_N$ the total number of partitions (source-plans) over $\mathcal{A}_N$. Then:

1. $P_N^1 = 1$ (obviously) and

$$P_N^n = C_{N-1}^0 P_{N-1}^{n-1} + C_{N-1}^1 P_{N-2}^{n-2} + \ldots + C_{N-1}^{N-n} P_{n-1}^{n-1}, \quad \forall N \geq 2, \forall n \in \overline{2,N}.$$

2. $\mathcal{S}_0 = 1$, by convention, and

$$\mathcal{S}_N = C_{N-1}^0 \mathcal{S}_{N-1} + C_{N-1}^1 \mathcal{S}_{N-2} + \ldots + C_{N-1}^{N-2} \mathcal{S}_1 + C_{N-1}^{N-1} \mathcal{S}_0, \quad \forall N \geq 1.$$

Here, $C_N^n \overset{def}{=} \dfrac{N!}{n!(N-n)!}$ is the combinations number of n taken over N ($n \in \overline{0,N}$).

PROOF

1. The rationale for this step is inductive, according to $n \in \overline{1,N}$. First, let $n = 1$. Then the only partition with one single cluster is $\mathcal{A}_N$ itself and, thus $P_N^1 = 1$. Moreover, if $n = N$, then $P_N^N = 1$ as well, because,this time, the partition consists of exactly N singleton clusters.
   Let $n = 2$. In this case, only two clusters share all members of $\mathcal{A}_N$. If one of them includes $m \in \overline{1,N-1}$ members of $\mathcal{A}_N$, the other one includes $N - m$ members. The cases $m = 0$ and $m = N$ are removed since they were considered when $n = 1$. Then, we will show next that

$$P_N^2 = C_N^1 + C_N^2 + \ldots + C_N^{\left[\frac{N}{2}\right]-1} + 2^{(N\%2-1)} C_N^{\left[\frac{N}{2}\right]}$$

where $\lfloor a \rfloor$ is the integer part of $a \in \mathcal{R}$ and $N\%p$ is the reminder of N when divided by p (here, $p = 2$).

a. When $m = 1$, only $C_N^1 = N$ singleton clusters (or, equivalently, only $C_N^{N-1} = N$ clusters with $N - 1$ members) are available.

b. For $m \geq 1$, only C_N^m clusters with m members (or, equivalently, only C_N^{N-m} clusters with $N - m$ members) are available. Each cluster points to a partition, in this case. In order to avoid repetitions (i.e., not to count partitions that have been already considered), m cannot overpass $\left\lfloor \frac{N}{2} \right\rfloor$. Moreover,

b1. If N is even, then the two clusters with $\left\lfloor \frac{N}{2} \right\rfloor$ members each can constitute only $\frac{1}{2} C_N^{N/2}$ distinct partitions.

b2. If N is odd, then one cluster has $\frac{N-1}{2}$ members and the other one has $\frac{N+1}{2}$ members; in this case, $\left\lfloor \frac{N}{2} \right\rfloor = \frac{N-1}{2}$, so that the number of distinct partitions constituted by this couple of clusters is $C_N^{\frac{N-1}{2}} = C_N^{\left\lfloor \frac{N}{2} \right\rfloor}$.

Only half of $C_N^{\left\lfloor \frac{N}{2} \right\rfloor}$ partitions are considered when N is even, because all combination of clusters starting from the $\left(\frac{1}{2} C_N^{\left\lfloor \frac{N}{2} \right\rfloor} + 1 \right)$-th have been considered previously. More specifically, if $A_1 = \left\{ a_j \right\}_{j \in J}$ and $A_2 = \left\{ a_k \right\}_{k \in \overline{1,N} \setminus J}$ (where $J \subset \overline{1,N}$ is an arbitrary subset), then the combination $B_1 = \left\{ a_j \right\}_{j \in I}$ and $B_2 = \left\{ a_k \right\}_{k \in \overline{1,N} \setminus I}$, where $I = \overline{1,N}, J$, is identical with $\{A_1, A_2\}$, because $A_1 = B_2$ and $A_2 = B_1$. This cannot happen when N is odd, since the cardinals of J and $I = \overline{1,N} \setminus J$ are different.

In conclusion,

$$
\begin{aligned}
P_N^2 &= C_N^1 + C_N^2 + \ldots + C_N^{\left\lfloor \frac{N}{2} \right\rfloor - 1} + 2^{(N\%2-1)} C_N^{\left\lfloor \frac{N}{2} \right\rfloor} = C_N^0 + C_N^1 + C_N^2 + \ldots + C_N^{\left\lfloor \frac{N}{2} \right\rfloor - 1} + 2^{(N\%2-1)} C_N^{\left\lfloor \frac{N}{2} \right\rfloor} - 1 = \\
&= \begin{cases} 2\left[C_N^0 + C_N^1 + \ldots \frac{1}{2} C_N^{\frac{N}{2}} \right] / 2 - 1, & \text{when } N \text{ is even} \\ 2\left[C_N^0 + C_N^1 + \ldots \frac{1}{2} C_N^{\frac{N-1}{2}} \right] / 2 - 1, & \text{when } N \text{ is odd} \end{cases} = \\
&= \frac{1}{2} \begin{cases} C_N^0 + C_N^1 + \ldots C_N^{\frac{N}{2}} + C_N^{\frac{N}{2}+1} + \ldots + C_N^N - 2, & \text{when } N \text{ is even} \\ C_N^0 + C_N^1 + \ldots C_N^{\frac{N-1}{2}} + C_N^{\frac{N+1}{2}} + \ldots + C_N^N - 2, & \text{when } N \text{ is odd} \end{cases} = \\
&= \frac{2^N - 2}{2} = 2^{N-1} + -1 = C_{N-1}^0 + C_{N-1}^1 + \ldots + C_{N-1}^{N-2} + C_{N-1}^{N-1} - 1 = \\
&= C_{N-1}^0 P_{N-1}^1 + C_{N-1}^1 P_{N-2}^1 + \ldots + C_{N-1}^{N-2} P_1^1
\end{aligned}
$$

In these manipulations, the identity $C_N^m = C_N^{N-m}$ was used. The concluding identity is thus verified for $n = 2$.

Consider now the general case, $n \in \overline{3,N}$ (where $N \geq 3$). When we are computing P_N^n, the main problem is how to avoid repetitions in partition counting. This can be solved by an appropriate rationale, that we detail hereafter.

From the start, the set $\mathcal{A}_N$ can be cut into two parts: one containing $(N-1)$ agents, say $\mathcal{A}_{N-1}$, and another one including the remaining agent, say a. It will be shown later that the index of the agent a is not important. So $\mathcal{A}_N = \mathcal{A}_{N-1} \cup \{a\}$. We can interpret this union as follows: to the existent set $\mathcal{A}_{N-1}$ is added a new agent a, in order to constitute the whole agents set $\mathcal{A}_N$. We know that $\mathcal{A}_{N-1}$ can generate P_{N-1}^{n-1} distinctive partitions (with $(n-1)$ clusters each). It follows that $\mathcal{A}_N$ can generate $P_{N-1}^{n-1} = C_{N-1}^0 P_{N-1}^{n-1}$distinctive partitions (with n clusters each, because the cluster $[a\}$ just has been added). However, these are not all possible partitions off $\mathcal{A}_N$, since a can be clustered not only with itself, but also with another members of $\mathcal{A}_{N-1}$. Thus, the next step is to extract one member from $\mathcal{A}_{N-1}$, say b, and to clustering it with a. Denote the remaining set by $\mathcal{A}_{N-2}(b)$. Now $\mathcal{A}_N = \mathcal{A}_{N-2}(b) \cup \{a,b\}$. In this case, $\mathcal{A}_{N-2}(b)$ can generate P_{N-2}^{n-1} distinctive partitions (the number of clusters for each partition is still $(n-1)$, because a and b construct together one single cluster). Since the cluster $\{a,b\}$ never appeared in the partitions counted before extracting b (so far, a was never clustered with any member of $\mathcal{A}_{N-1}$), the P_{N-2}^{n-1} partitions of $\mathcal{A}_N$ considered by adding $\{a,b\}$ to each partition of $\mathcal{A}_{N-2}(b)$ are different from the previous P_{N-1}^{n-1}. But, this time, any member of $\mathcal{A}_{N-1}$ can be clustered with a, so one generates $C_{N-1}^1 P_{N-2}^{n-1}$ new distinctive partitions of $\mathcal{A}_N$, because there are C_{N-1}^1 possibilities to extract one member from $\mathcal{A}_{N-1}$.

The argument continues by putting back b inside $\mathcal{A}_{N-1}$ and by extracting two members instead, provided that this operation doesn't empty $\mathcal{A}_{N-1}$. Denote these members by b and c. The remaining set is now $\mathcal{A}_{N-3}(b,c)$(assumed non-empty) and it can generate P_{N-3}^{n-1} distinctive partitions (still with $(n-1)$ clusters, obviously). Any of these partitions, together with the cluster $\{a,b,c\}$, is a new partition of $\mathcal{A}_N$, not counted before (because a was never clustered with any couple of members from $\mathcal{A}_{N-1}$). The new generated partitions of $\mathcal{A}_N$ are in number of $C_{N-1}^2 P_{N-3}^{n-1}$, because there are C_{N-1}^2 possibilities to extract two members from $\mathcal{A}_{N-1}$.

By repeating this procedure, where the number of members extracted from $\mathcal{A}_{N-1}$ is each time incremented by 1, finally, the maximum number of members available for extraction is $(N-n)$. In this case, the remaining set has only $(n-1)$ members and it generates one single partition ($P_{n-1}^{n-1} = 1$) with $(n-1)$ clusters (which are singletons). But the number of new partitions generated for $\mathcal{A}_N$ is $C_{N-1}^{N-n} P_{n-1}^{n-1} = C_{N-1}^{N-n}$, because there are C_{N-1}^{N-n} possibilities to extract $(N-n)$ members from $\mathcal{A}_{N-1}$.

Why does $\mathcal{A}_{N-1}$ not depend actually on a? Because reinserting a inside $\mathcal{A}_{N-1}$, extracting another member, say b, and repeating the argument before, any partition of $\mathcal{A}_N$ that can be generated now has been already counted when a was clustered exactly with the same members as it is clustered now, in this partition.

Thus the total numbers of partitions with n clusters is exactly

$$P_N^n = C_{N-1}^0 P_{N-1}^{n-1} + C_{N-1}^1 P_{N-2}^{n-1} + \ldots + C_{N-1}^{N-n} P_{n-1}^{n-1}.$$

It is interesting to verify if $P_N^N = 1$. Indeed: $P_N^N = C_{N-1}^0 P_{N-1}^{N-1} = P_{N-1}^{N-1} = \ldots = P_1^1 = 1$.

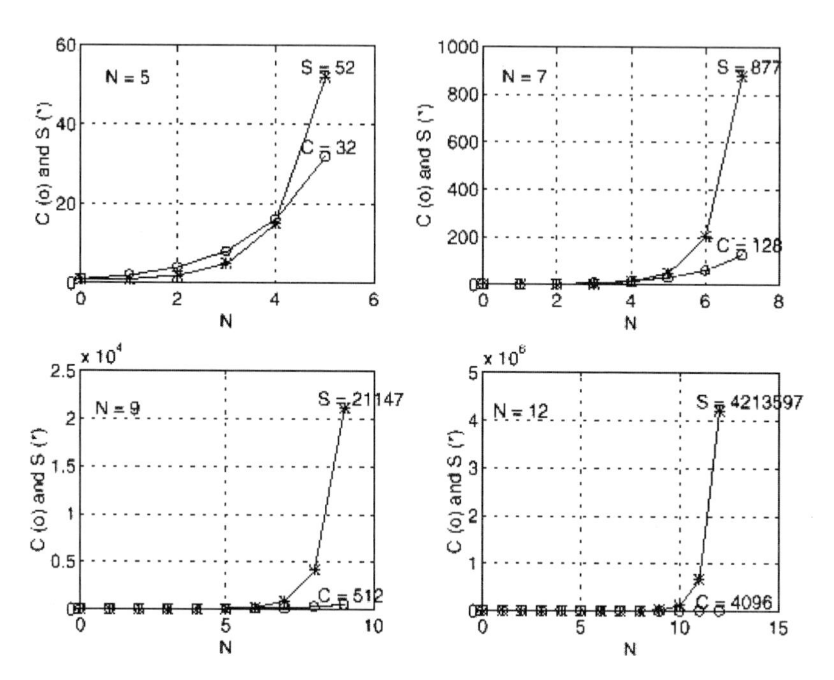

FIGURE A-1 Variation of partitions number ($S = \mathcal{S}_N$) and clusters number ($C = 2^N$).

2. The identity proved before is exceedingly useful in verifying the second identity of the theorem. More specifically, the identities below can be added:

$$
\left[\begin{array}{l|lcl}
n=1 & P_N^1 & = & 1 \\
n=2 & P_N^2 & = & C_{N-1}^0 P_{N-1}^1 + C_{N-1}^1 P_{N-2}^1 + \ldots + C_{N-1}^{N-n} P_{n-1}^1 + \ldots + C_{N-1}^{N-3} P_2^1 + C_{N-1}^{N-2} P_1^1 \\
n=3 & P_N^3 & = & C_{N-1}^0 P_{N-1}^2 + C_{N-1}^1 P_{N-2}^2 + \ldots + C_{N-1}^{N-n} P_{n-1}^2 + \ldots + C_{N-1}^{N-3} P_2^2 \\
\vdots & & & \\
n=n & P_N^n & = & C_{N-1}^0 P_{N-1}^n + C_{N-1}^1 P_{N-2}^{n-1} + \ldots + C_{N-1}^{N-n} P_{n-1}^{n-1} \\
\vdots & & & \\
n=N-1 & P_N^{N-1} & = & C_{N-1}^0 P_{N-1}^{N-2} + C_{N-1}^1 P_{N-2}^{N-1} \\
n=N & P_N^N & = & C_{N-1}^0 P_{N-1}^{N-1}
\end{array}\right.
$$

Thus,

$$
\mathcal{S}_N = \sum_{n=1}^{N} P_N^n = C_{N-1}^0 \sum_{n=1}^{N-1} P_{N-1}^n + C_{N-1}^1 \sum_{n=1}^{N-2} P_{N-2}^n + \ldots + C_{N-1}^{N-m} \sum_{n=1}^{m-1} P_{m-1}^n + \ldots + C_{N-1}^{N-2} P_1^1 + 1 \Leftrightarrow
$$

$$
\Leftrightarrow \mathcal{S}_N = C_{N-1}^0 \mathcal{S}_{N-1} + C_{N-1}^1 \mathcal{S}_{N-2} + \ldots + C_{N-1}^{N-m} \mathcal{S}_{m-1} + \ldots + C_{N-1}^{N-2} \mathcal{S}_1 + C_{N-1}^{N-1} \mathcal{S}_0 .
$$

This concludes the proof.

Note: Actually, the expression of $\mathcal{S}_N$ generalizes the well-known Newton's identity:

$$2^N = \left(1+1\right)^N = C_N^0 + C_N^1 + \ldots + C_N^N .$$

The number of all distinctive partitions that can cover the agents set is increasingly faster with N than the number of all distinctive clusters (2^N), if $N > 4$. This can be observed in the four graphics of Figure A–1 where N varies among the sets: $\overline{1,5}$, $\overline{1,7}$, $\overline{1,9}$, $\overline{1,12}$.

Appendix B Properties of Composition between Fuzzy Relations

The main properties of the composition operation between fuzzy relations are summarized by Theorem 2. In order to prove this result, two interesting and helpful lemmas are proven before. They show interesting similarities between numerical operations where the operators max and min are involved and same operations using the summation and, respectively, the product instead.

LEMMA 1
Let $\{a_n\}_{n\in\overline{1,N}}$ and $\{b_m\}_{m\in\overline{1,M}}$ be two finite sets of arbitrarily chosen real numbers. Then the following inequality holds:

$$\min\{\max_{n\in\overline{1,N}}\{a_n\}, \max_{m\in\overline{1,M}}\{b_m\}\} \le \max_{n\in\overline{1,N}}\max_{m\in\overline{1,M}}\{\min\{a_n, b_m\}$$

PROOF
Since the two sets are finite, there exists a pair of numbers (a_{n_0}, a_{m_0}) so that $a_{n_0} = \max_{n\in\overline{1,N}}\{a_n\}$ and $b_{m_0} = \max_{m\in\overline{1,M}}\{b_m\}$. Then, the conclusion follows immediately by observing that

$$\min\left\{\max_{n\in\overline{1,N}}\{a_n\}, \max_{m\in\overline{1,M}}\{b_m\}\right\} = \min\{a_{n_0}, b_{m_0}\} \le \max_{n\in\overline{1,N}}\max_{m\in\overline{1,M}}\{\min\{a_n, b_n\}\}$$

because

$$\min\left\{a_n, b_n\right\} \le \max_{n\in\overline{1,N}}\max_{m\in\overline{1,M}}\left\{\min\{a_n, b_n\}\right\}, \quad \forall n\in\overline{1,N}, \ \forall m\in\overline{1,M}$$

If the operators are replaced as mentioned, then the corresponding well-known equality below holds:

$$\left[\sum_{n=1}^{N} a_n\right]\left[\sum_{m=1}^{M} b_m\right] = \sum_{n=1}^{N}\sum_{m=1}^{M} a_n b_m ..$$

As it will be observed also in the further results, in general, the equality symbol between expressions with sums and products is replaced by an inequality symbol between expressions using max and min operators instead, but the equality symbol could be unchanged (as in Lemma 2 below).

LEMMA 2
Let $\{a_n\}_{n\in\overline{1,N}}$ and $\{b_n\}_{n\in\overline{1,N}}$ be two finite sets of arbitrarily chosen real numbers and $c \in \mathfrak{R}$, a constant. Then the following identity holds:

$$\max_{n\in 1,N} \min\{a_n,b_n,c\} = \max_{n\in 1,N} \min\{\min\{a_n,b_n\},c\} = \min\{\max_{n\in 1,N} \min\{a_n,b_n\},c\}.$$

PROOF

The proof uses an induction argument, following N.

For $N = 1$, the equality

$$\max \min\{a_1,b_1,c\} = \min\{a_1,b_1,c\} = \min\{\min\{a_1,b_1\},c\} = \min\{\max \min\{a_1,b_1\},c\}$$

is obviously verified and, thus, the concluding identity holds. It is based on the fact that min (or max) applied to a string of numbers can be computed recursively, by considering only two arguments simultaneously.

Consider now the case $N = 2$. Then, we have to prove that the identity below holds:

$$\max\{\min\{a_1,b_1,c\}, \min\{a_1,b_1,c\}\} = \min\{\max\{\min\{a_1,b_1\}, \min\{a_2,b_2\}\},c\}.$$

Obviously, if the constant c is $\leq$ or $\geq$ than both $\min\{a_1,b_1\}$ and $\min\{a_2,b_2\}$, then the identity is verified. If $\min\{a_1,b_1\} \leq c \leq \min\{a_2,b_2\}$,then the left term is c; but the right term is also c. A similar argument holds if $\min\{a_1,b_1\} \geq c \geq \min\{a_2,b_2\}$. These are all possibilities here.

Suppose that the concluding equality of lemma holds for a fixed $N \geq 2$. Then, when $N \to N + 1$, one can evaluate the expression below:

$$\max_{n\in 1,N} \min\{a_n, b_n, c\} = \max\{\max_{n\in 1,N} \min\{a_n,b_n,c\},\min\{a_{N+1},b_{N+1},c\}\}.$$

The induction hypothesis applied to the first N terms of max operator leads to the identity below:

$$\max_{n\in 1,N} \min\{a_n,b_n,c\} = \max\{\min\{\max_{n\in 1,N} \min\{a_n,b_n\},c\},\min\{a_{N+1},b_{N+1},c\}\}.$$

Next, if the case $N = 2$ is used, it follows that

$$\max_{n\in 1,N} \min\{a_n,b_n,c\} = \min\{\max\{\max_{n\in 1,N} \min\{a_n,b_n\},\min\{a_{N+1},b_{N+1}\}\},c\} = \min\{\max_{n\in 1,N} \min\{a_n,b_n\},c\},$$

which concludes the proof.

If, again, the operators are replaced as mentioned, then the corresponding obvious next identity holds:

$$\sum_{n=1}^{N} ca_n b_n = c\sum_{n=1}^{N} a_n b_n.$$

Both results above show the computation rules when operators max and min are inverted with each other.

THEOREM 2

Let $\mathcal{Q}$ and $\mathcal{R}$ be two binary fuzzy relations and $\mathcal{M}_{\mathcal{Q}}$, $\mathcal{M}_{\mathcal{R}}$ their $N \times N$ membership matrices, respectively. Denote by $\mathcal{C}$ the composition between $\mathcal{Q}$ and $\mathcal{R}$, that is $\mathcal{C} = \mathcal{Q} \circ \mathcal{R}$. Then $\mathcal{M}_{\mathcal{C}} = \mathcal{M}_{\mathcal{Q}} \circ \mathcal{M}_{\mathcal{R}}$ (fuzzy product) and

1. If $\mathcal{Q}$ and $\mathcal{R}$ are reflexive relations, $\mathcal{C}$ is reflexive as well.
2. If $\mathcal{Q}$ and $\mathcal{R}$ are symmetric relations, $\mathcal{C}$ is symmetric as well.
3. If $\mathcal{Q}$ and $\mathcal{R}$ is a transitive relation, then $\mathcal{C}$ is transitive as well.

Proof

The fact that the corresponding matrix of $\mathscr{C}$ is the fuzzy product between $\mathscr{M}_{\mathscr{Q}}$ and $\mathscr{M}_{\mathscr{R}}$ is well known in fuzzy sets theory. Denote by $(\mathscr{M}_{\mathscr{T}})_{i,j}$, $\forall i,j \in \overline{1,N}$ the elements of the matrix $\mathscr{M}_{\mathscr{T}}$, where $\mathscr{T} \in \{\mathscr{Q},\mathscr{R},\mathscr{C}\}$.

Assume that $\mathscr{Q}$ and $\mathscr{R}$ are both fuzzy reflexive relations. This means that the main diagonals of the corresponding matrices $\mathscr{M}_{\mathscr{Q}}$ and $\mathscr{M}_{\mathscr{R}}$ are unitary. Compute the elements of $\mathscr{M}_{\mathscr{C}}$ the main diagonal in $\mathscr{M}_{\mathscr{C}}$:

$$(\mathscr{M}_{\mathscr{C}})_{i,j} = \max_{n\in \overline{1,N}} \min\{(\mathscr{M}_{\mathscr{Q}})_{i,j},(\mathscr{M}_{\mathscr{R}})_{i,j}\}=1, \quad \forall i \in \overline{1,N},$$

since $\min\{(\mathscr{M}_{\mathscr{Q}})_{i,j},(\mathscr{M}_{\mathscr{R}})_{i,j}\}=1, \ \forall i \in \overline{1,N}$, due to the reflexivity of the original relations.

Assume that $\mathscr{Q}$ and $\mathscr{R}$ are both fuzzy symmetric relations. This means that the corresponding matrices $\mathscr{M}_{\mathscr{Q}}$ and $\mathscr{M}_{\mathscr{R}}$ are symmetric. Compute the elements of $\mathscr{M}_{\mathscr{C}}$, according to this property:

$$(\mathscr{M}_{\mathscr{C}})_{i,j} = \max_{n\in \overline{1,N}} \min\{(\mathscr{M}_{\mathscr{Q}})_{i,n},(\mathscr{M}_{\mathscr{R}})_{n,j}\} = \max_{n\in \overline{1,N}} \min\{(\mathscr{M}_{\mathscr{Q}})_{n,j},(\mathscr{M}_{\mathscr{R}})_{j,n}\}$$

$$= \max_{n\in \overline{1,N}} \min\{(\mathscr{M}_{\mathscr{R}})_{i,n}, (\mathscr{M}_{\mathscr{Q}})_{n,j}\} = (\mathscr{M}_{\mathscr{C}})_{i,j}, \quad \forall i,j \in \overline{1,N}$$

Thus, the matrix $\mathscr{M}_{\mathscr{C}}$ is symmetric as well, i.e., $\mathscr{C}$ is a symmetric fuzzy relation.

Suppose now that the fuzzy relation $\mathscr{Q} = \mathscr{R}$ is transitive. In matrix terms this property is equivalent with $\mathscr{M}_{\mathscr{R}} \geq^{\bullet} \left(\mathscr{M}_{\mathscr{R}} \circ \mathscr{M}_{\mathscr{R}}\right) = \mathscr{M}_{\mathscr{C}}$. We have to prove that $\mathscr{C} = \mathscr{R} \circ \mathscr{R}$ preserves the transitivity of $\mathscr{R}$. More specifically, any of the following equivalent inequality must hold:

$$\mathscr{M}_{\mathscr{C}} = \left(\mathscr{M}_{\mathscr{R}} \circ \mathscr{M}_{\mathscr{R}}\right) \geq^{\bullet} \left(\mathscr{M}_{\mathscr{R}} \circ \mathscr{M}_{\mathscr{R}}\right) \circ \left(\mathscr{M}_{\mathscr{R}} \circ \mathscr{M}_{\mathscr{R}}\right) = \left(\mathscr{M}_{\mathscr{C}} \circ \mathscr{M}_{\mathscr{C}}\right) \Leftrightarrow$$

$$\Leftrightarrow (\mathscr{M}_{\mathscr{C}})_{i,j} = \left(\mathscr{M}_{\mathscr{R}} \circ \mathscr{M}_{\mathscr{R}}\right)_{i,j} \geq \max_{n\in \overline{1,N}} \min\{ \left(\mathscr{M}_{\mathscr{R}} \circ \mathscr{M}_{\mathscr{R}}\right)_{i,n}, \left(\mathscr{M}_{\mathscr{R}} \circ \mathscr{M}_{\mathscr{R}}\right)_{n,j}\}, \quad \forall i,j \in \overline{1,N} \Leftrightarrow$$

$$\Leftrightarrow \max_{n\in \overline{1,N}} \min\{(\mathscr{M}_{\mathscr{R}})_{i,n},(\mathscr{M}_{\mathscr{R}})_{n,j}\} \geq$$

$$\geq \max_{n\in \overline{1,N}} \min\{\max_{p\in \overline{1,N}} \min\{(\mathscr{M}_{\mathscr{R}})_{i,p},(\mathscr{M}_{\mathscr{R}})_{p,n}\}, \max_{q\in \overline{1,N}} \min\{(\mathscr{M}_{\mathscr{R}})_{n,q},(\mathscr{M}_{\mathscr{R}})_{q,j}\}\}, \quad \forall i,j \in \overline{1,N}$$

In order to prove this inequality, start from the right term. According to Lemma 1, one can derive the following inequality:

$$\max_{n\in \overline{1,N}} \min\{\max_{p\in \overline{1,N}} \min\{(\mathscr{M}_{\mathscr{R}})_{i,p},(\mathscr{M}_{\mathscr{R}})_{p,n}\}, \max_{q\in \overline{1,N}} \min\{(\mathscr{M}_{\mathscr{R}})_{n,q},(\mathscr{M}_{\mathscr{R}})_{q,j}\}\} \leq$$

$$\leq \max_{n\in \overline{1,N}} \max_{p\in \overline{1,N}} \max_{q\in \overline{1,N}} \min\{\min\{(\mathscr{M}_{\mathscr{R}})_{i,p},(\mathscr{M}_{\mathscr{R}})_{p,n}\}, \min\{(\mathscr{M}_{\mathscr{R}})_{n,q},(\mathscr{M}_{\mathscr{R}})_{q,j}\}\} = \tag{B.1}$$

$$= \max_{n\in \overline{1,N}} \max_{p\in \overline{1,N}} \max_{q\in \overline{1,N}} \min\{(\mathscr{M}_{\mathscr{R}})_{i,p},(\mathscr{M}_{\mathscr{R}})_{p,n},(\mathscr{M}_{\mathscr{R}})_{n,q},(\mathscr{M}_{\mathscr{R}})_{q,j}\} =$$

$$\max_{n\in \overline{1,N}} \max_{p\in \overline{1,N}} \max_{q\in \overline{1,N}} \min\{\min\{(\mathscr{M}_{\mathscr{R}})_{p,n},(\mathscr{M}_{\mathscr{R}})_{n,q}\}(\mathscr{M}_{\mathscr{R}})_{i,p},(\mathscr{M}_{\mathscr{R}})_{q,j}\}, \quad \forall i,j \in \overline{1,N}$$

where the last term was obtained from previous one by computing min recursively and by inverting the order of max operators (which, obviously, is always possible).

Assume that $i, j, p, q \in \overline{1,N}$ are fixed. Then $(\mathcal{M}_{\mathcal{R}})_{i,p}$ and $(\mathcal{M}_{\mathcal{R}})_{q,j}$ are constant when n varies in the range $\overline{1,N}$. This remark allows us to use Lemma 2 for expressing the identity below:

$$\max_{n\in\overline{1,N}} \min\{\min\{(\mathcal{M}_{\mathcal{R}})_{p,n},(\mathcal{M}_{\mathcal{R}})_{n,q}\},(\mathcal{M}_{\mathcal{R}})_{i,p},(\mathcal{M}_{\mathcal{R}})_{q,j}\} =$$

$$= \min\{\max_{n\in\overline{1,N}} \min\{(\mathcal{M}_{\mathcal{R}})_{p,n},(\mathcal{M}_{\mathcal{R}})_{n,q}\},(\mathcal{M}_{\mathcal{R}})_{i,p},(\mathcal{M}_{\mathcal{R}})_{q,j}\} =$$

$$= \min\{ \left(\mathcal{M}_{\mathcal{R}} \circ \mathcal{M}_{\mathcal{R}}\right)_{p,q},(\mathcal{M}_{\mathcal{R}})_{i,p},(\mathcal{M}_{\mathcal{R}})_{q,j}\}.$$

Now, recall that $\mathcal{R}$ is transitive and thus

$$\min\{ \left(\mathcal{M}_{\mathcal{R}} \circ \mathcal{M}_{\mathcal{R}}\right)_{p,q},(\mathcal{M}_{\mathcal{R}})_{i,p},(\mathcal{M}_{\mathcal{R}})_{q,j}\} \le \min\{(\mathcal{M}_{\mathcal{R}})_{p,n},(\mathcal{M}_{\mathcal{R}})_{n,q}\}.$$

Return to (B.1) above. This inequality is now equivalent with

$$\max_{n\in\overline{1,N}} \min\{\max_{p\in\overline{1,N}} \min\{(\mathcal{M}_{\mathcal{R}})_{i,p},(\mathcal{M}_{\mathcal{R}})_{p,n}\}, \max_{q\in\overline{1,N}} \min\{(\mathcal{M}_{\mathcal{R}})_{n,q},(\mathcal{M}_{\mathcal{R}})_{q,j}\}\} \le$$

$$\le \max_{p\in\overline{1,N}} \max_{q\in\overline{1,N}} \min\{(\mathcal{M}_{\mathcal{R}})_{i,p},(\mathcal{M}_{\mathcal{R}})_{p,q}(\mathcal{M}_{\mathcal{R}})_{q,j}\}, \ \forall i,j \in \overline{1,N} \tag{B.2}$$

In this inequality, all terms depending on $n \in \overline{1,N}$ have disappeared and one single element of matrix $\mathcal{M}_{\mathcal{R}}$ was specified instead. The same argument can be invoked once again to cancel the presence of $p \in \overline{1,N}$ or $q \in \overline{1,N}$ in (B.2). Thus, the following inequality is also equivalent with (B.1) and (B.2):

$$\max_{n\in\overline{1,N}} \min\{\max_{p\in\overline{1,N}} \min\{(\mathcal{M}_{\mathcal{R}})_{i,p},(\mathcal{M}_{\mathcal{R}})_{p,n}\}, \max_{q\in\overline{1,N}} \min\{(\mathcal{M}_{\mathcal{R}})_{n,q},(\mathcal{M}_{\mathcal{R}})_{q,j}\}\} \le$$

$$\le \max_{q\in\overline{1,N}} \min\{(\mathcal{M}_{\mathcal{R}})_{i,q},(\mathcal{M}_{\mathcal{R}})_{q,i}\} = \left(\mathcal{M}_{\mathcal{R}} \circ \mathcal{M}_{\mathcal{R}}\right)_{i,j} = (\mathcal{M}_{\mathcal{C}})_{i,j}, \ \forall i,j \in \overline{1,N}$$

This completes the proof.

4

Neural Network Applications for Group Technology and Cellular Manufacturing

Nallan C. Suresh
State University of New York at Buffalo
University of Groningen

4.1 Introduction

Recognizing the potential of artificial neural networks (ANNs) for pattern recognition, researchers first began to apply neural networks for group technology (GT) applications in the late 1980s and early 1990s. After a decade of effort, neural networks have emerged as an important and viable means for pattern classification for the application of GT and design of cellular manufacturing (CM) systems. ANNs also hold considerable promise, in general, for reducing complexity in logistics, and for streamlining and synergistic regrouping of many operations in the supply chain. This chapter provides a summary of neural network applications developed for group technology and cellular manufacturing.

Group technology has been defined to be, in essence, a broad philosophy that is aimed at (1) identification of part families, based on similarities in design and/or manufacturing features, and (2) systematic exploitation of these similarities in every phase of manufacturing operation [Burbidge, 1963; Suresh and Kay, 1998].

Figure 4.1 provides an overview of various elements of group technology and cellular manufacturing. It may be seen that the identification of part families forms the first step in GT/CM. The formation of part families enables the realization of many synergistic benefits in the design stage, process planning stage, integration of design and process planning functions, production stage, and in other stages downstream.

In the design stage, classifying parts into families and creating a database that is easily accessed during design results in:

- Easy retrieval of existing designs on the basis of needed design attributes
- Avoidance of “reinvention of the wheel” when designing new parts

0-8493-0592-6/01/$0.00+$.50

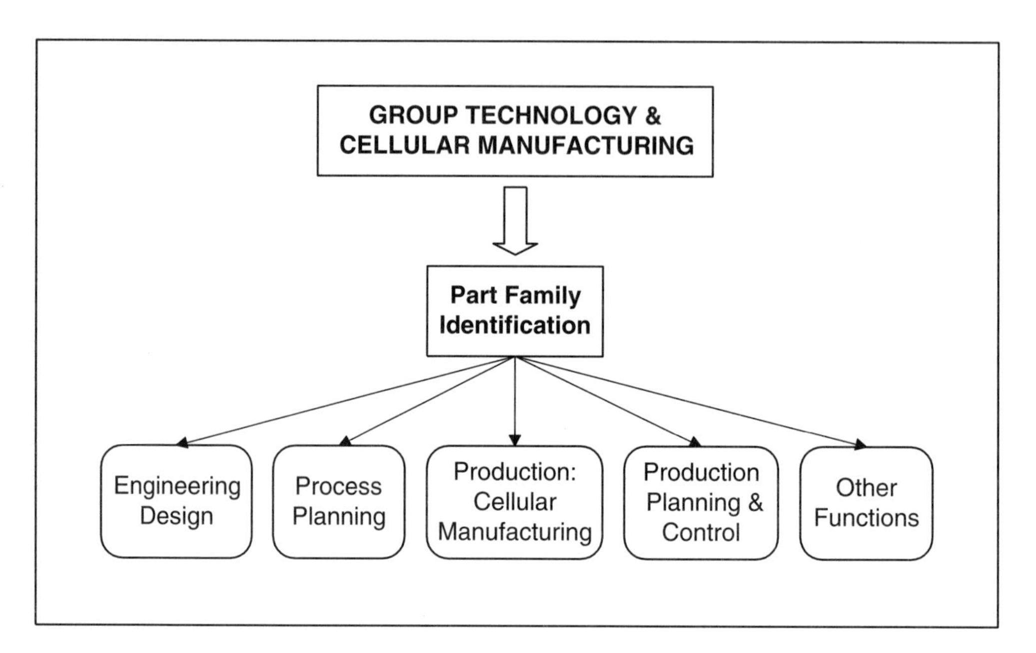

FIGURE 4.1 Elements of GT/CM. (From Suresh, N.C. and Kay, J.M. (Eds.), 1998, *Group Technology and Cellular Manufacturing: State-of-the-Art Synthesis of Research and Practice,* Kluwer Academic Publishers, Boston. With permission.)

- Countering proliferation of new part designs
- Reduction in developmental lead times and costs
- Better data management, and other important benefits.

Likewise, in the downstream, production stage, part families and their machine requirements form the basis for the creation of manufacturing cells. Each cell is dedicated to manufacturing one or more part families. The potential benefits from (properly designed) cellular manufacturing systems include:

- Reduced manufacturing lead times and work-in-process inventories
- Reduced material handling
- Simplified production planning and control
- Greater customer orientation
- Reduced setup times due to similarity of tool requirements for parts within each family
- Increased capacity and flexibility due to reduction of setup times, etc.

For implementing GT and designing cells, early approaches relied on *classification and coding systems,* based on the premise that part families with similar designs will eventually lead to identification of cells. Classification and coding systems involve introducing codes for various design and/or manufacturing attributes. A database is created and accessed through these "GT codes." This offers several advantages, such as design rationalization and variety reduction and better data management, as mentioned above. But the codification activity involves an exhaustive scrutiny of design data, possible errors in coding, and the necessity for frequent recoding. The need for classification and coding systems has also been on the decline due to advances in database technologies, especially the advent of relational databases.

Therefore, in recent years, cell design methods have bypassed the cumbersome codification exercise. They have relied more on a direct analysis of part routings, to identify parts with similar routings and machine requirements. Part families and machine families are identified simultaneously by manipulating part-machine incidence matrices.

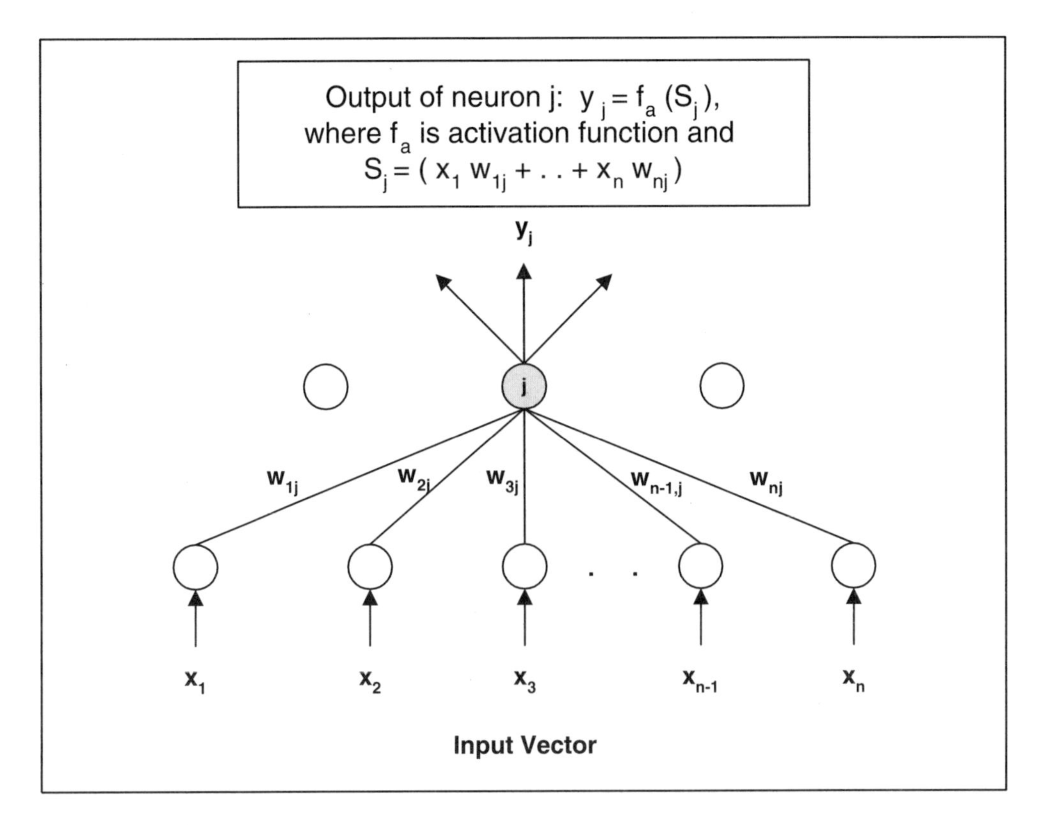

FIGURE 4.2 Neural computation.

The application of neural networks for GT/CM has undergone a similar evolution. As described below, early efforts for utilizing ANNs for GT/CM were devoted to identification of part families based on design and manufacturing process features, while much of the later efforts have been devoted to the use of neural networks for part-machine grouping based on direct analysis of part routings.

The objective of this chapter is to provide a systematic, and state-of-the-art overview of various neural network architectures developed to support group technology applications. A taxonomy of this literature is provided, in addition to a summary of the implementation requirements, pros and cons, computational performance and application domain for various neural network architectures.

4.2 Artificial Neural Networks

Artificial neural networks have emerged in recent years as a major means for *pattern recognition*, and it is this particular capability that has made ANNs a useful addition to the tools and techniques applicable for group technology and design of cellular manufacturing systems.

ANNs are "massively parallel, interconnected networks of simple processing units (neurons), and their hierarchical organizations and connections which interact with objects in the real world along the lines of biological nervous systems" [Kohonen, 1984]. The basic elements of a neural network are the processing units (neurons), which are the nodes in the network, and their connections and connection weights.

The operation of a neural network is specified by such factors as the propagation rule, activation/transfer function, and learning rule. The neurons receive weighted input values, which are combined into a single value. This weighted input is transformed into an output value through a nonlinear *activation function*. The activation function could be a hard limiter, sigmoidal nonlinearity or a threshold logic limit. This neuro-computing process is illustrated in Figure 4.2.

In a neural network, the nodes respond to information converging from other layers via the connections. The connection weights represent almost all the stored information in a network, and these weights are updated in response to new information entering the system. The learning rule specifies how the weights are to be updated in response to new information. For further details on basics of neural networks, readers are referred to works such as Wasserman [1989] and McClelland and Rumelhart [1988]. It must be stressed that all the above networks, though based on massive parallelism, are all still simulated using conventional, sequential computing, awaiting the development of neuro-computing hardware in the future.

Among the many properties of ANNs, their pattern recognition capability is of foremost relevance in the context of GT/CM. Unlike traditional artificial intelligence (AI) methods, employing logic and rule-driven procedures for pattern recognition, ANNs are adaptive devices that recognize patterns more through experience. Neural networks also have the ability to learn complex patterns and to generalize the learned information faster. They have the ability to work with incomplete information. Compared to rule-driven expert systems, neural networks are applicable when [Burke, 1991]

- The rules underlying decisions are not well understood
- Numerous examples of decisions are available
- A large number of attributes describe the inputs.

In contrast to traditional, statistical clustering, ANNs offer a powerful classification option when [Burke, 1991]

- The input generating distribution is unknown and probably non-Gaussian
- Estimating statistical parameters can be expensive and/or time consuming
- Nonlinear relationships, and noise and outliers in the data may exist
- On-line decision making is required.

Neural networks are characterized by parallelism: instead of serially computing the most likely classification, the inputs, outputs, as well as internal computations are performed in parallel. The internal parameters (weight vectors) are typically adapted or trained during use. In addition to this ability to adapt and continue learning, neural network classifiers are also nonparametric and make weaker assumptions regarding underlying distributions.

Based on the direction of signal flow, two types of neural networks can be identified. The first type of architecture is the *feedforward* network, in which there is unidirectional signal flow from the input layers, via intermediate layers, to an output stage. In the *feedback* network, signals may flow from the output of any neuron to the input of any neuron.

Neural networks are also classified on the basis of the type of learning adopted. In *supervised learning*, the network is trained, so that the inputs, as well as information indicating correct outputs, are presented to the network. The network is also "programmed" to know the procedure to be applied to adjust the weights. Thus, the network has the means to determine whether its output was correct and the means to apply the learning law to adjust its weights in response to the resulting errors. The weights are modified on the basis of the errors between desired and actual outputs in an iterative fashion.

In *unsupervised learning*, the network has no knowledge of what the correct outputs should be, since side information is not provided to convey the correct answers. As a series of input vectors are applied, the network clusters the input vectors into distinct classes depending on the similarities. An *exemplar vector* (representative vector) is used to represent each class. The exemplar vector, after being created, is also updated in response to a new input that has been found to be similar to the exemplar. As all inputs are fed to the network, several exemplars are created, each one representing one cluster of vectors. *Combined unsupervised–supervised learning* first uses unsupervised learning to form clusters. Labels are then assigned to the clusters identified and a supervised training follows.

Many types of neural network models have been developed over the years. The taxonomy of neural network models proposed by Lippmann [1987] is widely used in the literature. This classifies ANNs first

TABLE 4.1 Pattern Classification Based on Design and Manufacturing Features

	Supervised Learning		Unsupervised Learning				
Application Area	Back-Propagation	Hopfield	Competitive Learning	Interactive Activation	Kohonen's SOFM	ART1 and Variants	Fuzzy ART
Facilitate Classification and Coding	•						
Kaparthi & Suresh [1991]							
Design Retrieval Systems							
Kamarthi et al. [1990]	•						
Venugopal & Narendran [1992]		•					
Part Family Formation							
Kao & Moon [1990, 1991]	•						
Moon & Roy [1992]	•						
Chakraborty & Roy [1993]	•						
Liao & Lee [1994]						•	
Chung & Kusiak [1994]	•						
Support GT-Based Design Process							
Kusiak & Lee [1996]	•						

into those that accept binary-valued inputs and those accepting continuous-valued inputs. Secondly, these are classified on the basis of whether they are based on supervised or unsupervised training. These are further refined into six basic types of classifiers. However, within ART networks, with the emergence of Fuzzy ART, which accepts continuous values, and other developments, this taxonomy requires revision.

4.3 A Taxonomy of Neural Network Application for GT/CM

The application of neural networks for GT/CM can be classified under several major application areas along with the types of neural network used within each context. Reviewing the literature, three broad application areas for neural networks can be seen in the context of group technology: (1) pattern classification (part family formation) based on design and manufacturing features; (2) pattern classification (part-and-machine family formation) from part–machine incidence matrices; and (3) other classification applications such as part and tool grouping, which are also useful in the context of flexible manufacturing systems (FMS).

Within each of the above application areas, a wide range of networks have been applied, and we classify them into the schemes shown in Table 4.1 and Table 4.2. A taxonomy of neural networks and fuzzy set methods for part–machine grouping was also provided by Venugopal [1998]. The sections below are based on the three broad application areas mentioned above.

4.3.1 Pattern Classification Based on Design and Manufacturing Features

The application of neural networks based on design and manufacturing features can be placed within the contexts of part family identification, engineering design, and process planning blocks shown in Figure 4.1. Based on a review of ANNs developed for these application areas, they may be classified further into four subcategories. These include the use of neural networks primarily to

1. Facilitate classification and coding activity
2. Retrieve existing designs based on features required for a new part
3. Form part families based on design and/or manufacturing features
4. Support GT-based design and process planning functions.

TABLE 4.2 Pattern Classification Based on Part–Machine/Tool Matrix Elements

	Supervised Learning		Unsupervised Learning					
Application Area	Back-Propagation	Hopfield	Competitive Learning	Interactive Activation	Kohonen's SOFM	ART1 and Variants	Fuzzy ART	Other Models
Block Diagonalization								
Jamal [1993]	•							
Malave & Ramachandran [1991]			•					
Venugopal & Narendran [1992a, 1994]			•		•	•		
Chu [1993]			•					
Malakooti & Tang [1995]			•					
Moon [1990a, 1990b]				•				
Moon & Chi [1992]				•				
Currie [1992]				•				
Lee et al. [1992]					•			
Kiang, Hamu & Tam [1992]					•			
Kiang, Kulkarni & Tam [1995]					•			
Kulkarni & Kiang [1995]					•			
Kusiak & Chung [1991]						•		
Dagli & Huggahalli [1991]						•		
Kaparthi & Suresh [1992, 1994]						•		
Dagli & Sen [1992]						•		
Kaparthi, Cerveny & Suresh [1993]						•		
Liao & Chen [1993]						•		
Dagli & Huggahalli [1995]						•		
Chen & Chung [1995]						•		
Burke & Kamal [1992, 1995]							•	
Suresh & Kaparthi [1994]							•	
Kaparthi & Suresh [1994]							•	
Kamal & Burke [1996]							•	
Capacitated Cell Formation								
Rao and Gu [1994, 1995]						•		
Suresh, Slomp & Kaparthi [1995]							•	
Sequence-Dependent Clustering								
Suresh, Slomp & Kaparthi [1999]							•	
Part-Tool Matrix Elements								
Arizono et al. [1995]								•

Traditionally, the identification of part families for group technology has been via a classification and coding system which, as stated earlier, has generally given way to more direct methods that analyze process plans and routings to identify part families. Neural network applications for GT/CM have undergone a similar evolution. Table 4.1 presents a classification of the literature based on the above four categories. Table 4.1 also categorizes them under various supervised and unsupervised network categories. As seen in the table, most of the methods developed for this problem are based on supervised neural networks, especially the feedforward (back-propagation) network.

The work of Kaparthi and Suresh [1991] belongs to the first category. This study proposed a neural network system for shape-based classification and coding of rotational parts. Given the fact that classification and coding is a time-consuming and error-prone activity, a back-propagation network was designed to generate shape-based codes, for the Opitz coding system, directly from bitmaps of part drawings. The network is first trained, using selected part samples, to generate geometry-related codes of the Opitz coding system. The examples demonstrated pertained to rotational parts, but extension to

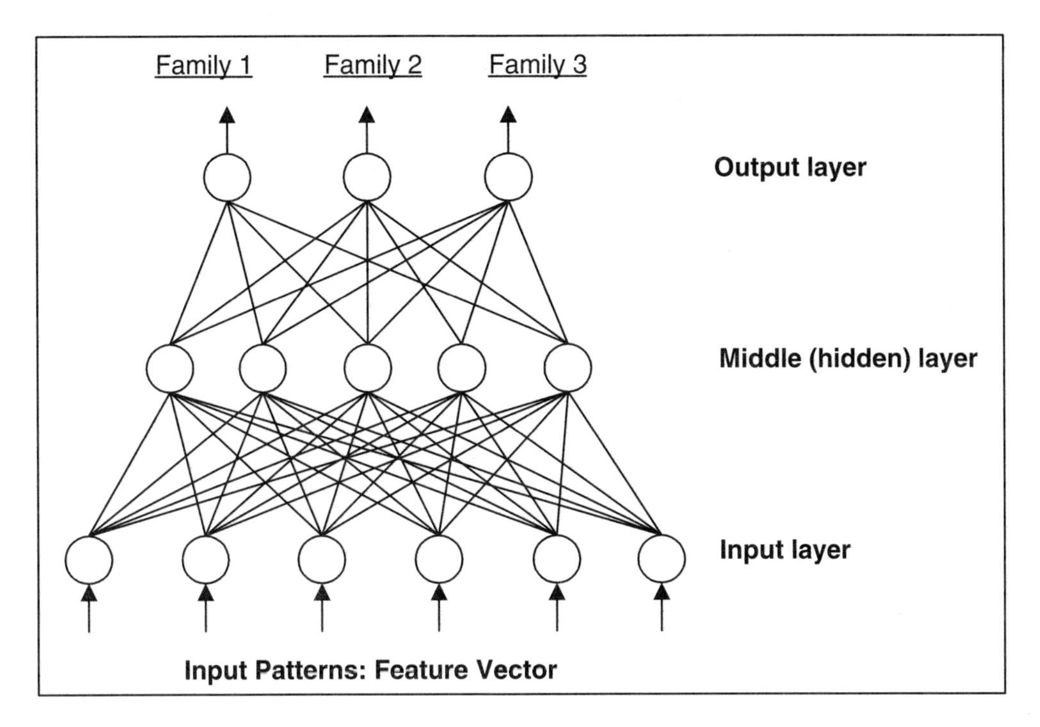

FIGURE 4.3 Feedforward network.

prismatic parts was seen to be feasible. The network was viewed as an element of a computer-aided design (CAD) system, serving to facilitate design procedures in general, and to retrieve existing designs and foster standardization and variety reduction among parts.

Works such as Kamarthi et al. [1990] and Venugopal and Narendran [1992b] have addressed the problem of design retrieval. Kamarthi et al. [1990] used the feedforward neural network model trained with the back-propagation algorithm. It was shown that neural networks can be effectively utilized for design retrieval even in the presence of incomplete or inexact design data. Venugopal and Narendran [1992b] applied a Hopfield network to model design retrieval systems in terms of a human associative memory process. Test cases involved both rotational and nonrotational parts.

The third category of methods involve the use of neural networks for forming part families based on design and/or manufacturing features. Instead of resorting to the laborious part classification and coding activity, these methods are aimed at clustering directly from part features presented to the networks. Almost all these methods have utilized the three-layer feedforward network, with part features forming the input. The network classifies the presented features into families and helps assign new parts to specific families. The basic mode of operation is as follows.

First, design features are identified to cover design attributes of all the parts. Features are design primitives or low-level designs, along with their attributes, qualifiers and restrictions which affect functionality or manufacturability. Features can be described by form (size and shape), precision (tolerances and finish) or material type. The feature vector can either be extracted from a CAD system or codified manually based on part features. Almost all the works have considered the feature vectors as binary-valued inputs (with an activation value of one if a specified feature is present and a value of zero otherwise). However, future implementations are expected to utilize continuous-valued inputs.

The neural network is constructed as shown in Figure 4.3. The processing units (neurons) in the input layer correspond to all the part features. The number of neurons in the input layer equals the number of features codified. The output layer neurons represent the part families identified. The number of neurons in the output layer equals the expected (or desired) number of families. The middle, hidden layer provides a nonlinear mapping between the features and the part families. The number of neurons required for the middle layer is normally determined through trial and error. Neural networks are at times criticized for the arbitrariness, or absence of guidelines for the number of neurons to be used in middle layers.

The input binary-valued vector is multiplied by the connection weight w_{ip}, and all the weighted inputs are summed and processed by an activation function $f'_\iota(\mathrm{net}_{pi})$. The output value of the activation function becomes the input for a neuron in the next layer. For more details on the basic operation of three-layer feedforward networks and back-propagation learning rule, the reader is referred to standard works of Wasserman [1989] and McClelland and Rumelhart [1988].

The net input, net_{pi}, to a neuron i, from a feature pattern p, is calculated as

$$\mathrm{net}_{pi} = \Sigma_j \, w_{ij} \, a_{pj} \qquad \text{Equation (4.1)}$$

$$a_{pi} = f'_\iota(\mathrm{net}_{pi}) = 1/[1 + \exp(-\mathrm{net}_{pi})] \qquad \text{Equation (4.2)}$$

where a_{pi} is the activation value of processing unit p from pattern p. This is applied for processing units in the middle layer. The net input is thus a weighted sum of the activation values of the connected input units (plus a bias value, if included). The connection weights are assigned randomly in the beginning and are modified using the equation

$$\Delta w_{ij} = \varepsilon \, \delta_{pi} \, a_{pj} \qquad \text{Equation (4.3)}$$

These activation values are in turn used to calculate the net inputs and activation values of the processing units in the output layer using Equations 4.1 and 4.2. Next, the activation values of the output units are compared with desired target values during training. The discrepancy between the two is propagated backwards using

$$\delta_{pi} = (\mathrm{t}_{pi} - \mathrm{a}_{pi}) \, f'_i(\mathrm{net}_{pi}) \qquad \text{Equation (4.4)}$$

For the middle layer, the following equation is used to compute discrepancy:

$$\delta_{pi} = f'_\iota(\mathrm{net}_{pi}) \, \Sigma_j \, \delta_{pk} \, w_{ki} \qquad \text{Equation (4.5)}$$

With these discrepancies, the weights are adjusted using Equation 4.3. Based on the above procedure, Kao and Moon [1991] presented a four-phased approach for forming part families, involving (1) seeding phase, (2) mapping phase, (3) training phase, and (4) assigning phase. In the seeding phase, a few very distinct parts are chosen from the part domain to identify basic families. In the mapping phase, these parts are coded based on their features. The network is trained in the training phase utilizing the back-propagation rule. In the assigning phase, the network compares a presented part with those with which it was trained. If the new part does not belong to any assigned family, a new family is identified.

Moon and Roy [1992] developed a feature-based solid modelling scheme for part representations. Again, part features are input to a three-layer feedforward network and classified into part families by the output layer. The training proceeds along conventional lines using predetermined samples of parts and their features. The network can then be used to classify new parts. The new part needs to be represented as a solid model, and the feature-extraction module presents the features to the input layer of the neural network. The system was found to result in fast and consistent responses. This basic procedure is followed in other works shown in Table 4.1.

A fourth application area of neural networks is to support engineering functions more generally. Several researchers have approached the design process in fundamental terms as a mapping activity, from a function space, to a structure space and eventually to a physical space. The design activity is based on an associative memory paradigm for which neural networks are ideally suited. Works such as Coyne and

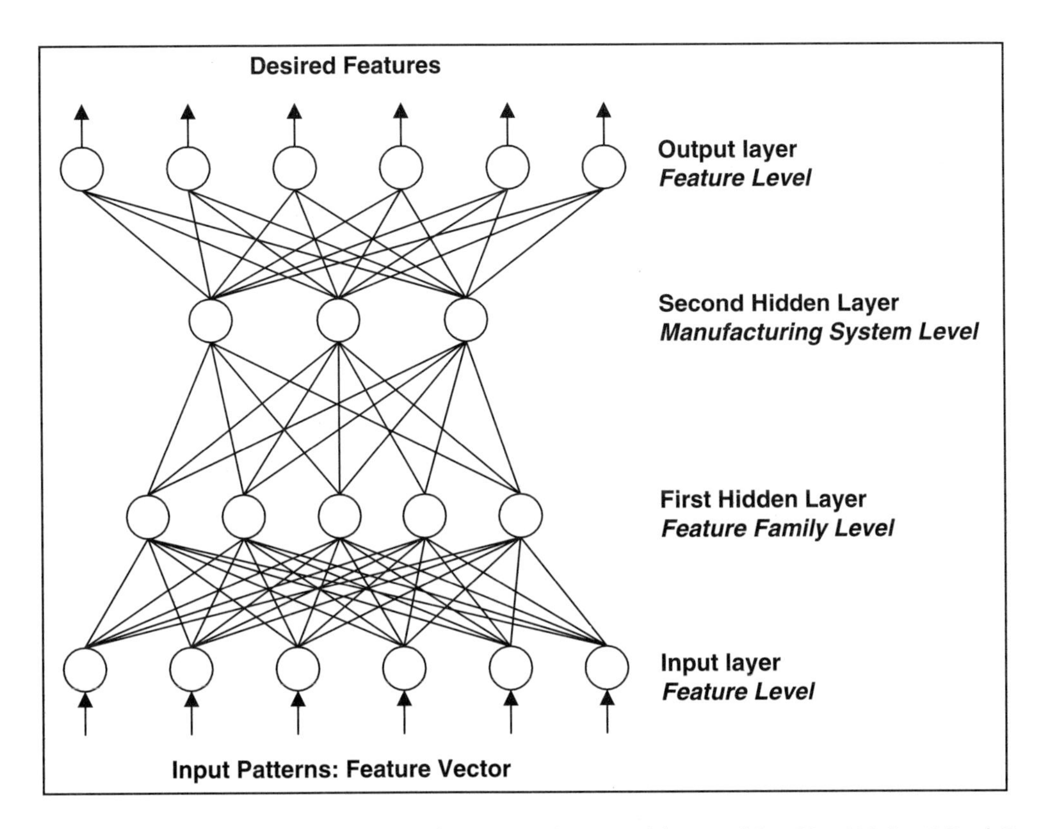

FIGURE 4.4 ANN for design of components for GT/CM. (From Kusiak, A. and Lee, H., 1996, *Int. J. Prod. Res.*, 34(7): 1777–1790. With permission.)

Postmus [1990] belong to this type of application, which falls somewhat outside the realm of group technology. Similarly, neural networks are beginning to be applied in the context of computer-aided process planning (CAPP) in fundamental terms. The reader is referred to Zhang and Huang [1995] for a review of these closely related works.

A promising new application of neural networks is to utilize them to support GT-based design activity, within a *concurrent engineering* framework. A related aim is *design for manufacturability* (DFM), to ensure that new designs can be produced with ease and low cost, using existing manufacturing resources of a firm.

The procedure developed by Kusiak and Lee [1996] represents a promising new approach in this context. It utilizes a back-propagation network to design components for cellular manufacturing, keeping a concurrent engineering framework in mind. It utilizes two middle layers, as shown in Figure 4.4.

The inputs for the network include design features of a component, extracted from a CAD system. After passing through the input layer, the feature vectors are fed into the first hidden layer, referred to as *the feature family layer.* The second hidden layer corresponds to manufacturing resources. The output layer yields *desired features* to fully process the part within a manufacturing cell. Thus the system provides immediate feedback for the designer regarding producibility and potential manufacturing problems encountered with a proposed new design.

The procedure consists of three phases: (1) formation of feature families, (2) identification of machine resources (cells), and (3) determination of a set of features required.

The number of neurons in input layer corresponds to n, the total number of features derived for parts from a CAD system. The inputs are in the form of a binary n-dimensional vector. The number of neurons in the first hidden layer (feature family level) equals the number of feature families. Features of all existing parts are collected and a feature–part incidence matrix is formed. The hidden layer neurons, representing feature families, are connected only to related input neurons that represent the various sets of features.

The number of neurons in the second hidden layer (manufacturing system level) equals the number of machine resources to process the feature families. The selection of manufacturing resources (which may be machines, tools, fixtures) is made based on the features. This may be performed in two phases: first, an operation (e.g., drilling) may be selected based on a feature, followed by selection of a machine resource to perform the operation. Thus, for a feature family, a set of machines and tools is identified and grouped together into a machine cell.

The number of neurons in the output layer (feature level) equals the number of features, like the input layer. Each neuron in the second hidden layer is connected to the output layer neurons. The output layer neurons are connected to the machine neurons only when the features can be performed on the machine. Given this network structure, the following sequence of operation takes place:

1. The binary feature vector for a part, extracted from CAD system, is input to the first layer.
2. A feature family is identified by computing

$$S_j = \Sigma_i \, w_{ij} \, x_i \qquad \text{Equation (4.6)}$$

$$y_j = \max\,(\,S_1,..\,S_p) \qquad \text{Equation (4.7)}$$

   where x_i is the input to neuron i, S_j is the output of neuron j, w_{ij} is the connection weight between an input and hidden layer neuron, and y_j is the output value of the activation function.
3. Next, machine resources to process the selected feature family are identified by computing

$$S_k = \Sigma_j \, w_{kj} \, y_i \qquad \text{Equation (4.8)}$$

$$y_k = f_a(S_k) = 1\,/\,(\,1 + \exp(-\,(S_k + \Phi_k\,)) \qquad \text{Equation (4.9)}$$

   where Φ_k is the activation threshold of a sigmoidal transfer function and w_{kj} is the connection weight between the hidden layer neurons.
4. Next, features that can be processed by the current manufacturing system are identified by computing

$$S_l = \Sigma_k \, w_{lk} \, y_k \qquad \text{Equation (4.10)}$$

$$y_l = f_a(S_l) = 1\,/\,(\,1 + \exp(-\,(S_l + \Phi_l)) \qquad \text{Equation (4.11)}$$

Thus, the network processes the input features and generates desired features as an output, based on current manufacturing resources and their capabilities, fostering design for manufacturability and concurrent engineering.

It is very likely that further development of neural networks for design and process planning functions will be along these lines.

4.3.2 Pattern Classification Based on Part–Machine Incidence Matrices

The *part–machine grouping problem* is one of the first steps in the design of a cellular manufacturing system. It was first identified by Burbidge [1963, 1989] as *group analysis* within the context of production flow analysis (PFA). A description of the overall cell formation problem, referred to here as the *capacitated cell formation problem,* may be found in Suresh et al. [1995], Rao and Gu [1995], and other works. The larger problem involves the consideration of several real-world complexities such as part demand volumes, presence of multiple copies of machines of each type, alternate routings, the need to balance work loads among various cells, and ensuring flexibility.

In the part–machine grouping problem, a binary-valued incidence matrix is first extracted from the routings. A value of one in this matrix denotes the requirement of a particular machine by a part; a value

Parts	Machine								
	6	3	7	1	5	9	8	4	2
9	.	.	1	.	.	1	1	.	.
4	1	.	.	.	1	.	.	1	.
1	.	1	.	1	.	.	.	.	1
5	.	.	.	.	1	.	.	1	.
8	.	.	1	.	.	1	1	.	.
2	.	1	.	1	.	.	1	.	1
3	.	1	.	1	.	.	.	.	.
6	1	.	.	.	1	.	.	1	.
7	.	.	.	.	.	.	1	.	.

Parts	Machine								
	5	4	6	3	1	2	8	7	9
4	1	1	1	.	.	.	.	.	.
6	1	1	1	.	.	.	.	.	.
5	1	1		.	.	.	.	.	.
2	.	.	.	1	1	1	1	.	.
1	.	.	.	1	1	1	.	.	.
3	.	.	.	1	1	.	.	.	.
9	.	.	.	.	.	.	1	1	1
8	.	.	.	.	.	.	1	1	1
7	.	.	.	.	.	.	1	.	.

FIGURE 4.5 (a) Part–machine matrix (zeros shown as "." for clarity). (b) Part–machine matrix in block diagonal form.

of zero indicates that a machine is not required by the part. Figure 4.5a shows a matrix of machine requirements for nine parts, processed on nine machines. Part–machine grouping involves a reorganization of the rows and columns of this matrix so that *a block diagonal form* is obtained. Figure 4.5b shows a block diagonal form of the matrix obtained through a traditional algorithm such as the *rank order clustering* (ROC2) method of King and Nakornchai [1982]. These diagonal blocks are, ideally, mutually exclusive groups of parts and machines, with no overlapping requirements. In Figure 4.5b, it may be noted that for part 2 one operation has to be performed outside the "cell," resulting in an "inter-cell movement." Each of these blocks is potentially a manufacturing cell, and is subjected to further evaluation as a feasible cell in subsequent steps in cell formation.

The intractability of this subproblem, despite its simplicity, has led to the development of numerous heuristic procedures over the years. Chu (1989) classified the literature under the categories of (1) array-based methods, (2) hierarchical clustering techniques, (3) nonhierarchical clustering methods, (4) mathematical programming formulations, (5) graph-theoretic approaches, and (6) other, heuristic methods. However, neural networks, as we see below, have emerged as a powerful method, especially for rapid clustering of large, industry-size data sets.

4.3.2.1 Evolution of Neural Network Methods for Part–Machine Grouping

A wide variety of neural network methods have been applied, and Table 4.2 summarizes the literature based on the type of network used and the specific application. It may be seen that most of these methods (unlike those seen earlier, in Table 4.1) are based on *unsupervised* learning methods. This may be understood given the fact that given a set of part routings, patterns of similar routings are not always known completely *a priori* and, from a practical standpoint, unsupervised networks are much more desirable.

Jamal [1993] represents the sole example so far for the application of a supervised network for part–machine grouping. The three-layer feedforward network is used, and the rows of a part–machine incidence matrix form the input to the network. The outputs, obtained by propagating the inputs via the middle layer, are compared with target (desired) values based on a training sample. The errors are measured and propagated back toward the input layer, and the weights of the interconnections are iteratively modified in order to reduce the measured error, in the customary manner.

The practical limitations and inflexibility of supervised, back-propagation systems have encouraged the development of numerous unsupervised methods, which take advantage of the natural groupings that may exist within a data set. Unsupervised methods do not require training and supervised prior learning, and they also have the capability of processing large amounts of input data.

Unsupervised neural networks can be classified as (1) competitive learning model; (2) interactive activation and competitive learning model; (3) methods based on adaptive resonance theory (ART); (4) Kohonen's self-organizing feature maps (SOFM); or (5) Fuzzy ART method, which is an extension of the ART methods.

4.3.2.2 Competitive Learning Model

Competitive learning models use a network consisting of two layers — an input layer and an output layer — which are fully connected, as shown in Figure 4.6. First, the weight vectors are initialized using small random or uniform values. The input vector, $\boldsymbol{x}$, is one row of the part–machine incidence matrix. The output for each node in the output layer is computed as the weighted sum of the inputs and weight vectors in the customary manner.

The output node with the largest net input, j^* is selected as the winning node. In this "winner-take-all" approach, the weight vector associated with the winning node, $\boldsymbol{w}(j^*)$ is updated as $\boldsymbol{w}'(j^*) = \boldsymbol{w}(j^*) + g\,\{\boldsymbol{x} - \boldsymbol{w}(j^*)\}$, where g is a learning rate which assumes values between zero and one.

Malave and Ramachandran [1991], Venugopal and Narendran [1992a, 1994] and Chu [1993] simulated the competitive learning model and reported good results on a few, relatively small problems. Malave and Ramachandran [1991] utilized a modified version of the Hebbian learning rule for the competitive learning algorithm. Malakooti and Yang [1995] modified competitive learning algorithm by using generalized Euclidean distance, and a momentum term in the weight vector updating equation to improve stability of the network.

The competitive learning algorithm, with its simple structure, emulates the k-means clustering algorithm. This network is known to be very sensitive to the learning rate. Instead of indicating the need for a new group for significantly different parts (machines), the model tends to force an assignment to one of the existing groups [Venugopal, 1998]. Adaptive resonance theory (ART) networks, developed later, extend the competitive learning methods by introducing additional properties of stability and vigilance, as we see below.

4.3.2.3 Interactive Activation and Competition Model

In the interactive activation and competition (IAC) model, the processing units are organized into several pools of neurons. Each pool represents specific characteristics of the problem. In each pool, all the units are mutually inhibitory. Between pools, units may have excitatory connections. The model assumes that these connections are bi-directional. For more details of this network, the reader is referred to McClelland and Rumelhart [1988].

Moon [1990a, 1990b] applied the interactive activation and competition network, with three pools of neurons: one for parts, one for machines, and one for part instances. The entries in the machine similarity matrix were used as connection weights among units in the machine pool. Similarly, entries in the part-similarity matrix were used as connection weights among units in the part pool. The method was illustrated using two small examples, but it was also envisioned that larger problems can be solved through the parallel processing capability of ANN.

The above network was generalized further in Moon and Chi [1992]. This network utilizes connection weights based on similarity coefficients. Operation sequences were also considered while computing the similarities among machines, and other generalizing features such as alternate process plans. Currie [1992] used the activation levels as similarity measure of parts and then used bond-energy algorithm [McCormick et al., 1972] to reorganize the activation level matrix (similarity matrix) to find part families. Like competitive learning, interactive activation models are also precursors to ART networks, which are described below.

4.3.2.4 Kohonen's Self-Organizing Feature Map Model

The self-organizing feature map (SOFM) network was developed by Kohonen [1984]. The unique feature of the SOFM network is the use of a two-dimensional output layer (Kohonen layer). It utilizes the same competitive learning framework described above and, when a winning output node is selected, its weight

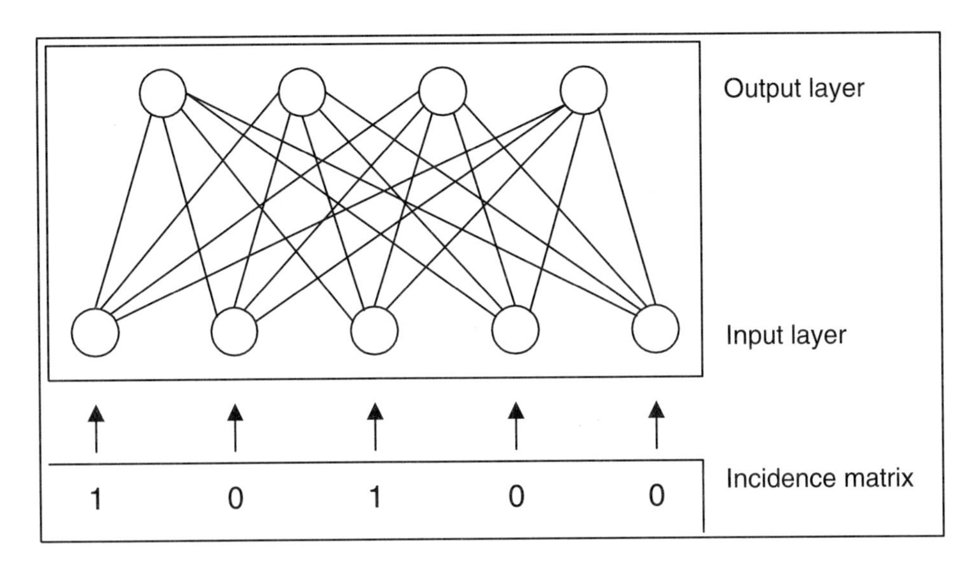

FIGURE 4.6 ART networks for part–machine grouping.

vectors are updated as mentioned above. However, the weights of nearby nodes, within a specified neighborhood, are also updated using the same learning rule. The size of the neighborhood is made to decrease progressively. The net result is that eventually each output node has an associated topological relationship with other nodes in the neighborhood. The SOFM essentially implements a clustering algorithm that is similar to the classical k-means clustering algorithm [Lippmann, 1987].

Venugopal and Narendran [1992a] simulated and compared SOFM model with competitive learning algorithm and ART1. Using small example problems, they demonstrated the applicability of all three networks for the part–machine grouping problem. The ART1 method seemed to perform well for some problems, while SOFM performed better for others. The SOFM model was also shown to be applicable and promising by Lee et al. [1992], Kiang, Hamu and Tam [1992], Kiang et al. [1995] and Kulkarni and Kiang [1995]. Based on these later studies, it appears that SOFM, along with the Fuzzy ART network, which is described below, are among the most effective networks for part-machine grouping.

4.3.2.5 Adaptive Resonance Theory (ART) Model

Adaptive resonance represents an advancement over competitive learning and interactive activation models. This model uses the same two-layer architecture shown in Figure 4.6, but introduces a vigilance measure and stability properties. This model implements a clustering algorithm that is similar to the sequential *leader algorithm* [Lippmann, 1987].

Adaptive resonance theory (ART) has led to a series of networks for unsupervised learning and pattern recognition. Among these, ART1 [Carpenter and Grossberg, 1987] is the earliest development. The inputs for ART1 network are still the binary-valued rows of the part–machine incidence matrix. As the inputs are presented to the network, the model selects the first input as belonging to the first family. The first neuron in the output layer is made to represent this family. The weight vector associated with this neuron becomes the *exemplar* (representative vector) for the first family. If the next input is similar to this vector, within a specified *vigilance threshold* (ρ), then it is treated as a member of the first group. The weights connected to this group are also updated in light of the new input vector. If the new input is not similar to the exemplar, it becomes the exemplar for a new group, associated with the second neuron in the output layer. This process is repeated for all inputs. This same process is followed in all ART networks, including Fuzzy ART, which represents the latest development in the series. The specific steps involved are explained in Section 4.3.2.6. For a more detailed description of ART1 for this application, the reader is referred to Dagli and Huggahalli [1991], Kaparthi and Suresh [1992] and other works shown under ART1 in Table 4.2.

Dagli and Huggahalli [1991] analyzed the performance of ART1 and encountered the *category proliferation problem,* which is frequently encountered with ART networks. This problem refers to a proliferation in the number of clusters formed due to contraction of the exemplar vector, especially when varying norms (i.e., varying number of ones in the input vector) are present. To overcome this problem, Dagli and Huggahalli [1991] suggested presenting the inputs in decreasing order of the number of ones, and to update the weights in a different way.

In Kaparthi and Suresh [1992], it was shown that ART1 network can effectively cluster large, industry-size data sets. The data sets tested included a matrix of 10,000 parts and 100 machine types — by far the largest data set considered in the literature — which was clustered in about 60 seconds on a mainframe computer. The low execution times are due to the fact that it is a *leader algorithm*, which does not require the entire part–machine matrix to be stored and manipulated.

The category proliferation problem was encountered again in Kaparthi and Suresh [1992], and their solution was to process the data set in *reverse notation,* i.e., the zeros and ones were reversed in the part–machine incidence matrix to increase the norm. In Kaparthi et al. [1993], use of reverse notation with ART1 was justified further based on several test cases.

Dagli and Sen [1992] investigated the performance of the improvements for ART1 mentioned in Dagli and Huggahalli [1991] on larger problems (1200×200 and 2000×200 part–machine incidence matrices). Dagli and Huggahalli [1995] have suggested a procedure to determine near-optimal value of vigilance parameter and supplementary procedures to improve the solutions. Venugopal and Narendran [1992a, 1994] used the ART1 model, along with competitive learning and SOFM, for small-sized problems, and demonstrated the applicability of ART1 for this problem.

4.3.2.6 Fuzzy ART Model

The Fuzzy ART network was introduced by Carpenter et al. [1991]. It represents an improvement over ART1. It incorporates fuzzy logic and can handle both analog and binary-valued inputs. In addition, it uses a different learning law and permits a fast-commit–slow-recode option.

Like other ART networks, Fuzzy ART is also based on unsupervised learning. No training is performed initially to provide correct responses to the network. The network is operated as a leader algorithm. As each part routing is read, the network clusters each routing into a distinct class. An *exemplar vector* (a representative vector) is created and maintained for each new class. If a new input is found to be similar (within a specified limit, referred to as vigilance threshold) to an existing exemplar, the input is classified under the category of that exemplar. The matching exemplar is also updated in the light of the new input. If a new input is not similar to any of the existing exemplars, it becomes a new exemplar. Thus, the routings information is scanned one row at a time, without having to store the entire data in memory. After all inputs are fed to the network, several exemplars are created, each representing one part family.

For the part–machine grouping problem, as demonstrated in Suresh and Kaparthi [1994], Fuzzy ART solutions tend to be superior to traditional algorithms such as BEA and ROC2 as well as those of ART1, and ART1 using reverse input notation of Kaparthi and Suresh [1992]. The execution times were also found to be much less than traditional algorithms, making them particularly suitable to cluster large, industry-sized data sets. This study was performed on large data sets, and the robustness of the algorithm was also tested by randomly reordering the inputs and presenting them several times, in a *replicated clustering experimental framework,* which is desirable for evaluating leader algorithms.

Burke and Kamal [1992, 1995] and Kamal and Burke [1996] showed that Fuzzy ART is a viable alternative that is superior to ART1 and several traditional algorithms. But they concluded that *complement coding* of inputs, an option recommended for Fuzzy ART, did not result in superior solutions, at least for the binary matrices involved in part–machine grouping.

Fuzzy ART is operated as follows. It utilizes the two-layer structure shown in Figure 4.6. Each upper layer is made to correspond to one class (part family identified). Associated with each of these neurons is an exemplar vector. The neurons required in this recognition layer are dynamically created whenever a new class is identified. The number of neurons required in the upper layer is situation-dependent, and also need not be known beforehand. The lower, input layer neurons interact with the routings input. The number of neurons in the input layer equals the number of machine types. The steps involved in Fuzzy ART are as follows:

Step 1. Initialization
 Connection weights: $w_{ij}(0) = 1$ [i = 0 to (N –1), j = 0 to (M – 1)]. Select values for choice parameter, $\alpha > 0$; learning rate, $\beta \in [0, 1]$; and vigilance parameter, $\rho \in [0, 1]$.
Step 2. Read new input vector **I** consisting of binary or analog elements (binary values have been used so far).
Step 3. For every output node j, compute choice function
 $(T_j) = [\,|\mathbf{I} \wedge \mathbf{w}_j|\,] / [\alpha + |\mathbf{w}_j|\,]$ for nodes $j = 0$ to $(M - 1)$, where $\wedge$ is the fuzzy AND operator, defined as $(\mathbf{x} \wedge \mathbf{y}) = \min(x_i, y_i)$
Step 4. Select best-matching exemplar (maximum output value): $T_\theta = \max_j \{T_j\}$
Step 5. Resonance test (i.e., degree of similarity with best-matching exemplar)
 If similarity = $[\,|\mathrm{I} \wedge \mathbf{w}_\theta| / |\mathrm{I}|\,] \geq \rho$ go to learning step 7; else go to the next, step 6.
Step 6. Mismatch reset: Set $T_\theta = -1$ and go to step 4.
Step 7. Update best-matching exemplar (learning law):
 $\mathbf{w}_\theta^{new} = \beta\,(\mathbf{I} \wedge \mathbf{w}_\theta^{old}) + (1 - \beta)\,\mathbf{w}_\theta^{old}$
Step 8. Repeat: go to step 2.

Consider the results of applying Fuzzy ART (with $\alpha = 0.5$, $\beta = 0.1$, and $\rho = 0.7$) to the matrix shown in Figure 4.5a. Each row is presented to the network starting with the input vector of part 9. The clustering results are shown in Figure 4.7. The sequence of operation is as follows:

- When the first input vector (part 9) is presented, it is coded as belonging to cluster 1, and the first neuron in the output layer is made to identify this class.
- Next, part 4 is coded as belonging to a new, cluster 2, since its vector is quite dissimilar to the exemplar of neuron 1. Likewise, part 1 is clustered as belonging to a new, family 3, since its vector is dissimilar to the exemplars of both the first and second neurons.
- Next, part 5 vector is classified as cluster 2 based on the level of similarity with the class-2 exemplar; the class-2 exemplar is also updated as a result of this step.
- Part 8 is coded as a class-1 vector, while parts 2 and 3 get coded as cluster-3 vectors; similarly, parts 6 and 7 are clustered within families 2 and 1, respectively.

Thus, after all the parts are processed, three clusters have been identified with Fuzzy ART. In order to present these results in the traditional form, a separate routine was written to reconstruct the matrix in the traditional, block diagonal form. The reconstructed matrix is shown in Figure 4.8. It is seen to be identical, in terms of part and machine families, to the one in Figure 4.5b using the traditional, ROC2 algorithm.

Figure 4.7 also shows the solutions obtained with ART1, and ART1 with inputs in reverse notation (referred to as ART1/KS in Suresh and Kaparthi [1994]). ART1/KS results in a solution identical to Fuzzy ART for this case. However, it is seen that ART1 algorithm results in four clusters, which is attributable to the category proliferation problem.

In Fuzzy ART, category proliferation can be dampened with slower learning in step 7, with $\beta < 1$. A further modification to Fuzzy ART involves the *fast-commit–slow-recode* option. This involves a fast update of the exemplar in the first occurrence, i.e., a new exemplar is made to coincide with an input vector (using a β value of one) in the first instance, but subsequently the updating process is dampened (using a β value less than one). This change is likely to improve clustering efficiency. Fuzzy ART thus provides additional means to counter the category proliferation problem, and is likely to yield better solutions than ART1. Replicated clustering experiments based on large data sets [Suresh and Kaparthi, 1994] show that Fuzzy ART solutions do tend to be superior to those of algorithms such as ROC2 as well as those of ART1 and ART1 with the reverse notation.

Parts	Input Vectors	Clusters Assigned Using Fuzzy ART ($\alpha = 0.5$, $\beta = 0.1$, and $\rho = 0.7$)	ART1/KS ($\rho = 0.7$)	ART1 ($\rho = 0.7$)
9	. . 1 . . 1 1 . .	1	1	1
4	1 . . . 1 . . 1 .	2	2	2
1	. 1 . 1 1	3	3	3
5	. . . . 1 . . 1 .	2	2	2
8	. . 1 . . 1 1 . .	1	1	1
2	. 1 . 1 . . 1 . 1	3	3	3
3	. 1 . 1	3	3	3
6	1 . . . 1 . . 1 .	2	2	4
7	. . . . . . 1 . .	1	1	1

FIGURE 4.7 Clustering with Fuzzy ART, ART1/KS, and ART1. (From Suresh, N.C. and Kaparthi, S., 1994, *Int. J. Prod. Res.*, 32(7): 1693–1713. With permission.)

Parts	7 9 8	6 5 4	3 1 2
9	1 1 1	. . .	. . .
8	1 1 1	. . .	. . .
7	. . 1	. . .	. . .
4	. . .	1 1 1	. . .
5	. . .	. 1 1	. . .
6	. . .	1 1 1	. . .
1	. . .	. . .	1 1 1
2	. . 1	. . .	1 1 1
3	. . .	. . .	1 1 .

FIGURE 4.8 Reconstructed matrix. (From Suresh, N.C. and Kaparthi, S., 1994, *Int. J. Prod. Res.*, 32(7): 1693–1713. With permission.)

4.3.3 Sequence-Dependent Clustering

We next consider the use of Fuzzy ART for sequence-dependent clustering based on the recent work of Suresh et al. [1999]. In an earlier section, it was mentioned that Moon and Chi [1992] utilized an interactive activation network in which the connection weights were based on similarity coefficients. While computing the similarity coefficients among machines, operation sequences were also considered, along with other generalizing features such as alternate process plans. However, here we consider the application of neural networks to discern clusters based on sequences by themselves.

In the traditional part–machine incidence matrix, assuming binary values, sequence information is ignored. To include sequence information, a *sequence-based incidence matrix* is introduced, as shown in Table 4.3. The elements of the matrix are general integer values, representing the serial number of the requirement in the part's operation sequence. A value of zero indicates non-requirement of a machine type. For instance, for part 5 , the row vector is "1 4 2 3 5 0 0 0 0 0 0 0 0 0 0" (the zeroes are shown as "." for clarity). This specifies that part 5 requires machine type 1 for its first operation, machine type 3 for its second operation, machine type 4 for its third operation, etc. Machine types 6 through 15 are not required, as indicated by the zeros under these machine types.

TABLE 4.3 Sequence-Based Incidence Matrix

	Machine Type														
Part	1	2	3	4	5	6	7	8	9	10	11	12	13	14	15
1	2	1	3	5	4	.	.	.	.	.	.	.	.	.	.
2	3	1	2	4	5	.	.	.	.	.	.	.	.	.	.
3	1	3	2	5	4	.	.	.	.	.	.	.	.	.	.
4	1	2	4	5	3	.	.	.	.	.	.	.	.	.	.
5	1	4	2	3	5	.	.	.	.	.	.	.	.	.	.
6	.	.	.	.	.	1	3	4	2	5	.	.	.	.	.
7	.	.	.	.	.	1	3	4	2	5	.	.	.	.	.
8	.	.	.	.	.	1	3	2	5	4	.	.	.	.	.
9	.	.	.	.	.	2	1	4	3	5	.	.	.	.	.
10	.	.	.	.	.	2	1	3	5	4	.	.	.	.	.
11	.	.	.	.	.	.	.	.	.	.	1	3	2	5	4
12	.	.	.	.	.	.	.	.	.	.	1	2	5	3	4
13	.	.	.	.	.	.	.	.	.	.	2	3	1	4	5
14	.	.	.	.	.	.	.	.	.	.	1	2	4	5	3
15	.	.	.	.	.	.	.	.	.	.	1	3	2	5	4

Note: Zeros shown as "." for clarity.
Source: Suresh, N.C. et al., 1999, *Int. J. Prod. Res.*, 37(12): 2793–2816. With permission.

TABLE 4.4 Precedence Matrices for Parts 5 and 3

					Part 5												Part 3						
M/c	1	2	3	4	5	6	7	8	..	14	15	M/c	1	2	3	4	5	6	7	8	..	14	15
1	.	1	1	1	1	.	.	.	..	.	.	1	.	1	1	1	1	.	.	.	..	.	.
2	.	.	.	.	1	.	.	.	..	.	.	2	.	.	.	1	1	.	.	.	..	.	.
3	.	1	.	1	1	.	.	.	..	.	.	3	.	1	.	1	1	.	.	.	..	.	.
4	.	1	.	.	1	.	.	.	..	.	.	4	.	.	.	.	.	.	.	.	..	.	.
5	.	.	.	.	.	.	.	.	..	.	.	5	.	.	.	1	.	.	.	.	..	.	.
6	.	.	.	.	.	.	.	.	..	.	.	6	.	.	.	.	.	.	.	.	..	.	.
7	.	.	.	.	.	.	.	.	..	.	.	7	.	.	.	.	.	.	.	.	..	.	.
8	.	.	.	.	.	.	.	.	..	.	.	8	.	.	.	.	.	.	.	.	..	.	.
9	.	.	.	.	.	.	.	.	..	.	.	9	.	.	.	.	.	.	.	.	..	.	.
10	.	.	.	.	.	.	.	.	..	.	.	10	.	.	.	.	.	.	.	.	..	.	.
11	.	.	.	.	.	.	.	.	..	.	.	11	.	.	.	.	.	.	.	.	..	.	.
12	.	.	.	.	.	.	.	.	..	.	.	12	.	.	.	.	.	.	.	.	..	.	.
13	.	.	.	.	.	.	.	.	..	.	.	13	.	.	.	.	.	.	.	.	..	.	.
14	.	.	.	.	.	.	.	.	..	.	.	14	.	.	.	.	.	.	.	.	..	.	.
15	.	.	.	.	.	.	.	.	..	.	.	15	.	.	.	.	.	.	.	.	..	.	.

Source: Suresh, N.C. et al., 1999, *Int. J. Prod. Res.*, 37(12): 2793–2816. With permission.

When a part routing is read from the routings database, a binary-valued *precedence matrix* is created. The precedence matrix is a mapping of the operation sequence. Table 4.4 shows the precedence matrices for parts 5 and 3. This binary-valued, machine type by machine type matrix is specified such that, in a given row, the ones indicate all machine types visited subsequently. Consider the precedence matrix for part 5 in Table 4.4. Part 5 requires the machine types in the following order: < 1 3 4 2 5 >. For the row corresponding to machine type 3, machine types 4, 2, and 5 have ones, as follower machines.

Part 3 requires the same set of machine types, but in a slightly different order: < 1 3 2 5 4 >. Comparing the matrix elements for parts 3 and 5, it is seen that there are eight matching ones, out of a total ten nonzero elements each. In traditional clustering, the similarity coefficient would have been equal to one. However, given the few backtracks, part 3 is not totally similar in terms of machine types required *and* material flows, to part 5. The similarity coefficient specified (ρ), and a similarity measure computed determine if parts 3 and 5 may be grouped under the same family. Thus, this measure determines

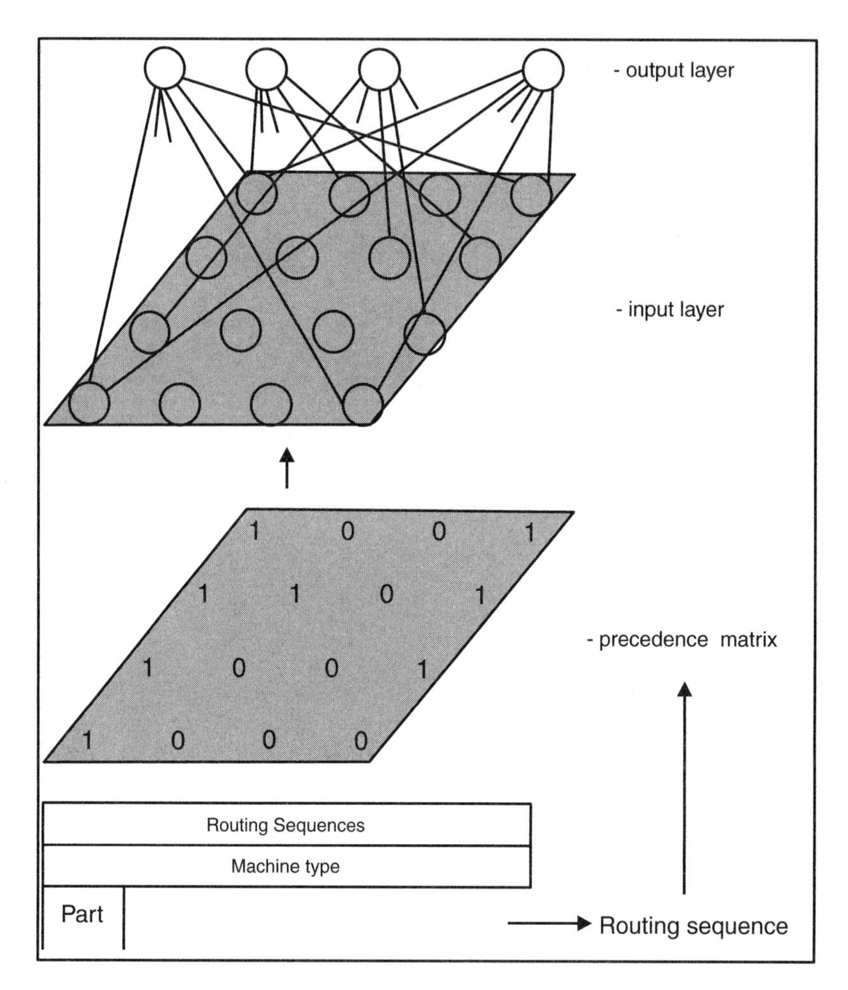

FIGURE 4.9 Fuzzy ART for sequence-based part–machine grouping. (From Suresh, N.C. et al., 1999, *Int. J. Prod. Res.*, 37(12):2793–2816. With permission.)

similarity in machine requirements *and* the extent of backtracks that may be tolerated among two parts, to be grouped within the same family.

The network structure also has to be modified slightly. In the input layer, *a matrix of neurons* is introduced, as shown in Figure 4.8. When a new part routing is read from the database, a precedence matrix containing the sequence information generated. The precedence matrix interacts with the neurons in the two-dimensional lower layer. Otherwise, the steps involved with Fuzzy ART remain practically the same as with non-sequence-based clustering. The basic network connections and properties are unaltered by this modification.

Experiments with past data sets, and replicated clustering with larger data sets, confirm the usefulness of this network for sequence-based clustering. Further details on the operation of this network are provided in Suresh et al. [1999].

4.3.4 Capacitated Cell Formation

Going beyond the part–machine grouping subproblem, researchers have attempted to utilize neural networks for the larger, capacitated cell formation problem. The larger problem requires the consideration of part demand volumes, machine capacities, multiple machines of each type, alternate process plans, etc.

It appears that neural networks by themselves cannot adequately address all of these issues, and they necessarily have to be utilized with other tools, within a decision support or expert system framework. Rao and Gu [1994, 1995] illustrate the use of a neural-network-based expert system that considers additional constraints such as duplicate machines of each type and capacity constraints. A three-layer network is considered that forms an interesting adaptation of ART networks for such constrained grouping.

In Suresh et al. [1995], a Fuzzy ART network is utilized for rapid clustering of parts and machine types, followed by the application of a goal programming formulation for multi-objective cell formation that considers part demands, capacity constraints, duplicate machines, alternate routings, etc. This method represents possibly a new application area for neural networks, namely *pattern detection and reduction of problem sizes* prior to the application of traditional mathematical programming methods. It is likely that neural networks will be utilized for this purpose for many other applications as well.

4.3.5 Pattern Classification Based on Part–Tool Matrix Elements

Finally, neural networks are beginning to be utilized in other, related grouping applications as well. Arizono et al. [1995] presented a stochastic neural network to solve the part–tool grouping problem for flexible manufacturing systems (FMS). Deterministic neural network models do not have the capability to escape from local optimal solution. Stochastic neural network models attempt to avoid local optimal solutions. Stochastic neural network models include the Boltzmann machine model and Gaussian machine model. Arizono et al. [1995] formulated the part–tool grouping problem as a 0–1 integer programming problem, and utilized a recurrent network, stochastic network along the lines of the Boltzmann machine.

4.4 Conclusions

In summary, artificial networks have emerged as a useful addition to the set of tools and techniques for the application of group technology and design of cellular manufacturing systems. The application of neural networks for the part–machine grouping problem, in particular, have produced very encouraging results. Neural networks also hold considerable promise, in general, for pattern classification and complexity reduction in logistics, and for streamlining and synergistic regrouping of many operations in the supply chain.

This chapter provided a taxonomy of the literature, tracing the evolution of neural network applications for GT/CM. A concise summary of the workings of several supervised and unsupervised networks for this application domain was provided along with a summary of their implementation requirements, pros and cons, computational performance, and domain of applicability. Along with the contents of this chapter, and the references cited, the interested reader is referred to Zhang and Huang [1995] for a review of artificial neural networks for other, closely related areas in manufacturing.

References

Arizono, I., Kato, M., Yamamoto, A. and Ohta, H., 1996, A new stochastic neural network model and its application to grouping parts and tools in flexible manufacturing systems, *Int. J. Prod. Res.*, 33(6): 1535-1548.

Burbidge, J.L., 1963, Production flow analysis, *Prod. Eng.*, 42:742.

Burbidge, J.L., 1989, *Production Flow Analysis*, Clarendon Press, Oxford, UK.

Burke, L.I., 1991, Introduction to artificial neural systems for pattern recognition, *Comp. Ops. Res.* 18(2): 211-220.

Burke, L.I. and Kamal, S., 1992, Fuzzy ART and cellular manufacturing, *ANNIE 92 Conf. Proc.*, 779-784.

Burke, L.I. and Kamal, S., 1995, Neural networks and the part family/machine group formation problem in cellular manufacturing: A framework using Fuzzy ART, *J. Manuf. Sys.*, 14(3): 148-159.

Carpenter, G.A. and Grossberg, S., 1987, A massively parallel architecture for a self-organizing neural pattern recognition machine, *Computer Vision, Graphics and Image Processing*, 37: 54-115.

Carpenter, G. A., Grossberg, S. and Rosen, D. B., 1991, Fuzzy ART: Fast stable learning and categorization of analog patterns by an adaptive resonance system, *Neural Networks*, 4: 759-771.

Chakraborty, K. and Roy, U., 1993, Connectionist models for part-family classifications, *Comp. Ind. Eng.*, 24(2): 189-198.

Chen, S.-J. and Cheng, C.-S., 1995, A neural network-based cell formation algorithm in cellular manufacturing, *Int. J. Prod. Res.*, 33(2): 293-318.

Chu, C.H., 1989, Clustering analysis in manufacturing cell formation, *OMEGA: Int. J. Mgt. Sci.*, 17: 289-295.

Chu, C.H., 1993, Manufacturing cell formation by competitive learning, *Int. J. Prod. Res.*, 31(4): 829-843.

Chung, Y. and Kusiak, A., 1994, Grouping parts with a neural network, *J. Manuf. Systems*, 13(4): 262-275.

Coyne, R.D. and Postmus, A.G., 1990, Spatial application of neural networks in computer aided design, *Artificial Intelligence in Engineering*, 5(1), 9-22.

Currie, K.R., 1992, An intelligent grouping algorithm for cellular manufacturing, *Comp. Ind. Eng.*, 23(1): 109-112.

Dagli, C. and Huggahalli, R., 1991, Neural network approach to group technology, in *Knowledge-Based Systems and Neural Networks*, Elsevier, New York, 213-228.

Dagli, C. and Huggahalli, R., 1995, Machine-part family formation with the adaptive resonance theory paradigm, *Int. J. Prod. Res.*, 33(4): 893-913.

Dagli, C. and Sen, C.F., 1992, ART1 neural network approach to large scale group technology problems, in *Robotics and Manufacturing: Recent Trends in Research, Education and Applications*, 4, ASME Press, New York, 787-792.

Jamal, A.M.M., 1993, Neural network and cellular manufacturing, *Ind. Mgt. Data Sys.*, 93(3): 21-25.

Kamal, S. and Burke, L.I., 1996, FACT: A new neural network based clustering algorithm for group technology, *Int. J. Prod. Res.*, 34(4): 919-946.

Kamarthi, S.V., Kumara, S.T., Yu, F.T.S. and Ham, I., 1990, Neural networks and their applications in component design data retrieval, *J. Intell. Manuf.*, 1(2), 125-140.

Kao, Y. and Moon, Y.B., 1991, A unified group technology implementation using the back-propagation learning rule of neural networks, *Comp. Ind. Eng.*, 20(4): 425-437.

Kaparthi, S. and Suresh, N.C., 1991, A neural network system for shape-based classification and coding of rotational parts, *Int. J. Prod. Res.*, 29(9): 1771-1784.

Kaparthi, S. and Suresh, N.C., 1992, Machine-component cell formation in group technology: Neural network approach, *Int. J. Prod. Res.*, 30(6): 1353-1368.

Kaparthi, S., Suresh, N.C. and Cerveny, R., 1993, An improved neural network leader algorithm for part–machine grouping in group technology, *Eur. J. Op. Res.*, 69(3): 342-356.

Kaparthi, S. and Suresh, N.C., 1994, Performance of selected part-machine grouping techniques for data sets of wide ranging sizes and imperfection, *Decision Sci.*, 25(4): 515-539.

Kiang, M.Y., Hamu, D. and Tam, K.Y., 1992, Self-organizing map networks as a clustering tool — An application to group technology, in V.C. Storey and A.B. Whinston (Eds.), *Proc. Second Annual Workshop on Information Technologies and Systems*, Dallas, TX, 35-44.

Kiang, M.Y., Kulkarni, U.R. and Tam, K.Y., 1995, Self-organizing map network as an interactive clustering tool — An application to group technology, *Decision Support Sys.*, 15(4): 351-374.

King, J.R. and Nakornchai, V., 1982, Machine-component group formation in group technology: Review and extension, *Int. J. Prod. Res.*, 20(2): 117.

Kohonen, T., 1984, *Self-Organisation and Associative Memory*, Springer-Verlag, Berlin.

Kosko, B., 1992, *Neural Networks and Fuzzy Systems*, Prentice-Hall International, Englewood Cliffs, NJ.

Kulkarni, U.R. and Kiang, M.Y., 1995, Dynamic grouping of parts in flexible manufacturing systems — A self-organizing neural networks approach, *Eur. J. Op. Res.*, 84(1): 192-212.

Kusiak, A. and Chung, Y., 1991, GT/ART: Using neural networks to form machine cells, *Manuf. Rev.*, 4(4): 293-301.

Kusiak, A. and Lee, H., 1996, Neural computing based design of components for cellular manufacturing, *Int. J. Prod. Res.*, 34(7): 1777-1790.

Lee, H., Malave, C.O. and Ramachandran, S., 1992, A self-organizing neural network approach for the design of cellular manufacturing systems, *J. Intell. Manuf.*, 3: 325-332.

Liao, T.W. and Chen, J.L., 1993, An evaluation of ART1 neural models for GT part family and machine cell forming, *J. Manuf. Sys.*, 12(4): 282-289.

Liao, T.W. and Lee, K.S., 1994, Integration of feature-based CAD system and an ART1 neural model for GT coding and part family forming, *Comp. Ind. Eng.*, 26(1): 93-104.

Lippmann, R.P., 1987, An introduction to computing with neural nets, *IEEE ASSP Mag.*, 4-22.

Malakooti, B. and Yang, Z., 1995, A variable-parameter unsupervised learning clustering neural network approach with application to machine-part group formation, *Int. J. Prod. Res.*, 33(9): 2395-2413.

Malave, C.O. and Ramachandran, S., 1991, A neural network-based design of cellular manufacturing system, *J. Intell. Mfg.*, 2: 305-314.

McClelland, J.L. and Rumelhart, D.E., 1988, *Explorations in Parallel Distributed Processing: A Handbook of Models, Programs and Exercises*, MIT Press, Cambridge, MA.

McCormick, W.T., Schweitzer, P.J. and White, T.W., 1972, Problem decomposition and data reorganization by a clustering technique, *Op. Res.*, 20(5), 992-1009.

Moon, Y.B., 1990a, An interactive activation and competition model for machine-part family formation in group technology, *Proc. Int. Joint Conf. on Neural Networks*, Washington, D.C., vol. 2, 667-670.

Moon, Y.B., 1990b, Forming part families for cellular manufacturing: A neural network approach, *Int. J. Adv. Manuf. Tech.*, 5: 278-291.

Moon, Y.B., 1998, Part family identification: New pattern recognition technologies, in N.C. Suresh and J.M. Kay, (Eds.), *Group Technology and Cellular Manufacturing: State-of-the-Art Synthesis of Research and Practice*, Kluwer Academic Publishers, Boston.

Moon, Y.B. and Chi, S.C., 1992, Generalised part family formation using neural network techniques, *J. Manuf. Sys.*, 11(3): 149-159.

Moon, Y.B. and Roy, U., 1992, Learning group technology part families from solid models by parallel distributed processing, *Int. J. Adv. Mfg. Tech.*, 7: 109-118.

Rao, H.A. and Gu, P., 1994, Expert self-organizing neural network for the design of cellular manufacturing systems, *J. Manuf. Sys.*, 13(5): 346-358.

Rao, H.A. and Gu, P., 1995, A multi-constraint neural network for the pragmatic design of cellular manufacturing systems, *Int. J. Prod. Res.*, 33(4): 1049-1070.

Suresh, N.C. and Kaparthi, S., 1994, Performance of Fuzzy ART neural network for group technology cell formation, *Int. J. Prod. Res.*, 32(7): 1693-1713.

Suresh, N.C. and Kay, J.M. (Eds.), 1998, *Group Technology and Cellular Manufacturing: State-of-the-Art Synthesis of Research and Practice*, Kluwer Academic Publishers, Boston.

Suresh, N.C., Slomp, J. and Kaparthi, S., 1995, The capacitated cell formation problem: a new hierarchical methodology, *Int. J. Prod. Res.*, 33(6): 1761-1784.

Suresh, N.C., Slomp, J. and Kaparthi, S., 1999, Sequence-dependent clustering of parts and machines: A Fuzzy ART neural network approach, *Int. J. Prod. Res.*, 37(12): 2793-2816.

Venugopal, V., 1998, Artificial neural networks and fuzzy models: New tools for part-machine grouping, in N.C. Suresh, and J.M. Kay, (Eds.), *Group Technology and Cellular Manufacturing: State-of-the-Art Synthesis of Research and Practice*, Kluwer Academic Publishers, Boston.

Venugopal, V. and Narendran, T.T., 1992a, A neural network approach for designing cellular manufacturing systems, *Advances in Modelling and Analysis*, 32(2): 13-26.

Venugopal, V. and Narendran, T.T., 1992b, Neural network model for design retrieval in manufacturing systems, *Comp. Industry*, 20: 11-23.

Venugopal, V. and Narendran, T.T., 1994, Machine-cell formation through neural network models, *Int. J. Prod. Res.*, 32(9): 2105-2116.

Wasserman, D., 1989, *Neural Computing — Theory and Practice,* Van Nostrand Reinhold.

Zhang, H.-C. and Huang, S.H., 1995, Applications of neural networks in manufacturing: A state-of-the-art survey, *Int. J. Prod. Res.*, 33(3), 705-728.

5

Application of Fuzzy Set Theory in Flexible Manufacturing System Design

A. Kazerooni
University of Lavisan

K. Abhary
University of South Australia

L. H. S. Luong
University of South Australia

F. T. S. Chan
University of Hong Kong

5.1 Introduction

In design of a flexible manufacturing system (FMS), different combinations of scheduling rules can be applied to its simulation model. Each combination satisfies a very limited number of *performance measures* (PM). *Evaluation of scheduling rules* is an inevitable task for any scheduler. This chapter explains a framework for evaluation of scheduling using *pair-wise comparison, multi-criterion decision-making techniques, and fuzzy set theory.*

Scheduling criteria or performance measures are used to evaluate the system performance under applied scheduling rules. Examples of scheduling criteria include *production throughput, makespan, system utilization, net profit, tardiness, lateness, production cost, flow time,* etc. Importance of each performance measure depends on the objective of the production system. More commonly used criteria were given by Ramasesh [1990].

Based on the review of the literature on FMS production scheduling problems by Rachamadugu and Stecke [1988] and Gupta et al. [1990], the most extensively studied scheduling criteria are *minimization of flow time* and *maximization of system utilization.* However, some authors found some other criteria to be more important. For example, Smith et al. [1986] observed the following criteria to be of most importance:

0-8493-0592-6/01/$0.00+$.50

- Minimizing lateness/tardiness
- Minimizing makespan
- Maximizing system/machine utilization
- Minimizing WIP (work in process)
- Maximizing throughput
- Minimizing average flow time
- Minimizing maximum lateness/tardiness

Hutchison and Khumavala [1990] stated that production rate (i.e., the number of parts completed per period) dominates all other criteria. Chryssolouris et al. [1994] and Yang and Sum [1994] selected total cost as a better overall measure of satisfying a set of different performance measures.

One of the most important considerations in scheduling FMSs is the right choice of appropriate criteria. Although the ultimate objective of any enterprise is to maximize the net present value of the shareholder wealth, this criterion does not easily lend itself to operational decision making in scheduling [Rachamadugu and Stecke 1994]. An example of conflict in these objectives is minimizing WIP and average flow time necessitates lower system utilization. Similarly, minimizing average flow time necessitates a high maximum lateness, or minimizing makespan can result in higher mean flow time. Thus, most of the above listed objectives are mutually incompatible, as it may be impossible to optimize the system with respect to all of these criteria. These considerations indicate that a scheduling procedure that does well for one criterion, is not necessarily the best for some others. Furthermore, a criterion that is appropriate at one level of decision making may be unsuitable at another level. These issues raise further complications in the context of FMSs due to the additional decision variables including, for example, *routing*, *sequencing alternatives*, and AGV (automatic guided vehicle) selections.

Job shop research uses various types of criteria to measure the performance of scheduling algorithms. In FMS studies usually some performance measures are considered more important than the others such as *throughput time*, *system output*, and *machine utilization* [Rachamadugu and Stecke 1994]. This is not surprising, since many FMSs are operated as dedicated systems and the systems are very capital-intensive. However, general-purpose FMSs operate in some ways like job shops in the manner that part types may have to be scheduled according to customer requirements. In these systems due-date-related criteria such as *mean tardiness* and *number of tardy parts* are important too.

But from a scheduling point of view, all criteria do not possess the same importance. Depending on the situation of the shop floor, importance of criteria or performance measures varies over the time. Virtually no published paper has considered performance measures bearing different important weights. They have evaluated the results by considering the same importance for all performance measures.

5.2 A Multi-Criterion Decision-Making Approach for Evaluation of Scheduling Rules

Scheduling rules are usually involved with combination of different decision rules applied at different decision points. Determination of the best scheduling rule based on a single criterion is a simple task, but decision on an FMS is made with respect to different and usually conflicting criteria or performance measures. The simple way to consider all criteria at the same time is assigning a weight to each criterion. It can be defined mathematically as follows [Hang and Yon 1981]: Assume that the decision-maker assigns a set of important weights to the attributes, $W = \{w_1, w_2, \ldots, w_m\}$. Then the most preferred alternative, X^*, is selected such that

$$X^* = \left\{ X_i \,|\, \max_i \sum_{j=1}^{m} w_j x_{ij} \Big/ \sum_{j=1}^{m} w_j \right\}, i = 1, \ldots, n \qquad \text{Equation (5.1)}$$

where x_{ij} is the outcome of the ith alternative (X_i) related to the jth attribute or criterion. In the evaluation of scheduling rules, x_{ij} is the simulation result of the ith alternative related to the jth performance measure or criterion and w_j is the important weight of the jth performance measure. Usually the weights of performance measures are normalized so the $\Sigma w_j = 1$. This method is called simple additive weighting (SAW) and uses the simulation results of an alternative and regular arithmetical operations of multiplication and addition.

The simulation results can be converted to new values using fuzzy sets and through building membership functions. In this method, called modified additive weighting (MAW), x_{ij} from Equation 5.1 is converted to the membership value mvx_{ij}, which is the simulation results for the ith alternative related to the jth performance measure. Therefore, x_{ij} in Equation 5.1 is replaced with its membership value mvx_{ij}.

$$X^* = \left\{ X_i \mid \max_i \sum_{j=1}^{m} w_j mvx_{ij} \Big/ \sum_{j=1}^{m} w_j \right\}, i = 1, \ldots, n \qquad \text{Equation (5.2)}$$

Considering the objectives, $A_1, A_2, \ldots, A_m$, each of which associated with a fuzzy subset over the set of alternatives $X = [X_1, X_2, \ldots, X_n]$, the decision function D(x) can be denoted, in terms of fuzzy subsets, as [Yager 1978]

$$D(x) = A_1(x) \cap A_2(x) \cap, \ldots, \cap A_m(x)\; x \in X \qquad \text{Equation (5.3)}$$

or

$$D(x) = \min\{A_1(x) A_2(x), \ldots, A_m(x)\}\; x \in X \qquad \text{Equation (5.4)}$$

$D(x)$ is the degree to which x satisfies the objectives, and the solution, of course, is the highest $\{D(x) \mid x \in X\}$. For unequal important weights αi associated with the objectives, Yager represents the decision model D as follows:

$$D(x) = A_1^{a_1}(x) \cap A_2^{a_2}(x) \cap, \ldots, \cap A_m^{a_m}(x)\; x \in X \qquad \text{Equation (5.5)}$$

$$D(x) = \min\left\{ A_j^{a_j}(x) \mid j = 1, \ldots, m \right\} x \in X \qquad \text{Equation (5.6)}$$

This method is also called max–min method. For evaluation of scheduling rules, *objectives* are performance measures, *alternatives* are combinations of scheduling rules and α_j is the weight of the jth performance measure w_j. For this model, the following process is used:

1. Select the smallest membership value of each alternative X_i related to all performance measures and form $D(x)$.
2. Select the alternative with the highest member in D as the optimal decision.

Another method, the max–max method, is similar to the MAW in the sense that it also uses membership functions of fuzzy sets and calculates the numerical value of each performance measure via multiplying the value of the corresponding membership function by the weight of the related performance measure. This method determines the value of an alternative by selecting the maximum value of the performance measures for that particular alternative, and mathematically is defined as

$$X^* = \left\{ X_i \mid \max_i \left(\max_j \left(w_j mvx_{ij} \right) \right) i = 1, \ldots, n \text{ and } j = 1, \ldots, m \right\} \qquad \text{Equation (5.7)}$$

5.3 Justification of Representing Objectives with Fuzzy Sets

Unlike ordinary sets, fuzzy sets have gradual transitions from membership to nonmembership, and can represent both very vague or fuzzy objectives as well as very precise objectives [Yager 1978]. For example, when considering net profit as a performance measure, earning $200,000 in a month is not simply earning twice as much as $100,000 for the same period of time. With $100,000 the overhead cost can just be covered, while with $200,000 the research and development department can benefit as well. Membership functions can show this kind of vagueness. The membership functions play a very important role in multi-criterion decision-making problems because they not only transform the value of outcomes to a nondimensional number, but also contain the relevant information for evaluating the significance of outcomes. Some examples of showing outcomes with membership values are depicted in Figure 5.1.

5.4 Decision Points and Associated Rules

Evaluation of scheduling rules always involves the evaluation of a combination of different decision rules applied at different decision points. Some decision points are explained by Montazeri and Van Wassenhove [1990], Tang et al. [1993], and Kazerooni et al. [1996] that are general enough for most of the simulation models; however, depending on the complication of the model, even more decision points can be considered. A list of these decision points (DPi) is as follows:

DP1. Selection of a Routing.
DP2. Parts Select AGVs.
DP3. Part Selection from Input Buffers.
DP4. Part Selection from Output Buffers.
DP5. Intersections Select AGVs.
DP6. AGVs Select Parts.

The rules of each decision point can have different important weights, say AGV selection rules SDS (shortest distance to station), CYC (cyclic), and RAN (random). In a general case, a scheduling rule can be a combination of p decision rules, and the possible number of these combinations, n, depends on the number of rules at each level or decision point. A combination of scheduling rules can be shown as $rule_1$ $/rule_2$ / . . . $/rule_p$ in which rule_k is a decision rule applied at DP_k $1 \le k \le p$. This combination of rules is one of the possible combinations of rules. If three rules are assumed for each decision point, the number of possible combinations would be 3^p. Each combination of rules, namely an alternative, is denoted by c_i, whose simulation result for performance measure j is shown by x_{ij} and the related membership value by mvx_{ij}, where i varies from 1 to n and j varies from 1 to m. Πwc_i is the product of important weights of the rules participated in c_i.

5.5 A Hierarchical Structure for Evaluation of Scheduling Rules

As described previously, evaluation of scheduling rules depends on the important weight of performance measures and decision rules applied at decision points. Figure 5.2 shows a hierarchical structure for evaluation of scheduling rules. There are m performance measures and six decision points. The number of decision points can be extended, and depends on the complexity of the system under study.

Regarding the hierarchical structure of Figure 5.2, the mathematical equation of different multi-criterion decision-making (MCD) methods are reformulated and customized for evaluation of scheduling rules as follows:

SAW method:

$$D = \max_i \left(\Pi wc_i \left(\sum_{j=1}^{m} w_j \times s_j \times x_{ij} \right) \right) \quad i = 1, \ldots, n \qquad \text{Equation (5.8)}$$

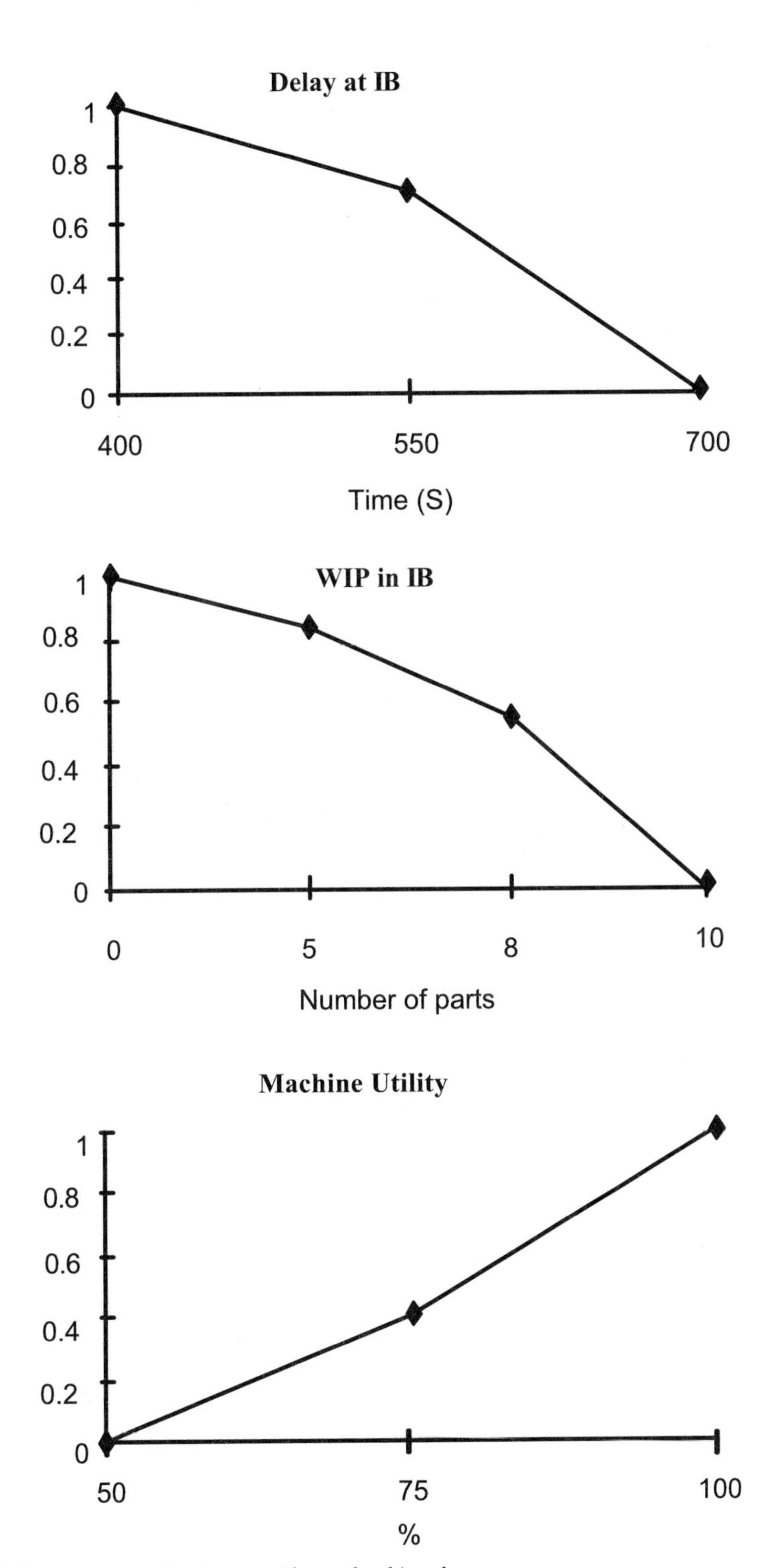

FIGURE 5.1 Some examples of outcomes with membership values.

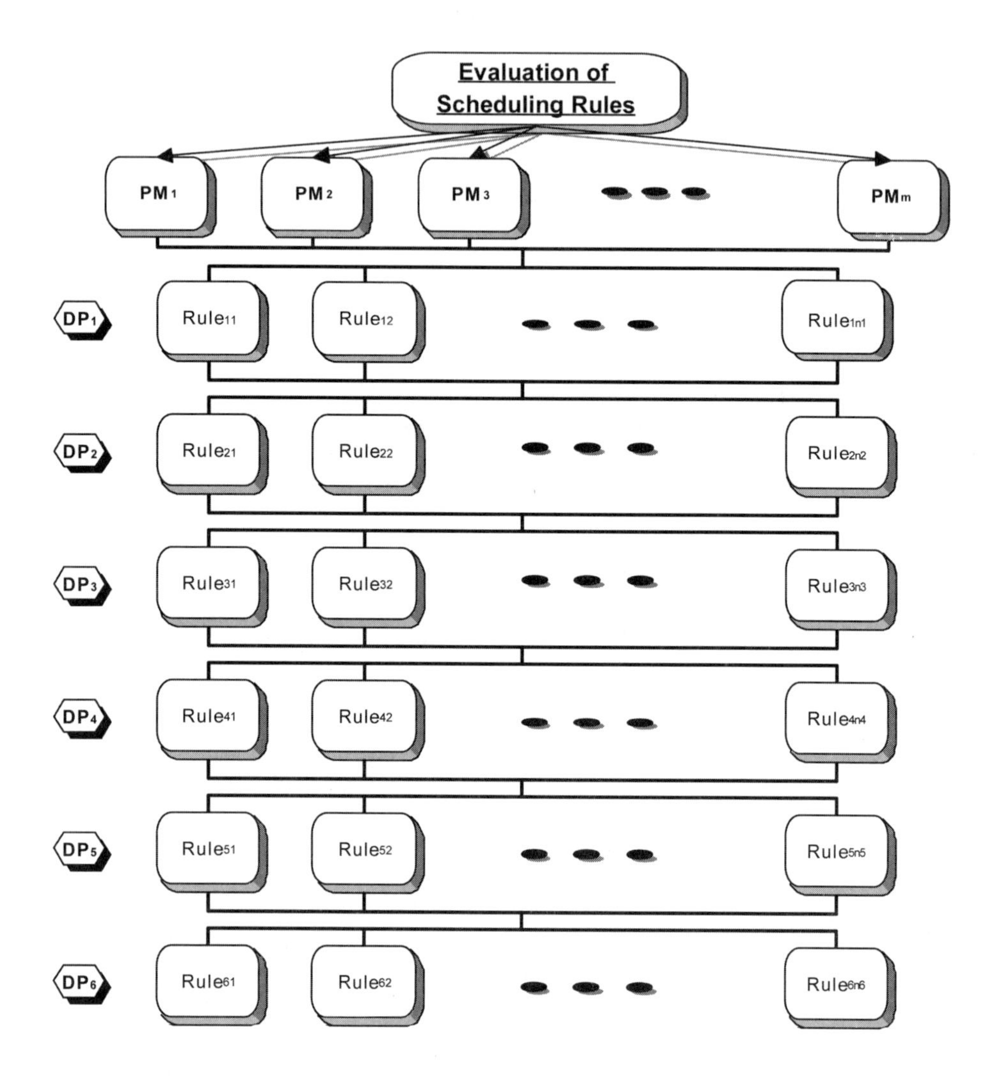

FIGURE 5.2 Hierarchical structure for evaluation of scheduling rules.

where $\begin{cases} s_j = 1 & \text{if the PM}_j \text{ is to be maximized} \\ s_j = -1 & \text{if the PM}_j \text{ is to be minimized} \end{cases}$

MAW method:

$$D = \max_i \left(\Pi wc_i \left(\sum_{j=1}^{m} w_j \times mvx_{ij} \right) \right) \quad i = 1, \ldots, n \qquad \text{Equation (5.9)}$$

Max–Min method:

$$D = \max_i \left(\min_j \left(mvx_{ij} \left(w_j / \Pi wc_i \right) \right) \right) \quad i = 1,\ldots,n,\ j = 1,\ldots,m \qquad \text{Equation (5.10)}$$

Max–Max method:

$$D = \max_i \left(\max_j \left(\Pi wc_i \times w_j \times mvx_{ij} \right) \right) i = 1,\ldots,n, j = 1,\ldots,m \qquad \text{Equation (5.11)}$$

where it is assumed that $\Sigma w_j = 1$. The value inside the outermost parenthesis of each of the above equations shows the overall scores of all scheduling rules with respect to the related method.

5.5.1 Important Weight of Performance Measures and Interval Judgment

The first task of the decision-maker is to find the important weight of each performance measure. Saaty [1975, 1977, 1980, 1990] developed a procedure for obtaining a ratio scale of importance for a group of m elements based upon paired comparisons. Assume there are m objectives and it is desired to construct a scale and rate these objectives according to their importance with respect to the decision, as seen by the analyst. The decision-maker is asked to compare the objectives in pairs. If objective i is more important than objective j then the former is compared with the latter and the value a_{ij} from Table 5.1 shows how objective i dominates objective j (if objective j is more important than objective i, then a_{ji} is assigned). The values a_{ij} and a_{ji} are inversely related:

$$a_{ji} = 1/a_{ij} \qquad \text{Equation (5.12)}$$

When the decision-maker cannot articulate his/her preference by a single scale value that serves as an element in a comparison matrix from which one drives the priority vector, he/she has to resort to approximate articulations of preference that still permit exposing the decision-makers underlying preference and priority structure. In this case, an interval of numerical values is associated with each judgment, and the pairwise comparison is referred to as an interval pairwise comparison or simply interval judgment [Saaty and Vargas 1987; Arbel 1989; Arbel and Vargas 1993]. A reciprocal matrix of pairwise comparisons with interval judgment is given in Equation 5.13 where l_{ij} and u_{ij} represent the lower and upper bounds of the decision-maker's preference, respectively, in comparing element i versus element j using comparison scale (Table 5.1). When the decision-maker is certain about his/her judgment, l_{ij} and u_{ij} assume the same value. Justifications for using interval judgments are described by Arbel and Vargas [1993].

$$[A] = \begin{bmatrix} 1 & [l_{12}, u_{12}] & \cdots & [l_{1m}, u_{1m}] \\ \left[1/u_{12}, 1/l_{12}\right] & 1 & \cdots & l_{2m}, u_{2m} \\ \vdots & \vdots & \vdots & \vdots \\ \left[1/u_{1m}, 1/l_{1m}\right] & \left[1/u_{2m}, 1/l_{2m}\right] & \cdots & 1 \end{bmatrix} \qquad \text{Equation (5.13)}$$

A preference programming is used to find the important weight of each element in matrix $[A]$, Equation 5.13 [Arbel and Vargas 1993].

5.5.2 Consistency of the Decision-Maker's Judgment

In the evaluation of scheduling rules process, it is necessary for the decision-maker to find the consistency of his/her decision on assigning intensity of importance to the performance measures. This is done by first constructing the matrix of the lower limit values $[A]_l$ and the matrix of the upper limit values $[A]_u$, Equation 5.14 below, then calculating the consistency index, CI, for each of the matrices:

TABLE 5.1 Intensity of Importance in the Pair-Wise Comparison Process

Intensity of Importance Value of a_{ij}	Definition
1	Equal importance of i and j
2	Between equal and weak importance of i over j
3	Weak importance of i over j
4	Between weak and strong importance of i over j
5	Strong importance of i over j
6	Between strong and demonstrated importance of i and j
7	Demonstrated importance of i over j
8	Between demonstrated and absolute importance of i over j
9	Absolute importance of i over j

$$[A]_l = \begin{bmatrix} 1 & [l_{12}] & \cdots & [l_{1m}] \\ \left[\frac{1}{l_{12}}\right] & 1 & \cdots & [l_{2m}] \\ \vdots & \vdots & 1 & \vdots \\ \left[\frac{1}{l_{1m}}\right] & \left[\frac{1}{l_{2m}}\right] & \cdots & 1 \end{bmatrix} \quad [A]_u = \begin{bmatrix} 1 & [u_{12}] & \cdots & [u_{1m}] \\ \left[\frac{1}{u_{12}}\right] & 1 & \cdots & [u_{2m}] \\ \vdots & \vdots & 1 & \vdots \\ \left[\frac{1}{u_{1m}}\right] & \left[\frac{1}{u_{2m}}\right] & \cdots & 1 \end{bmatrix}$$

Equation (5.14)

Saaty [1980] suggests the following steps to find the consistency index for each matrix in Equation 5.14:

1. Find the important weight of each performance measure (w_i) for the matrix:
   a. Multiply the m elements of each row of the matrix by each other and construct the column-wise vector (X_i):

$$\{X_i\} = \left\{\prod_{j=1}^{m} A_{ij}\right\} \qquad i = 1,\ldots, m \qquad \text{Equation (5.15)}$$

   b. Take the nth root of each element X_i and construct the column-wise vector $\{Y_i\}$:

$$\{Y_i\} = \left\{\sqrt[n]{X_i}\right\} \qquad i = 1,\ldots, m \qquad \text{Equation (5.16)}$$

   c. Normalize vector $\{Y_i\}$ by dividing its elements by the sum of all elements (ΣY_i) to construct the column-wise vector w_i:

$$\{w_i\} = \left\{\frac{Y_i}{\sum_{i=1}^{m} Y_i}\right\} \qquad i = 1,\ldots, m \qquad \text{Equation (5.17)}$$

TABLE 5.2 The Random Index (RI) for the Order of Comparison Matrix

m	1	2	3	4	5	6	7	8	9	10	11	12
RI	0	0	0.58	0.9	1.12	1.24	1.32	1.41	1.45	1.49	1.51	1.58

2. Find vector $\{F_i\}$ by multiplying each matrix of Equation 5.14 by $\{w_i\}$:

$$\{F_i\} = \sum_{j=1}^{m} (A_{ij} \times w_j) \qquad i = 1,\ldots, m \qquad \text{Equation (5.18)}$$

3. Divide F_i by w_i to construct vector $\{Z_i\}$:

$$\{Z_i\} = \left\{\frac{F_i}{w_i}\right\} \qquad i = 1,\ldots, m \qquad \text{Equation (5.19)}$$

4. Find the maximum eigenvalue (λ_{max}) for the matrix by dividing the sum of elements of $\{Z_i\}$ by m:

$$\lambda_{max} = \frac{\sum_{i=1}^{m} Z_i}{m} \qquad \text{Equation (5.20)}$$

5. Find the consistency index (CI) = $(\lambda_{max} - m)/(m - 1)$ for the matrix.
6. Find the random index (RI) from Table 5.2 for m = 1 to 12.
7. Find the consistency ratio (CR) = (CI)/(RI) for each matrix. Any value of CR between zero and 0.1 is acceptable.

5.5.3 Advantages and Disadvantages of Multi-Criterion Decision-Making Methods

Results of evaluation of scheduling rules depend on the selected MCD method. Each MCD method has some advantages and disadvantages, as follows:

SAW Method: This method is easy to use and needs no membership function, but for evaluation of scheduling rules it can be applied only to those areas in which performance measures are of the same type and of close magnitudes. Even the graded values will not be indicative of the real difference between two successive values. This method is not appropriate for evaluation of scheduling rules, because the preference measure whose values prevail over the other performance measures' values is selected as the best rule regardless of its poor values in comparison with those of the other performance measures.

Max–Max Method: In this method, membership functions are used to interpret the outcomes. Its disadvantage is that only the highest value of one performance measure determines which combination of scheduling rules is selected, regardless of poor values of other performance measures.

Max–Min Method: Like max–max method, membership functions are used to interpret the outcomes. Sometimes max–min method will lead to bad decisions. For example, in a situation where a combination of scheduling rules leads to a poor value for one performance measure but extremely satisfactory values for the other ones, the method rejects the combination. The advantage of this method is that the selected combination of rules does not lead to a poor value for any performance measure.

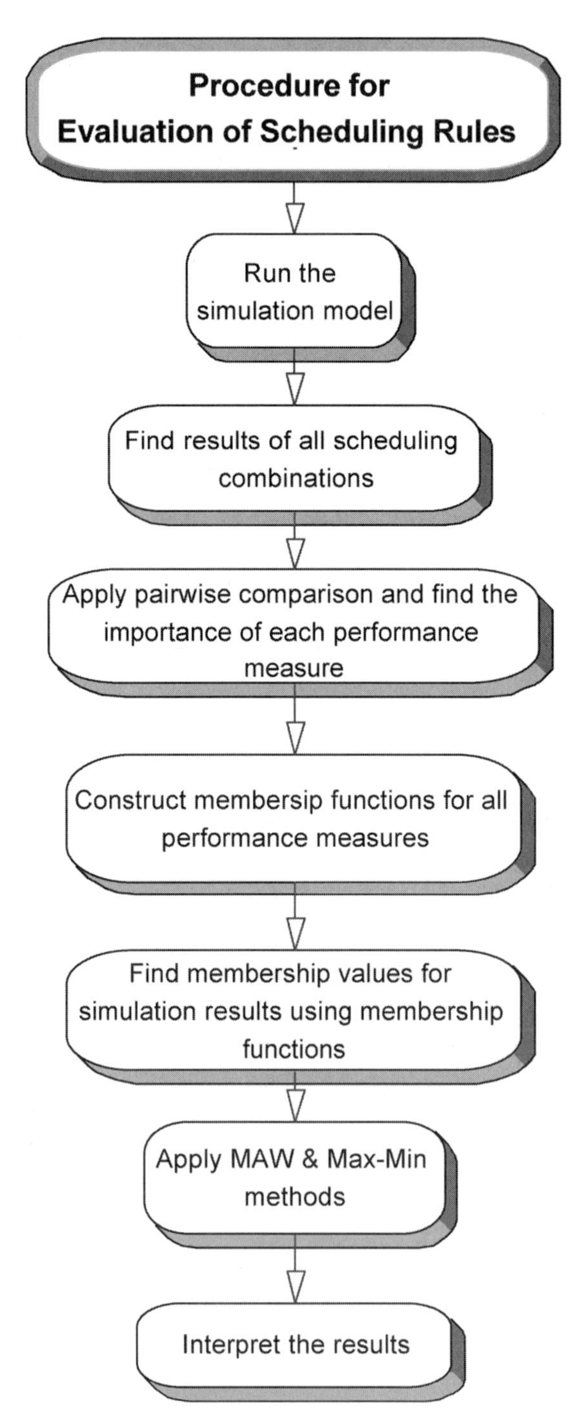

FIGURE 5.3 Procedure for evaluation of scheduling rules.

MAW Method: Like the two immediately previous methods, membership functions are used to interpret the outcomes. This method does not have the shortcomings of the three preceding methods; however, it does not guarantee non-poor values for all performance measures.

The procedure for evaluation of scheduling rules is depicted in Figure 5.3.

5.6 A Fuzzy Approach to Operation Selection

Due to the flexibility of FMSs, alternative operations are common. The real-time scheduling schemes use simulation techniques and dispatching rules to schedule parts in real time. The advantage of this kind of scheduling over analytical approaches is its capability to consider any unexpected events in the simulation model. This section presents a fuzzy approach to a real-time operation selection. This approach uses membership functions to find the share of each goal in the final decision rule.

5.6.1 Statement of the Problem

Production management and scheduling problems in an FMS are more complex than those problems in job shops and transfer lines. Each machine is versatile and capable of holding different tools to perform different operations; therefore, several alternative part types can be manufactured at any given time. An operation is capable of being performed on a number of alternative machines with possibly different process times. Therefore, production is continuous even under unexpected events such as breakdown of machines. Alternative operations give more flexibility to control the shop floor [Wilhelm and Shin 1985]. An alternative operation could be used if one workstation is overloaded. One machine may be preferred to another for a particular operation due to its queue length, job allocation cost, set-up time, and/or processing time. These objectives appear to be particularly important because of substantial investment required to install an FMS, and using the system in the most efficient way should compensate for this investment. Consequently, an FMS is viewed as a dynamic system.

The problem is to investigate the influence of a multi-criterion operation selection procedure on the performance of an FMS. The solution to the problem via the application of fuzzy set theory is explained below.

5.6.2 The Main Approach

Each traditional operation-selection rule, sometimes known as next-station-selection rule, is aimed at a particular objective and considers one attribute. For example, the work in queue (WINQ) rule searches the queues of next stations to find a queue with the least total work content. In other words, this rule finds the queue with the minimum total processing time of all waiting jobs. This will assure that flow time will be reduced. The number of parts in queue (NINQ), sometimes known as shortest number in queue (SNQ) rule searches the queues of next stations to find a queue with the minimum number of jobs. This will assure the reduction in number of blockage of queues. The lowest utilization of local input buffer (LULIB) rule searches all queues and selects the one with the lowest utilization level. This will provide a balance between utilization of queues and consequently workstations.

Production scheduling in an automated system is a multi-criterion task. Therefore, a more complicated operation-selection rule that can satisfy more objectives, by considering more attributes of queues, should be designed. For example, using the designed rule, a queue can be selected such that it has

- Fewest jobs
- Least total work content
- Lowest utilization level
- Lowest job allocation rate

To achieve this, a decision rule (an operation-selection rule) should be defined to include all of the above criteria. This decision rule should be able to react to the changes in the system in real time and has to be consistent with a multiple criteria approach. It can be a combination of traditional rules to consider all attributes of the queues.

There are several ways to combine the operation-selection rules. The simplest one is to apply the rules one after another. For example, one can give highest priority to those queues with least jobs; if there is any tie (i.e., two or more queues with the same length), then the queues with highest priority are searched for the least total work content until the last rule selects the queue of highest priority. This approach cannot be so effective, because most of the time a queue is selected after the application of the first or the second rule, leaving no reason for applying the other rules.

To overcome the above-mentioned difficulty, the problem is approached by constructing "a compromise of traditional rules" such as WINQ, SNQ, and LULIB. For this purpose, a multi-criterion approach combining all the above-mentioned criteria is used in conjunction with fuzzy set theory to model a compromise of the criteria and to achieve a balance between elementary (traditional) rules.

5.6.3 Operation-Selection Rules and Membership Functions

Bellman and Zadeh (1970) pointed out that in a fuzzy environment, goals and constraints normally have the same nature and can be presented by fuzzy sets on a set of alternatives, say X whose elements are denoted by x. Let C_k be the fuzzy domain delimited by the kth constraint ($k = 1, \ldots, l$) and G_i the fuzzy domain associated with the ith goal ($i = 1, \ldots, n$). When goals and constraints have unequal importance, membership functions can be weighted by x-dependent coefficients α_k and β_i such that

$$\forall x \in X, \sum_{k=1}^{l} \alpha_k(x) + \sum_{i=1}^{n} \beta_i(x) = 1 \qquad \text{Equation (5.21)}$$

Bellman and Zadeh then called the fuzzy set D, a fuzzy decision on X

$$D = \left(\bigcap_{k=1 \text{ to } l} C_k \right) \cap \left(\bigcap_{i=1 \text{ to } n} G_i \right) \qquad \text{Equation (5.22)}$$

whose membership function can be defined as

$$\mu_D(x) = \sum_{k=1}^{l} \alpha_k(x) \mu_{C_k}(x) + \sum_{i=1}^{n} \beta_i(x) \mu_{G_i}(x) \qquad \text{Equation (5.23)}$$

where μ_D is the membership of alternative x with respect to all goals and constraints, μ_{G_i} is the membership of alternative x to goal i, and μ_{C_k} is the membership of alternative x to constraint k. In the alternative operation problem no constraint is considered as a separate function, i.e., the coefficients α_k do not exist, therefore, the first term of Equation 5.21 and Equation 5.23 would vanish, i.e.,

$$\mu_D(x) = \sum_{i=1}^{n} \beta_i(x) \mu_{G_i}(x) \qquad \text{Equation (5.24)}$$

and

$$\sum_{i=1}^{n} \beta_i(x) = 1 \qquad \text{Equation (5.25)}$$

which means β_i's are normalized. Non-normalized weight of goals can be normalized by using the following equation:

$$\beta_i = \frac{w_i}{\sum_{i=1}^{n} w_i} \qquad \text{Equation (5.26)}$$

where w_i is the non-normalized weight of goal i.

Based on Equations 5.23 through 5.26, the following equation can be derived:

$$\mu_D(j) = \sum_{i=1}^{n} \left(\frac{(w_i)}{\sum_{i=1}^{n} w_i} \times \mu_i(j) \right) \times S(j) \qquad \text{Equation (5.27)}$$

where j = 1, … , m alternative operations
i = 1, … , n goals
$\mu_i(j)$ = membership of operation j to goal i
$\mu_D(j)$ = membership of operation j to all goals
w_i = weight of goal i

$$\frac{S(j)}{\mu_D(j)} = \begin{cases} 0 & \text{if } \mu_i(j) \text{ is negative} \\ 1 & \text{if } \mu_i(j) \text{ is non-negative} \end{cases}$$

In Equation 5.27, $S(j)$ will assure that a job will not be sent to a full queue (blocked queue). Finally, the queue with Max$\{\mu_D(j)\}$ will be selected.

To clarify Equation 5.27, let us consider a case where there are two different workstations on which a job can be processed. To find the better option, the following goals are to be achieved:

1. Minimizing number of jobs in queue.
2. Minimizing job allocation cost.
3. Minimizing lead time through the system.
4. Balancing machine utilization.

Not all the above goals can be achieved together, but it is desirable to achieve them to the greatest degree possible.

For the above goals, four fuzzy membership functions should be built to evaluate $\mu i(j)$. Membership functions have to be built to evaluate the contribution of alternative operations to the goals. First, to find the contribution of all alternatives to goal 1, the function can be defined as follows:

$$\mu_1(j) = (QCAP_j - NIQ_j - SR_j) / QCAP_j \qquad \text{Equation (5.28)}$$

where $QCAP_j$ = capacity of queue j
$NIQj$ = number of jobs in queue j

$$SR_j = \begin{cases} 0 & \text{if workstation } j \text{ is idle} \\ 1 & \text{if workstation } j \text{ is busy} \end{cases}$$

Assume two queues, say q1 with 6 jobs and q2 with 9 jobs, with capacity of 10 and 15 jobs, respectively, and both queues are busy at the decision-making time. The values of $\mu_1(1)$ and $\mu_1(2)$, for q1 and q2, respectively, are calculated as follows:

$\mu_1(1)$ = (10-6-1)/10 = 0.3 and $\mu_1(2)$ = (15-9-1)/15 = 0.33

Therefore, q2 is preferred to q1 as far as goal 1 is concerned.

For goal 2, the following membership function is defined:

$$\mu_2(j) = (MAXJR - JAR_j)/(MAXJR - MINJR) \qquad \text{Equation (5.29)}$$

where $MAXJR$ = maximum job allocation rate of all alternative operations
JAR_j = job allocation rate of operation j
$MINJR$ = minimum job allocation rate of all alternative operations

Suppose the job allocation rate for q1 and q2 are $0.6/min and $0.7/min, respectively, while the maximum job allocation rate is $1.2/min and the minimum one is $0.4/min, respectively. The values $\mu_2(1)$ and $\mu_2(2)$ are calculated as follows:

$\mu_2(1)$ = (1.2-0.6)/(1.2-0.4) = 0.75 and $\mu_2(2)$ = (1.2-0.7)/(1.2-0.4) = 0.63

Therefore, q1 is preferred to q2 as far as goal 2 is concerned.

For goal 3, the following membership function can be defined in a similar way:

$$\mu_3(j) = (MAXPT - TPT_j)/(MAXPT - MINPT) \qquad \text{Equation (5.30)}$$

where $MAXPT$ = maximum possible total processing time of jobs in a queue
TPT_j = total processing time of queue j
$MINPT$ = minimum possible total processing time of jobs in a queue, i.e., zero

Assume maximum possible total processing time of jobs in q1 and q2 to be 300 and 450 minutes, respectively, and at the decision-making time the total processing time of six jobs in q1 is 180 minutes and total processing time of nine jobs in q2 is 250 minutes. The values of $\mu_3(1)$ and $\mu_3(2)$ are calculated as follows:

$\mu_3(1)$ = (300-180)/(300) = 0.4 and $\mu_3(2)$ = (450-250)/(450) = 0.44

Therefore, q2 is preferred to q1 as far as goal 3 is concerned.

It should be noted that the maximum total processing time of jobs in each queue depends on the dispatching rule applied to the input buffers and is determined by using pilot runs.

The membership value for goal 4 is:

$$\mu_4(j) = 1 - (AVGUTIL_j / 100) \qquad \text{Equation (5.31)}$$

where $AVGUTIL_j$ = average utilization level of workstation j. For example, if the utilization of workstation 1 is 60% and that of workstation 2 is 78%, then $\mu_4(1)$ and $\mu_4(2)$ are 0.4 and 0.22, respectively. This shows preference of q1 to q2 as far as goal 4 is concerned.

Assuming that $w_1 = 0.55$, $w_2 = 0.45$, $w_3 = 0.25$, and $w_4 = 0.05$, $\mu_D(1)$ and $\mu_D(2)$ are calculated as follows:

$$\mu_D(1)=\sum_{i=1}^{4}\left(\frac{(w_j)}{\sum_{i=1}^{4}w_i}\times\mu_i(1)\right)\times S(1)$$

$$=\left(\frac{0.55}{1.3}\times 0.3+\frac{0.45}{1.3}\times 0.75+\frac{0.25}{1.3}\times 0.4+\frac{0.05}{1.3}\times 0.4\right)\times 1=0.48$$

$$\mu_D(2)=\sum_{i=1}^{4}\left(\frac{(w_i)}{\sum_{i=1}^{4}w_i}\times\mu_i(2)\right)\times S(2)$$

$$=\left(\frac{0.55}{1.3}\times 0.33+\frac{0.45}{1.3}\times 0.63+\frac{0.25}{1.3}\times 0.44+\frac{0.05}{1.3}\times 0.22\right)\times 1=0.45$$

Therefore, q1 is selected, because its membership value to all goals ($\mu_D(1)$) is greater than that of q2 ($\mu_D(2)$).

5.7 Fuzzy-Based Part Dispatching Rules in FMSs

This section explains a fuzzy real-time routing selection and an intelligent decision-making tool using fuzzy expert systems for machine loading in an FMS. In the complex environment of an FMS, proper expertise and experience are needed for decision making. Artificial intelligence, along with simulation modeling, can help imitate human expertise to schedule manufacturing systems [Baid and Nagarur 1994]. Machine centers in FMSs have automatic tool-changing capability. This means that a variety of machining operations can be done on the same machine. This facility introduces alternative routings and operations in FMSs. Alternate routings give more flexibility to control the shop floor. Chen and Chung [1991] evaluated loading formulations and routing policies in a simulated environment and concluded that an FMS is not superior to a job shop if routing flexibility is not utilized.

One routing may be preferred to another due to the number of parts in its queues, total work in its queues, and/or processing time. These objectives appear to be particularly important because of substantial investment required to install an FMS, and using the system in the most efficient fashion should compensate for this investment.

On the basis of fuzzy logic, founded by Zadeh [1965], many control applications were developed. Some examples are control of subways, cranes, washing machines, cameras, televisions, and many other devices. The literature on scheduling FMSs includes a few works in which fuzzy logic and fuzzy sets have been used, especially in simulation applications.

Karwowski and Evans [1986], Watanabe [1990], Raoot and Rakshit [1991], and Dweiri and Meier [1996] found that a fuzzy decision-making approach to evaluating the traditional facilities layout planning and fuzzy facilities layout planning leads to better results than the other approaches. O'Keefe and Rao [1992], Grabot and Geneste [1994], and Petrovic and Sweeney [1994] concluded that from an analytical and computational point of view, the fuzzy knowledge-based approach was relatively simple and efficient.

Dispatching rules are extensively used in the literature, and tested by many authors. There is no consensus on an individual rule that outperforms all others regarding multiple performance measures. According to Montazeri and Van Wassenhove [1990]:

McCartney and Hinds [1981] tested three priority rules (FIFO (first in first out), SIO (shortest imminent operation time), and SLACK/RO (slack per number of remaining operations)) to assign workpieces to machines in an FMS consisting of seven machines and a transport system. They concluded that for average tardiness, SLACK/RO performs better than two other rules when due-dates are loose. Alternatively, when due-dates are tight, the SIO rule leads to better results than the two other rules.

Blackstone et al. [1982], in their survey of the state of the art of dispatching rules, mentioned that SIO was the best priority rule in minimizing mean flowtime (with a few very late parts) when

- The shop has no control over due-dates, or
- The shop has control over due-dates and due-dates are tight, or
- The shop has control over due-dates and due-dates are loose, and there is a great congestion in the shop.

They defined a tight due date as less than six times the total processing time. They recommended that when due dates are loose, a modified SIO rule that also takes account of due dates (e.g., SIO^X (truncated SIO)) or a due-date-oriented rule (e.g., SLACK (shortest remaining slack time)) performs better on the criterion of tardiness than the other rules.

Ballakur and Steudel [1984], in their review of the state of the art of job shop control systems, compared several dispatching rules and indicated that the SIO rule appears to be the best dispatching rule for relatively loose due-dates and moderate machine utilization.

Based upon the above summary, it can be concluded that when due dates are tight, SIO rule is more effective than due-date based rules even with respect to the tardiness measures of performance. However, when due dates are loose, slack-based rules perform better than SIO rule, and among these, SLACK/RO rule seems to be the most promising one. But the main problem is that, in a dynamic and real-time system in which the parts arrive at the system over a period of time and in an irregular way, the tightness of the due dates is hard to determine, as due-date tightness depends on two factors, K and Tpt_i, described in the following paragraphs.

A variety of division rules have been used for setting due dates in simulation studies. The reader can find a classified table of due-date setting policies by Ramasesh [1990]. Among the due-date setting policies, total work content (TWK) is the best approach to set the due dates internally. The TWK approach usually is set up as

$$d_i = At_i + K \times Tpt_i \qquad \text{Equation (5.32)}$$

where d_i = due date of part i
At_i = arrival time of part i
Tpt_i = total processing time of part i

The value of the multiplier K, assumed between 1 to 9 by most authors, determines the degree of tightness of the due dates. According to Equation 5.32, a higher value of K results in a lower degree of tightness, but this is not always true. In other words, there is another important concern related to the interdependency of the due-date tightness and the shop-load level or utilization. Studies by Baker [1984] and Elvers and Tabue [1983] show that these two dimensions are closely connected and their effects cannot be separated, at least with respect to due-date related measures of performance. The interdependency of the due-date tightness and the shop-load level can be stated as follows: *If* the shop-load level is high, due dates may be tight even when K = 7.0, although in most studies this value is used to represent moderate or loose due dates. Conversely, a value of K = 3.0 may not necessarily represent tight due dates *if* the shop-load levels are very low.

Scheduling in FMSs is different from scheduling in job shops or transfer lines. In FMSs, utilization level of each machine can be different. For example, if a machine is dedicated to packaging only and is used as the final process, its time could be very tight because all parts should be packed on this unique machine. In this case, shop-load level cannot show the utilization level of this particular machine and machine utilization level should be used instead of shop-load level. Using machine utilization level, due-date tightness of each machine's input buffer queue can be determined. In subsequent sections machine utilization levels are used to construct the fuzzy expert-based rules.

5.8 Fuzzy Expert System-Based Rules

5.8.1 Fuzzy Set, Fuzzy Logic

Fuzzy set theory, introduced by Zadeh [1965], is a concept that can bring the reasoning used by computers closer to that used by human beings. Fuzzy set theory was invented to deal with vague, imprecise, and uncertain problems. The lack of certain data is the reason for uncertainty in many problems. While a conventional or a crisp set has sharp boundaries (such as the set of all numbers greater than 10), the transition between membership and nonmembership in a fuzzy set is gradual, and the degree of membership is specified by a number between 1 (full member) and 0 (full nonmember). Figure 5.4 shows the distinction between a traditional set and a fuzzy set.

The rationale behind any fuzzy system is that truth-values (in fuzzy logic) or membership values (in fuzzy sets) are indicated by a value in the range [0.0, 1.0], with 0.0 representing absolute falseness and 1.0 representing absolute truth. For example, "the truth value of 0.85" may be assigned to the statement "the value of M is high" if M = 7; likewise, "the truth value of 0.35" may be assigned to the statement "the value of M is high" if M = 4. This statement could be translated into set terminology as "M is a member of the set of high values." This statement would be rendered symbolically with fuzzy sets as

$$m(\text{HIGH VALUES}(7))=0.85 \qquad \text{or} \qquad m(\text{HIGH VALUES}(4))=0.35$$

where m in this example is the membership function, operating on the fuzzy set of high values, and returning a value between 0.0 and 1.0.

5.8.2 Fuzzy Expert Systems

In the late 1960s to early 1970s, a special branch of artificial intelligence (AI) known as expert systems, began to emerge. It has grown dramatically in the past few years and it represents the most successful demonstration of the capabilities of AI [Darvishi and Gill 1990; Mellichamp et al. 1990; Naruo et al. 1990; Singh and Shivakumar 1990]. Expert systems are the first truly commercial application of work done in the AI field and have received considerable publicity. Due to the potential benefits, there is currently a major concentration in the research and development of expert systems compared to other efforts in AI.

Reflecting human expertise, much of the information in the knowledge base of a typical expert system is imprecise, incomplete, or not totally reliable. Due to this reason, the answer to a question or the advice rendered by an expert system is usually qualified with a "certainty factor" that gives the user an indication of the degree of confidence in the outcome of the system. To arrive at the certainty factor, the existing expert systems employ what are essentially probability-based methods. However, since much of the uncertainty in the knowledge base of a typical expert system are due to the fuzziness and incompleteness of data, rather than from its randomness, the computed values of the certainty factor frequently lack reliability. This is still one of the serious shortcomings of expert systems. By providing a single inferential system for dealing with the fuzziness, incompleteness, and randomness of information in the knowledge base, fuzzy logic furnishes a systematic basis for the computation of certainty factors in the form of fuzzy numbers. At this juncture, fuzzy logic provides a natural framework for the design of expert systems. Indeed, the design of expert systems may well prove to be one of the most important applications of fuzzy logic in knowledge engineering and information technology.

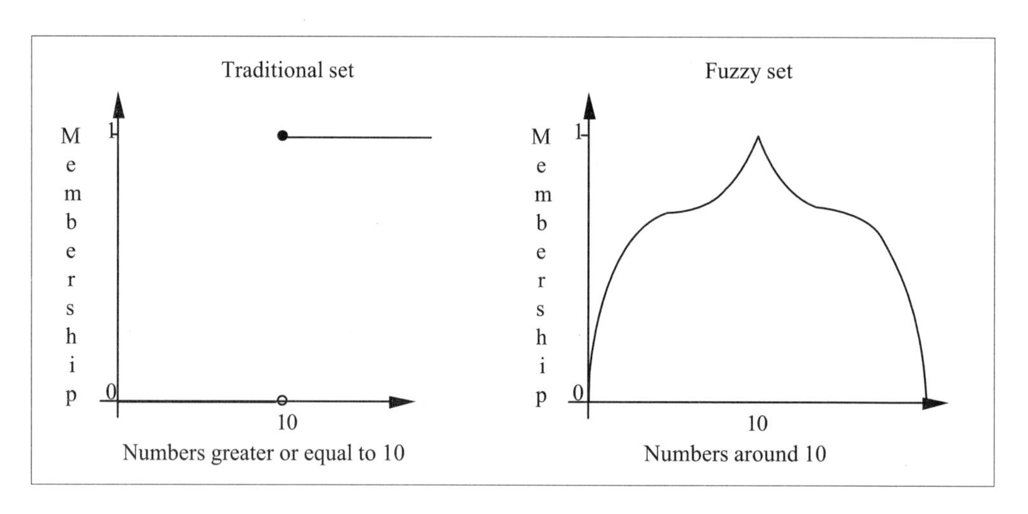

FIGURE 5.4 Traditional set vs. fuzzy set.

A *fuzzy expert system* is an expert system that applies reasoning to data via a number of *fuzzy membership functions* and *rules*, instead of Boolean logic. The rules in a fuzzy expert system are usually of a form similar to the following:

If 'x is low' *and* 'y is high' *then* 'z is medium'

where x and y are input variables (known data values), z is an output variable (a value to be computed), *low* is a membership function (fuzzy subset) defined on x, *high* is a membership function defined on y, and *medium* is a membership function defined on z. The antecedent (the rule's premise) describes to what degree the rule applies, and the conclusion (the rule's consequent) assigns a membership function to each of one or more output variables. Most tools for working with fuzzy expert systems allow for more than one conclusion per rule. The set of rules in a fuzzy expert system is known as the rule base or knowledge base. A closely related application is the use of fuzzy expert system in simulation and models that are intended to aid decision-making.

The general inference process proceeds in three (or four) steps, as follows:

1. Under fuzzification, the membership functions defined on the input variables are applied to their actual values, to determine the degree of truth for each rule premise. In other words, under defuzzification the values of the input variables are transferred into natural language like high, low, etc.
2. Under inference, the truth-value for the premise of each rule is computed and applied to the conclusion part of each rule. This results in one fuzzy subset to be assigned to each output variable for each rule. Usually only MIN or PRODUCT is used as inference rules. In MIN inferencing the output membership function is clipped off at a height corresponding to the rule premise's computed degree of truth (fuzzy logic AND). In PRODUCT inferencing the output membership function is scaled by the rule premise's computed degree of truth.
3. Under composition, all of the fuzzy subsets assigned to each output variable are combined together to form a single fuzzy subset for each output variable. Usually MAX or SUM are used as composition rules. In MAX composition, the combined output fuzzy subset is constructed by taking the point-wise maximum over all of the fuzzy subsets assigned to the output variable by the inference rule (fuzzy logic OR). In SUM composition, the combined output fuzzy subset is constructed by taking the point-wise sum over all of the fuzzy subsets assigned to the output variable by the inference rule.

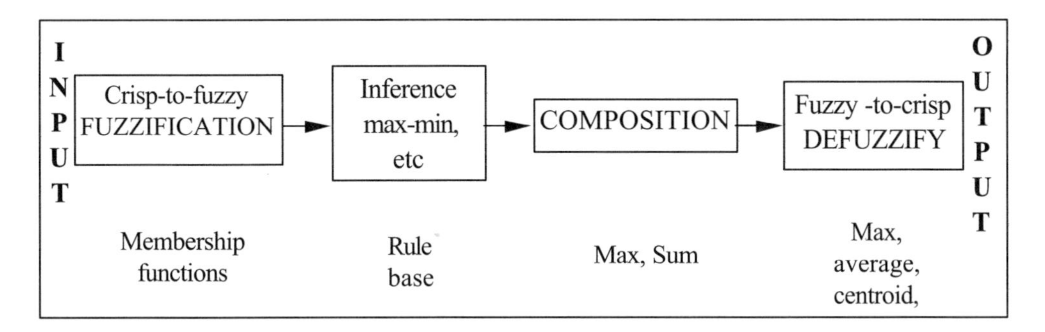

FIGURE 5.5 A typical fuzzy system.

4. Finally is the (optional) defuzzification that is used when it is useful to convert the fuzzy output-set to a crisp number. There are a number of defuzzification methods (at least 30). Two of the more common techniques are the CENTROID and MAXIMUM methods. In the centroid method, the crisp value of the output variable is computed by finding the variable value of the center of the gravity of the membership function for the fuzzy value. In the maximum method, one of the variable values at which the fuzzy subset has its maximum truth-value is chosen as the crisp value for the output variable. Figure 5.5 shows a typical fuzzy system.

5.8.3 Decision Points

In real-time scheduling, a decision point arises whenever seizing of one available production resource is necessary. A resource can be any individual production unit such as a single machine, a transport vehicle, an operator, or even a path of transportation.

Six decision points are considered herein for part processing, Figure 5.6. As a part arrives at the system, it should select a routing between several alternative routings (decision point 1). Then it requests for an AGV for traveling (decision point 2). All parts transferred to the input buffer queues will be selected by machines for processing (decision point 3). After completion of the process, parts reside in output buffer queues and compete to access idle AGVs (decision point 4). When more than one AGV tries to occupy an intersection, a decision is made to determine which AGV can proceed (decision point 5). When multiple parts are requesting the same idle AGV from different queues, the tie should be broken by a decision making to determine which queue is to be serviced first (decision point 6). These decision points are described below.

5.8.3.1 Selection of a Routing

After being released from the loading station, the part will select a routing as a fixed sequence of operations among alternative routings. A routing is selected such that the following goals are achieved:

- Minimizing number of blocked machines.
- Minimizing total processing time.
- Minimizing number of processing steps

Four different routing selection rules used herein are SNQ (or NINQ), STPT (shortest total processing time), WINQ, and FUZZY. In previous sections it was explained how fuzzy set can be employed for selecting alternative routing as an alternative operation selection rule (FUZZY rule). It must be noted that alternative routing is different from alternative operation.

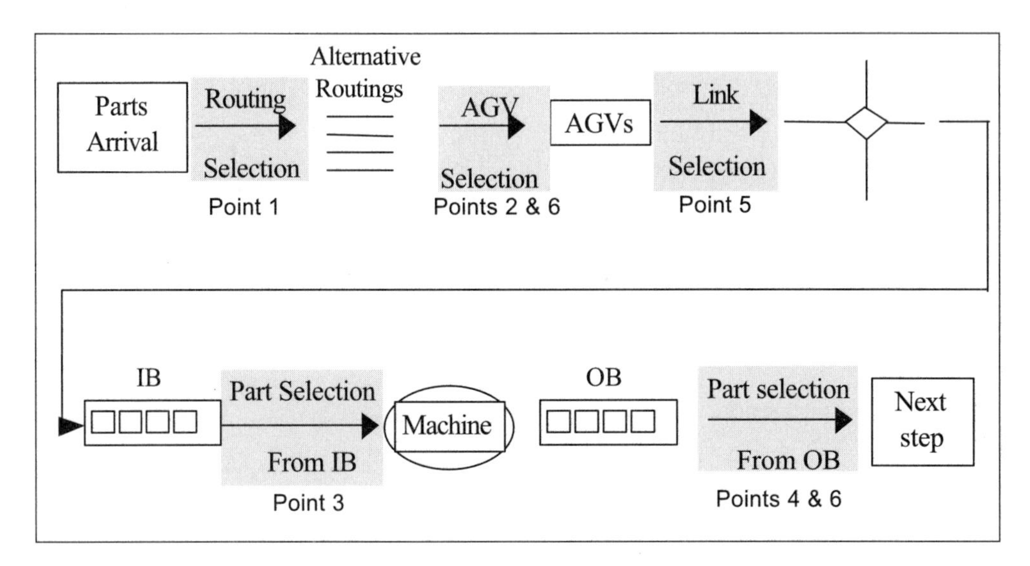

FIGURE 5.6 Decision points.

5.8.3.2 Parts Select AGVs

After selection of appropriate routing, the part will select an AGV to be transferred to the next station. The transporter selection rules, commonly used herein, include CYC, RAN, LDS (longest distance), SDS, and POR (preferred order).

5.8.3.3 Part Selection from Input Buffers

When a machine becomes idle and more than one part is waiting in its preceding queue, a part has to be selected to be processed next. The decision rules applied to this decision point include FIFO, SPT (shortest processing time), SRPT (shortest remaining processing time), SLRO (ratio of slack to remaining operation time), SIO, and FUZZY (fuzzy expert-based rules).

5.8.3.4 Part Selection from Output Buffers

When more than one part is waiting in one output buffer queue, they compete to obtain an idle AGV to be transported to the next station. The decision rules employed for this decision point include SPT, EDD (earliest due date), FIFO, SIO, and LPT (longest processing time).

5.8.3.5 Intersections Select AGVs

When more than one AGV tries to seize an intersection, the following rules (link selection rules) determine which one goes first: FCFS (first come first served), LCFS (last come first served), CLSD (closest distance), and FRTD (further distance).

5.8.3.6 AGVs Select Parts

When multiple parts are requesting the same AGV from different queues, the following steps are taken to break the tie:

1. The part with the highest priority is selected. All parts in output buffers of machines have the same priority. When the number of parts in the loading station queue is greater than or equal to ten, the highest priority is given to these parts rather than the parts sitting in the output buffers of machines.
2. When the number of parts in the loading queue is less than ten, the closest part to the AGV is selected.
3. If ties exist, select the part, which belongs to the queue of highest number of parts.

5.9 Selection of Routing and Part Dispatching Using Membership Functions and Fuzzy Expert System-Based Rules

In this section, the routing selection and the part dispatching rules based on fuzzy set and fuzzy logic are explained.

5.9.1 Routing Selection Using Fuzzy Set

Alternative routings are considered in advance for each incoming part. When a routing is selected for the part, it does not change during the simulation. Membership functions are employed to find the contribution of a routing to a goal. For each routing the following goals can always be set:

1. The possible minimum number of parts in queues of each routing will be zero while the maximum number will not reach Q_B.
2. The total possible minimum processing time of each routing for each part will be TPT_A, while the maximum time will not reach TPT_B.
3. The possible minimum number of processing steps of each routing will be P_A, while the maximum number of processing steps will not reach P_B.
4. The possible minimum total work in queues of any routing is zero, while the maximum total work in queues of any routing will not reach TWC_B.

To illustrate the method, an example of FMS with seven parts and six machines is considered according to Table 5.3.

To evaluate the contribution of a routing to a goal, a fuzzy membership function is set up. When goal 1 is considered for, say, part type 3, four alternative routings are to be observed. Routings number 1 and 3 have four operations on four different machines. Assuming the maximum length of a queue is 10, parts for routings number 1 and 3 is $Q_B = 41$ (= $4 \times 10 + 1$) and for routings number 2 and 4 is $Q_B = 31$ (= $3 \times 10 + 1$).

A simple function can be set up to evaluate the contribution of a routing to goal 1:

$$\mu_1(j) = (Q_B - Q_j)/(Q_B) \qquad \text{Equation (5.33)}$$

where Q_j = the number of parts in input buffers of routing *j*. The membership of routings is shown in Figure 5.7.

For goal 2, the following membership function is defined:

$$\mu_2(j) = (TPT_B - TPT_j)/(TPT_B - TPT_A) \qquad \text{Equation (5.34)}$$

where TPT_j = the total processing time needed for the part if routing *j* is selected.

For goals 3 and 4, similar membership functions can be set up as follows:

$$\mu_3(j) = (P_B - P_j)/(P_B - P_A) \qquad \text{Equation (5.35)}$$

$$\mu_4(j) = (TWC_B - TWC_j)/(TWC_B) \qquad \text{Equation (5.36)}$$

where P_j = the number of processing steps needed for the part if routing *j* is selected; TWC_j = the total work of parts in input buffers of routing j.

Each goal can be weighted using pairwise comparison [Saaty 1980]. The following formula is used to determine the final membership value of all alternatives.

TABLE 5.3 Operation Times (min) for Part Types and the Given Routings

Part Type	Machine No. 1	2	3	4	5	6	Routing No.
1	—	30	—	70	—	40	1
	40	—	55	—	50	—	2
	—	—	45	60	—	75	3
2	—	40	40	—	—	50	1
	—	70	—	60	50	—	2
	70	—	—	80	30	—	3
3	40	—	50	30	—	60	1
	—	50	—	—	70	80	2
	60	40	30	—	50	—	3
	—	50	—	30	40	—	4
4	30	—	—	40	20	—	1
	—	15	35	—	65	—	2
	—	20	—	35	15	—	3
5	—	40	—	70	—	—	1
	—	—	25	—	45	15	2
	15	25	35	—	—	—	3
6	—	75	—	40	35	—	1
	—	45	40	25	—	—	2
	—	30	25	—	40	—	3
7	—	—	—	—	65	50	1
	—	65	45	—	25	—	2
	—	25	—	15	—	40	3

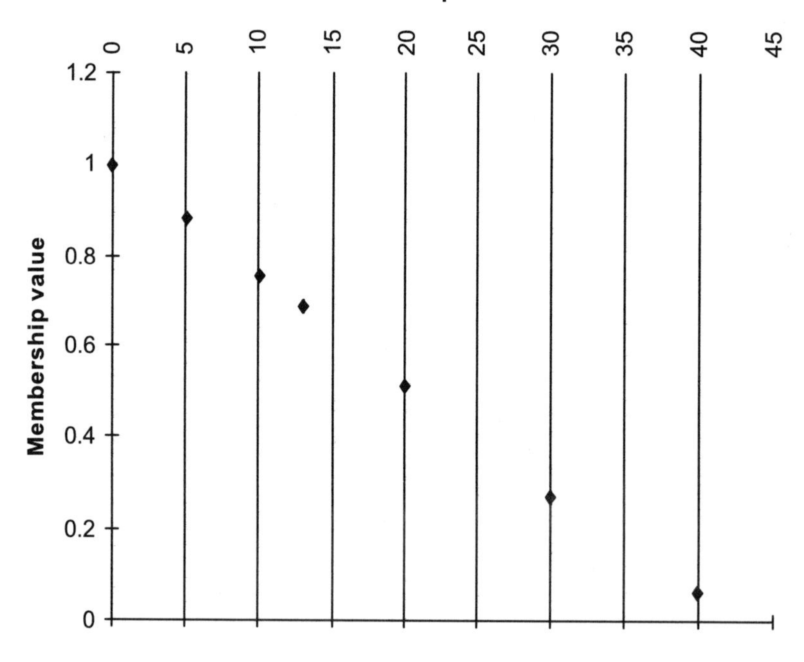

FIGURE 5.7 Membership function for goal 1.

$$\mu_D(j)=\sum_{i=1}^{n}\left(\frac{(w_i)}{\sum_{i=1}^{n} w_i}\times\mu_1(j)\right) \quad \text{Equation (5.37)}$$

where $j = , \ldots , m$ alternative routings
$i = , \ldots , n$ goals
$\mu_i(j)$ = membership of routing j to goal i
$\mu_D(j)$ = membership of routing j to all goals
w_i = weight of goal i

Finally, the routing with $\text{Max}\{\mu_D(j)\}$ will be selected.

5.9.2 Fuzzy Expert System-based Rule Approach for Dispatching of Parts

It was previously shown that selection of a scheduling rule heavily depends on the production system's due-date tightness, which itself depends on multiplier K and machine utilization. It was assumed that the due-date is assigned to the parts by the TWK policy.

Here, a fuzzy logic approach is proposed to determine the tightness of the production system. For this purpose the following rules are defined:

Rule 1: If multiplier K is *low*, and the machine utilization level is *high*, then the due-date tightness of the preceding queue is *high*.
Rule 2: If multiplier K is *high*, and the machine utilization level is *low*, then the due-date tightness of the preceding queue is *low*.

The fuzzy propositions of the antecedent of the rule 1 and 2 are represented by membership functions as described in Figures 5.8 and 5.9, respectively, and fuzzy propositions of the concluding parts are presented in Figure 5.10.

Mamdani's min–max method [1974] is used herein to perform fuzzy inference rules in which the membership functions of the concluding parts of the rules are cut at min $\mu_k(v)$ where $k = 1$ or 2 and v = K or L (Figures 5.11 and 5.12). The global conclusion is given by the maximum value of these graphs.

To demonstrate the implementation of the procedure, assume that multiplier K is set to 6.0 and at the moment that the rules are being applied, machine utilization level is 90%. The due-date tightness of the preceding queue can be found as follows:

1. Construct the graphical representation of membership functions to represent the fuzzy declaration, Figures 5.8 through 5.10.
2. Find the membership for each fuzzy proposition, Figures 5.11 and 5.12 (the left and the middle one).
3. Use min–max method to perform fuzzy inference rule, Figures 5.11 and Figure 5.12 (the right ones).

To determine the priority of parts in the queue using areas $S1$ and $S2$, a combination of SIO rule and SLACK/RO rule is used. The contribution of each rule to the final decision rule will depend on the areas $S1$ and $S2$. It can be seen from Figures 5.11 and 5.12 that when a machine center is completely tight, i.e., K = 1, and the machine center utilization level is 100%, $S1$ is equal to $A1$ and $S2$ is zero, as a consequence of which the final rule is determined to be SIO. While when the tightness of a machine center is completely loose, i.e., K = 9, and the machine center utilization level is close to 10%, $S2$ is close to $A2$ and $S1$ is close to zero, as a consequence of which the final rule is determined to be SLACK/RO. In general, the final

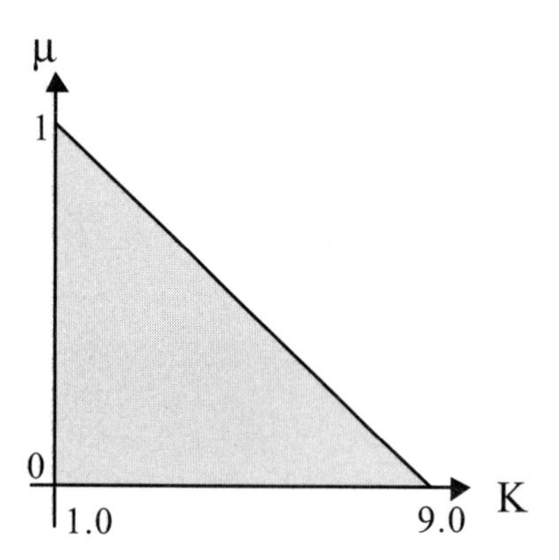

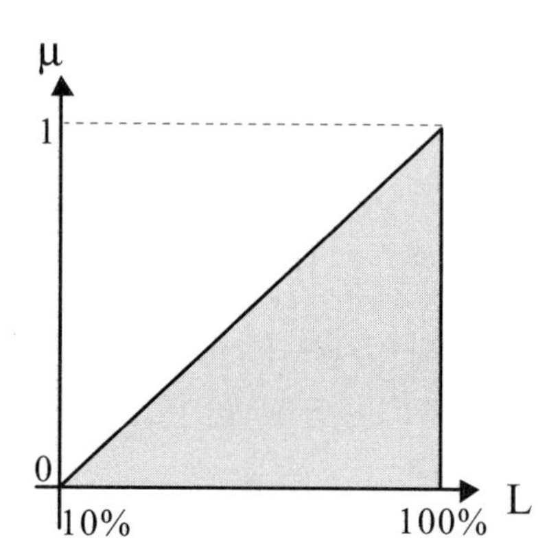

FIGURE 5.8 Membership functions for antecedent rule 1.

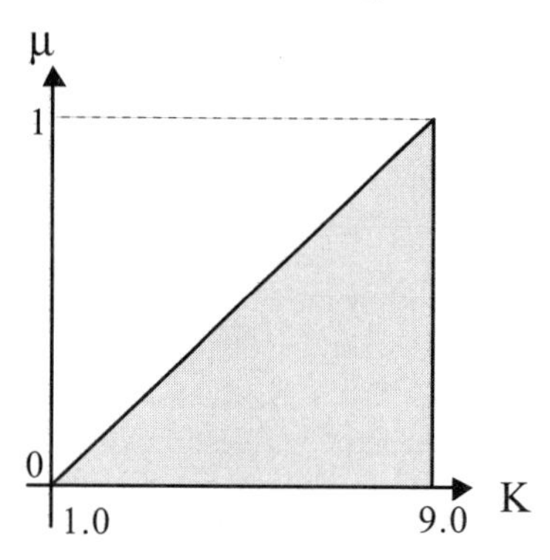

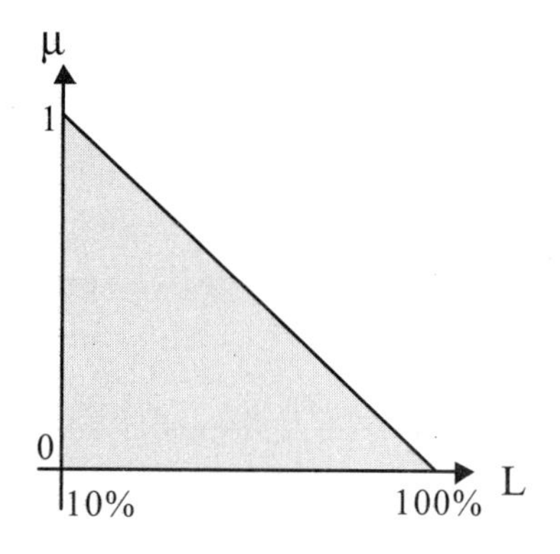

FIGURE 5.9 Membership functions for antecedent rule 2.

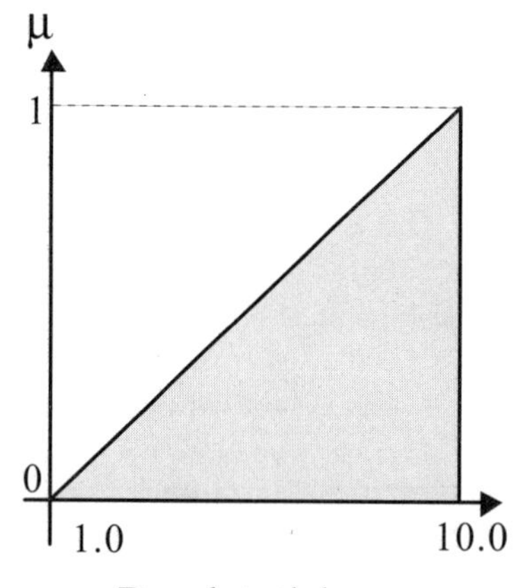

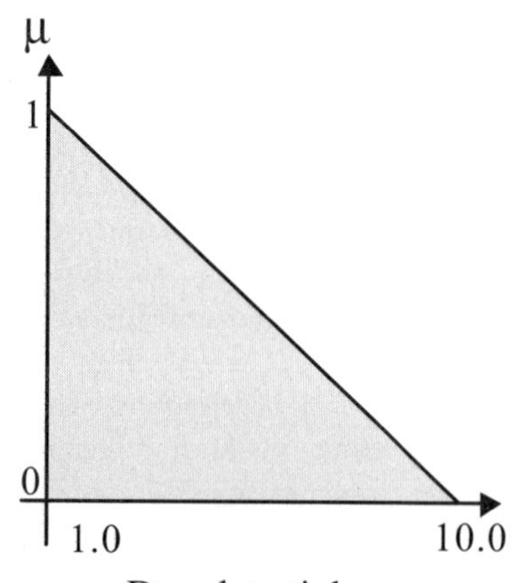

FIGURE 5.10 Conclusion modelling for rule 1 (left) and rule 2 (right).

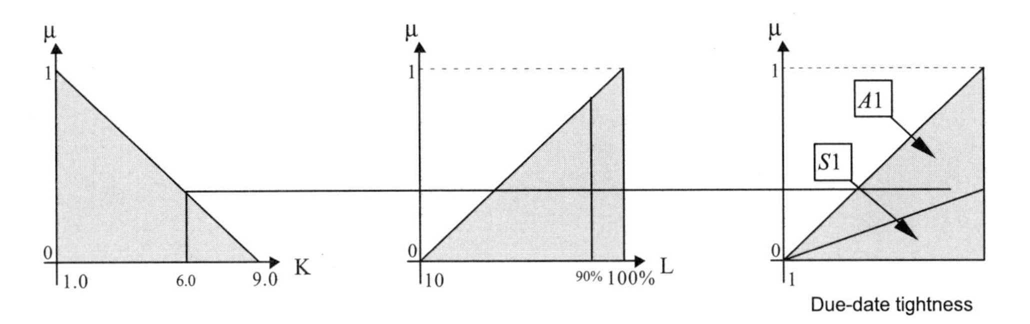

FIGURE 5.11 Conclusion modification for rule 1.

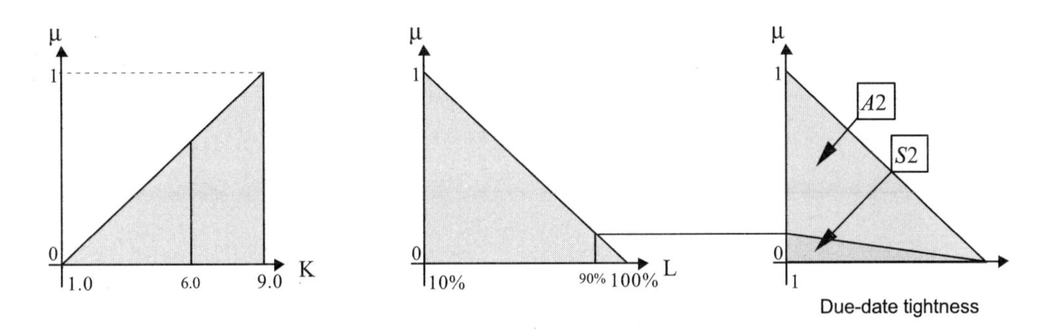

FIGURE 5.12 Conclusion modification for rule 2.

rule is calculated on the basis of both areas $S1$ and $S2$. For example, when K is 6 and the machine center utilization level is 90%, Figures 5.11 and 5.12, the final decision rule is determined as follows:

$$(\text{Priority with LVF})_{n},\ j = (S1)_j \times (\text{IOT})_{n,j} + (S2)_j \times (\text{SLRO})_{n,j} \qquad \text{Equation (5.38)}$$

where j = queue number
n = part number in the preceding queue of machine center j (or simply queue j)
LVF = low value first
$(\text{IOT})_{n,j}$ = imminent operation time of part n in queue j
$(\text{SLRO})_{n,j}$ = slack/remaining number of operation (SLACK/RO) of part n in queue j

It can be seen from Equation 5.38 that when $S2$ is zero, priority = IOT, i.e., SIO rule applies, and when $S1$ is zero then priority = SLRO, i.e., SLACK/RO rule applies.

Defining Terms

AGV	Automatic guided vehicle
AI	Artificial intelligence
CLSD	Closest distance
CYC	Cyclic
EDD	Earliest due date
FCFS	First come first served
FIFO	First in first out

FMS	Flexible manufacturing system
FRTD	Further distance
LCFS	Last come first served
LDS	Longest distance
LPT	Longest processing time
LULIB	Lowest utilization of local input buffer
MAW	Modified additive weighting
MCD	Multi-criterion decision making
NINQ	Number of parts in queue
PM	Performance measure
POR	Preferred order
RAN	Random
SAW	Simple additive weighting
SDS	Shortest distance to station
SIO	Shortest imminent operation time
SIO^x or SI^x	Truncated SIO
SLACK	Shortest remaining slack time
SLACK/RO	Slack per number of remaining operations
SLRO	Ratio of slack to remaining operation time
SNQ	Shortest number in queue
SPT	Shortest processing time
SRPT	Shortest remaining processing time
STPT	Shortest total processing time
TWK	Total work
WINQ	Work in queue
WIP	Work in process

References

Arbel, A. 1989, Approximate articulation of preference and priority derivation, *European Journal of Operational Research,* vol. 43, pp. 317-326.

Arbel, A. and Vargas, G. L. 1993, Preference simulation and preference programming: robustness issues in priority derivation, *European Journal of Operational Research,* vol. 69, pp. 200-209.

Baid, N. K. and Nagarur, N. N. 1994, An integrated decision support system for FMS: using intelligent simulation, *International Journal of Production Research,* vol. 32, no. 4, pp. 951-965.

Baker, K. R. 1984, Sequencing rules and due-date assignments in a job shop, *Management Science,* vol. 30, no. 9, pp. 1093-1104.

Ballakur, A. and Steudel, H. J. 1984, Integration of job shop control systems: a state-of-the-art review, *Journal of Manufacturing Systems,* vol. 3, no. 1, pp. 71-79.

Bellman, R. E. and Zadeh, L. A. 1970, Decision making in a fuzzy environment, *Management Science,* vol. 17, pp. 141-164.

Blackstone, J. H., Philips, D. T. and Hogg, G. L. 1982, A state of the art survey of dispatching rules for job shop operations, *International Journal of Production Research,* vol. 20, no. 1, pp. 27-45.

Chen, I. J. and Chung, C. H. 1991, Effects of loading and routing decisions on performance flexible manufacturing systems, *International Journal of Production Research,* vol. 29, pp. 2209-2225.

Chryssolouris, G., Dicke, K. and Moshine, L. 1994, An approach to real-time flexible scheduling, *International Journal of Flexible Manufacturing Systems,* vol. 6, pp. 235-253.

Darvishi, A. R. and Gill, K. F. 1990, Expert system design for fixture design, *International Journal of Production Research,* vol 28, no. 10, pp. 1901-1920.

Dweiri, F. and Meier, F. A. 1996, Application of fuzzy decision-making in facilities layout planning, *International Journal of Production Research,* vol. 34, no. 11, pp. 3207-3225.

Elvers, D. A. and Tabue, L. R. 1983, Time completion of various dispatching rules in job shops, *Omega,* vol. 11, no. 1, pp. 81-89.

Grabot, B. and Geneste, L. 1994, Dispatching rules in scheduling: a fuzzy approach, *International Journal of Production Research,* vol. 32, no. 4, pp. 903-915.

Gupta, Y. P., Gupta, M. C. and Bector, C. R. 1990, A review of scheduling rules in flexible manufacturing systems, *International Journal of Computer Integrated Manufacturing,* vol. 2, no. 6, pp. 356-377.

Hang, C. L. and Yon, K. 1981, *Multiple Attribute Decision Making,* Springer-Verlag, New York.

Hutchison, J. and Khumavala, B. 1990, Scheduling random flexible manufacturing systems with dynamic environments, *Journal of Operations Management,* vol. 9, no. 3, pp. 335-351.

Karwowski, W. and Evans, G. W. 1986, Fuzzy concepts in production management research: a review, *International Journal of Production Research,* vol. 24, no. 1, pp. 129-147.

Kazerooni, A., Chan, F. T. S., Abhary, K. and Ip, R. W. L. 1996, Simulation of scheduling rules in a flexible manufacturing system using fuzzy logic, *IEA-AIE96 Ninth International Conference on Industrial and Engineering Application of Artificial Intelligence and Expert System,* Japan, pp. 491-500.

Mamdani, E. H. 1974, Application of fuzzy algorithms for control of simple dynamic plant, *Proceeding of IEE,* vol. 121, no. 12, pp. 1585-1588.

McCartney, J. C. and Hinds, B. K. 1981, Interactive scheduling procedures for FMS, *Proceedings of 22nd International Machine Tool Design and Research Conference,* Manchester, U.K., pp. 47-54.

Mellichamp, J. M., Kwon, O. J. and Wahab, A. F. A. 1990, FMS designer: an expert system for flexible manufacturing system design, *International Journal of Production Research,* vol. 28, no. 11, pp. 2013-2024.

Montazeri, M. and Van Wassenhove, L. N. 1990, Analysis of scheduling rules for an FMS, *International Journal of Production Research,* vol. 28, no. 4, pp. 785-802.

Naruo, N., Lehto, M. and Salvendy, G. 1990, Development of a knowledge-based decision support system for diagnosing malfunctions of advanced production equipment, *International Journal of Production Research,* vol. 28, no. 12, pp. 2259-2276.

O'Keefe, R. M. and Rao, R. 1992, Part input into a flexible input flow system: an evaluation of look-ahead simulation and a fuzzy rule base, *International Journal of Flexible Manufacturing Systems,* vol. 4, pp. 113-127.

Petrovic, D. and Sweeney, E. 1994, Fuzzy knowledge-based approach to treating uncertainty in inventory, *Computer Integrated Manufacturing Systems,* vol. 7, no. 3, pp. 147-152.

Rachamadugu, R. and Stecke, K. E. 1988, Classification and review of FMS scheduling procedures, Working paper # 481 c, The University of Michigan, Ann Arbor.

Rachamadugu, R. and Stecke, K. E. 1994, Classification and review of FMS scheduling procedures, *Production Planning and Control,* vol. 5, no. 1, pp. 2-20.

Ramasesh, R. 1990, Dynamic job shop scheduling: a survey of simulation research, *Omega International Journal of Management Science,* vol. 18, no. 1, pp. 43-57.

Raoot, A. D. and Rakshit, A. 1991, A fuzzy approach to facilities lay-out planning, *International Journal of Production Research,* vol. 29, no. 4, pp. 835-857.

Saaty, T. L. 1975, Hierarchies and priorities-eigenvalue analysis, internal report, University of Pennsylvania, Wharton School, Philadelphia, PA.

Saaty, T. L. 1977, A scaling method for priorities in hierarchical structures, *Journal of Mathematical Psychology,* vol. 15, pp. 234-281.

Saaty, T. L. 1980, *The Analytic Hierarchy Process,* McGraw-Hill, New York.

Saaty, T. L. 1990, How to make a decision: the analytic hierarchy process, *European Journal of Operational Research,* vol. 48, no. 1, pp. 9-26.

Saaty, T. L. and Vargas, L. G. 1987, Uncertainty and rank order in the Analytic Hierarchy Process, *European Journal of Operational Research,* vol. 32, no. 3, pp. 107-117.

Singh, R. and Shivakumar, R. 1990, METEX — an expert system for machining planning, *International Journal of Production Research,* vol. 30, no. 7, pp. 1501-1516.

Smith, M. L., Ramesh, R., Dudeck, R. and Blair, E. 1986, Characteristic of US flexible manufacturing systems, *Computers and Industrial Engineering*, vol. 7, no. 3, pp. 199-207.

Tang, L. L., Yih, Y. and Liu, C. Y. 1993, A study on decision rules of scheduling model in an FMS, *Computers in Industry*, vol. 22, 1-13.

Watanabe, T. 1990, Job shop scheduling using fuzzy logic in a computer integrated manufacturing environment, *5th International Conference on System Research, Information and Cybernetics*, Baden-Baden, Germany, pp. 1-7.

Wilhelm, W. E. and Shin, H. M. 1985, Effectiveness of alternate operations in a flexible manufacturing system, *International Journal of Production Research*, vol. 23, no. 1, pp. 65-79.

Yager, R. R. 1978, Fuzzy decision making including unequal objectives, *Fuzzy Sets and Systems*, vol. 1, pp. 87-95.

Yang, K. K. and Sum, C. C. 1994, A comparison of job dispatching rules using a total cost criterion, *International Journal of Production Research*, vol. 32, no. 4, pp. 807-820.

Zadeh, L. A. 1965, Fuzzy sets, *Information and Control*, vol. 8, pp. 338-353.

For Further Information

Jamshidi, M., Vadaiee, N. and Ross T. J. 1993, *Fuzzy Logic and Control: Software and Hardware Applications*, Prentice-Hall, Englewood Cliffs, NJ.

Parsaei, H. R. 1995, *Design and Implementation of Intelligent Manufacturing Systems: From Expert Systems, Neural Networks, to Fuzzy Logic*, Prentice-Hall, Englewood Cliffs, NJ.

Chen, C. H. 1996, *Fuzzy Logic and Neural Network Handbook*, McGraw-Hill, New York.

6

Genetic Algorithms in Manufacturing System Design

L. H. S. Luong
University of South Australia

M. Kazerooni
Toosi University of Technology

K. Abhary
University of South Australia

6.1 Introduction

Batch manufacturing is a dominant manufacturing activity in many industries due to the demand for product customization. The high level of product variety and small manufacturing lot sizes are the major problems in batch manufacturing systems. The improvement in productivity is therefore essential for industries involved in batch manufacturing.

Group technology is a manufacturing philosophy for improving productivity in batch production systems and tries to retain the flexibility of job shop production. The basic idea of group technology (GT) is to divide a manufacturing system including parts, machines, and information into some groups or subsystems. Introduction of group technology into manufacturing has many advantages, including a reduction in flow time, work-in-process, and set-up time. One of the most important applications of group technology is cellular manufacturing system. A *cellular manufacturing system* is a manufacturing system that is divided into independent groups of machine cells and part families so that each family of parts can be produced within a group of machines. This allows batch production to gain economic advantages of mass production while retaining the flexibility of job shop methods. Wemmerlov and Hyer [1986] defined cellular manufacturing as follows:

> A manufacturing cell is a collection of dissimilar machines or manufacturing processes dedicated to a collection of similar parts and cellular manufacturing is said to be in place when a manufacturing system encompasses one or more such cells.

0-8493-0592-6/01/$0.00+$.50

PARTS

MACHINES	1	2	3	4	5	6	7	8	9	10	11	12	13	14	15	16	17	18	19	20	21	22	23	24	25	26	27	28	29	30	31	32	33	34	35	36	37	38	39	40
1				1													1		1			1	1						1											
2				1											1		1					1						1	1											
3							1																									1		1		1	1			
4							1																									1		1		1	1			
5							1																											1			1			
6									1												1						1												1	
7					1			1																	1					1								1		
8	1												1					1																						1
9	1										1		1			1		1																	1					
10		1																		1						1					1									
11									1					1										1			1												1	
12		1	1							1		1														1					1									
13		1	1							1		1								1											1									
14									1												1			1			1													
15	1										1					1		1																	1					
16	1										1		1			1		1																						
17		1	1							1		1								1						1					1									
18									1					1							1			1			1												1	
19	1												1			1		1																	1					
20	1												1			1		1																	1					
21																						1																		
22				1											1		1		1			1	1					1	1											
23							1																									1		1			1			
24				1											1		1		1			1	1					1	1											
25																																		1		1	1			
26					1	1		1																	1								1					1		
27							1																									1		1		1				1
28						1																			1					1			1							
29							1																													1	1			1
30					1	1		1																	1					1			1					1		

FIGURE 6.1 An initial machine–component matrix.

When some forms of automation are applied to a cellular manufacturing system, it is usually referred to as a flexible manufacturing system (FMS). These forms of automation may include numerically controlled machines, robotics, and automatic guided vehicles. For these reasons, FMS can be regarded as a subset of cellular manufacturing systems, and the design procedures for both cellular manufacturing systems and FMS are similar.

The benefits of cellular manufacturing system in comparison with the traditional functional layout are many, including a reduction in set-up time, work-in-process, and manufacturing lead-time, and an increase in product quality and job satisfaction. These benefits are well documented in literature. This chapter presents an integrated methodology for the design of cellular manufacturing systems using genetic algorithms.

6.2 The Design of Cellular Manufacturing Systems

The first step in the process of designing a cellular manufacturing system is called *cell formation.* Most approaches to cell formation utilize a machine-component incidence matrix, which is derived and oversimplified from the information included in the routing sheets of the parts to be manufactured. A typical machine-component incidence matrix is shown in Figure 6.1. The a_{ji}, which is the $(j,i)^{th}$ entry of this matrix, is 1 if the part i requires processing on machine j and a_{ji} is otherwise 0. Many attempts have been made to convert this form of matrix to a block diagonal form, as shown in Figure 6.2. Each block in Figure 6.2 represents a potential manufacturing cell. Not all incidence matrices can be decomposed to a complete block diagonal form. This problem can come from both **exceptional elements** and **bottleneck machines**. There are two possible ways to deal with exceptional elements. One way is to investigate alternative routings for all exceptional elements and choose a process route that does not need any machine from another cell. However, this solution cannot be achieved in most cases. Another way is subcontracting the exceptional elements to other companies. If there are not many exceptional elements, this way seems more reasonable, although it may incur extra handling costs and create problems with production planning and control.

In the presence of bottleneck machines, the system cannot be decomposed into independent cells, and some intercellular movements are inevitable. The impact of bottleneck machines on the system is increasing usage of material handling devices due to parts moving amongst the cells. Obviously a high number of intercellular movements will lead to an increase in material handling costs. Therefore, to decrease the

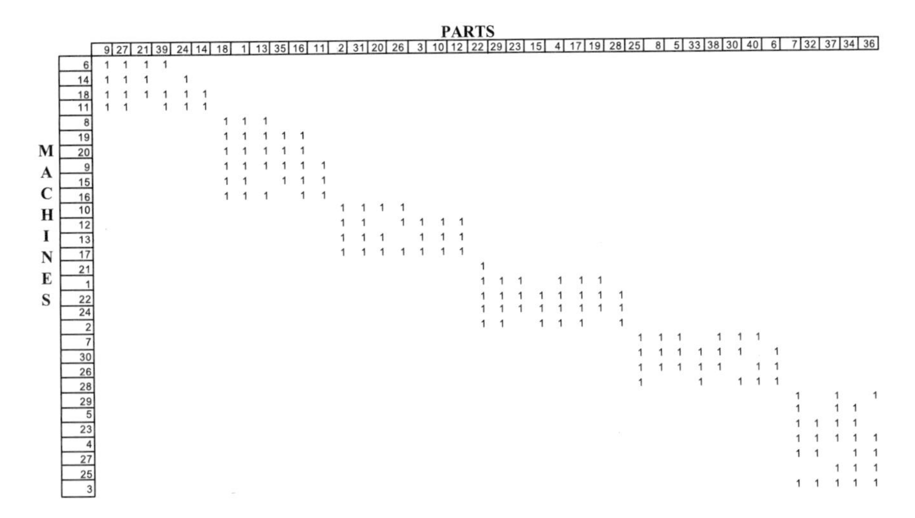

FIGURE 6.2 A block diagonal form (BDF) of machine–component matrix.

number of intercellular movements, some or all bottleneck machines should be duplicated. However, duplicating of bottleneck machines is not always economical. To justify which machine is to be duplicated, some subproblems including **clustering procedure, intracell layout,** and **intercell layout** of machines should be considered simultaneously in any attempt to optimize the design.

The above discussion indicates that the design of cellular manufacturing systems can be divided into two major stages: cell formation and system layout. The activities in the cell formation stage include constructing a group technology database of parts and their process routings, finding the most suitable routings among parts' alternative routings, grouping machines into machine groups, and forming parts into part families dedicated to the machine groups. In the system layout stage, the activities are selecting candidates for machine duplication, designing intercellular and intracellular layout, and detailed design.

As in any design process, the design of cellular manufacturing systems should take into consideration all relevant production parameters, design constraints, and design objectives. The relevant production parameters are process routings of parts, parts' production volume or annual demand, parts' alternative routings, processing time of each operation, and machine capacity or machine availability. There are also some constraints that should be considered while designing a cellular manufacturing system, such as minimum and/or maximum cell size, minimum and/or maximum number of cells, and maximum number of each machine type. In design optimization, there are many design objectives with regard to a cellular manufacturing system that can be considered individually or combinatorially. The design objectives may include minimizing intercellular movements, minimizing set-up time, minimizing machine load variation or maximizing machine utilization, and minimizing the system's costs. Some of these objectives can be conflicting. The goal of attaining all of these objectives, and at the same time satisfying the relevant design constraints, is a challenging task and may not be achievable because of conflicting objectives.

Many analytical, heuristic, cost-based and artificial intelligence techniques have been developed for solving the cell formation problem. Some examples are branch and bound method [Kusiak et al., 1991], nonlinear integer programming [Adil et al., 1996], cellular similarity [Luong, 1993], fuzzy technique [Lee et al., 1991], and simulated annealing [Murthy and Srinivasan, 1995]. There are also a number of review papers in this area. Waghodekar and Sahu [1983] provide an exhaustive bibliography of papers on group technology that appeared from 1928 to 1982. They also have classified the bibliography into four categories relating to both design and operational aspects. Another extensive survey with regard to different aspects of cellular manufacturing systems can be found in Wemmerlov and Hyer [1987]. Kusiak and Cheng [1991] have also reviewed some applications of models and algorithms for the cell formation

A review of current works in literature has revealed several drawbacks in the existing methods for designing cellular manufacturing systems. These drawbacks can be summarized as follows:

- Most methods work only with binary data or binary machine-component matrix. These approaches are far from real situations in industry, as they do not take all relevant production data into consideration in the design process. For example, production volumes, process sequences, processing times, and alternative routings are neglected in the majority of methods.
- Most methods are not able to handle design constraints such as minimum or maximum cell size or the maximum number of each machine type.
- Most methods are heuristic, and there is no optimization in the design process. Although many attempts have been made to optimize the design process using traditional optimization techniques such as integer programming, their scope of application is very limited as they can only deal with problems of small scale.

This chapter presents an integrated methodology for cellular manufacturing system design based on genetic algorithms. This methodology takes into account all relevant production data in the design process. Other features of this methodology include design optimization and the ability to handle design constraints such as cell size and machine duplication.

6.3 The Concepts of Similarity Coefficients

The basic idea of cellular manufacturing systems is to take the advantages of similarities in the process routings of parts. Most clustering algorithms for cell formation rely upon the concept of similarity coefficients. This concept is used to quantify the similarity in processing requirements between parts, which is then used as the basis for cell formation heuristic methods. This section introduces the concept of machine chain similarity (MCS) coefficient and part similarity coefficient that can be used to quantify the similarities in processing requirements for use in the design process. A unique feature of these similarity coefficients is that they take into consideration all relevant production data such as production volume, process sequences, and alternative routings in the early step of cellular manufacturing design.

6.3.1 Mathematical Formulation of the MCS Coefficient

The MCS_{ij}, which presents machine chain similarity between machines i and j, can be expressed mathematically as follows:

$$MCS_{ij} = \begin{cases} \dfrac{\sum_{l=1}^{M}\left(Min\left(\sum_{k=1}^{N} P_{il}^{k}, \sum_{k=1}^{N} P_{jl}^{k} \right)\right)}{\sum_{l=1}^{M}\sum_{k=1}^{N}\left(V_{kl} + V_{kl}^{'}\right)} & \text{if } i \neq j \\ 1 & \text{if } i = j \end{cases} \quad \text{Equation (6.1)}$$

where V_{kl} = volume of kth part moved out from machine l

$V_{kl}^{'}$ = volume of kth part moved in to machine l

N = number of parts

M = number of machines

$$P_{il}^{k} = \frac{\text{production volume for part } k \text{ moved between machines } i \text{ and } l \text{ if } i \neq 1}{\text{production volume for part } k \text{ moved between machines } i \text{ and } l \text{ if } i = 1}$$

or mathematically,

$$P_{il}^{k} = \begin{cases} \sum_{k=1}^{N}\sum_{l=1}^{G_k} C_l V_k & \text{if } i=1 \\ \sum_{k=1}^{N}\sum_{l=1}^{G_k} W_{il}^{k} V_k & \text{if } i \neq 1 \end{cases} \qquad \text{Equation (6.2)}$$

where $C_l = \begin{cases} 1 & \text{if } \{l = 1 \text{ or } l = G_k\}, \\ 2 & \text{otherwise} \end{cases}$

G_k = the last machine in processing route of part type k

V_k = production volume for part type k

W_{il}^{k} = number of trips that part type k makes between machines i and l, directly or indirectly

The extreme values for an MCS coefficient are 0 and 1. When the value of MCS_{ij} is 1, it means that all production volume transported in the system are moving between machines i and j. On the other hand, an MCS_{ij} with a value of zero means that there is no part transported between machines i and j whether *directly* or *indirectly*. In order to illustrate the concept of MCS coefficient, consider Table 6.1, which shows an example of five parts and six machines. The relationship between these six machines can be depicted graphically as in Figure 6.3. As can be seen from Figure 6.3, there is no direct part movement between machines M_2 and M_3. However, these two machines are indirectly connected together by machine M_6, implying that machines M_2 and M_3 can be positioned in the same cell. Consequently, the MCS coefficient for these machines is more than zero. On the other hand, if these two machines are in separate cells, then their MCS coefficient would be zero.

Table 6.2 is the production volume matrix showing the volume of parts transported between any pair of machines. The element a_{ij} in this table indicates the production volume transported between machines i and j ($i \neq j$), which has been calculated using Equation 6.2. For example:

$$a_{2,6} = \sum_{k=1}^{5} P_{2/6}^{k} = 1{*}150 \text{ (part 2)} + 3{*}70 \text{ (part 3)} = 360.$$

TABLE 6.1 Production Information for the Six-Machine/Five-Part Problem

	Parts				
	P_1	P_2	P_3	P_4	P_5
Production volume	100	150	70	150	160
Routing sequence	M_1-M_3-M_5-M_6	M_2-M_4-M_6	M_2-M_6-M_2-M_6	M_5-M_3-M_6-M_1-M_3	M_5-M_1-M_3

TABLE 6.2 Production Volume Transported between Pair of Machines

	M_1	M_2	M_3	M_4	M_5	M_6
M_1	720	0	560	0	410	250
M_2	0	360	0	150	0	360
M_3	560	0	810	0	410	400
M_4	0	150	0	300	0	150
M_5	410	0	410	0	510	250
M_6	250	360	400	150	250	760

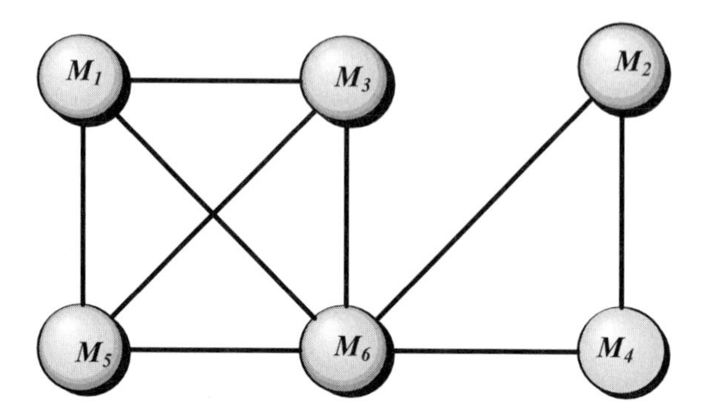

FIGURE 6.3 Graphical presentation of the example shown in Table 6.1.

It should be noted that the first term in the above calculation (part 2) is due to the indirect relationship between machines 2 and 6, while the second term indicates that there are three trips between these two machines. In the case of $i = j$, a_{ij} indicates the sum of parts transported *to and from* machine *i*. For example,

$$a_{1,1} = \sum_{k=1}^{5} P_{1/1}^{k} = 1{*}100 \text{ (part 1)} + 2{*}150 \text{ (part 4)} + 2{*}160 \text{ (part 5)} = 720.$$

Having computed all machine pair similarities, MCS coefficients for all machines can then be written in a MCS matrix (Table 6.3) in which element a_{ij} indicates the MCS coefficient between machines *i* and *j*. For example, the MCS coefficient between machines M_3 and M_6 can be computed as follows:

$$MCS_{M_3M_6} = \frac{\min(250,560)+\min(360,0)+\min(400,810)+\min(150,0)+\min(250,410)+\min(760,400)}{720+360+810+300+510+760} = 0.3757$$

Once the MCS matrix is obtained, it can be normalized by dividing all elements in the matrix by the largest element in that matrix (Table 6.4). In comparison with McAuley's similarity coefficient [1972], the results in Table 6.4 indicate that production volume and process sequence can make a significant difference in the pairwise similarity between machines.

6.3.2 Parts Similarity Coefficient

For each pair of parts, the parts similarity coefficient is defined as:

$$PS_{ij} = PS_{ij} \sum_{k=1}^{M} \min\left(N_{ki}, N_{kj}\right) \qquad \text{Equation (6.3)}$$

where PS_{ij} = the similarity between parts *i* and *j*
N_{ki} = k^{th} element of $\mathbf{MRV}_i$
N_{kj} = k^{th} element of $\mathbf{MRV}_j$
M = number of machines

MRV_i is the machine required vector for part *i*, which is defined as

$$\mathbf{MRV}_i = [N_{1i}, N_{2i}, N_{3i}, \ldots, N_{ki}, \ldots, N_{mi}] \qquad \text{Equation (6.4)}$$

where *k* is k^{th} machine and *m* is the total number of machines. N_{ki} is defined as follows:.

TABLE 6.3 The Initial MCS Matrix between a Pair of Machines

	Machines					
	M_1	M_2	M_3	M_4	M_5	M_6
M_1	1	0.0723	0.5144	0.0434	0.4277	0.3324
M_2		1	0.1040	0.1301	0.0723	0.2514
M_3			1	0.0434	0.4277	0.3757
M_4				1	0.0434	0.1300
M_5		SYMMETRIC			1	0.3324
M_6						1

TABLE 6.4 The Normalized MCS Matrix

	Machines					
	M_1	M_2	M_3	M_4	M_5	M_6
M_1	1	0.1404	1	0.0843	0.8315	0.6461
M_2		1	0.2022	0.2528	0.1405	0.4888
M_3			1	0.0843	0.8315	0.7303
M_4				1	0.0843	0.2528
M_5		SYMMETRIC			1	0.6461
M_6						1

$$N_{ki} = \begin{cases} N_{uki} & \text{if part } i \text{ meets machine } k \\ 0 & \text{if part } i \text{ does not meet machine } k \end{cases}$$

N_{uki} shows the frequency that part i travels *to and from* machine k multiplied by the production volume required for part i. For example, consider the problem of six machines and five parts shown in Table 6.1, and lets assume that the machines have been sequenced in the order of [M_2, M_3, M_5, M_1, M_4, M_6]; then the MRV for part 1 is [0, 200, 200, 100, 0, 100].

The MCS matrix and the parts similarity coefficients discussed above are used as the tools to identify the best routings of parts that yield the most independent cells. In addition, they are also used for clustering the machines and the parts into machine groups and part families, respectively. Figure 6.4 depicts the three major stages in the cell formation process, using the concept similarity coefficients and genetic algorithms (GA). The details of each stage are described in the following sections.

6.4 A Genetic Algorithm for Finding the Optimum Process Routings for Parts

The aim of a cellular manufacturing system design is minimizing the cost of the system. It can be gained by dividing the system into independent cells (machine groups and part families) to minimize the costs of material handling and set-up. Accordingly, in a case where there are alternative process routings for parts, it is therefore necessary to identify the combination of parts' process routings, which minimizes the number of intercellular movements, and consequently maximizes the number of independent cells. It has been shown [Kazerooni, Luong, and Abhary, 1995a and 1995b] that maximum clusterability of parts can be achieved by maximizing the number of zero elements (or number of elements below a certain threshold value) in the MCS matrix. A genetic algorithm for this purpose is described below.

6.4.1 Chromosome Representation for Different Routings

Suppose a problem including n parts in which each part can have d different alternative routings where $1 \leq d \leq p$, and p is the maximum number of alternative routings a part can possess. For such a problem

A GA-based algorithm to find the optimum process routings for parts

Aim: Minimizing the number of intercellular movements of parts.

Input: Normalized MCS matrix.

Objective: Maximize the number of zeros in the MCS matrix.

Output: A MCS matrix which represents the selected process routings for parts which yield the maximum number of independent cells.

↓

A GA-based algorithm to cluster machines into machine groups

Aim: Clustering machines into machine groups.

Input: An MCS matrix for the selected process routings for parts.

Objective: Maximize the similarity of adjacent machines in the optimized MCS matrix.

Output: A diagonal MCS matrix that represents groups of machines.

↓

A GA-based algorithm to cluster parts into part families

Aim: Clustering parts into part families.

Input: Diagonal MCS matrix and parts similarity coefficients.

Objective: Maximizing parts similarity coefficient of adjacent parts in the diagonal MCS matrix.

Output: Final machine-component matrix.

FIGURE 6.4 The three stages in the cell formation process.

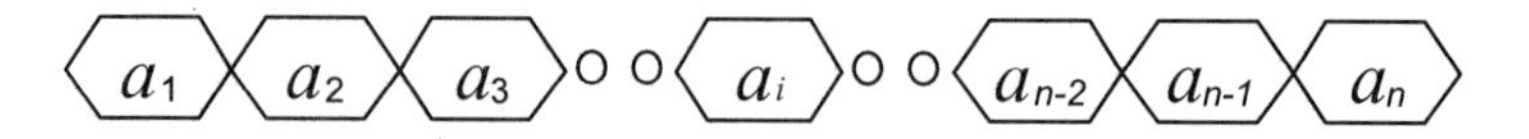

FIGURE 6.5 Chromosome representation of parts' process routings.

the following chromosome representation shown in Figure 6.5 is used. In Figure 6.5, a_i represents the selected process routing for part i, and can be any number between 1 and p for part i. However, all parts do not have the same number of routings and every number between 1 and p cannot be valid for all parts. To overcome such a drawback, the following procedure is done to validate the value of all genes regardless of the number of routings that the corresponding part has.

1. Set the counter i to 1.
2. Read p_i, the maximum number of routings that part i can have.

3. Read a_i, the value of gene i.
4. $a_i = \begin{cases} a_i & \text{if } a_i \le p_i \\ a_i - p_i & \text{if } a_i \ge p_i \end{cases}$
5. If $a_i \ge p_i$, go to step 4 =, otherwise increment i by one.
6. If $i > n$ (number of parts), stop, otherwise go to step 2.

With this procedure, for example if the value of the first gene in Figure 6.5 is 5 and there are only three different alternative routings for part 1, then the gene value is changed to 2 (i.e., 5 – 3). Using the above procedure, the gene values in the chromosome will be valid. It should be noted that if a part has only one process plan, it should not participate in the chromosome, because its process routing has been already specified.

6.4.1.1 The Crossover Operator

Since a gene in the chromosome can take any number between 1 and p, and repeated value for gene is allowed, any normal crossover technique such as two-point crossover or multiple-point crossover can be used in this algorithm.

6.4.1.2 The Fitness Function

The fitness function for this algorithm is to maximize the number of zeros in the MCS matrix.

6.4.1.3 The Convergence Policy

The entropic measure H_i, as suggested by Grefenstette [1987], is used for this algorithm. H_i in the current population can be computed using the following equation:

$$H_i = \frac{-\sum_{j=1}^{p} \left(\frac{n_{ij}}{SP}\right) * Log\left(\frac{n_{ij}}{SP}\right)}{Log(p)} \qquad \text{Equation (6.5)}$$

where n_{ij} is the number of chromosomes in which process plan j is assigned to part i in the current population, SP is the population size, and p is the maximum number of process plans for part i. The divergence (H) is then calculated as

$$H = \sum_{i=1}^{n} \frac{H_i}{n} \qquad \text{Equation (6.6)}$$

where n is the number of parts.

6.4.1.4 The Algorithm

A genetic algorithm, which is described below, has been developed to find those parts' routings, which yield the maximum number of zero elements (or number of elements below a certain threshold value) in the MCS matrix.

Step 1. Read the input data:
- Number of parts.
- Number of process plans for each part.
- L_n, threshold value to count small values in MCS matrix.

Step 2. Initialize the problem:
- Assign an integer number to each process plan for each part.
- Initialize the value of GA control parameters, including population size, crossover probability, low and high mutation probability, maximum generation, maximum number of process plans.

Step 3. Initialize the first population:

- Create the first population at random.
- Decode the chromosome and modify those process plans, which are more than the number of their corresponding maximum parts' process plans.
- For each member of the population, evaluate the fitness (i.e., the number of elements in the MCS matrix that have a value equal to zero or below a certain limit).

Step 4. Generate new population while termination criterion not reached:

- Select a new population from the old population according to a selection strategy. For this study, the tournament strategy has shown good performance.
- Apply two-cut-point crossover and low mutation with respect to their associated probability.
- For each member of the population evaluate the fitness value (number of zero in the corresponding MCS matrix).

Step 5. Measure the diversity of new population and applying high probability mutation if the population's diversity passes the threshold value.

Step 6. Evaluate the fitness of new population's members.

Step 7. Report the best chromosome.

Step 8. Go to Step 4, if maximum generation has not been reached.

6.5 A Genetic Algorithm to Cluster Machines into Machine Groups

The second step in the cell formation process is to group the machines in machine groups. As previously discussed, the objective here is to maximize the similarity of machines that belong to the same cell. A genetic algorithm described below has been developed for this purpose.

6.5.1 Chromosome Representation

The path representation method is used for chromosome representation. For example, the following sequence of machines [5 – 1 –7 –8 – 9 – 4 – 6 – 2 – 3] is represented simply as (5 1 7 8 9 4 6 2 3). In other words, machine 5 is the first and machine 3 is the last machine in the MCS matrix.

6.5.2 The Crossover Operator

A crossover technique called advanced edge recombination (AER) that has been developed based on the traditional edge recombination (ER) crossover [Whitley et al., 1989] is used for this algorithm. The AER crossover operator can be set up as follows:

1. For each machine *m*, make an edge list including all other machines connected to machine *m* in at least one of the parent.
2. Select the first machine of parent p_1 to be the current machine *m* of the offspring.
3. Select the connected edges to the current machine *m*.
4. Define a probability distribution over connected edges based on their similarity.
5. Select a new machine *m* from the previous *m* edge list, according to roulette wheel probability distribution. If previous *m* edge list has no member, selection is done at random from all nonselected machines.
6. Extract the current machine *m* from all edges list.
7. If the sequence is completed, stop, otherwise, go to step 3.

6.5.3 The Fitness Function

A cell is a set of machines with maximum similarity among the machines within the cell. Accordingly, the following term should be maximized:

$$\text{Max}\left(\sum_{i=1}^{k}\sum_{j=1}^{k}\sum_{c=1}^{C} MCS_{c(i,j)}\right) \text{ subject to } i \neq j \qquad \text{Equation (6.7)}$$

where k is the number of machines in each cell, C is the number of cells, and $MCS_{c(i,\,j)}$ is the MCS coefficient between machines i and j in cell c.

Since the number of cells and the number of machines within the cells are not known beforehand, the search space for Equation 6.5 can become very large. To reduce the search space, which can significantly affect the optimum result and computational time, consider a MCS matrix in which the rows (or columns) are permuted to achieve maximum similarity between any two adjacent machines. The machines with high similarity coefficients must therefore be placed close to each other. As a result, the following maximization is employed instead of Equation 6.5 to obtain maximum similarity between adjacent machines in the machine chain similarity matrix:

$$MMCS = \text{Max}\left[\left(\sum_{i=1}^{m}\sum_{j=1}^{m-1}\left(a_{i,j}, a_{i,j+1}\right)_{\min}\right) + \left(\sum_{j=1}^{m}\sum_{i=1}^{m-1}\left(a_{i,j}, a_{i+1,j}\right)_{\min}\right)\right] \qquad \text{Equation (6.8)}$$

where $a_{i,j}$ is the MCS coefficient between machine i and machine j. Equation 6.7 is used as the fitness function in this algorithm.

6.5.4 The Convergence Policy

The convergence policy used for this algorithm is the same as the policy used for finding the optimum process routings (Equations 6.5 and 6.6). The only differences are that in this case n_{ij} is the number of chromosomes in which machine j is assigned to gene i in the current population, and p is the number of machines.

6.5.5 The Replacement Strategy

The replacement strategy is that if either the first or the last chromosome of the new generation is better than the best of the old generation, then all the offsprings will form the new population. On the other hand, if the first and the last chromosome of the new population are not as fit as the best chromosome of the previous population, then the best chromosome of the previous population will form the next generation.

6.5.6 The Algorithm

Step 1. Read the input data.
Step 2. Initialize the GA control parameters.
Step 3. Set the generation counter, $G_i = 1$.
Step 4. Initialize the first population at random.
Step 5. Report the situation of the first population, including maximum, minimum, and average fitness value of the population.
Step 6. Increment the generation counter, $G_i = G_i + 1$.
Step 7. While the mating pool is not full:
    Step 7.1 Select two chromosomes using tournament strategy.

Step 7.2 Crossover the parent and generate corresponding offspring considering crossover probability.
Step 7.3 Apply the mutation to all genes of offspring considering mutation probability.
Step 7.4 Evaluate the population diversity; if it is less than threshold value, apply the high value mutation probability.
Step 7.5 Evaluate the fitness of new generated chromosomes.
Step 8. Replace the best chromosome of the old population with the first and last newly generated chromosome according to the replacement strategy.
Step 9. Report the statistical situation of the new generation.
Step 10. Record the chromosome and its associated fitness.
Step 11. Go to step 6 if the termination criteria are not met.

6.6 A Genetic Algorithm to Cluster Parts into Part Families

The third step in the process is the clustering of machines and components into a machine–component matrix. The objective function for this genetic algorithm is to maximize the similarity of adjacent parts in the machine–component matrix in which machines have been sequenced according to MCS coefficients as described in the previous section. This step will produce a machine-component matrix by which the cells, machine groups, and part families can be recognized.

6.6.1 Chromosome Representation

The path representation technique is used to represent the sequence of parts in the final machine–component matrix. For example, a sequence of [5 – 1 – 7 – 8 – 9 – 4 – 6 – 2 – 3] is represented as (5 1 7 8 9 4 6 2 3). In other words, part 5 is the first and part 3 is the last part in the machine–component matrix.

6.6.2 The Crossover Operator

The AER crossover technique described in the previous section is also used for this genetic algorithm.

6.6.3 The Fitness Function

The objective here is to rearrange parts to yield maximum parts similarity coefficients for adjacent parts in the machine–component matrix. This can be done by using the following equation:

$$\text{Maximize } (APS) = \text{Maximize} \left(\sum_{j=1}^{M} \sum_{i=1}^{N-1} \left(PS_{i,j} \right) \right) \qquad \text{Equation (6.9)}$$

where APS is the sum of any two adjacent parts' similarity coefficients in the machine–component matrix, and $PS_{i,j}$ is as defined by Equation 6.3.

6.6.4 The Convergence Policy

The convergence policy used for this algorithm is the same as the policy used for finding the optimum process routings (Equations 6.5 and 6.6). The only differences are that in this case n_{ij} is the number of chromosomes in which part j is assigned to gene i in the current population, and p is the number of parts.

6.6.5 The Replacement Strategy

The replacement strategy for this algorithm is the same as the strategy used for machine grouping previously described.

6.6.6 The Algorithm

Step 1. Read the input data.
Step 2. Enter the initial value for the GA control parameters.
Step 3. Set the generation counter, $G_i = 1$.
Step 4. Initialize the first population at random.
Step 5. Report the situation of the first population, including maximum, minimum, and average fitness value of the population.
Step 6. Increment the generation counter, $G_i = G_i + 1$.
Step 7. While the mating pool is not full:
  Step 7.1 Select two chromosomes using tournament strategy.
  Step 7.2 Crossover the parent and generate corresponding offspring considering crossover probability.
  Step 7.3 Apply the mutation to all genes of offspring considering mutation probability.
  Step 7.4 Evaluate the population diversity; if it is less than threshold value, apply the high value mutation probability.
  Step 7.5 Evaluate the fitness of new generated chromosomes.
Step 8. Replace the best chromosome of the old population with the first and last newly generated chromosome according to the replacement strategy.
Step 9. Report the statistical situation of new generation.
Step 10. Record the best chromosome and its associated fitness.
Step 11. Go to step 6, if the termination criteria are not met.

6.7 Layout Design

6.7.1 Machine Duplications

Layout design, which includes both intracellular and intercellular layout, is the next step after machines have been clustered into machine groups and parts in part families. This step often involves the duplication of some machines. Machine duplications can be classified into two categories: compulsory duplication and economic duplication. *Compulsory duplication* refers to those cases where the processing time carried out on a particular machine exceeds the availability of the machine. In those cases the number of machines to be duplicated can be easily calculated based on the required processing time and set-up time. *Economic duplication* of machines, on the other hand, aims to reduce intercellular movements of parts, resulting in a reduction in the total material handling cost.

In most cases of cell formation, it is not always possible to have perfectly independent cells. This is due to some exceptional parts that need one or more machines outside the cells to which they belong. The machines associated with these exceptional parts are bottleneck machines, which become the source of intercellular moves. To eliminate intercellular moves, one or more machines need to be duplicated in one or more appropriate cells. However, machine duplication is often expensive, and is justified only if the savings achieved by the reduction in intercell movements outweighs the initial expenditure for purchasing the extra machines. It is therefore necessary to correctly identify those machines whose duplications are feasible economic options. Finally, it should be noted that layout design could not be considered as a hierarchical process because the duplication of machines will change the number of machines in the system, and therefore will alter the layout design.

6.7.2 Methodology for Layout Design

The objective of the layout design can be mathematically expressed as follows:

$$\text{Minimize } (mh\_cost + dup\_cost) \qquad \text{Equation (6.10)}$$

where mh_cost = material handling cost
dup_cost = duplication cost

The iterative procedure for layout design and machine duplication process is as follows:

Step 1. Determine the machine groups and the part families.
Step 2. Calculate the number of machines of each type (compulsory duplication).
Step 3. For the current number of machines, optimize the machines' layout with respect to system handling costs, using the genetic algorithm described in Section 6.8.
Step 4. Evaluate the system cost (Sc) including material handling cost and machine duplication cost.
Step 5. Determine the bottleneck machines.
Step 6. Duplicate the bottleneck machine, which is required by the most number of machines.
Step 7. For the current number of machines, optimize the machines' layout with respect to system handling costs, using the genetic algorithm described in Section 6.8.
Step 8. Evaluate the system cost including material handling cost and machine duplication cost (Sc_{New}).

$$\text{Step 9.} \begin{cases} \text{if } Sc_{New} \text{ is less than } Sc_{Old} & \text{the bottleneck machine is duplicated} \\ \text{otherwise} & \text{the bottleneck machine cannot be duplicated} \end{cases}$$

Step 10. If there are any other bottleneck machines go to step 5.
Step 11. The layout is the final layout.

6.8 A Genetic Algorithm for Layout Optimization

This section presents a genetic algorithm for the optimization of layout design for a given cell composition.

6.8.1 The Chromosome Representation

In this algorithm, a factory site is divided into k grids (positions) where $k \geq m$ (m = number of machines). This enables the calculation of geometric distances between machines. In this representation, the length of the chromosome is equal to the number of machines. The location of a gene in the chromosome indicates the position of a machine on the grids, while the value of the gene indicates the machine. As a result, the chromosome consists of a set of integer string; each indicates a machine number and its position number. For example, Figure 6.6 shows a chromosome representation for nine machines and its corresponding positions on the grids. It is worth mentioning that, if the number of positions is more than the number of machines, some dummy machines are used to fill the chromosome. For example, if a site comprises of 16 positions and there are only 14 machines, then machines 15 and 16 are created as dummy machines.

6.8.2 The Crossover Operator

The AER crossover technique previously described is also used for this genetic algorithm. However, in this algorithm the string ordering has a two-dimensional aspect, and accordingly any gene's neighborhood is not restricted to one previous gene and one next gene. It can be seen from Figure 6.7 that the neighborhood of genes depends on their position in the site. The AER crossover operator can be set up as follows:

1. For each machine m, make an edge list including all other machines connected to machine m in at least one of the parent.
2. Select the first machine of parent p_1 to be the current machine m of the offspring.
3. Select the connected edges to the current machine m.
4. Define a probability distribution over connected edges based on their similarity.

Gene position	*1*	*2*	*3*	*4*	*5*	*6*	*7*	*8*	*9*
Gene value	9	3	4	2	7	5	8	6	1

9	3	4
2	7	5
8	6	1

FIGURE 6.6 Chromosome representation (top) and its corresponding interpretation (bottom).

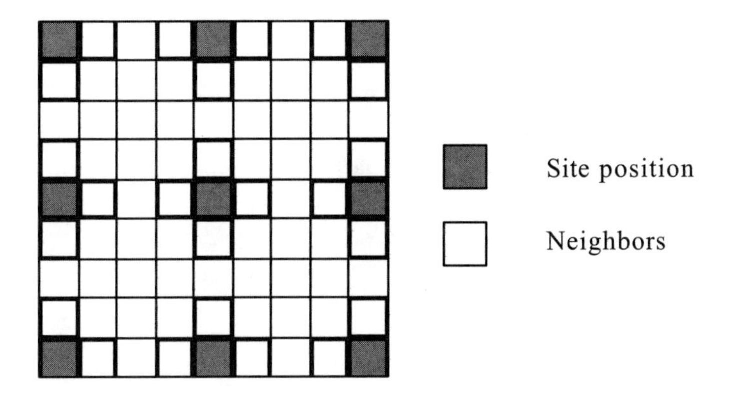

FIGURE 6.7 Genes neighborhood for two-dimensional problems.

5. Select a new machine m from the previous m edge list, according to roulette wheel probability distribution. If the previous m edge list has no member, selection is done at random from all nonselected machines.
6. Extract the current machine m from all edges list.
7. If the sequence is completed, stop, otherwise, go to step 3.

6.8.3 The Fitness Function

The transportation cost for a cellular manufacturing system can be defined as:

$$TTC = \sum_{n=1}^{nf} \sum_{i=1}^{pf_n} \sum_{l=1}^{M} \sum_{j=1}^{M} d_{l,j} \times CT_{l,j}^{i} \times YPV_i \times X_{l,j}^{i} \qquad \text{Equation (6.11)}$$

where CT_{lj}^{i} = the unit transportation cost for part i to be moved from machine l to machine j per unit of distance ($/m)

X_{lj}^{i} = is a safety factor in this genetic algorithm, for placing the machines, belong to the same cell, in adjacent positions

$$X_{lj}^{i} = \begin{cases} >1 & \text{if part } i \text{ needs machine } j \text{ after meeting machine } l \text{ in different cells} \\ 1 & \text{if part } i \text{ needs machine } j \text{ after meeting machine } l \text{ in the same cell} \\ 0 & \text{otherwise} \end{cases}$$

TTC = total transportation cost ($/year)
$d_{l,j}$ = the distance between machine j and machine l (m)
YPV_i = the production volume required for part i (unit/year)
pf_n = number of parts in n^{th} part family
nf = number of part families

The facility layout problem aims at minimizing the cost of transportation between the various machines in different cells. Since the fitness function in a genetic algorithm is a measure of goodness of the solution to the objective function, the following fitness function, which is an inverse of the TTC, is employed:

$$Fitness = TTC^{-\text{Log}\,(TTC)} \qquad \text{Equation (6.12)}$$

6.8.4 The Replacement Strategy

The replacement strategy for this algorithm is the same as the strategy used for machine grouping previously described.

6.8.5 The Algorithm

Step 1. Read input data, including the number of machines, parts, cells, number of machines and parts in each cell, the processing routes and production volume of each part.
Step 2. Enter GA parameters (including crossover and mutation probability, population size, number of generation, and seed number).
Step 3. Set the generation counter, $G_i = 1$.
Step 4. Initialize the first population at random.
Step 5. Report the situation of the first population, including maximum, minimum, and average fitness value of the population.
Step 6. Increment the generation counter, $G_i = G_i + 1$.
Step 7. While the mating pool is not full:
- A. Select two chromosomes using tournament or roulette-wheel strategy.
- B. Crossover the parent and generate corresponding offspring considering crossover probability.
- C. Apply the mutation to all genes of offspring considering mutation probability.
- D. Evaluate the fitness of the new generated chromosomes.

Step 8. Replace the best chromosome of the old population with the first and last newly generated chromosomes according to the replacement strategy.
Step 9. Report the statistical situation of the new generation.
Step 10. Record the best chromosome and its associated fitness.
Step 11. Go to step 6 if the termination criteria are not met.

6.9 A Case Study

This case study consists of 30 parts and 17 machines with the relevant production data shown in Table 6.5.

6.9.1 Finding the Best Alternative Routings

It can be seen from Table 6.5 that there are alternative routings for several parts. The first step toward a cellular manufacturing system design is to select a process plan for each part. This can be done using

TABLE 6.5 Production Data for a Case Study That Includes 30 Parts and 17 Machines

Part No.	P.V. x 100	Routing No.	Process Sequences					Part No.	P.V. x 100	Routing No.	Process Sequences					
1	110	1	11	6	13	15	4	14	125	1	4	11	6	13	12	10
		2	6	13	1	8				2	10	5	9	4	12	6
		3	14	7	3	15	2	15	100	1	12	10	5	9		
2	95	1	11	13	12	6	10			2	8	1	5	6	11	
		2	11	15	13	12				3	4	12	6	13	10	
3	100	1	4	15	12	6				4	10	5	9	4	3	
		2	17	1	9			16	105	1	7	3	2	15		
		3	6	13	1	8		17	95	1	14	7	3	15	2	11
4	95	1	11	6	13	12		18	105	1	1	5	6	11		
5	130	1	11	6	13	12	10			2	14	7	3	15	2	
		2	8	1	5	6	11	19	110	1	11	6	13	12	10	
6	130	1	11	6	13	4	12			2	8	1	5	6	11	
		2	11	15	13	12		20	100	1	11	12	5	7	8	
		3	14	7	13	15	2			2	17	1	9			
		4	11	15	5	6	13	21	100	1	11	6	13	12		
7	130	1	8	1	5	6				2	12	10	5	9		
		2	11	15	5	6		22	100	1	8	4	16	5		
		3	8	5	4					2	12	10	5	9		
8	80	1	16	8	5	4		23	90	1	17	1	9			
9	130	1	14	7	3	15	2			2	12	10	5	9		
		2	8	1	5	6	9	24	100	1	14	7	3	15	2	
10	105	1	8	1	5	6	11	25	90	1	17	1	8	9		
		2	11	6	13	4	12			2	11	15	13	12		
		3	11	15	13	12				3	1	5	6	11		
		4	11	6	13	12	10	26	100	1	10	5	9	13	12	
11	115	1	16	8	5	4				2	14	7	3	15	2	
		2	8	1	5	6	11	27	90	1	11	6	12	10		
12	120	1	11	12	5	7				2	5	9	13	12		
		2	17	1	9			28	90	1	11	6	13	12	10	
13	105	1	14	7	3	15	2	29	110	1	14	7	15	2		
		2	13	12	10	5	9			2	11	12	5	7	8	
								30	120	1	5	16	8	4		

TABLE 6.6 The Best Alternative Routings for Each Part

Part No.	1	2	3	4	5	6	7	8	9	10	11	12	13	14	15
Routing No.	3	1	2	1	1	3	3	1	1	4	1	2	1	1	3
Part No.	16	17	18	19	20	21	22	23	24	25	26	27	28	29	30
Routing No.	1	1	2	1	2	1	1	1	1	1	2	1	1	1	1

the genetic algorithm discussed previously. The GA simulation is shown in Figure 6.8, and the corresponding optimum routing for each part is shown in Table 6.6. The MCS matrix for the selected process routings is shown in Table 6.7.

6.9.2 Machine Grouping

The next step in the cell formation process is to group machines into a diagonal MCS matrix using genetic algorithm as previously discussed. The objective here is to maximize the MCS coefficients of adjacent machines. The simulation result for the genetic algorithm can be seen in Figure 6.9, and the corresponding diagonal MCS matrix is shown in Table 6.8.

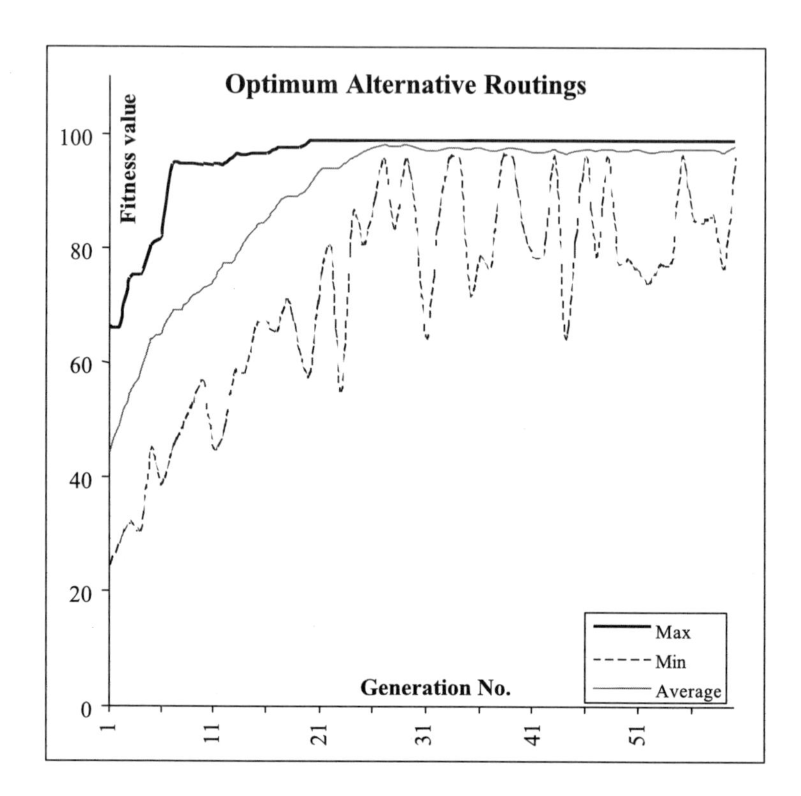

FIGURE 6.8 The GA simulation to find parts' optimum process routings.

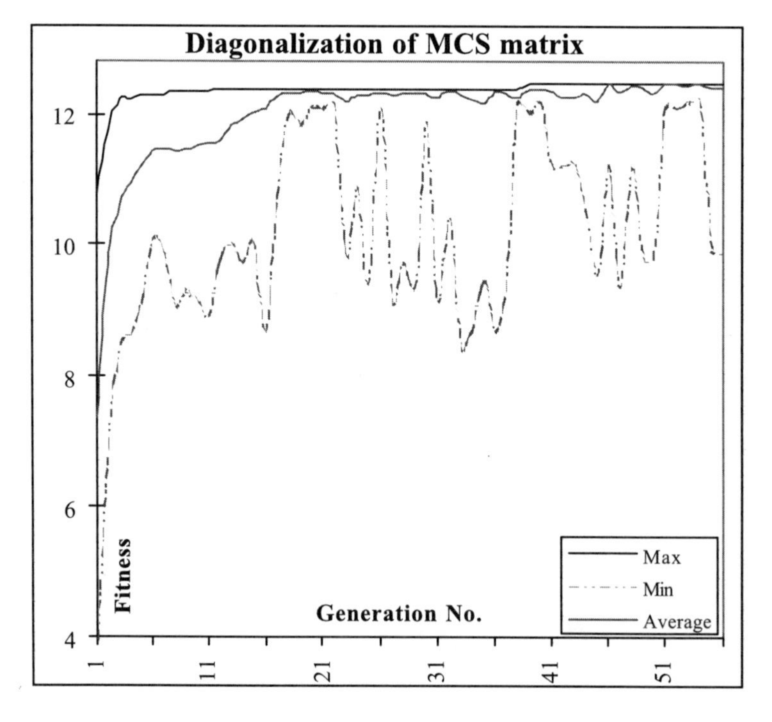

FIGURE 6.9 The GA simulation for machine grouping.

6.9.3 Clustering Parts into Part Families Using Parts' Similarity Coefficients

After the sequencing of machines as indicated in the diagonal MCS matrix, parts are then rearranged to produce maximum parts' similarity between any adjacent parts in the final machine–component matrix using the genetic algorithm. The GA simulation result is depicted in Figure 6.10, and the final machine–component matrix can be seen in Table 6.9. The results of the cell formation process are summarized in Table 6.10.

6.9.4 Layout Design

Figure 6.11 shows the factory site for positioning the 17 machines in this case study. The site has been divided into 24 positions to cater to compulsory and economic machine duplications. In this case, there is no need for compulsory machine duplication, and the layout of the 17 machines can be optimized using the genetic algorithm previously described. The resultant layout is shown in Figure 6.12. This result has been achieved after 155 generations using a population size of 120. The material handling cost for this layout is $12,765 per year.

The next step is to consider machine duplication in order to minimize the number of intercellular movements (economic machine duplication). Table 6.11 indicates the number of intercellular movements due to each bottleneck machine. It can be seen from this table that machine 13 creates the most number of intercellular movements. Therefore, this should be the first machine to be considered for duplication in cell 4. There is now a total of 18 machines (instead of 17 machines), and the layout for these 18 machines needs to be optimized again, using the same GA as in the case of 17 machines. Figure 6.13 shows the layout for 18 machines. The difference in material handling cost between this layout and the previous layout is $1845 per year, which is $580 more than duplication cost of machine 13. As a result, the duplication of machine 13 is justified. Next, machine 4 is selected to be duplicated in cell 3, resulting in a total of 19 machines. The genetic algorithm is run again, and the optimum layout for these 19 machines is shown in Figure 6.14. The difference in material handling cost between this layout and the previous layout (for 18 machines) is $956, which is $193 less than the duplication cost of machine 4. As a result, this duplication is not economically justified. For the same reason, machines 11 and 8 cannot be economically duplicated. Therefore, the final layout design is as indicated in Figure 6.13, with a total of 18 machines.

6.10 Conclusion

This chapter has presented an integrated methodology cellular manufacturing system design that covers both the cell formation process and cellular layout. The methodology is based on a new concept of similarity coefficients, and the use of genetic algorithms as an optimization tool. In comparison with previous works, this methodology takes into account in the design process all relevant production data such process routings and production volumes of parts. The issue of the interaction between layout procedure and machine duplication to justify economic machine duplications is also addressed.

The GA-based optimization tool used in this work is not only robust and able to handle large-scale problems, but also parallel in nature, and hence can reduce computational time significantly. Another advantage of the genetic algorithm is that it is independent of the objective function and the number constraints.

Defining Terms

Bottleneck machine: A machine that is required by parts from different part families.

Clustering procedure: A procedure for clustering machines into machine groups and parts into part families.

Exceptional element: A part that needs one or more machines from different cells in its process routing.

Intercell layout: Arrangement of machines for the whole cellular manufacturing system, which may include many machine cells.

Intracell layout: Arrangement of machines within a cell.

TABLE 6.7 The MCS Matrix for the Selected Process Routings

	Machines																
	1	2	3	4	5	6	7	8	9	10	11	12	13	14	15	16	17
1	1	0	0	0.004	0.004	0	0	0.016	0.073	0	0	0	0	0	0	0.004	0.073
2	0	1	0.194	0.01	0	0.015	0.244	0	0	0.015	0.036	0.015	0.038	0.225	0.244	0	0
3	0	0.194	1	0.004	0	0.009	0.194	0	0	0.009	0.03	0.009	0.032	0.189	0.194	0	0
4	0.004	0.01	0.004	1	0.094	0.057	0.01	0.094	0.004	0.057	0.053	0.057	0.057	0.029	0.01	0.076	0.004
5	0.004	0	0	0.094	1	0.01	0	0.094	0.004	0.01	0.006	0.01	0.01	0	0	0.076	0.004
6	0	0.015	0.009	0.057	0.01	1	0.015	0.01	0	0.199	0.212	0.235	0.218	0.036	0.015	0.01	0
7	0	0.244	0.194	0.01	0	0.015	1	0	0	0.015	0.036	0.015	0.038	0.225	0.244	0	0
8	0.016	0	0	0.094	0.094	0.01	0	1	0.016	0.01	0.006	0.01	0.01	0	0	0.076	0.016
9	0.073	0	0	0.004	0.004	0	0	0.016	1	0	0	0	0	0	0	0.004	0.073
10	0	0.015	0.009	0.057	0.01	0.199	0.015	0.01	0	1	0.19	0.199	0.195	0.36	0.015	0.01	0
11	0	0.036	0.03	0.053	0.006	0.212	0.036	0.006	0	0.19	1	0.212	0.225	0.058	0.036	0.006	0
12	0	0.015	0.009	0.057	0.01	0.235	0.015	0.01	0	0.199	0.212	1	0.218	0.036	0.015	0.01	0
13	0	0.038	0.032	0.057	0.01	0.218	0.038	0.01	0	0.195	0.225	0.218	1	0.06	0.038	0.01	0
14	0	0.225	0.189	0.029	0	0.036	0.225	0	0	0.036	0.058	0.036	0.06	1	0.225	0	0
15	0	0.244	0.194	0.01	0	0.015	0.244	0	0	0.015	0.036	0.015	0.038	0.225	1	0	0
16	0.004	0	0	0.076	0.076	0.01	0	0.076	0.004	0.01	0.006	0.01	0.01	0	0	1	0.004
17	0.073	0	0	0.004	0.004	0	0	0.016	0.073	0	0	0	0	0	0	0.004	1

TABLE 6.8 The Diagnonal MCS Matrix for Machine Grouping

	Machines																
	17	9	1	16	8	5	4	10	12	6	13	11	14	15	2	7	3
17	1	0.0727	0.0727	0.0041	0.0165	0.0041	0.0041										
9	0.0727	1	0.0727	0.0041	0.0165	0.0041	0.0041										
1	0.0727	0.0727	1	0.0041	0.0165	0.0041	0.0041										
16	0.0041	0.0041	0.0041	1	0.0759	0.0759	0.0759	0.0103	0.0103	0.0103	0.0103	0.0057					
8	0.0165	0.0165	0.0165	0.0759	1	0.0938	0.0938	0.0103	0.0103	0.0103	0.0103	0.0057					
5	0.0041	0.0041	0.0041	0.0759	0.0938	1	0.0938	0.0103	0.0103	0.0103	0.0103	0.0057					
4	0.0041	0.0041	0.0041	0.0759	0.0938	0.0938	1	0.0572	0.0572	0.0572	0.0572	0.0526	0.029	0.0103	0.0103	0.0103	0.0043
10				0.0103	0.0103	0.0103	0.0572	1	0.1992	0.1992	0.1951	0.1901	0.0364	0.0146	0.0146	0.0146	0.0087
12				0.0103	0.0103	0.0103	0.0572	0.1992	1	0.2349	0.2184	0.212	0.0364	0.0146	0.0146	0.0146	0.0087
6				0.0103	0.0103	0.0103	0.0572	0.1992	0.2349	1	0.2184	0.212	0.0364	0.0146	0.0146	0.0146	0.0087
13				0.0103	0.0103	0.0103	0.0572	0.1951	0.2184	0.2184	1	0.2253	0.0602	0.0384	0.0384	0.0384	0.0325
11				0.0057	0.0057	0.0057	0.0526	0.1901	0.212	0.212	0.2253	1	0.0581	0.0364	0.0364	0.0364	0.0304
14							0.029	0.0364	0.0364	0.0364	0.0602	0.0581	1	0.2246	0.2246	0.2246	0.1892
15							0.0103	0.0146	0.0146	0.0146	0.0384	0.0364	0.2246	1	0.2438	0.2438	0.194
2							0.0103	0.0146	0.0146	0.0146	0.0384	0.0364	0.2246	0.2438	1	0.2438	0.194
7							0.0103	0.0146	0.0146	0.0146	0.0384	0.0364	0.2246	0.2438	0.2438	1	0.194
3							0.0043	0.0087	0.0087	0.0087	0.0325	0.0304	0.1892	0.194	0.194	0.194	1

TABLE 6.9 The Final Machine–Component Matrix

	Parts																													
	23	3	12	20	25	8	7	11	30	22	27	28	4	21	10	19	5	14	15	2	29	6	1	9	13	18	24	26	17	16
17	90	100	120	100	90																									
9	90	100	120	100	90																									
1	180	200	240	200	180																									
16						80		115	240	200																				
8					180	160	130	230	240	100																				
5						160	260	230	120	100																				
4						80	130	115	120	200								125	100											
10											90	90			105	110	130	125	100	95										
12											180	180	95	100	210	220	260	250	200	190										
6											180	180	190	200	210	220	260	250	200	190										
13											180	180	190	200	210	220	260	250	200	190		260								
11											90	90	95	100	105	110	130	250		95									95	
14																				190	110	130	110	130	105	105	100	100	95	
15																					220	260	220	260	210	210	200	200	190	105
2																					110	130	110	130	105	105	100	100	190	210
7																					220	260	220	260	210	210	200	200	190	105
3																							220	260	210	210	200	200	190	210

TABLE 6.10 Machine Groups and Associated Part Families for the Case Study

Cell No.	Machines	Parts
1	14,2,15,7,3	29,13,24,9,1,26,18,17,16,6
2	11,6,12,13,10	21,4,27,2,10,5,28,14,15
3	16,8,5,4	7,22,11,30,8
4	17,1,9	25,20,12,23,3

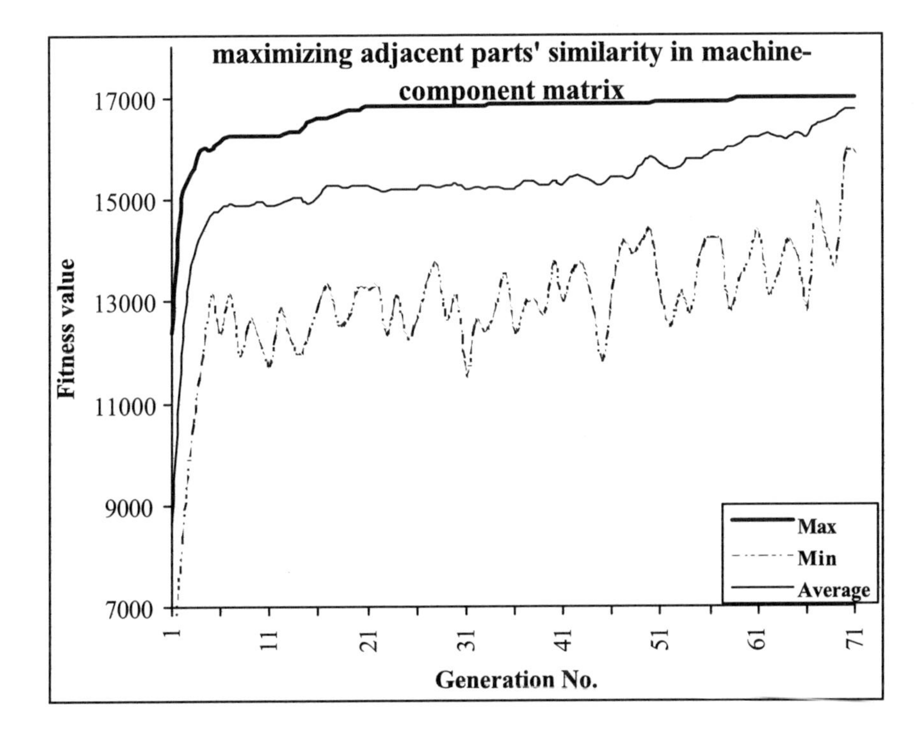

FIGURE 6.10 The GA simulation for maximizing adjacent parts' similarity in the machine–part matrix.

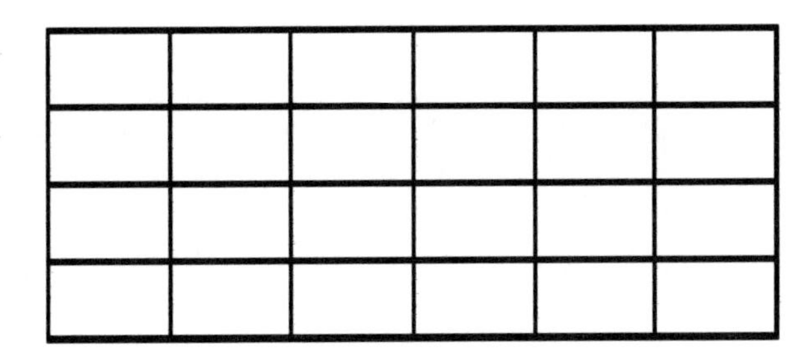

FIGURE 6.11 The site of a factory divided into 24 positions of equal size.

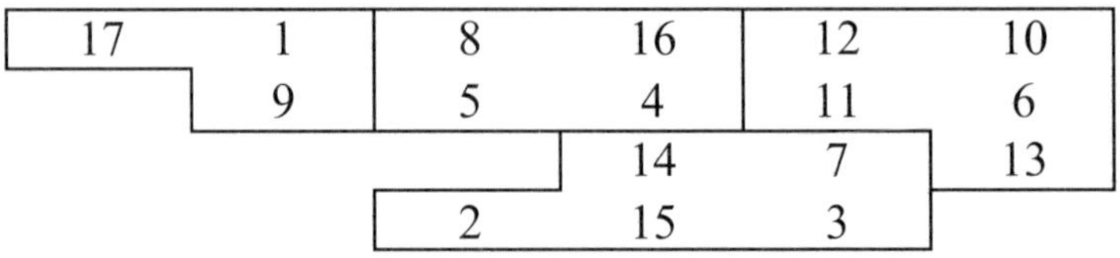

17 1 8 16 12 10
9 5 4 11 6
14 7 13
2 15 3

FIGURE 6.12 The first layout design with no machine duplication.

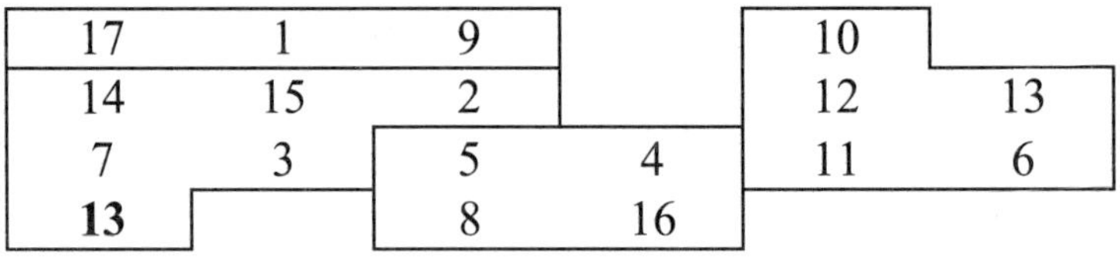

17 1 9 10
14 15 2 12 13
7 3 5 4 11 6
13 8 16

FIGURE 6.13 The layout for 18 machines, after the duplication of machine 13 in cell 4.

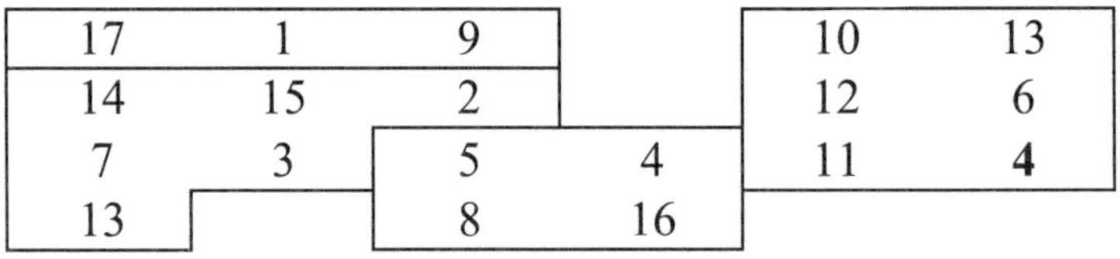

17 1 9 10 13
14 15 2 12 6
7 3 5 4 11 **4**
13 8 16

FIGURE 6.14 The new layout after duplication of machine 4 in cell 3, giving a total of 19 machines.

TABLE 6.11 Number of Intercellular Movements Created by the Bottleneck Machines

Machine no.	8	4	13	11
From cell no.	2	2	3	3
Duplicated in cell no.	1	3	4	4
For part no.	25	5,14	6	17
Number of intercellular movements	180	225	260	95

References

Adil, G.K., D. Rajamani, and D. Strong, 1996, Cell Formation Considering Alternate Routings, *International Journal of Production Research,* vol. 34, no. 5, pp. 1361-1380.

Grefenstette, J.J., 1987, Incorporating Problem-Specific Knowledge into Genetic Algorithms in *Genetic Algorithms and Simulated Annealing,* Davis, L. (Ed.), Morgan Kaufman.

Kazerooni, M., L.H.S. Luong, and K. Abhary, 1995a, Cell Formation Using Genetic Algorithms, *International Journal of Flexible Automation and Integrated Manufacturing,* vol. 3, pp. 219-235.

Kazerooni, M., L.H.S. Luong, and K. Abhary, 1995b, Machine Chain Similarity, A New Approach to Cell Formation, *Proceedings of the 11th International Conference on Computer Aided Production Engineering,* London, pp. 97-105.

Kusiak, A., W. Boe, and C.H. Cheng, 1991, Designing Cellular Manufacturing Systems: Branch-and-Bound Approaches, *IIE Transaction,* vol. 25, no. 4, pp. 46-56.

Kusiak, A. and C.H. Cheng, 1991, Group Technology: Analysis of Selected Models and Algorithms, *PED-Design, Analysis, and Control of Manufacturing Cells ASME,* vol. 53, pp. 99-114.

Lee, S., C. Zhang, and H.P. Wang, 1991, Fuzzy Set-Based Procedures for Machine Cell Formation, *PED-Design, Analysis, and Control of Manufacturing Cells ASME,* vol. 53, pp. 31-45.

Luong, L.H.S., 1993, A Cellular Similarity Coefficient Algorithm for the Design of Manufacturing Cells, *International Journal of Production Research,* vol. 31, no. 8, pp. 1757-1766.

McAuley, J., 1972, Machine Grouping for Efficient Production, *The Production Engineer,* vol. 51, no. 2, pp. 53-57.

Murthy, C.V.R. and G. Srinivasan, 1995, Fractional Cell Formation on Group Technology, *International Journal of Production Research,* vol. 33, no. 5, pp. 1323-1337.

Waghodekar, P.H. and S. Sahu, 1983, Group Technology: A Research Bibliography, *OPSEARCH,* vol. 20, no. 4, pp. 225-249.

Wemmerlov, U. and N.L. Hyer, 1986, Procedures for the Part Family/Machine Group Identification Problem in Cellular Manufacturing, *Journal of Operations Management,* vol. 6, no. 2, pp. 125-147.

Wemmerlov, U. and N.L. Hyer, 1987, Research Issues in Cellular Manufacturing, *International Journal of Production Research,* vol. 25, no. 3, pp. 413-431.

Whitley, D., T. Strakweather, and D.A. Fuquay, 1989, Scheduling Problems and Traveling Salesman: The Genetic Edge Recombination, *Proceedings of the Third International Conference on Genetic Algorithms,* Los Altos, CA, pp. 133-140.

For Further Information

A more complete discussion on genetic algorithms, including extensions and related topics, can be found in *The Handbook of Genetic Algorithms* by L. Davis, Van Nostrand Reinhold, New York, 1991 and *Genetic Algorithms + Data Structure = Evolution Programs* by Z. Michalewicz, Springer-Verlag, New York, 1994.

The *International Journal of Production Research* frequently publishes articles on advances in cellular manufacturing systems.

7

Intelligent Design Retrieving Systems Using Neural Networks

C. Alec Chang
University of Missouri – Columbia

Chieh-Yuan Tsai
Yuan-Ze University

7.1 Introduction

Design is a process of generating a description of a set of methods that satisfy all requirements. Generally speaking, a design process model consists of the following four major activities: analysis of a problem, conceptual design, embodiment design, and detailing design. Among these, the conceptual design stage is considered a higher level design phase, which requires more creativity, imagination, intuition, and knowledge than detail design stages. Conceptual design is also the phase where the most important decisions are made, and where engineering science, practical knowledge, production methods, and commercial aspects are brought together. During conceptual design, designers must be aware of the component structures, such as important geometric features and technical attributes that match a particular set of functions with new design tasks.

Several disciplines, such as variant design, analogical design, and case-based design, have been explored to computerize the procedure of conceptual design in CAD systems. These techniques follow similar problem-solving paradigms that support retrieval of an existing design specification for the purpose of adaptation. In order to identify similar existing designs, the development of an efficient design retrieval mechanism is of major concern. Design retrieval mechanisms may range from manual search to computerized identification systems based on tailored criteria such as targeted features. Once a similar design is identified, a number of techniques may be employed to adapt this design based upon current design goals and constraints. After adapting the retrieved design, a new but similar artifact can be created.

7.1.1 Information Retrieval Systems vs. Design Retrieving Systems

An information retrieval system is a system that is capable of storage, retrieval, and maintenance of information. Major problems have been found in employing traditional information retrieval methods for component design retrieval. First, these systems focus on the processing of textual sources. This type of design information would be hard to describe using traditional textual data.

Another major problem with using traditional information retrieving methods is the use of search algorithms such as Boolean logic. In a typical Boolean retrieval process, all matched items are returned,

0-8493-0592-6/01/$0.00+$.50

and all nonmatched documents are rejected. The component design process is an associative activity through which "designers retrieve previous designs with similar attributes in memory," not designs with identical features for a target component.

7.1.2 Group Technology-Based Indexing

Group technology (GT) related systems such as Optiz codes, MICLASS, DCLASS, KK-3, etc., and other tailored approaches are the most widely used indexing methods for components in industry. While these methods are suitable as a general search mechanism for an existing component in a database, they suffer critical drawbacks when they are used as retrieval indexes in the conceptual design task for new components.

Lately, several methods have been developed to fulfill the needs for component design such as indexing by skeleton, by material, by operation, or by manufacturing process. However, indexing numbers chosen for these design retrieving systems must be redefined again and again due to fixed GT codes for part description, and many similar reference designs are still missed. In the context of GT, items to be associated through similarity are not properly defined.

7.1.3 Other Design Indexing

Several researchers also experiment with image-bitmap-based indexing methods. Back-propagation neural networks have been used as an associative memory to search corresponding bitmaps for conceptual designs. Adaptive resonance theory (ART) networks are also explored for the creation of part families in design tasks (Kumara and Kamarthi, 1992). Other researchers also propose the use of neural networks with bitmaps for the retrieval of engineering designs (Smith et al., 1997). However, these approaches are not proper tools for conceptual design tasks because bitmaps are not available without a prototype design, and a prototype design is the result of a conceptual design. The limitations in hidden line representation as well as internal features also make them difficult to use in practice.

7.1.4 Feature-Based Modeling

A *part feature* is a parameter set that has specified meanings to manufacturing and design engineers. Using proper classification schemes, part features can represent form features, tolerance features, assembly features, functional features, or material features. Comprehensive reviews on feature-based modeling and feature recognition methods can be found in recent papers (Allada, 1995). There are important works related to feature mapping processes that transform initial feature models into a product model (Chen, 1989; Case et al., 1994; Lim et al., 1995; Perng and Chang, 1997; Lee and Kim, 1998; and Tseng, 1999).

7.2 Characteristics of Intelligent Design Retrieval

There is no doubt that design is one of the most interesting, complicated, and challenging problem-solving activities that human beings could ever encounter. Design is a highly knowledge-intensive area. Most of the practical problems we face in design are either too complex or too ill defined to analyze using conventional approaches. For the conceptual design stage of industrial components, we urgently need a higher level ability that maps processes from design requirements and constraints to solution spaces. Thus, an intelligent design retrieving system should have the characteristics detailed in the following subsections.

7.2.1 Retrieving "Similar" Designs Instead of Identical Designs

Most designers start the conceptual design process by referring to similar designs that have been developed in the past. Through the process of association to similar designs, designers selectively retrieve reference designs, defined as existing designs that have similar geometric features and technological attributes.

They then modify these referenced designs into a desired design. Designers also get inspiration from the relevant design information.

7.2.2 Determining the Extent of Reference Corresponding to Similarity Measures

Design tasks comprise a mixture of complicated synthesis and analysis activities that are not easily modeled in terms of clear mathematical functions. Defining a clear mathematical formula or algorithm to automate design processes could be impractical. Thus, methods that retrieve "the" design are not compatible with conceptual design tasks.

Moreover, features of a conceptual design can be scattered throughout many past designs. Normally designers would start to observe a few very similar designs, then expand the number of references until the usefulness of design references diminishes. An intelligent design retrieving system should be able to facilitate the ability to change the number of references during conceptual design processes.

7.2.3 Relating to Manufacturing Processes

An integrated system for CAD/CAPP/CAM includes modules of object indexing, database structure, design retrieving, graphic component, design formation, analysis and refinement, generation for process plan, and finally, process codes to be downloaded. Most computer-aided design (CAD) systems are concentrated on the integration of advanced geometric modeling tools and methods. These CAD systems are mainly for detailed design rather than conceptual design. Their linking with the next process planning stage is still difficult. An intelligent design retrieving system should aim toward a natural linking of the next process planning and manufacturing stages.

7.2.4 Conducting Retrieval Tasks with a Certain Degree of Incomplete Query Input

Currently, users are required to specify initial design requirements completely and consistently in the design process utilization CAD systems. During a conceptual design stage, designers usually do not know all required features. Thus a design retrieving system that relies on complete query input would not be practical. It is necessary to provide designers with a computer assistant design system that can operate like human association tasks, using incomplete queries to come up with creative solutions for the conceptual design tasks.

7.3 Structure of an Intelligent System

There have been some studies to facilitate design associative memory, such as case-based reasoning, artificial neural networks, and fuzzy set theory. As early as two decades ago, Minsky at MIT proposed the use of frame notion to associate knowledge, procedural routines, default contents, and structured clusters of facts. Researchers have indicated that stories and events can be represented in memory by their underlying thematic structures and then used for understanding new unfamiliar problems.

CASECAD is a design assistance system based on an integration of case-based reasoning (CBR) and computer-aided design (CAD) techniques (Maher and Balachandran, 1994). A hybrid intelligent design retrieval and packaging system is proposed utilizing fuzzy associative memory with back-propagation neural networks and adaptive resonance theory (Bahrami et al., 1995). Lin and Chang (1996) combine fuzzy set theory and back-propagation neural networks to deal with uncertainty in progressive die designs.

Many of these presented methods do not integrate associative memory with manufacturing feature-based methods. Others still use GT-based features as their indexing methods and suffer the drawbacks inherited from GT systems. These systems try to use a branching idea to fulfill the need for "similarity" queries. This approach is not flexible enough to meet the need in conceptual design tasks.

7.3.1 Associative Memory for Intelligent Design Retrieval

According to these recent experiences, the fuzzy ART neural network can be adopted as a design associative memory in our intelligent system. This associative memory is constructed by feeding all design cases from a database into fuzzy ART. After the memory has been built up, the query of a conceptual design is input for searching similar reference designs in an associative way. By adjusting the similarity parameter of a fuzzy ART, designers can retrieve reference designs with the desired similarity level. Through the process of computerized design associated memory, designers can selectively retrieve qualified designs from an immense number of existing designs.

7.3.2 Design Representation and Indexing

Using a DSG or CSG indexing scheme, a raw material with minimum covered dimension conducts addition or subtraction Boolean operations with necessary form features from the feature library ψ. Based on either indexing scheme, design case d_k can be represented into a vector format $\vec{d}_k$ in terms of form features from ψ. Accordingly, this indexing procedure can be described as

$$d_k \mapsto \vec{d}_k \equiv [\pi_F(k,1),\ldots,\pi_F(k,i),\ldots,\pi_F(k,M)] \quad \text{Equation (7.1)}$$

where $\pi(k,i) \in [0,1]$ is a membership measurement associated with the appearance frequency of form feature $i \in \psi$ in design case k and M is the total number of form features.

After following the similar indexing procedure, all design cases in vector formats are stored in a design database A:

$$A=\{d_1,\ldots,d_k,\ldots,d_N\} \quad \text{Equation (7.2)}$$

where N is the total number of design cases.

The query construction procedure can be represented as

$$q \mapsto \vec{q} \equiv [\pi_F(c,1),\ldots,\pi_F(c,i),\ldots,\pi_F(c,M)] \quad \text{Equation (7.3)}$$

where $\pi_F(c,i) \in [0,1]$ is a membership measurement defined in Equation 7.1 for conceptual design c.

7.3.3 Using a Fuzzy ART Neural Network as Design Associative Memory

Introduced as a theory of human cognitive information processing, fuzzy art incorporates computations from fuzzy set theory into the adaptive resonance theory (ART) based models (Carpenter et al. 1991; Venugopal and Narendran, 1992). The ART model is a class of unsupervised as well as adaptive neural networks. In response to both analog and binary input patterns, fuzzy ART incorporates an important feature of ART models, such as the pattern matching between bottom-up input and top-down learned prototype vectors. This matching process leads either to a resonant state that focuses attention and triggers stable prototype learning or to a self-regulating parallel memory search. This makes the performance of fuzzy ART superior to other clustering methods, especially when industry-size problems are applied (Bahrami and Dagli, 1993; Burke and Kamal, 1992).

Mathematically, we can view a feature library as a universe of discourse. Let R be a binary fuzzy relation in $\psi \times \psi$ if

$$R=\{(x,y),\pi_R(x,y)|(x,y) \in \psi \times \psi\} \quad \text{Equation (7.4)}$$

where $\pi_R(x,y) \in [0,1]$ is the membership function for the set R.

7.4 Performing Fuzzy Association

After the fuzzy feature relation has been defined, a feature association function is activated for a query vector and design vectors (Garza and Maher, 1996; Liao and Lee, 1994). To combine the fuzzy feature relation into vectors, operating a composition operation to them is necessary. Through max–min composition, a new query vector and design vectors contain not only feature-appearing frequency but also associative feature information. Specifically, proposed fuzzy feature association procedure, ***FFA***, can be defined as

$$\underline{FFA}\left[\vec{a},R\right]\rightarrow\vec{a}^{\,new}. \qquad \text{Equation (7.5)}$$

where $\vec{a}$ is the vector, R is the fuzzy feature relation, and $\vec{a}^{\,new}$ is the modified vector containing association information.

By implementing max–min composition, the ***FFA***[] can be accomplished as

$$\vec{a}^{\,new}\equiv[\pi_{F\circ R}(k,1),\ldots,\pi_{F\circ R}(k,y),\ldots,\pi_{F\circ R}(k,M)] \qquad \text{Equation (7.6)}$$

where $\pi_{F\circ R}(k,y)=\vee_x[\pi_F(k,x)\wedge\pi_R(x,y)]=\max\text{-}\min_{x\in\psi}[\pi_F(k,x),\pi_R(x,y)]$. Therefore, fuzzy feature association for design vectors and query vector can be conducted as

$$\underline{FFA}\left[\vec{d}_k,R\right]\rightarrow\vec{d}_k^{\,new}. \qquad \underline{\forall k\in A} \qquad \text{Equation (7.7)}$$

Fuzzy ART cluster vectors are based on two separate distance criteria, choice function and match function. To categorize input patterns, the output node j receives input pattern $\mathbf{I}$ in the form of a *choice function*, T_j, which is defined as $Tj(\mathrm{I}) = |\ |\mathbf{I}\wedge\mathbf{w}_j|/(\alpha+|\mathbf{w}_i|)$ where $\mathbf{w}_j$ is an analog-valued weight vector associated with cluster j and is a choice parameter that is suggested to be close to zero. The fuzzy AND operator $\wedge$ is defined by $(\mathbf{p}\wedge\mathbf{q})_i\equiv\min(p_i,q_i)$ and where the norm $|\cdot|$ is defined by $|\mathbf{p}|\equiv\sum_i|p_i|$. The system makes a *category choice* when at most one node can become active at a given time. The output node, J, with the highest value of T_j is the candidate to claim the current input pattern. For node J to code the pattern, the *match function* should exceed the vigilance parameter. That is, $|\mathbf{I}\wedge\mathbf{w}_j|/|\mathbf{I}|\geq\rho$, where the vigilance parameter ρ is a constant, $0\leq\rho\leq1$. Once the search ends, the weight vector, $\mathbf{w}_J$, of the winning node J learns according to the equation

$$\mathbf{w}_J^{(\mathrm{new})}=\beta\left(\mathbf{I}\wedge\mathbf{w}_J^{(old)}\right)+\left(1-\beta\right)\mathbf{w}_J^{(\mathrm{old})} \qquad \text{Equation (7.8)}$$

To perform associative searching, designers specify a desired similarity parameter ρ and sequentially feed the design vectors evaluated from Equation 7.1 into fuzzy ART to construct the geometric associative memory of achieve design cases. By varying the similarity parameter from 0 to 1, the similarity level of design cases in fuzzy ART can be adjusted.

7.5 Implementation Example

A database of industrial parts is used in this chapter to demonstrate the proposed system (Figure 7.1). There are 35 form features defined for this database, as shown in Table 7.1.

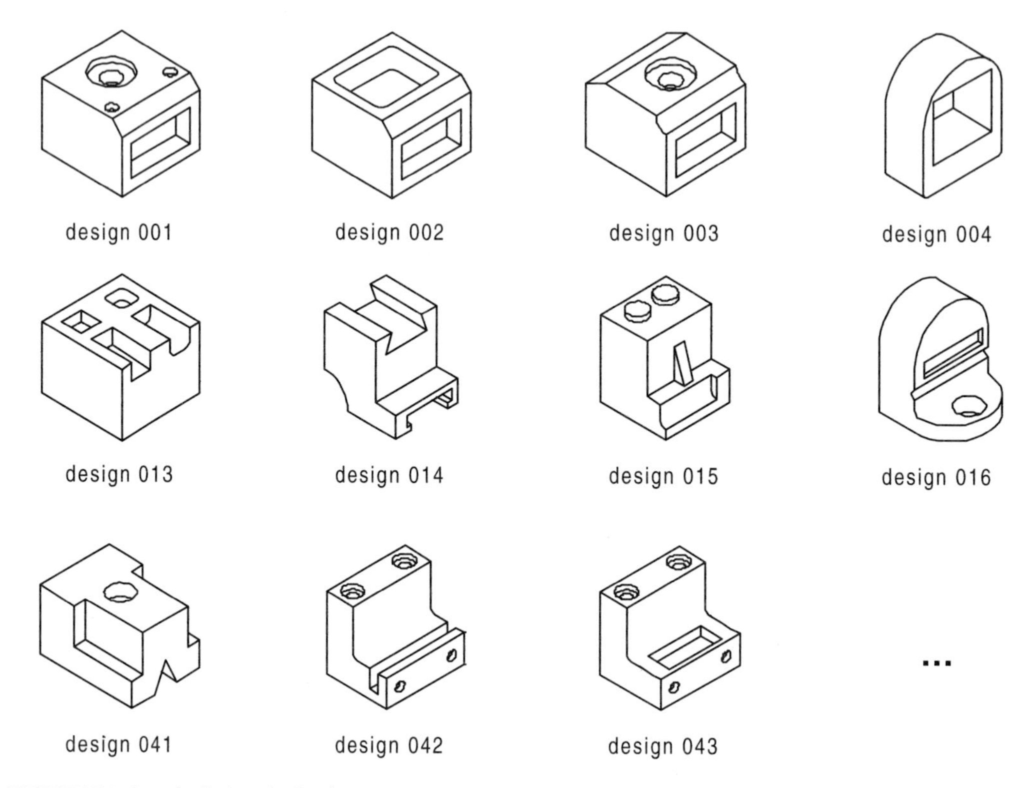

FIGURE 7.1 Sample designs in database.

TABLE 7.1 Sample Features for Prismatic Parts

	Name		Name
1	Hole thru	19	Slot blind cylinder
2	Hole blind flat bottomed	20	T slot
3	Hole blind conic bottomed	21	Dove tail slot
4	Bore thru	22	Step thru
5	Bore blind flat	23	Step blind
6	Bore blind conic	24	Step thru round

7.5.1 Constructing Geometric Memory

Using the DSG coding scheme and the predefined feature library, these designs are coded in the design coding module first (Chang and Tsai, 1997; Chen, 1989). After completing the coding process, a set of normalized arrays based on the largest number of same features is obtained and stored in the existing part database, as shown in Table 7.2.

TABLE 7.2 Sample Arrays with Normalized Feature Codes for Current Designs

Feature		Design												
		2	3	4	5	6	7	8	9	10	11	12	13	14	1
1		0	0	0	0	0	0	0	0	0	0	0	0	0	
2	0.	0	0	0	0	0	0	0	0.25	0	0	0	0	0	
3		0	0	0	0	0	0	0	0	0	0	0	0	0	
4		0	0	0	0	0.5	0	0	0	0	0	0	0	0	
5	0.2	0	0.25	0	0	0	0	0	0	0	0	0	0	0	
6		0	0	0	0	0	0	0	0	0	0	0	0	0	

7.5.2 Generating a Design Description Vector

Conceptual design 48 shown in Figure 7.2 is provided as an implementation example for this proposed system. The feature, HOLE-THRU, is selected first from the feature library. This design has four through holes; thus the first number in the input feature array is a "4." Then, as there are two blind holes with flat bottom, a "2" is registered as the second number of the input feature array, and so forth. The complete array can be shown as A = {4, 2, 2, 3, 2, 2, 3, 2, 2, 0, 0, 0, 0, 0, 0, 0, 0, 0, 0, 0, 0, 0, 0, 0, 0, 0, 0, 0, 0, 1, 1, 1, 0, 0, 0, 0}.

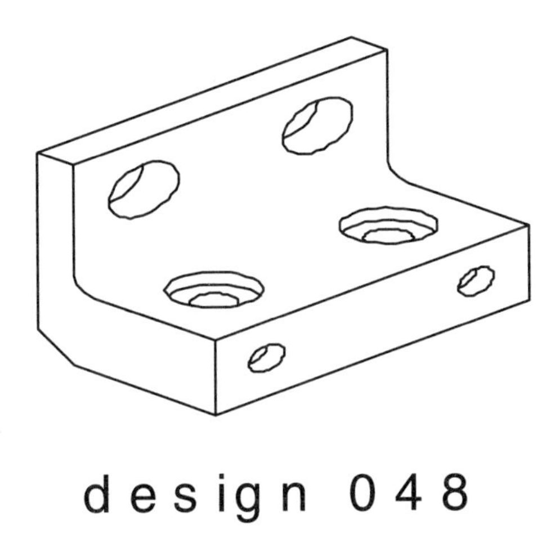

FIGURE 7.2 A conceptual design.

After being normalized by the largest number of this input array, which is "4," the input array for the system is I = {1, 0.5, 0.5, 0.75, 0.5, 0.5, 0.75, 0.5, 0.5, 0, 0, 0, 0, 0, 0, 0, 0, 0, 0, 0, 0, 0, 0, 0, 0, 0, 0, 0, 0, 0.25, 0.25, 0.25, 0, 0, 0, 0}.

7.5.3 Retrieving Reference Designs

One of the main advantages of the proposed design retrieving system is that users can easily control the number of retrieved designs. By adjusting the vigilance parameter, the user can get the proper number of similar designs. Designs retrieved at similarity parameter = 0.5 are 31, 37, 38, 43, 44, and 45 as shown in Figure 7.3.

7.5.4 Similarity Parameter to Control the Similarity of References

To perform associative searching, designers specify a desired similarity parameter and sequentially feed the design vectors evaluated from Equation 7.1 into fuzzy ART to construct the geometric associative memory of achieve design cases. Fuzzy ART automatically clusters design cases into design categories, based on their geometric likeness. When the query vector depicted in Equation 7.7 is input into the fuzzy ART, design cases are claimed as "retrieved" if they are in the design category having the highest similarity to the query. This searching task is expressed as

$$\underline{\mathbf{\mathit{FFA}}}\,[\{\vec{d}_k^{\,new} \mid k \in A\}, \vec{q}^{\,new}, \rho] \rightarrow \{k \mid k \in B\} \qquad \text{Equation (7.9)}$$

where $\underline{\mathbf{\mathit{FFA}}}$[ ] is a fuzzy ART neural network, $\rho = [0,1]$ is the similarity parameter of $\underline{\mathbf{\mathit{FFA}}}$, and B is the set of reference design retrieved from A.

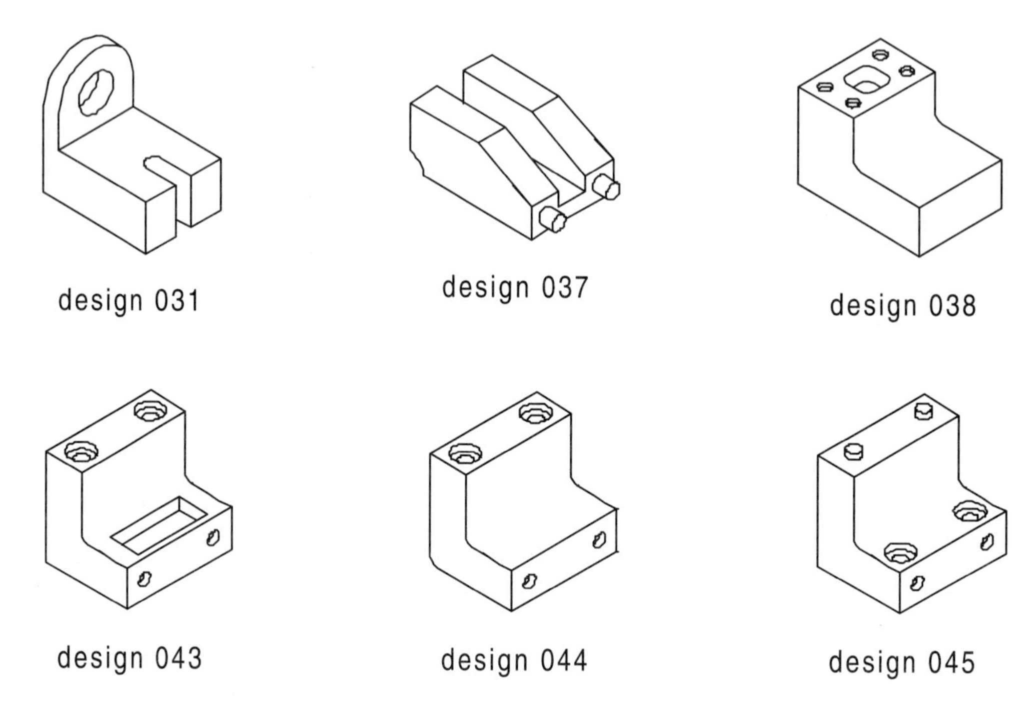

FIGURE 7.3 Retrieve reference designs for the conceptual design 048 using similarity = 0.5.

By varying the similarity parameter from 0 to 1, the similarity level of design cases in fuzzy ART can be adjusted. When adapting a higher value of similarity parameter, designers tend to retrieve fewer designs, but the designs received have higher similarity. When adapting a lower value of similarity parameter, designers often receive a longer list of designs, but with lower similarity.

After receiving similar designs, designers decide whether retrieved reference designs are suitable. If they are, designers can adapt and modify these designs into a new design. Otherwise, they can request the fuzzy ART ***FFA***[ ] again by using a different similarity parameter until satisfactory designs are retrieved.

7.5.5 Testing Robustness

When the user wants to retrieve a design that exists in the existing parts database, a well-designed retrieval system should have the ability to quickly retrieve that existing design. A new design, which is identical to Design 13 but not known beforehand, is fed into the neural-based retrieval module. The experiment result shows that no matter how the vigilance parameter is changed, from 0.001 to 0.999, the user will always receive Design 13 as a design candidate.

A designer may not always remember all the details of a design. Some of the information may be missed or neglected. Therefore, a design retrieving system should be able to retrieve a design based on slightly incomplete coding. Experiments also show that even with some noisy or lost information imbedded, a similar or exact design can still be retrieved.

7.5.6 Comparison with GT-Based Methods

The GT-based indexing approach considers the geometric and technological information of a design. However, because the procedure of coding and classification is completed simultaneously, users are not allowed to change the number of retrieved designs. That is, whenever a design is assigned a unique code according to the geometric and technological rules, the classification is also completed. This makes the number and similarity of retrieved designs unchangeable. Also, inaccurate and incomplete queries are not allowed in GT-based methods.

In the proposed method, the tasks described are solved separately, while in GT-based methods they are all merged together. The separated procedures provide the ability to change the similarity and number of retrieved designs. Also, the proposed associative models can relieve the problem of incomplete and inaccurate query/input.

7.5.7 Comparison with Hopfield Neural Network Approach

In comparison to the work of Venugopal and Narendran (1992), who use a Hopfield neural network to conduct design retrieval, the proposed system provides users more flexibility in choosing retrieved reference designs. The major disadvantage of their design retrieving system is that only one design case can be retrieved at a time, due to a mathematical property of the Hopfield neural network. In many practical situations, however, users want to receive several reference designs instead of only one. In the proposed system, users simply adjust a similarity parameter, and a list of reference designs with the desired similarity will be received. Thus, users have more flexibility when using the proposed system.

7.5.8 Comparison with Adaptive Resonance Theory (ART) Systems

In comparison to three published research works that utilize adaptive resonance theory (ART1) for design retrieval, the proposed method shows better results. Bitmap images of engineering designs can be adapted in the research to represent a design. One major disadvantage of using image-based indexing is that the disappearance of hidden features and internal lines is inevitable. Also, constructing an image-based query may be very cumbersome and time consuming. Liao and Lee (1994) utilize a feature-based indexing system for GT grouping and classification. In their research, only appearance or disappearance of form features is considered. However, ignoring the appearance frequency of a specific form feature could dramatically reduce the capability to discriminate between retrieved designs, especially as the design database grows. Using the proposed fuzzy ART neural network, the system is capable of dealing with the appearance frequency of a specific form feature, while keeping the advantage of adaptive resonance theory.

Acknowledgments

This work is partially supported by National Science Foundation Grant No. DMI-9900224 and National Science Council 89-2213-E-343-002.

Defining Terms

Conceptual design: M. J. French presents a four-stage model for engineering design process: analysis of problem, conceptual design, embodiment of schemes, and detailing for working drawings. In the conceptual design stage, designers generate broad solutions in the form of schemes that solve specified problems. This phase makes the greatest demands for engineering science and all related knowledge.

Bitmap: A bitmap file is an image data file that generally encodes a gray-level or color image using up to 24 bits per pixel.

Group technology (for industrial parts): An approach that groups parts by geometric design attributes and manufacturing attributes. Groups of parts are then coded with a predetermined numbering system, such as Optiz codes or MICLASS codes, etc.

References

Allada, V. and S. Anand, (1995) Feature-based modelling approaches for integrated manufacturing: state-of-the-art survey and future research directions, *International Journal of Computer Integrated Manufacturing*, 8(6):411-440.

Bahrami, A. and C.H. Dagli, (1993) From fuzzy input requirements to crisp design, *International Journal of Advanced Manufacturing Technology*, 8:52-60.

Burke, L. and S. Kamal, (1995) Neural networks and the part family/machine group formation problem in cellular manufacturing: a framework using fuzzy ART, *Journal of Manufacturing Systems*, 14(3):148-159.

Carpenter, G.A., S. Grossberg, and D.B. Rosen, (1991) Fuzzy ART: fast stable learning and categorization of analog patterns by an adaptive resonance system, *Neural Networks*, 4(6):759-771.

Case, K., J.X. Gao, and N.N.Z. Gindy, (1994) The implementation of a feature-based component representation for CAD/CAM integration, *Journal of Engineering Manufacture*, 208(B1):71-80.

Chang, C.A. and C.-Y. Tsai, (1997) Using ART1 neural networks with destructive solid geometry for design retrieval systems, *Computers in Industry*, 34(1):27-41.

Chen, C.S. (1989) A form feature oriented coding scheme, *Computers and Industrial Engineering*, 17(1-4):227-233.

Garza, A.G. and M.L. Maher, (1996) Design by interactive exploration using memory-based techniques, *Knowledge-Based Systems*, 9:151-161.

Kumara, S.R.T. and S.V. Kamarthi, (1992) Application of adaptive resonance networks for conceptual design, *Annals of the CIRP*, 41(1):213-216.

Lee J.Y. and K. Kim, (1998) A feature based approach to extracting machining features, *Computer-Aided Design*, 30(13):1019-1035.

Liao, T.W. and K.S. Lee, (1994) Integration of a feature-based CAD system and an ART1 neural network model for GT coding and part family forming, *Computers and Industrial Engineering*, 26(1):93-104.

Lim, S.S., et al., (1995) Multiple domain feature mapping: a methodology based on deep models of features, *Journal of Intelligent Manufacturing*, 6(4):245-262.

Lin, Z.-C. and H. Chang, (1996) Application of fuzzy set theory and back-propagation neural networks in progressive die design, *Journal of Manufacturing Systems*, 15(4):268-281.

Maher, M.L. and M.B. Balachandran, (1994) A multimedia approach to case-based structural design, *ASCE Journal of Computing in Civil Engineering*, 8(3):137-150.

Perng, D.-B. and C.-F. Chang, (1997) A new feature-based design system with dynamic editing, *Computers and Industrial Engineering*, 32(2):383-397.

Smith, S.D.G., et al., (1997) A deployed engineering design retrieval system using neural networks, *IEEE Transactions on Neural Networks*, 8(4):847-851.

Tseng, Y.-J., (1999) A modular modeling approach by integrating feature recognition and feature-based design, *Computers in Industry*, 39(2):113-125.

Venugopal, V. and T.T. Narendran, (1992) Neural network model for design retrieval in manufacturing systems, *Computers in Industry*, 20(1):11-23.

For Further Information

A good introduction to engineering design is presented in *Engineering Design Methods: Strategies for Product Design* by Nigel Cross. A recent survey about part/machine classification and coding can be found in Part B of *Group Technology and Cellular Manufacturing: State-of-the-Art Synthesis of Research and Practice*, edited by Nallan C. Suresh and John M. Kay. *Introduction to Artificial Neural Systems* by Jacek M. Zurada is a thorough, easy-to-read introductory text with plenty of numerical examples. Detlef Nauck, Frank Klawonn, and Rudolf Kurse present a more recent introduction to fuzzy neural systems in *Foundations of Neuro-Fuzzy Systems.* Several chapters in Part II of *Associative Neural Memories: Theory and Implementation*, edited by Mohamad H. Hassoun, present essential discussion about artificial associative neural memory models. To track progress in related areas, readers should refer to future publications of technical journals cited in references.

III

Process Planning and Scheduling

8

Soft Computing for Optimal Planning and Sequencing of Parallel Machining Operations

Yuan-Shin Lee*
North Carolina State University

Nan-Chieh Chiu
North Carolina State University

Shu-Cherng Fang
North Carolina State University

Abstract

Parallel machines (mill-turn machining centers) provide a powerful and efficient machining alternative to the traditional sequential machining process. The underutilization of parallel machines due to their operational complexity has raised interests in developing efficient methodologies for sequencing the parallel machining operations. This chapter presents a mixed integer programming model for the problems. Both the genetic algorithms and tabu search methods are used to find an optimal solution. Testing problems are randomly generated and computational results are reported for comparison purposes.

8.1 Introduction

Process planning transforms design specifications into manufacturing processes, and computer-aided process planning (CAPP) uses computers to automate the tasks of process planning. The recent introduction of parallel machines (mill-turn machining centers) can greatly reduce the total machining cycle time required by the conventional sequential machining centers in manufacturing a large batch of mill-turn parts [13, 14]. In this chapter, we consider the CAPP for this new machine tool.

*Dr. Lee's work was partially supported by the National Science Foundation (NSF) CAREER Award (DMI-9702374). E-mail: yslee@cos.ncsu.edu

0-8493-0592-6/01/$0.00+$.50

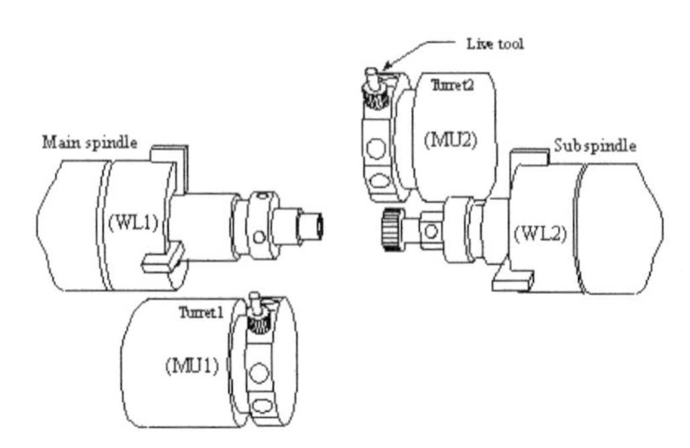

FIGURE 8.1 An example of a parallel machine equipped with two turrets (MUs) and two spindles (WLs). (From Lee, Y.-S. and Chiou, C.-J., *Computers in Industry*, vol. 39, 1999. With permission.)

One characterization of parallel machines is based on the location of the cutting tools and workpiece. As shown in Figure 8.1, a typical parallel machine is equipped with a main spindle, a subspindle (or work locations), and two or more turrets (or machining units), each containing several cutting tools. For a given workpiece to be machined on parallel machines, the output of the CAPP generates a set of precedent operations needed for each particular workpiece to be completed. A major issue to be resolved is the sequencing of these precedent operations.

The objective is to find a feasible operation sequence with an associated parallel machining schedule to minimize the total machining cycle time. Because of the relatively new trend of applying parallel machines in industrial manufacturing, only a handful of papers are found on sequencing machining operations for parallel machines [3, 22]. The combinatorial nature of sequencing and the complication of having precedence constraints make the problem difficult to solve.

A definition of such parallel machines can be found in [11, 22]:

Definition 1 (Workholding Location (WL)): WL refers to a workholding location on a machine tool.

Definition 2 (Machining Unit (MU)): MU refers to a toolholding location on a machine tool.

Definition 3 (Parallel Machine $P(I, L)$): $P(I, L)$ is a machine tool with $I(>1)$ MUs and $L(\geq 1)$ WLs with the capability of activating i cutting tools ($I \geq i \geq 1$) on distinct MUs, in parallel, either for the purpose of machining a single workpiece, or for the purpose of machining, in parallel, l workpieces ($L \geq l > 1$) being held on distinct WLs.

The necessary and sufficient condition for a machine tool to be parallel is $I > 1$. However, for a parallel machine to perform machining in sequential operations, we can simply set $i = 1$ and $l = 1$.

A mixed integer programming model will be introduced in Section 8.2 to model the process of parallel machining. Such a model, with only five operations, can easily result in a problem with 300 variables and 470 constraints. This clearly indicates that sequencing the parallel machining operations by using conventional integer programming method could be computationally expensive and inefficient [4]. An alternative approach is to apply random search heuristics. To determine an optimal operation sequence, Veeramani and Stinnes employed a tabu search method in computer-aided process planning [19]. Shan et al. [16] applied Hopfield neural networks to sequencing machining operations with partial orders. Yip-Hoi and Dutta [22] explored the use of genetic algorithms searching for the optimal operation sequences. Usher and Bowden [20] proposed a coding strategy that took into account a general scenario of having multiple parents in precedence relations among operations. Other reported searching strategies can also be found in Usher and Bowden [20].

This chapter is organized as follows. In Section 8.2, a mixed integer program for parallel operation sequencing is presented. In Section 8.3, a genetic-based algorithm for sequencing parallel machining operations with precedence constraints is proposed. A new crossover operator and a new mutation operator designed for solving the order-based sequencing problem are included. Section 8.4 presents a tabu search procedure to solve the operations sequencing problem for parallel machines. Sections 8.5 and 8.6 detail the computational experiments on using both the proposed genetic algorithm and tabu search procedure. To compare the quality of solutions obtained by the two methods, a random problem generator is introduced and further testing results are reported in Section 8.7. Concluding remarks are given in Section 8.8.

8.2 A Mixed Integer Program

The problem of sequencing parallel machining operations was originated from the manufacturing practice of using parallel machines, and so far there is no formal mathematical model for it. In this section, we propose a mixed integer program to model the process of sequencing parallel operations on parallel machines.

The proposed mixed integer program seeks the minimum cycle time (completion time) of the corresponding operation sequence for a given workpiece. The model is formulated under the assumptions that each live tool is equipped with only one spindle and the automated tool change time is negligibly small. Consider a general parallel machine with I MUs and L WLs. The completion of a workpiece requires a sequence of J operations which follows a prescribed precedence relation. Let K denote the number of time slots needed to complete the job. Under the parallel setting, $K \leq J$, because some time slots may have two operations performed in parallel. In case $I = L$, the process planning of a parallel machine with I MUs and L WLs can be formulated as a mixed integer program. The decision variables for the model are defined as follows:

$$x_{ijl}^{k} = \begin{cases} 1 & \text{if operation } j \text{ is performed by } MU_i \text{ on } WL_l \text{ in the } k\text{th time slot,} \\ 0 & \text{if not applicable,} \end{cases}$$

$$a_{ij} = \begin{cases} \text{processing time of operation } j \text{ performed by } MU_i, \\ +\infty \quad \text{if not applicable,} \end{cases}$$

s_{ijl}^{k} = starting time of operation j performed by MU_i on WL_l in the kth time slot. Define $s_{ijl}^{k} = 0$, if $k=1$; $s_{ijl}^{k} = +\infty$ for infeasible i, j, k, l; and $s_{ijl}^{k} = +\infty$ if $\sum_i x_{ijl}^{k} = 0$ for all j, k, l, i.e., for any particular operation j on WL_l in the kth time slot, if no MU is available then the starting time is set to be $+\infty$,

f_{ijl}^{k} = completion time of operation j performed by MU_i on WL_l in the kth time slot and define $f_{ijl}^{k} = +\infty$ for infeasible i, j, k, l.

For example, let 1–3–2–6–4–7–8–5 be a feasible solution of a sequence of eight operations required for the completion of a workpiece. Then x_{261}^{4} indicates that the fourth time slot (or the fourth operation being carried out) in the feasible solution was performed by applying MU_2 and WL_1 on operation 6. Denote $\delta(\cdot)$ as the Dirac delta function. For any particular operation j, with its corresponding starting time s_{ijl}^{k} and completion time f_{ijl}^{k}, no other operation $j', j' \neq j$, at any other time slot $k', k' \neq k$, can be scheduled between $[s_{ijl}^{k}, f_{ijl}^{k}]$, i.e., either $s_{ijl}^{k} \geq f_{ij'l}^{k'}$ or $f_{ijl}^{k} \leq s_{ij'l}^{k''}$, for $j' \neq j$, and $k' < k$ or $k < k''$. Thus, for a feasible schedule, the following conditions are required:

$$\delta(s_{ijl}^{k} - f_{ij'l}^{k'}) = \begin{cases} 1 & \text{if } s_{ijl}^{k} - f_{ij'l}^{k'} \geq 0, \\ 0 & \text{if } s_{ijl}^{k} - f_{ij'l}^{k'} < 0, \end{cases}$$

$$\delta(s_{ij'l}^{k''} - f_{ijl}^{k}) = \begin{cases} 1 & \text{if } s_{ij'l}^{k''} - f_{ijl}^{k} \geq 0, \\ 0 & \text{if } s_{ij'l}^{k''} - f_{ijl}^{k} < 0. \end{cases}$$

With the above definitions, a mixed integer program for sequencing parallel operations is formulated as

$$\min \alpha, \qquad \text{Equation (8.1)}$$

$$f_{ijl}^{K} \leq \alpha, \; i = 1,\ldots, I, \; j = 1,\ldots, J, \; l = 1,\ldots, L, \qquad \text{Equation (8.2)}$$

$$\sum_{j=1}^{J}\sum_{l=1}^{L} x_{ijl}^{k} \leq 1, i = 1,\ldots, I, k = 1,\ldots, K, \quad \text{Equation (8.3)}$$

$$\sum_{i=1}^{I}\sum_{j=1}^{J} x_{ijl}^{k} \leq 1, k = 1,\ldots, K, l = 1,\ldots, L, \quad \text{Equation (8.4)}$$

$$\sum_{i=1}^{I}\sum_{k=1}^{K}\sum_{l=1}^{L} x_{ijl}^{k} = 1, j = 1, \ldots, J, \quad \text{Equation (8.5)}$$

$$\sum_{i=1}^{I}\sum_{j=1}^{J}\sum_{l=1}^{L} x_{ijl}^{k} \leq 2, k = 1,\ldots, K, \quad \text{Equation (8.6)}$$

$$\left(\sum_{i=1}^{I}\sum_{l=1}^{L}\sum_{k'=1}^{k-1} x_{ihl}^{k'}\right) \geq \left(\sum_{i=1}^{I}\sum_{l=1}^{L} x_{ijl}^{k}\right), \forall k, (h, j), \quad \text{Equation (8.7)}$$

where (h, j) is a precedence relation on operations,

$$f_{ijl}^{k} = s_{ijl}^{k} + a_{ij}, \quad \text{for feasible } i, j, k, l, \quad \text{Equation (8.8)}$$

$$s_{ijl}^{k} = \max\left(\max_{\substack{k'=1\ldots k-1,\\ l=1,\ldots L\\ j'\neq j}} [x_{ij'l}^{k'} f_{ij'l}^{k'}], \quad \left[\sum_{i'=1}^{I}\sum_{k'=1}^{k-1}\sum_{l=1}^{L} x_{i'hl}^{k'} f_{i'hl}^{k'}\right]\right), \quad \text{Equation (8.9)}$$

for feasible i, j, k, l, with (h, j) being a precedence relation on operations and $0 \cdot \infty = 0$,

$$\delta(s_{ijl}^{k} - f_{ij'l}^{k'}) + \delta(s_{ij'l}^{k''} - f_{ijl}^{k}) = 1, \quad \text{for feasible } i, j, k, l, \quad \text{with } j' \neq j, k' < k, k < k'', \quad \text{Equation (8.10)}$$

$$x_{ijl}^{k} = 0 \quad \text{or} \quad 1, \forall i, j, k, l, \quad \text{and} \quad \alpha \geq 0. \quad \text{Equation (8.11)}$$

The objective function 8.1 is to minimize the total cycle time (completion time). Constraint 8.2 says that every operation has to be finished in the cycle time. Constraint 8.3 ensures that each MU can perform at most one operation in a time slot. Constraint 8.4 ensures that each WL can hold at most one operation in a time slot. Constraint 8.5 ensures that each operation is performed by one MU on one WL in a particular time slot. Constraint 8.6 is the parallel constraint which ensures that at most two operations can be performed in one time slot. Constraint 8.7 ensures that in each time slot, the precedence order of operations must be satisfied. Constraint 8.8 denotes the completion time as the sum of the starting time and the processing time. Constraint 8.9 ensures the starting time of operation *j* cannot be initialized until both (i) an MU is available for operation *j* and (ii) operation *j*'s precedent operations are completed. Constraint 8.10 ensures that no multiple operations are performed by the same MU in the same time slot. Constraint 8.11 describes the variables assumption.

The combinatorial nature of the operation sequencing problem with precedence constraints indicates the potential existence of multiple local optima in the search space. It is very likely that an algorithm for solving the above mixed integer program will be trapped by a local optimum. The complexity of the problem is also an issue that needs to be considered. Note that each of the variables x_{ijl}^{k}, s_{ijl}^{k}, and f_{ijl}^{k} has multiple indices. For a five-operation example performed on a 2-MU, 2-WL parallel machine, given that both MUs and one WL are available for each operation, there are $50 \times 3 = 150$ variables

($i \times j \times k \times l = 2 \times 1 \times 5 \times 5 = 50$ for each variable) under consideration. To overcome the above problems, we explore the idea of using "random search" to solve the problem.

8.3 A Genetic-Based Algorithm

A genetic algorithm [8, 12] is a stochastic search that mimics the evolution process searching for optimal solutions. Unlike conventional optimization methods, GAs maintain a set of potential solutions, i.e., a population of individuals, $P(t) = \{x_1^t, \ldots, x_n^t\}$, in each generation t. Each solution x_i^t is evaluated by a measurement called *fitness value*, which affects its likelihood of producing offspring in the next generation. Based on the fitness of current solutions, new individuals are generated by applying genetic operators on selecting individuals of this generation to obtain a new and hopefully "better" generation of individuals. A typical GA has the following structure:

1. Set generation counter $t = 0$.
2. Create initial population $P(t)$.
3. Evaluate the fitness of each individual in $P(t)$.
4. Set $t = t + 1$.
5. Select a new population $P(t)$ from $P(t - 1)$.
6. Apply genetic operator on $P(t)$.
7. Generate $P(t + 1)$.
8. Repeat steps 3 through 8 until termination conditions are met.
9. Output the best solutions found.

8.3.1 Applying GAs on the Parallel Operations Process

The proposed genetic algorithm utilizes Yip-Hoi and Dutta's single parent precedence tree [22]. The outline of this approach is illustrated in Figure 8.2. An initial population is generated with each chromosome representing a feasible operation sequence satisfying the precedence constraints. The genetic operators are then applied. After each generation, a subroutine to schedule the operations in parallel

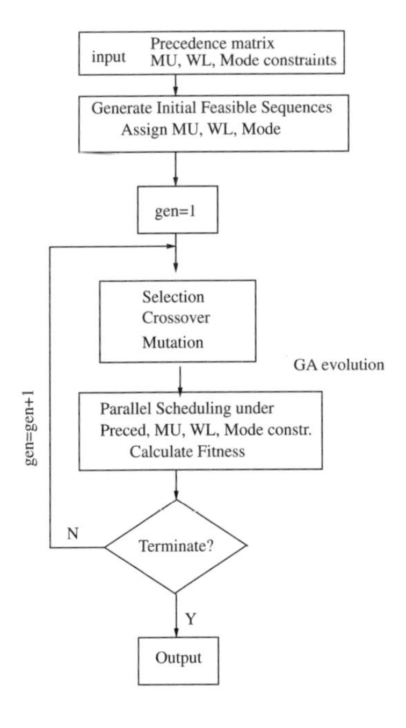

FIGURE 8.2 Flow chart for parallel operations implementing GAs.

TABLE 8.1 The Precedence Constraint Matrix P

$$
\begin{array}{cc}
 & \text{level} \rightarrow \;\; 1 \quad 2 \quad 3 \\
P = & \begin{array}{c} \text{op1} \\ \text{op2} \\ \text{op3} \\ \text{op4} \\ \text{op5} \end{array} \begin{pmatrix} M & 0 & 0 \\ M & 0 & 0 \\ 0 & 1 & 0 \\ 0 & 1 & 0 \\ 0 & 0 & 3 \end{pmatrix}
\end{array}
$$

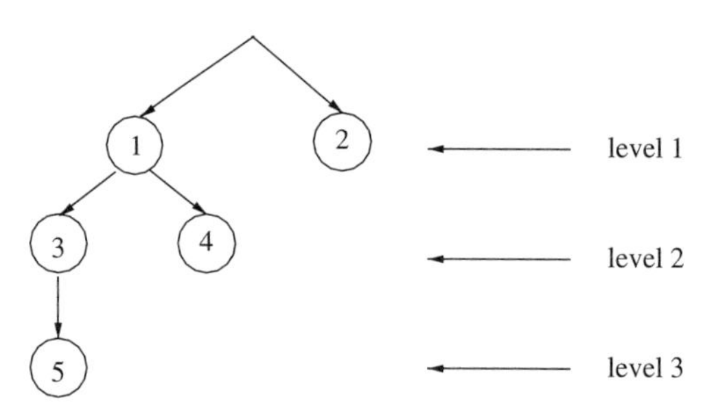

opE = [1 1 2 2 3]: earliest possible op order

opL = [2 5 4 5 5]: latest possible op order

FIGURE 8.3 A five-operation precedence tree.

according to the assignments of MU and WL is utilized to find the minimum cycle time and its corresponding schedule.

8.3.1.1 Order-Based Representations

The operation sequencing in our problem has the same nature of the traveling salesman problem (TSP). More precisely, the issue here is to find a Hamiltonian path of an asymmetric TSP with precedence constraints on the cities. Thus, we adopt a TSP path representation [12] to represent a feasible operation sequence. For an eight-operation example, an operation sequence (tour) 1–3–2–4–6–8–7–5 is represented by [1 3 2 4 6 8 7 5]. The approach is similar to the ordered-based representation discussed in [5], where each chromosome represents a feasible operation sequence, each gene in the chromosome represents an operation to be scheduled, and the order of the genes in the chromosomes is the order of the operations in the sequence.

8.3.1.2 Representation of Precedence Constraints

A precedence constraint is represented by a precedence matrix P. For the example, with five operations (Figure 8.3), the operations occupy three levels. A 5×3 matrix (Table 8.1) P is constructed with each row representing an operation and each column representing a level. Each element $P_{i,j}$ assigns a predecessor of operation i which resides at level j, e.g., $P_{3,2} = 1$ stands for "operation 3 at level 2 has a precedent operation 1." The operations at level 1 are assigned with a large value M. The initial population is then generated based on the information provided by this precedence matrix.

8.3.1.3 Generating Initial Population

The initial population is generated by two different mechanisms and then the resulting individuals are merged to form the initial population. We use the five-operation example to explain this work.

In the example, operation 1 can be performed as early as the first operation (level 1), and as late as the second (= total nodes − children nodes) operation. Thus, the earliest and latest possible orders are opE = [1 1 2 2 3] and opL = [2 5 4 5 5], respectively. This gives the possible positions of the five operations in determining a feasible operating sequence (see Figure 8.3). Let $pos(i, n)$ denote the possible locations of operation i in the sequence of n operations, $lev(i)$ denote the level of operation i resides, and $child(i)$ denote the number of child nodes of operation i. Operation i can be allocated in the following locations to ensure the feasibility of the operation sequence, $lev(i) \leq pos(i, n) \leq n - child(i)$. The initial population was generated accordingly to ensure its feasibility. A portion of our initial population was generated by the "level by level" method. Those operations in the same level are to be scheduled in parallel at the same time so that their successive operations (if any) can be scheduled as early as possible and the overall operation time (cycle time) can be reduced. To achieve this goal, the operations in the same level are scheduled as a cluster in the resulting sequence.

8.3.1.4 Selection Method

The roulette wheel method is chosen for selection, where the average fitness (cycle time) of each chromosome is calculated based on the total fitness of the whole population. The chromosomes are selected randomly proportional to their average fitness.

8.3.2 Order-/Position-Based Crossover Operators

A crossover operator combines the genes in two parental chromosomes to produce two new children. For the order-based chromosomes, a number of crossover operators were specially designed for the evolution process. Syswerda proposed the order-based and position-based crossovers for solving scheduling problem with GAs [17]. Another group of crossover operators that preserve orders/positions in the parental chromosomes was originally designed for solving TSP. The group consists of a partially-mapped crossover (PMX) [9], an order crossover (OX) [6], a cycle crossover (CX) [15] and a commonality-based crossover [1]. These crossovers all attempt to preserve the orders and/or positions of parental chromosomes as the genetic algorithm evolves. But none of them is able to maintain the precedence constraints required in our problem. To overcome the difficulty, a new crossover operator is proposed in Section 8.3.3.

8.3.3 A New Crossover Operator

In the parallel machining operation sequencing problem, the ordering comes from the given precedence constraints. To maintain the relative orders from parents, we propose a new crossover operator that will produce an offspring that not only inherits the relative orders from both parents but also maintains the feasibility of the precedence constraints.

The Proposed Crossover Operator

Given parent 1 and parent 2, the child is generated by the following steps:

Step 1. Randomly select an operation in parent 1. Find all its precedent operations. Store all the operations in a set, say, *branch*.

Step 2. For those operations found in Step 1, store the locations of operations in parent 1 as $location_1$. Similarly, find $location_2$ for parent 2.

Step 3. Construct a $location_c$ for the child, $location_c(i) = \min\{location_1(i), location_2(i)\}$ where i is a chosen operation stored in *branch*. Fill in the child with operations found in Step 1 at the locations indicated by $location_c$.

Step 4. Fill in the remaining operations as follows:
If $location_c = location_1$, fill in remaining operations with the ordering of parent 2, else if $location_c = location_2$, fill in remaining operations with the ordering of parent 1, else ($location_c \neq location_1$ and $location_c \neq location_1$), fill in remaining operations with the ordering of parent 1.

TABLE 8.2 The Proposed Crossover Process on the Eight-Operation Example

	The Proposed Crossover	
parent 1	[1 3 2 6 4 7 8 5]	
parent 2	[1 3 6 7 2 4 5 8]	
step 1:	[x x x x x x x 5]	randomly choose 5, *branch* = {1, 2, 5}
step 2:	[1 x 2 x x x x 5]	$location_1$ = {1, 3, 8}
	[1 x x x 2 x 5 x]	$location_2$ = {1, 5, 7}
step 3:	[1 x 2 x x x 5 x]	$location_c$ = {1, 3, 7}
step 4:	[1 3 2 6 4 7 5 8]	

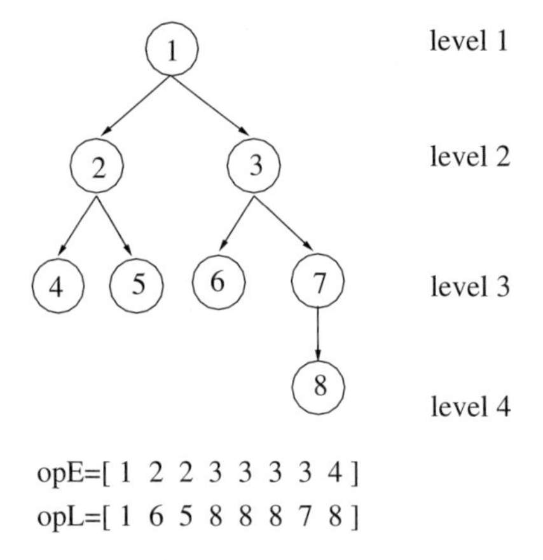

FIGURE 8.4 An eight-operation example.

Table 8.2 shows how the operator works for the eight-operation example (Figure 8.4). In step 1, operation 5 is randomly chosen and then traced back to all its precedent operations (operations 1 and 2), together they form *branch* = {1, 2, 5}. In step 2, find the locations of operations 1, 2, 5 in both parents, and store them in $location_1$ = {1, 3, 8} and $location_2$= {1, 5, 7}. In step 3, the earliest locations for each operation in {1, 2, 5} to appear in both parents is stored as $location_c$ = {1, 3, 7}. Fill in the child with {1, 2, 5} at the locations given by $location_c$ = {1, 3, 7} while at the same time keeping the precedence relation unchanged. In step 4, fill in the remaining operations {3, 6, 4, 7, 8} following the ordering of parent 1. The crossover process is now completed with a resulting child [1 3 2 6 4 7 5 8] that not only inherits the relative orderings from both parents but also satisfies the precedence constraints.

To show that the proposed crossover operator always produces feasible offspring, a proof is given as follows. Let T_n denote a precedence tree with n nodes and $D := \{(i, j): i < j, \forall(i, j) \in T_n\}$ denote the set of all precedent nodes. Thus, if $i < j$ in both parent 1 and parent 2 then both $location_1(i) < location_1(j)$, and $location_2(i) < location_2(j)$. Let $\{i_1, \ldots, i_k\}$ denote the chosen operations in step 1, we know $location_1(i_1) < location_1(i_2) < location_1(i_k)$, and $location_2(i_1) < location_2(i_2) < \cdots < location_2(i_k)$.

In step 3, $location_c(i_l) = \min_{l=1}^{k}\{location_1(i_l), location_2(i_l)\}$ is defined to allocate the location of chosen operation i_l in the child. We claim that the resulting child is always feasible. Otherwise, there exists a precedent pair $(i_l, i_m) \in D$ such that $location_c(i_l) > location_c(i_m)$, for $l, m = 1, \ldots, k$. However, this cannot be true. Because $i_l < i_m$ is given, we know that $location_1(i_l) < location_1(i_m)$, and $location_2(i_l) < location_2(i_m)$. This implies that $location_c(i_l) = \min\{location_1(i_l), location_2(i_l)\} < \min\{location_1(i_m), location_2(i_m)\}$. Thus, if $(i_l, i_m) \in D$, then $location_c(i_l) < location_c(i_m)$. This guarantees the child to be a feasible sequence after applying the proposed crossover operator.

TABLE 8.3 The Proposed Mutation Process on the Eight-Operation Example

	The Proposed Mutation	
parent	[1 3 2 6 4 7 8 5]	
step 1:	[1 **3** |**2 6 4**| **7** 8 5]	operations 3, 7 chosen, *branch* = {3, 2, 6, 4, 7}
step 2, 3:	[1 **2** |3 6 4| **7 8 5**]	mutate operations 2 and 3
child	[1 2 3 6 4 7 8 5]	

8.3.4 A New Mutation Operator

The mutation operators were designed to prevent GAs from being trapped by a local minimum. Mutation operators carry out local modification of chromosomes. To maintain the feasible orders among operations, some possible mutations may (i) mutate operations between two independent subtrees or (ii) mutate operations residing in the same level. Under this consideration, we develop a new mutation operator to increase the diversity of possible mutations that can occur in a feasible sequence.

The Proposed Mutation Operator

Given a parent, the child is generated by the following steps:

Step 1. Randomly select an operation in the parent, and find its immediate precedent operation. Store all the operations between them (including the two precedent operations) in a set, say, *branch*.

Step 2. If the number of operations found in step 1 is less than or equal to 2, (i.e., not enough operations to mutate), go to step 1.

Step 3. Let m denote the total number of operations ($m \geq 3$). Mutate either $branch(1)$ with $branch(2)$ or $branch(m-1)$ with $branch(m)$ given that $branch(1) \nprec branch(2)$ or $branch(m-1) \nprec branch(m)$, where "$\nprec$" indicates there is no precedence relation.

Table 8.3 shows how the mutation operator works for the example with eight operations. In step 1, operation 7 is randomly chosen, with its immediate precedent operation 3 from Figure 8.4 to form *branch* = {3, 2, 6, 4, 7}. In step 3, mutate operation 2 with 3 (or 4 with 7) and produce a feasible offspring, child = [1 2 3 6 4 7 8 5].

For the parallel machining operation sequencing problem, the children generated by the above mutation process are guaranteed to keep the feasible ordering. Applying the proposed mutation operator in the parent chromosome results in a child which is different from its parental chromosome by one digit. This increases the chance to explore the search space. The proposed crossover and mutation operators will be used to solve the problems in Sections 8.5 and 8.7.

8.4 Tabu Search for Sequencing Parallel Machining Operations

8.4.1 Tabu Search

Tabu search (TS) is a heuristic method based on the introduction of adaptive memory to guide local search processes. It was first proposed by Glover [10] and has been shown to be effective in solving a wide range of combinatorial optimization problems. The main idea of tabu search is outlined as follows.

Tabu search starts with an initial feasible solution. From this solution, the search process evaluates the "neighboring solutions" at each iteration as the search progresses. The set of neighboring solutions is called the *neighborhood* of the current solution and it can be generated by applying certain transformation to current solution. The transformation that takes the current solution to a new neighboring solution is called a *move*. Tabu search then explores the best solution in this neighborhood and makes the best available move. A move that brings a current solution back to a previously visited solution is called a *tabu* move. In order to prevent cycling of the search procedure, a first-in first-out *tabu list* is created to record tabu moves. Each time a move is executed, it is recorded as a tabu move. The status of this tabu move will last for a given number of iterations called *tabu tenure.* A tabu move is freed from the tabu list when it reaches tabu tenure. A strategy called *aspiration criteria* is introduced to override the tabu status of moves once a better solution is encountered. The above tabu conditions and aspiration criteria are the backbone of the tabu search heuristic. Together they make up the bases of the short-term memory process. The implementation of the short-term memory process is what differentiates tabu search from local search (hill climbing/descending) techniques. Tabu search follows the greatest improvement move in the neighborhood. If such a move is not available, the least nonimprovement move will be performed in order to escape from the trap of a local optimum. Local search, on the other end of the spectrum, always searches for the most improvement for each iteration.

A more refined tabu search may employ the intermediate-term and long-term memory structures in the search procedure to reach regional intensification and global diversification. The intermediate-term memory process is implemented by restricting the search within a set of potentially best solutions during a particular period of search to *intensify* the search. The long-term memory process is invoked periodically to direct the search to less explored regions of the solution space to *diversify* the search.

In this section, a tabu search with a short-term recency-based memory structure is employed to solve the operations sequencing problem. The neighborhood structure, move mechanism, data structure of tabu list, aspiration criteria, and the stopping rule in the tabu search are described as follows.

8.4.2 Neighborhood Structure

The neighborhood structure in the proposed tabu search is determined based on the precedence constraints among operations. There are $n(n-1)/2$ neighbors considered for each given operation sequence with n operations. This idea is illustrated with an example. Suppose we have a sequence [1 2 3 4 5 6 7 8]; starting from operation 1, there are seven potential exchanges, operations 2 to 8, to be examined for an admissible move. Then for operation 2, there remains six potential exchanges of operations. This process is repeated for every operation. Excluding the reverse exchange of any two previously exchanged operations, there are $(8 \times 7)/2 = 28$ potential exchanges. These 28 potential exchanges make up all the neighboring solutions of sequence [1 2 3 4 5 6 7 8]. Note that not all of the 28 exchanges lead to admissible moves. A precedence check is required for choosing an admissible move. The best available neighbor is the sequence in the neighborhood that minimizes the cycle time without resulting a tabu move.

8.4.3 Move Mechanism

An intuitive type of move for the operation sequencing is to exchange two nonprecedent operations in an operation sequence, so that each operation occupies the location formerly occupied by the other. To perform such a move that transforms one feasible operation sequence to another, the exchange of operations needs to satisfy the following two constraints to guarantee an admissible move.

1. *The location constraint:* For a given precedence tree, the location of each operation in an operation sequence is determined by the location vectors *opE* and *opL* [4]. An exchange of operations i and j is admissible only if operation i is swapped with operation j, provided j currently resides between locations $opE(i)$ and $opL(i)$. Similarly, operation j can only be swapped with operation i, provided i currently resides between locations $opE(j)$ and $opL(j)$. Consider the precedence tree in Figure 8.4 and a given feasible operation sequence [1 3 6 7 8 2 4 5]. The earliest and latest possible locations for operations 2 and 3 are determined by $[opE(2), opL(2)] = [2, 6]$ and $[opE(3), opL(3)] = [2, 5]$, respectively (Figure 8.4). An exchange of operations 2 and 3 is not legitimate in this case because the current location of operation 2, the sixth operation in the sequence [1 3 6 7 8 2 4 5], is not a feasible location for operation 3 (i.e., $[opE(3), opL(3)] = [2, 5]$). Operation 3 can only be placed between the second and the fifth locations, a swap of operations 2 and 3 will result in

an infeasible operation sequence [1 2 6 7 8 3 4 5]. Consider another operation sequence [1 2 3 4 5 6 7 8], whereas the exchange of operation 2 and 3 is admissible because both operations are swapped to feasible locations. This results in a new feasible operation sequence [1 3 2 4 5 6 7 8].

2. *The precedence constraint:* Even after an "admissible" exchange, a checkup procedure is needed to ensure the precedence constraint is not violated. Consider the above operation sequence [1 2 3 4 5 6 7 8]. Operations 6 has possible locations between the third and the eighth in the sequence, i.e., $[opE(6), opL(6)] = [3, 8]$. Similarly, $[opE(8), opL(8)] = [4, 8]$. Although both operations are in permissible locations for exchange, the exchange of operations 6 and 8 incurs an infeasible operation sequence [1 2 3 4 5 8 7 6] which violates the precedence constraint (Figure 8.4) of having operation 7 before operation 8.

8.4.4 Tabu List

The type of elements in the tabu list L_{TABU} for the searching process is defined in this section. The data structure of L_{TABU} takes the form of an $n \times n$ matrix, where n is the number of operations in the parallel machining process. A move consists of a pair (i, j) which indicates that operations (i, j) are exchanged. The element $L_{\text{TABU}}(i, j)$ keeps track of the number of times the two operations are exchanged and the prohibition of a tabu move is thus recorded. A fixed tabu tenure is used in the proposed tabu search. The status of a tabu move is removed when the value of $L_{\text{TABU}}(i, j)$ reaches the tabu tenure.

8.4.5 Aspiration Criterion and Stopping Rule

An aspiration criterion is used to override the tabu status of moves once better solutions are encountered. The aspiration criterion together with the tabu conditions are the means used by a tabu search to escape from the trap of local optima. A fixed aspiration level is adopted in the proposed tabu search. The stopping rule used here is based on the maximum number of iterations. The search process terminates at a given maximum number of iterations. The best solution found in the process is the output of the tabu search.

8.4.6 Proposed Tabu Search Algorithm

To outline the structure of the proposed tabu search algorithm, we denote

n	as the number of operations,
$maxit$	as the maximum iteration number,
k, p	as the iteration counters,
$opseq$	as the input operation sequence,
$f(opseq)$	as the objective value (machining cycle time) of sequence $opseq$,
$opbest$	as the current best (with minimum value of f) operation sequence,
$N_B(opseq)$	as the neighborhood of $opseq$,
$opseq(p)$	as the pth operation sequence in $N_B(opseq)$,
$M(opseq(p))$	as the move to sequence $opseq(p)$,
L_{TABU}	as the tabu list.

The steps of the proposed tabu search algorithm are depicted as follows. The flow chart of the algorithm is shown in Figure 8.5.

1. $k, p \leftarrow 1$, $L_{\text{TABU}} \leftarrow O_{n\times n}$ (zero matrix).
2. Start with $opseq$. Update $f^* \leftarrow f(opseq)$.
3. Choose an admissible move $M(opseq(p)) \in N_B(opseq)$.
4. If $M(opseq(p))$ is not tabu,
   if $f(opseq(p)) < f^*$,
   store the tabu move, update L_{TABU},
   $opbest \leftarrow opseq(p)$, $f^* \leftarrow f(opseq(p))$.

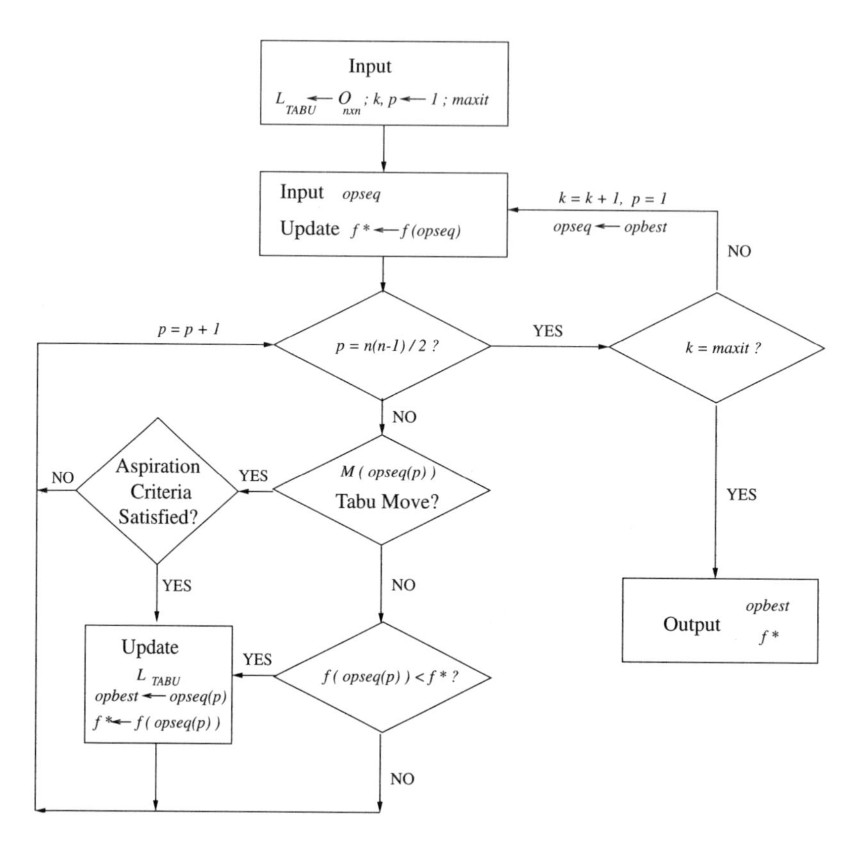

FIGURE 8.5 Tabu search flow chart.

Else, check aspiration level.
if $f(opseq(p)) < f^*$, update L_{TABU},
$opbest \leftarrow opseq(p), f^* \leftarrow f(opseq(p))$.

5. If $p < n(n-1)/2$, set $p \leftarrow p + 1$ and Go to step 3.
6. $opseq \leftarrow opbest$.
   If $k = maxit$, STOP.
   Else $k \leftarrow k + 1$ and $p \leftarrow 1$, Go to step 2.

The performance of the proposed tabu search will be investigated in Sections 8.6 and 8.7.

8.5 Two Reported Examples Solved by the Proposed GA

In this section, two examples are used to test our genetic algorithm with the proposed crossover and mutation GA operators. The computation experiments were conducted on a Sun Ultra1 workstation, and all programs were written in MATLAB environment.

In both examples, there are three machining constraints [22]:

1. *The precedence constraints.* All the scheduled operations must follow a prescribed precedence relation.
2. *The cutter* (MU), *spindle* (WL) *allocation constraint.* In the examples, both MUs are accessible to all operations. Due to a practical machining process, each operation can only be performed on one specific spindle (WL).
3. *The mode conflict constraint.* Operations with different machining modes, such as milling/drilling and turning, cannot be performed on the same spindle.

TABLE 8.4 Operation Times and Resource for the 18 Operations

Volume	Operation Type	Operation Number	WL_1	WL_2	MU_1	MU_2	Mode	Machining Time
MV1	External Turning	1	1	0	1	1	1	20
MV2	Center Drilling	2	1	0	1	1	1	5
MV3	External Turning	3	1	0	1	1	1	28
MV4	External Turning	4	1	0	1	1	1	10
MV5	External Grooving	5	1	0	1	1	1	12
MV6	Flat Milling	6	0	1	1	1	2	22
MV7	Flat Milling	7	0	1	1	1	2	22
MV8	Radial Center Drill	8	1	0	1	1	2	5
MV9	Radial Center Drill	9	1	0	1	1	2	5
MV10	Axial Drilling	10	1	0	1	1	1	44
MV11	External Threading	11	0	1	1	1	1	12
MV12	External Grooving	12	0	1	1	1	1	8
MV13	Radial Drilling	13	0	1	1	1	2	10
MV14	Radial Drilling	14	0	1	1	1	2	10
MV15	Internal Turning	15	0	1	1	1	1	32
MV16	Boring	16	1	0	1	1	1	14
MV17	Internal Turning	17	0	1	1	1	1	29
MV18	Boring	18	0	1	1	1	1	18

Note: Mode 1: turning. Mode 2: milling/drilling.

In the proposed GA, the MU, WL assignments are attached to the operation sequence as separate vectors. Together they go through the selection, crossover, and mutation processes. After each generation, a best solution is then rescheduled with a repair algorithm to avoid the mode conflict.

The first example of 18 operations is taken from [22] with a new labeling for the operations (see Figure 8.6 and Table 8.4). For a run with 100 generations and a population size of 25, the proposed GA takes about 2 to 3 CPU-minutes to complete (Table 8.5). The reported optimal cycle time for this example is 154 time units [22]. Using our proposed genetic algorithm, Figure 8.7 shows the resulting Gantt chart of the optimal sequence of machining operations. In the chart, there are two workpieces held on the two spindles of the parallel machines: a new workpiece on WL_1 and a partially completed "old" piece on WL_2 from the last cycle of operations. For this partially completed workpiece, operations 1, 2, 3, 4, 5, 8, 9, 10, and 16 had been performed before it was transformed to WL_2.

At time 0, the machine started working on the new workpiece. Operations 1 and 2 were performed simultaneously by MU_1 and MU_2 on WL_1. At time 20, MU_1 moved to WL_2 to perform operations 7 and 6 on the "old" workpiece. At the same time, MU_2 kept performing operations 10, 16 on the new workpiece

TABLE 8.5 CPU Time of the Eight Examples

GA	Gen.	Pop.	CPU Time
10	100	25	0.64
	100	100	1.36
18	100	50	2.48
	100	100	4.94
20	100	50	2.88
	100	100	5.70
26	100	50	4.02
	100	100	8.16
30	100	50	4.46
	100	100	9.53
40	100	100	12.75
	100	150	19.56
50	100	100	17.43
	100	150	25.52
60	100	100	22.35
	100	150	32.11

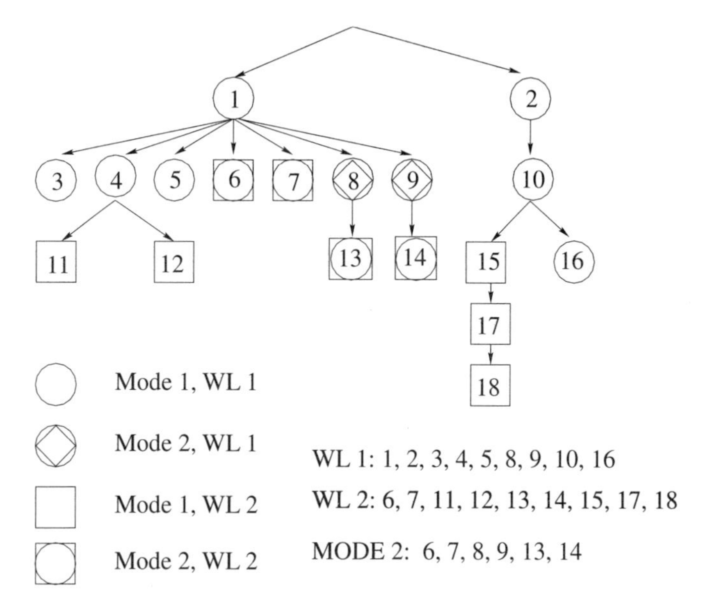

FIGURE 8.6 The precedence tree of an 18-operation example.

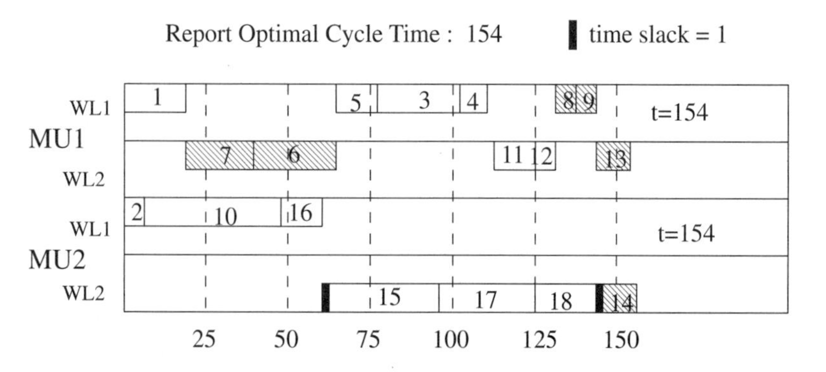

FIGURE 8.7 Optimal cycle time for the 18-operation example.

TABLE 8.6 A Precedence Matrix *P*

$P =$ level →	1	2	3	4
op1	M	0	0	0
op2	0	1	0	0
op3	0	1	0	0
op4	0	0	2	0
op5	0	0	2	0
op6	0	0	3	0
op7	0	0	3	0
op8	0	0	0	7

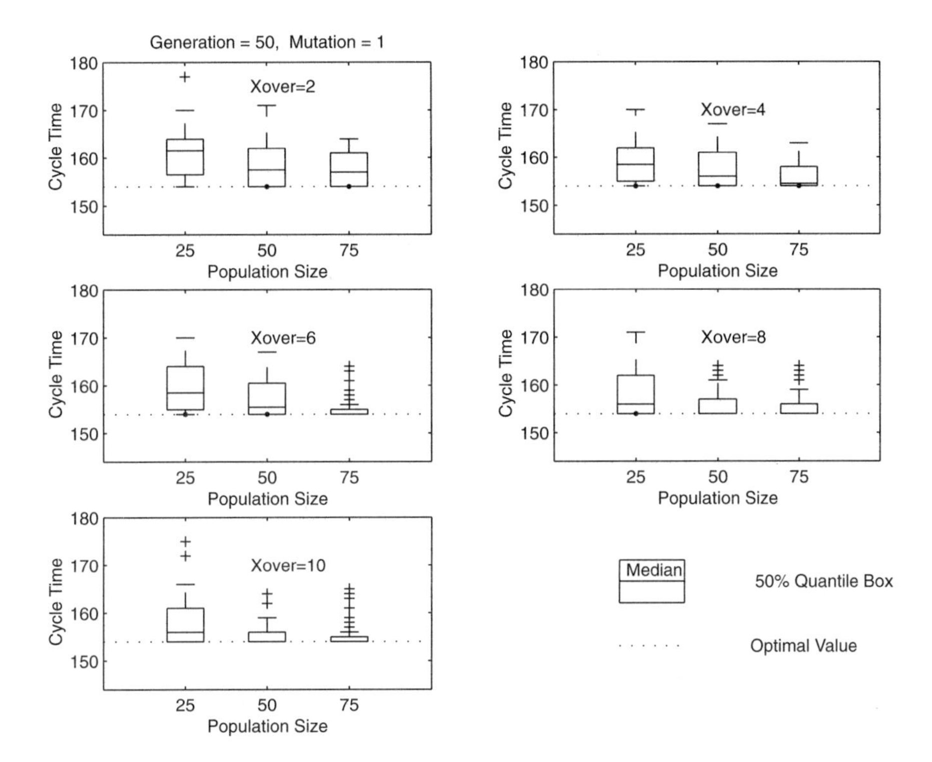

FIGURE 8.8 18-operation example with different number of crossovers (mutation = 1, generations = 50).

(from time 5 to time 63). There is a slack time of one unit for MU_2 at this point (time 64) to avoid the mode conflict between operation 6 and 15. Both operations belong to different machining modes and thus cannot be performed on the same spindle at the same time. Starting from time 64, MU_1 moved back to perform operations 5, 3 and 4 on the new workpiece at WL_1 and then on operations 11, 12, 8, 9 and 13 at their respective locations while MU_2 moved to WL_2 for operations 15, 17, 18 and 14. Notice that, from Figure 8.7, another slack time of one unit occurs at time 143 to ensure the precedence constraint between operations 9 and 14 is not violated. The cycle time was calculated as the maximum operations time occurring on both MUs, which is 154 time units as shown in Figure 8.7. Our proposed GA indeed found an optimal solution with a cycle time of 154 (Table 8.4) as reported.

To analyze the overall quality of the proposed genetic algorithm, we tested the algorithm with different parameters. Figure 8.8 shows the result by taking 50 generations and one mutation at each generation.

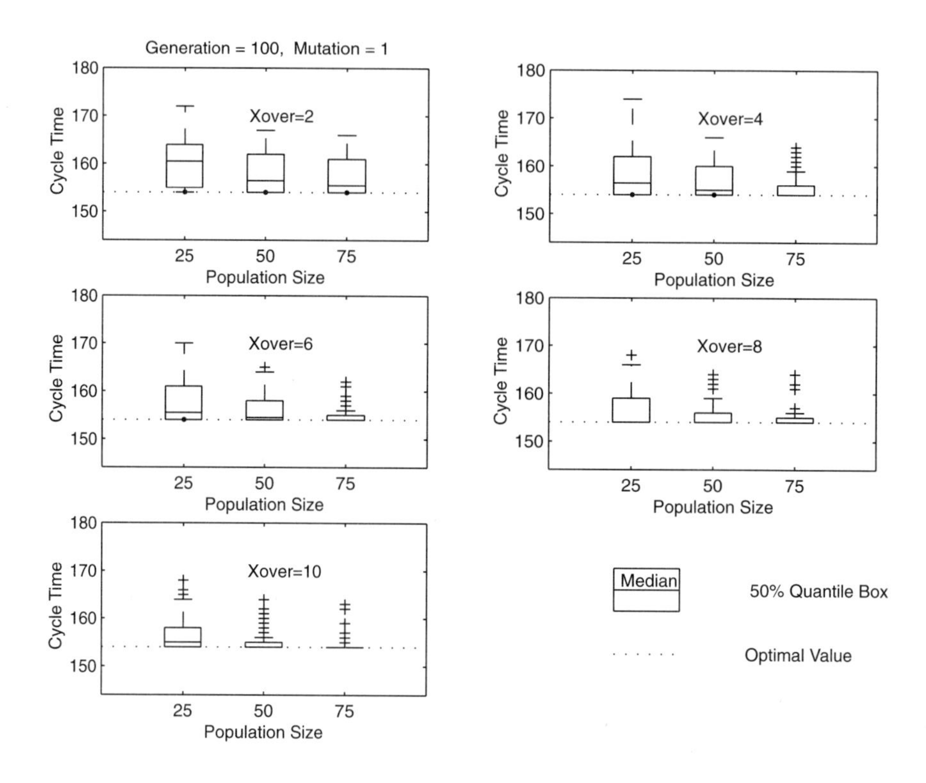

FIGURE 8.9 18-operation example with different number of crossovers (mutation = 1, generations = 100).

Population sizes were set to be 25, 50, and 75, respectively. Each subplot in the figure represents a simulation with the number of generations, mutations, and crossovers fixed. The number of crossovers were then increased from 2 to 10 to analyze the performance of the proposed genetic algorithm. We ran100 times for each setting. The result was plotted as a 50% quantile box, as shown in Figure 8.8. Each box contains 100 data points for the final cycle time in a run of 50 generations. The solid line inside each box is the median of the 100 points. The upper boundary denotes the 75% quantile and the lower boundary for 25% quantile. In Figure 8.8, the symbol "+" represents the outliner in the simulation and the dotted line represents the reported optimum. From Figure 8.8, we observe that a larger population or more crossovers led to a better result. Figure 8.9 shows the effect when the generation increased from 50 to 100 for each run. The proposed GA had more runs converged to the optimum with more crossovers and a larger population. On the other hand, when the crossovers increased from eight to ten, there was no significant improvement for higher population sizes (50, 75), as shown in Figure 8.9.

To observe how mutation affects the performance of the proposed GA, the number of mutations was increased from one to two in each of the above settings (see Figures 8.10 and 8.11). From Figures 8.10 and 8.11, it is noticed that mutations of two did not significantly improve the performance. The above simulation verifies the fact that crossover operators play a major role in a GA's evolution process. The result also shows that the proposed GA can locate the optimal solution of the problem.

The second example of 26 operations is taken from [7] (see Figure 8.12). For a run with 100 generations and a population size of 25, the proposed GA takes about 4 to 5 CPU-minutes to complete. The reported optimal cycle time for the proposed 26-operation example is 5410 time units. The proposed GA also found an optimal solution of that value. A further investigation on the reported 18- and 26-operation examples is given in Figures 8.13 and 8.14. They both confirmed the best solutions reported in [22].

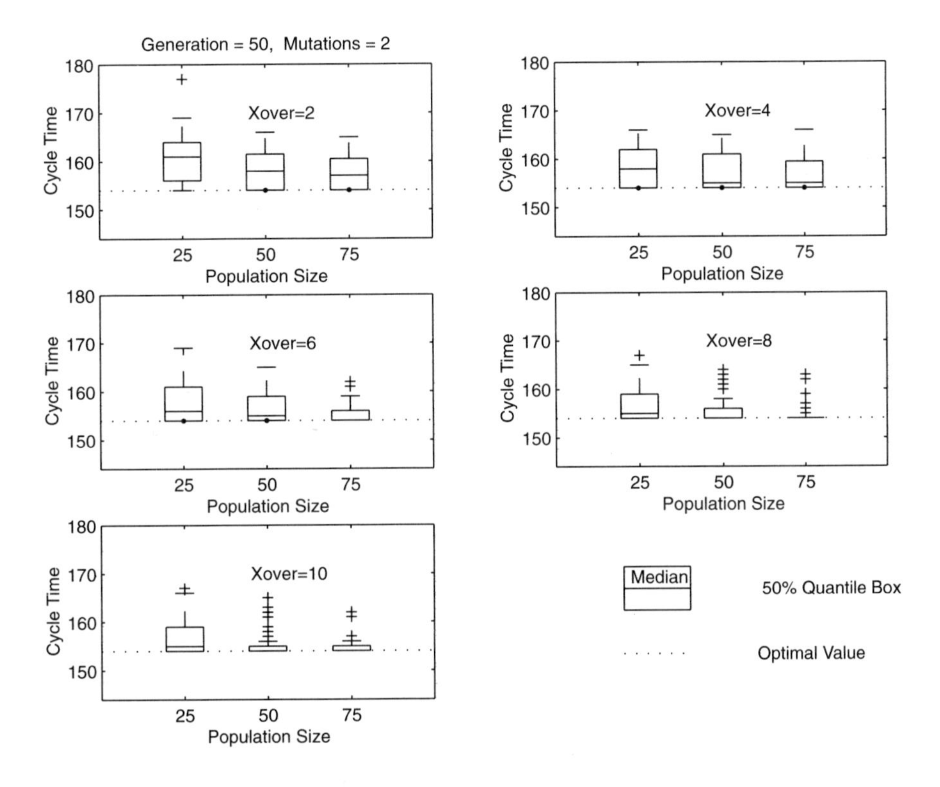

FIGURE 8.10 18-operation example with different number of crossovers (mutation = 2, generations = 50).

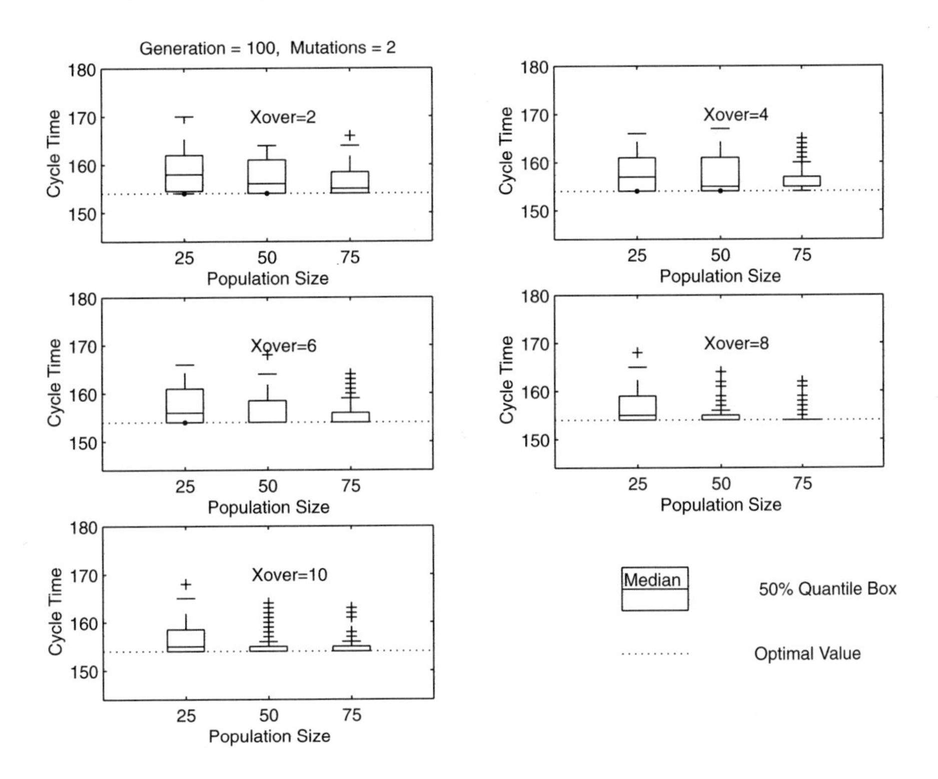

FIGURE 8.11 18-operation example with different number of crossovers (mutation = 1, generations = 100).

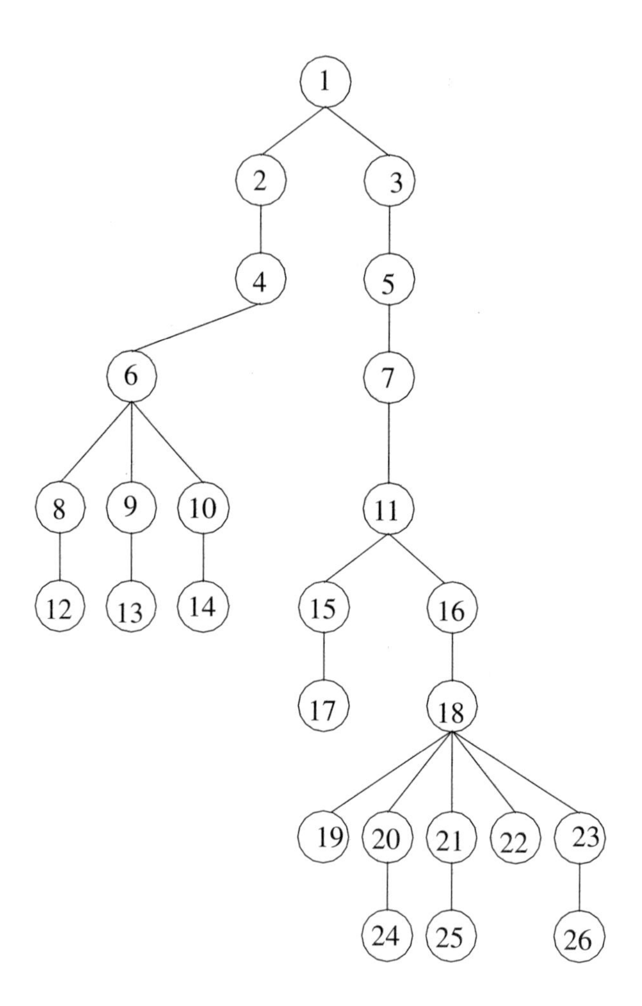

FIGURE 8.12 The precedence tree of 26-operation example.

8.6 Two Reported Examples Solved by the Proposed Tabu Search

In this section, computational results of the proposed tabu search are presented. The experiments were done on a Motorola PowerStack (133 MHz) workstation, and all programs were written in MATLAB environment. The two reported examples described in Section 8.5 will be tested.

For each example, ten feasible operation sequences were randomly generated as the seeds. The proposed tabu search algorithm took each seed as an initial solution to make four runs. The first run stopped at the 100th iteration, the second run stopped at the 200th iteration, the third run stopped at the 400th iteration, and the fourth run stopped at the 600th iteration. These runs were carried out independently. The details are given below.

8.6.1 The 18-Operation Example

For the 18-operation example (Table 8.4 and Figure 8.6), the reported best schedule found by genetic algorithms [4, 22] has a cycle time of 154. The proposed tabu search finds an improved schedule with a cycle time of 153. The corresponding solutions are given in Figure 8.15. Notice that the two time slacks in the genetic algorithm solution were eliminated by the proposed tabu search (Figure 8.15).

The detailed results obtained by tabu search are shown in Figure 8.16(a). For the run with 100 iterations, the best (only one) seed ended up with a cycle time of 156 and the worst seeds (five) with 178. For the

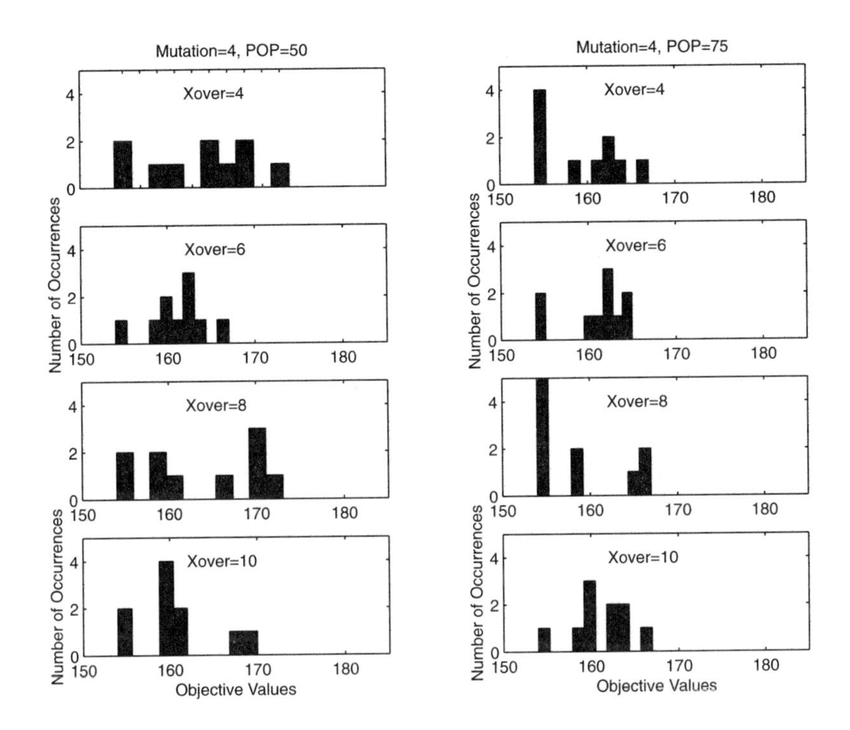

FIGURE 8.13 Histogram of 18-operation example; each box consists of ten runs.

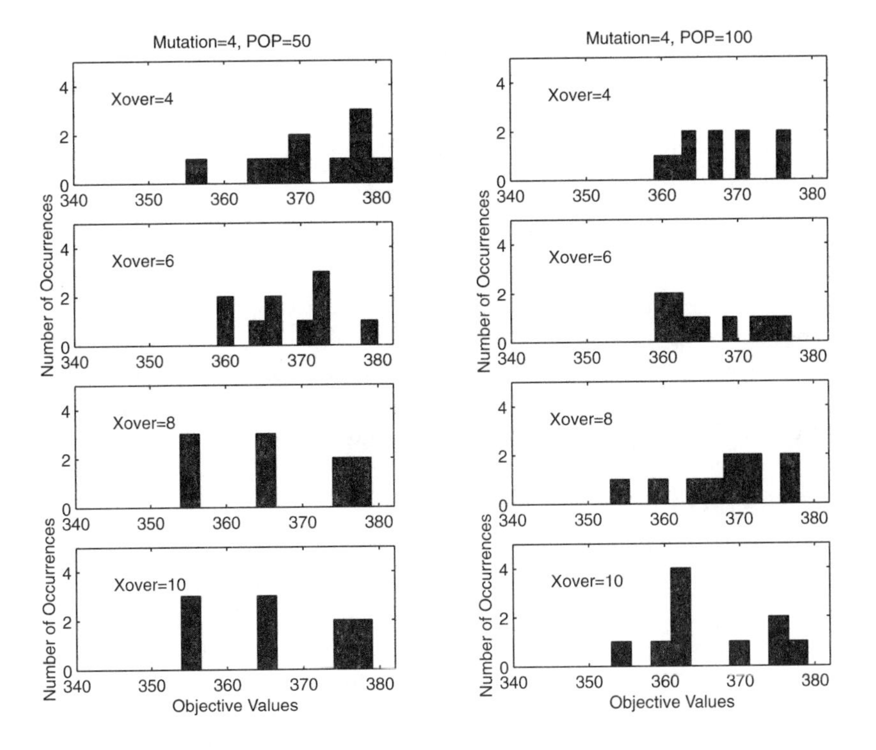

FIGURE 8.14 Histogram of 26-operation example; each box consists of ten runs.

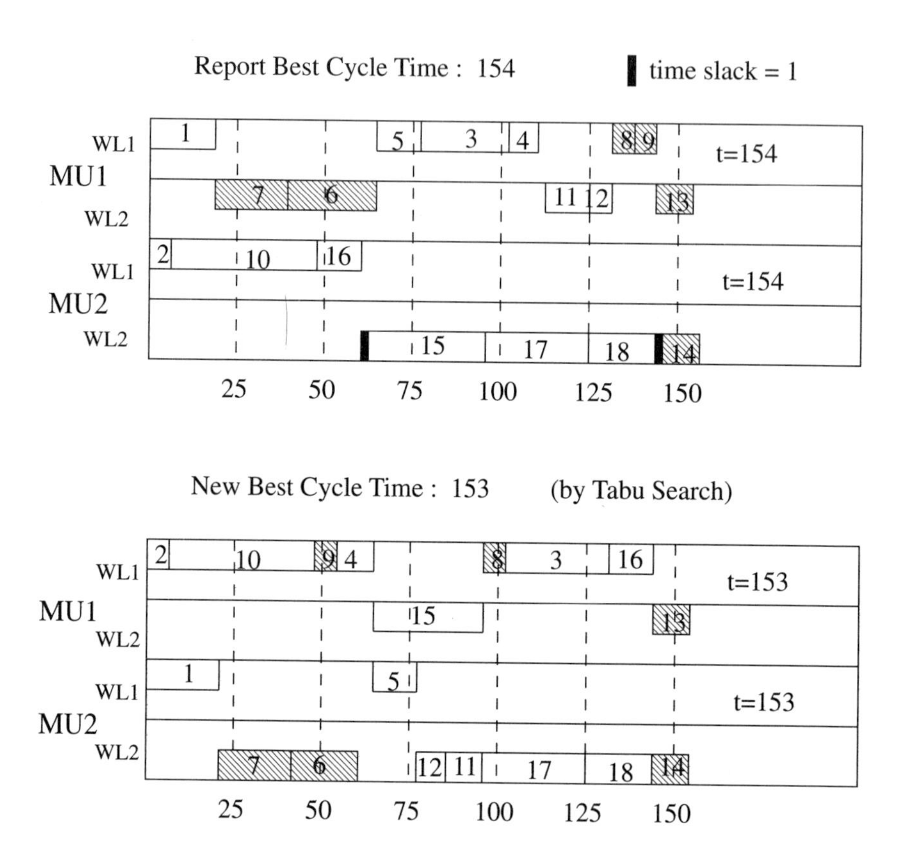

FIGURE 8.15 A new best schedule for the 18-operation example.

run with 200 iterations, an overall best schedule with a cycle time of 153 was found by one seed. Although a seed ended up with a cycle time of 183, the results obtained in this run are in general better than before. This "converging to good objective values" phenomenon is more obvious for both the runs with 400 and 600 iterations. The worst case of these two runs ended with a cycle time of 168. The overall solution quality improves as we take more iterations. This is not necessarily true for individual seeds in different runs. However, solutions with high quality can be generated if the tabu search algorithm runs long enough.

The results were also obtained by using the genetic algorithm approach. The genetic algorithm [4] was run with two different population sizes of 50 and 75, respectively. The results with different mutation and crossover parameters are shown in Figure 8.16(b). The best schedule in all runs ended with a cycle time of 154. This is the same with the results as reported by [4, 22]. As expected, keeping a large population size improves the solution quality at the cost of computation efforts.

8.6.2 The 26-Operation Example

For the 26-operation example (Figure 8.12), the reported best schedule found by genetic algorithms [3, 7] has a cycle time of 5410. The proposed tabu search also found a schedule of 5410 (Figure 8.17).

The detail results obtained by tabu search are shown in Figure 8.18(a). For the run with 100 iterations, the best seeds (two) ended up with a cycle time of 5410 and the worst seed (one) with 5660. For the runs with 200 and 400 iterations, the best seeds (seven) ended with a cycle time of 5410. Although the worst seed (one) in the run with 200 iterations ended with 6029, the results obtained in this run in general are better than those obtained in the run with 100 iterations. As mentioned earlier, all four runs were carried out independently; this explains why fewer seeds (five) ended up with a cycle time of 5410 in the run with 600 iterations as compared to the results in the runs with 200 and 400 iterations.

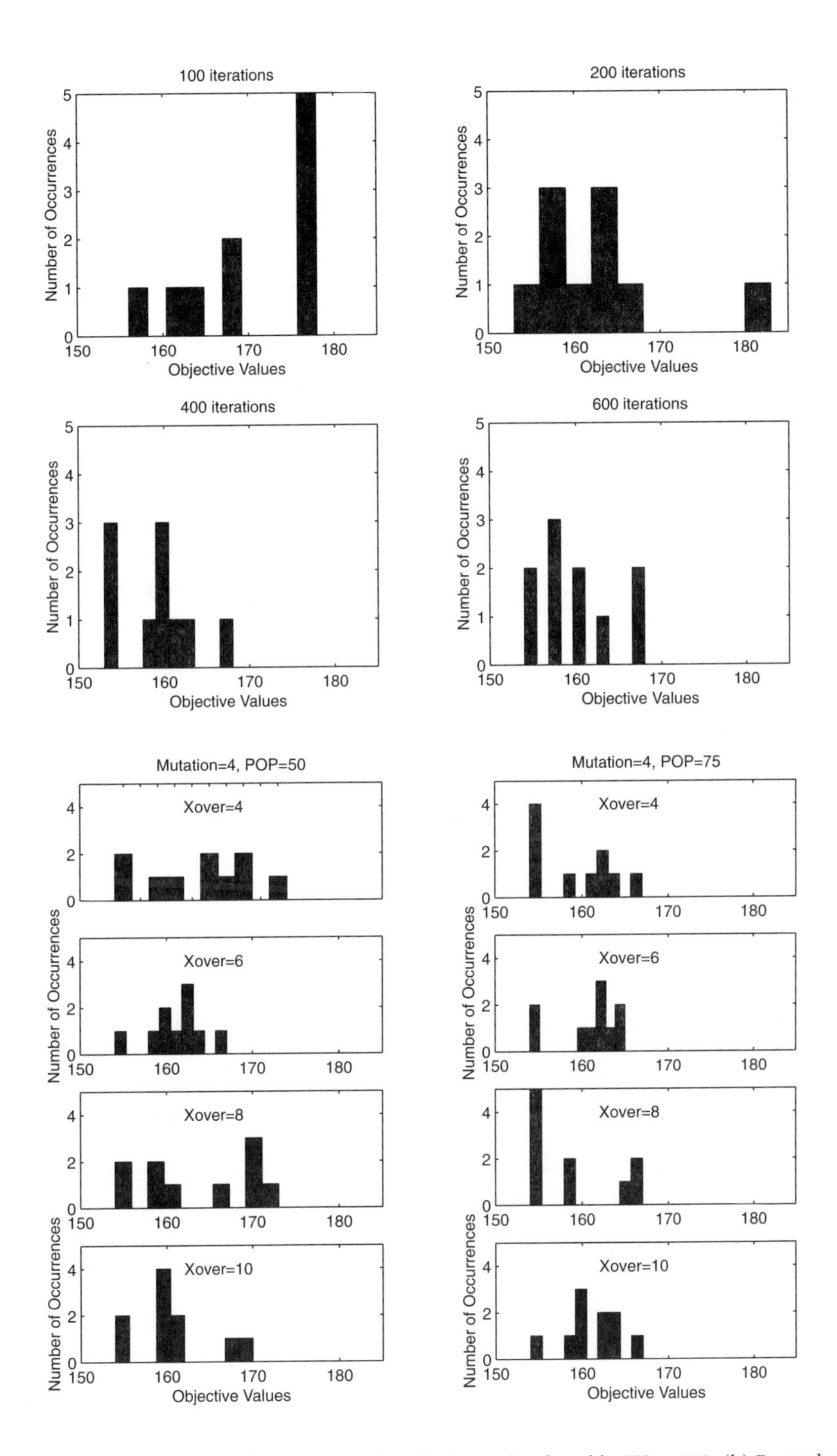

FIGURE 8.16 (a) Histograms of the 18-operation example. Best value found by TS = 153. (b) Best value found by GA = 154.

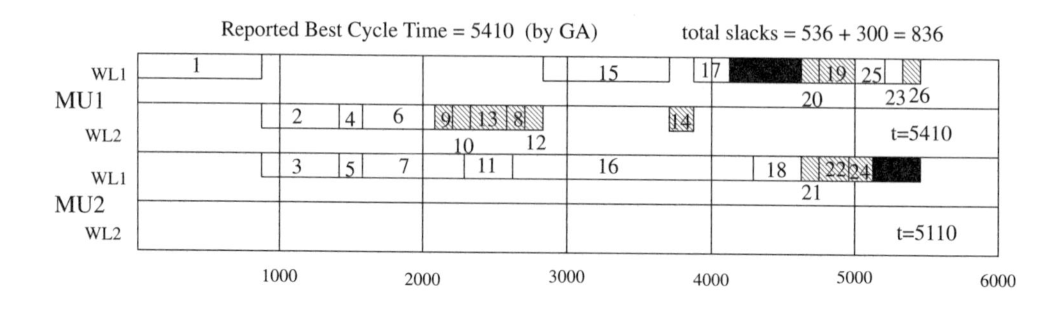

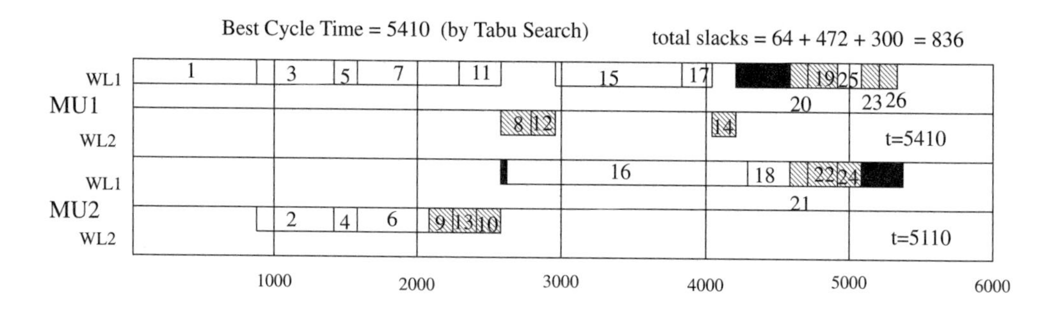

FIGURE 8.17 Best schedules for the 26-operation example.

The results were also obtained by using the genetic algorithm approach. The genetic algorithm was run with two different population sizes of 50 and 100, respectively. The results with different mutation and crossover parameters are shown in Figure 8.18(b). Seven out of eight runs ended with a cycle time of 5410. This is the same results as reported by [4, 22].

The above results have shown that the proposed tabu search is capable of finding and even improving the best known solutions. To further test the overall quality of the proposed tabu search, a random problem generator is needed to produce a set of representative test problems. The design of such a random problem generator and a comparison of the outcomes by applying the proposed tabu search and the existing genetic algorithm to the test examples will be presented in the next section.

8.7 Random Problem Generator and Further Tests

To fully explore the capability of the proposed genetic algorithm and tabu search, a problem generator has been designed with the following guiding rules as observed in literature and practice:

1. 30 to 50% of total operations are accessible to spindle 1, and the rest to spindle 2.
2. Both cutters are accessible to all operations.
3. 30 to 50% of total operations are of one machining mode [22].

Examples are randomly generated in the following manner:

Step 1: Set L = maximum number of levels in the precedence tree. In other words, the resulting tree has at most L levels.

Step 2: At each level, two to six operations are randomly generated and each operation is assigned to, at most, one predecessor.

Step 3: Once the precedence tree is generated, the spindle location, cutter location, mode constraint, and machining time of each operation are randomly assigned.

Six testing problems with 10, 20, 30, 40, 50, and 60 operations, respectively, were generated and are shown in Figures 8.19 to 8.24.

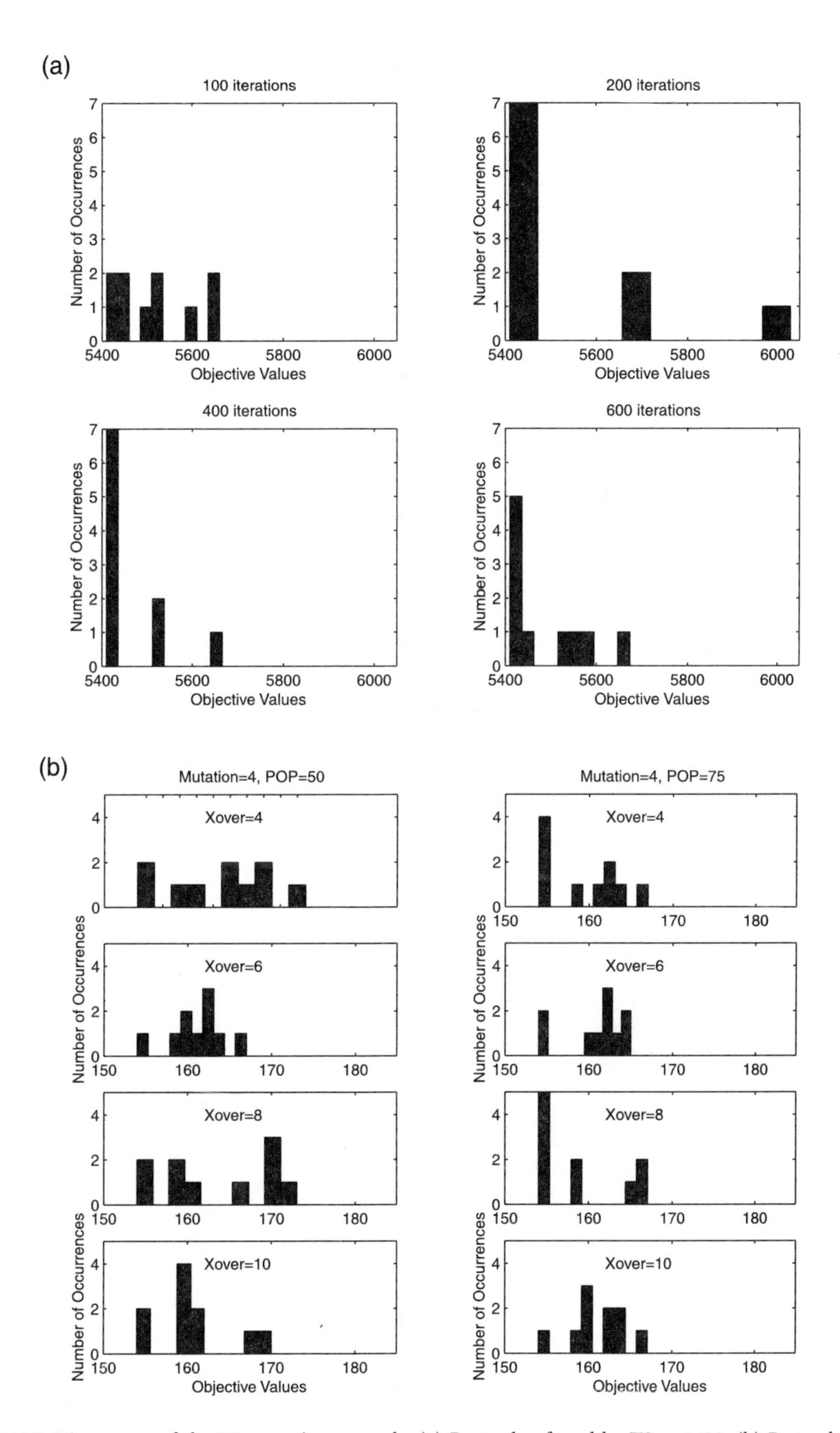

FIGURE 8.18 Histograms of the 26-operation example. (a) Best value found by TS = 5410. (b) Best value found by GA = 5410.

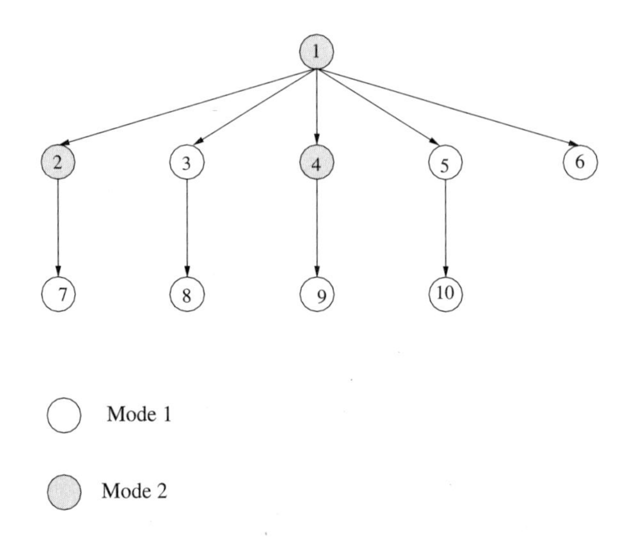

FIGURE 8.19 10-operation problem.

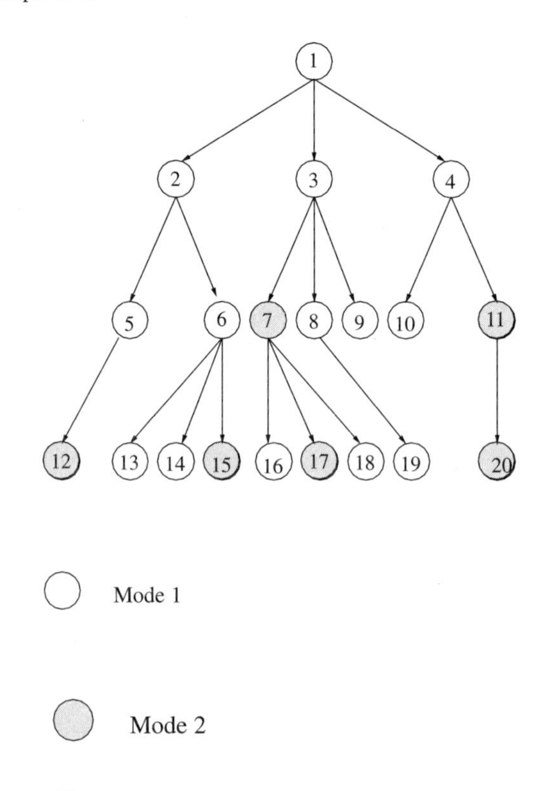

FIGURE 8.20 20-operation problem.

8.7.1 Performance Comparison

Similar to the setting of Section 8.6, results obtained by the proposed tabu search and genetic algorithms are reported in this section.

For the test problem with ten operations, tabu search located its best solution with a cycle time of 93 in each of the four runs (with 100, 200, 400, and 600 iterations), as shown in Figure 8.25(a). On the other hand, genetic algorithm found its best solution with a cycle time of 94 (Figure 8.25(b)). Therefore,

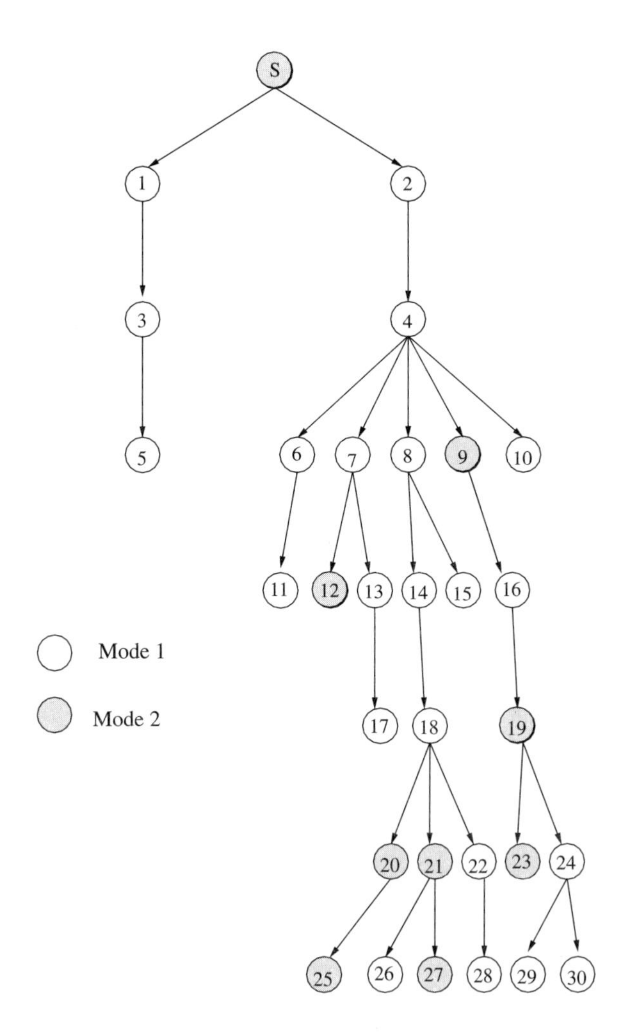

FIGURE 8.21 30-operation problem.

the tabu search had a 1% improvement over the genetic algorithm in this test problem. For such small problems, the proposed tabu search clusters tightly in the vicinity of its best solutions as the number of iterations increases.

For the test problem with 20 operations, tabu search found its best solution with a cycle time of 194 (Figure 8.26(a)). Notice that the solutions converge toward this value as the number of iterations increases. In particular, all ten seeds converge to this value for the run with 600 iterations. On the other hand, genetic algorithm found its best cycle time of 195 (Figure 8.26(b)), which is 0.5% worse.

For the test problem with 30 operations, tabu search found its best solution with a cycle time of 340 for the runs with 200, 400, and 600 iterations (Figure 8.27(a)). On the other hand, genetic algorithm found its best cycle time of 353 (Figure 8.27(b)), which is 4% higher.

For the test problem with 40 operations, tabu search found its best cycle time of 385 (Figure 8.28(a)). The convergence of solutions in the vicinity of the 385 value was observed. This convergent behavior of the tabu search was consistently observed in the experiments. On the other hand, genetic algorithm found its best cycle time of 391 when the population size was increased to 150 (Figure 8.28(b)). The proposed tabu search has a 1.5% improvement over the genetic algorithm.

For the test problem with 50 operations, tabu search found its best cycle time of 499 (Figure 8.29(a)), while genetic algorithm found its best of 509 (Figure 8.29(b)). A 2% improvement was achieved by the proposed

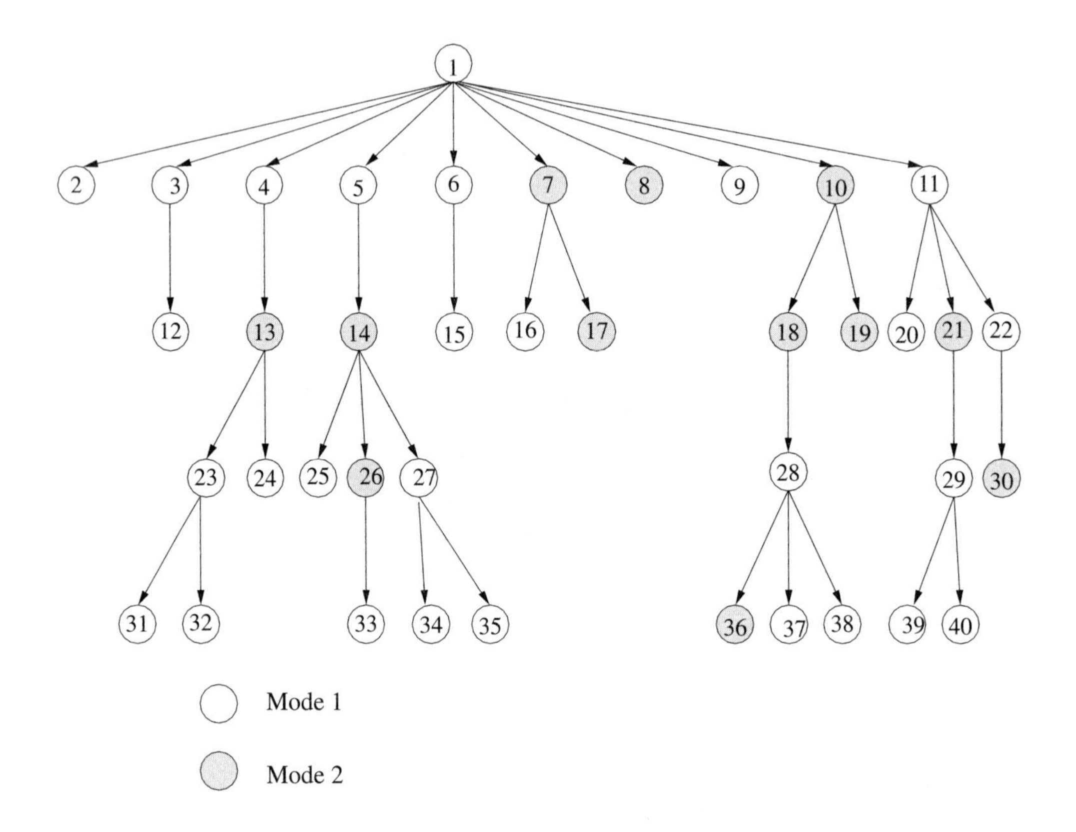

FIGURE 8.22 40-operation problem.

tabu search. Similarly, for the test problem with 60 operations, tabu search found its best of 641 (Figure 8.30(a)), which represents a 1.2% improvement over the genetic algorithm's best cycle time of 643 (Figure 8.30(b)).

The comparison of the two methods is summarized in Table 8.7. Notice that the differences range from 0.00 to 3.82%. Tables 8.8 and 8.5 record the CPU time required by the tabu search procedure and the existing genetic algorithm, respectively. For the small size problems (10-, 20-, and 30-operations), the proposed tabu search consumed less CPU time than genetic algorithms in generating high quality solutions. As the problem size increases from 30 to 40 operations, a dramatic increase of the CPU time for the tabu search was observed (Table 8.8) and the proposed genetic algorithm consumed less CPU time for large size problems. This is due to the $O(n^2)$ neighborhood structure implemented in the tabu search procedure, which requires a complete search of the entire $O(n^2)$ neighborhood in each iteration. Nevertheless, the tabu search procedure always generates better quality solutions than the existing genetic algorithms in this application.

8.8 Conclusion

In this chapter, we presented our study of the optimal planning and sequencing for parallel machining operations. The combinatorial nature of sequencing and the complication of having precedence and mode constraints make the problem difficult to solve with conventional mathematical programming methods. A genetic algorithm and a tabu search algorithm were proposed for finding an optimal solution.

A search technique for generating a feasible initial population and two genetic operators for order-based GAs were proposed and proved to generate feasible offsprings. An analysis on the proposed GA was performed with different parameters. A random problem generator was devised to investigate the overall solutions quality. The experiments showed that the proposed genetic algorithm is capable of finding high quality solutions in an efficient manner.

A tabu search technique was also proposed to solve the problem. The proposed tabu search outperforms the genetic algorithms in all testing cases, although the margin is small.

Additional work on designing better neighborhood structure to reduce the CPU time and on implementing the intermediate-term intensification and long-term diversification strategies in the tabu search process may further enhance the solution quality.

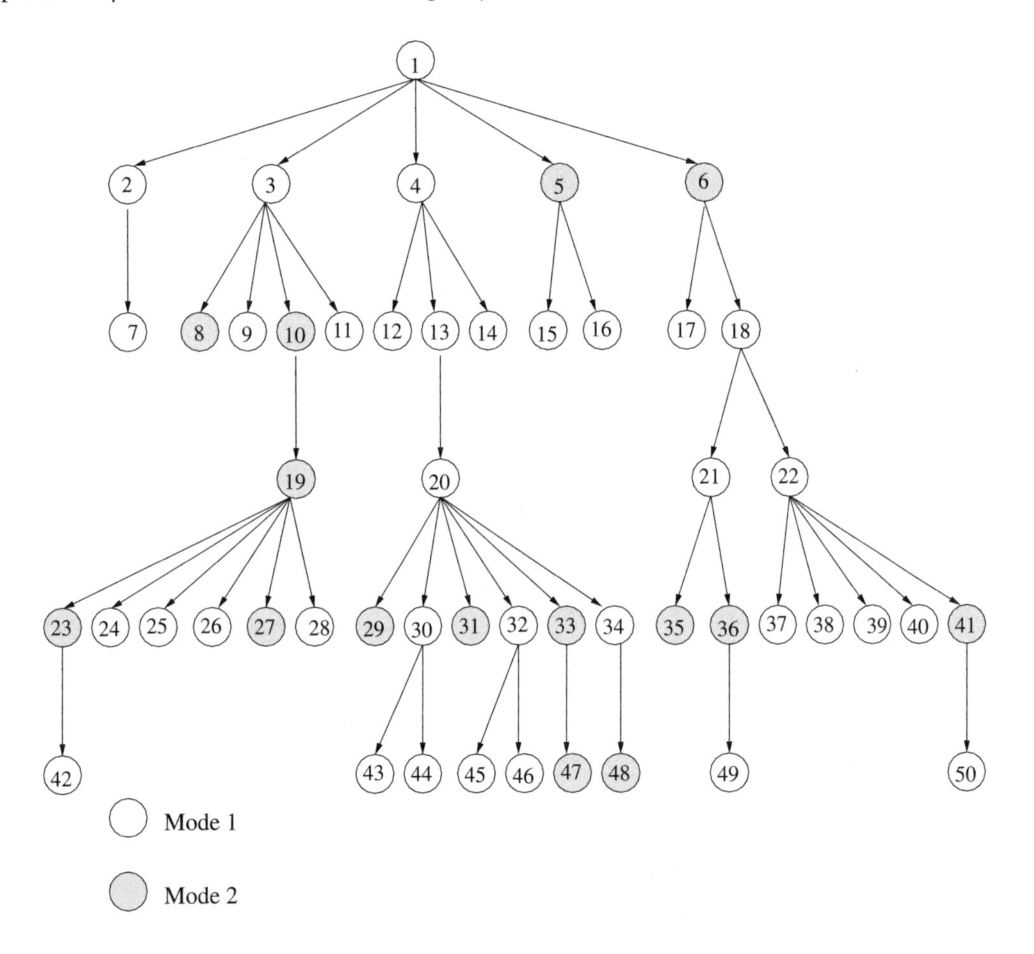

FIGURE 8.23 50-operation problem.

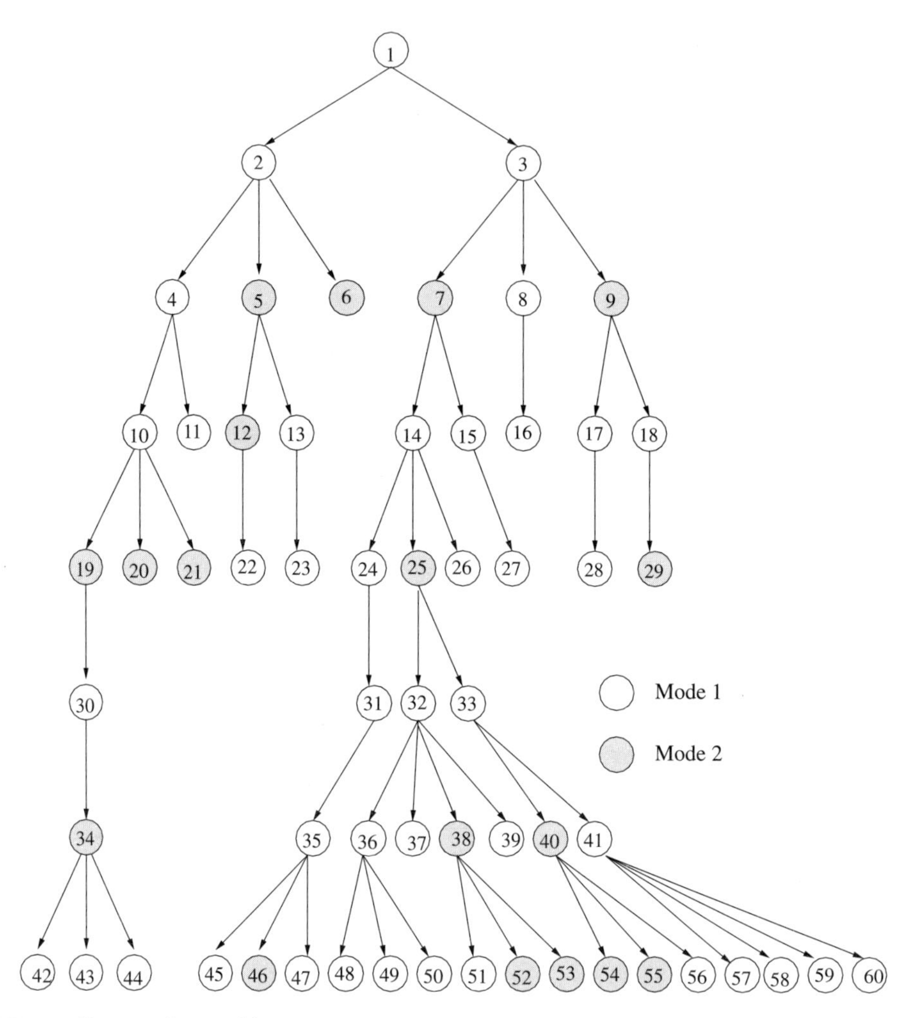

FIGURE 8.24 60-operation problem.

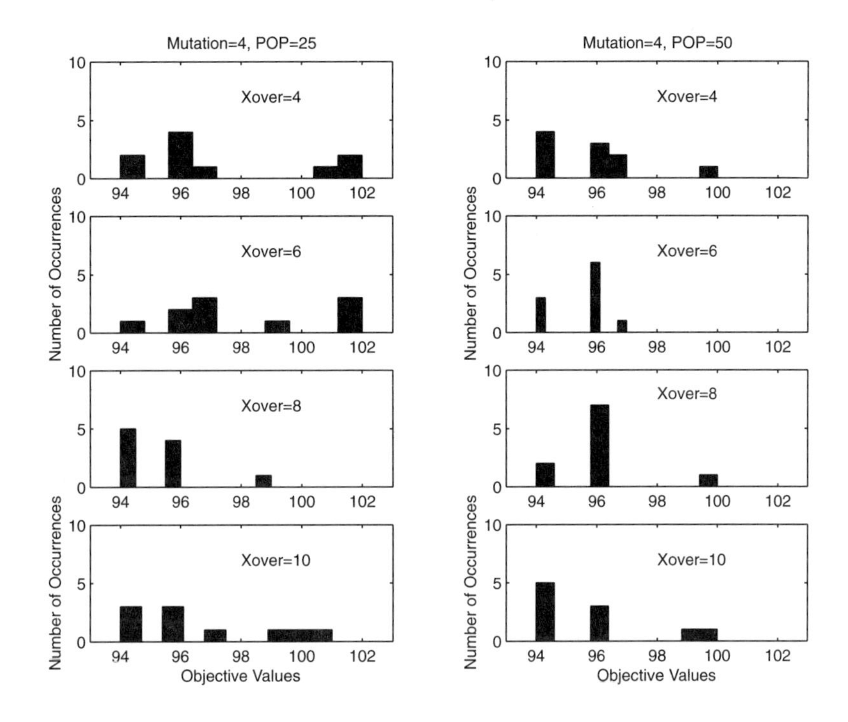

FIGURE 8.25 A 10-operation example, best solution found by GA = 94.

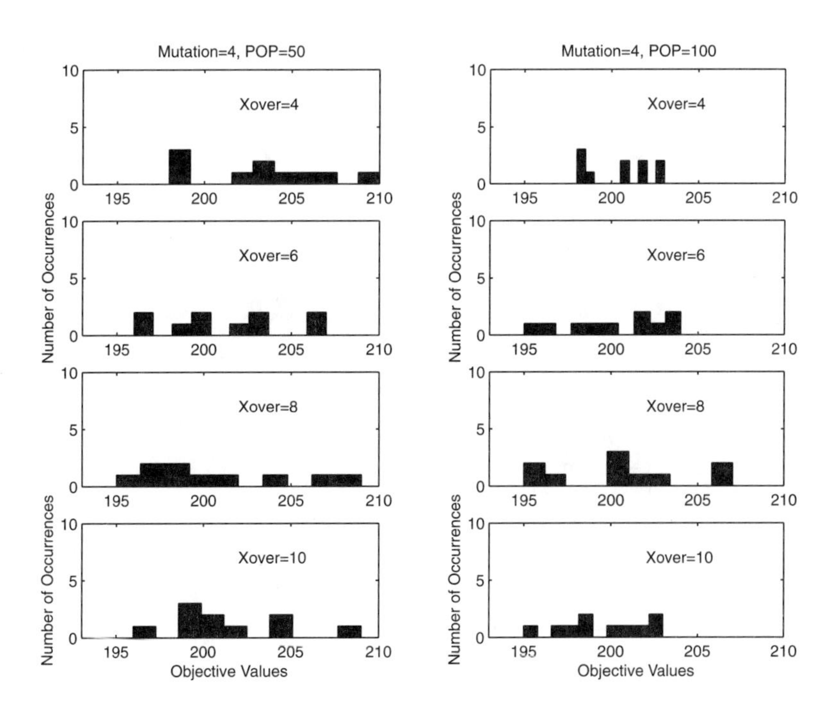

FIGURE 8.26 A 20-operation example, best solution found by GA = 195.

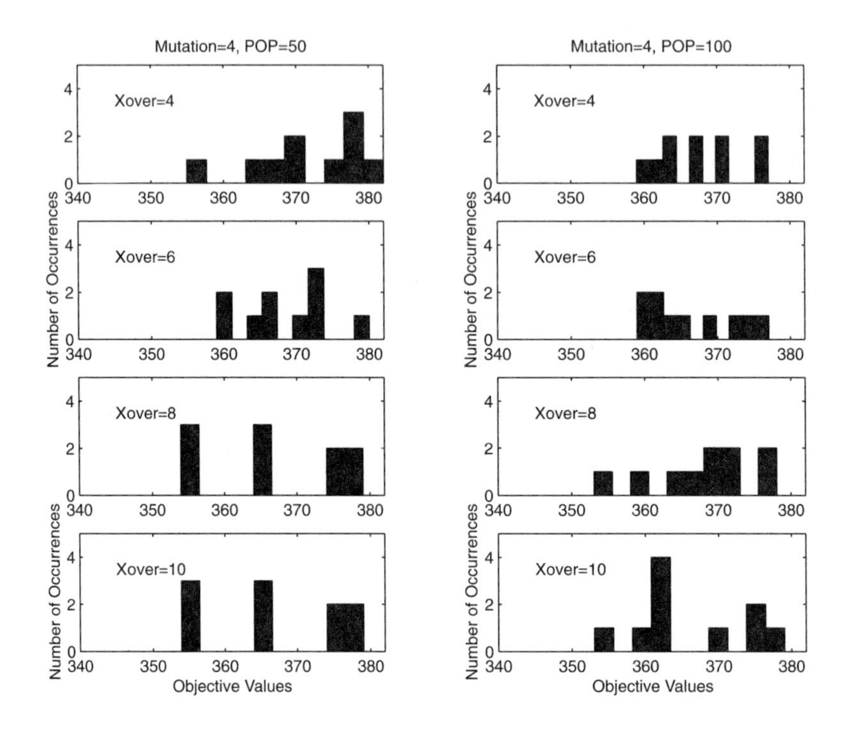

FIGURE 8.27 A 30-operation example, best solution found by GA = 353.

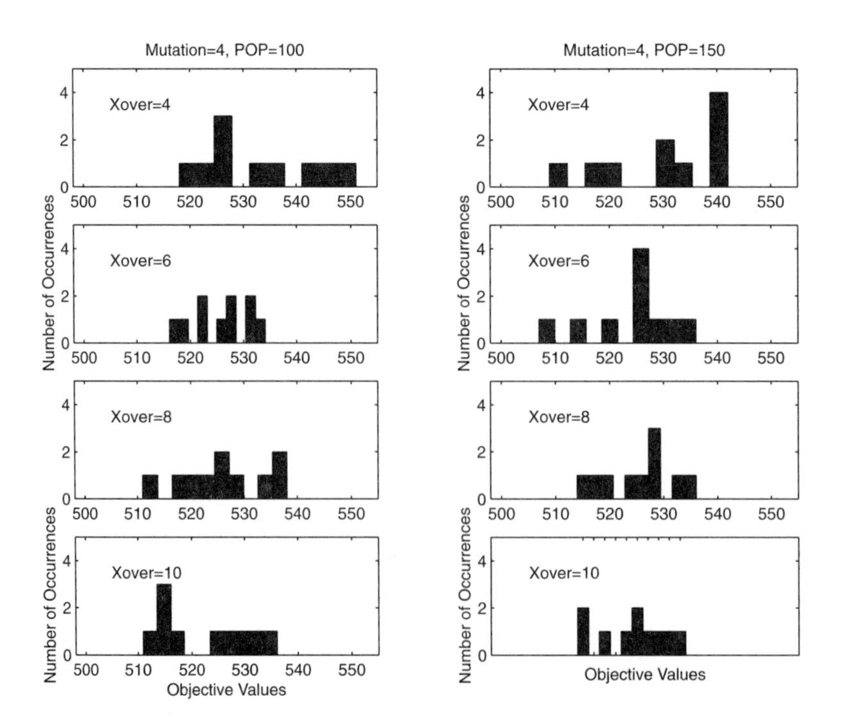

FIGURE 8.28 A 40-operation example, best solution found by GA = 391.

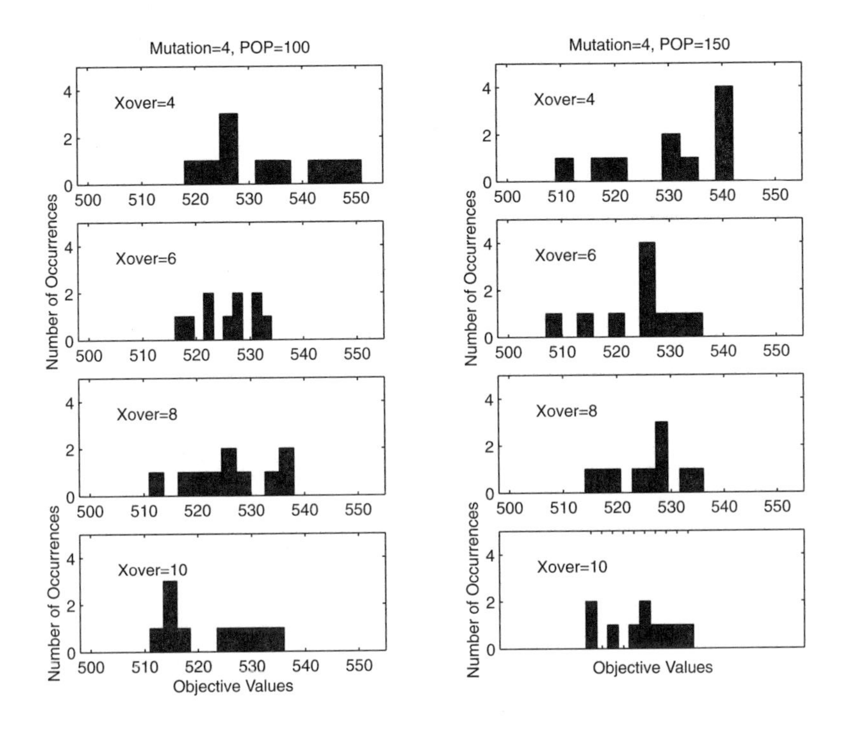

FIGURE 8.29 A 50-operation example, best solution found by GA = 509.

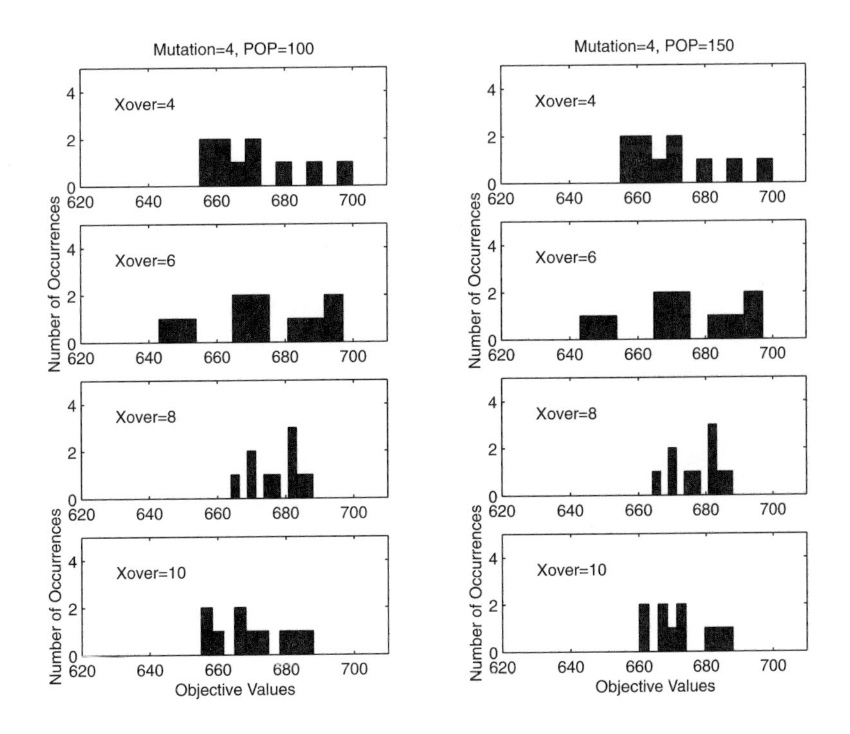

FIGURE 8.30 A 60-operation example, best solution found by GA = 643.

TABLE 8.7 Difference Between the Best Objective Values Found by TS and GA

Operations	*a* Best of TS	*b* Best of GA	Difference in % $\|a - b\|/\min(a, b)$
10	93	94	1.08
18	153	154	0.65
20	194	195	0.52
26	5410	5410	0.00
30	340	353	3.82
40	385	391	1.56
50	501	509	2.00
60	641	643	0.31
average			1.24%

TABLE 8.8 CPU Time (in Hours) of the Tabu Search (TS) Implementation

TS	Iterations			
	100	200	400	600
Operations	CPU Time			
10	0.07	0.14	0.27	0.39
18	0.34	0.68	1.36	2.04
20	0.53	1.05	2.19	3.31
26	0.53	1.09	2.20	3.22
30	1.07	2.16	4.24	6.39
40	3.38	7.10	14.60	22.12
50	7.07	14.12'	28.23	42.35
60	8.74	17.59	35.19	52.78

References

1. S. Chen and S. Smith, Commonality and genetic algorithms, Technical Report, Carnegie Mellon University, CMU-RI-TR-96-27, December, 1996.
2. Y. -S. Lee and C. -J. Chiou, Unfolded projection approach to machining non-coaxial parts on mill-turn machines, *Computers in Industry,* vol. 39, no. 2, 1999, pp. 147–173.
3. N.-C. Chiu, Sequencing Parallel Machining Process by Soft Computing Techniques, Ph.D. Dissertation, Graduate Program in Operations Research, North Carolina State University, Raleigh, Fall 1998.
4. N.-C. Chiu, S.-C. Fang and Y.-S. Lee, Sequencing parallel machining process by genetic algorithms, *Computers and Industrial Engineering,* vol. 36, no. 2, 1999, pp. 259–280.
5. L. Davis, *Handbook of Genetic Algorithms,* Van Nostrand Reinhold, New York, 1991.
6. L. Davis, Applying adaptive algorithms to epistatic domains, *Proceedings of the International Joint Conference of Artificial Intelligence,* 1985, pp. 162–164.
7. D. Dutta, Y.-S. Kim, Y. Kim, E. Wang and D. Yip-Hoi, Feature extraction and operation sequencing for machining on mill-turns, *ASME Design Engineering Technical Conference,* DETC97/CIE-4276, (CD), 1997.
8. M. Gen and R. Cheng, *Genetic Algorithms and Engineering Design,* Wiley, New York, 1997.
9. D. Goldberg and R. Lingle, Alleles, loci, and the TSP, in *Proceedings of the First International Conference on Genetic Algorithms,* J. J. Grefenstette (Ed.), Lawrence Erlbaum Associates, Hillsdale, NJ, 1985, pp. 154–159.
10. F. Glover and M. Laguna, *Tabu Search,* Kluwer Academic Publishers, Boston, 1997.
11. J.B. Levin and D. Dutta, On the effect of parallelism on computer-aided process planning, *Computers in Engineering,* vol. 1, ASME 1992, pp. 363–368.

12. Z. Michalewicz, *Genetic Algorithms + Data Structures = Evolution Algorithms,* Springer-Verlag, New York, 3rd ed., 1996.
13. P. Miller, Lathes turn to other tasks, *Tooling & Production,* March, 1989, pp. 54–60.
14. K.H Miska, Driving tools turn on turning centers, *Manufacturing Engineering,* May, 1990, pp. 63–66.
15. I.M. Oliver, D.J. Smith and J.R.C. Holland., A study of permutation crossover operators on the traveling salesman problem, in *Proceedings of the Second International Conference on Genetic Algorithms,* J.J. Grefenstette (Ed.), Lawrence Erlbaum Associates, Hillsdale, NJ, 1987, pp. 224–230.
16. X.H. Shan, A.Y.C. Nee and A.N. Poo, Integrated application of expert systems and neural networks for machining operation sequencing, in *Neural Networks in Manufacturing and Robotics,* Y.C. Shin, A.H. Abodelmonem and S. Kumara (Eds.), PED-vol. 57, ASME 1992, pp. 117–126.
17. G. Syswerda, Scheduling optimization using genetic algorithms, in *Handbook of Genetic Algorithms,* L. Davis (Ed.), Van Nostrand Reinhold, New York, 1991.
18. J. Váncza and A. Márkus, Genetic algorithms in process planning, *Computers in Industry,* vol. 17, 1991, pp. 181–194.
19. D. Veeramani and A. Stinnes, A hybrid computer-intelligent and user-interactive process planning framework for four-axis CNC turning centers, *Proceedings of the 5th Industrial Engineering Research Conference,* 1996, pp. 233–237.
20. J.M. Usher and R.O. Bowden, The application of genetic algorithms to operation sequencing for use in computer-aided process planning, *Computers and Industrial Engineering,* vol. 30, no. 4, 1996, pp. 999–1013.
21. D. Yip-Hoi and D. Dutta, An introduction to parallel machine tools and related CAPP issues, *Advances in Feature Base Manufacturing,* J.J. Shah, M. Mäntylä and D.S. Nau (Eds.), Elsevier, New York, 1994, pp. 261–285.
22. D. Yip-Hoi and D. Dutta, A genetic algorithm application for sequencing operations in process planning for parallel machining, *IIE Transaction,* vol. 28, no. 1, 1996, pp. 55–68.

9

Application of Genetic Algorithms and Simulated Annealing in Process Planning Optimization

Y. F. Zhang
National University of Singapore

A. Y. C. Nee
National University of Singapore

9.1 Introduction

Process planning represents the link between engineering design and shop floor manufacturing. More specifically, it is the function within a manufacturing facility that establishes the processes and process parameters to be used in order to convert a piece-part from its original form to a final form that is specified on a detailed engineering drawing. For machining, process planning includes determination of the specific operations required and their sequence, which include tooling, fixturing, and inspection equipment. In some cases, the selection of depth of cut, feed rate, and cutting speed for each cut of each operation is also included in the process planning for a part. Due to the advent of computer technology and great market need for short production lead time, there has been a general demand for computer-aided process planning (CAPP) systems to assist human planners and achieve the integration of computer-aided design (CAD) and computer-aided manufacturing (CAM). Moreover, the emerging product design practice employing the design for manufacturing (DFM) philosophy also needs a CAPP system to generate the best process plan for a given part for manufacturing evaluation. In general, a CAPP system is required to be able to provide the following supporting roles to design and manufacturing:

1. Capable of conducting manufacturability assessment on a given part and generating modification suggestions for the poor design features.

0-8493-0592-6/01/$0.00+$.50

2. Capable of generating the best process plan for a given part based on available machining resources in the shop floor. The term "best" refers to a plan that satisfies a predetermined criterion.
3. Capable of generating alternative process plans to suit production scheduling needs and/or changes in the shop floor status.

Over the last 20 years, there have been numerous attempts to develop CAPP systems for various parts manufacturing domains, such as rotational and prismatic parts. The levels of automation among the reported systems range from interactive, variant, to generative [Alting and Zhang, 1989]. The discussion in this chapter focuses on the generative type as it is the most advanced and preferred. In general, the main characteristics of those reported generative CAPP systems, in terms of various decision-making activities in process planning, can be summarised as follows:

1. **Machining features recognition** — The workpiece, both the raw material and finished part, is generally given in a solid model representation. The difference between the finished part and the raw materials represents the volumes that need to be removed. The total removal volumes are extracted and decomposed into individual volumes, called machining features (e.g., holes and slots), which can be removed by a certain type of machining process (e.g., drilling and milling).
2. **Operations selection** — For each machining feature, an operation or a set of operations (e.g., roughing and finishing operations) is selected.
3. **Machines and cutters selection** — For each operation, a machine and a cutter are assigned to it for execution, based on shop floor resources.
4. **Set-up plan generation** — A *set-up* refers to any particular orientation of the workpiece on the machine table together with a fixture arrangement where a number of operations can be carried out. The tool approach direction (TAD) for each operation is determined. A set-up is determined based on the commonality of TAD and fixture arrangement for several operations.

Most reported CAPP systems carry out decision-making activities 2 through 4 in a linear manner, and each is treated as an isolated deterministic problem (see Figure 9.1(a)). Such systems can only generate a feasible process plan by going through one or several iterations of decision-making due to possible conflicts among the different selection results. As a result, they found little industrial acceptance, as the plans generated are far from optimal. The main reasons behind this are

1. For each decision-making problem, the feasible solutions are most likely many but one. Selecting only one solution from each problem leads to premature reduction of the overall solution space and hence the optimal selection combination may well be lost on the way to the final solution.
2. The decision-making problems are interrelated rather than independent of each other. The linear decision-making strategy may create various conflicts that result in many iterations.

In conclusion, these decision-making problems must be considered simultaneously in order to achieve an optimal process plan for a given part (see Figure 9.1(b)). Recently, the authors have noted that some CAPP approaches have been designed to tackle the problems mentioned above by partially integrating the decision-making activities [Chen and LeClair, 1994; Chu and Gadh, 1996; Yip-Hoi and Dutta, 1996; Hayes, 1996; Gupta, 1997; Chen et al., 1998]. However, full integration has not yet been achieved.

When all the decision-making problems are considered simultaneously, process planning becomes a combinatorial problem. Even for a problem with a reasonable number of machining features, it is impossible to check every possible solution. Efficient search algorithms must be developed to find the optimal or near-optimal solution.

In this chapter, a novel algorithm that models process planning in an optimisation perspective is presented. The application of genetic algorithms and simulated annealing to solve the process planning model are discussed together with a case study.

(a) Process planning in a linear manner.

(b) Process planning in a concurrent manner.

FIGURE 9.1 Two typical CAPP approaches.

9.2 Modeling Process Planning Problems in an Optimization Perspective

9.2.1 The Process Planning Domain

The process planning domain for a CAPP system is defined by the types of parts, machining features, and machining environment it deals with. The discussion focuses on prismatic parts. The geometry of a part is represented as a solid model created using a CAD software. The raw material is assumed to be a pre-machined prismatic block that just encloses the part (a convex hull). The basic machining features, which can be used to construct the most commonly encountered prismatic parts, are shown in Figure 9.2. Some of these simple features can be merged together to form complex intersecting machining features. As for the machining environment domain, a CAPP system should be flexible enough to handle common parts in traditional job shops. In this discussion, the job shop information in terms of its machining capability and current status is treated as user input through an open architecture, to update the currently available machining resources (machines, tools) along with their technological attributes such as the dimensions

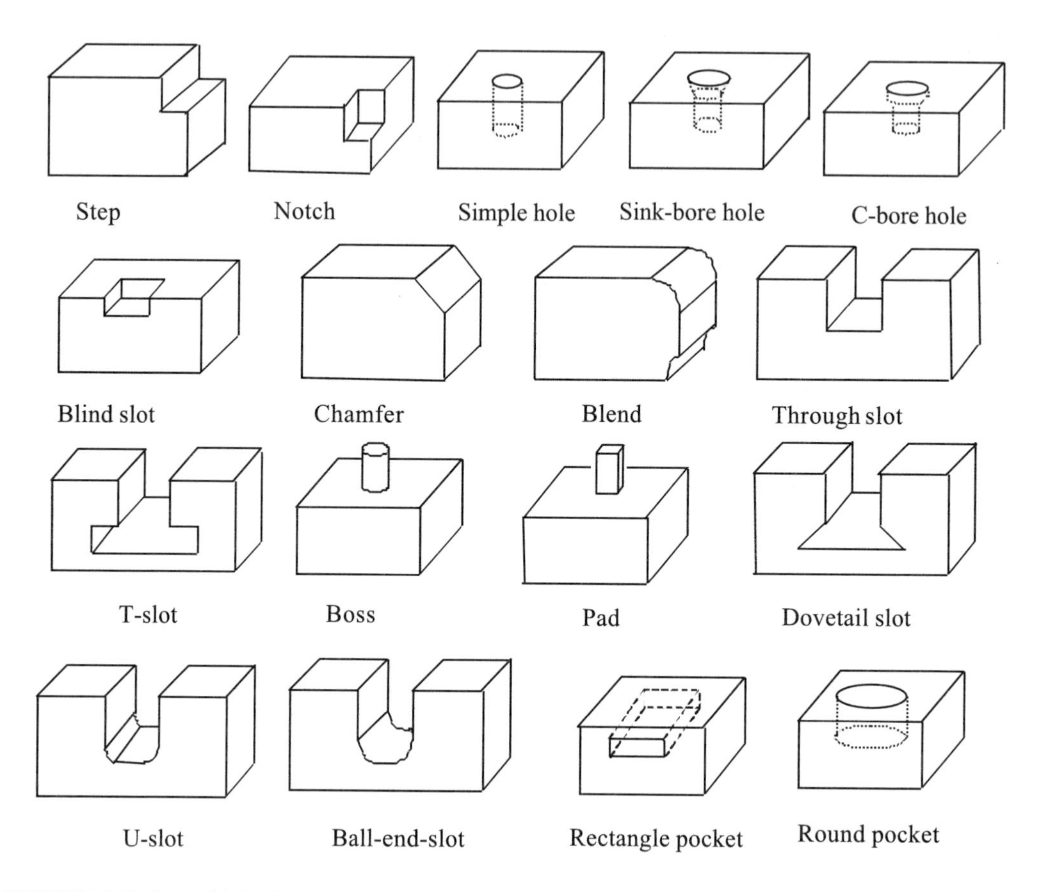

FIGURE 9.2 Basic machining features.

limits, achievable accuracy and surface finish. The machines used include conventional machines (e.g., horizontal and vertical milling, drilling, and grinding) and CNC machining centers.

9.2.2 The Process Planning Problem

Given a part and a set of manufacturing resources in a job shop, the process planning problem can be generally defined as follows:

1. **Operations selection** — For each machining feature, determine one or several operations required. This includes the selection of machines, tools (cutters), TADs, and fixtures based on the feature's geometric and technological specification and available machining resources. This process can also be categorized into two substages: operation type selection and operation method selection. An operation type (OPT) refers to a general machining method without concerning any specific machines and tools (e.g., end-milling and boring). An operation method (OPM), on the other hand, refers to the machine and tool to be used to execute an OPT. It is a common practice for human planners to make decisions over these two stages simultaneously.
2. **Operations sequencing** — Determine the sequence of executing all the OPMs required for the part so that the precedence relationships among all the operations are maintained. Set-ups can then be generated by clustering the neighboring OPMs that share the same machine and fixture arrangement.

It is obvious that decisions made in steps 1 and 2 may contradict each other. Therefore, the decision-making tasks in 1 and 2 must be carried out simultaneously in order to achieve an optimal plan. For

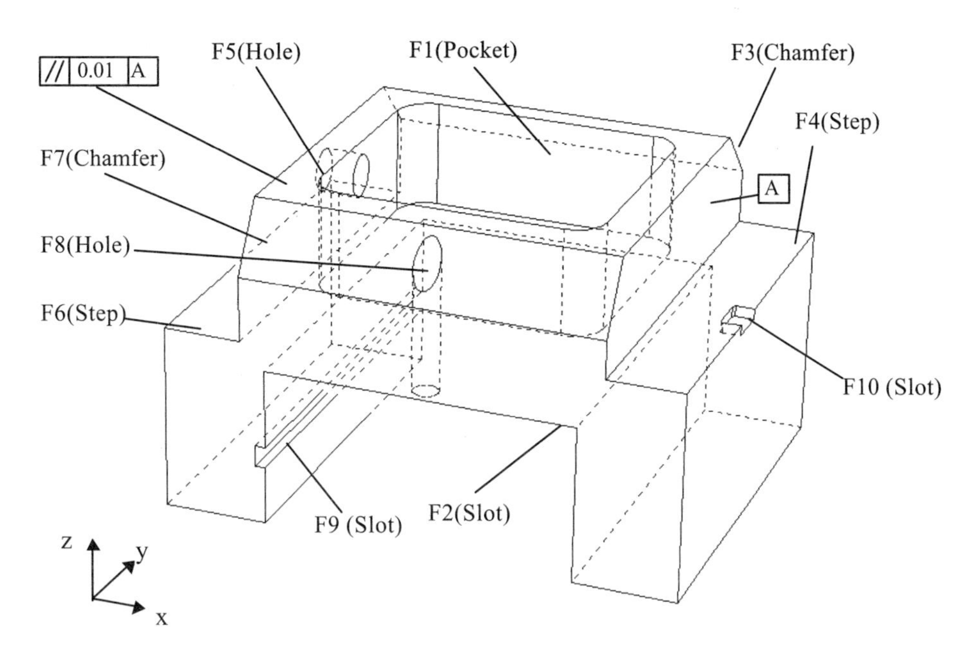

FIGURE 9.3 A machined block.

each of the decision-making tasks described above, process planning knowledge at both global and local levels is required, which is described in the following subsections.

9.2.2.1 OPT Selection

OPT selection is carried out by mapping each machining feature to one or a set of OPTs, based on feature's type, dimensions, tolerance, and surface finish. Feasible solutions can be multiple. For example, the OPT for a through hole of 20 mm in diameter without any tolerance and surface finish requirements is by drilling only, where as the OPT for the same hole but with a surface finish of 3.2 μm can be either (drilling, reaming) or (drilling, boring).

9.2.2.2 Machine (M) and Cutter (T) Selection

For each OPT, a machine and cutter can be selected from the available machined and cutters in the job shop, based on the geometry and accessibility of the feature to which the OPT is applied. Similarly, feasible solutions can be multiple. If no such machine or suitable cutter can be found to perform the OPT, it is eliminated. For example, an end-milling operation is selected for F1 (pocket) in Figure 9.3. A feasible end-mill cutter must have sufficient length to reach the bottom of the pocket and its radius is no more than the corner radius of the pocket.

9.2.2.3 TAD Selection

A TAD for an OPM refers to an unobstructed path along which the tool can approach the feature to be machined. Similarly, feasible TADs for an OPT(M/T) can be multiple. For a prismatic part, six possible TADs, i.e., the six normal directions of a prismatic block ($\pm x$, $\pm y$, $\pm z$), are assumed. For a cutter acting on a feature alone, its theoretical TADs are fixed. However, interference may occur when considering the part and tool geometry. One of the approaches to check such interference is to simulate the movement of the cutter in a solid mode [Ma, 1999]. The solid model of a tool is moved from a predefined path toward the feature along its theoretical TADs. After reaching the feature, it is moved to cover the entire feature. During this simulation, any TAD that causes interference is discarded. If an OPT(M/T) does not have any feasible TADs, it is discarded. Referring to the part shown in Figure 9.3, although drilling a through-hole has two TADs in theory, a drill can only approach F5 along "$+x$."

9.2.2.4 Fixture Selection

Fixtures are mechanical devices used to secure a workpiece on the machine table during machining. For machining of prismatic parts, modular fixtures and vices are the most common fixture devices [Nee et al., 1995]. For an OPT(M/T/TAD), the workpiece geometry is checked to see if a particular fixture can be applied. A vice needs two parallel faces vertical to the machine table while a modular fixture needs three orthogonal faces based on the 3–2–1 principle. Since the workpiece geometry is a result of the OPT sequence, fixture selection can be done during operations sequencing.

9.2.2.5 Precedence Relationships between Operations

Although a possible process plan can be a permutation generated from the OPM pool, it is considered valid only if none of the precedence relationships (PRs) between operations caused by geometrical and technological consideration need is violated. In other words, these PRs have to be identified to check if a randomly generated sequence is valid. In general, the PRs between operations can be identified from the following:

1. **PRs between the realisation of machining features** — These can be derived from process constraints such as fixture constraint, datum dependency, and good machining practices. Some typical process constraints considered are as follows:
   - **Fixture constraint:** A PR between two features exists when machining one feature first may cause another to be unfixturable. An example is given in Figure 9.3 where F2 must be milled before F1, F3, F4, F6, and F7. Otherwise, the supporting area for milling F2 is not sufficient.
   - **Datum dependency:** When two features have a dimensional or geometrical tolerance relationship, the feature containing the datum should be machined first. For example, F4 needs to be machined before F6 since the side face of F4 is the datum of F6 (see Figure 9.3).
   - **Parent–child dependency:** When a feature (A) must be produced before a tool can access another feature (B), A is called the parent of B. For example, F9 can only be machined after F2 is produced (see Figure 9.3).
   - **Avoiding cutter damage:** For two intersecting features A and B, if machining B after the creation of A may cause cutter damage, B should be machined first. Referring to Figure 9.3, if F8 is a blind hole, it should be drilled before F7 in order to avoid cutter damage. However, if F8 is a through hole, this PR does not hold if "+ z" is chosen as the TAD for drilling F8, provided that the slot F2 is not too deep
   - **Better machining efficiency:** For two intersecting features A and B, machining either one first may partially create the other. If machining A has a larger material removal rate (e.g., an end-mall cutter with a larger diameter) than machining B, A should be machined first. Referring to Figure 9.3, the machining of F4 and F10 is intersected. As F10 can only be machined using an end-mill cutter with a much smaller diameter than that for F4, F4 should be machined before F10.
2. **PRs among a set of OPTs for machining a feature** — For every set of OPTs obtained through mapping from a feature, there exists a fixed PR among those OPTs, i.e., roughing operations come before finishing operation (e.g., drilling comes before reaming, milling comes before grinding, etc.).

Since the PRs generated based on the above considerations may contradict each other, they can be categorized into two types: the *hard* PRs and *soft* PRs. The part cannot be manufactured if any of the hard PRs are violated; while the part can still be manufactured even if several soft PRs are violated, although the practice is less preferred. For example, for the part in Figure 9.3, the PR of F2 → F9 is a hard one, while the PR of F4 → F10 is a soft one. By categorizing the PRs into hard and soft ones, a feasible process plan can be achieved when conflicts between the hard PRs and soft PRs arise. Generally, a well-designed part will not present any conflicts between hard PRs.

9.2.2.6 Relationships between OPMs and the Sequence of OPMs

Although operation methods selection and operation methods sequencing need to be carried out simultaneously, in actual operation execution, one is carried out before the other, without considering the result of the latter. To maintain the concurrency between these two activities, the validity of the result from the latter must be checked and ensured to satisfy the result from the former. Depending on which one (OPMs selection or OPMs sequencing) is executed first, the validity of OPMs and a sequence of OPMs is discussed as follows:

1. **Validity of a sequence of OPMs (OPMs selection first)** — The validity of a sequence of OPMs refers to the PRs among the OPMs are satisfied. The influence of selecting OPMs first on the validity of a sequence of OPMs can be seen from the variation of PRs caused by the selected OPMs. In other words, some PRs are invariant regardless of the OPMs selection, while others are the results of the OPMs selection. Referring to Figure 9.3, if F8 is a blind hole, it should be drilled before F7 as discussed previously. However, if F8 is a through hole, this PR does not hold if "+z" is chosen as the TAD for drilling F8, provided that the slot F2 is not too deep. This indicates that sometimes the validity of a PR depends on the selection of TAD or OPMs. These kinds of PRs are designated as *conditional* PRs and the conditions are attached to them. When OPMs are selected first, the conditional PRs can be validated.
2. **Validity of OPMs (OPMs sequencing first)** — The validity of an operation method refers to the feasibility of applying its M/T/TAD under particular circumstances. For instance, an OPT(M/T/TAD) certainly depends on the availability of the M and T, which are naturally ensured during the OPMs selection phase. When a sequence of OPMs (OPM templates, actually) is selected first, some of the TADs identified earlier for the OPMs may become invalid. Referring to Figure 9.3, F8 (through-hole) has two alternative TADs, i.e., "+z" and "–z." If OPM(F7) precedes OPM(F8) in a preselected sequence, OPM(F8) with "–z" will be invalid. This indicates that sometimes the validity of a TAD or OPM depends on the sequence of OPMs. These kinds of TADs are designated as *conditional* TADs and the conditions are attached to them. When a sequence of OPMs is selected first, the conditional TADs can be validated, as well as the OPMs they belong to.

9.2.2.7 Grouping Effect on the Validity of OPMs

In addition, there are situations where a group of OPMs are needed for producing a feature. In that case, the validity of an OPM may depend on its neighbouring OPM. For instance, a through hole can be produced through (centre-drilling, drilling, reaming) from a solid part. Although drilling may have two opposite TADs in theory, in the OPM selection for drilling, it must have the same TAD as centre-drilling in the same group.

9.2.3 The Process Plan Evaluation Criterion

In the last section, the solution space for a process planning problem is described based on discussions on various selection stages and constraints. It is obvious that there can be many feasible solutions for a given process planning problem. Therefore, there is a need to find the best plan. In order to achieve this, an evaluation criterion must be defined.

Currently, the most commonly used criteria for evaluating process plans include shortest processing time and minimum processing cost. Since the detailed information on tool paths and machining parameters is not available at the stage where only operation methods and their sequences are determined, the total processing time cannot be accurately calculated for plan evaluation. On the other hand, the processing cost can be approximated. Generally, the processing cost for fabricating a part consists of two parts, i.e., cost due to the usage of machines and tools and cost due to set-ups (machine change, set-up change on the same machine, and tool change). For the cost due to machine usage, a fixed cost is assumed every time a particular machine or a tool is used. For the cost due to set-ups, a fixed cost is assumed when a particular set-up is required. Based on these assumptions, five cost factors (machine usage, tool usage, machine change, set-up change, and tool change) for a process plan can be derived as follows:

1. Machine usage cost (*MC*)

$$MC = \sum_{i=1}^{n} MCI_i \qquad \text{Equation (9.1)}$$

   where n is the total number of OPMs in the process plan and MCI_i is the machine cost index for using machine i, a constant for a particular machine.

2. Tool usage cost (*TC*)

$$TC = \sum_{i=1}^{n} TCI_i \qquad \text{Equation (9.2)}$$

   where TCI_i is the tool cost index for using tool i, a constant for a particular tool.

3. Machine change cost (*MCC*): a machine change is needed when two adjacent operations are performed on different machines.

$$MCC = MCCI \times \sum_{i=1}^{n} \Omega(M_{i+1} - M_i) \qquad \text{Equation (9.3a)}$$

$$\Omega(M_i - M_j) = \begin{cases} 1 \text{ if } M_i \neq M_j \\ 0 \text{ if } M_i = M_j \end{cases} \qquad \text{Equation (9.3b)}$$

   where *MCCI* is the machine change cost index, a constant, and is the ID of the machine used for operation i.

4. Set-up change cost (*SCC*): a set-up change is needed when two adjacent OPMs performed on the same machine have different part orientations. Generally, when a part is placed on a machine table, a TAD along which an OPM approaches a feature dictates the orientation of the part. Therefore, TADs are used to represent the required orientations of the part.

$$SCC = SCCI \times \sum_{i=1}^{n-1} \left(\left(1 - \Omega\left(M_{i+1} - M_i\right)\right) \times \Omega\left(TAD_{i+1} - TAD_i\right) \right) \qquad \text{Equation (9.4)}$$

   where *SCCI* is the set-up change cost index, a constant.

5. Tool change cost (*TCC*): a tool change is needed when two adjacent OPMs performed on the same machine use different tools.

$$TCC = TCCI \times \sum_{i=1}^{n-1} \left(\left(1 - \Omega\left(M_{i+1} - M_i\right)\right) \times \Omega\left(T_{i+1} - T_i\right) \right) \qquad \text{Equation (9.5)}$$

   where *TCCI* is the tool change cost index, a constant.

The sum of these five cost factors approximates the total processing cost. On the other hand, these cost factors can also be used either individually or collectively as a cost compound based on the requirement and the data availability of the job shop. For example, if a process plan with minimum number of set-up changes is required, the evaluation criterion is set as *SCC*. This criterion setting provides much flexibility required by different job shops.

9.2.4 An Optimization Process Planning Model

Based on the above discussion, the solution space for a process planning problem can be explicitly formed by all the OPTs required, the available OPMs for each OPT, possible sequences, and various constraints. A modelling algorithm to obtain this solution space is described as follows:

Algorithm: process planning modelling

Input: A set of features, job shop information (machines and tools).
Output: Number of OPTs required, OPMs (M/T/TAD) for each OPT, precedence relationships between OPTs.

1. For each feature, find all the OPTs that can achieve its attributes (shape, dimensions, tolerances, and surface finish), based on shop level machining capabilities. The resulting OPTs for a feature can be expressed as one or several sets of (OPT_1, OPT_2, . . . , OPT_k), where k is the total number of OPTs required in a set. It is worth mentioning that different sets of OPTs for a feature may have different number of OPTs. For example, a blind hole can have two sets of OPTs: (centre-drill, drill, ream) and (centre-drill, mill). To achieve a uniform representation, the concept of a "dummy" OPT (D-OPT) is introduced, which is defined as an operation incurring no machining cost and having no effects in terms of machine, set-up, and tool changes. By adding the dummy operation into the sets that have fewer operations, each set will have the same number of OPTs. For the above example, the two sets of OPTs for producing a hole can be expressed as (centre-drill, drill, ream) and (centre-drill, D-OPT, mill). Finally, the total number of OPTs required for the part is determined.
2. Identify all the PRs among features as well as among the OPTs in each set for a feature. Convert the PRs between two features to the PRs between their OPTs according to the following rule:

IF	$F(X) \rightarrow F(Y)$
AND	$OPT(i) \in F(X)$
AND	$OPT(j) \in F(Y)$
THE	$OPT(i) \rightarrow OPT(j)$

   As a result, both *hard* and *soft* PRs are identified. The hard and soft PRs are compared pair-wise. If a conflict arises, the respective soft PR in the respective conflict is deleted.
3. For each OPT, find all the combinations of machines (Ms) and tools (Ts) with which it can be executed, based on machine-level machine capabilities.
4. For each combination of M and T, find all the feasible TADs.
5. Attach a condition to a TAD if it depends on a particular sequence or the assignment of a TAD to a particular OPT. Similarly, attach a condition to a PR if it depends on the assignment of a TAD to a particular OPT.

End of algorithm process planning modelling.

The output from the above algorithm is a fixed number of OPTs required for fabricating a part and along with each OPT, one or several sets of (M/T/TAD). The PRs among the OPTs are also explicitly represented. To further illustrate this novel process planning modelling technique, the part shown in Figure 9.4 is used as an example. It is constructed by adding several machining features to a chuck–jaw in order to pose certain difficulty for process planning. The dimensions and tolerances are also shown in Figure 9.4. The machining process starts from a rectangular block. It is assumed that the block is pre-machined to a size of $160 \times 50 \times 70$. Therefore, the part consists of 14 features as shown in the figure. This part is assumed to be machined in a job shop, in which the available machines are: one three-axis conventional vertical milling machine (M1, $MCI = 35$), one three-axis CNC vertical milling machine (M2, $MCI = 70$), one drill press (M3, $MCI = 10$), one conventional horizontal milling machine (M4,

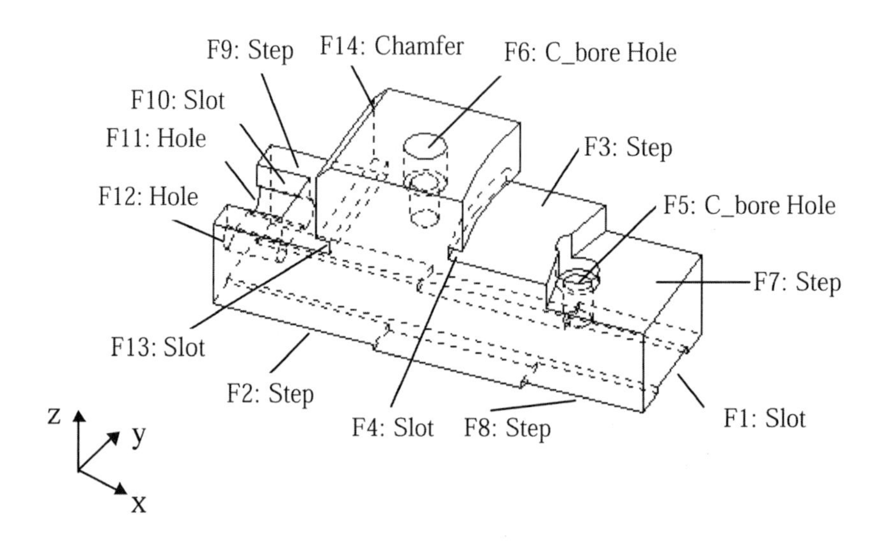

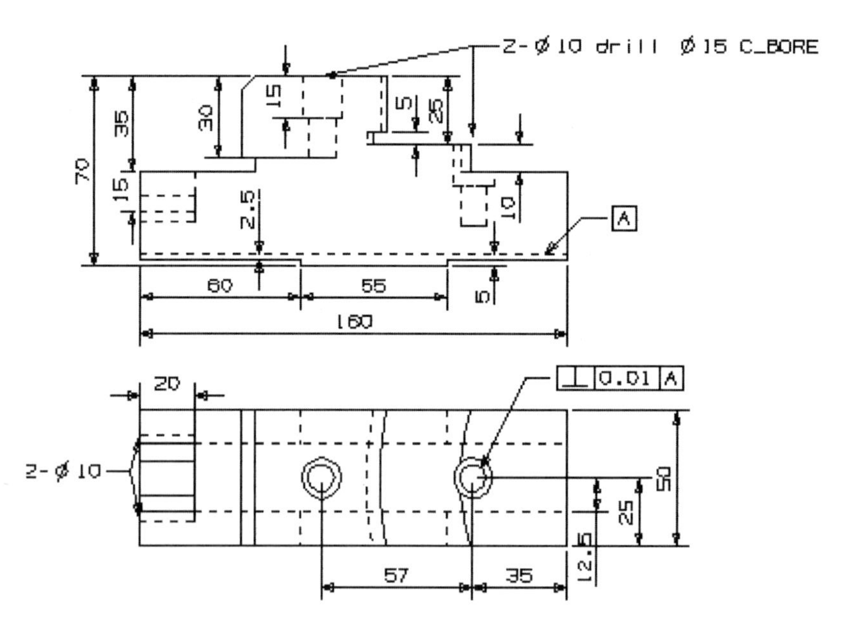

FIGURE 9.4 An example part and specifications.

MCI = 50), and one CNC horizontal milling machine (M5, *MCI* = 85). It is assumed that all the machines are capable of accommodating the part. The available cutting tools with their dimensions as well as their cost indices are shown in Table 9.1. The other cost indices are assumed to be *MCCI* = 110, *SCCI* = 90, and *TCCI* = 20. The process planning modelling process for this part is described as follows:

1. **OPTs** — Based on the shop floor machining capability, the identified OPTs are shown in column 3 in Table 9.2. It can be seen that only F5 and F6 need three OPTs (centre-drilling, drilling, and end-milling) each, while the remaining features need one OPT each (the two holes, F11 and F12, do not need centre-drilling, as there is no special tolerance requirement). The total number of OPTs required to fabricate this part is 18.

TABLE 9.1 Cutting Tools and Their Cost Indices (*TCI*) Used in the Job Shop

Tool	Type (Diameter, Flute Length)	TCI	Tool	Type (Diameter, Flute Length)	TCI
T1	End_mill (20, 30)	10	T10	Centre_drill	2
T2	End_mill (30, 50)	10	T11	Angle_cutter (40, 45°)	10
T3	End_mill (15, 20)	10	T12	Drill (70, 100)	5
T4	End_mill (40, 60)	12	T13	Drill (8, 30)	6
T5	Side_mill (100, 40)	8	T14	Drill (10, 35)	3
T6	T-slot_cutter (30, 15[a])	16	T15	T-slot)_cutter (20, 5[a])	6
T7	Drill (20, 55)	3	T16	Drill (5, 30)	3
T8	Drill (30, 50)	3	T17	Drill (15, 50)	4
T9	Drill (50, 80)	4	T18	Side-angle_mill (50, 45°)	10

[a] For T-slot_cutter, the second parameter in the bracket represents the width of the cutter.

TABLE 9.2 OPTs and Possible OPMs (M/T/TAD) for the Part in Figure 9.4

OPT	Feature	Type	Machine	Tool	TADs
OPT1	F1	Milling	M1, M2	T1, T3	$+z$
OPT2	F2	Milling	M1, M2	T1, T2, T3, T4	$+z$
			M4, M5	T5	$+x$
OPT3	F3	Milling	M1, M2	T1, T2, T4	$-z$
OPT4	F4	Milling	M1, M2	T15	$-z$
OPT5	F5	Center_drilling	M1, M2, M3, M4, M5	T10	$-z$
OPT6	F5	Drilling	M1, M2, M3, M4, M5	T14	$-z$
OPT7	F5	Milling	M1, M2	T3	$-z$
OPT8	F6	Center_drilling	M1, M2, M3, M4, M5	T14	$-z$
OPT9	F6	Drilling	M1, M2, M3, M4, M5	T14	$-z$
OPT10	F6	Milling	M1, M2	T3	$-z$
OPT11	F7	Milling	M1, M2	T1, T2, T3, T4	$-z$
OPT12	F8	Milling	M1, M2	T1, T2, T3, T4	$+z$
			M4, M5	T5	$-x$
OPT13	F9	Milling	M1, M2	T2, T4	$+x, -z$
			M4, M5	T5	$+x, -z$
OPT14	F10	Milling	M1, M2	T3	$+x$
OPT15	F11	Drilling	M1, M2, M3, M4, M5	T14	$+x$
OPT16	F12	Drilling	M1, M2, M3, M4, M5	T14	$+x$
OPT17	F13	Milling	M1, M2	T15	$-z$
OPT18	F14	Milling	M1, M2	T11	$+x, -z$
			M4, M5	T18	$+x, -z$

2. **PRs** — Firstly, the PRs between features are identified. F1 should be produced after F2 and F8 for better machining efficiency. F4 should be produced after F3 due to parent–child relationship. F5 should be produced after F3, otherwise the drill (T14) would not be able to reach the bottom of F5. On the other hand, F5 should also be produced after F1 as F1 is the datum feature of F5. F7 should be produced after F5 to avoid cutter damage. F7 should also be produced after F1 to ensure sufficient supporting area for machining F1. F11 and F12 should be drilled before F10 to avoid cutter damage. F14 should be machined after F9 due to parent–child relationship. Similarly, F13 should be machined after F9. These PRs between features are then converted to PRs between their respective OPTs. Secondly, the PRs between OPTs for every individual feature are identified, where only features needing multiple OPTs are concerned. As a result, OPT5 should proceed OPT6 and OPT6 proceed OPT7 for F5; OPT8 proceed OPT9 and OPT9 proceed OPT10 for F6. Finally, these two sets of PRs between OPTs are merged, as shown in Table 9.3.

TABLE 9.3 Precedence Relationships between OPTs for the Part in Figure 9.4

OPTS	Predecessors	OPTs	Predecessors
OPT1	OPT2, OPT12	OPT10	OPT8, OPT9
OPT2	Nil	OPT11	OPT1, OPT5, OPT6, OPT7
OPT3	Nil	OPT12	Nil
OPT4	OPT3	OPT13	Nil
OPT5	OPT1, OPT3	OPT14	OPT15, OPT16
OPT6	OPT1, OPT3, OPT5	OPT15	Nil
OPT7	OPT1, OPT3, OPT5, OPT6	OPT16	Nil
OPT8	Nil	OPT17	OPT13
OPT9	OPT8	OPT18	OPT13

3. **Ms and Ts** — The suitable machines and tools are selected to perform each OPT based on the consideration of the shape-producing capability of the machine, tool dimension, and the respective feature geometry. The results are shown in column 4 and column 5 of Table 9.2. If an OPT has more than one row of Ms and Ts in the table, the possible combinations of Ms and Ts are derived among the alternatives in the same row. For example, the possible combinations (M/T) for OPT18 are M1/T11, M2/T11, M4/T18, and M5/T18.
4. **TADs** — The suitable TADs for each OPT (M/T) are identified as shown in column 6 of Table 9.2. The representation of TADs is based on the coordinate system (*x–y–z*) shown in Figure 9.4. Similarly, the possible combinations (M/T/TAD) for every OPT are derived among the alternatives in the same row.
5. **Conditional TADs and PRs** — For this example, there are no conditional TADs or conditional PRs.

Up to this point, an explicit solution space for the example is successfully developed. The next step is to choose an objective function (cost compound) and "to identify a combination of (M, T, TAD) for every OPT and put them into an order that does not violate any precedence relationships between any two OPTs while achieving the least cost compound." Based on this definition, there could be more than one optimal solution for a given process planning problem, and each has the same cost compound but different operation methods and/or sequence order.

For a part needing n OPTs and if OPT-i has $m(i)$ alternatives, the total number of feasible process plans (K) for the part is expressed as

$$K - \left(n! - \kappa\right) \times \prod_{i=1}^{n} m\left(i\right) \qquad \text{Equation (9.6)}$$

where κ is the number of invalid sequences. Obviously, to find the best plan by enumerating all the feasible process plans is a NP-hard problem. Therefore, optimization search algorithms should be used to find an optimal or near-optimal solution.

9.2.5 Set-Up Plan Generation

Upon solving the process planning model using a search method, the output is a set of OPMs (M/T/TAD) in a sequence. A set-up plan can then be generated by grouping the neighboring OPMs, which share the same machine and TAD, into the same set-up. The details are given as follows:

Algorithm: set-up plan generation

Input: All the OPMs (M/T/TAD) in a sequence.
Output: Set-ups in sequence; each includes a set of OPMs in sequence.

1. Start with the first set-up with the first OPM in the total OPMs sequence as its first OPM. This set-up is called the *current set-up* and the first OPM the *current OPM.*

2. If the current OPM is the last OPM in the total OPMs sequence, a solution has been found.
3. Check the OPM next to the current OPM in the total OPMs sequence. If it shares the same M and TAD with the current OPM, group it into the current set-up and this OPM becomes the current OPM. Otherwise, close the current set-up and start a new set-up with this OPM as its first OPM. This OPM becomes the current OPM and the new set-up becomes the current set-up.
4. Go to 2.

End algorithm set-up plan generation

9.3 Applying a Genetic Algorithm to the Process Planning Problem

The genetic algorithm (GA) is a stochastic search technique based on the mechanism of natural selection and natural genetics. It starts with a set of random solutions called *population,* and each individual in the population is called a *chromosome.* The chromosomes evolve through successive iterations, called *generations,* during which the chromosomes are evaluated by means of measuring their *fitness.* To create the next generation, new chromosomes called *offspring* are formed by two genetic operators, i.e., *crossover* (merging two chromosomes) and *mutation* (modifying a chromosome). A new generation is formed by *selection,* which means that chromosomes with higher fitness values have higher probability of being selected to survive. After several generations, the algorithm converges to the best chromosome, which hopefully represents the optimal or near-optimal solution to the problem. Recently, GA has been applied to many combinatorial problems such as job shop scheduling and the travelling salesman problem. To formulate a GA for process planning, the following tasks need to be carried out:

1. Develop a process plan representation scheme
2. Generate the first population
3. Set up a fitness function
4. Develop a reproduction method
5. Develop GA operators (crossover and mutation)
6. Develop GA parameters (population size, crossover and mutation rates, and stopping criterion)

9.3.1 Process Plan Representation

A string is used to represent a process plan. For an *n*-OPT process planning problem, the string is composed of *n* segments. Each segment contains a M/T/TAD combination from the alternatives of an OPT and its order in the string. This representation scheme can cover all the solution space due to the selection of machines, tools, TADs, and sequences among the OPTs.

9.3.2 Initial Population

The strings in the initial population need to be randomly generated, and at the same time they must satisfy the PRs. An algorithm to randomly generate a valid string is described as follows:

Algorithm: generate a string

Input: All the alternative M/T/TADs for the required OPT, PRs.
Output: A valid string.

1. Start with an empty sequence.
2. Select (at random) an OPT among those that have "no predecessors" and append it to the end of the sequence. Randomly select a set of M/T/TAD for the OPT from all its alternative OPMs.
3. Delete the OPT just handled from the OPT space, as well as from the PRs.
4. If there are OPTs that have not been handled, go to 1.

End of algorithm generate a string

9.3.3 Fitness Function

A fitness function in a GA acts as an objective function in an optimization process. The fitness function used here is a cost compound constructed using the cost factors in Equations 9.1 through 9.5. Therefore, the lower the fitness of a string is, the better the quality of the process plan represented by the string.

9.3.4 Reproduction

The reproduction operator works in two steps. First, it applies "elitism" by copying the string solution having the lowest cost value, thus keeping the cost function nonincreasing. Second, it uses the "roulette wheel" method for the reproduction of the remaining string solutions.

9.3.5 Crossover and Mutation Operators

The crossover operator is developed to change the order between the OPTs in a string solution. Through crossover, the strings obtained from reproduction are mated at a given probability (crossover rate P_c). To ensure that a crossover will not result in any violation of PRs and each OPT appears in the offspring once and only once, a cyclic crossover operator is developed, which is described as follows:

Algorithm: crossover of two strings

Input: string-1 and string-2.
Output: offspring-1 and offspring-2.

1. Determine a cut point randomly from all the positions of a string. Each string is divided into two parts, the left side and the right side, according to the cut point.
2. Copy the left side of string-1 to form the left side of offspring-1. The right side of offspring-1 is constructed by the OPTs in the right side of string-1 and their sequence is reorganized according to the order of these OPTs in string-2.
3. Copy the left side of string-2 to form the left side of offspring-2. The right side of offspring-2 is constructed by the OPTs in the right side of string-2 and their sequence is reorganized according to the order of OPTs in string-1.

End of algorithm crossover of two strings

This crossover algorithm is illustrated with an example shown in Figure 9.5, in which string-1 and string-2 are under crossover operation. The cut point is chosen between positions 3 and 4. The left side of string-1, OPT4–OPT1–OPT2, is used to form the left side of offspring-1. The order of the rightside of string-1, OPT5–OPT7–OPT8–OPT3–OPT6–OPT9, is adjusted according to the order of these OPTs in string-2 to form the right side of offspring-1. By doing so, the order among the OPTs in both string-1 and string-2 is maintained in offspring-1. Offspring-2 is formed similarly.

Three mutation operators are developed to change the machine, tool, and TAD used for any OPT in a string. To improve the searching efficiency, process planning heuristics are incorporated into these mutation operators. The three mutation operators have the same mechanism except the subject to be changed. Therefore, only machine mutation is described here.

Machine mutation is used to change the machine (current machine) to perform an OPT if more than one machine can be applied. To reduce the total machine changes, machine mutation does not stop at the selected OPT. Instead, the machine alternatives for the remaining OPTs are also checked. If the current machine is used by any of the remaining OPTs and at the same time the OPT has the selected machine in its alternatives, the selected machine is selected to replace the current machine for the OPT. By doing this, the number of machine changes can be reduced intuitively. The algorithm is as follows:

Algorithm: machine mutation

Input: A string in a generation, all machine alternatives for all OPTs.
Output: A new string.

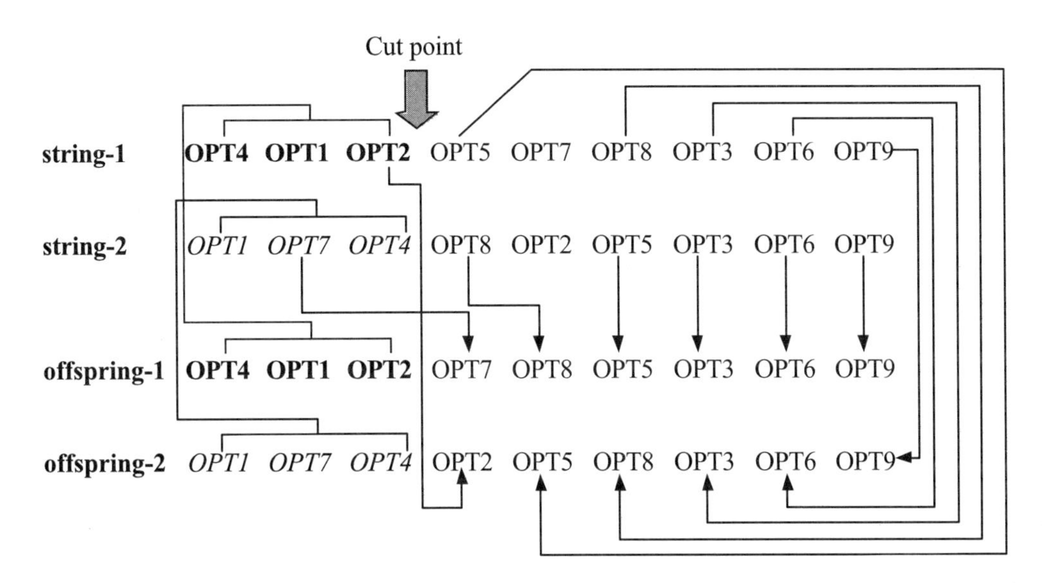

FIGURE 9.5 An example of applying the cyclic crossover for changing OPT sequence.

1. Select an OPT randomly in the string and use a predetermined probability (machine mutation rate P_m^m) to determine if the machine for the OPT needs to be changed.
2. Randomly select a machine (M-b) from all the alternative machines of the OPT to replace the current machine (M-a).
3. Identify all the other OPTs in the string in which the current machine used is M-a. If any one of these OPTs has M-b as an alternative, assign M-b to replace M-a.

End of algorithm machine mutation

This mutation is illustrated with an example shown in Figure 9.6. It can be seen that OPT3 is selected for mutation where M-a is the current machine. M-c is randomly selected from M-c and M-e and assigned to OPT3 to replace M-a. It is then found that M-a is also the current machine used by OPT1, OPT4, OPT5, and OPT2. Among these OPTs, OPT1, OPT4, and OPT5 have M-c as one of their alternative machines. Therefore, M-c is assigned to OPT1, OPT4, and OPT5 to replace M-a. Compared with single point mutation, this group mutation can accelerate the convergence speed of the GA. By utilizing this mechanism, mutation is not an arbitrary action only to introduce new schema but "intends" to introduce better schemata into the gene pool.

Through crossover operation, the solution space formed by possible sequences can be traversed, or in other words, operations sequence randomization can be performed. On the other hand, the solution space formed by the alternative machines, tools, and TADs for OPTs is traversed through machine mutation, followed by tool mutation and TAD mutation. Therefore, the concurrent decision making mechanism is realized in this GA, i.e., operations routing and sequencing are carried out simultaneously. It is worth mentioning that at the end of tool and TAD mutation processes, the resulted M/T and M/T/TAD combinations need to be checked against the available M/T/TAD combinations. If a match cannot be found, the selected T or TAD is abandoned and the respective mutation process (tool or TAD) is repeated until such a match is found. Moreover, since the sequence among the OPTs is selected first, the resulting TAD from TAD mutation needs to be validated against the sequence if it is a *conditional* TAD.

9.3.6 GA Parameters

After a GA's mechanism has been devised, its control parameters must be selected. These parameters include population size, crossover rate, mutation rate, and stopping criterion:

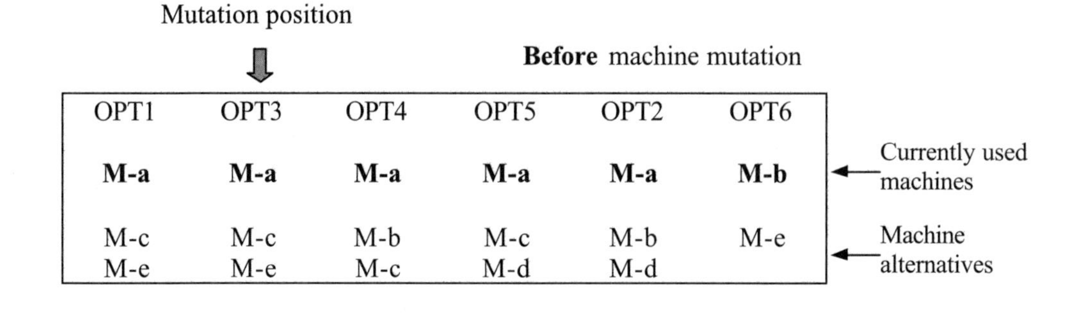

FIGURE 9.6 An example of machine mutation with six OPTs.

1. **Population size** — In principle, the population size should be sufficiently large such that the population associated with the solution space can be adequately represented. A larger population, however, needs larger computation cost in terms of memory requirement and computation time. The authors' experiments show that a population size of 50 works well for most of the process planning problems tested.
2. **Crossover rate P_c** — Most GA literature suggests that crossover rate should be set between 0.5 and 1.0. In the present GA formulation, crossover is equivalent to a change of operation sequence that should therefore be vigorously performed to traverse more points in the solution space. Based on this principle, P_c is set as 0.7.
3. **Mutation rate** — Generally, mutation acts as a background operator that provides a small amount of random search. Its purpose is to maintain diversity within the population and inhibit premature convergence due to loss of information. Mutation rate is often in the range of [0.001, 0.1]. However, in the GA for process planning, the three mutation operations should play a role similar to the crossover operation since the extended solution space due to the availability of alternative machines, tools, and TADs must be adequately traversed in the optimization process. Therefore, mutation should be performed with vigor equal to the crossover. Through trial and error, the three mutation rates are set equally as 0.6.
4. **Stopping criterion** — Since GA is a stochastic search method, it is quite difficult to formally specify a convergence criterion, as it is often observed that the fitness of a population may remain static for a number of generations before a superior string is found. For the GA developed, a number of numerical simulation experiments have been conducted based on different stopping criteria including a fixed number of generations and a fixed number of generations in which no improvement is gained. It is observed that all the tested cases (number of OPTs less than 40) achieved satisfactory results after 8000 generations. Therefore, 8000 generations is chosen as the stopping criterion.

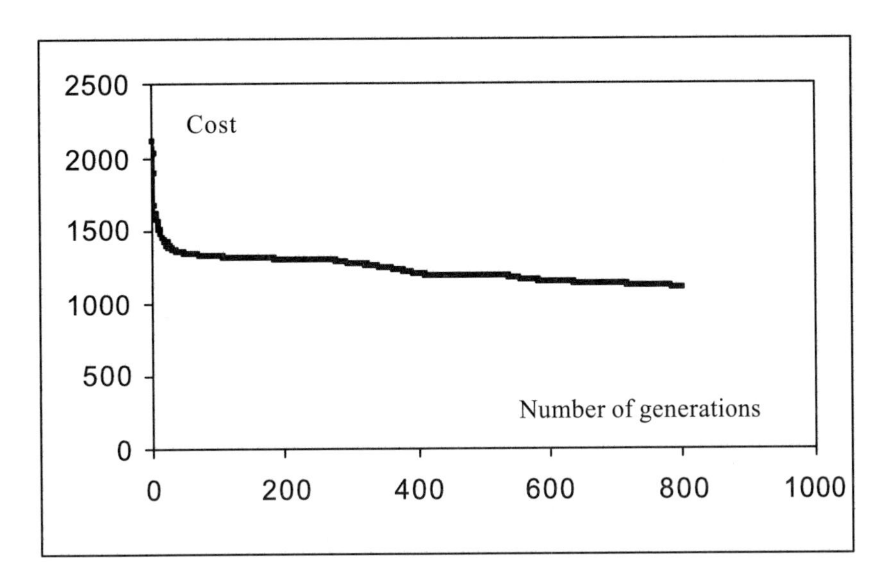

FIGURE 9.7 The mean performance curve of the GA over 50 trials (setting 1).

9.3.7 A Process Planning Example Using the GA

The part shown in Figure 9.4 is used as an example to test the search capability of the developed GA. The job shop information (machines, tools, and all the cost indices) is described earlier and the required OPTs and their machine, tool, and TAD alternatives are shown in Table 9.2 and the PRs between the OPTs in Table 9.3. The GA is coded in C and runs on a HP C200 workstation.

In order to test the capability of the GA for generating optimal plans as well as dealing with the dynamic characteristics of job shops, process planning for the part is carried out under two different settings. Setting 1 assumes that all the facilities are available, while setting 2 assumes that M1 is down. For each setting, 50 trials (same set up, different initial plan) were conducted. The objective function (fitness) used is the sum of all the cost factors described in Equations 9.1 through 9.5. The results are summarized as follows

1. *Setting-1: All machines and tools are available.* Over the 50 trials, the minimum cost of the founded plans is 1098, the maximum cost is 1218, and the average cost is 1120.3. Using a state-space search method, it was found that a process plan with a cost of 1098 is the best known so far. The GA found the optimal plans (cost = 1098) 28 times out of the 50 trials (56%). Figure 9.7 illustrates the convergence characteristics of the optimization curve (cost vs. number of generations) that was produced by averaging the 50 trials (for each generation in a trial, the best process plan is selected). It can be seen that the performance curve is nonincreasing, which is the result of applying "elitism" in the reproduction. One of the process plans with a cost of 1098 is shown in Table 9.4 with its number of machine changes, set-up changes, and tool changes.
2. *Setting-2: M1 is down.* Looking at the solutions for setting 1, it was found that in all the optimal plans M1 is used to perform all the OPTs. This result matches with intuition well, as M1 can perform all the OPTs and is the cheapest among the available machines. In this setting, M1 is assumed to be down in order to make the planning problem more complicated. The OPTs and their alternative machines, tools, and TADs for this setting is the same as for setting 1 (see Table 9.2), except that all the entries related to M1 are excluded. The PRs between the OPTs are the same as in setting 1. The best solutions found have a cost of 1598. This is also the minimum cost found by the state-space search algorithm so far. The average cost of the founded plans over the

TABLE 9.4 One of the Best Process Plans Found by the GA and SA Algorithm (Setting 1)

OPT	M	T	TAD	Summary
OPT2	M1	TE	+z	
OPT12	M1	T3	+z	
OPT1	M1	T3	+z	
OPT15	M1	T14	+x	Total Cost: 1098
OPT16	M1	T14	+x	
OPT14	M1	T3	+x	
OPT13	M1	T2	−z	No. of machine changes: 0
OPT3	M1	T2	−z	
OPT8	M1	T10	−z	
OPT5	M1	T10	−z	No. of set-up changes: 2
OPT9	M1	T14	−z	
OPT6	M1	T14	−z	No. of tool changes: 8
OPT18	M1	T11	−z	
OPT17	M1	T15	−z	
OPT4	M1	T15	−z	
OPT10	M1	T3	−z	
OPT7	M1	T3	−z	
OPT11	M1	T3	−z	

50 trials is 1635.7. The worst plan among the 50 solutions has a cost of 1786. The GA found the best plans (cost = 1598) 27 times out of 50 trials (54%). One of the best plans found is shown in Table 9.5. Compared with the best plans found for setting 1, this plan employs more machines (M2 and M3) instead of simply replacing M1 by M2 in the best plans for setting 1, as this would result in a plan with a cost of 1728. This demonstrates that the GA algorithm is able to handle changes in terms of machining resources to accommodate the dynamic nature of job shops.

The GA solution time for both settings is about 5 minutes. Most of the trials conducted reach their respective final solution within 6000 generations. This suggests that the stopping criterion of 8000 generations is sufficient, if not too conservative. The success rate of finding the best known solution in both settings is more than 50%. Even those nonoptimal solutions found in the trials are considered acceptable, according to the machinists. This indicates that for a given process planning problem, a good solution, optimal or near-optimal, can be found by running the GA several times.

9.4 Applying Simulated Annealing to the Process Planning Problem

Compared with GA, a simpler and more direct search method is the gradient descent method. The idea is to start with a random (but valid) solution and then consider small changes to it. Only those changes leading to a smaller cost are accepted. This is repeated until no changes can be made that lead to a cost reduction. The problem with this approach is that the solution is often trapped in local minimum. To solve this problem, the principle of simulated annealing (SA) can be incorporated into the gradient descent algorithm.

The principle of simulated annealing is illustrated in Figure 9.8, where a ball is initially positioned randomly along the landscape. It is equally likely to end up in either A or B. Assuming that the ball initially sits in valley A, if the system is slightly agitated, there is a greater likelihood of the ball moving to the bottom of A than jumping to valley B. If, on the other hand, the system is vigorously agitated, the likelihood of the ball moving from B to A is roughly as great as that of moving from A to B. Therefore, to move the ball to the lowest valley in a landscape, the best strategy would be to start with a vigorous shaking and then gradually slow down as the system works its way toward a global minimum solution. This idea is similar to that in metallurgy, where the low energy state of a metal is reached by melting it,

TABLE 9.5 One of the Best Process Plans Found by the GA and SA Algorithm (Setting 2)

OPT	M	T	TAD	Summary
OPT3	M2	T2	−z	
OPT13	M2	T2	−z	
OPT18	M2	T11	−z	
OPT2	M2	T3	+z	
OPT12	M2	T3	+z	Total Cost: 1598
OPT1	M2	T3	+z	
OPT5	M3	T10	−z	No. of machine changes: 2
OPT8	M3	T10	−z	
OPT6	M3	T14	−z	No. of set-up changes: 3
OPT9	M3	T14	−z	
OPT15	M3	T14	+x	No. of tool changes: 4
OPT16	M3	T14	+x	
OPT14	M2	T3	+x	
OPT10	M2	T3	−z	
OPT7	M2	T3	−z	
OPT11	M2	T3	−z	
OPT17	M2	T15	−z	
OPT4	M2	T15	−z	

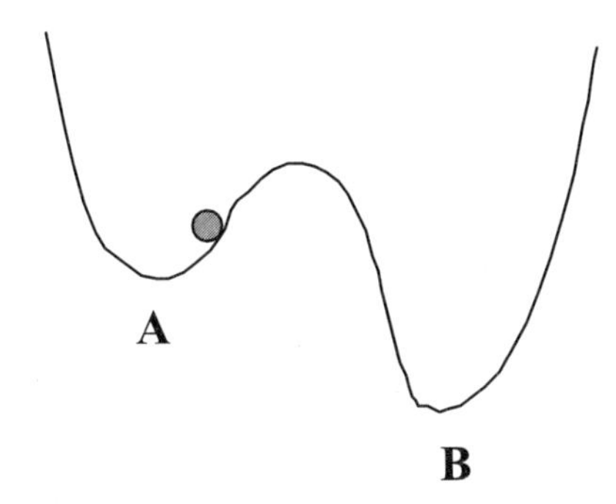

FIGURE 9.8 Principle of simulated annealing.

then slowly reducing its temperature until a stable and low energy configuration is achieved. In the annealing process, the relationship between the temperature (T) and a possible state change from one energy state (E_1) to another (E_2) can be described by the Boltzman's expression [Abe, 1990]:

$$P = e^{-(E_2 - E_1)/T} \qquad \text{Equation (9.7)}$$

where P is the probability that a system, in a state with energy E_1 and in an environment at temperature T, when excited by random thermal fluctuations, will move to a state with energy E_2. This means that the higher the temperature is, the greater the chance of the system moving to a higher energy state. At very high temperatures all moves will be accepted, but at very low temperatures only moves leading to a lower energy state will occur. These moves to higher energy states are necessary to escape from local minima; but the number of the upward jumps needs to be slowly reduced so that an escape from the global minimum eventually becomes unlikely. This characteristic is what gives SA the advantage over gradient descent and other "greedy" search methods, i.e., those that only allow moves to lower energy states.

Let a randomly generated but valid process plan be the initial configuration with the plan's cost as its energy, the general algorithm of applying SA to search for the best process plan can be outlined as follows:

1. Randomly generate a feasible plan (OPM_1, OPM_2, . . . , OPM_n), called the *current-plan.*
2. Start from the initial temperature $T = T_0$, while not reaching the final temperature T_{lowest}

{

a) Make a random change to the *current-plan*, let *temp-plan* be the plan after the change

b) Check to make sure that *temp-plan* is valid. Otherwise, go back to a)

c) Calculate the costs of *current-plan* (E_1) and *temp-plan* (E_2)

d) Determine whether to replace the *current-plan* with the *temp-plan*

d) Repeat a) to c) until a criterion is satisfied

e) Reduce the temperature to a new T according to a cooling schedule

}

For implementing this generic algorithm, the following tasks need to be carried out:

1. Develop a method for generating the first process plan randomly.
2. Develop the types of random changes to be applied to a process plan.
3. Develop a criterion for determining whether to accept or reject a change.
4. Determine the initial temperature and the final temperature.
5. Develop a cooling schedule.
6. Develop a criterion for stopping further changes under a temperature.

9.4.1 A Method for Generating the Initial Process Plan

The initial process plan needs to be randomly generated, and at the same time it must satisfy all the PRs. The algorithm used in the GA to randomly generate a valid string is used here. The generated process plan is called the *current-plan*.

9.4.2 Random Changes to a Process Plan

There are two kinds of changes that can be applied to a process plan, i.e., *sequence change* and *method change:*

1. **Sequence change:** Two OPMs are randomly selected and their positions in the plan are swapped. It is clear that such an exchange may result in an invalid sequence because the PRs (including conditional PRs and TADs) may be violated. Therefore, the selected OPMs will be checked to satisfy the precedence constraints before proceeding to the next step. Otherwise, it will go back to select the two OPMs again, which are to be swapped.

2. **Method change:** An OPM is randomly selected and replaced by randomly selecting one of the alternative OPMs of its respective OPT, if there is one. The validity of the TAD in the new OPM needs to be checked if it is a conditional TAD. A new OPM needs to be selected if the TAD is invalid.

During a single cycle of applying changes to a process plan, only one type of change is used, which is randomly selected. The process plan after a change is called the *temp-plan*.

9.4.3 A Criterion for Accepting or Rejecting a Change

The costs of the *current-plan* (E_1) and *temp-plan* (E_2) are first calculated based on the defined objective function. The algorithm for accepting or rejecting a change is as follows:

Algorithm: accepting or rejecting a change

Input: current-plan (E_1), temp-plan (E_2)

Output: current-plan (by accepting or rejecting a change)

IF $E_1 > E_2$

Accept the change (let the temp-plan be the current-plan)

ELSE

Generate a random number, X $(0<X<1)$

IF $X < e^{\frac{E_1 - E_2}{T}}$

Accept the change (let the temp-plan be the current-plan)

ELSE

Reject the change (let the current-plan remain)

END IF

END IF

End of algorithm accepting or rejecting a change

9.4.4 Initial and Final Temperature

The initial temperature T_0 should be chosen sufficiently high, so that even in the worst cost increase after a random change, the possibility given by Boltzman's expression is high enough to lead to further randomization of the plan. If all the cost factors described in Equations 9.1 through 9.5 are included in the objective function, three times the largest cost index among *MCCI, SCCI*, and *TCCI* is considered as the worst cost increase after a random change, and the possibility is set at 0.75. According to Boltzman's expression, T_0 is calculated as

$$T_0 = 10 \times Max\ (MCCI,\ SCCI,\ TCCI) \qquad \text{Equation (9.8)}$$

On the other hand, the final temperature T_{lowest} should be chosen low enough so that only changes leading to a lower cost are accepted. If all the cost factors described in Equations 9.1 through 9.5 are included in the objective function, the least cost increase is the minimum cost index among *MCCI, SCCI*, and *TCCI* and the possibility is set as 0.02. According to Boltzman's expression, T_{lowest} is calculated as

$$T_{lowest} = 0.25 \times Min(MCCI,\ SCCI,\ TCCI) \qquad \text{Equation (9.9)}$$

In cases where the objective cost function does not include all five cost factors, T_0 and T_{lowest} can be determined using the above heuristics similarly.

9.4.5 A Fast Cooling Schedule

The cooling schedule affects the opportunities that the configuration eventually settles in the globally lowest energy state. If the temperature is lowered too quickly, the configuration tends to settle in a local minimum. On the other hand, if the temperature is lowered too slowly, it will take too much computation time to converge to a final solution. In this application, the fast simulated annealing schedule suggested by Szu [1986] is used as

$$\frac{T_t}{T_0} = \frac{1}{1+t} \qquad \text{Equation (9.10)}$$

where, $t = 1, 2, \ldots.$

9.4.6 A Criterion for Stopping Further Changes under a Temperature

The time required to reach quasi-equilibrium under a temperature is another important parameter of the cooling schedule. In theory, every state must be visited infinitely often, in order to reach absolute convergence. Practically, this is not feasible. It is suggested that the number of calculations under a temperature should be sufficiently large to reach a good solution, but not so long as little information is provided for the time spent. In this research, simulation experiments were conducted on four commonly used criteria and the results were compared to identify the best one. The four criteria are as follows:

1. A fixed number of accepted changes under each temperature.
2. A fixed number of rejected changes under each temperature.
3. A fixed number of accepted changes, together with a fixed cost ratio between the *temp-plan* and the *current-plan.*
4. A fixed number of accepted changes, together with a fixed cost ratio between *temp-plan* and the average cost of all accepted plans.

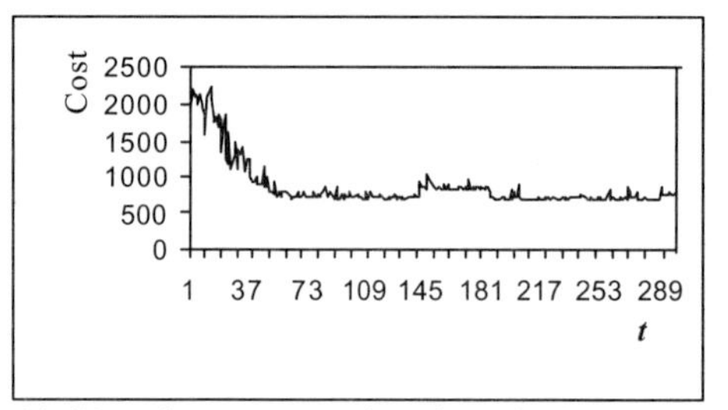

(a) SA performance curve based on criterion-1

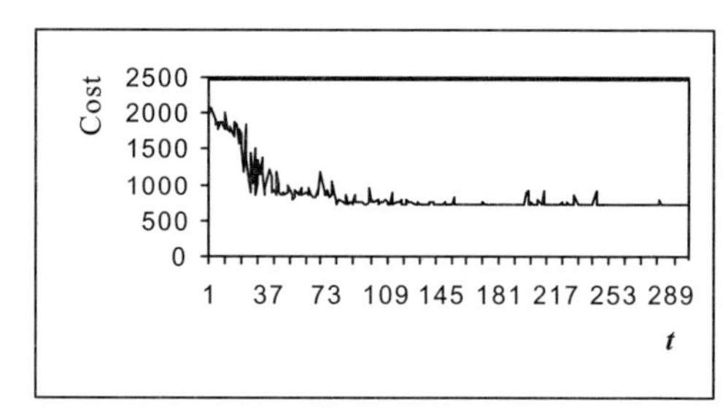

(b) SA performance curve based on criterion-2

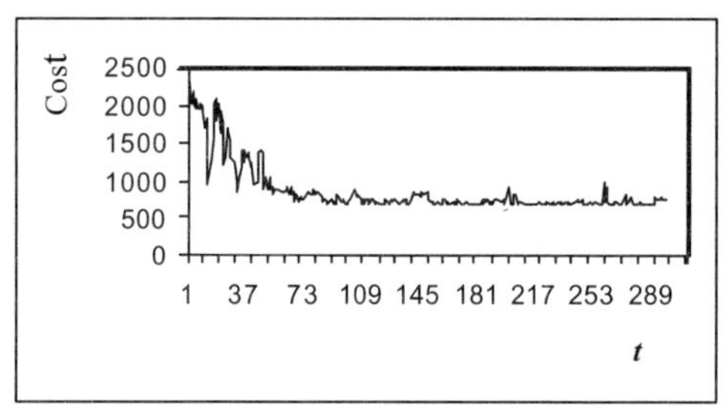

(c) SA performance curve based on criterion-3

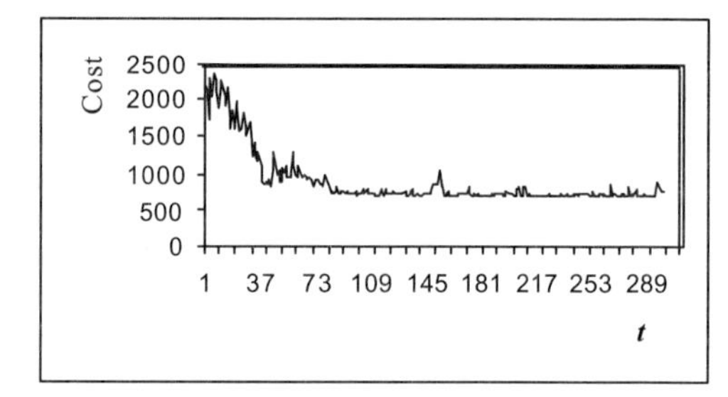

(d) SA performance curve based on criterion-4

FIGURE 9.9 Simulation results under different stopping criteria. (a) SA performance curve based on criterion 1. (b) SA performance curve based on criterion 2. (c) SA performance curve based on criterion 3. (d) SA performance curve based on criterion 4.

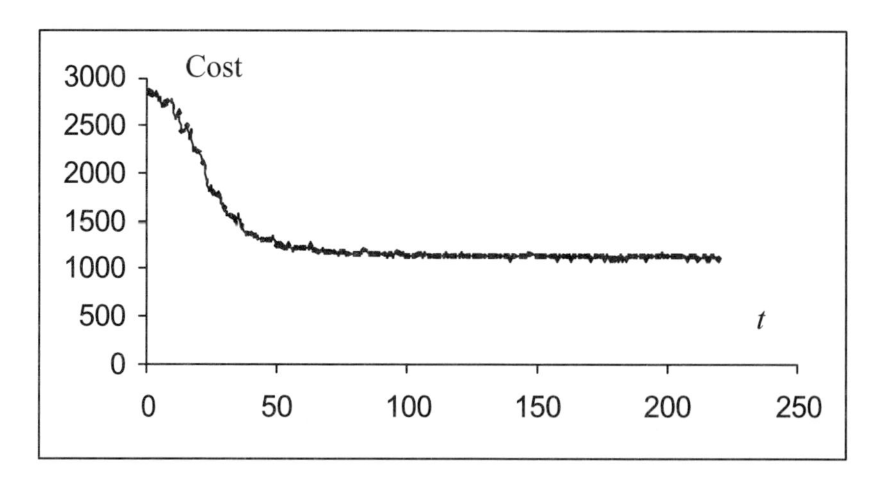

FIGURE 9.10 The mean performance curve of the SA algorithm over 50 trials (setting 1).

The results of one such experiment are given in Section 9.4.7, in which the case used by Zhang et al. [1997] was used. All the SA algorithms using the four criteria achieved the optimal result. The costs of the final plan under each temperature based on different criteria are shown in Figure 9.9. It is obvious that criterion 2 is the most preferred as Figure 9.9(b) clearly shows that the final stage of the annealing is slightly more stable compared with other graphs. On the other hand, the SA algorithm using criterion 2 converged faster than the other three. Similar conclusion has been reached based on other experiments, although for some cases, the difference of using different criteria is not obvious. Therefore, a fixed number of rejected changes is chosen as the stopping criterion under each temperature.

9.4.7 A Process Planning Example Using the SA Algorithm

The part shown in Figure 9.4 is used as an example to test the capability of the developed SA algorithm. The job shop information (machines, tools, and all the cost indices) is described earlier; the required OPTs and their machine, tool, and TAD alternatives are shown in Table 9.2 and the PRs between the OPTs in Table 9.3. The SA algorithm is coded in C and runs on a HP C200 workstation.

In order to compare the performance of the SA algorithm and the GA described earlier, the SA algorithm was run under the same two settings as for the GA. For each setting, 50 trials (same set up, different initial plan) were conducted. The objective function used is the same as for the GA, i.e., the sum of all the cost factors described in Equations 9.1 through 9.5. Therefore, T_0 is set at $10MCCI$, i.e., 1100, and T_{lowest} is set at $0.25TCCI$, i.e., 5. Based on Equation 9.10, the SA algorithm will be run from $t = 0$ to $t = 219$. The calculation results are summarized as follows:

1. *Setting 1: All machines and tools are available.* Over the 50 trials, the minimum cost of the founded plans is 1098, the maximum cost is 1168, and the average cost is 1116.4. The SA algorithm found the optimal plans (cost = 1098) 34 times out of the 50 trials (68%). Figure 9.10 illustrates the convergence characteristics of the optimization curve (cost vs. number of generations) that was produced by averaging the 50 trials (for each temperature in a trial, the last process plan is selected). One of the process plans with a cost of 1098 is the same as the one shown in Table 9.4.
2. *Setting 2: M1 is down.* The OPTs and their alternative machines, tools, and TADs for this setting are the same as for setting 1 (see Table 9.3) except that all the entries related to M1 are excluded. The PRs between the OPTs are the same as setting 1. The best solutions found have a cost of 1598. The average cost of the plans found over the 50 trials is 1623.4. The worst plan among the 50 solutions has a cost of 1726. The SA algorithm found the best plans (cost = 1598) 30 times out of 50 trials (60%). One of the best plans found is the same as the one shown in Table 9.5.

The SA solution time for both settings is about 25 seconds, which is much less than the GA solution time. Most of the trials conducted reach their respective final solution well before the lowest temperature is reached. This suggests that the stopping criterion for the SA is sufficient, if not conservative. The success rate of finding the best known solution in both settings is also more than 50% (slightly better than that of the GA). Similarly, those nonoptimal solutions found in the trials are considered acceptable, according to the machinists. This indicates that for a given process planning problem, a good solution, optimal or near optimal, can be found by running the SA algorithm several times.

9.5 Comparison between the GA and the SA Algorithm

Both the GA and the SA algorithm described earlier are stochastic search algorithms. They are devised to handle the same problem model, i.e., the concurrent process planning model. Although it is difficult to come to a comprehensive conclusion, these two developed algorithms can be compared based on the following aspects:

1. **Traversing the solution space** — Both the GA and the SA algorithm traverse the overall solution space by incorporating sequence and operation-method changes into the seed process plan. In the

GA, a sequence change is achieved by applying the cyclic crossover operation to two strings (plans). Although the validity of the sequence of the new offspring is ensured, the cyclic crossover may lose possible sequence space as the sequence of the OPTs in the offspring follows the sequence of OPTs in its predecessors, i.e., the original strings. The random sequence changes incorporated in the SA algorithm, on the other hand, do not have this problem. In the GA, the operation-method change is achieved by randomly changing the machine, tool, and TAD individually. The operation-method space can be fully traversed but the efficiency may be affected as the randomly formed M/T/TAD may not be valid and a number of iterations may be required before a valid change is reached. In the SA algorithm, operation-method change is carried out by randomly selecting a valid M/T/TAD from the alternatives. The operation-method space can also be fully traversed.

2. **Escaping from local minima** — Both the GA and the SA algorithm have incorporated mechanisms to overcome the local minimum problem. In the GA, this is achieved by keeping a pool of solutions in each generation, so that even a currently poor solution, which could lead to an optimal one eventually, is kept. In the SA algorithm, on the other hand, this is achieved by changing the temperature from high to low, which affects the decision on whether to accept or reject a change, so that even a poorer solution could be accepted, especially at a high temperature in order to escape from local minima. In theory, both mechanisms can enable their respective algorithms to achieve the global optimal solution.
3. **Setting algorithm parameters** — Both the GA and the SA algorithm have the problem of parameter setting. Heuristics and trial-and-error methods are used for setting the parameters for both algorithms. It was found that setting parameters for the GA is particularly difficult.
4. **Performance measures** — Based on the present limited case studies, both the algorithms are able to achieve the known best solution for any of those process planning problems used in the case studies in more than 50% of the trials. However, for a single trial, there is no guarantee that an optimal solution can be found. It was also found that the GA generally takes much more computation time for a single trial than the SA algorithm, although both of the computation time (less than 10 minutes) taken are acceptable. This may be due to the conservative setting of 8000 generations as the stopping criterion for the GA.

In summary, both the GA and SA algorithms are feasible for solving the process planning model by finding a process plan with high quality. On the other hand, it can be concluded that the SA algorithm can outperform the GA in terms of solution time.

9.6 Conclusions

Process planning is an optimization problem with interrelated decision-making activities and multiple constraints. This chapter describes a process planning approach that integrates the tasks of routing and sequencing for obtaining a globally optimal process plan for a machined part (prismatic). A GA and a SA-based algorithm have been developed to search for the globally optimal solution based on the proposed process planning model. Results in the case studies suggest that the modelling method is able to cover the solution space and the search methods developed are effective. The approach employed has advantages over previous approaches in the following aspects:

1. The system is developed based on a customisable job shop environment, such that a user can modify the manufacturing resource database to suit his/her needs. This makes the system more flexible as well as realistic compared to the approaches in which a fixed machining environment is assumed.
2. By concurrently considering the selection of machines, tools, TADs for each operation type and the sequence among the operations, together with the constraints of PR, conditional PR, and validity of operation methods, the resulting process plan model successfully retains the entire solution space. This makes it possible to find a globally optimal process plan.

3. GA and SA have been successfully applied to search for the globally optimal solutions in the process planning model, with the incorporation of process planning heuristics into their operation procedures.

In addition, this novel concurrent process planning modelling scheme can provide a flexible base for the integration of process planning and job scheduling. Based on the scheduling statistics (e.g., machine utilisation rate) for a set of process plans generated by the method introduced in this chapter, a control measure can be easily applied to the process plan model to generate another set of process plans for another trial of scheduling. For example, if a machine is overutilised, it can be deleted from the alternative machines used for those OPTs that have other alternatives. If a machine is underutilised, it can be assigned to those OPTs that have it as an alternative, as the only machine that can be used.

A limitation of the process planning modelling approach is that the dynamic aspects of the fixture constraints are not considered. First, only modular fixture elements are considered, and it is assumed that a fixture solution for an OPM can be found if the part under the TAD has sufficient base area and clamping does not cause too much deformation. Second, it is assumed that two OPMs sharing the same machine and TAD also share the same fixture solution, hence no set-up change is required. These assumptions may affect the flexibility on fixture device choices and overconstrain the solution space. Currently, work is underway to consider other fixture devices and include the fixture solution as another attribute of OPMs, in addition to M, T, and TAD.

Defining Terms

CAD Computer-aided design
CAPP Computer-aided process planning
CAM Computer-aided manufacturing
CNC Computer numerical control
D-OPT Dummy OPT
DFM Design for manufacture
OPM Operation method, an operation to be executed under a specific machine, cutter, and tool approach direction combination
OPT Operation type, an operation in general without relating to any machine, cutter, or tool approach direction
PR Precedence relationship
TAD Tool approach direction

References

Abe, S., 1990, *Simulated Annealing and Boltzman's Machine*, Wiley, New York.

Alting, L. and H. C. Zhang, 1989, Computer-aided process planning: the state-of-the-art survey, *International Journal of Production Research*, vol. 27, no. 4, pp. 553-585.

Chen, C. L. P. and S. R. LeClair, 1994, Integration of design and manufacturing: solving set-up generation and feature sequencing using an unsupervised-learning approach, *Computer Aided Design*, vol. 26, no. 1, pp. 59-75.

Chen, J., Y. F. Zhang, and A. Y. C. Nee, 1998, Set-up planning using Hopfield net and simulated annealing, *International Journal of Production Research*, vol. 36, no. 4, pp. 981-1000.

Chu, C. C. P. and R. Gadh, 1996, Feature-based approach for set-up minimisation of process design from product design, *Computer Aided Design*, vol. 28, no. 5, pp. 321-332.

Gupta, S. K., 1997, Using manufacturing planning to generate manufacturability feedback, *Journal of Mechanical Design*, vol. 119, pp. 73-80.

Hayes, C. C., 1996, P^3: a process planner for manufacturability analysis, *IEEE Transactions on Robotics and Automation*, vol. 12, no. 2, pp. 220-234.

Ma, G. H., 1999, An Automated Process Planning System for Prismatic Parts, M. Eng. thesis, National University of Singapore, Republic of Singapore.

Nee, A. Y. C., K. Whybrew, and A. Senthil Kumar, 1995, *Advanced Fixture Design for FMS*, Springer-Verlag, London.

Szu, H., 1986, Fast simulated annealing, *Proceedings of the American Institute of Physics Conference on Neural Computing*, pp. 420-425.

Yip-Hoi, D. and D. Dutta, 1996, A genetic algorithm application for sequencing operations in process planning for parallel machining, *IIE Transactions*, vol. 28, pp. 55-68.

Zhang, F., Y. F. Zhang, and A. Y. C. Nee, 1997, Using genetic algorithms in process planning for job shop machining, *IEEE Transactions on Evolutionary Computation*, vol. 1, no. 4, pp. 278-289.

10

Production Planning and Scheduling Using Genetic Algorithms

Runwei Cheng
Ashikaga Institute of Technology

Mitsuo Gen
Ashikaga Institute of Technology

10.1 Introduction

Production scheduling problems concern the allocation of limited resources over time to perform tasks to satisfy certain criteria. Resources can be of a very different nature, for example, manpower, money, machines, tools, materials, energy, and so on. Tasks can have a variety of interpretations from machining parts in manufacturing systems up to processing information in computer systems. A task is usually characterized by some factors, such as ready time, due date, relative urgency weight, processing time, resource consumption, and so on. Moreover, a structure of a set of tasks, reflecting precedence constraints among them, can be defined in different ways. In addition, different criteria that measure the quality of the performance of a schedule can be taken into account.

Many scheduling problems from manufacturing industries are characterized as combinatorial optimization problems subject to highly complex constraints, which are very difficult to solve by conventional optimization techniques. This has led to the recent interest in using genetic algorithms to address the problem. In the following sections, we explain how to solve them with genetic algorithms, including resource-constrained project scheduling, parallel machine scheduling, job-shop scheduling, multistage process planning, and part loading scheduling problem.

10.2 Resource-Constrained Project Scheduling Problem

The problem of scheduling activities under resource and precedence restrictions with the objective of minimizing the project duration is referred to as a resource constrained project scheduling problem in literature [Baker, 1974]. The basic type of the problem can be stated as follows: A project consists of a number of interrelated activities. Each activity is characterized by a known duration and given resource requirements. Resources are available in limited quantities but renewable from period to period. There is no substitution between resources and activities cannot be interrupted. A solution is to determine the start times of activities with respect to the precedence and resource constraints so as to optimize the objective.

0-8493-0592-6/01/$0.00+$.50

The problem can be stated mathematically as follows:

$$\min \quad t_n \qquad \text{Equation (10.1)}$$

$$\text{s.t.} \quad t_j - t_i \geq d_i, \quad \forall j \in S_i \qquad \text{Equation (10.2)}$$

$$\sum_{t_i \in A_{t_i}} r_{ik} \leq b_k, \quad k = 1, 2, \ldots, m \qquad \text{Equation (10.3)}$$

$$t_i \geq 0, \quad i = 1, 2, \ldots, n \qquad \text{Equation (10.4)}$$

where t_i is the starting time of activity i, d_i the duration (processing time) of activity i, S_i the set of successors of activity i, r_{ik} the amount of resource k required by activity i, b_k the total availability of resource k, A_{t_i} the set of activities in process at time t_i, and m the number of different resource types. Activities 1 and n are dummy activities that mark the beginning and end of the project. The objective is to minimize total project duration. Constraint 10.2 ensures that none of the procedence constraints are violated. Constraint 10.3 ensures that the amount of resource k used by all activities does not exceed its limited quantity in any period.

The earliest attempts were made to find an exact optimal solution to the problem by using standard solution techniques of mathematical programming. Because the resource-constrained project scheduling problem is NP-hard, for large projects, the size of the problem may render optimal methods computationally impracticable. In such cases, the problem is most amenable to heuristic problem solving, using fairly simple scheduling rules capable of producing reasonably good suboptimal schedules [Alvarez-Valdés and Tamarit, 1989]. Most heuristic methods known so far can be viewed as priority dispatching rules that assign activity priorities in making sequencing decisions for resolution of resource conflicts according to either temporally related heuristic rules or resource-related heuristic rules.

In essentials, the problem consists of the following two basic issues: (i) to determine the processing order of activities without violating the precedence constraints and (ii) subsequently to determine the start time for each activity without violating the resource constraint. How to determine the order of activities is critical to the problem, because if the order of activities is determined, a schedule then can be easily constructed with some determining procedures according to the order.

Cheng and Gen [1998] have proposed a hybrid genetic algorithm to the resource-constrained project scheduling problem. The basic idea of the approach is to (i) use genetic algorithms to evolve an appropriate processing order of activities and (ii) use a fit-in-best procedure to calculate the start times of activities. Their study focuses on how to handle the precedence constraint existing in the problem. A new encoding method is proposed, which is essentially capable of representing all feasible permutations of activities for a given instance.

10.2.1 Priority-Based Encoding

The key issue of the problem is to find an appropriate processing order of activities. This is a permutation problem in nature. Due to the existence of precedence constraints among activities, an arbitrary permutation may yield an infeasible processing order. Making an encoding that can treat the precedence constraint efficiently is a critical step and conditions all subsequent steps. A priority-based encoding method is proposed by Cheng and Gen to handle this difficulty, which is based on the concepts of a *directed acyclic graph* (DAG) model.

A sample project can be represented with a directed acyclic graph. A directed acyclic graph $G = (V, A)$ consists of a set of nodes V representing activities and a set of directed edges A representing the precedence constraints among activities. The terms *node* and *activity* will be used interchangeably in the following sections. For a given directed graph, a *topological sort* is a linear ordering of all its nodes such that for any directed edge $(u, v) \in A$, node u appears before node v in the ordering. In other words, a topological sort corresponds to a feasible ordering of activities, that is, a feasible solution. Cheng and Gen suggest a new encoding method, *priority-based encoding,* which is capable of representing all possible topological sort for a given instance.

Recall that a gene contains two kinds of information: *locus,* the position of the gene located within the structure of a chromosome, and *allele,* the value the gene takes. Here, the position is used to denote an activity ID and the value is used to denote the priority associated with the activity, as shown in Figure 10.1. The value of a gene is an integer exclusively within $[1, n]$. The larger the integer, the higher the priority.

A one-pass procedure is used to generate a topological sort from a chromosome: to determine activities from left to right. When making a decision for a position, several activities may compete for the position and the one with the highest priority wins the position. The encoding does not explicitly represent a topological sort for a given DAG. It just contains some message for resolution of conflicts. A topological sort can be uniquely determined according to the encoding. Any changes in priorities usually result in a different topological sort. Therefore, this encoding is essentially capable of representing all possible topological sort for a given DAG.

Let us see how to generate a topological sort from the encoding. Consider the example given in Figure 10.2. An array $A[\bullet]$ is used to store the generated topological sort. At the beginning, $A[1] = 1$. Three activities, 2, 3, and 4, compete for $A[2]$. Their priorities as defined in above encoding are 7, 1, and 6, respectively. Activity 2 wins the position because it has the highest priority. After fixing $A[2] = 2$, the candidates for the next position, $A[3]$, are activities 3, 4, and 5. Activity 4 wins for the position and fixs $A[3] = 4$. Repeat these two steps: (i) construct the set of candidates for current position and (ii) select the highest-priority activity, until we obtain a topological sort, as shown in Figure 10.3

The tricky part, of course, is how to find a set of eligible nodes. The following definitions and theorems give us a better understanding about how to make such a set and how the procedure works.

A *partial topological sort* is the one under development, which just contains the first t $(t < |V|)$ nodes with fixed orders. Let $PS_t \subset V$ be the set of nodes corresponding to a given partial topological sort, where

1	2	3	4	5	6	7
3	7	1	6	4	5	2

position : activity ID
value : priority of activity

FIGURE 10.1 Priority-based encoding.

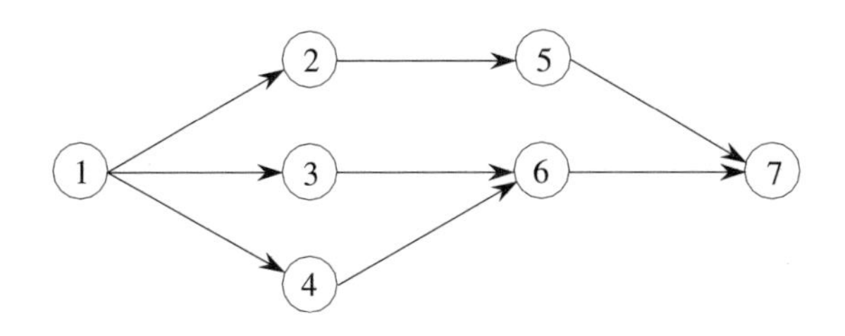

FIGURE 10.2 Network representation of a project.

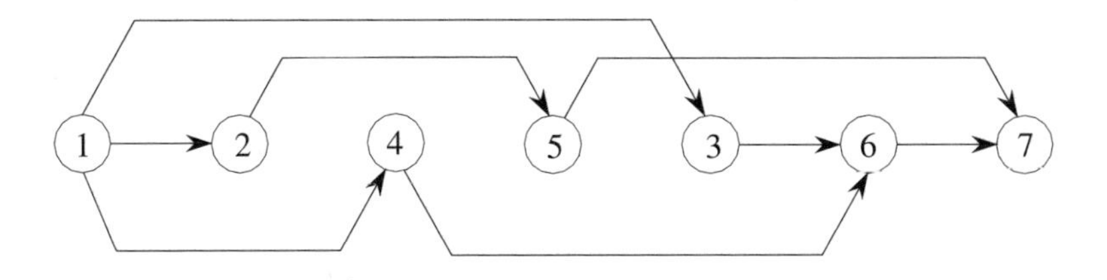

FIGURE 10.3 The topological sort of the DAG shown in Figure 10.2.

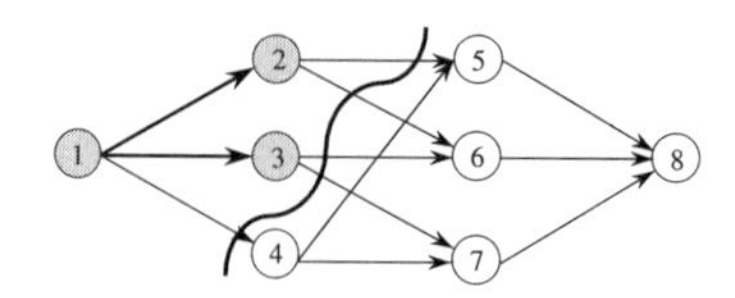

FIGURE 10.4 Partial topological sort, cut, and eligible nodes.

the subscript t denotes the cardinal number of the set, that is, $|PS_t| = t$. Let $C(PS_t, V - PS_t) = \{(i, j) \mid i \in PS_t, j \in V - PS_t\}$ be the cut of the directed graph with respect to the given partial topological sort. Then we have the following lemma:

Lemma 1 (eligible node) *For a given partial topological sort with nodes PS_t, a node $j \in V - PS_t$ is eligible if and only if we can have a set $S(j) = \{(i, j) \mid (i, j) \in A\}$ of edges incident to j such that $S(j) \subseteq C(PS_t, V - PS_t)$.*

Proof. For a given node $j \in V - PS_t$, if there is an edge (x, j) incident to j, the node x is a parent node of j. If all such edges belong to the cut $C(PS_t, V - PS_t)$, it means that all the parent nodes of j belong to the set PS_t, that is, they are the sorted node, therefore, the node j is eligible.

Assume that there is an eligible node j and not all the edges incident to j belong to the cut. That is, $|C(PS_t, V - PS_t) \cup S(j)| > |C(PS_t, V - PS_t)|$. Then at least one of its parent node belongs to set $V - PS_t$, that is, at least one of its parent node is not the sorted node. This is a contradiction to the definition of eligible node. ■

Theoretically, we can check whether a node is eligible with the lemma, but it is usually not easy for programming to check if a set is a subset of others. The following theorem provides a criterion to determine an eligible node.

Theorem 1 (criterion of eligible node) *For an eligible node $j \in V - PS_t$, let $S_t(j)$ be a proper subset of cut $C(PS_t, V - PS_t)$ containing all edges incident to j, we have $|S_t(j)| = d^{IN}(j)$.*

Proof. For an eligible node, we have, according to Lemma 1, $S_t(j) \cap S(j) = S(j)$ and $S_t(j) \cup S(j) = S(j)$. Because $|S(j)| = d^{IN}(j)$, we prove the theorem. ■

Now we can identify an eligible node simply by checking whether the number of edges incident to it in the cut equals its indegree. This criterion is easy for programming. Let us consider the example given in Figure 10.4. The partial topological sort is $PS_3 = \{1, 2, 3\}$ and the cut contains the directed edges $C(PS_3, V - PS_3) = \{(1, 4), (2, 5), (2, 6), (3, 6), (3, 7)\}$. Node 6 is an eligible one because its indegree $d^{IN}(6) = 2$ and the two edges incident to node 6 belong to the cut. Node 5 is a free node because its indegree is 2 and only one edge incident to it belongs to the cut. Its other parent node is 4, which is an eligible one but not a sorted one.

10.2.1.1 Procedure of Topological Sort

The basic idea of the topological sort procedure is, at each step as the procedure progresses, to (i) identify the set of eligible nodes with Theorem 1, (ii) remove the one with the highest priority from the set, and (iii) fix the removed node in the partial topological sort.

Let t be the iteration index of the procedure. Let V be the set of all node. Let Q_i be the set of all direct successors of activities i. Let $PS[\bullet]$ be the array for storing topological sort. Let $CUT[i]$ be the number of edges incident to node i in cut. Let S_t be the set of eligible nodes at step t. The procedure for generating a topological sort from a chromosome is given as below:

procedure: topological sort

step 1: (initialization)

$t \leftarrow 1$ (iteration index)
$PS[t] \leftarrow 1$ (initial topological sort)
$S_t \leftarrow Q_1$ (initial priority queue)
$CUT[i] \leftarrow 1, \forall i \in Q_1$ (initial number of edges in the cut)

$CUT[i] \leftarrow 0, \forall i \in V - Q_1$

step 2: (termination test)

If $PS[t] = n$, go to step 6; otherwise $t \leftarrow t + 1$, continue.

step 3: (fixing the tth node)

Remove the highest priority node i^* from priority queue S_t and put it in array $PS[t]$.

step 4: (cut set update) $CUT[i] \leftarrow CUT[i] + 1, \forall i \in Q_{i*}$

step 5: (eligible node set update)

For all $i \in Q_{i*}$, if $CUT[i] = d^{IN}(i)$, then put i in priority queue S_t. Go back to step 2.

step 6: (topological sort)

Return a complete topological sort $PS[\bullet]$.

10.2.2 Genetic Operators

Genetic search is implemented through genetic operators and directed by selection pressure. Usually, a crossover operator is used as the main genetic operator, and the performance of a genetic system depends heavily on it; a mutation operator is used as a background operator, which produces spontaneous random changes in various chromosomes.

Gen and Cheng [1997] proposed an alternative approach to design genetic operators: one operator is designed to perform a widespread search to explore the area beyond local optima; the other is designed to perform an intensive search to hunt for an improved solution. Two kinds of search approaches, the intensive search and the widespread search, form the mutual complementary components of genetic search. With this approach, the crossover operator and mutation operator play the same important role in the genetic search.

10.2.2.1 Position-Based Crossover

The nature of the proposed encodings can be viewed as a kind of permutation encodings. A number of recombination operators have been investigated for permutation representation. The position-based crossover operator proposed by Syswerda [1991] was adopted, shown in Figure 10.5. Essentially, it takes some genes from one parent at random and fills vacuum positions with genes from the other parent by a left-to-right scan.

10.2.2.2 Swap Mutation

The swap mutation operator used here simply selects two positions at random and swaps their contents, as shown in Figure 10.6.

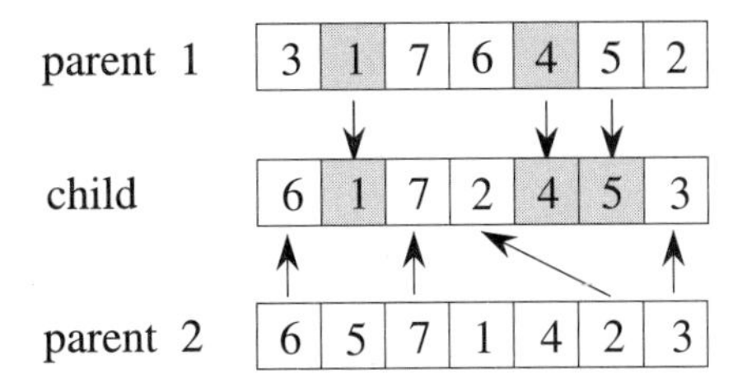

FIGURE 10.5 The position-based crossover operator.

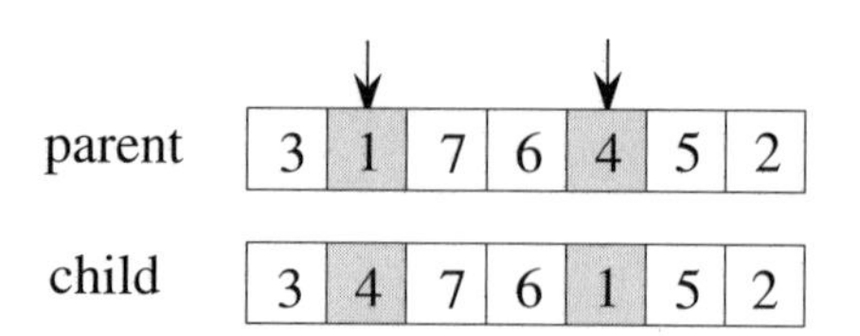

FIGURE 10.6 The swap mutation operator.

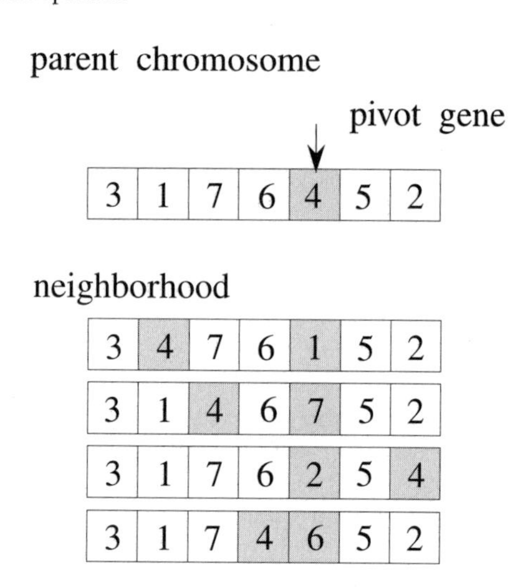

FIGURE 10.7 The incumbent chromosome and its neighborhood.

10.2.2.3 Local Search-Based Mutation

Local search methods seek improved solutions to a problem by searching in the neighborhood of an incumbent solution. The implementation of local search requires an initial incumbent solution, the definition of a neighborhood for an incumbent solution, and a method for choosing the next incumbent solution. The idea of hunting for an improved solution by making a small change can be used in mutation operator. A *neighborhood* of a chromosome is then defined as a set of chromosomes generated by pairwise interchanges. For a pair of genes, one is called *pivot*, which is fixed for a given neighborhood, and the other is selected at random, as shown in Figure 10.7. For a given neighborhood, a chromosome is called a *local optima* if it is better than any other chromosomes according to the fitness value. The size of a neighborhood affects the quality of the local optima. There is a clear trade-off between small and large neighborhoods: if the number of neighbors is larger, the probability of finding a good neighbor may be higher, but looking for it takes more time.

10.2.3 Evaluation and Selection

During each generation, chromosomes are *evaluated*, using some measure of fitness. The following four major steps are included in the evaluation phase: (i) convert chromosomes to topological sorts, (ii) generate schedules from the topological sorts, (iii) calculate objective values for each schedule, and (iv) convert objective values into fitness values.

Because a topological sort gives a feasible order of activities, we construct a schedule by selecting the activities in order of their appearance in the topological sort and scheduling them one at a time as early as resource availabilities permit. Let i be the iteration index of the procedure. Let V be the set of all node. Let P_i be the set of all direct predecessors of activities i. Let $PS[\bullet]$ be the array for storing topological sort. Let σ_j and ϕ_j be start and finish times associated with activity j. Let $b_k[l]$ be the array for storing available amount of resource k in time l. Let d_j be the duration associated with activity j. Let r_{jk} be the consumption of resource k associated with activity j. The procedure for determining start and finish times of each activity from a given topological sort is given below:

procedure: start and finish times of activities

step 1: (initialization)

$i \leftarrow 1$ (iteration index)
$j \leftarrow PS[i]$ (initial activity)
$\sigma_j \leftarrow 0,\ \phi_j \leftarrow 0$ (start and finish times for initial activity)
$b_k[l] \leftarrow b_k,\ l = 1, 2, \ldots, \Sigma_{j=1}^{n} d_j,\ k = 1, 2, \ldots, m$ (initial resources)

step 2: (termination test)

If $i = n$, go to step 5; otherwise $i \leftarrow i + 1$, continue.

step 3: (start and finish times)

$j \leftarrow PS[i]$
$\sigma_j^{\min} \leftarrow \max\{\phi_l \mid l \in P_j\}$
$\sigma_j \leftarrow \min\{t \mid t \geq \sigma_j^{\min},\ b_k[l] \leq r_{jk},\ l = t, t+1, \ldots, t+d_j,\ k = 1, 2, \ldots, m\}$
$\phi_j \leftarrow \sigma_j + d_j$

step 4: (available resources update)

$b_k[l] \leftarrow b_k[l] - r_{jk},\ l = t, t+1, \ldots, t+d_j,\ k = 1, 2, \ldots, m$
go back to step 2.

step 5: (stop)

return σ_j and ϕ_j.

Because we use the measure of project duration, the finish time of the last activity is the objective value. Since we deal with a minimization problem, we have to convert the original objective value to a fitness value in order to ensure that the fitter individual has a larger fitness value.

Let v_k be the kth chromosome in the current generation, $g(v_k)$ be the fitness function, $f(v_k)$ be the objective value, that is, the project duration, $f_{\max}$ and $f_{\min}$ be the maximum and minimum values of the objective values in current generation. The transformation is given as follows:

$$g(v_k) = \frac{f_{\max} - f(v_k) + \gamma}{f_{\max} - f_{\min} + \gamma} \qquad \text{Equation (10.5)}$$

where γ is a positive real number that is usually restricted within the open interval (0, 1). The purpose of using it is twofold: (1) to prevent Equation 10.5 from zero division and (2) to make it possible to adjust the selection behaviors from fitness-proportional selection to pure random selection. When the difference of fitness among chromosomes is relatively large, the selection is fitness-proportional; when the difference becomes too small, the selection tends to pure random among relatively competitive chromosomes.

The *roulette wheel* approach was adopted as the selection procedure that is one of the fitness-proportional selection. The *elitist selection* method was combined with this approach in order to preserve the best chromosome in the next generation and overcome the stochastic errors of sampling. With the elitist selection, if the best individual in the current generation is not reproduced into the new generation, one individual is randomly removed from the new population and the best one added to the new population.

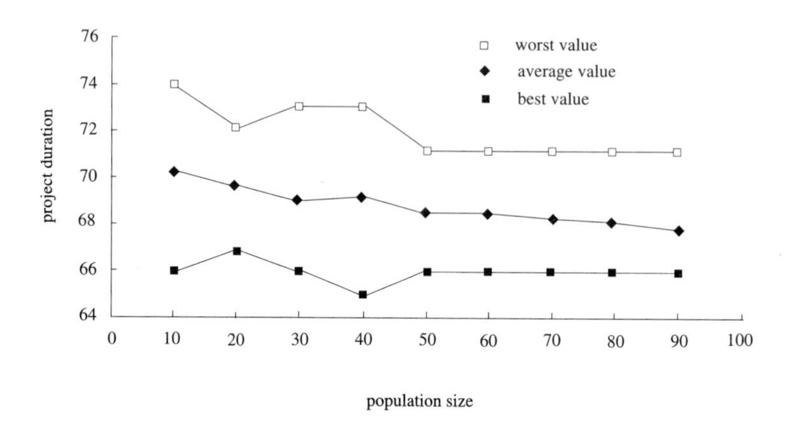

FIGURE 10.8 Comparison on the best value, worst value, and average values of objective under different *pop_size*.

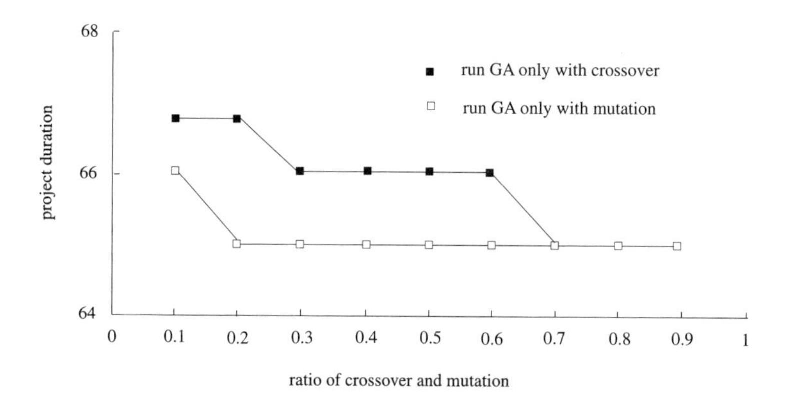

FIGURE 10.9 The best values over 200 random runs under different ratios of crossover and mutation.

10.2.4 Experimental Results

To investigate how population size affects the performance of genetic algorithms, experiments on parameter tuning were conducted. Fix maximum generation as 100, crossover and mutation ratios as 0.1, respectively. Under the condition of lower ratio of crossover and mutation, population size becomes one of the leading factors for the performance of genetic algorithms. Population size was varied from 10 to 100. Figure 10.8 shows the best, worst, and average values of objective over 100 random runs for each parameter setting. From the results we can see that when *pop_size* is larger than 50, any increase of it has no significant influence on the performance of the genetic algorithm.

Comparison between crossover and mutation operators was also performed to confirm which plays a more important role in the genetic search. Genetic algorithms were tested in the following two cases: (i) fix mutation ratio as 0 and vary crossover ratio from 0.1 to 0.9; (ii) fix crossover ratio as 0 and vary mutation ratio from 0.1 to 0.9. In both cases, fix *max_gen* = 100 and *pop_size* = 20. The best values of objective function over 200 random runs for each different parameter setting are given in Figure 10.9, From the results we can see that the chance for obtaining an optimal solution is much higher when running genetic algorithms with mutation only than when running with crossover only. The results reveal that mutation plays a critical role in this genetic system, contradicting conventional beliefs. In conventional genetic algorithms, crossover is used as the main operator and mutation is just used as a subsidiary means. Although the mechanism of swap mutation is very simple, it provides the

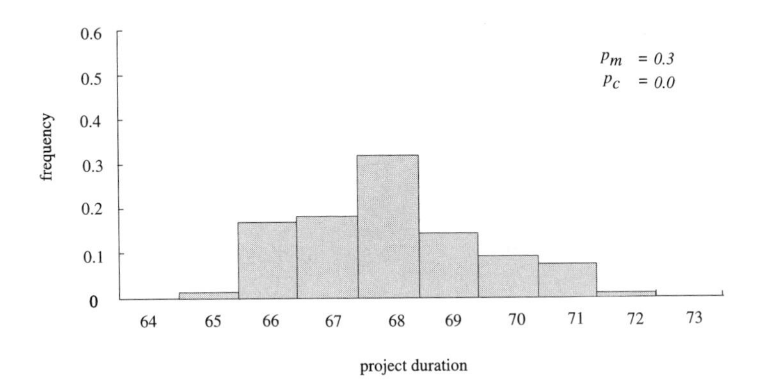

FIGURE 10.10 Solution distribution over 200 runs with the swap mutation only.

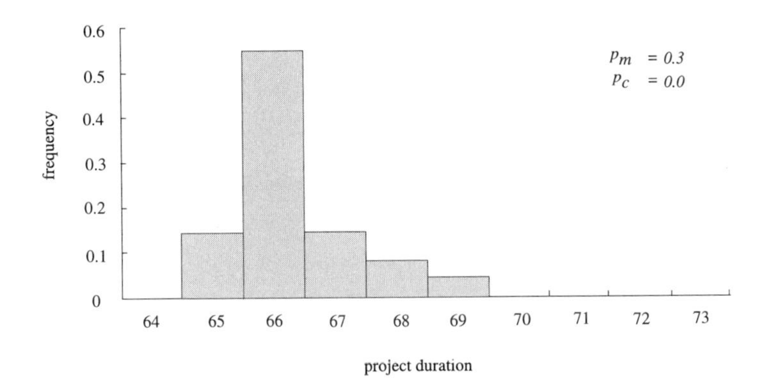

FIGURE 10.11 Solution distribution over 200 runs with the local search-based mutation only.

exploitation on the neighborhood of a given chromosome. This is the reason that mutation can have a high probability of obtaining the optimal solution.

To demonstrate the behavior of local search-based mutation in the genetic search, a comparative experiment was conducted by running with the swap mutation only and running with the local search-based mutation only. To give a fair comparison, for the cases of swap mutation, population size was fixed as 50, maximum generation was fixed as 200; for the case of local search-based mutation, population size was fixed as 20, maximum generation was fixed as 100. For each case, the ratio was fixed as 0.3. The size of the neighborhood for local search-based mutation was fixed as 6 so that the total number of examined chromosomes for each case was nearly the same. The solution distribution over 200 runs is given in Figure 10.10 for the case where only the swap mutation was used and in Figure 10.11 for the case where only the local search-based mutation was used. It is easy to see that the local search-based mutation has a significant impact on the performance of the genetic algorithms.

10.3 Parallel Machine Scheduling Problem

The machine scheduling problem is a rich and promising field of research with applications in manufacturing, logistics, computer architecture, and so on. The parallel machine scheduling problem is concerned with how to construct a schedule of a set of jobs on several machines in order to ensure the execution of all jobs in a reasonable amount of time. All machines are assumed to be identical such that the processing time of a job is independent of the machine. A job is characterized by a processing time and a weight. Each job can be processed by at most one machine at a time, while each machine can process at most one job at a time. Each finished job will free a machine and leave the system. A due date is associated with each job [Cheng and Sin, 1990].

The objective for machine scheduling problem can be broadly classified into regular or nonregular measures. In recent years, scheduling research involving nonregular performance measures has received much attention in response to increasing competitive pressure in domestic and international markets. There are two nonregular performance measures commonly used in machine scheduling problems: minsum measure and minmax measure. The *minsum measure* attempts to minimize the sum of weighted absolute deviations of job completion times about the due date to reduce the aggregate disappointment of customers; the *minmax* measure attempts to minimize the maximum weighted absolute deviation of job completion times about the due date to reduce the maximum disappointment of customers.

Cheng and Gen examined the minmax weighted absolute lateness scheduling problem, which is NP-complete even for a single machine problem, and developed a hybrid genetic algorithm to solve the problem [Cheng, 1997]. There are two essential issues to be dealt with for all kinds of parallel machine scheduling problems: job partition among machines and job sequence within each machine. In Cheng and Gen's method, the genetic algorithm is used to evolve the job partition and a heuristic procedure is used to adjust the job permutation to push each chromosome climb to its local optima.

Consider the following parallel machine scheduling problem [Li and Cheng, 1990]: There are m ($m < n$) identical parallel machines and n independent jobs with known weights $w_1, w_2, \ldots, w_n$ as well as processing times $p_1, p_2, \ldots, p_n$. The jobs are immediately available for processing and can be processed by any one of m machines. No job can be preempted once its processing has begun. Given an unrestricted due date d, i.e.,

$$d \geq \sum_{j=1}^{n} p_j$$

which is common to all jobs.

It is easy to verify that an optimal schedule for the problem has no idle time between jobs. Let Π denote the set of feasible schedules without idle times between jobs. For a given schedule $\sigma \in \Pi$, let c_j be the completion time of job j under the schedule σ for $j = 1, 2, \ldots, n$, and $f(\sigma)$ denote the corresponding objective function value. The problem is to minimize the maximum weighted absolute lateness as follows:

$$\min_{\sigma \in \Pi} f(\sigma) = \max\{w_j |c_j - d|; j = 1, \ldots, n\} \qquad \text{Equation (10.6)}$$

10.3.1 Dominance Condition

A common way to determine jobs order on a machine is to establish some dominance properties among jobs. A *dominance property* gives precedence relations among jobs in an optimal schedule. It is usually called an *a posteriori* precedence relation because it is not a part of the original problem statement. This kind of precedence relations will be used to build a fast heuristic algorithm for job sequencing on a machine.

Let J be the set of jobs. A job belongs to either an early job set or a tardy job set. The early job set is defined as $E = \{j \mid c_j \leq d \text{ and } j \in J\}$ and the tardy job set is defined as $T = \{j \mid c_j > d \text{ and } j \in J\}$. For a given schedule, a job is called a *dominant job* if it has the maximal weighted absolute lateness; that is, if job i is the dominant job, then we have $w_i |c_i - d| = \max_{j \in J} \{w_j \mid c_j - d|\}$. A machine is called a *dominant machine* if it processes the dominant job.

Property 1 For a pair of jobs i and j in the tardy job set T, if $w_i \geq w_j$, then job i precedes job j in at least one optimal schedule.

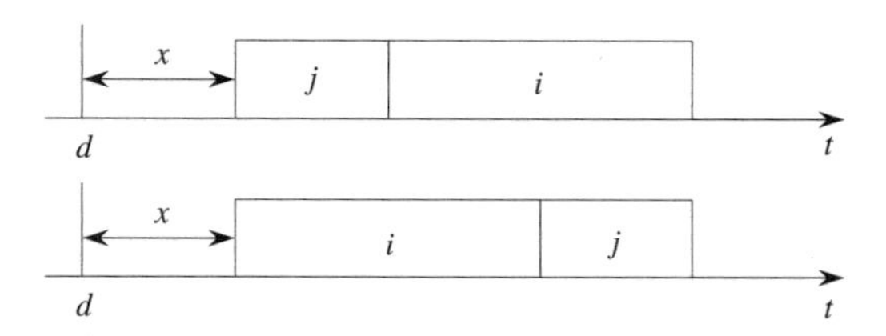

FIGURE 10.12 Two possible orders for jobs i and j in the tardy job set.

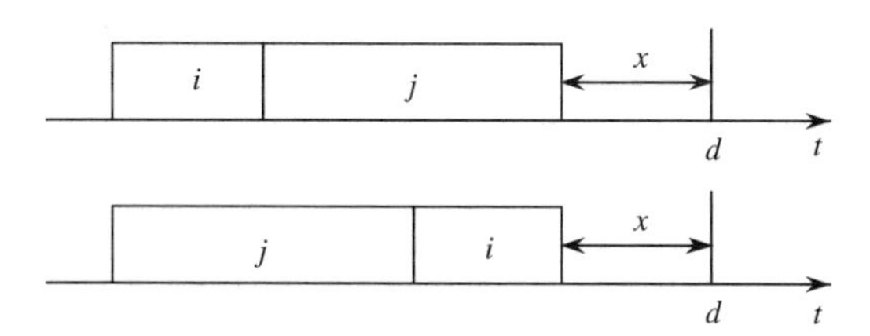

FIGURE 10.13 Two possible orders for jobs i and j in the early job set.

Proof. Consider two possible orders of jobs i and j as shown in Figure 10.12, where x denote the total length of jobs scheduled in set T until now. We have

$$\frac{w_i(p_i + p_j + x)}{w_j(p_i + p_j + x)} = \frac{w_i}{w_j} \geq 1 \qquad \text{Equation (10.7)}$$

Therefore, job i must precede job j. ■

This property gives the fact that a weightier job precedes a lighter job in the tardy job set T in at least one optimal schedule. In other words, jobs in the set T are in nonincreasing order of weights.

How to order jobs in the early job set E is not as simple as in the tardy job set T. A job is characterized by two factors: weight and processing time. For given two jobs i and j, there are four basic patterns of ordering relations of them.

1. $w_i \geq w_j$ and $p_i \geq p_j$
2. $w_i \geq w_j$ and $p_i < p_j$
3. $w_i < w_j$ and $p_i \geq p_j$
4. $w_i < w_j$ and $p_i < p_j$

We just need to examine pattern 1 and pattern 2, because for the precedence relation of a given pair of jobs, pattern 4 describes the same matter as the pattern 1, and pattern 3 describes the same matter as the pattern 2.

Property 2 For a pair of jobs i and j in the early job set E, if $w_i \geq w_j$ and $p_i < p_j$, then job j must precede job i in at least one optimal schedule.

Proof. Consider two possible orders of job i and job j shown in Figure 10.13, where x denotes the total length of jobs scheduled in the set E until now. We have

$$\frac{w_i(p_j + x)}{w_j(p_i + x)} \geq \frac{w_i(p_j + x)}{w_j(p_j + x)} = \frac{w_i}{w_j} \geq 1 \qquad \text{Equation (10.8)}$$

Therefore, job j precedes job i. ■

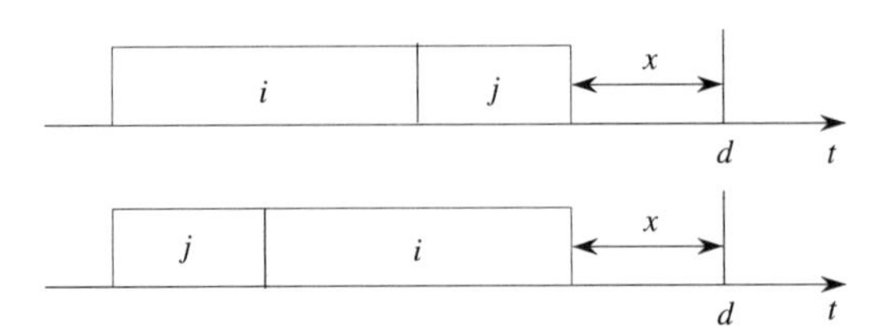

FIGURE 10.14 Two possible orders for jobs i and j in the early job set.

This property gives the fact that a longer and lighter job precedes a shorter and weightier job in the set E in at least one optimal schedule.

PROPERTY 3 For a pair of jobs i and j in the early job set E, if $w_i \geq w_j$, $p_i \geq p_j$ and $w_i/p_i \geq w_j/p_j$, then job j precedes job i.

PROOF. Consider two possible orders of job i and job j shown in Figure 10.14. We have

$$\frac{w_i(p_j + x)}{w_j(p_i + x)} \geq \frac{w_i p_j + w_j x}{w_j(p_i + x)} \geq \frac{w_j p_i + w_j x}{w_j(p_i + x)} = 1 \qquad \text{Equation (10.9)}$$

Therefore, job j must precede job i in an optimal schedule. ■

This property gives one of the necessary conditions that a shorter and lighter job precedes a longer and weightier job in the set E in at least one optimal schedule.

PROPERTY 4 For a pair of jobs i and j of set E associated with $w_i \geq w_j$, $p_i \geq p_j$ and $w_i/p_i < w_j/p_j$, if the total length of scheduled jobs x in the set E is

$$x \geq \frac{w_j p_i - w_i p_j}{w_i - w_j}$$

then job j precedes job i in at least one optimal schedule.

PROOF. Consider two possible orders of job i and job j shown in Figure 10.14. If job j precedes job i in set E, we have

$$\frac{w_i(p_j + x)}{w_j(p_i + x)} \geq 1 \qquad \text{Equation (10.11)}$$

It implies that

$$x \geq \frac{w_j p_i - w_i p_j}{w_i - w_j} \qquad \text{Equation (10.12)}$$

■

This property gives another necessary condition that a shorter and lighter job precedes a longer and weightier job in an optimal schedule. From the above properties we know that there exists at least one optimal schedule where jobs in the set T of the dominant machine are in nondecreasing order of weights while jobs in the set E of the dominant machine are in nonincreasing order of weights for most cases. The exception may occur if a weightier and longer job does not satisfy the condition given in Property 4.

PROPERTY 5 For a given optimal schedule, there exist two dominant jobs on the dominant machine: one is in the set T and the other is in the set E.

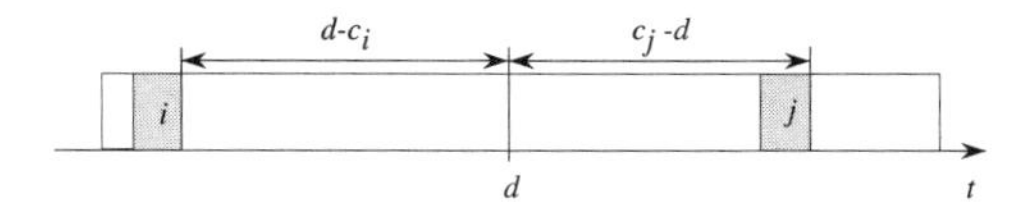

FIGURE 10.15 Two dominant jobs i and j in an optimal schedule.

Proof. Suppose that for a given optimal schedule, there is only one dominant job. Without loss of generality, assume that job $i \in E$ is the dominant job. Then there must exist a job $j \in T$ that dominates the set T, and $w_j\,|c_j - d\,| < w_i\,|c_i - d|$ as shown in Figure 10.15. If we delay all jobs with the amount of δ, which is determined by the equation

$$\delta = \frac{w_i|c_i - d| - w_j|c_j - d|}{w_i + w_j} \qquad \text{Equation (10.13)}$$

then these two jobs i and j have the equal value of absolute lateness, and the objective value of the given schedule is reduced by the amount of $w_i\delta$. This is a contradiction to the precondition of optimal schedule. Therefore, we prove the property.

Property 6 There exists at least an optimal schedule where the following condition holds true for the dominant machine:

$$\sum_{i\in E} p_i \geq \sum_{j\in T} p_j \qquad \text{Equation (10.14)}$$

Proof. There are four possible patterns in which the dominant jobs may occur on the dominant machine.

1. One job is in the head of E while thc other is in the tail of T.
2. One job is in the head of E while the other is in the middle of T.
3. One job is in the middle of E while the other is in the tail of T.
4. One job is in the middle of E while the other is in the middle of T.

For case 1 and case 3 as shown in Figure 10.16, suppose that the total length of T is larger than the length of E. Then we can make a much better schedule by moving the rightmost one of T to the leftmost of E. It means that the given schedule is not an optimal one. This is a contradiction to the precondition of optimal schedule. For case 2 and 4 as shown in Figure 10.16, without loss of generality, assume that if the rightmost jobs k and m are removed from the tardy job set T, the two sets have nearly the same length. In such a case, we can make a new schedule by putting job m at the leftmost of E and job k at the position following job k so that the total length of E is larger than T without increasing the maximal absolute lateness. Therefore, we prove the property. ■

10.3.2 Hybrid Genetic Algorithms

Genetic algorithms have proved to be a versatile and effective approach for solving optimization problems. Nevertheless, there are many situations where the simple genetic algorithms do not perform particularly well. Various methods of *hybridization* have been proposed. A common way is to incorporate local optimization as an add-on extra to the basic loop of genetic algorithms. With the hybrid approach, genetic algorithms are used to perform global exploration among the population while heuristic methods are used to perform local exploitation around chromosomes. Because of the complementary properties of genetic algorithms and heuristics, the hybrid approach often outperforms either method operating alone.

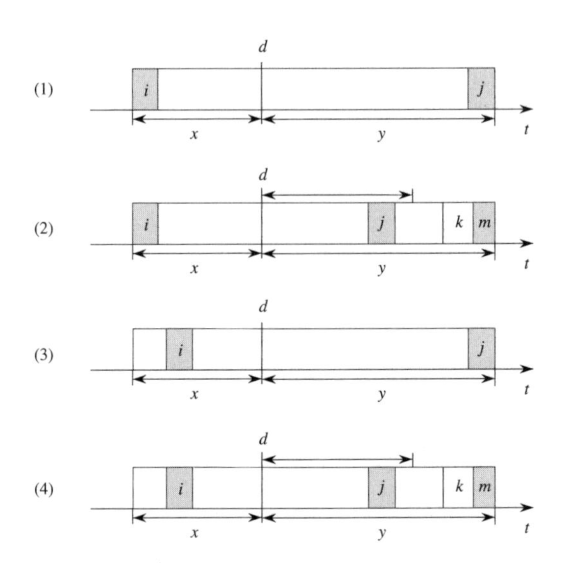

FIGURE 10.16 Four possible patterns of dominant jobs.

Let $P(t)$ and $C(t)$ be parents and offspring, respectively, in current generation t. The hybrid genetic algorithms procedure is described as follows:

procedure: hybrid genetic algorithms
  begin
    $t \leftarrow 0$;
    initialize $P(t)$;
    evaluate $P(t)$;
    while (not termination condition) **do**
    begin
      recombine $P(t)$ to yield $C(t)$;
      locally climb $C(t)$;
      evaluate $C(t)$;
      select $P(t+1)$ from $P(t)$ and $C(t)$;
      $t \leftarrow t+1$;
    end
  end

10.3.2.1 Representation

As we know, there are two essential issues to be dealt with for the parallel machine scheduling problems: (i) partition jobs to machines (combination nature) and (ii) sequence jobs for each machine (permutation nature). The problem involves both permutation and combination components. By using the dominance conditions given in the last section, job sequence on machines can be easily determined; therefore, the problem reduces to a combinatorial problem of finding the best way of partitioning jobs into machines. We then just need to encode the component of *job partition* into a chromosome.

Let $J = \{1, 2, \ldots, n\}$ be the set of jobs and $M = \{1, 2, \ldots, m\}$ be the set of machines. A chromosome consists of n genes and each gene takes an integer number from the set M. Thus, the position of a gene represents the job ID while the value of the gene represents a machine ID. Figure 10.17 illustrates the encoding method with a simple example of three machines and nine jobs. In this example, jobs 2, 3, and 9 are processed by machine 1, jobs 6 and 7 are processed by machine 2, and jobs 1, 4, 5, and 8 are processed by machine 3.

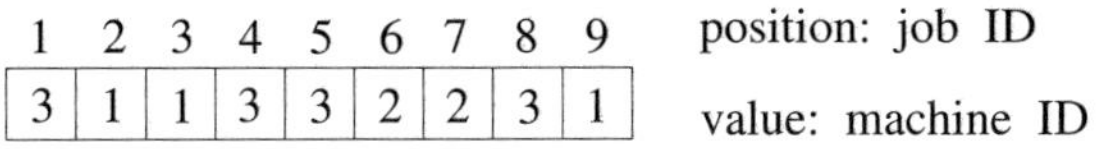

1	2	3	4	5	6	7	8	9	position: job ID
3	1	1	3	3	2	2	3	1	value: machine ID

FIGURE 10.17 Illustration of the proposed encoding method.

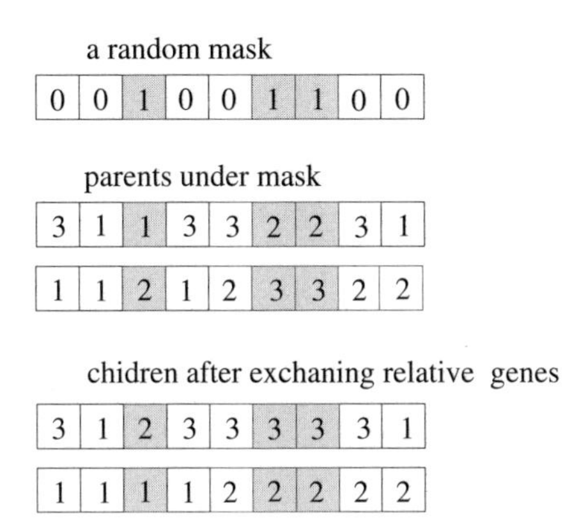

FIGURE 10.18 Illustration of uniform crossover operation.

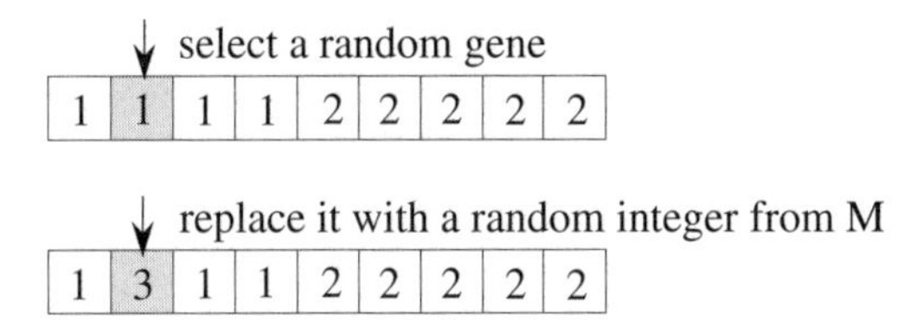

FIGURE 10.19 Illustration of random replacement mutation operation.

10.3.2.2 Genetic Operators

The uniform crossover operator was used to produce offspring. This method first generates a random mask and then exchanges relative genes under the mask between parents. A mask is simply a binary string with the same size of chromosome, and the parity of each bit in the mask determines, for each corresponding bit in an offspring, the parent from which it will receive that bit [Syswerda, 1989]. It is easy to see that the crossover operator can adjust the job partition among machines.

Observing the second child in Figure 10.18, we can see that machine 3 disappears from the chromosome: no job is assigned to machine 3. Such disappearance certainly produces a bad schedule. To enable genetic search to recover a missing machine for a chromosome, a *random replacement* mutation operator is used. The mutation operator first selects a gene randomly and then replaces it with a random integer from the set M, as shown in Figure 10.19. Essentially, the mutation operator can cause a gene either to recover or disappear and can adjust the job partition among machines.

10.3.2.3 Job Sequence

The heuristic procedure is used to sequence jobs on each machine, which is based on the dominance conditions of the properties from 1 to 4.

Let $d - x$ denote the start time of the earliest job and $d + y$ the completion time of the latest job that has been scheduled on a machine, as shown in Figure 10.20. Let J be the set of jobs, let E be the early job set, let $E[i]$ be the ith element in E, and let T be the tardy job set. The heuristic procedure works as follows:

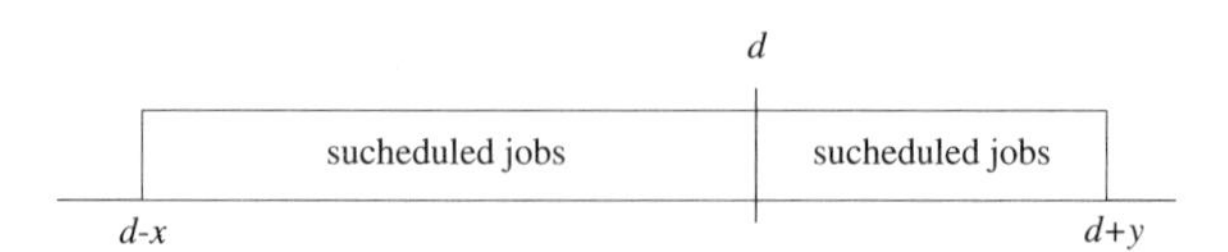

FIGURE 10.20 The start and completion times for scheduled jobs.

procedure: job sequencing

input: a set of jobs J on a machine;
output: a sequence of jobs in the set E and the set T;
begin
  sort all jobs in J on nonincreasing order of weight;
  save the ordered job into array $S[\bullet]$;
  $x \leftarrow 0,\ y \leftarrow 0$;
  $i \leftarrow 0$;
  while $(i \leq |J|)$ **do**
    $p \leftarrow p_{s[i]}$;
    if $(y + p > x)$
      then put job i in the head of the set E;
      $x \leftarrow x + p$;
      if $p_{E[2]} \geq p_{E[1]}$ and $w_{E[1]}p_{E[2]} > w_{E[2]}p_{E[1]}$ and
        $y < (w_{E[1]}p_{E[2]} - w_{E[2]}p_{E[1]})/(w_{E[2]} - w_{E[1]}) - p_{E[2]}$
      then swap $E[1]$ and E[2];
      end
    else
      put job i in the tail of the set T;
      $y \leftarrow y + p$;
    end
    $i \leftarrow i + 1$;
  end
end

10.3.2.4 Evaluation and Selection

Two major steps are involved in the evaluation phase: (i) calculate the objective values for each chromosome and (ii) convert the objective values to fitness values.

For a given chromosome, the job sequence on each machine is determined by the heuristic procedure described above. In the schedule obtained from the heuristic procedure, there may exist only one dominant job for the dominant machine. According to Property 5, we can further slide each job properly to yield a better schedule. Then we calculate the objective value of maximal absolute lateness for the chromosome.

Since the problem is a minimization problem, we have to convert the original objective value to a fitness value in order to ensure that a fitter individual has a larger fitness value. This is done by a transformation through the following fractional linear function:

$$eval(v_t) = \frac{1}{f(v_t)}, \qquad t = 1, 2, \ldots, pop\_size \qquad \text{Equation (10.15)}$$

where $eval(v_t)$ is the fitness function for the tth chromosome and $f(v_f)$ is the objective function value.

The roulette wheel selection method was used as the basic selection mechanism to reproduce the next generation based on the current enlarged population. The *elitist* method was combined with it in order to preserve the best chromosome in the next generation and overcome the stochastic errors of sampling.

TABLE 10.1 Results of the Random Test Problems

Problems	Size	Heuristic	Genetic Algorithms		
			Best	Worst	Average
1	$j30 \times m5$	38.50	32.00	35.20	33.52
2	$j50 \times m5$	100.15	94.50	102.86	98.59
3	$j50 \times m10$	49.09	38.77	43.31	40.91
4	$j60 \times m5$	117.78	114.55	120.00	117.37
5	$j60 \times m10$	132.00	125.36	140.00	133.59
6	$j70 \times m8$	88.00	84.00	93.17	87.06
7	$j70 \times m10$	150.86	142.22	156.80	147.88
8	$j80 \times m5$	292.50	284.39	304.00	292.43
9	$j80 \times m10$	97.07	92.00	99.47	95.75
10	$j80 \times m12$	98.18	92.31	97.78	94.72
11	$j80 \times m14$	81.67	73.85	82.13	76.84
12	$j90 \times m7$	150.00	145.38	151.20	148.47
13	$j90 \times m11$	77.50	71.11	78.24	74.05
14	$j100 \times m20$	113.75	106.36	116.31	110.61

10.3.3 Experimental Results

Because there is no benchmark problem available for this problem, the proposed hybrid method was tested on randomly generated problems. A total of 14 test problems were randomly generated with sizes ranging from 30 jobs and 5 machines to 100 jobs and 20 machines. The basic setting of parameters for the hybrid genetic algorithm is the same: $pop\_size = 50$, $p_c = 0.4$, $p_m = 0.4$, and $max\_gen = 500$. The results over 20 runs for each problem are given in Table 10.1, where *heuristic* stands for Li and Cheng's heuristic [1993]. The results show that the genetic algorithm outperforms the heuristic.

10.4 Job-Shop Scheduling Problem

In the job-shop scheduling problem, we are given a set of jobs and a set of machines. Each machine can handle at most one job at a time. Each job consists of a chain of operations, each of which needs to be processed during an uninterrupted time period of a given length on a given machine. The purpose is to find a schedule—that is, an allocation of the operations to time intervals on the machines—that has a minimum duration required to complete all jobs.

This problem is one of the best-known difficult combinatorial optimization problems. During the last three decades, the problem has captured the interest of a significant number of researchers, and many solution methods have been proposed, ranging from simple and fast dispatching rules to sophisticated branch-and-bound algorithms, but no efficient solution algorithm has been found yet for solving it to optimality in polynomial time. This has led to the recent interest in using genetic algorithms to address it.

The problem of how to adapt genetic algorithms to the job-shop scheduling problems is very challenging but frustrating. During the past decade, two important issues have been extensively studied. One is how to encode a solution of the problem into a chromosome so as to ensure that a chromosome will correspond to a feasible solution. The other issue is how to enhance the performance of genetic search by incorporating traditional heuristic methods.

10.4.1 Basic Approaches

The essence of the job-shop scheduling problem is to find out a permutation of operations on each machine subject to precedence constraints in order to minimize the makespan. If such a permutation can be determined, a solution can then be easily derived with a problem-specific procedure. A general approach for applying genetic algorithms to the job-shop scheduling problem is to (i) use genetic algorithms to evolve an appropriate permutation and (ii) use a heuristic method to construct a solution subsequently according to the permutation.

Because genetic algorithms are not well suited for fine tuning of structures close to optima, various methods of hybridization have been suggested to compensate for this shortcoming. The hybridization methods for the job-shop scheduling problem can be classified into the following three categories: (i) adapted genetic operators, (ii) heuristic-featured genetic operators, and (iii) hybrid genetic algorithms. The first approach is to revise or invent genetic operators so as to meet the feature of a given encoding representation. The second approach is to create new genetic operators inspired from conventional heuristics. The third approach involves hybridizing conventional heuristics into the main loop of genetic algorithms where possible.

10.4.2 Encodings

During the past decade, the following ten representations have been proposed for the job-shop scheduling problems [Cheng et al., 1996]:

- Operation-based representation
- Preference list-based representation
- Topological sort-based representation
- Random key representation
- Priority rule-based representation
- Job-based representation
- Machine-based representation
- Job pair relation-based representation
- Disjunctive graph-based representation
- Completion time-based representation

10.4.2.1 Operation-Based Representation

For an n-job m-machine problem, a chromosome contains $n \times m$ genes and each gene stands for one operation. All operations for a job are encoded with the same symbol and then interpreted according to order of occurrence in a given chromosome [Gen et al., 1994]. For example, for a three-job, three-machine instance, a chromosome takes the following form: [2 1 1 3 2 2 3 1 3]. Each job appears in the chromosome exactly three times and each repeating (each gene) does not indicate a concrete operation of a job but refers to a unique operation that is context-dependent.

10.4.2.2 Preference List-Based Representation

For an n-job m-machine problem, a chromosome is formed of m subchromosomes, each for one machine. Each subchromosome is a string of symbols with length of n, and each symbol identifies an operation that has to be processed on the relevant machine [Davis, 1985]. For example, for a three-job, three-machine instance, a chromosome takes the following form: [(2 3 1) (1 3 2) (2 1 3)]. Each subchromosome does not describe an operation sequence on a machine. It is a *preference list*. The actual schedule is deduced from the chromosome through a simulation, which uses the preference lists to determine the schedule, that is, the operation that appears first in the preference list will be chosen.

10.4.2.3 Topological Sort-Based Representation

This representation is based on the concept of a topological sort of a directed acyclic graph, derived from a disjunctive graph [Cheng, 1997]. The *disjunctive graph model* is a useful tool to represent an instance of the job shop scheduling problem. In a disjunctive graph, there are two kinds of arcs: conjunctive arc and disjunctive arc. The *conjunctive arcs* represent precedence constraints among operations for a job; the *disjunctive arcs* represent operation sequence on a machine. A disjunctive arc can be settled by either of its two possible orientations. A schedule is constructed to settle the orientations of all disjunctive arcs so as to determine the sequences of operations on the same machines. The topological sort of a digraph formed by all conjunctive arcs defines a unique feasible complete orientation on the disjunctive arcs. A topological sort can be represented by a string such as the following, for a three-job, three-machine instance: [2 3 2 1 1 2 3 1 3]. The order for a given digit describes the precedence constraints among operations for a given job, while the order between different digits determines a unique feasible complete orientation, and therefore the operation sequence on machines.

10.4.2.4 Random Key Representation

This representation encodes a solution with *random number.* These values are used as sort *keys* to decode the solution. For an n-job m-machine scheduling problem, each gene (any random key) consists of two parts: an integer in set $\{1,2,\cdots,m\}$ and a fraction generated randomly from (0.1). The integer part of any random key is interpreted as the machine assignment for that job. Sorting the fractional parts provides the job sequence on each machine [Bean, 1994]. For example, for a three-job, three-machine instance, a chromosome is represented as follows: [1.34 1.09 1.88 2.66 2.91 2.01 3.23 3.21 3.44]. It is easy to see that the job sequences given above may violate the precedence constraints. Therefore, a pseudocode is used for handling precedence constraints.

10.4.2.5 Priority Rule-Based Representation

A chromosome is encoded as a sequence of dispatching rules for job assignment and a schedule is constructed with priority dispatching heuristic based on the sequence of dispatching rules [Dorndorf and Pesch, 1995]. Genetic algorithms here are used to evolve a better sequence of dispatching rules. Let 1 denote SPT rule, 2 denote LPT rule, 3 denote MWR rule, and 4 denote LWR rule. A chromosome can be represented as follows: [1 2 2 1 4 4 2 1 3]. Giffler and Thompson algorithms are then used to build a schedule according to the sequence of priority rule [Baker, 1974].

10.4.2.6 Job-Based Representation

A chromosome consists of a list of n jobs, and a schedule is constructed according to the sequence of jobs [Holsapple et al., 1993]. For a three-job, three-machine instance, a chromosome is simply represented as [2 3 1]. For a given sequence of jobs, all operations of the first job in the list are scheduled first, and then the operations of the second job in the list are considered. The first operation of the job under treatment is allocated the best available processing time for the corresponding machine the operation requires, and then the second operation, and so on until all operations of the job are scheduled. The process is repeated with each of the jobs in the list considered in the appropriate sequence.

10.4.2.7 Machine-Based Representation

A chromosome is encoded as a sequence of machines and a schedule is constructed with *shifting bottleneck heuristic* based on the sequence [Dorndorf and Pesch, 1995]. The shifting bottleneck heuristic sequences the machines one by one, successively, each time taking the machine identified as a bottleneck among the machines not yet sequenced. After each new machine is sequenced, all previously established sequences are locally reoptimized [Adams et al., 1988]. For a three-job, three-machine instance, a chromosome is represented simply as [1 3 2]. It is used as the machine sequence for shifting bottleneck heuristic. Genetic algorithms are used to evolve those chromosomes to find out a better sequence of machines.

10.4.2.8 Job Pair Relation-Based Representation

A schedule is encoded by a binary matrix. The matrix is determined according to the precedence relation of a pair of jobs on corresponding machines [Nakano and Yamada, 1991]. This method is perhaps the most complex one among all representations and is highly redundant. Besides its complexity, with this representation, the chromosomes produced either by initial procedure or by genetic operations are illegal in general.

10.4.2.9 Disjunctive Graph-Based Representation

A chromosome, encoded by a binary string, is used to represent an order list of disjunctive arcs [Tamaki and Nishikawa, 1992]. For a three-job, three-machine instance, a chromosome is given as follows: [0 0 1 1 0 1 0 1 1]. It is easy to see that an arbitrary chromosome may not correspond to a perfect settlement of orientations of all disjunctive arcs, which means that an arbitrary permutation of the method may yield a infeasible schedule. Therefore, the chromosome is not used to represent a schedule but only used as a decision preference. A critical path-based procedure was used to deduce a schedule. During the process of deduction, when conflict occurs on a machine, the corresponding bit of the chromosome is used to settle the processing order of the two operations.

10.4.2.10 Completion Time-Based Representation

A chromosome is an ordered list of completion time of operations [Yamada and Nakano, 1992]. For a three-job, three-machine instance, a chromosome is represented as follows: [c_{111} c_{122} c_{133} c_{211} c_{223} c_{232} c_{312} c_{321} c_{333}], where c_{jir} denotes the completion time for operation i of job j on machine r. It is easy to know that such representation is not suitable for most genetic operators because it will yield an illegal schedule. Yamada and Nakano designed a special crossover operator for it.

10.4.2.11 Summary

For ease of explanation, we first define two types of literal strings: pure and general. A *pure literal string encoding* consists of distinctive symbols; a *general string encoding* allows a symbol to repeat a prescribed number. In other words, duplication is prohibited in a pure literal string, whereas the same symbol can coexist in a general literal string. Among the ten encoding methods discussed above, the job-based encoding, the machine-based encoding, the random key encoding, and the completion time-based encoding are the pure literal string; the operation-based encoding, the preference list-based encoding, the topological sort-based encoding, and the priority rule-based encoding are the general literal string.

As we know, there are two types of order relations in the job-shop scheduling problem: (i) the operation sequence on each machine and (ii) the precedence constraints among operations for a job. The first one must be determined by a solution method, while the second must be maintained in a schedule. Accordingly, several different ways have been used to handle the order relations. One is that the information with respect to both operation sequence and precedence constraints is preserved in the encoding simultaneously. In such a case, all crossover methods for literal permutation encodings are not directly applicable. A chromosome may be either *infeasible* in a sense that some precedence constraints are violated, or *illegal* in a sense that the repetitions of some symbols are not equal to the prescribed numbers. For a pure literal string encoding, illegality means that some symbols are repeated more than once while other symbols get lost. For a general literal string encoding, illegality means that some symbols may appear more than necessary, others less than necessary. Special attention must be given to how to handle infeasibility and illegality when designing a new crossover operator for such kinds of encodings so as not to disorder precedence relations. Among literal string encodings, operation-based encoding is the only type that keeps both information of operation sequence and precedence constraints in a chromosome.

In another situation, only the information with respect to operation sequence is encoded in the encodings, while a special decoder procedure or schedule builder procedure is used to resolve the precedence constraints. In such a case the chromosome is one type of literal permutation, with attention directed to preventing a crossover operator from producing illegal offspring. Among literal string encodings, preference list-based encoding, job-based encoding, and machine-based are of this type.

In the last case, neither order relation, but some guild information, is encoded in the chromosome. Such encoding is a pure permutation of literal strings. Although the duplication of some genes results in illegality, as the above one, the order of genes has no direct correspondence to the operation sequences of the jobs. The priority rule-based encoding is of this type.

10.4.3 Adapted Genetic Operators

During the past two decades, various crossover operators have been proposed for literal string encodings, such as partial-mapped crossover (PMX), order crossover (OX), position-based crossover, order-based crossover, cycle crossover (CX), linear order crossover (LOX), sequence exchange crossover, job-based order crossover, partial schedule exchange crossover, substring exchange crossover, and so on [Gen and Cheng, 1999]. Note that there are two basic considerations when designing crossover operators for literal string encodings: (i) to make fewer changes when crossing over so as to inherit the parents' features as much as possible—all variations of two-cut-point crossover operator belong to this class—and (ii) to make more changes when crossing over so as to explore new patterns of permutation and thereby enhance the search ability [Cheng et al., 1999]. All variations of uniform crossover belong to this class.

10.4.3.1 Partial-Mapped Crossover (PMX)

Partial-mapped crossover was proposed by Goldberg and Lingle [1985]. It can be viewed as a variation of two-cut-point crossover by incorporating a special repairing procedure to resolve possible illegitimacy. PMX has the following major steps: (i) Select two cut-points along the string at random. The substrings defined by the two cut-points are called the *mapping sections.* (ii) Exchange two substrings between parents to produce proto-children. (iii) Determine the *mapping relationship* between two mapping sections. (iv) Legalize offspring with the *mapping relationship.*

10.4.3.2 Order Crossover (OX)

Order crossover was proposed by Davis [1985]. It can be viewed as a kind of variation of PMX that uses a different repairing procedure. OX has the following major steps: (i) Select a substring from one parent at random. (ii) Produce a proto-child by copying the substring into the corresponding positions as they are in the parent. (iii) Delete all the symbols from the second parent, which are already in the substring. The resulting sequence contains the symbols the proto-child needs. (ii) Place the symbols into the unfixed positions of the proto-child from left to right according to the order of the sequence to produce an offspring.

10.4.3.3 Position-Based Crossover

Position-based crossover was proposed by Syswerda [1991]. It is essentially a kind of uniform crossover for literal permutation encodings incorporated with a repairing procedure. Uniform crossover operator was proposed for bit-string encoding by Syswerda [1989]. It first generates a random mask and then exchanges relative genes between parents according to the mask. A crossover mask is simply a binary string with the same size of chromosome. The parity of each bit in the mask determines, for each corresponding bit in an offspring, which parent it will receive that bit from. Because uniform crossover will produce illegal offspring for literal permutation encodings, position-based crossover uses a repairing procedure to resolve the illegitimacy. Position-based crossover has the following major steps: (i) Select a set of positions from one parent at random. (ii) Produce a proto-child by copying the symbols on these positions into the corresponding positions of the proto-child. (iii) Delete the symbols that are already selected from the second parent. The resulting sequence contains only the symbols the proto-child needs. (iv) Place the symbols into the unfixed positions of the proto-child from left to right according to the order of the sequence to produce one offspring.

10.4.3.4 Order-Based Crossover

Order-based crossover was also proposed by Syswerda [1991]. It is a slight variation of position-based crossover in which the order of symbols in the selected position in one parent is imposed on the corresponding ones in the other parent.

10.4.3.5 Cycle Crossover (CX)

Cycle crossover was proposed by Oliver et al. [1987]. Like the position-based crossover, it takes some symbols from one parent and the remaining symbols from the other parent. The difference is that the symbols from the first parent are not selected randomly and only those symbols are selected that define a cycle according to the corresponding positions between parents. CX works as follows: (i) Find the cycle that is defined by the corresponding positions of symbols between parents. (ii) Copy the symbols in the cycle to a child with the corresponding positions of one parent. (iii) Determine the remaining symbols for the child by deleting those symbols that are already in the cycle from the other parent. (iv) Fulfill the child with the remaining symbols.

10.4.3.6 Linear Order Crossover (LOX)

Falkenauer and Bouffouix [1991] proposed a modified version of order crossover, the linear order crossover. Order crossover tends to transmit the relative positions of genes rather than the absolute ones. In the order crossover, the chromosome is considered to be circular since the operator is devised for the TSP. In the job-shop problem, the chromosome cannot be considered to be circular. For this reason they developed a variant of the OX called Linear Order Crossover (LOX), where the chromosome is considered linear instead of circular. The LOX works as follows: (i) Select sublists from parents randomly. (ii) Remove $sublist_2$ from parent p_1 leaving some "holes" (marked with h) and then slide the holes from the extremities toward the center until they reach the cross section. Similarly, remove $sublist_1$ from parent p_2 and slide holes to cross section. (iii) Insert $sublist_1$ into the holes of parent p_2 to form the offspring o_1 and insert $sublist_2$ into the holes of parent p_1 to form the offspring o_2.

The crossover operator can preserve both the relative positions between genes and the absolute positions relative to the extremities of parents as much as possible. The extremities correspond to the high- and low-priority operations.

10.4.3.7 Subsequence Exchange Crossover

Kobayashi et al. [1995] proposed a subsequence exchange crossover method. A job sequence matrix is used as encodings. For an n-job m-machine problem, the encoding is an $m \times n$ matrix where each row specifies an operation sequence for each machine. A subsequence is defined as a set of jobs that are processed consecutively on a machine for both parents but not necessarily in the same order. The methods has the following two major steps: (i) Identify subsequences, one for each parent. (ii) Exchange these subsequences machine by machine among parents to create offspring.

Because it is difficult to maintain the precedence relation among operations in either initial population or offspring by use of the job sequence matrix encoding, a Giffler and Thompson algorithm is used to carefully adjust job orders on each machine to resolve the infeasibility and to convert offspring into active schedules.

10.4.3.8 Job-Based Order Crossover

Ono et al. [1996] gave a variation of their subsequence exchange crossover method, called job-based order crossover, by relaxing the requirement that all jobs in the subsequence must be processed consecutively. The job-based order crossover is also designed for the encoding of job-sequence matrix. It has the following major steps: (ii) Identify the sets of jobs from parents, one set for one machine. (ii) Copy the selected jobs of the first parent onto the corresponding positions of the first child machine by machine. Do the same thing for the second child. (iii) Fulfill the unfixed position of the first child by the not-selected jobs from left to right according to the order as they appear in the second parent. Do the same thing for the second child.

10.4.3.9 Partial Schedule Exchange Crossover

Gen et al. [1994] proposed a partial schedule exchange crossover for an operation-based encoding. They consider partial schedules to be the natural building blocks and intend to use such crossover to maintain building blocks in offspring in much the same manner as Holland described. The method has the following major steps: (i) Identify a partial schedule in one parent randomly and in the other parent accordingly. (ii) Exchange the partial schedules to generate proto-offspring. (iii) Determine the missed and exceeded genes for the proto-offspring. (iv) Legalize offspring by deleting exceeded genes and adding missed genes. The partial schedule is identified with the same job in the head and tail of the partial schedule.

10.4.3.10 Substring Exchange Crossover

Cheng and Gen gave another version of partial schedule exchange crossover, called substring exchange crossover [Cheng, 1997]. It can be viewed as a kind of adaptation of two cut-points crossover for general literal string encodings. It contains the following major steps: (i) First, select two cut-points along the string at random. Exchange two substrings defined by the two cuts between two parents to produce proto-children. (ii) Determine the missed and exceeded genes for each proto-child by making a comparison between two substrings. (iii) Legalize the proto-children by replacing the exceeded genes with the missed genes in a random way.

10.4.3.11 Mutation

It is relatively easy to make some mutation operators for permutation representation. During the last decade, several mutation operators have been proposed for permutation representation [Gen and Cheng, 1997]. *Inversion mutation* selects two positions within a chromosome at random and then inverts the substring between these two positions. *Insertion mutation* selects a gene at random and inserts it in a random position. *Displacement mutation* selects a substring at random and inserts it in a random position. *Reciprocal exchange mutation* selects two positions at random and then swaps the genes on these positions. *Shift mutation* first chooses a gene randomly and then shifts it to a random position right or left from the gene's position.

10.4.4 Heuristic-Featured Genetic Operators

Inspired by some successful heuristic methods, several heuristic-featured genetic operators have been created for the job-shop scheduling problems. The kernel procedures within this kind of genetic operator are heuristic methods. A method can be identified as a crossover simply because it will merge two parents to produce two offspring, and as a mutation because it will alter certain amount of genes of one parent to produce offspring.

10.4.4.1 Giffler and Thompson Algorithm-Based Crossover

Giffler and Thompson algorithm is a tree search approach. At each step, it essentially identifies all processing conflicts (the operations competing for the same machine), and an enumeration procedure is used to resolve these conflicts. Inspired by the idea of the Giffler and Thompson algorithm, Yamada and Nakano [1992] proposed a special crossover operator, which is essentially a kind of one-pass procedure but not with a tree search approach. When generating offspring, at each step it identifies all processing conflicts in the manner of Giffler and Thompson, and then chooses one operation from the conflict set of operations according to one of their parents' schedules [Yamada and Nakano, 1992].

10.4.4.2 Neighborhood Search-Based Mutation

Cheng and Gen proposed a mutation inspired by neighbor search technique [Cheng, 1997]. Many definitions may be considered for the neighborhood of a schedule. For a permutation representation, the neighborhood for a given chromosome can be considered as the set of chromosomes (schedules) transformable from a given chromosome by exchanging the positions of λ genes (randomly selected and nonidentical genes). A chromosome (schedule) is said to be λ-*optimum* if it is better than any others in the neighborhood according to some measure.

10.4.5 Hybrid Genetic Algorithms

The role of local search in the context of genetic algorithms has been receiving serious consideration, and many successful applications are strongly in favor of such a hybrid approach. Because of the complementary properties of genetic algorithms and conventional heuristics, the hybrid approach often outperforms either method operating alone. The hybridization can be done in a variety of ways, such as (i) incorporating heuristics into initialization to generate well-adapted initial population. In this way, a hybrid genetic algorithm with elitism can be guaranteed to do no worse than the conventional heuristic does. Other ways include (ii) incorporating heuristics into the evaluation function to decode chromosomes to schedules, and (iii) incorporating local search heuristic as an add-on extra to the basic loop of genetic algorithm, working together with mutation and crossover operators, to perform quick and localized optimization in order to improve offspring before returning it to be evaluated.

10.4.5.1 Combining Genetic Algorithm with Local Search

A common form of hybrid genetic algorithm is the combination of local search and genetic algorithm. Genetic algorithms are used to perform global exploration among population to escape from local optima, while local search methods are used to perform local exploitation around chromosomes to conduct fine-tuning.

Local search in this context can be thought of as being analogous to a kind of learning that occurs during the lifetime of an individual string. With the hybrid method, a offspring will pass on its traits acquired during its learning (local optimization) to future offspring through common crossover. This phenomenon is call *Lamarckian evolution* [Kennedy, 1993]. Therefore, this hybrid approach can be viewed as the combination of Darwin's evolution and Lamarck's evolution.

10.4.5.2 Combining Genetic Algorithm with Giffler and Thompson Method

In this hybrid approach, the genetic algorithm is used to evolve an operation sequence of jobs or some information for producing a sequence of operation, and Giffler and Thompson algorithm is used to deduce an active schedule from the encoding. Giffler and Thompson method can be considered as the common basis of all priority rule-based heuristics. Essentially, Giffler and Thompson method is an enumerative tree search. Embedded with the genetic algorithms, it is used as a one-pass heuristic in order to resolve the precedence constraints and to convert each offspring into an active schedule.

10.4.5.3 Combining Genetic Algorithm with Bottleneck Shifting Heuristic

In this hybrid approach, genetic algorithm is used to evolve a sequence of machines and shifting bottleneck heuristic is used to deduce a schedule from the encoding of machine sequence [Dorndorf and Pesch, 1995]. The shifting bottleneck heuristic is based on the classic idea of giving priority to bottleneck machines. Different measures of bottleneck quality of machines will yield different sequences of bottleneck machines. The quality of the schedules obtained by the shifting bottleneck heuristic heavily depends on the sequence of bottleneck machines. In this method, genetic algorithms are used to evolve those chromosomes to find out a better sequence of machines for the shifting bottleneck heuristic.

10.4.6 Discussion

During the past decade, genetic job-shop scheduling has become a hot topic. This is because the problem exhibits all aspects of constrained combinatorial problems and serves as a paradigm for testing any new algorithmic ideas.

Note that a solution of genetic algorithms is not necessarily a solution of a given problem, but it may contain the necessary information for constructing a solution to a given problem. As we know, there are two types of order relations in the job-shop scheduling problem: (i) the operation sequence on each machine and (ii) the precedence constraints among operations for a job. The first one must be determined by a solution method; the second must be maintained in a schedule. If these two types of order relations are mingled together in the chromosomes, it will impose a heavy burden on genetic operators as to how to handle the infeasibility and the illegality of offspring. How to represent a solution to a job-shop scheduling problem is a well-studied issue. Because of the existence of precedence constraints among operations, the permutation of operations usually corresponds to infeasible solutions. A common way suggested by many researchers is to cope with the trouble of precedence constraints in a special schedule builder. Then the job-shop scheduling problem is treated as a permutation problem: genetic algorithms are used to evolve an appropriate permutation of operations on each machine and a schedule builder is used to produce a feasible solution according to both the permutation and the precedence requirements.

Most encoding methods for the job-shop scheduling problems are of general literal strings type in the sense that a symbol can be repeated in a chromosome. Among them, the preference list-based encoding, consisting of several substrings and one for one machine, is often used in the studies. Because each substring of the encoding is a pure literal string, genetic operators are usually applied to each substring. In this way, the general literal string is treated essentially as several pure literal strings, and many genetic operators developed for pure literal strings can then be used directly. This multiple use of genetic operators will increase the amount of computation greatly as the problem size increases. It is then hoped that genetic search ability can be increased accordingly. Note that multiple genetic operators are independently used in each substring; one consequence of such multiple use is that precedence constraints among operations may be violated. This is why we need a builder to readjust the order of operations to obtain a feasible solution. The trade-off among the complexity of encoding and builder should thus be taken into account.

A trend in genetic job-shop scheduling practice is to incorporate local search techniques into the main loop of a genetic algorithm to convert each offspring into an active schedule. In general, feasible permutation of operations corresponds to the set of semiactive schedules. As the size of the problem increases, the size of the set of semiactive schedules will become larger and larger. Because genetic algorithms are not very good at fine tuning, search within the huge space of semiactive schedules will make genetic algorithms less effective. We know that the optimal solution to the job-shop scheduling problem lies within the set of active schedules, which is much smaller than the set of semiactive schedules. By using a permutation encoding, we have no way to confine genetic search within the space of active schedule. One possible way would be to leave genetic search in the whole search space of semiactive schedules while we convert each chromosome into an active schedule by an add-on extra procedure, the schedule builder. One way to do this is to incorporate the Giffler and Thompson algorithm into genetic algorithms to perform the conversion. The Giffler and Thompson algorithm originally was a kind of enumeration method. In most of such hybrid approaches, it is used in a one-pass heuristic way to readjust the operation order in order to resolve the precedence constraints, and to convert each offspring into an active schedule.

The research on how to adapt the genetic algorithms to the job-shop scheduling problem provides very rich experiences with the constrained combinatorial optimization problems. All of the techniques developed for the problem are very useful for other scheduling problems in modern flexible manufacturing systems and other difficult-to-solve combinatorial optimization problems.

10.5 Multistage Process Planning

Mulistage process planning (MPP) problems are common among manufacturing systems. The problem, in general, provides a detailed description of manufacturing capabilities and requirements for transforming a raw stock of materials into a completed product through a multistage process. In modern manufacturing systems, the problem can be classified into two tpes: variant MPP and generative MPP [Chang and Wysk, 1985]. In the variant MPP, a family of components of similar manufacturing requirements are grouped together and coded using a coding scheme. When a new component is required to be manufactured, its code is mapped onto one of these component families and the corresponding process plan is retrieved. In the generative MPP, the new process plan is automatically generated according to some decision logic. The decision logic interacts with some automatic manufacturing feature recognition machanisms and corresponding machining requirements are derived by utilizing optimization techniques. The generative MPP problem is now receiving more attention, since it can accommodate both general and automated process planning models, which is particularly important in flexible manufacturing systems.

The generative MPP problem lends itself naturally to optimization techniques to find the best process plan among numerous alternatives given a criterion such as minimum cost, minimum time, maximum quality, or a combination of these criteria. The implicit enumeration of all these alternatives can be formulated as network flow, but with the increase of the problem scale, it has the difficulty known as *dimension explosion*. Zhou and Gen [1997] proposed a genetic algorithm to deal with the MPP problem.

10.5.1 Problem Description

The MPP system usually consists of a series of machining operations, such as turning, drilling, grinding, finishing, and so on, to transform a part into its final shape or product. The whole process can be divided

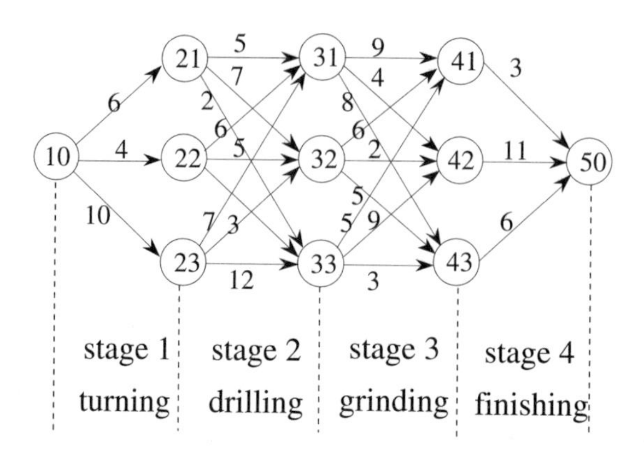

FIGURE 10.21 Flow network for a simple MPP problem.

into several stages. At each stage there is a set of similar manufacturing operations. The MPP problem is to find the optimal process planning among all possible alternatives given certain criteria. Figure 10.21 shows a simple MPP problem.

For an n-stage MPP problem, let s_k be the state at stage k, $x_k \in D_k\ (s_k)$, $k = 1, 2, \ldots, n$ be the decision variable to determine which state to choose at stage k, obviously $D_k(s_k)$, $k = 1, 2, \ldots, n$. Then the MPP problem can be formulated as follows:

$$\min_{\substack{x_k \in D_k(s_k) \\ k=1,2,\ldots,n}} V(x_1, x_2, \ldots, x_n) = \sum_{k=1}^{n} v_k(s_k, x_k) \qquad \text{Equation (10.16)}$$

where $v_k(s_k, x_k)$ represents the criterion to determine x_k under state s_k at stage k. The problem can be rewritten as a dynamic recurrence expression. Let $f_k(x_k, s_k)$ be the optimal process plan from stage k to the last stage n, then the dynamic recurrence expression of the problem can be formulated as follows:

$$f_k(x_k, s_k) = \min_{x_k \in D_k(s_k)} \{v_k(s_k, x_k) + f_{k+1}(x_{k+1}, s_{k+1})\}, \quad k = 1,2,\ldots,n-1.$$

The problem can be approached by the shortest path method or dynamic programming. Obviously, with increased problem scale, many stages and states must be considered, which will greatly affect the efficiency of any solution method.

If there are multiple objectives to be treated simultaneously, the problem has the following formulation:

$$\begin{aligned} \min V_1(x_1, x_2, \ldots, x_n) &= \sum_{k=1}^{n} v_k(s_k, x_k) \\ \min V_2(x_1, x_2, \ldots, x_n) &= \sum_{k=1}^{n} v_k(s_k, x_k) \\ &\cdots \\ \min V_p(x_1, x_2, \ldots, x_n) &= \sum_{k=1}^{n} v_k(s_k, x_k) \\ \text{s.t.} \quad x_k \in D_k(s_k), \quad k &= 1,2,\ldots,n \end{aligned} \qquad \text{Equation (10.17)}$$

where p is the number of the objectives.

Obviously, it is impossible to transform the problem into its equivalent dynamic recurrence expression. As to small-scale problems, the problem can be solved by some traditional multiple objective optimization

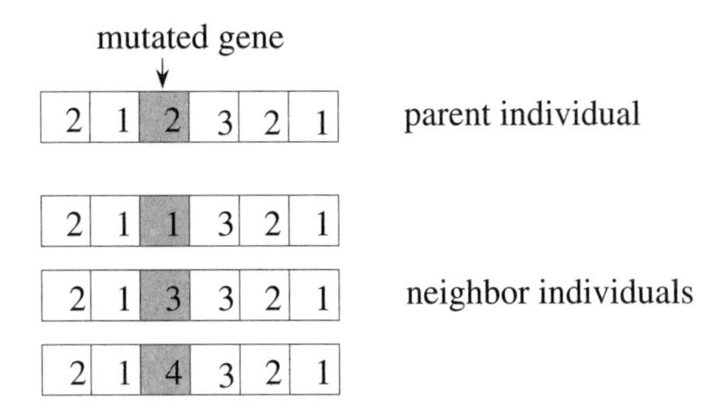

FIGURE 10.22 Illustration of the neighborhood search-based mutation.

techniques. However, if the problem scale increases, it will become difficult to deal with even in the case of a single objective, let alone the case of multiple objective.

10.5.2 Genetic Algorithm Approach

10.5.2.1 Genetic Representation

The MPP problem can be encoded as a permutation of states. In such an encoding, the position of a gene indicates the stage and the value of the gene indicates a state for an operation chosen at that stage. For example, [3 1 2] is an encoding for a feasible solution to the instance given in Figure 10.21, which represents the processing plan as $10 \rightarrow 23 \rightarrow 31 \rightarrow 42 \rightarrow 50$. This encoding method is able to generate all feasible individuals through genetic operations such as crossover or mutation. Initial population is generated randomly with the number of all possible states in the corresponding stage.

10.5.2.2 Genetic Operation

Only neighborhood search-based mutation was used in Zhou and Gen's [1997] method. A neighbor to a chromosome is formed by replacing the mutated gene with all its possible states. Figure 10.22 shows an example for the mutation operation.

10.5.2.3 Evaluation

In a multiobjective context, criteria are usually conflicting with each other in nature, and the concept of optimal solution gives place to the concept of nondominated solutions (or efficient solutions, or Pareto optimal solutions, or noninferior solutions), for which no improvement in any objective function is possible without sacrificing at least one of the other objective functions.

When applying a genetic algorithm in a multiobjective context, a crucial issue is how to evaluate the credit of chromosomes and then how to represent the credit as fitness values. Zhou and Gen [1997] used a weighted-sum approach, which assigns weights to each objective function, combines the weighted objectives into a single objective function, and take its reciprocal as the fitness value:

$$\text{eval}(x) = \frac{1}{\Sigma_{i=1}^{p} \lambda_i V_i(x)} \qquad \text{Equation (10.18)}$$

where p is the number of objectives, λ_i $(i = 1,2,\ldots,p)$ is the weight coefficient, which is determined by the *adaptive weights approach* proposed by Gen and Cheng [1999]. In the adaptive weights approach, weights will be adaptively adjusted according to the current generation in order to obtain a search pressure toward the positive ideal point. Because some useful information from the current population is used to readjust the weights at each generation, the approach gives a selection pressure toward the positive ideal point.

TABLE 10.2 Comparison among Different Approaches to the MPP Problems

Problem	Number of Stages	Number of Nodes	CPU Time OK (s)	CPU Min.	Time Av.	SP (s) %
1	7	24	3.01	0.03	0.04	100
2	7	27	2.84	0.03	0.04	100
3	8	38	4.31	0.06	0.09	100
4	9	37	4.25	0.06	0.12	100
5	10	47	4.48	0.09	0.27	100
6	11	53	4.58	0.34	0.45	100
7	12	63	4.69	0.32	0.48	100
8	13	72	5.08	0.61	1.03	100
9	14	79	5.19	0.49	1.01	90
10	15	89	5.35	0.97	1.28	80

Notes: OK: out-of-kilter algorithm using SAS/OR program; *SP*: state permutation encoding by GA, *pop_size*: 100; Min.: the minimal CPU time in all 20 runs; Av.: the average CPU time in all 20 runs; %: the frequency of obtaining the optimal solution in all 20 runs.

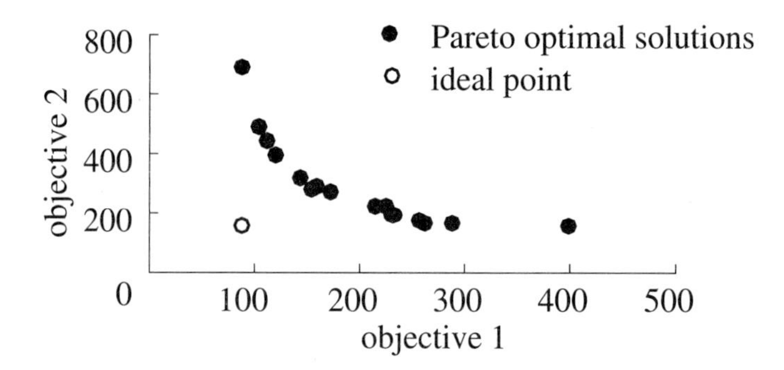

FIGURE 10.23 Illustration of the multicriteria MPP problem.

10.5.4 Numerical Examples

To test the performance of a genetic algorith, ten problems were randomly generated. The results, given in Table 10.2, show that Zhou and Gen's [1997] genetic algorithm is more efficient than the traditional method in obtaining the optimal solutions.

An instance of the problem with two objectives, 15 stages, and 89 nodes is randomly generated. Two attributes, processing time and processing cost, are integers distributed uniformly over [1, 50] and [1, 100], respectively. The ideal point is (91, 159). By using the adaptive weight method as the fitness function, the results are plotted in Figure 10.23.

Compared with the enumeration of all possible Pareto optimal solutions, which takes 3840 seconds CPU time, 85% results obtained by genetic algorithm are Pareto optimal solutions, and it only takes 2.36 seconds CPU time on average results.

10.6 Part Loading Scheduling Problem

Flexible forging machines are used to process a variety of parts according to production requirements [Kusiak, 1990]. Generally, tools and fixtures are heavy, and it takes a considerable amount of time to change them. When processing a variety of parts, the production cost is greatly affected by a parts loading schedule. In order to reduce the production cost, it is very important to minimize the total changeover cost of parts. The Parts Loading Scheduling problem belongs to the class of NP-hard. Tsujimura and Gen [1999] proposed a genetic algorithm to address this problem.

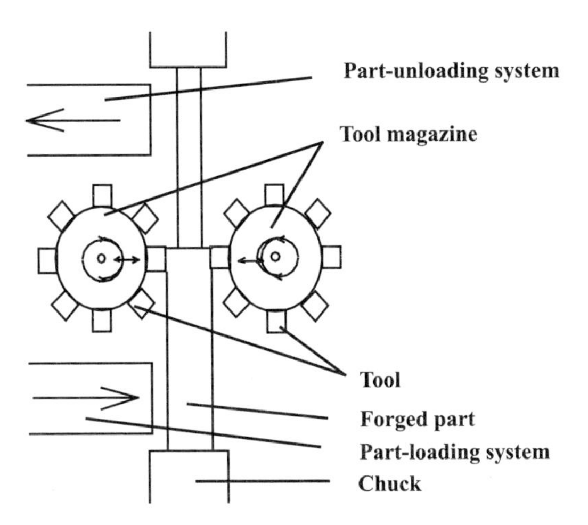

FIGURE 10.24 Flexible forging machine.

10.6.1 Problem Description

The flexible forging machine, as shown in Figure 10.24, has the following four automatic options: (i) part loading, (ii) chuck rotation and horizontal movement, (iii) tool magazine rotation and oscillation, and (iv) part loading and unloading. Reduction of changeover costs is the major concern for making a processing schedule. There are the four types of changeover costs: (i) the cost for changing parts, (ii) the cost of NC programs, (iii) the cost for changing chucks holding parts, and (iv) the cost for changing tools.

The changeover cost imposed by loading and unloading parts is dictated by the hardware design, and its reduction is beyond operational control. These costs are sequence-dependent. Before a new part (or a batch of parts) is forged, in general it is necessary to change same tools stored in the tool magazines.

The parts loading scheduling problem can be considered as a single-machine scheduling problem with sequence-dependent changeover costs and precedence constraints. It is equivalent to the traveling salesman problem. The problem can be formulated as the following mixed-integer programming problem:

$$\min \quad z = \sum_{t=1}^{n} \sum_{j=1}^{n} c_{ij} x_{ij} \qquad \text{Equation (10.19)}$$

$$\text{s.t.} \quad \sum_{i=1}^{n} x_{ij} = 1, \quad j = 1, 2, \ldots, n \qquad \text{Equation (10.20)}$$

$$\sum_{j=1}^{n} x_{ij} = 1, \quad i = 1, 2, \ldots, n \qquad \text{Equation (10.21)}$$

$$u_i - u_j + n x_{ij} \leq n - 1, \quad i = 2, 3, \ldots, n, \; j = 2, 3, \ldots, n, \quad i \neq j \qquad \text{Equation (10.22)}$$

$$x_{ij} = 0 \quad \text{or} \quad 1, \quad i, j = 1, 2, \ldots, n \qquad \text{Equation (10.23)}$$

$$u_i \geq 0, \quad i = 1, 2, \ldots, n \qquad \text{Equation (10.24)}$$

where n is the number of parts c_{ij}, the changeover cost from part i to part j, u_i the nonnegative variable, and $x_{ij} = 1$ if part i immediately precedes part j; otherwise, 0. The objective is to minimize the total changeover cost. Constraint (10.23) eliminates subschedules generated by solving the assignment problem 10.20–10.22.

10.6.2 Genetic Algorithm Approach

10.6.2.1 Representation Scheme

Path encoding is used to represent a parts loading schedule [Grefenstette et al., 1985]. Let p_i be the ith part to be processed, and a chromosome s_k can be represented as $s_k = (p_1\ p_2\ \cdots p_n)$.

10.6.2.2 Evaluation Function

The evaluation function eval(s_k) is defined as follows:

$$\text{eval}(s_k) = \frac{1}{\sum_{i=1}^{n} (c(p_i, p_{i+1}) + c(p_n, p_1))} \qquad \text{Equation (10.25)}$$

where $c(p_i, p_{i+1})$ is the changeover cost from part p_i to part p_{i+1}.

10.6.2.3 Genetic Operators

Cycle crossover (CX) is employed, which essentially takes some symbols from one parent and the remaining symbols from the other parent [Oliver et al., 1987]. The swap mutation operation is used to introduce some new gene materials into the population.

10.6.3 Measure of Diversity

To prevent genetic search from falling into a local optima in the evolution process, a measure was introduced to monitor the diversity of the population. The total diversity of a population is the sum of the locus diversity, as shown in Figure 10.25.

The locus diversity H_i of the ith locus of a chromosome is defined as follows:

$$\begin{aligned} H_i &\equiv -\sum_{p \in P} d_{ip} \ln d_{ip} \\ d_{ip} &= \frac{\text{the number of occurrence of part } p \text{ at locus } i}{\text{the size of population}} \end{aligned} \qquad \text{Equation (10.26)}$$

where P is the set of parts that should be processed.

The locus diversity H_i is derived from the concept of *information entropy*, and used as a measure of the diversity of the ith locus of chromosomes in the current population. H_i approaches the maximum value (ln *pop_size*) when each part appears uniformly at the locus in the population; conversely, it approaches the minimum value 0 when a part appears in that locus at most chromosomes in the current population.

The locus diversity H_i is evaluated by comparing with the threshold value. Tsujimura and Gen [1999] used ln 2 as the threshold value. If $H_i \leq \ln 2$, the locus i has low diversity. Compare every loca with the threshold value, and if the number of loca with locus diversity below the threshold value is greater than or equal to n/a, the diversity of the chromosome population is too low. Then, for such a low-diversity population, the diversity of the population must be increased. The parameter a is the control parameter, which takes an integer value from the closed interval [2, 5]. The value of a is decided by the relation between the population size and the number of parts in P, and the larger the value of a, the higher the probability of improvement.

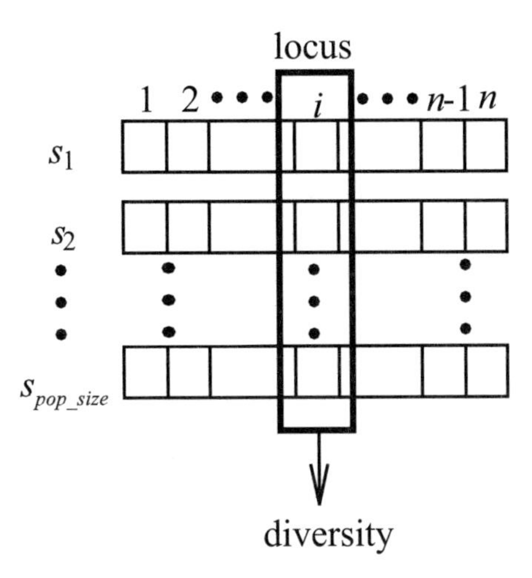

FIGURE 10.25 Diversity measure at a locus.

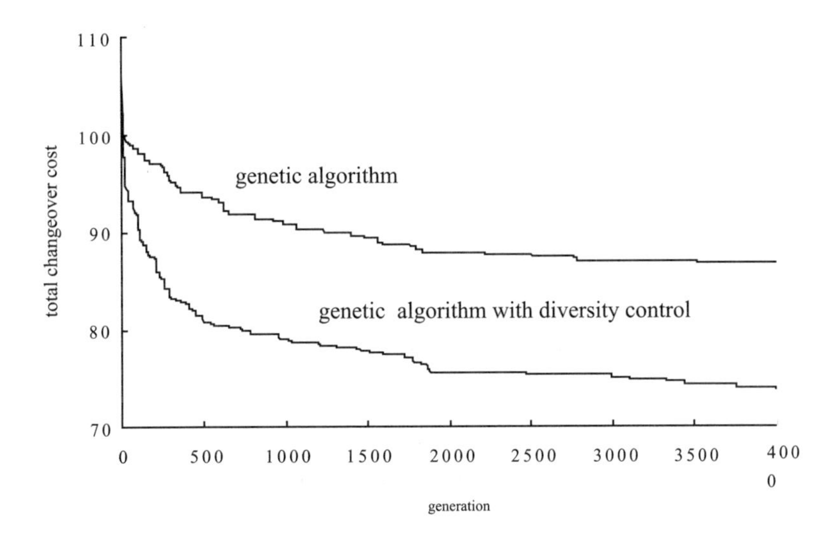

FIGURE 10.26 Evolutionary convergence process.

10.6.3.1 Procedure of Diversity Improvement

The improvement of the population diversity is performed according to the following procedure: (i) Select m chromosomes from the population; m is a random integer number in $[\frac{pop\_size}{x} \quad pop\_size - 1]$. (ii) In each of the chromosomes selected, exchange genes among the loca that have lower locus diversities than the threshold value. Repeat the procedure until population diversity becomes high enough.

10.6.4 Numerical Experiments

A ten-part parts-loading scheduling problem with symmetric cost data is given in Table 10.3. The optimal solution obtained by using the branch-and-bound method is 1 – 2 – 6 – 7 – 9 – 10 – 8 – 5 – 4 – 3 and the total changeover cost is 73.

Results were compared between a simple genetic algorithm and a genetic algorithm with diversity control. Parameter setting for both cases are population size $pop\_size = 10$, maximum generation $max\_gen = 4000$, crossover probability $p_c = 0.6$, mutation probability $p_m = 0.01$, and control parameter $a = 3$.

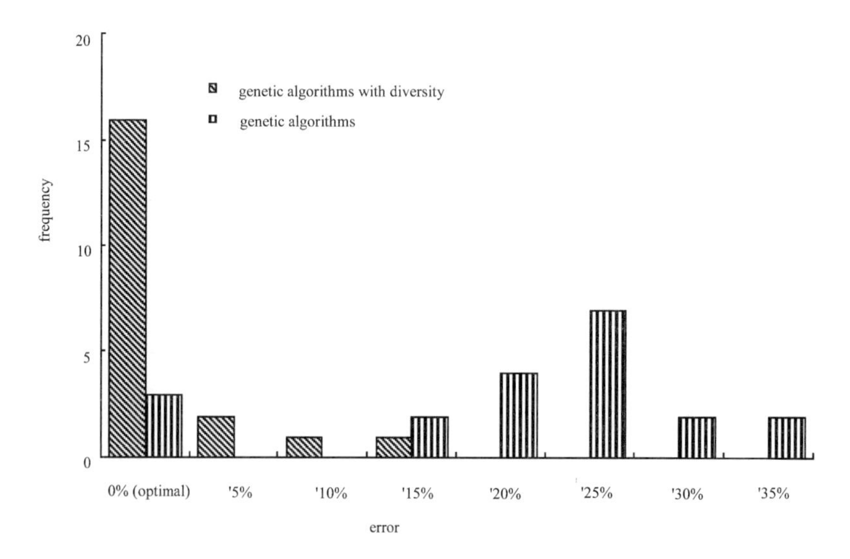

FIGURE 10.27 Distribution of the errors.

TABLE 10.3 Ten-Part Instance for a Parts Loading Scheduling Problem

		i									
		1	2	3	4	5	6	7	8	9	10
	1	—	8	5	9	12	14	12	16	17	22
	2	8	—	9	15	17	8	11	18	14	22
	3	5	9	—	7	9	11	7	12	12	17
	4	9	15	7	—	3	17	10	7	15	18
j	5	12	17	9	3	—	8	10	6	15	15
	6	14	8	11	17	8	—	9	14	8	16
	7	12	11	7	10	10	9	—	8	6	11
	8	16	18	12	7	6	14	8	—	11	11
	9	17	14	12	15	15	8	6	11	—	10
	10	22	22	17	18	15	16	11	11	10	—

The evolutionary convergence processes are shown in Figure 10.26. Distribution of the errors of 20 runs is given in Figure 10.27. The results imply that the procedure with diversity control can have both better convergent ability and better quality solution than a simple genetic algorithm.

References

Adams, J., E. Balas, and D. Zawack, 1988. The shifting bottleneck procedure for job shop scheduling, *Flexible Manufac. Sys.*, vol. 34, no. 3, pp. 391–401.

Alvarez-Valdés, R. and J. Tamarit, 1989. Heuristic algorithms for resource constrained project scheduling: A review and an empirical analysis, in *Advances in Project Scheduling*, Ed. R. Slowinski and J. Weglarz, pp. 113–134, Elsevier Science Publishers, Amsterdam.

Baker, K., 1974. *Introduction to Sequencing and Scheduling*, John Wiley & Sons, New York.

Bean, J., 1994. Genetic algorithms and random keys for sequencing and optimization, *ORSA J. Comput.*, vol. 6, no. 2, pp. 154–160.

Chang, T. C. and R. A. Wysk, 1985. *An Introduction to Automated Process Planning Systems*, Prentice-Hall, Englewood Cliffs, NJ.

Cheng, R., 1997. A study on genetic algorithms-based optimal scheduling techniques, Ph.D. dissertation, Tokyo Inst. of Tech., Japan.

Cheng, R. and M. Gen, 1998. An evolution program for the resource constrained project scheduling problem, *Comp. Integr. Manuf.*, vol. 11, no. 3, pp. 274–287.

Cheng, T. and C. Sin, 1990. A state-of-the-art review of parallel-machine scheduling research, *Eur. J. Oper. Res.*, vol. 47, pp. 271–292.

Cheng, R., M. Gen, and Y. Tsujimura, 1996. A tutorial survey of job-shop scheduling problems using genetic algorithms: Part I. Representation, *Comp. Ind. Eng.*, vol. 30, no. 4, pp. 983–997.

Cheng, R., M. Gen, and Y. Tsujimura, 1999. A tutorial survey of job-shop scheduling problems using genetic algorithms: Part II. Hybrid genetic search strategies, *Comp. Ind. Eng.*, vol. 36, no. 2.

Davis, L., 1985. Job shop scheduling with genetic algorithms, in *Proc. 1st ICGA*, Ed. J. Grefenstette, pp. 136–140, Lawrence Erlbaum Associates, Hillsdale, NJ.

Dorndorf, U. and E. Pesch, 1995. Evolution based learning in a job shop scheduling environment, *Comp. Opns. Res.*, vol. 22, no. 1, pp. 25–40.

Falkenauer, E. and S. Bouffoix, 1991. A genetic algorithm for job shop, in *Proc. 1991 IEEE Int. Conf. Robot. Auto.*, pp. 824–829.

Gen, M. and R. Cheng, 1997. *Genetic Algorithms and Engineering Design,* John Wiley & Sons, New York.

Gen, M. and R. Cheng, 1999. *Genetic Algorithms and Engineering Optimization,* John Wiley & Sons, New York.

Gen, M., Y. Tsujimura, and E. Kubota, 1994. Solving job-shop scheduling problem using genetic algorithms, in *Proc. 16th Int. Conf. Comp. Indus. Eng.*, Ed. M. Gen, and T. Kobayashi, pp. 576–579, Ashikaga, Japan.

Goldberg, D. and R. Lingle, 1985. Alleles, loci and the traveling salesman problem, in *Proc. 1st ICGA*, Ed. Grefenstette, pp. 154–159, Lawrence Erlbaum Associates, Hillsdale, NJ.

Grefenstette, J., R. Gopsl, B. J. Rosmaita and D. Van Gucht, 1985. Genetic algorithms for the traveling salesman problem, in *Proc. 1st ICGA*, Ed. J. Grefenstette, pp. 160–168, Lawrence Erlbaum Associates, Hillsdale, NJ.

Holsapple, C., V. Jacob, R. Pakath, and J. Zaveri, 1993. A genetics-based hybrid scheduler for generating static schedules in flexible manufacturing contexts, *IEEE Trans. Sys. Man. Cyber.*, vol. 23, pp. 953–971.

Kennedy, S., 1993. Five ways to a smarter genetic algorithm, *AI Expert*, pp. 35–38.

Kobayashi, S., I. Ono, and M. Yamamura, 1995. An efficient genetic algorithm for job shop scheduling problems, in *Proc. 6th ICGA*, Ed. L. J. Eshelman, pp. 506–511, Morgan Kaufmann, San Francisco.

Kusiak, A., 1990. *Intelligent Manufacturing System*, Prentice-Hall, Englewood Cliffs, NJ.

Li, C. and T. Cheng, 1993. The parallel machine min-max weighted absolute lateness scheduling problem, *Naval Res. Log.*, vol. 41, pp. 33–46.

Nakano, R. and T. Yamada, 1991. Conventional genetic algorithms for job-shop problems, in *Proc. of 4th ICGA*, pp. 477–479, Eds. Belew and Booker, Morgan Kaufmann, San Mateo, CA.

Oliver, I., D. Smith, and J. Holland, 1987. A study of permutation crossover operators on the traveling salesman problem, in *Proc. 2nd ICGA*, Ed. J. Grefenstette, 224–230, Lawrence Erlbaum Associates, Hillsdale, NJ.

Ono, I., M. Yamamura, and S. Kobaysshi, 1996. A genetic algorithms for job-based order crossover, in *Proc. 1996 IEEE ICEC*, Ed. D. Fogel, IEEE Press, Japan.

Panwalkar, S. and W. Iskander, 1977. A survey of scheduling rules, *Opns. Res.*, vol. 25, pp. 45–61.

Syswerda, G., 1989. Uniform crossover in genetic algorithms, in *Proc. 3rd ICGA*, Ed. J. Schaffer, pp. 2–9, Morgan Kaufmann, San Mateo, CA.

Syswerda, G., 1991. Scheduling optimization using genetic algorithms, in *Handbook of Genetic Algorithm*, Ed. L. Davis, pp. 332–349, Van Nostrand Reinhold, New York.

Tamaki, H. and Y. Nishikawa, 1992. A paralleled genetic algorithm based on a neighborhood model and its application to the jobshop scheduling, in *Proc. PPSN II*, pp. 573–582, Eds. Männer and Manderick, Elsevier Science, Amsterdam.

Tsujimura, Y. and M. Gen, 1999. Parts loading scheduling in flexible forging machine using advanced genetic algorithm, *J. Intel. Manuf.*, vol. 10, no. 2, pp. 149–159.

Yamada, T. and R. Nakano, 1992. A genetic algorithm applicable to large-scale job-shop problems, in *Proc. PPSN II,* pp. 281–290. Eds. Männer and Manderick, Elsevier Science, Amsterdam.

Zhou, G. and M. Gen, 1997. Evolutionary computation on multicriteria production process planning problem, in *Proc. 1997 ICEC,* Ed. B. Porto, pp. 419–424, IEEE Press, Indianapolis, IN.

IV

Manufacturing Process Monitoring and Control

11

Neural Network Predictive Process Models: Three Diverse Manufacturing Applications

Sarah S. Y. Lam
State University of New York at Binghamton

Alice E. Smith
Auburn University

11.1 Introduction to Neural Network Predictive Process Models

In a broad sense, predictive models describe the functional relationship between input and output variables of a data set. When dealing with real-world manufacturing applications, it is usually not an easy task to precisely define the set of input variables that potentially affect the output variables for a particular process. Oftentimes, this is further complicated by the existence of interactions between the variables. Even if these variables can be identified, finding an analytical expression of the relationship may not always be possible. The process of selecting the analytical expression and estimating the parameters of the selected expression could be very time-consuming.

Neural networks, a field that was introduced approximately 50 years ago, have been getting more attention over the past 15 years. There are a number of survey papers that summarize some of the applications of neural networks. Udo [1992] surveys within the manufacturing domain, which covers resource allocation, scheduling, process control, robotic control, and quality control. Zhang and Huang [1995] provide a good overview of many manufacturing applications. Hussain [1999] discusses a variety of applications in chemical process control. One of the advantages of neural network modeling is its ability to learn relationships through the data itself rather than assuming the functional form of the relationship. A neural network is known as a universal approximator [Hornik et al., 1989; Funahashi, 1989]. It can model any relationship to any degree of accuracy given that there are sufficient data for modeling. It can tolerate noisy and incomplete data representations. Moreover, it can dynamically adjust to new process conditions by continuous training. Through an iterative learning process, a neural network extracts information from the training set and stores the information in its weight connections. After a

0-8493-0592-6/01/$0.00+$.50

network is trained, it can then be used to provide predictions for new inputs. But how good is the network when it is used to make predictions on data that are not used to train the network?

Being an empirical modeling technique, validating the network is at least as important as constructing the network. Theoretically speaking, an infinite number of data points should be used to validate and evaluate the performance of a network. However, this is not feasible in practice. In order to maximally leverage the available data, resampling methods such as *cross validation* and *group cross validation* can be used to validate the network [Twomey et al., 1995; Lam et al., 2000; Lam and Smith, 1998; Coit et al., 1998]. These validation methods are more appealing than the traditional *data splitting* approach, especially when the data are sparse. They allow the construction of the network based upon the entire data set but also allow the evaluation of the network using all the data that are available [Efron, 1982; Wolpert, 1993]. Hence, these methods ensure the extraction of as much information from the available data as possible for developing an *application network,* ensuring the best possible prediction performance. The trade-off of using these resampling methods is the incurred computational expense of developing multiple *validation networks* in order to infer the performance of the application network [Twomey and Smith, 1998].

This chapter discusses some applications where neural networks have been used successfully as predictive process models. These applications have relatively sparse data sets; therefore, resampling methods are used to validate the application networks. More specifically, the manufacturing processes covered include: (1) a ceramic slip casting process, (2) an abrasive flow machining process, and (3) a chemical oxidation process. These examples are real-world engineering problems where designed experiments were conducted for the first and the last examples to supplement production data.

11.2 Ceramic Slip Casting Application

The slip-casting process is used to produce ceramic ware in many complicated shapes — such as bowls, statues, and sinks — that cannot be achieved through conventional pressing. However, this flexibility does not come without a price. It is generally more difficult to achieve a desired level of product quality in the presence of many controllable and uncontrollable process factors. Basically, the manufacturing of ceramic ware consists of the following steps:

1. Preparing the liquid clay (a slurry, or slip)
2. Casting the slip in a plaster mold for a specified duration
3. Removing the mold
4. Air drying the cast piece
5. Spray gazing the dried piece
6. Firing the gazed piece in a kiln
7. Inspecting the finished product.

The slip is prepared by mixing clay powder with a suspending liquid. Deflocculants are added to the slurry to provide stability and density, and binders are added to ensure that the resulting cast is strong enough to be handled. This slip is then poured into a plaster mold and stays there for a specified time period in order to form a solid product. The liquid in the slip is absorbed into the mold through capillary action, resulting in a solid cast inside the mold. When it is estimated (by the slip cast operators) that the cast has reached the desired wall thickness, it is then removed from the mold, air dried, glazed, and fired to produce a finished product [Adams, 1988].

Slip casting largely determines the quality of the final product. If the slip casting process takes too long, the cast will be too dry and may result in cracks. On the other hand, if it does not allow sufficient time period for the slip to cast, the cast piece will be too wet and may result in instabilities. These defects are manifest in the subsequent steps of the manufacturing process. Defects that are found before the ware is fired can often be repaired. For defects that cannot be repaired, the material can be recovered, but the considerable labor and overhead are still irretrievably lost. Most defects that are found after firing result in a complete loss of the defective piece. The proportion of defective pieces due to casting

TABLE 11.1 Slip Casting Process Parameters

Input	Parameter	Definition
1	Plant temperature (°F)	The temperature of the plant.
2	Relative humidity (%)	The humidity level of the plant.
3	Cast time	The time duration that the liquid slip is left in the mold before draining.
4	Sulfate (SO_4) content	The proportion of soluble sulfates in the slip.
5	Brookfield–10 RPM	Viscosity of the slip at 10 revolutions per minute.
6	Brookfield–100 RPM	Viscosity of the slip at 100 revolutions per minute.
7	Initial reading	Initial viscosity (taken at 3 1/2 minutes).
8	Build up	Change in viscosity from initial reading (taken after 18 minutes)
9	20 minute gelation	Thixotropy (viscosity vs. time).
10	Filtrate rate	The rate at which the slip filtrates.
11	Slip cake weight	Approximation of the cast rate without considering a mold.
12	Cake weight water retention	Moisture content of the cake. See slip cake weight.
13	Slip temperature	The temperature of the slip.

Source: Lam, S. S. Y., Petri, K. L. and Smith, A. E., Prediction and Optimization of a Ceramic Casting Process Using A Hierarchical System of Neural Networks and Fuzzy Logic, *IIE Transactions,* 32(1), 83–91, 2000.

imperfections can approach as much as 30%. This figure obviously poses a significant problem that affects the efficiency and profitability of these manufacturing firms.

The primary causal factor for cast fractures and/or deformities is the distribution of moisture content inside the cast before firing. When the moisture differential, or *moisture gradient,* inside the wall of the cast is too steep, it results in stress differences. It is the stress differences that cause the piece to deform and eventually fracture. In order to minimize the possibility of fractures or deformities, the moisture content should be as uniform as possible. In other words, the moisture gradient should be close to zero in order to have a good cast. Another important measure is *cast rate,* which is actually the thickness of the cast (in inches) achieved during a set time in the mold. A larger cast rate is more desirable because it indicates an increase in production efficiency.

The quality of the cast (moisture gradient) and the cast rate depend on the slip conditions, the ambient conditions in the plant, and the plaster mold conditions. As the age of the mold increases, the capillary action of the mold degrades, which causes an increase in the required casting time. Ambient conditions also have a significant effect on casting time. Molds that are cast under hot, dry conditions (i.e., near a kiln) require less casting time than molds cast under cooler, wetter conditions (i.e., near the building exterior). The variance of ambient conditions across the plant can be a significant problem in ceramic casting facilities. Two ambient variables (the plant temperature and humidity), ten slip property variables, and the cast time are identified as significant factors for determining the cast quality and the cast efficiency. These variables are summarized in Table 11.1.

11.2.1 Neural Network Modeling for Slip Casting Process

Two separate networks were developed, one to predict the moisture gradient and the other to predict the cast rate. The dual models' approach was motivated partly because of the uneven distribution of the data. However, the main motivation behind this was the undefined underlying relationship between the process variables and the unknown interactive behavior of the variables.

Two separate data sets (consisting of production data supplemented by experimental data) were collected and obtained by the plant technicians. There were 367 observations for the moisture gradient model and 952 observations for the cast rate model. Both of these models have the same set of input variables as illustrated in Table 11.1. The network architectures, training parameters, and stopping criteria were finalized through experimentation and examination of preliminary networks. An ordinary back-propagation algorithm was used because of its simplicity and its documented ability as a continuous function approximator [Hornik et al., 1989; Funahashi, 1989]. The final application network architecture

TABLE 11.2 MAE and RMSE for the Moisture Gradient Network

Network	MAE	RMSE
Application network	0.0025	0.0045
Validation networks	0.0036	0.0061

TABLE 11.3 MAE and RMSE for the Cast Rate Network

Network	MAE	RMSE
Application network	0.0156	0.0207
Validation networks	0.0167	0.0243

for the moisture gradient model had 13 inputs, two hidden layers with 27 hidden units in each layer, and a single output. The output unit represents the moisture gradient of the cast piece. Similarly, the architecture for cast rate model had the same input variables, two hidden layers with 11 hidden units in each layer, and a single output that represents the cast rate of the slip casting process.

In order to assure that the plant engineers and supervisors have faith in the models, and that the neural network predictions were accurate, a five-fold group cross-validation method was used to validate the two application networks. This resampling method allows all the data to be used for both construction and validation of the neural network prediction models. This approach divided the available data into five mutually exclusive groups. Five validation networks were constructed, using the same parameters as the application network described above, where each used four groups of data to train the validation network and test it on the hold-back group. Each validation network had a different hold-back group as a test set. The errors of the five tests provide an estimate of the generalization ability of the application network.

Tables 11.2 and 11.3 summarize the mean absolute errors (MAE) and root mean squared errors (RMSE) of the validation and application networks for the moisture gradient network and the cast rate network. Typically, the error measures on the validation networks are not as good as those on the application network. This observation is obvious because the errors of the application network were calculated by resubstituting the training data back into the model, whereas the errors of the validation networks were obtained using the five hold-back test sets.

Figure 11.1 shows typical cross-validation network predictions for the moisture gradient network, and Figure 11.2 shows a similar graph for the cast rate network. By comparing these figures, it appears that the cast rate network performs better than the moisture gradient network. The predictions of the cast rate application network are accurate over the entire range of the process variables. On the other hand, the predictions of the moisture gradient network are fairly accurate, except for large values of the target. This can be explained by the skewed distribution of the data (367 observations) where almost 90% percent of the data have values of moisture gradient less than 0.01 (see Figure 11.3). Also, there was considerable human error possible in the moisture gradient measurement. However, despite the imperfection of the performance of the moisture gradient network, it provides adequate precision for use in the manufacturing plant.

11.3 Abrasive Flow Machining Application

Abrasive flow machining (AFM) was originally developed for deburring aircraft valve bodies. It has many applications in the aerospace, automotive, electronic, and die-making industries. The product spectrum includes turbine engines, fuel injector nozzles, combustion liners, and aluminum extrusion dies. It is a special finishing process that is used to deburr, polish, or radius surfaces of critical components. However, it is not a mass material removal process. AFM removes small quantities of material by a viscous, abrasive-laden, semi-solid grinding media flowing under pressure through or across a workpiece. The AFM process acts in a manner similar to grinding or lapping where the extruded abrasive media gently hones edges and surfaces. The abrasive action occurs only in areas where the media flow is restricted. The passageways that have the greatest restriction will experience the largest grinding forces and the highest deburring ability.

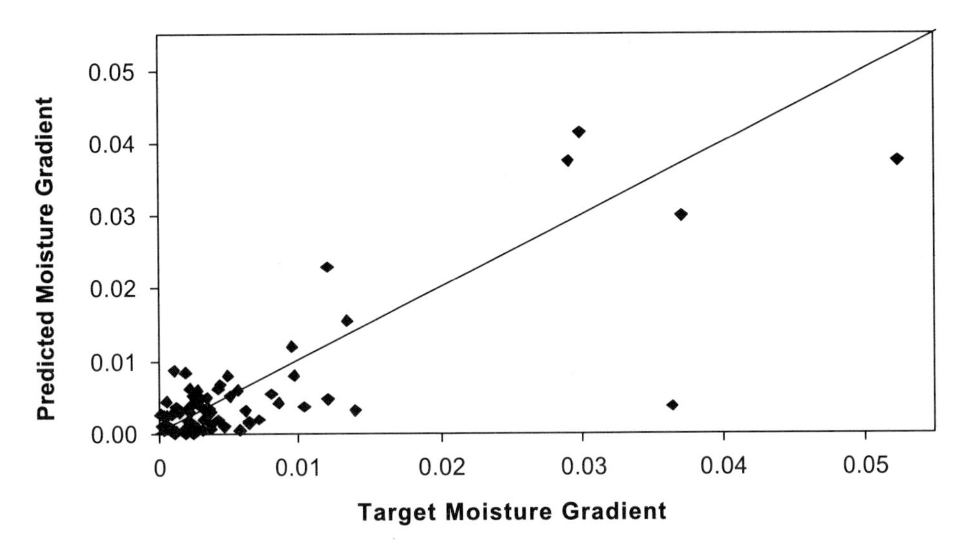

FIGURE 11.1 Performance of a typical cross-validation network for moisture gradient.

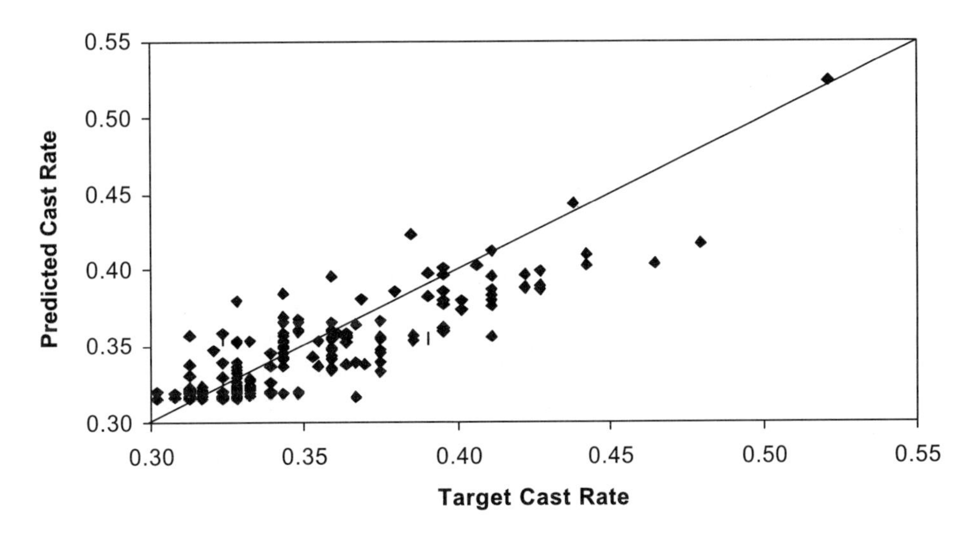

FIGURE 11.2 Performance of a typical cross-validation network for cast rate.

AFM can process many selected passages on a single workpiece or multiple parts simultaneously. Generally, the media is extruded through or over the workpiece with motion, usually in both directions. It is particularly useful when applied to workpieces containing internal passageways that are inaccessible using conventional deburring and polishing tools [Rhoades, 1991]. However, AFM has not been widely used because of the lack of theoretic support for the complex behavior of the process. In order to understand the process, a large range of process parameters such as extrusion pressure, media viscosity, media rheology, abrasive size and type, and part geometry must be taken into consideration.

11.3.1 Engine Manifolds

An air intake manifold is a part of automotive engines (see Figure 11.4), consisting of 12 cylindrical "runners" (the shaded parts in Figure 11.4) through which air flows. These runners have complex

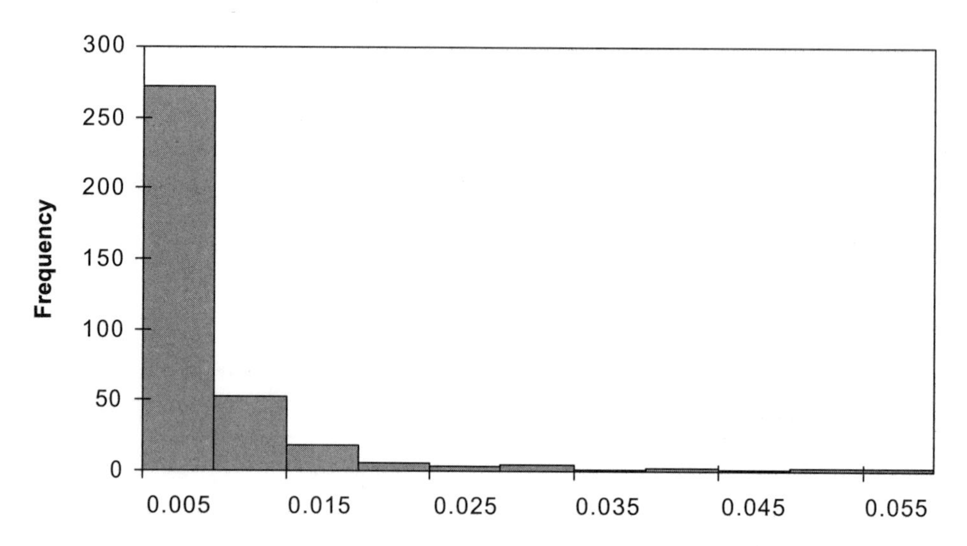

FIGURE 11.3 Data distribution of moisture gradient.

geometries. The manifold is attached to the throttle body in the engine through a large hole (middle front of Figure 11.4). Engine manifolds are typically sand-cast aluminum and are too complex to be economically machined by conventional methods. The sand-cast cavities have rough and irregular surfaces that retard air flow, particularly at the passage walls. This imperfect finish has a significant detrimental impact on the performance, fuel efficiency, and emissions of automotive engines [Smith, 1997].

AFM can finish sand-cast manifolds so that the interior passages are smoother and more uniform, and can achieve more precise air flow specifications. This can increase engine horsepower and improve vehicle performance. However, the AFM process is not currently economical for mass production of manifolds. Currently, in order to AFM engine manifolds, technicians preset values for the volume of media that will be extruded through the manifold and the hydraulic extrusion pressure based on their judgment and experience. After this operation is finished, they clean the manifold, air-dry the part and test the outgoing air flow. If specifications have not been made, they will repeat the process until the manifold achieves the desired air flow requirements. On the other hand, if the manifold is overmachined, the part is scrapped.

11.3.2 Process Variables

The outgoing air flow rate (the air flow rate achieved after the AFM process) of an engine manifold depends on the part characteristics, the AFM machine settings, the media conditions, and the ambient conditions in the plant. These process variables are categorized as follows:

1. Incoming part characteristics:
   - Weight
   - Surface roughness inside the throttle body orifice
   - Air flow rate through each paired runner
   - Variability of the air flow rate among the six pairs of runners
   - Diameter of the throttle body orifice.
2. AFM machine settings:
   - Volume of media extruded through the manifold
   - Hydraulic extrusion pressure

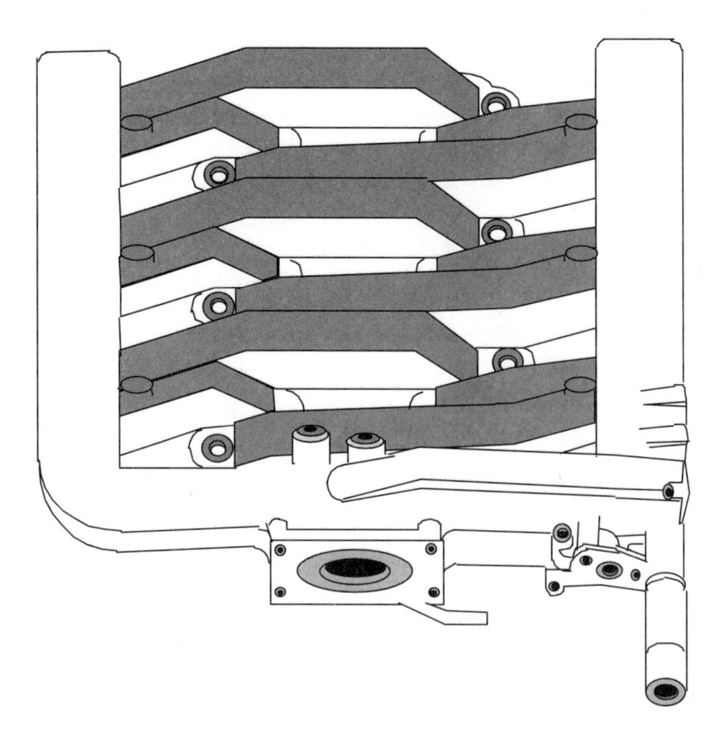

FIGURE 11.4 Drawing of an engine manifold.

- Media flow rate through the manifold
- Number of cycles of the AFM machine piston.

3. Media conditions:
   - Temperature of the media
   - Temperature of the manifold
   - Number of parts machined prior to the current part in a work day
   - Sequence of production during the day.

There is considerable variability among the shipped manifolds due to the limitations of the sand casting process. The volume of media extruded and the extrusion pressure are preset by the operator depending on his judgment of when the manifold reaches air flow specifications. These two machine settings determine the number of cycles needed and the media flow rate for the process. Media condition is extremely important because it relates to the cutting ability of the media. The media starts new with an amount of abrasive grit and no impurities. Over time, impurities enter into the media from the metal being machined and the grit becomes less abrasive and contaminated. This has a profound impact on the AFM process. However, measurement of media condition during processing is impossible. Another change in media condition occurs daily because the behavior of the media depends partially on its temperature. At the beginning of a day, the media is cold and a relatively higher machining ability can be achieved. However, after repeated processing, it becomes heated and hence lowers the AFM ability. The sequence of production and the number of parts machined prior to the current part are very crude approximations to this heating effect. Time of production was divided into five periods beginning in the morning (period 1) and ending with the work day (period 5). Each part was assigned to one of these periods depending on its time of production. The number of parts machined prior to the current part is simply a counter for another measure of the changing characteristics of the media during a work day.

11.3.3 Neural Network Modeling for Abrasive Flow Machining Process

Production data were collected, and after processing, 58 observations were left for analysis. The data set consisted of static information as well as dynamic information. The extrusion pressure, volume of media extruded, media flow rate, and media temperature were collected on a per-cycle basis during the machining process. These dynamic process variables were used to derive the statistics median, average, gradient, range, and standard deviation. Using these additional variables along with the static information, a first-order stepwise regression model was constructed to predict the outgoing average air flow rate of the manifolds. The critical process variables are as follows:

- Average air flow rate before AFM
- Average hydraulic extrusion pressure
- Median of media flow rate
- Range of media temperature
- Range of part temperature
- Standard deviation of volume of media extruded
- Number of parts machined prior to the current part

Regression results showed that these seven process variables can explain 87.00% of the variance of the outgoing average air flow rate. These variables were then used as inputs to the neural network. A static neural network model was developed in favor of a dynamic one because it is simpler to construct and to operate.

The neural network for predicting the outgoing average air flow rate of the engine manifolds was created using the cascade-correlation paradigm. A cascade-correlation learning algorithm was used because it learns very quickly and the network determines its own topology and its own size [Fahlman and Lebiere, 1990]. This algorithm begins with no hidden neurons, with only direct connections from the input units to the output units. Then, hidden neurons are added one at a time. The purpose of each new hidden neuron is to predict the current remaining output error in the network. Unlike the traditional backpropagation learning algorithm, hidden neurons are allowed to have connections from the preexisting hidden neurons along with connections from the input units. Figure 11.5 illustrates a typical cascade architecture.

The training parameters and the maximum number of epochs were selected through experimentation and examination of preliminary networks. The final network architecture had seven inputs, one hidden layer with ten hidden units, and a single output. The output unit represents the outgoing average air flow rate.

Since the data set was small (only 58 observations), a cross-validation method was used to evaluate the performance of the application network. This approach was chosen in favor of a group cross-validation approach because it provides a better estimate of the performance of the final network. Fifty eight validation networks were built using the same parameters as the application network with an upper bound of ten hidden units. This upper bound, which is the same as the number of hidden units for the application network, limits the number of hidden units chosen by the cascade-correlation algorithm for the validation networks. Each validation network was trained on 57 observations and tested on the hold-out observation. Each network had a different hold-out observation as its test set. The test set errors obtained from all 58 cross-validated networks provides an estimate of the generalization ability of the application network.

The final network was able to predict the outgoing average air flow rate with a mean absolute error of 0.0972 (0.0569%) and a root mean squared error of 0.1358 (0.0794%). The R-squared (coefficient of determination) value of the final network, which was estimated by the validation networks, is 0.8741, which proves to be somewhat superior to the regression model. Figure 11.6 provides a visual relationship between the predicted and the actual observed outgoing average air flow rate of the cross-validation networks. Although the predictions of the application network were expected to be fairly accurate, the precision level of the model was still not high enough for use in the plant. Additional data have to be

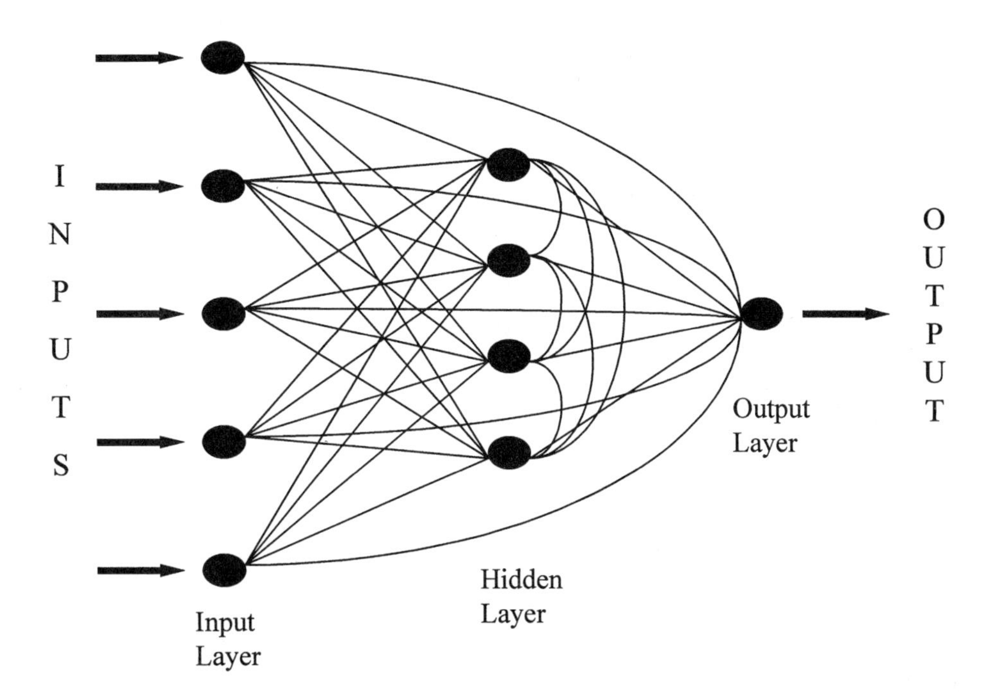

FIGURE 11.5 A typical cascade-correlation neural network.

gathered, and more information on the condition of the media will have to be collected in order to improve the model.

11.4 Chemical Oxidation Application

The cyclohexane oxidation process is used to produce cyclohexanol and cyclohexanone. These products are then used to generate caprolactam and adipic acid, which are eventually consumed as prime raw materials for the nylon-6 and nylon-6,6 polymerization processes. Conventionally, cyclohexane oxidation is the reaction between liquid cyclohexane and air. The reaction can be controlled in an experimental setting [Tekie et al., 1997], where the following parameters can be adjusted/changed in order to obtain measures of the volumetric mass transfer coefficient (k_La):

- Mixing speed of the reactor, which is used to mix the reactants, x_1
- Pressure of the feed gas, x_2
- Temperature of the reaction environment, x_3
- Liquid height of cyclohexane that is above the bottom of the reactor, x_4
- Type of feed gas (nitrogen or oxygen)
- Type of reactor (gas-inducing reactor, GIR, or surface-aeration reactor, SAR).

A central composite experimental design was carried out to collect data for the oxidation process of liquid cyclohexane. Based on the design of experiments data, the study of the effects of these variables on k_La values helped to develop a quadratic response surface model [Tekie et al., 1997] given in Equation 11.1:

$$\text{In}\left(k_La\right)=\beta_0+\sum_{i=1}^{4}\beta_i x_i+\sum_{i=1}^{4}\beta_{ii}x_i^2+\alpha_1\exp\left(\gamma_1\left(x_1+\gamma_2\right)^2\right)+\alpha_2\exp\left(\gamma_3\left(x_1+3\right)\left(4-x_4\right)\right)$$

Equation (11.1)

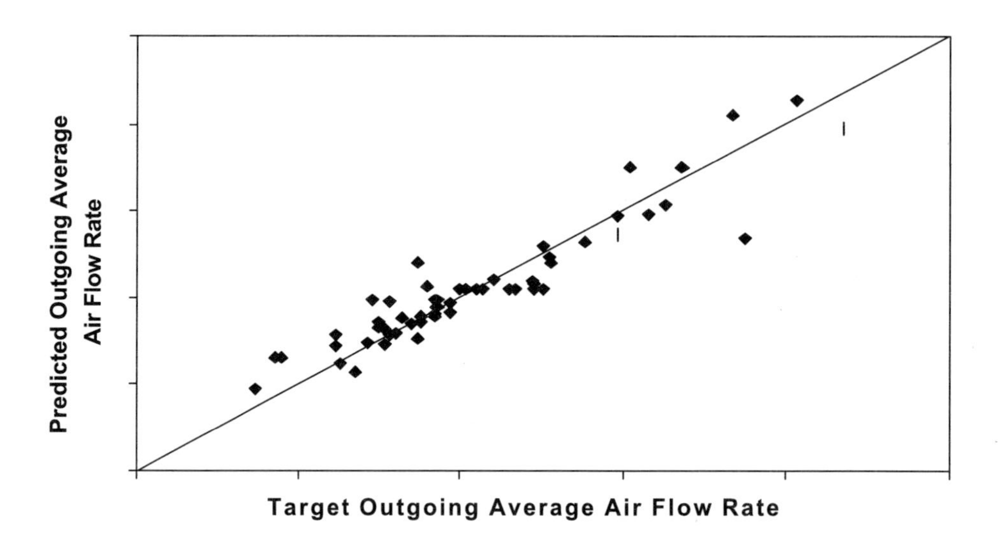

FIGURE 11.6 Performance of cross-validation networks for the cascade application network.

However, this model can only be applied to a specific combination of gas type and reactor type. For the experiments carried out by Tekie et al., there were a total of four different combinations of gas and reactor. Therefore, the authors developed four different quadratic response surface models, each with a different set of parameters (i.e., β's, α's, and γ's) for each model. This approach, however, cannot be used to generalize to a larger set of gas/reactor combinations without reestimating the equation parameter set. Furthermore, their approach required a considerable amount of effort to fine-tune the functional form and the parameters.

11.4.1 Neural Network Modeling for Chemical Oxidation Process

The use of neural networks is motivated because of their accommodation of nonlinearities, interactions, and multiple variables. Moreover, a generalized approach for predicting the volumetric mass transfer coefficient can be achieved while considering the operating variables (temperature, pressure, mixing speeds, and liquid heights) along with gas types and reactor types as input variables. This approach can eventually be used to optimize the oxidation process while considering variations in gas and reactor.

The data set collected by Tekie et al. [1997] was used for construction of the networks. After processing and purifying the data, the final set had 296 observations. The training parameters, network architectures, and stopping points were identified via preliminary experiments. An ordinary backpropagation algorithm was used. The final application network had six inputs, one hidden layer with eight hidden units, and one output. The output unit represents the volumetric mass transfer coefficient, k_La. This network was developed using the entire data set of 296 observations. Similar to the example from ceramic slip casting, a group cross-validation method was used to validate the application network. More specifically, a 74 group cross-validation approach, which required constructing 74 validation networks, was carried out. First, the entire data set was randomly divided into 74 disjoint groups of data (each group had four observations). Then each validation network was trained using 73 groups of data and tested on the hold-back group. Combining all 74 test sets exhausts the original data set.

Table 11.4 summarizes the error measures and the R-squared values of the application network and the validation networks. The R-squared value of the application network, which was calculated by the test set errors of the validation networks, is 0.9773. Therefore, the application network can explain about 98% of the variation of the volumetric mass transfer coefficient.

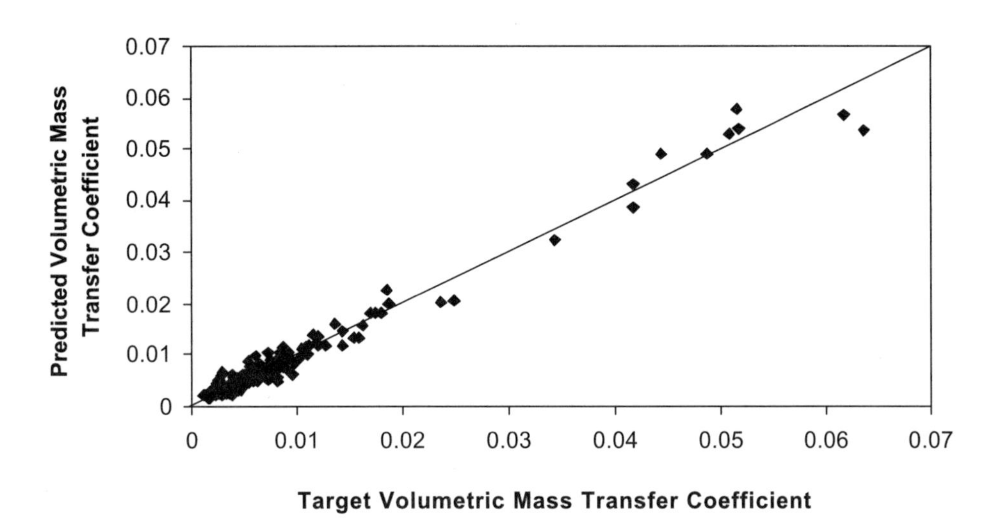

FIGURE 11.7 Performance of validation networks for the oxidation process.

TABLE 11.4 MAE, RMSE, and R-Squared Values for the Oxidation Network

Network	MAE	RMSE	R-Squared
Application network	0.00067	0.00106	0.9856
Validation networks	0.00079	0.00133	0.9773

Figure 11.7 describes the relationship between the predicted and the target volumetric mass transfer coefficient for the validation networks. It shows that the application network is expected to provide very accurate predictions.

11.5 Concluding Remarks

Three diverse neural network predictive process models were presented using real-world engineering problems. The ceramic slip casting application was part of a larger project for which only the neural network development was discussed in this chapter. The final product of this project has been implemented at a major U.S. manufacturer of sanitary ware to improve the product quality and the production efficiency [Lam et al., 2000]. The second example is a collaboration between a machining company, two automotive companies and the University of Pittsburgh. This project is currently underway, and this chapter discussed some of the more recent results in the model-building phase of the abrasive flow machining process. The last example demonstrates a more generalized approach to model the chemical oxidation process of cyclohexane. This approach does not require the construction of multiple functional models for each combination of gas and reactor and can be used for multiple reactor and gas combinations.

Acknowledgment

Alice E. Smith gratefully acknowledges the support of the U.S. National Science Foundation CAREER grant DMI 9502134.

Defining Terms

Application network: The network that is ultimately used for the manufacturing application.

Cross-validation: Also known as leave-one-out cross-validation. This approach requires the construction of n validation networks, where each is built using n-1 observations and tested on the left-out point. Combining all the test sets exhausts the original n observations.

Data splitting: Requires splitting the available data into two sets: a training set and a testing set, where the training set is used to train the network and the testing set is used to evaluate the performance of the network.

Group cross-validation: Also known as leave-k-out cross-validation. This approach is similar to cross-validation, except that k observations are held back each time.

Validation networks: Defined as the additional networks constructed to infer the performance of the application network.

References

Adams, E. F., 1988, *Introduction to the Principles of Ceramic Processing*, John Wiley & Sons, New York.

Coit, D. W., Jackson, B. T. and Smith, A. E., 1998, Static Neural Network Process Models: Considerations and Case Studies, *International Journal of Production Research*, 36 (11), 2953-2967.

Efron, B., 1982, *The Jackknife, the Bootstrap and Other Re-Sampling Plans*, SIAM, Philadelphia.

Fahlman, S. E. and Lebiere, C., 1990, The Cascade-Correlation Learning Architecture, *Advances in Neural Information Processing Systems 2* (D. S. Touretzky, Ed.), Morgan Kaufmann, San Mateo, CA, 524-532.

Funahashi, K., 1989, On the Approximate Realization of Continuous Mappings by Neural Networks, *Neural Networks*, 2, 183-192.

Hornik, K., Stinchcombe, M. and White, H., 1989, Multilayer Feed-Forward Networks are Universal Approximators, *Neural Networks*, 2, 359-366.

Hussain, M. A., 1999, Review of the Applications of Neural Networks in Chemical Process Control — Simulation and Online Implementation, *Artificial Intelligence in Engineering*, 13, 55-68.

Lam, S. S. Y. and Smith, A. E., 1998, Process Control of Abrasive Flow Machining Using a Static Neural Network Model, *Intelligent Engineering Systems through Artificial Neural Networks: Smart Engineering Systems: Neural Networks, Fuzzy Logic, Evolutionary Programming, Data Mining and Rough Sets*, Vol. 8 (C. H. Dagli et al., Eds.), ASME Press, 797-802.

Lam, S. S. Y., Petri, K. L. and Smith, A. E., 2000, Prediction and Optimization of a Ceramic Casting Process Using a Hierarchical System of Neural Networks and Fuzzy Logic, *IIE Transactions*, 32(1), 83–91.

Rhoades, L. J., 1991, Abrasive Flow Machining: A Case Study, *Journal of Materials Processing Technology*, 28, 107-116.

Smith, S. C., 1997, Four-Door Party Animals, *Car and Driver*, 42 (9), 46-56.

Tekie, Z., Li, J. and Morsi, B. I., 1997, Mass Transfer Parameters of O_2 and N_2 in Cyclohexane under Elevated Pressures and Temperatures: A Statistical Approach, *Industrial and Engineering Chemistry Research*, 36 (9), 3879-3888.

Twomey, J. M. and Smith, A. E., 1998, Bias and Variance of Validation Methods for Function Approximation Neural Networks Under Conditions of Sparse Data, *IEEE Transactions on Systems, Man, and Cybernetics, Part C*, 28, 417-430.

Twomey, J. M., Smith, A. E. and Redfern, M. S., 1995, A Predictive Model for Slip Resistance Using Artificial Neural Networks, *IIE Transactions*, 27, 374-381.

Udo, G. J., 1992, Neural Networks Applications in Manufacturing Processes, *Computers and Industrial Engineering*, 23 (1-4), 97-100.

Wolpert, D. H., 1993, Combining Generalizers by Using Partitions of the Learning Set, *1992 Lectures in Complex Systems* (L. Nadel and D. Stein, Eds.), SFI Studies in the Sciences of Complexity, Lect. Vol. V, Addison-Wesley, Reading, MA.

Zhang, H. C. and Huang, S. H., 1995, Applications of Neural Networks in Manufacturing: A State of the Art Review, *International Journal of Production Research*, 33, 705-728.

12

Neural Network Applications to Manufacturing Processes: Monitoring and Control

Hyung Suck Cho
Korea Advanced Institute of Science and Technology (KAIST)

12.1 Introduction

The nature of today's manufacturing systems is changing with greater speed than ever and is becoming tremendously sophisticated due to rapid changes in their environments that result from customer demand and reduced product life cycle. Accordingly, the systems have to be capable of responding to the rapid changes and solving the complex problems that occur in various manufacturing steps. The monitoring and control of manufacturing processes is one of the important manufacturing step that requires the capabilities described in the above.

Monitoring of the process state is comprised of three major steps carried out on-line: (i) the process is continuously monitored with a sensor or multiple sensor; (ii) the sensor signals are conditioned and preprocessed so that certain features and peaks sensitive to the process states can be obtained; (iii) by pattern recognition based on these, the process states are identified. Control of the process state is usually meant for feedback control, and is comprised of the following steps: (i) identifying the dynamic characteristics of the process, (ii) measuring the process state, (iii) correcting the process operation, observing the resulting product quality, and comparing the observed with the desired quality. It is noted that in the last step, the observed state needs to be related to product quality.

Normal operation of the above-mentioned steps should not be interrupted and needs to be carried out with little human intervention, in an unmanned manner if possible. To this end the process with this capability should be equipped with such functionalities as storing information, reasoning, decision making, learning, and integration of these into the process. In particular, the learning characteristic is a unique feature of the ANN. Neural networks are not programmed; they learn by example. Typically, a

0-8493-0592-6/01/$0.00+$.50

neural network is presented with a training set consisting of a group of examples that can occur during manufacturing processes and from which the neural network can learn. One typical example is to measure the quality-related variable of the process state and identify the product quality based on these measured data. The use of artificial neural networks (ANN) is apparently a good solution to make manufacturing processes truly intelligent and autonomous. The reason is that the networks possess most of the above functionalities along with massively computing power.

Utilizing such functionalities, ANNs have quite recently established themselves as the versatile algorithmic and information processing tool for use in monitoring and control of manufacturing process. In most manufacturing processes, the role of the artificial neural network is to perform signal processing, pattern recognition, mapping or approximation system identification and control, optimization and *multisensors data fusion.* In more detail, the ANNs being used for manufacturing process applications are able to exhibit the ability to

1. Generalize the results obtained from known situations to unforeseen situations.
2. Perform classification and pattern recognition from a given set of measured data.
3. Identify the uncertainties associated with the process dynamics.
4. Generate control signal based upon inverse model learning.
5. Predict the quality from the measured process state variables.

Due to such capabilities, there has been widespread recognition that the ANNs are an artificial intelligence (AI) technique that has the potential of improving the product quality, increasing the effect events in production, increasing autonomity and intelligence in manufacturing lines, reducing the reaction time of manufacturing systems, and improving system reliability. Therefore, in recent years, an explosion of interest that has occured in the application of ANNs to manufacturing process monitoring and control.

The purpose of this chapter is to provide the newest information and state-of-the-art technology in neural-network-based manufacturing process monitoring and control. Most applications are widely scattered over many different monitoring and control tasks but, in this chapter, those related to product quality will be highlighted. Section 12.2 reviews basic concept methodologies, and procedures of process monitoring and control. In this section the nature of the processes is discussed to give reasons and justification for applying the neural networks. Section 12.3 deals with the applications of neural networks in monitoring various manufacturing processes such as welding, laser heat treatment, and PCB solder joint inspection. Section 12.4 treats neural-network-based control and discusses the architecture of the control system and the role of the network within the system. Various manufacturing processes including machining, arc welding, semiconductor, and hydroforming processes are considered for networks applications. Finally, perspectives of future applications are briefly discussed and conclusions are made.

12.2 Manufacturing Process Monitoring and Control

In this chapter, we will treat the problems associated with monitoring and control of manufacturing processes but confine ourselves only to product quality monitoring and control problems. Furthermore, we will consider only on-line monitoring and control schemes.

12.2.1 Manufacturing Process Monitoring

Product quality of most processes cannot be measurable in an on-line manner. For instance, weld quality in the arc welding process depends on a number of factors such as the weld pool geometry, the presence of cracks and void, inclusions, oxide films, and the metallographic conditions. Among these factors, the weld pool geometry is of vital importance, since this is directly correlated to weld strength of the welded joint. The weld pool size representative of weld strength is very difficult to measure, since the weld pool formed underneath the weldment surface represents complex geometry and is not exposed from the outside. This makes it very difficult to assess the weld quality in an on-line manner. Due to this reason, direct quality monitoring is extremely difficult. Thus, one needs to resort to finding some process state variables that can represent the product quality. In the case of arc welding, the representative variable is the temperature spatially distributed over the weld pool surface, since formation of the weld pool geometry is directly affected by heat input. In this situation, the weld quality can be indirectly assessed by measuring the surface temperature.

Two methodologies of assessing product quality, are considered. One is the *direct method*, in which the quality variables are the monitoring variables. The other is the *indirect method*, which utilizes the measured state variable as measures of the quality variables. In this case, several prerequisite steps are required to design the monitoring system, since the relationship between product quality and process condition is not known *a priori*. In fact, it is very difficult to understand the physics involved with this issue. The prerequisite steps treat the issues, which include (i) relating the product quality with the process state variables, (ii) selection of sensors that accurately measure the state variables, (iii) appropriate instrumentation, and (iv) correlation of the obtained process state data to quality variables. The procedure stated here casts itself a heavy burden in monitoring of process condition problem. Once this relationship is clearly established, the quality monitoring problem can be replaced by a process state monitoring problem.

Figure 12.1 illustrates the general procedure of evaluating product quality from measurement of process variables and/or machine condition variables. This procedure requires a number of activities that are performed by the sensing element, signal interpretation elements, and quality evaluation unit. The sensors may include multiple types having different principles of measurement or multiples of one type. In using sensors of different types, sensing reliability becomes very important in synthesizing the information needed to estimate the process condition or product quality. The reliability may change relative to one another. This necessitates careful development of a synthesis method. In reality, in almost all processes whose quality cannot be measured directly, multisensor integration/fusion is vital to characterize the product quality; for instance weld pool geometry in arc welding, nugget geometry in resistance spot welding, hardened layer thickness in laser hardening, etc. This is because, under complex physical processing or varying process conditions, a single sensor alone may not adequately provide the information required to make reliable decisions on product quality or process condition. In this case, sensor fusion or integration is effective, since the confidence level of the information can be enhanced by fusion/integration of the multiple sensor domain. This multiple sensor approach is similar to the method a human would use to monitor a manufacturing process by using his own multiple senses, and processing the information about a variety of state variables that characterize the process. Since measurement of process variables is performed by several sensing devices, i.e., more sensor-based information is considered, the uncertainty and randomness involved with the process measurement may be drastically reduced.

The two typical methods used to evaluate product quality handle information differently. One makes use of the raw signal directly, the other uses features extracted from the raw signal. In the case of using the raw signal, indicated in a dotted arrow, the amount of data can be a burden on tasks for clustering and pattern recognition. On the other hand, the feature extraction method is very popular, since it allows analysis of data in lower dimensional space and provides efficiency and accuracy in monitoring. Usually, the features are composed of the compressed data due to the reduction of dimensionality, which is postulated to be much smaller than the dimensionality of the data space. The feature values used could be of entirely different properties, depending upon monitoring applications. For example, in most industrial inspection problems adopting machine vision technique, image features such as area, center of gravity, periphery, and moment of inertia of the object image are frequently used to characterize the shapes of the object under inspection. In some complicated problems, the number of features used has to be as many as 20 in order to achieve successful problem solution. On the contrary, in some simple problems one single feature may suffice to characterize the object. Monitoring the machine conditions frequently employs time and frequency domain features of the sensor signal such as mean variance, kurtosis, crest factor, skewness, and power in a specified frequency band.

The selection of features is often not an easy task and needs an expert to work with characteristics of the signal/data. Furthermore, computation of feature values may often constitute a rather cumbersome

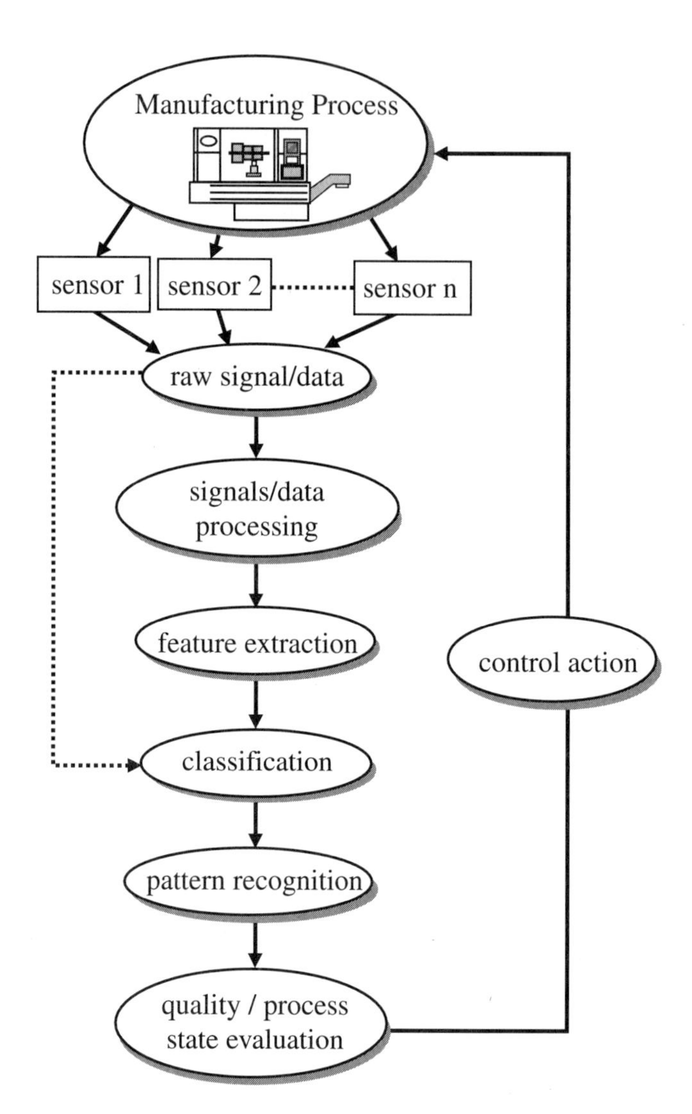

FIGURE 12.1 A general procedure for quality monitoring.

task. It is therefore important to obtain features that shows high sensitivity to product quality or a quality-related process variable and low sensitivity to process variation and uncertainty. It is equally important to obtain the fewest but the best combination of features in order to reduce the computational burden and increase efficiency in clustering. This can ensure better performance of the monitoring system, while reducing the monitoring cost.

When the choice of features is appropriately made, and their values are calculated, the next task is to find the similarity between the feature vector and the quality variables or process conditions, that is, to perform the classification task. If the feature vector is denoted by x, finding the similarity mathematically is to find the relationship R;

$$R{:}i(\underset{\sim}{x}) \rightarrow C \; (C = 1 \text{ or } 2, \text{ or } \ldots \text{ or, } m) \qquad \text{Equation (12.1)}$$

where C denotes the number assigned specifically to a class category and has m categories of classification. In the above equation, the category number C is assumed to be preassigned to represent the quality variables or process conditions. The operator i that yields the relationship expressed in Equation 12.1 is called the classifier.

A large number of classifiers have been developed for many classification problems. Depending upon the nature of the problem, the classifier needs to differ in its discriminating characteristics, since there is no universal classifier that can be effectively used for a large class of problems. In fact, it is observed from the literature that a specific method works for a specific application. Frequently used conventional classifiers include K-nearest mean, minimum distance, and the Bayes approach. This topic will be revisited in detail.

There are several important factors that affect classification accuracy, including the distribution characteristics of data in feature data space, and the degree of similarity between patterns. The set of extracted features yields the sets of pattern vectors to the classifier, and the vector components then are represented as the classifier input. The pattern vectors thus formed must be separable enough to discriminate each pattern that uniquely belongs to the corresponding category. This implies that we compute feature transformation such that the spread of each class in the output feature space is maximized. Therefore, the classifier should be designed in such a way that the designed methodology is insensitive to the influence of the above factors.

12.2.2 Manufacturing Process Control

Most of manufacturing processes suffer from the drawback that their operating parameters are usually preset with no provision for on-line adjustments. The preset values should be adjusted when process parameters are subject to change and external disturbances are present, as is usually the case in manufacturing process. As discussed previously, the manufacturing process is time-varying, highly nonlinear, complex, and of uncertain nature. Unlike nonlinearity and complexity, variability and uncertainty can be decreased if they are the result of some seemingly controllable factors such as incorrect machine setting, inconsistent material dimension and composition, miscalibration, and degradation of process machine equipment. Reducing the effect of these factors would improve process conditions, and therefore product quality. However, these controllable factors usually cannot be measurable in an on-line manner, and thus these effects cannot be easily estimated. This situation requires on-line adjustment or control of the operating parameters in response to the environment change, which in turn needs reliable, accurate models of the processes. This is due to the fact that, unless the process characteristics are exactly known, the performance of a control system that was designed based on such uncertainty may not be guaranteed to a satisfactory level.

A general feedback control system consists of a controller, an actuator, a sensor, and a feedback element that feeds the measured process signal to the controller. The role of the controller is to adjust its command signal depending upon the error characteristics. Therefore, performance of the controller significantly affects the overall performance of the control system for the manufacturing process. Equally important is the performance of the actuator and sensor to be used for control. Unless these are suitably designed or selected, the control performance would not be guaranteed, even though the controller was designed in a manner best reflecting the process characteristics.

For controller design of the manufacturing process, the greatest difficulty is that an accurate model of the process dynamics often does not exist. Lack of the physical models makes the design of a process controller difficult, and it is virtually impractical to use the conventional control methodologies. In this situation, these are two widely accepted methods of designing process controllers. One is to approximate the exact mathematical model dynamics by making some assumptions involved with the process mechanism and phenomena. As shown in Figure 12.2, the process model thus approximately obtained can be utilized for the design of the conventional controllers, which include all the model-based control schemes such as adaptive, optimal, predictive, robust, and time-delay control, etc. The advantage of the approach using the model dynamics is that the analytical method in design is possible by enabling us to investigate the effects of the design parameters. The disadvantage is that the control performance may not be satisfactory when compared with the desirable performance of the ideal case, since the controller is designed based upon an approximate model. Furthermore, when changes in the process characteristics occur with time, the designed controller may be further deteriorated.

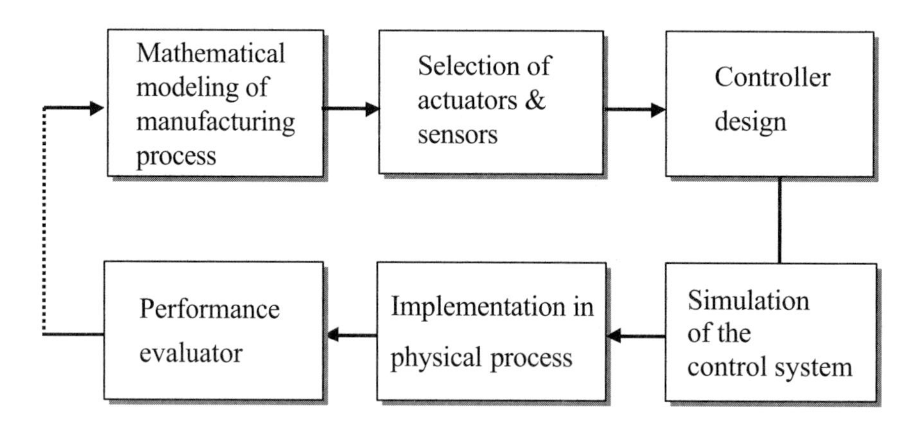

FIGURE 12.2 A feedback procedure for the design of a process controller.

The other widely accepted approach is based on an experimental trial-and-error method that uses heuristics of human operators rather than a mathematically based algorithm. In this case, human operators design the controller, making use of their own knowledge and past experience on the control action based upon observation of dynamic characteristics. The control actions of a human operator are generated from the inference of rules from which he formulates his knowledge. Accordingly, the performance of the control largely depends upon how broad and deep his knowledge of the process dynamic characteristic is and how well he can construct the appropriate rule base utilizing his knowledge and experience. As can be perceived, reliable control performance may not be guaranteed with a human operator's observation and experience alone, when the characteristics of manufacturing processes are uncertain and time-varying in nature.

12.3 Neural Network-Based Monitoring

In the previous sections we noted that monitoring requires identification or estimation of the characteristic changes of a manufacturing process based on the evaluation of a process signature without interrupting normal operations. In doing so, a series of tasks is performed, such as signal processing, feature extraction, feature selection and integration, classification, and pattern recognition. In some cases, a complete process model describing the functional relationship between process variables must be extracted. Some typical problems that arise in the conventional monitoring task may be listed as follows:

- Inability to learn and self-organize signals or data
- Inefficiency in solving complex problems
- Robustness problem in the presence of noise
- Inefficiency in handling the large amount of signals/data required

In any process, disturbances of some type arise during manufacturing. For example, in welding processes, there are usually some variation in incoming material and material thickness, variation in the wire feeder, variation in gas content, and variation in the operating conditions such as weld voltage and current. When some of these process parameter changes occur, the result is variation in monitoring signal. This situation requires adaptation of the monitoring strategy, signal processing, and feature extraction and selection by analyzing the changes in the signature of the incoming signal and incoming information on the observed phenomena. The conventional method, however, cannot effectively respond to these changing, real process variations. In contrast to this, a neural network has the capability of testing and selecting the best configuration of standard sensors and signal processing methods. In addition, it has a learning capability that can adapt and digest changes in the process.

Normally, it is not easy to directly measure product quality from sensors, as mentioned previously. Indirectly measuring a single measurement may suffice to give some correlation to the quality. However, the relationship between the quality variables and the measured variables is normally quite complex, being also subjected to the dependency of some other parameters. Furthermore, in some other cases, single sensor measurement may not provide a good solution, and thus multiple measurements may be required. This situation calls for a neural network role that has the capability to self-organize signals or data and fuse them together.

A robustness problem in the presence of signal noise and process noise is one of the major obstacles to achieving high quality in monitoring performance. In general, process noise has either long-term or short-term characteristics. For instance, in machining processes, if vibration from the ground is coming into the machine processing the materials, and lasts continuously for some time, it can be said to be a long-term noise. If it continues only for a short time and intermittently, it may be regarded as a short-term noise. A neural network can handle the short-term noise without difficulty due to its generalization characteristics; it provides monitoring performance that is almost immune to the process noise. Such a neural network easily takes the roles of association, mapping, and filtering of the incoming information on the observed phenomena.

Finally, a monitoring task requires a tremendous amount of signal/data to process. Handling this large volume of data is not a difficult task for the neural network, since it possesses the capability of a high-speed paralleled computation. And, if necessary, it has the ability to compress the data in an appropriate way.

The foregoing discussions imply that the role of networks is to provide generality, robustness, and reliability to the monitoring. When they are embedded in the monitoring system, the system is expected to work better, especially under operating conditions with uncertainty and noise. Even in such conditions the embedded system should be able to effectively extract feature of the measured signals, test and select the extracted features and, if necessary, integrate them to obtain better correlation to the quality-related variables. In addition, it should effectively classify the collected patterns and recognize each pattern to identify the quality variables.

The neural networks often used for monitoring and control purpose are shown in Figure 12.3. In this figure, the neural networks are classified in terms of the learning paradigm. These different types of the networks are used according to domain of problem characteristics and application area. Specifically, problem characteristics to be considered include ability of on-line monitoring, time limitation of classification and recognition robustness to uncertainty, and range of process operations. Even if one classifier works well in some problem and/or application area, it may not be effectively applied to some others because any single network does not process general functionality that can handle all types of complexity involved with the processes. For this reason, integration of two or more networks has become popular in monitoring and control of manufacturing processes.

12.3.1 Feature Selection Method

This issue concerns relating the feature vector to classification and recognition. Important input features can be selected in various ways within the neural network domain. The method introduced herein is based upon a multilayer perceptron with sigmoidal nonlinearity whose structure is shown in Figure 12.4(a).

The first method [Sokolowski and Kosmol, 1996] utilizes the concept of *weight pruning*, which can determine the importance of each input feature. The method starts with selection of a certain weight of an already trained network. This selected weight is then set to zero while the network processes a complete set of input feature vectors. Due to this change, the error will occur as follows:

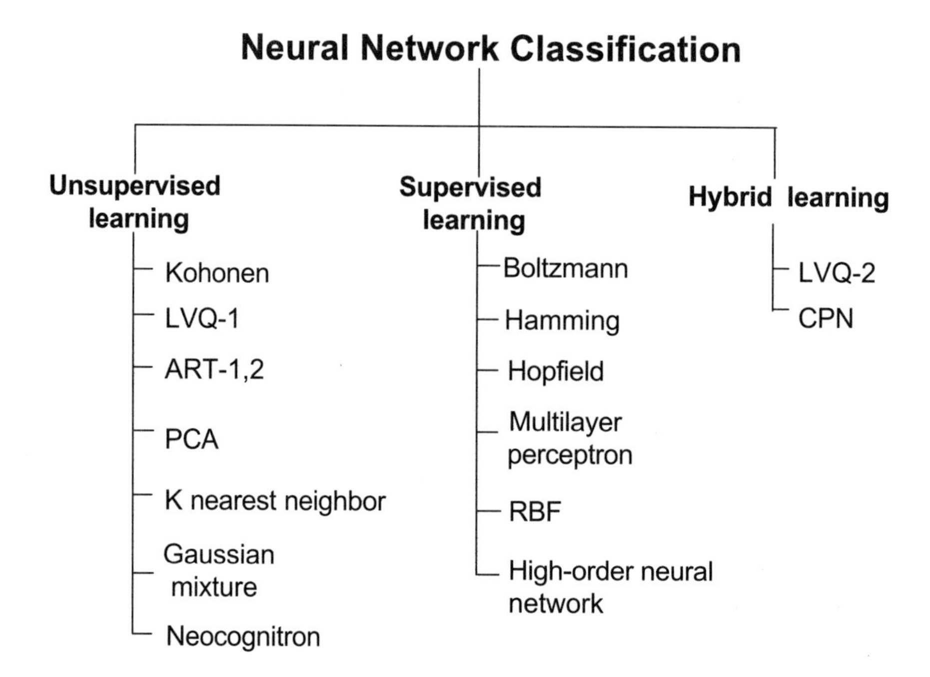

FIGURE 12.3 The neural networks frequently used for classifiers, identifiers, and controllers.

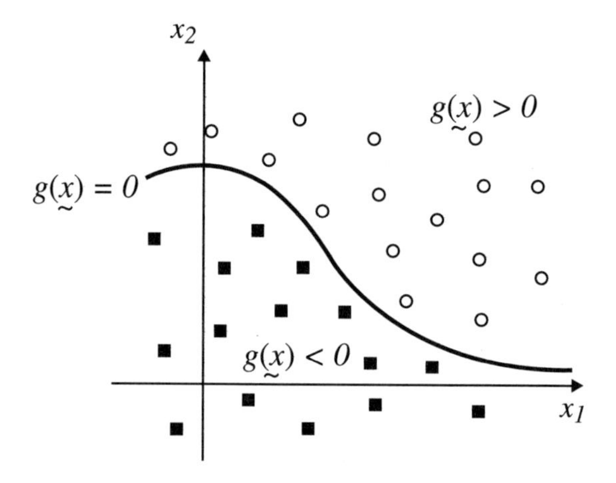

FIGURE 12.4 A discriminant function defined in a two-dimensional space.

$$E_s \max \| d_k^j - o_k^j \| \text{ for } k = 1, 2, \ldots M, j = 1, 2 \ldots, N \qquad \text{Equation (12.2)}$$

where the subscript j refers to the jth input vector, d_k^j is the desired output of the kth neuron in the output layer, is the actual output of the kth neuron for the jth input vector, M is the number of the output neurons, and N is the number of input vectors.

If the error does not exceed a prescribed maximum value E_s, the contribution of the weight omitted in the calculation to obtain the actual output is considered to be less important.

For each weight that satisfies E_k E_s the following total RMS error is calculated by

$$E_T = \frac{1}{NM}\sqrt{\sum_{k=1}^{M}\sum_{j=1}^{N}\left(d_k^j - o_k^j\right)^2} \qquad \text{Equation (12.3)}$$

After checking this weight its previous value is restored and another weight is tested. The procedure of weight pruning continues until the elimination of a weight leads to E_s error above the prescribed value.

The second method is referred to as weight sum method [Zurada, 1992]. In this method, the sensitivity of each input feature to total error is evaluated based on the sum of absolute values of the weight, which is defined by

$$\left\|w_{kj}\right\| = \sum_{k=1}^{M}\left|w_{kj}\right| \qquad (j=1, 2 \ldots, N) \qquad \text{Equation (12.4)}$$

where j is the jth input feature, and w_{kj} is the weight parameter related to the jth input. If the sum of the weight values $||w_{kj}||$ is below a prescribed value, the input can be discarded from further consideration, implying that the important input features can be removed.

12.3.2 Classification Method

With an appropriate set of input feature thus selected, the next task in monitoring is to perform classification and recognition. The goal of pattern classification is to assign input patterns to partition the multidimensional space spanned by the selected features into decision regions that indicate to which any belongs. Good classification performance therefore requires selection of an effective classifier, e.g., a type of neural network, in addition to selection of effective features. The network should be able to make good use of the selected features with limited training, memory, and computing power.

Figure 12.3 summarizes various types of neural networks popularly used for pattern classification. The Hopfield net, Hamming net, and Carpenter–Grossberg classifier have been developed for binary input classification, while the perceptron, Kohonen self-organizing feature maps, and radial basis function network have been developed for analog inputs. The training methods used with these neural networks include supervised learning, unsupervised learning, and hybrid learning (unsupervised + supervised). In the supervised classifier the desired class for given data is provided by a teacher. If an error in assigning correct classification occurs, the error can be used to adjust weight so that the error decreases. The multilayer perceptron and radial basis function classifiers are typical of this supervised learning.

In learning without supervision, the desired class is not known *a priori*, thus explicit error information cannot be used to adjust network behavior. This means that the network must discover for itself dissimilarity between patterns based upon observation of the characteristics of input patterns. The unsupervised learning classifiers include the Kohonen feature map, learning vector quantizer with a single layer and ART-1 and ART-2. Classifiers that employ unsupervised/supervised learning first form clusters by using unsupervised learning with unlabeled input patterns and then assign labels to the cluster using a small amount of training input patterns in the supervised manner. The supervised learning corrects the sizes and locations of the cluster to yield an accurate classification. The primary advantage of this classifier is that it can alleviate the effort needed to collect input data by requiring a small amount of training data. The classifiers that belong to this group are the learning vector quantizer (LVQ1 and 2) and feature map.

The role of the neural network classifiers is to characterize the decision boundaries by the computing elements or neurons. Lippmann [1989] divided various neural network classifiers into four broad groups according to the characteristics of decision boundaries made by neural network classifiers. The first group is based on *probabilistic distributions* such as probabilistic or Bayesian classifiers. These types of neural networks can learn to estimate probabilistic distributions such as Gaussian or Gaussian mixture distributions by using supervised learning. The second group is classifiers with *hyper-plane* decision boundaries. Nodes form a weighted sum of the inputs and pass this sum through a sigmoid nonlinearity. The group includes multilayer perceptrons, Boltzmann machines, and high-order nets. The third group has complex boundaries that are created from *kernel function* nodes that form overlapping receptive fields. Kernel function nodes use a kernel function, as shown in the figure, which provides the strongest output when the input is near a node's centroid. Kernel function indicates that the node output peaks when the input is near the centroid of the node and then falls off monotonically as the *Euclidean distance* between input and the centroid of a node increases. Classifications are made by high-level nodes that form functions from weighted sums of outputs of kernel function nodes. These type of neural network classifiers are based on the cerebellar model articulation controller (CMAC), radial basis function classifier. The fourth group is *exemplar classifiers,* which perform classification based on the identity of the training examples, or exemplars, that are nearest to the input, similar to the kernel function nodes. Exemplar nodes compute the weighted Euclidean distance between inputs and node centroids. Centroids correspond to previously given labeled training examples or to cluster center and called prototypes. These classifiers includes k-nearest neighbor classifiers, the feature map classifiers, the learning vector quantizer (LVQ), restricted coulomb energy (RCE) classifiers, and adaptive resonance theory (ART). These four group classifiers provide similar low error rate. But their characteristics for real world problems are different.

Let us illustrate the role of the neural network classifier in classification by illustrating a basic classification problem. Suppose that the input components of a classifier are denoted by an n-dimensional vector x. This then can be represented by a point in n-dimensional Euclidean space E'' called pattern space. An illustration is presented for the case of two-dimensional spaces, $n = 2$, in Figure 12.4.

In the figure, $g(x)$ is called the *discriminant function,* which can discriminate the decision boundary. The function $g(x)$ shown here is not a straight line dividing the pattern space, and represents an arbitrary curved line. This problem is called a *nonlinearly separable classification problem.* The pattern x belongs to the ith category if and only if

$$g_i(\underset{\sim}{x}) > g_j(\underset{\sim}{x});\ i, j = 1, 2\ (i \neq j) \qquad \text{Equation (12.5)}$$

Therefore, within the region, the ith discriminant function will have the largest value. When a monitoring problem is complex and highly nonlinear, adaptive *nonparametric* neural network *classifiers* have an advantage over the conventional methodologies. They take role of determining the decision surface $g(\underset{\sim}{x})$ in multidimensional space defined by the input feature vectors.

Determining the function depends upon which classifier is used, and which domain of the training data is considered for classification. Depending upon the problem characteristics and domain, the classifier, its structure, and the learning algorithm need to be carefully chosen. Once these are chosen, the next task is to provide the network with the capability of good classifications. To design such a classifier, the development of neural network classifiers must go through two major phases: training phase and test phase.

12.4 Quality Monitoring Applications

There have been tremendous research efforts in monitoring of manufacturing processes. A majority of the research deals with tool condition, machine process condition, and fault detection and diagnosis, which are not directly related to product quality. In contrast, not many studies deal with product quality monitoring. This is mainly due to the fact that direct quality monitoring is extremely difficult and that even correlating the process variables with it is not an easy task. Table 12.1 summarizes types of quality monitoring in various manufacturing processes, including types of sensor signals, neural networks, and quality variables used for monitoring. Some neural network applications will be summarized below for turning end milling, grinding, and spot welding processes.

TABLE 12.1 Types of Sensor Signals and Neural Networks in Monitoring

Process	Sensor Signal	Neural Network	Quality Variable
Turning	Force	Perceptron	Surface finish
	Diffraction image	Perceptron	Surface finish
Grinding	Image	RBF	Surface finish
	Wheel velocity grinding, depth	Perceptron	Grinding burn
	AE	RBF	Surface finish
Milling	Acoustic wave, spindle variation, cutting force	Perceptron	Surface finish, bore tolerance
Spot welding	Weld resistance	Perceptron	Weld nugget geometry
	Current	Perceptron	Quality factor
	Electrode force	LVQ	Strength, indentation
Arc welding	Weld current	Perceptron	Weld pool geometry
	Acoustic wave	Perceptron	Acceptability of weld
	Temperature	Perceptron	Weld pool geometry
	Vision image	Perceptron	Weld pool geometry
	Welding current, arc voltage	Perceptron	Weld pool geometry
	Ultrasonic sound	Perceptron	Weld defects (void, crack)
Pipe welding	Optical image	Kohonen	Weld pool geometry
PCB solder joint	CCD image	LVQ	Surface dimension
IC fabrication	Chamber pressure	Perceptron	Plasma etching fault detection
	DC bias, reflected RF power	Perceptron	Oxide thickness
	Etch time, gas flow rate, RF power pressure	Perceptron	Oxide thickness
Autoclave curing	Pressure, the 1st and 2nd holding temperature	Perceptron	Laminate thickness, void size
Web	CCD image	LVQ2	Surface roughness
Steel casting	Temperature	Time-series and spatial	Breakout
Steel types inspection	Vision (capture of spark)	Perceptron	Steel types
Metal forging	Ram load and velocity	Perceptron	Final shape, microscopic properties
Wire EDM	Pulse width, wire tension	Perceptron	Surface roughness
Color printing	Color	Perceptron	Desired color codes
Light wave inspection	CCD image	Perceptron, counter-propagation	Light ware defect
Tapping	Cutting force	Perceptron, RBF	Thread quality
Riveting	AE	Perceptron, Kohonen	Crack growth
Laser surface hardening	Temperature	Perceptron	Layer thickness

12.4.1 Tapping Process

Tapping is an important machining process that produces internal threads and requires relatively low cutting speed and effective cooling. Several malfunctions may occur in the process, including tap-hole misalignment, tap wear, tap-hole mismatch. With such malfunctions, the machine produces threads of undesirable quality such as hole undersize, hole oversize, eccentricity of the hole, and so on. To monitor

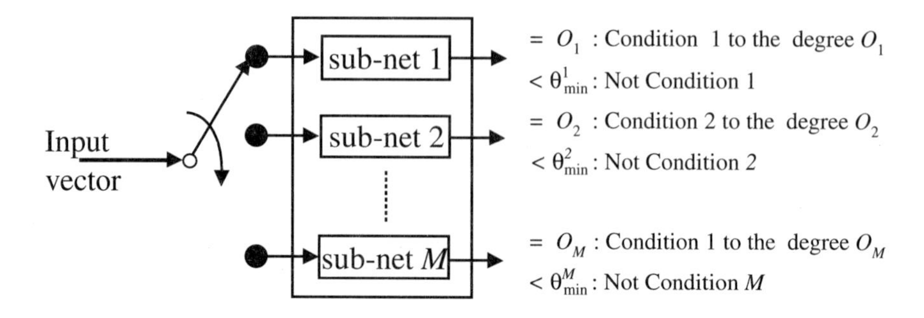

(a) The proposed neural architecture

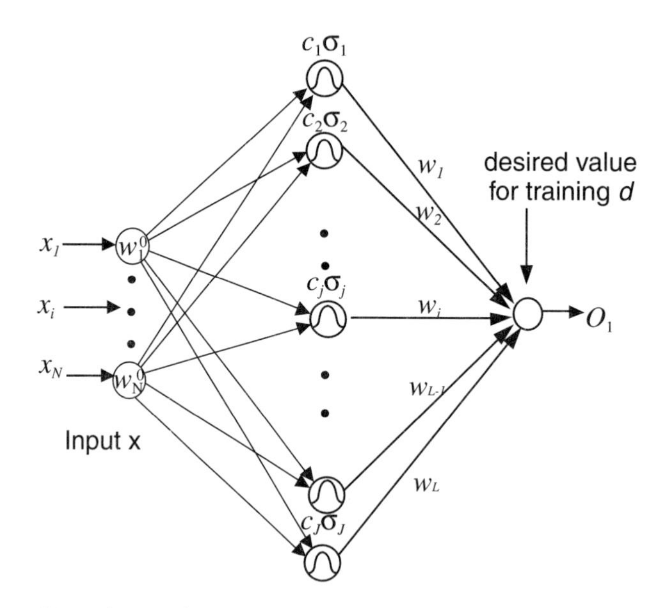

(b) Information-gain-weighted RBF as the sub-net

FIGURE 12.5 A neural network schematic proposed for tapping process monitoring. (a) The proposed neural architecture. (b) Information-gain-weighted RBF as the sub-net.

these conditions, a network-based monitoring system [Chen et al., 1996] has been developed. This system utilizes a dynamometer that measures tapping torque, thrust force, and lateral force. The network used here is composed of M subnetworks, as shown in Figure 12.5(a), where M denotes the number of categories to be classified. Each subnetwork is essentially the information-gain-weighted radial basis function, which accepts an input vector (x_1, x_2, ..., x_8) and produces one output. The input vector composed of eight nodes of the input layer of the RBF is extracted from the dynamometer, including x_1 = peak of torque, x_2 = mean of torque, x_3 = variance of torque, x_4 = mean of torque in retraction stroke, x_5 = mean of thrust force, x_6 = covariance of torque and force, x_7 = correlation of torque and thrust force, x_8 = correlation of torque and thrust force in retraction stroke. The w_i° assigned to each node represents the information gain-weighted value, which is learned based on information available at the signal/index evaluation stage. The information gains can always be updated when new data are available.

The weight parameters w_i° ($i = 1, 2, \ldots, N$) are calculated from entropy theory. According to the results given in the reference, the total information gain of the index x_k about the process is obtained by

$$G_\Omega(x_k) = \sum_{i=1}^{R} G_{c_i}(x_k) \qquad \text{Equation (12.6)}$$

In the above equation, $Gc_i(x_k)$ indicates the information gain of the index x_k about the class c_i, $\Omega = \{c_j; i = 1,2, \ldots, R\}$ is the class space with c_i representing the ith class and R the total number of classes. Equation 12.6 essentially implies that the larger the $G_\Omega(x_k)$ is, the more significant the index x_k is to the tapping process.

The classification is obtained by the proposed RBF when the minimum threshold θ_{min} of each sub-network is set to 0.2 based on statistical distribution obtained during training. These results are compared with the conventional RBF. The results indicate that, in the case of undersize and misalignment classes, the proposed system discerns the patterns much more clearly than the conventional one.

12.4.2 Solder Joint Monitoring

Solder joints of printed circuit board have various shapes as soldering conditions (amount of solder paste cream, heating condition, heat profile, etc.) are changed. In the aspect of classification problem, even though solder joints belong to a set of the same soldering quality, the shapes are not identical to each other, but vary to a certain degree. This makes it difficult to define a quantitative reference of solder joint shape as a soldering quality. In recent years, artificial neural network (ANN) approaches have been applied to solder joint inspection due to learning capability and nonlinear classification performance. Among many neural network approaches, the LVQ neural network classifier [Kim and Cho, 1995] is one example of such neural network applications for solder joint inspection.

A three-color tiered illumination system to acquire the shape of a solder joint is shown in Figure 12.6, which consists of three colored circular lamps (red, green, blue), a CCD camera, a color image processing board, and a PC with a display monitor. The lamps are coaxially tiered in the sequence of green, red, and blue upward from the bottom of the inspection surface. The three color lamps illuminate the solder joint surface with different incident angles: the blue lamp to 20°, the red lamp to 40°, the green lamp to 70°. With the help of the three colors of lamp with different incident angles, we can acquire color patterns of three different slope surfaces in an image at a time. Figure 12.6(a) shows a typical image of solder joints on the PCB under the illumination system. The color patterns of the specular surface on solder joints are generated according to surface slope.

Utilizing the acquired image of the solder joint surface, a neural architecture for solder joint inspection is adopted as shown in Figure 12.6(b). The color intensities at each pixel in the stored color image are used as the input data of the LVQ-1 neural network. An input of LVQ-1 neural network is represented by

$$\underset{\sim}{x}_c = \{x(l, i, j) \mid l = 1, 2, 3, \quad i = 1, 2, \ldots, n, \quad j = 1, 2, \ldots, m\} \qquad \text{Equation (12.7)}$$

where the subscript c indicates the class of the input data, $\underset{\sim}{x}_c(l, i, j)$ means an intensity value at the (i, j) pixel in the l color frame, the l is an index to represent each color frame (1 = the red, 2 = the green, 3 = blue), the (i, j) indicates a pixel position, and the (n, m) indicates the size of a window image. The dimension of input nodes is the same as that of the input image.

Output nodes are fully connected to all input nodes by the weight vectors. The number of output nodes in a competitive layer is set to 10. A set of the weight vectors between the kth output node and the input nodes is defined as

$$\underset{\sim}{w}_k = \{w_k(l, i, j) \mid l = 1, 2, 3, \quad i = 1, 2, \ldots, n, \quad j = 1, 2, \ldots, m\} \qquad \text{Equation (12.8)}$$

where all subscripts are the same as those of the input image. The input value of the kth output node is the Euclidean distance between the input data and the weights, which is expressed by

$$o_k = \underset{\sim}{w}_k - \underset{\sim}{x}_c = \sum_{i=1}^{n}\sum_{j=1}^{m}\sum_{l=1}^{3}\left\| w_k(l,i,j) - x_c(l,i,j)\right\|. \qquad \text{Equation (12.9)}$$

In the self-clustering module, the input value of the output node is a similarity measure for clustering that indicates the degree of resemblance between an input image and the weight vectors. The update of weight vectors are based on competitive learning, called *winner-take-all learning*. Only one node of the nearest weights to the current images is selected as the winner and has a chance to update its weight vector by

$$\underset{\sim}{w}_{win}(t+1) = \underset{\sim}{w}_{win}(t) + \eta(t)\left(\underset{\sim}{x}_c(t) - \underset{\sim}{w}_{win}(t)\right) \qquad \text{Equation (12.10)}$$

where $\eta(t)$ is the learning rate, and is initially set to 0.25 and decreases to 0.05 as t increases. During the training procedure, the weights of each output node are to be updated toward resembling the members in its own cluster. After the training procedure, a supervised learning technique is adopted to increase classification performance. Only if an input data is misclassified into a different class should the weight vector of the class be updated. For example, consider an input data, x_q belonging to q^{th} class. If it is misclassified into the cluster labeled as the cth class, then update the synaptic weights of the cluster as follows:

$$\underset{\sim}{w}_{win}(t+1) = \underset{\sim}{w}_{win}(t) + \eta(t)\left(\underset{\sim}{x}_q(t) - \underset{\sim}{w}_{win}(t)\right). \qquad \text{Equation (12.11)}$$

To evaluate the performance of the proposed neural network classifier, the classification is made for the test data that are not used for the training. The clustering result is compared to that of the expert inspector. The total classification success rate is found to be 93.1%. The proposed neural network has a simple structure that facilitates the use of raw intensity data of each pixel as its input data. It relieves the burden of performing a large number of experiments to find optimal visual features, and will also help to initially find the input feature space before designing a suitable classifier.

12.4.3 Pipe Welding Process

In a high-frequency electric resistance pipe welding process, the hot roll coils are progressively formed into cylinder shapes in several stages while high-frequency current is applied with contact tips to both edges of the formed metal to be joined. Applied current flowing between the adjacent surfaces of the edges metal heats and melts a small volume of metal along the edge by making the best use of the skin effect and proximity effects of high-frequency currents. When the molten metal from both adjacent edges runs together by the action of the squeeze rolls and cools down by cold water, a weld is produced.

The three typical shapes of bead in high-frequency electric resistance welding (HERW) are shown in Figure 12.7(a). Under insufficient heat input, the concave shape appears on the top bead while the slope of the bead is steep. The concave shape is produced when the molten metal diminishes between the base metal due to the squeezing force. A cold weld is a main defect from the lack of heat input. Under an optimum heat input condition, the concave shape disappears. In this case, the molten steel appears on the top of the bead and the smooth shape of the bead can be achieved. Under an excessive heat condition, the molten steel hangs over the top of the bead and the height of the bead becomes unstable while the bottom width of the bead increases with heat input. Penetration is one of the main defects from excessive heat input due to the inclusion of impure particles in the weld pool.

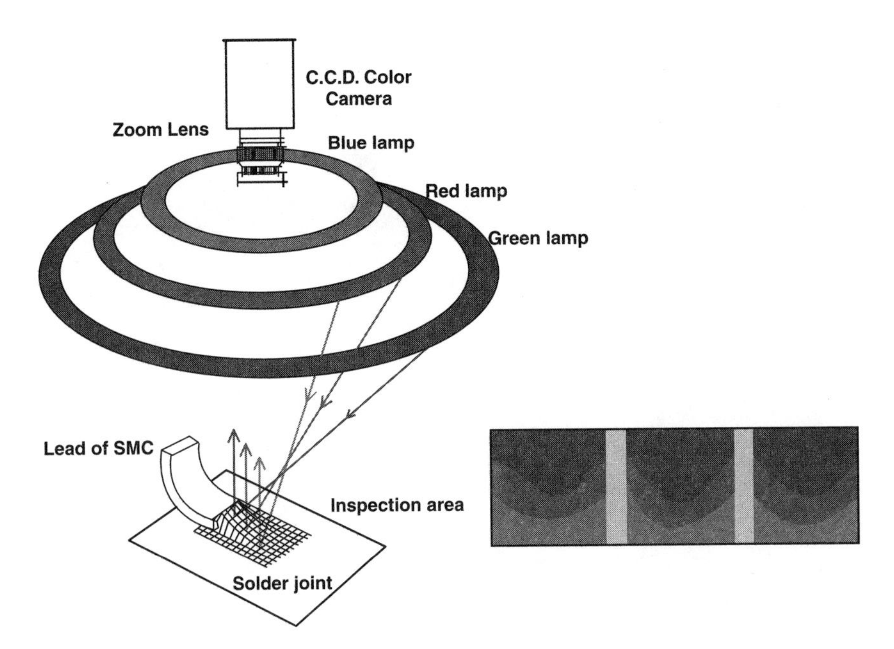

(a) Images of solder joint and 3-color ring illumination system

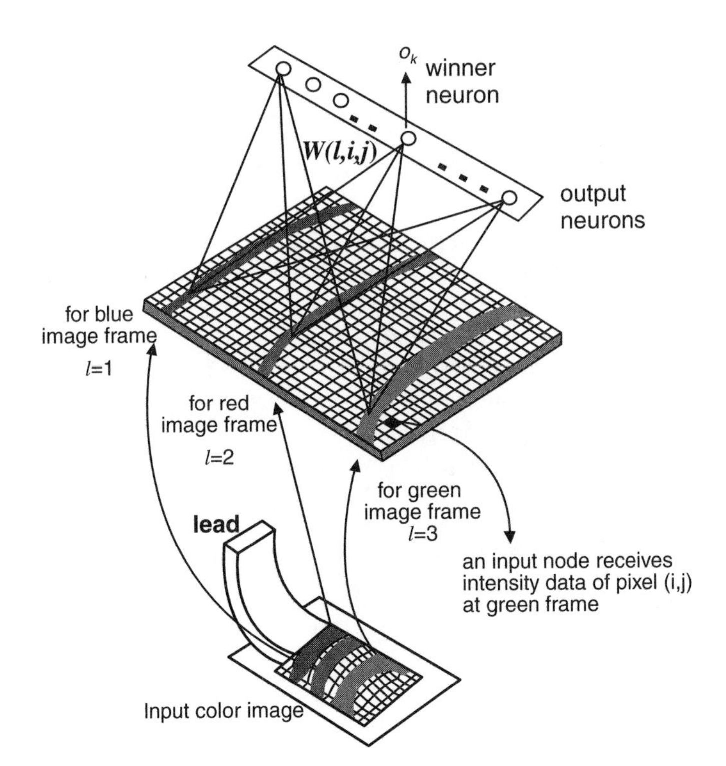

(b) The neural structure of LVQ for solder joint inspections

FIGURE 12.6 The solder joint image and LVQ architecture. (a) Images of solder joint and three-color ring illumination system. (b) The neural structure of LVQ for solder joint inspections.

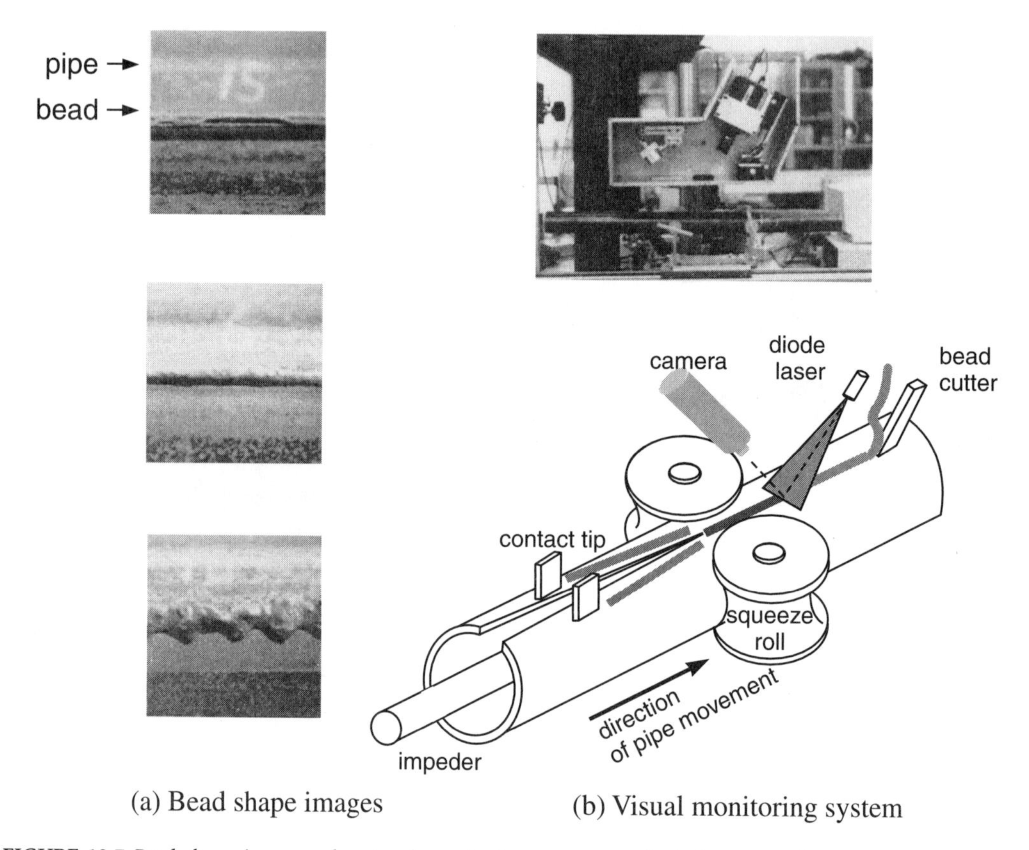

FIGURE 12.7 Bead shape image and a visual monitoring system in a high-frequency electric resistance welding process. (a) Bead shape images. (b) Visual monitoring system.

A visual bead shape monitoring system [Ko et al., 1994], as shown in Figure 12.7(b), was designed to acquire an image of the bead shape. It consists of a CID (charge injection device) camera, a laser with a cylindrical lens, fine mechanical alignment stage sets, filters, air spray, and cooling system. The monitoring system operates on a principle of triangulation that has been used to obtain visual information on the layout of bead shape. The Kohonen SOFM for bead shape classification is shown in Figure 12.8. The input layer and competitive layer consist of 180 nodes which corresponds to horizontal pixel and 12 nodes in the form of one dimensional array, respectively.

During the training procedure, the weight vectors of the network form a model of the input pattern space in terms of so-called prototypical feature patterns. A Kohonen network can segment an input bead image similar to the training pattern by comparing it with all trained weight vectors. For this classification procedure, a winning weight vector $\underset{\sim}{w}_{win}$ that is close to a given input pattern x is selected by

$$\|\underset{\sim}{w}_{win} - \underset{\sim}{x}\| = \min \|\underset{\sim}{w}_k - \underset{\sim}{x}\|, \quad _k = 1, 2, \ldots, N \qquad \text{Equation (12.12)}$$

where N is the number of output node.

The weights of winning node and its neighborhood nodes are updated by

$$\underset{\sim}{w}_{win}(t+1) = \underset{\sim}{w}_{win}(t) + \Lambda_{win}(t)\eta(t)\left(\underset{\sim}{x}(t) - \underset{\sim}{w}_{win}(t)\right) \qquad \text{Equation (12.13)}$$

where $\Lambda_{win}(t)$ is the neighborhood relation and is a learning rate. The initial value of the learning rate is 0.5 and gradually decreases to 0.01 with time. The extent of the neighborhood relation is set initially

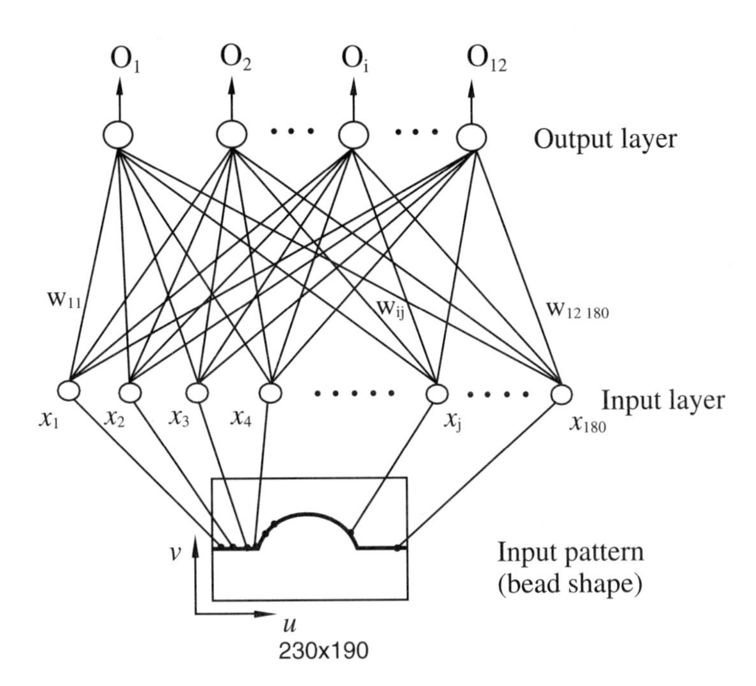

FIGURE 12.8 The NN architecture for bead shape classification.

at half the output node and is made to decrease in every 1000 iterations. Total number of iterations is 10,000.

To examine classification performance of the neural network, the clustering results are compared to those of the expert operator. The classification success rate for the whole image is found to be 98.05%. These results reveal that the relationship between heat input condition and bead shape has been successfully implemented on the neural network. If the welding condition is changed, a similar neural bead shape classifier can be designed and trained using the experimentally obtained bead shape. This monitoring method faciliates adjustment of welding condition according to the current state of the bead shape. This SOFM neural network approach is more efficient for monitoring current welding conditions than the geometrical feature-based method for real HERW process, considering accuracy, experimental cost, and implementation time.

12.4.4 Laser Surface Hardening Process

A laser surface hardening process is shown in Figure 12.9(a). The surface layer is heated above its austenite transformation temperature but below the melting temperature by high-power laser beam energy traveling along a specified direction. As a result, the microstructure previously composed of ferrite and pearlite is transformed into the austenite phase. At this high temperature, the microstructure of the substrate material is homogenized by carbon diffusion. After being heated, the material is rapidly cooled, resulting in a hard martensite phase due to the self-quenching effect caused by the rapid thermal conduction from the surface into the bulk substrate material.

Due to its high energy density and power controllability, high-power laser material hardening has been widely used in industrial processes. The difficulty involved with this process is that the hardening quality, in this case layer thickness, cannot be easily assessed during the process. To assess the coating or hardening thickness during the processing, a surface temperature measurement is used, which can represent the corresponding layer thickness. Figure 12.9(b) gives a schematic of the measurement arrangement and a neural network training procedure [Woo and Cho, 1998]. The neural network used is a multilayer perceptron and it adopts the error backpropagation algorithm. The input data used for training includes surface temperatures and operating conditions such as input powers and laser travel speed, and its outputs

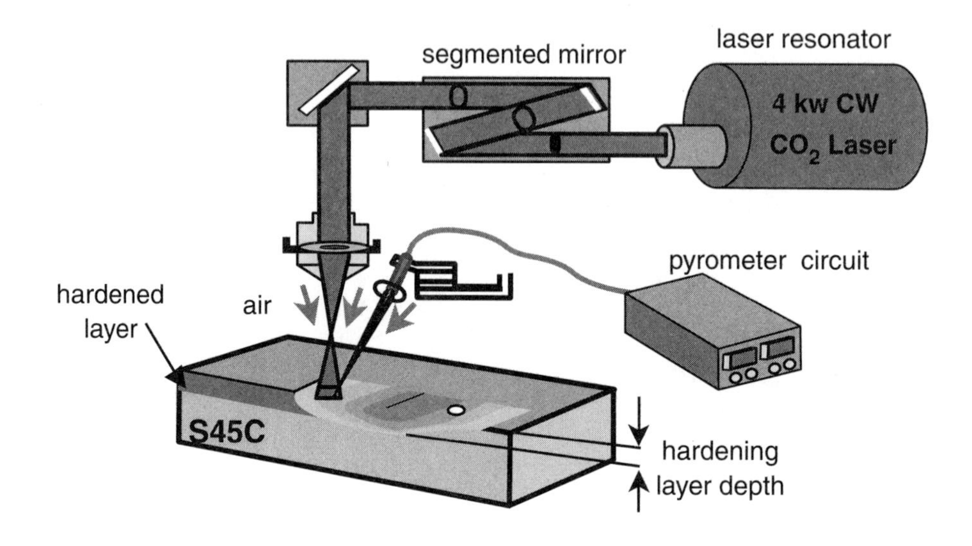

(a) A laser surface hardening process

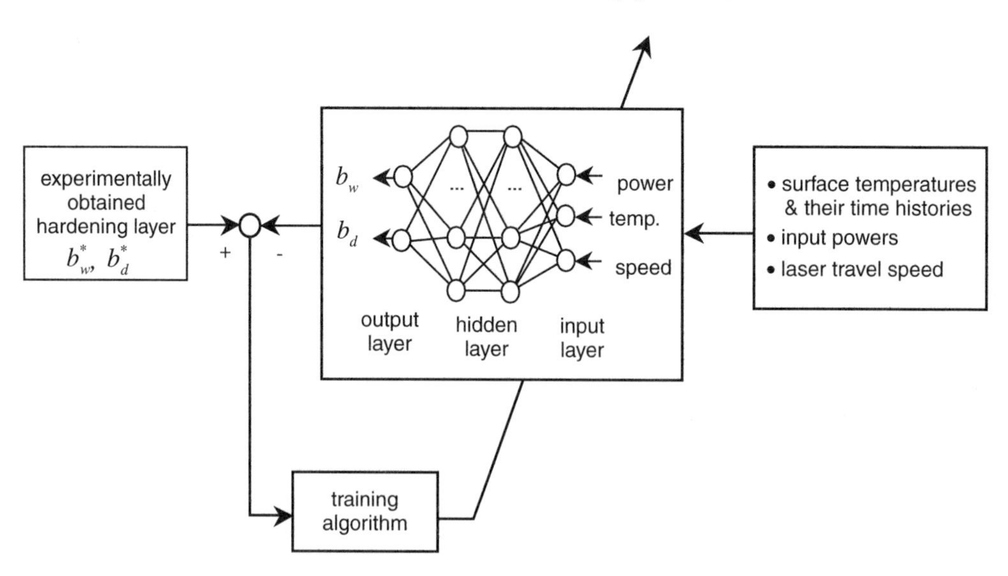

(b) The layer geometry estimation using a neural network model

FIGURE 12.9 The neural network architecture. (a) A laser surface hardening process. (b) The layer geometry estimation using a neural network model.

generate the hardening depth (b_d) and width (b_w). In the error backpropagation algorithm, the weight of weights is given in an iterative manner as

$$\begin{aligned} w_{lk}(t+1) &= w_{lk}(t) + \eta\delta_l(t)o_l(t) + \alpha\Delta w_{lk}(t) \\ \delta_l(t) &= \left(o_l^d - o_l\right)\dot{f}_l\left(net_l(t)\right) \end{aligned} \qquad \text{Equation (12.14)}$$

where w_{lk} is the weight linking the lth node in output layer and k^{th} node in the hidden layer; o_l and o_l^d are the output and the desired output of lth node, respectively, is the activation function; (·) derivative of ( ); and t, η, and α are the number of training iterations, the learning rate concerned with the speed of convergence of the error, and the momentum rate to avoid the oscillating phenomena, respectively. In the above, $net_l\,(t)$ is given by

$$net_l(t) = \sum_{k=0}^{K} w_{lk}(t) o_k(t) \quad (l = 1, 2, \ldots, L) \qquad \text{Equation (12.15)}$$

where L is the total number of neurons in the last layer. Hidden layer weights are adjusted according to

$$\delta_k(t)\, \dot{f}(net_k(t)) \sum_{l=0}^{L} \delta_l(t) \cdot w_{lk}(t) \quad (k = 1, 2, \ldots, K)$$
$$w_{kj}(t+1)\; w_{kj}(t) + \eta \cdot \delta_k(t) \cdot o_j(t) \quad (j = 1, 2, \ldots, J) \qquad \text{Equation (12.16)}$$

where J is the number of the neuron of the jth preceding hidden layer, w_{kj} is the weights linking the kth node of the last hidden layer and j^{th} node of the proceeding layer. In this application, learning and momentum gains are chosen to be 0.7 and 0.5, respectively. The network estimation is proven to yield fairy accurate results; the largest deviation is 10%.

12.5 Neural Network-Based Control

The conventional control approach lacks the ability to learn the property of the unknown system to self-organize the features of the system response, and to make an appropriate decision, based upon the learned process state, regarding how to generate the control signal so as to drive the control system to reach a designed state. The neural networks overcome the deficiency of the conventional control methodologies with their learning, self-organization, and decision-making abilities. This is why the network-based control is called one of *intelligent control.* It should be noted that "intelligent" here does not imply better system performance by intelligent control than by conventional control. In fact, for a system with its dynamic characteristics completely known there may be no need to utilize an intelligent control technique. This is because the conventional technique in this case provides better control performance and reliability in implementation.

In reality, the dynamic characteristics of most of the manufacturing processes are not exactly known. For this reason, neural networks come into play to learn the process characteristics and generate appropriate controller output based upon this learning to meet a certain specification of control performance. In carrying out these tasks the networks are embedded into the system in several forms.

The first form is to use the networks as an aid to the conventional controller. For example, the process modeled under certain assumptions can be controlled by one of the conventional controls such as the optimal control, adaptive, and so on, as shown in Figure 12.10(a). The objective of the control is to make y follow the desired value y_d by aid of a neural network that estimates the uncertain part of the process. In this case, an approximately modeled process has to be identified on-line. The network's role here is to identify uncertain process parameters and provide the estimated parameters to the controller.

The second form is to use a neural network as a direct tuner to obtain the desired gains of the conventional controllers such as PI, PD, and PID, as shown in Figure 12.10(b). The control objective here is to self-tune the controller gain in such a way that the output y follows the desired value y. The network is trained here to minimize the error function with respect to the weight value w_{ij},

$$\frac{\partial E}{\partial w_{ij}} = f\left(\frac{\partial K}{\partial w_{ij}}\right) \qquad \text{Equation (12.17)}$$

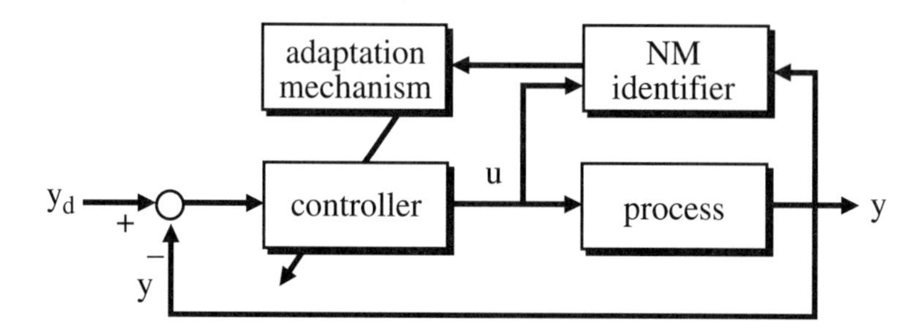

(a) A neural identifier combined with an adaptive controller

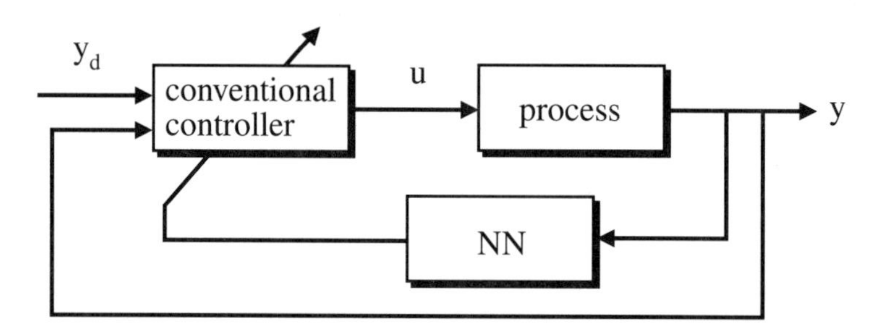

(b) A gain-tuning neural network controller

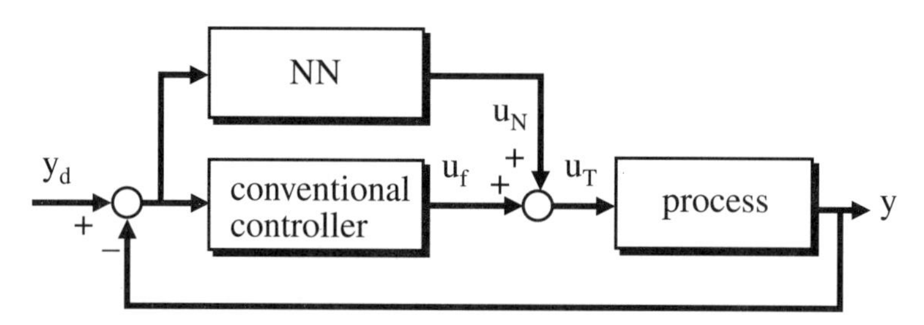

(c) A feed forward neural controller combined with a conventional feedback controller

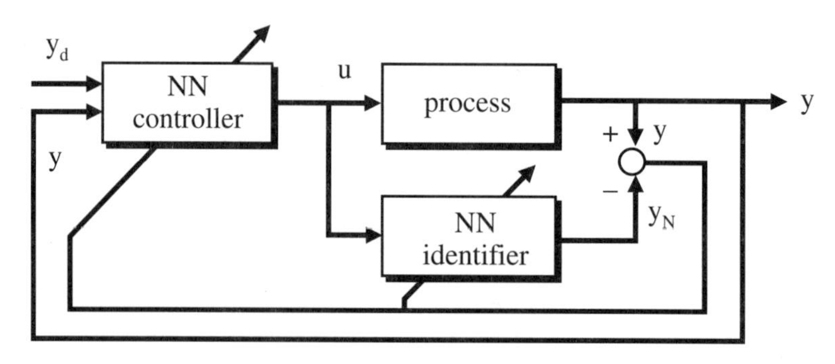

(d) A neural controller combined with a neural identifier

Figure 12.10 Various neural network based monitoring and control schemes. (a) A neural identifier combined with an adaptive controller. (b) A gain-tuning neural network controller. (c) A feedforward neural controller combined with a conventional feedback controller. (d) A neural controller combined with a neural identifier.

where E is the squared error function defined by

$$E = \frac{1}{2}(y_d - y)^2. \qquad \text{Equation (12.18)}$$

In the above, the K consists of the gain values if a PID controller is used,

$$K = K_p, K_i, K_d \qquad \text{Equation (12.19)}$$

and the function f indicates that the gain values are the explicit functions of the network weight values w_{ij}.

The third form is a form of the neural network controller combined with a simple conventional controller, shown in Figure 12.10(c). In this control structure, the conventional controller is used to provide the network controller with stability of the system response, which may be needed at a stage of initial learning. The network is used here to learn the *inverse model* of the process, that is, the relationship that relates the desired output y_d to the corresponding control action u. In this control scheme, the neural network is so trained to minimize the squared error function E defined by

$$E = \frac{1}{2}\left(u_T - u_N\right)^2. \qquad \text{Equation (12.20)}$$

In Equation 12.20, the resultant control effort u_T consists of u_N, the control part generated by the network, and u_f, that of the feedback signal.

$$\frac{\partial E}{\partial w_{ij}} = f\left(\frac{\partial u_N}{\partial w_{ij}}, u_f\right). \qquad \text{Equation (12.21)}$$

Equation 12.21 shows the gradient of the error with respect to the weights; $\partial E/\partial w_{ij}$ is a function of the u_N/w_{ij} and u_f.

The final form, shown in Figure 12.10(d), is an efficient usage of the multiple networks in control as well as identification of the process, a neural controller and a neural identifier. In this case, learning signal and convergence are the most important factors for speed and robustness of the system response. Often, finding learning signals can be a difficult task. In the block diagram, the neural network identifier estimates the unknown process model based upon the measured output y and control input u. The network's role herein is to make the estimated process model M_N follow the actual process model M as close as possible by carrying out on-line identification of the process.

$$M_N \rightarrow M \;\; \forall \;\; u \in U, y \in Y \qquad \text{Equation (12.22)}$$

where u belongs to admissible operating input range U and Y is the range of the corresponding output y.

Thus, the network is trained to minimize the error function E defined by

$$E = \frac{1}{2}\left(y_N - y\right)^2$$

and Equation (12.23)

$$\frac{\partial E}{\partial w_{ij}} = f\left(\frac{\partial y}{\partial w_{ij}}\right).$$

The other block representing the neural network controller is designed based upon the inverse model approach, as explained in Figure 12.10(c). That is to say, upon utilization of the identification result, the control input in the forward loop is generated such that the actual output y follows the desired value y_d as closely as possible. Thus, the network is so trained that the control error is modified in order to generate the desired output y_d,

$$E = \frac{1}{2}\left(u - u_N\right)^2 \qquad \text{Equation (12.24)}$$

and

$$\frac{\partial E}{\partial w_{ij}} = f\left(\frac{\partial u_N}{\partial w_{ij}}, \frac{\partial y}{\partial u}\right) \qquad \text{Equation (12.25)}$$

where the partial derivative y/u is included to account for generating the network training signal.

There have been tremendous research efforts in control of manufacturing processes. Table 12.2 summarizes the types of neural network control in various manufacturing processes. Some neural network applications to machining, arc welding, semiconductor fabrication, hydroforming, and hot plate rolling processes will be summarized in the following sections.

12.6 Process Control Applications

12.6.1 Machining Process

In machining, adaptive control has been viewed as a very promising strategy to adapt on-line the process parameters to widely varying process conditions. To effectively achieve this, a process model is needed for the feedback control of the process. Figure 12.11 shows a neural network based control system developed for this purpose [Azouzi and Guillot, 1996]. Two network models are used here for estimation and control of the quality variables, the dimensional deviation DD and the surface finish, Ra. The process neural model here is to provide a mathematical relationship for parameters needed to the optimizer, which minimizes the machining cost. This model estimates the DD and Ra, which implies that $\hat{x}(t) = [\hat{DD}, \hat{Ra}]$ learns the quality model output $\underset{\sim}{x}(t) = [DD, Ra]$.

In this model, a hybrid network model is adopted, which consists of a Kohonen feature map and a multilayer perceptron. The motivation of using this type of the network structure is to avoid the memory degradation due to distributed learning and process in most existing feedforward neural networks. Often, the feedforward neural model correctly represents the process behavior only in the vicinity of the most operational process input, partly forgetting the process behavior in other regions. As shown in Figure 12.12, a Kohonen network is used as a two-dimensional network tuned to a variety of input patterns through unsupervised learning. This network divides the multilayer perceptron network (MLP) into N specified clusters. Based upon this network, information flow and storage in the MLP is directed to a specified cluster. The neurons of the output layer in P-network accept their incoming activation signal weighted by the K-network. Let P_j be the weighting parameter of the jth cluster (node) in the P-network. Then, P_j is essentially the percent contribution of the jth cluster to the activation signal of the output neurons. The input to the kth output neuron in the lth layer is given by

$$net_k = \sum_{j=1}^{h} P_j \left\{ w_{kj}^{l-1} o_j^{l-1} \right\} \quad (k = 1, 2, \ldots\ldots, m) \qquad \text{Equation (12.26)}$$

TABLE 12.2 Types of Neural Networks in Control

Process	Control Input	Controller	Network	Quality Variable
Extrusion	Ram velocity	NN controller	Perceptron	Micro structure of the material
Semiconductor (etching)	Oxygen flow, pressure	NN controller	Perceptron	Etch rate, anisotropy
Arc welding	Heat power	NN controller	Perceptron	Weld bead width, weld depth
Steel making (galvanizing)	Temperature		RBF + perceptron	Percentage of iron at the surface
Hot plate mill	Servo valve current	NN controller + PD	RBF, perceptron	Slab width
Turning	Feed rate Cutting speed	NN controller	Perceptron + Kohonen	Surface finish, dimensional accuracy
	Feed	NN based tuning	ART 2	Machining efficiency, surface finish
	Feed	NN controller + PD	Perceptron	Machining efficiency
	Cutting Force	NN controller	Perceptron	Surface finish
Plastic injection molding	Inlet flow rate	NN controller	Perceptron	Product defect
Hydroforming	Servo valve current	NN controler	CMAC	Forming dimension, wrinkling

where h is the number of the neurons in layer l-1, w_{kj}^{l-1} is the weight linking the kth node in the output layer with the jth node in layer l-1, and o_j^{l-1} is the output of the jth node in layer l-1. The normalized P_j is estimated by

$$P_j = \frac{\exp(\lambda_j)}{\sum_{j=1}^{h} \exp(\lambda_j)} \quad (j = 1, 2, \ldots.., h) \qquad \text{Equation (12.27)}$$

where the λ_j is given by

$$\lambda_j = \frac{\mu_1}{1 + \exp\left(\mu_2 d_j^2\right)} \qquad \text{Equation (12.28)}$$

In Equation 12.28 above, d_j is the Euclidean distance between the input pattern and the weight w_j linking j^{th} node in the hidden layer with input nodes in input layer, so that

$$d_j \| \underset{\sim}{w}_j - \underset{\sim}{x} \| \quad j = 1, 2, \ldots.., h \qquad \text{Equation (12.29)}$$

And μ_1 and μ_2 are positive constants that are used to control the relative importance of the cluster. It is noted that the winning cluster min $\|\underset{\sim}{w}_i - \underset{\sim}{x}\|$ and its neighbors are allowed to contribute significantly to the activation of the output neuron. The training of each network in this hybrid architecture is carried out using a set of special data. The Kohonen network is trained independently using a winner-take-all learning method, while the quasi-Newton method is used to determine the weights and thresholds of the MLP. Using the network architecture and the learning method described in the above, a series of simulation works was conducted for various conditions for turning operation of AISI 108 steel parts.

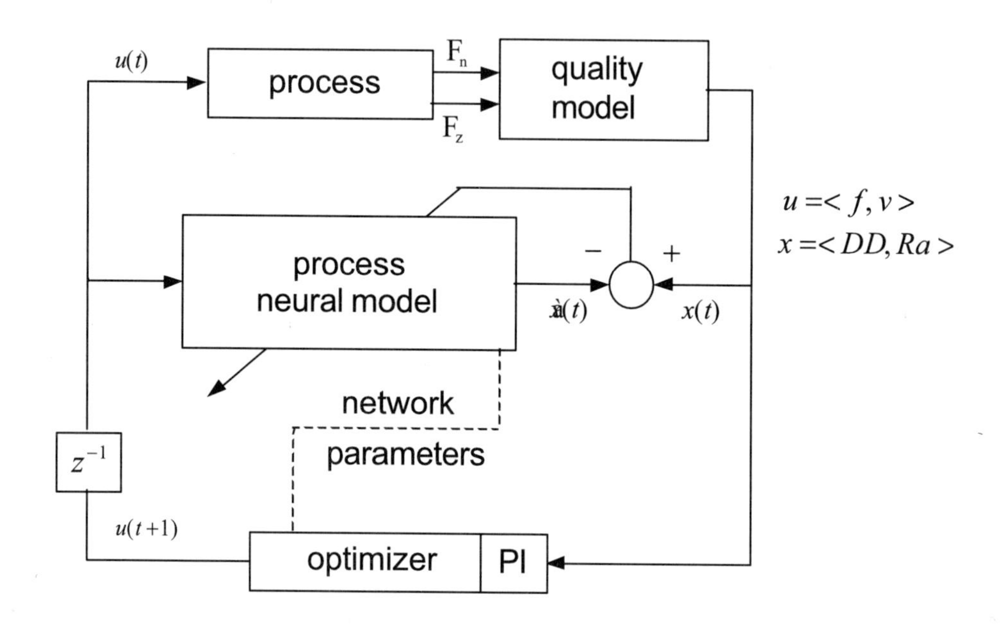

(a) The proposed synthesis scheme

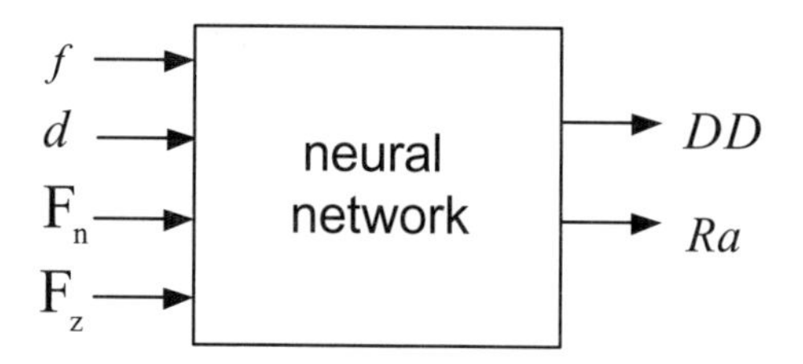

(b) The K-P network-based process model

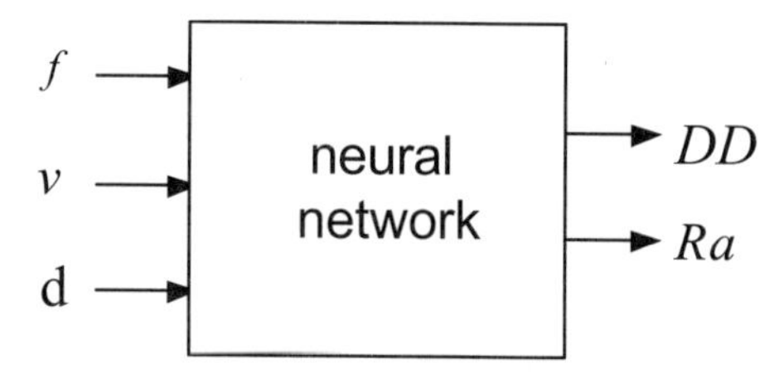

(c) The MLP-based quality model

FIGURE 12.11 The proposed neurocontrol scheme. (a) The proposed synthesis scheme. (b) The K-P network-based process model. (c) The MLP-based quality model.

The results are given in terms of the machining cost *L*, not in terms of dimensional deviation (*DD*) and surface finish (*Ra*). According to this work the optimized cost indicative of the *DD* and *Ra* shows 35% improvement in process performance by use of this proposed controller as compared to the well-known adaptive control with constraints (ACC).

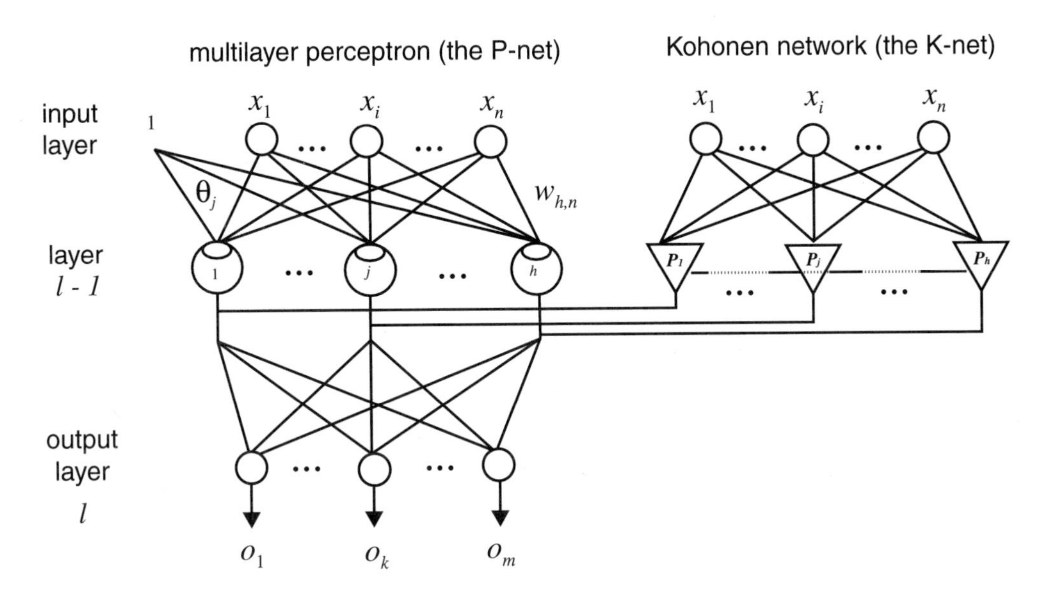

FIGURE 12.12 Schematic of the proposed K-P network.

12.6.2 Arc Welding Process

The objective of automated welding is to produce welds of high strength. To achieve this a number of research efforts have been made; one such effort is on-line control of the weld geometry such as width and penetration. The weld strength, indicative of quality, is usually represented by geometry of the weld pool, its width and penetration. These geometrical variables are not readily measurable during welding, and therefore an alternative that measures surface temperatures near the torch area has been used, since surface temperature distribution is indicative of the weld pool geometry. As shown in Figure 12.13(a), this temperature measurement is utilized to estimate an instantaneous weld pool size [Lim and Cho, 1993], which otherwise is not attainable. The control system to regulate the weld pool size is shown in Figure 12.13(b). In the figure, *PS* indicates the weld pool size and T_i denotes the temperature of the ith location. The system is basically a feedback error learning adopting two neural networks, one for estimation of the weld pool size and one for a feedback forward controller. The networks are a multilayer perceptron and the error back propagation method is utilized for training these. This architecture essentially utilizes the inverse dynamics of the welding process, and therefore, the total input u_T is given by

$$u_T(t) = u_N(t) + u_f(t) \qquad \text{Equation (12.30)}$$

where u_N is the network generated control signal and u_f is the feedback control signal. In fact, u_f can be any of the conventional controllers. The weights of the neural network controller are corrected according to the following weight adaptation equations:

$$\begin{aligned} w_{lk}(t+1) &= w_{lk}(t) + \eta\delta_1(t)o_k(t) \\ \delta_1(t) &= u_f(t)\dot{f}(net_l(t)) \end{aligned} \qquad \text{Equation (12.31)}$$

where w_{lk} is the weight linking the lth node in the output layer and the kth node in the last hidden layer adjacent to the output layer, f is the activation function, $(\cdot)$ derivative of ( ). In the above, $net_l(t)$ is given by Equation 12.15, and hidden layer weights are adjusted by Equation 12.16.

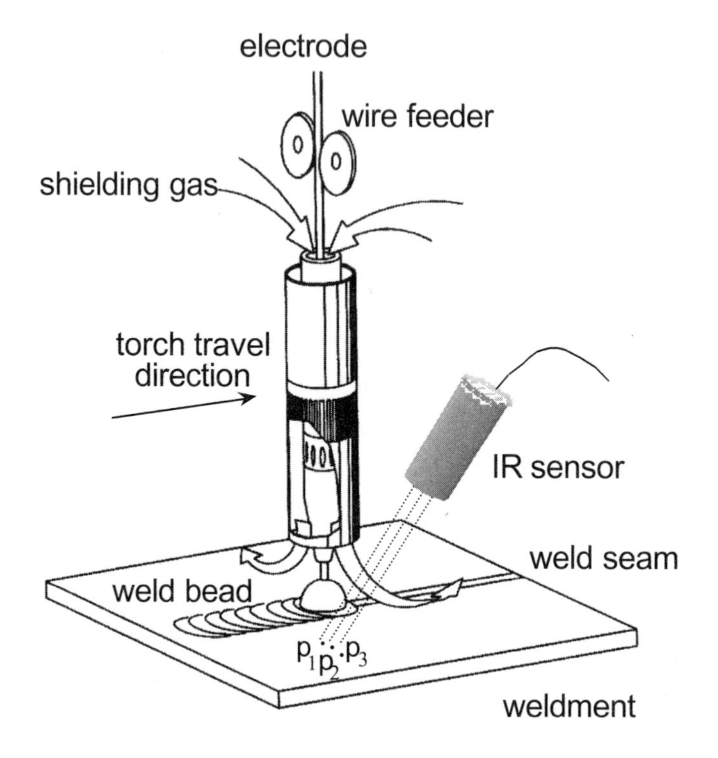

(a) Schematic description of GMA welding process

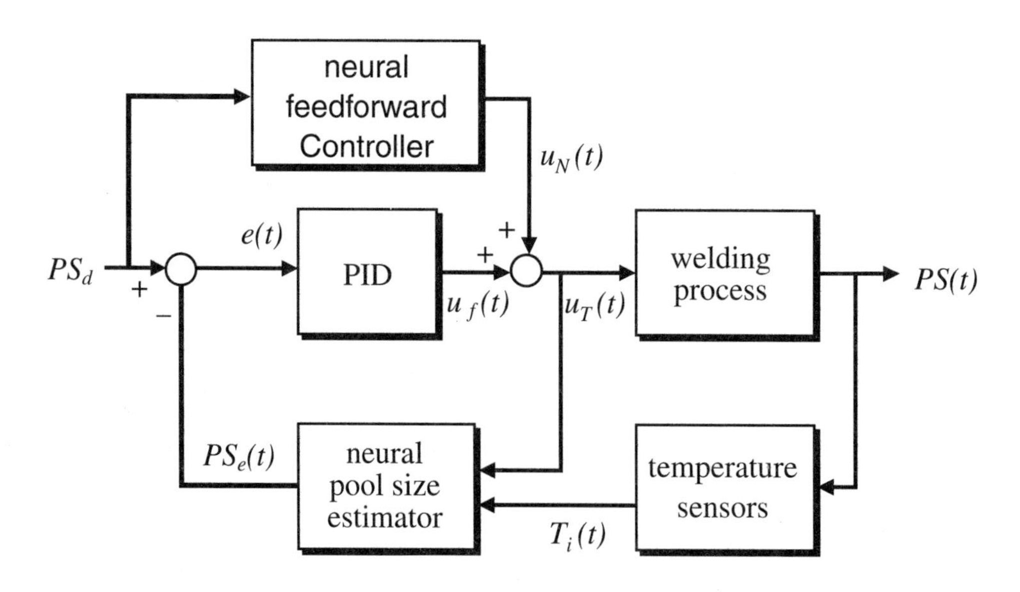

(b) A weld pool control system

FIGURE 12.13 Temperature sensing and control system for the GMA welding process. (a) Schematic description of GMA welding process. (b) A weld pool control system.

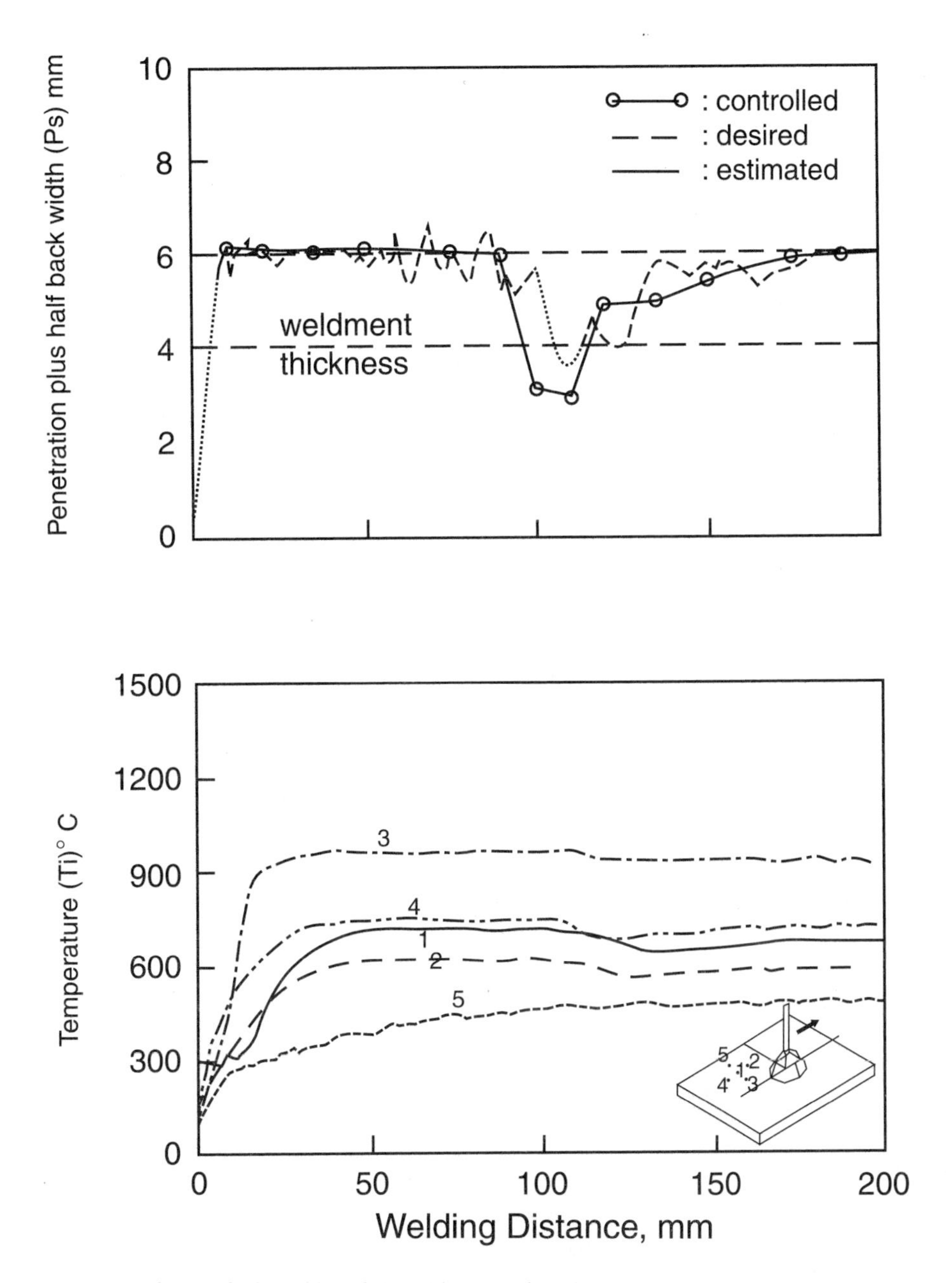

FIGURE 12.14 Neural control of a weld pool size with a neural pool.

To validate the capability of this control architecture, a series of welding experiments is performed with consideration of external disturbances. Two different network architectures are considered: $30 \times 50 \times 50 \times 1$ and $3 \times 25 \times 25 \times 1$ for the estimator and controller, respectively. An external disturbance torch travel speed is increased from 4 mm/s to 6 mm/s while the other input parameters are kept unchanged. In Figure 12.14, the experimental results of the welding control are shown together with responses of surfaces of surface temperatures at five locations. It can be seen that the pool size obtained by the neural control converges to a desired value, 6 mm. The controlled value, however, exhibits small fluctuations due to those of the estimation. The estimation results, however, denoted with a dotted line, indicate that the neural network estimates the actual pool size with satisfactory accuracy, which in this case is the controlled one.

12.6.3 Semiconductor Manufacturing

Semiconductor manufacturing processes typically exhibit complex interactions between multiple operating input signals and multiple output variables. In particular, reactive ion etching (RIE), as shown in Figure 12.15(a), is a highly nonlinear low-pressure form of plasma etching and remains a poorly understood process. The neural network control system [Stokes and May, 1996] employed here basically consists of an emulator (identifier) and an inverse neural model controller. The control system is illustrated in Figure 12.15(b). For simulation purpose, the process is assumed to be governed by a q-step ahead model described by

$$y(t+q) = f_N\left(\underset{\sim}{y}_t, \underset{\sim}{u}_{t+q-1}, \underset{\sim}{u}_{t-1}, \underset{\sim}{d}_t\right) \quad \text{Equation (12.32)}$$

where $\underset{\sim}{y}_t, \underset{\sim}{u}_{t+q-1}, \underset{\sim}{u}_{t-1}, \underset{\sim}{d}_t$ are the sampled output, input, and disturbance vectors, respectively. To generate the control input $u(t)$, the emulator is generated for the function in the q-step ahead predictive model given by Equation 12.32. The output of the process model and the resulting difference signal is used to tune an inverted neural model contained in the neural controller, which then generates the control input to drive the process. A series of simulations of the etch rate control are conducted with the emulator neural network 4-7-1 multilayer perceptron and the controller network a 1-8-4 structure. The etch rate was observed to converge quickly to the desired values, which are changed every 30 seconds.

12.6.4 Hydroforming Process

As shown in Figure 12.16(a), the hydroforming process does not require a die to form a sheet metal product of desired shape. Instead, a forming chamber takes this role by generating a hydraulic pressure against the sheet metal to be formed according to a prescribed schedule. The objective of the control is accurate tracking of the curve, pressure vs. the punch stroke, which ensures forming of high-quality products with no defects. A pressure control system based on cerebellar model articulation control (CMAC) [Park and Cho, 1990] is shown in Figure 12.16(b). In the diagram, p_d is the desired forming pressure, p_f is the actual forming pressure of the chamber, e is the differential signal between these two, u_T is the total control input consisting of the neural controller, u_N and the feedback controller PID, u_f.

The CMAC controller output is essentially the inverse dynamics of the forming process, namely,

$$u(t) = g\left(\underset{\sim}{p}_f(t)\right) \quad \text{Equation (12.33)}$$

where g represents the functional relationship between the u and p_f, and the chamber pressure vector at time step t consists of

$$p_f(t) = \{p_f(k+t), \ldots, p_f(1+t), p_f(t)\}. \quad \text{Equation (12.34)}$$

The g function is expressed by

$$g\left(\underset{\sim}{p}_f\right) = \sum_{i=1}^{N} a_i \cdot w_i. \quad \text{Equation (12.35)}$$

In Equation 12.35, a_i is the unipolar binary value and the weight w_i is given by

$$w_i(t+1) = w_i(t) + \Delta w_i \quad i = 1, 2, \ldots, N \quad \text{Equation (12.36)}$$

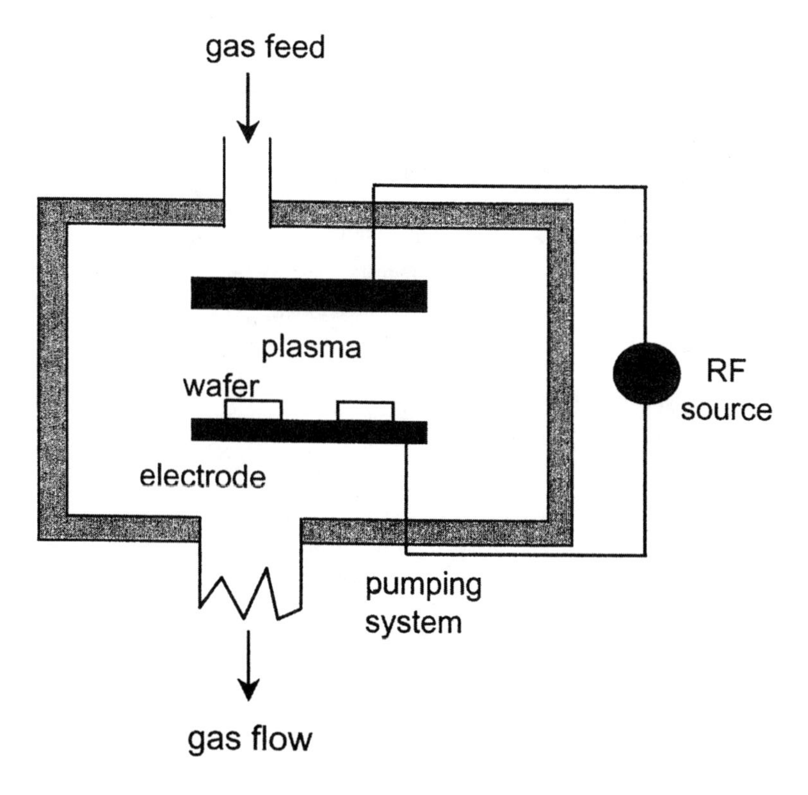

(a) Reactive ion etching process

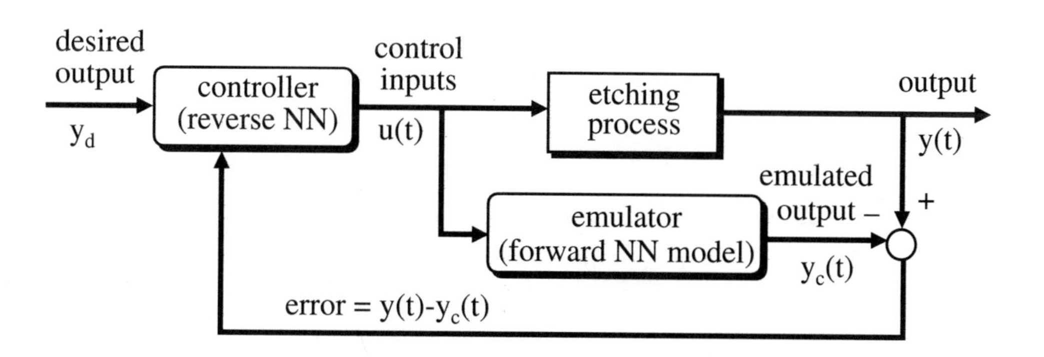

(b) Illustration of simple control scheme

FIGURE 12.15 Reactive ion etching process and the neural controller. (a) Reactive ion etching process. (b) Illustration of a simple control scheme.

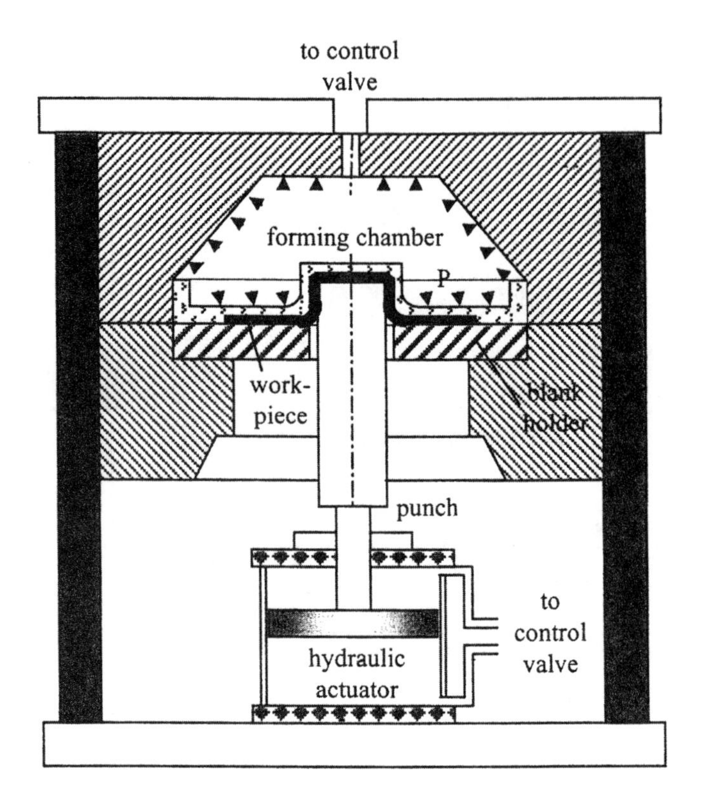

(a) A schematic of hydroforming process

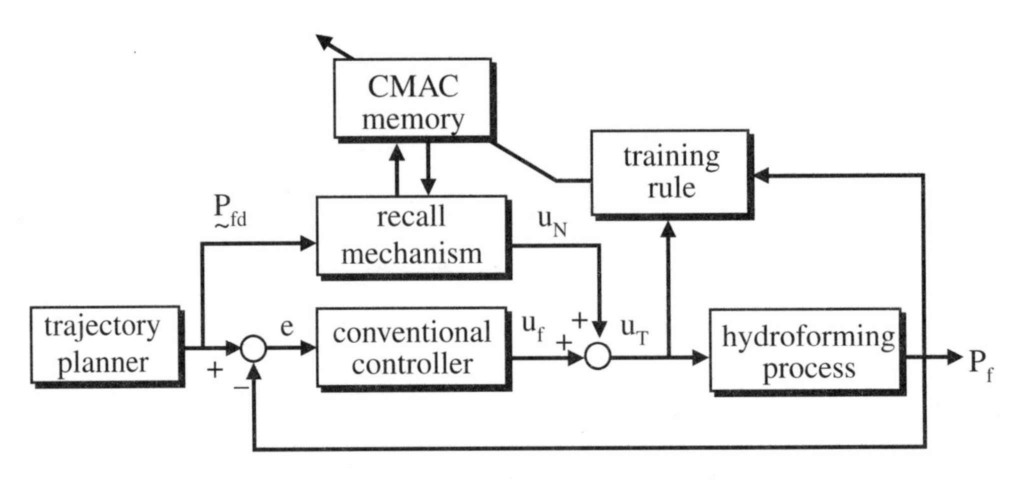

(b) The CMAC-based control system

FIGURE 12.16 The hydroforming process and the proposed pressure control system. (a) A schematic of the hydroforming process. (b) The CMAC-based control system.

where i is the memory address. The update rule for w_i is given by

$$\Delta w_i = \eta \cdot \frac{u(t) - g\left(\underset{\sim}{p}_f(t)\right)}{\gamma} \qquad \text{Equation (12.37)}$$

where γ is the generalization factor.

Utilizing the hydroforming machine, a series of experiments was carried out for various product shapes. The parameters of the CMAC used here such as learning rate η and generalization factor γ were set within a certain range. The controlled chamber pressure was found to approach the desired curve as the number of trials increased. In the long run, at the tenth trial, almost no difference was observed between the actual and desired chamber pressures. Several products produced by this control method show that control of the product quality in this manner can be used effectively for actual hydroforming process. The neural network application here is effective and highly recommendable, since the process is characterized by a highly nonlinear, uncertain, complex system, which may not achieve desired accuracy by conventional control techniques.

12.7 Conclusions

The current intense competition in markets requires manufacturers to produce high-quality products with lower cost and higher productivity. This stringent situation has created an issue of growing importance in the manufacturing community and evoked a necessity for reliable and robust monitoring and control systems with high performance that do not require any justification. Recently, artificial neural networks have emerged as an intelligent tool to approach this problem and are shown to offer great promise, especially in monitoring and control of manufacturing processes.

In this chapter the characteristics of the manufacturing processes were analyzed. According to this, the monitoring and control problems were identified and the use of artificial neural networks to solve them was justified. Types of sensor signals, network structures, and output variables for monitoring and control were surveyed for application domains covering a variety of processes. It has been pointed out that due to inherent functionalities of neural networks, they provide monitoring systems and network-based control systems with capabilities of handling time-varying parameters and uncertainty, and suppressing process noise and complexity involved with process phenomena better than the conventional techniques. These abilities will take the monitoring and control systems closer to a truly intelligent manufacturing system. However, there are problem domains that still need to further enhance the capabilities of the neural-based systems network. Current limitations or shortcomings of the neural networks are given in the following:

1. In identifying the processes, the networks are utilized like a transfer function which maps the input data into the output variables. The pitfall of this is that this method is just a black box approach, since the networks do not know what is going on with regard to the physical phenomena of the processes.
2. In pattern recognition and clustering, there still remains a limitation to the use of neural networks. Their performance here somewhat depends upon the statistical properties of sampled data. Accuracy clustering data located at the proximity of border lines of each group still need to be improved, since most manufacturing processes require nearly 100% correct judgment.
3. In process control, the networks assume roles in identification and/or generation of control signals based upon sensor data. As in many other problems, robustness and accuracy are the key issues that promote their implementation in real processes.

There are several other shortcomings not listed here, but all of these may be gradually solved when the networks are equipped with functionalities that accommodate the manufacturing-specific physical nature by enhancing the structure, learning algorithm, and convergency properties of the currently used networks. To approach and solve this, one must understand the characteristics of the manufacturing processes first and then design appropriate networks.

Defining Terms

Multisensor fusion: A technique for using a number of sensors to improve sensing accuracy by reducing the uncertainty of each sensor.

Weight pruning: A technique to prune an initially large structured network by weakening or eliminating certain synaptic weights in a selective and orderly fashion.

Hyper-plane: A plane is defined in a fictitious multidimensional space.

Kernel function: A simple function to estimate the probabilistic density.

Euclidean distance: A similarity measure expressed by the root-squared sum of the difference between two vectors.

Nonparametric classifier: Nonparametric methods do not need to use the parameter of a predefined model for classification, compared with parametric classifiers based on a model that can be completely described by the chosen mathematical function using the small and fixed number of parameters.

Intelligent control: A control method that does not rely on mathematical models of processes or plants but uses biologically inspired techniques and procedures.

Inverse model: A mathematical model of processes or plants to be controlled that yields the corresponding control/operating inputs for the given desired outputs.

References

Azouzi, R., and Guillot, M. 1996. Control and optimization of the turning process using a neural network, *Japan/USA Symposium on Flexible Automation ASME,* vol. 2, pp. 1437-1444.

Chen, Y., Li, X., and Orady, E. 1996. Integrated diagnosis using information-gain-weighted radial basis function network, *Computers and Ind. Engineering,* vol. 30, no. 2, pp. 243-255.

Chryssolouris, G., and Guillot, M. 1990. A comparison of statistical and AI approaches to the selection of process parameters in intelligent machining, *ASME Trans. J. Engineering for Industry,* vol. 112, pp. 122-131.

Grabec, I., and Kuljanic, E. 1994. Characterization of manufacturing processes based upon acoustic emission analysis by neural networks, *Annals of the CIRP,* vol. 43, pp. 77-80.

Khanchustambhan, R.G., and Zhang, G.M. 1992. A neural network approach to on-line monitoring of a turning process, *IEEE International Joint Conference on Neural Networks,* vol. 2, pp. 889-894.

Kim, J.H., and Cho, H.S. 1995. Neural network-based inspection of solder joints using a circular illumination, *Image and Vision Computing,* vol. 13, no. 6, pp. 479-490.

Ko, K.W., Cho, H.S., Kim, J.H., and Kong, W.I. 1998. A bead shape classification method using neural network in high frequency electric resistance weld, *Proc. of World Automation Congress.*

Javed, M.A., and Sanders, S.A.C. 1991. Neural networks based learning and adaptive control for manufacturing systems, *IEEE/RSJ International Workshop on Intelligent Robots and Systems,* pp. 242-246.

Lim, T.G., and Cho, H.S. 1993, A study on the estimation and control of weld pool sizes in GMA welding processes using multilayer perceptrons, Ph.D. thesis, *Korea Institute of Science and Technology.*

Lippmann, R.P. 1989. Pattern classification using neural networks, *IEEE Communication Magazine,* November, pp. 47-64.

Okafor, C., and Adetona, O. 1995. Predicting quality characteristic of end-milled parts based on multisensor integration using neural networks: individual effects of learning parameters and rules, *Journal of Intelligent Manufacturing,* vol. 6, pp. 389-400.

Park H.J., and Cho, H.S. 1990. A CMAC-based learning controller for pressure tracking control of hydroforming processes, *ASME Winter Annual Meeting,* Dollars, Texas.

Quero, J.M, Millan, R.L., and Franquelo, L.G. 1994. Neural network approach to weld quality monitoring, *International Conference on Industrial Electronics, Control and Instruments*, vol. 2, pp. 1287-1291.

Sokolowski, A., and Kosmol, J. 1996. Intelligent monitoring system designer, *Japan/USA Symposium on Flexible Automation*, vol. 2, pp. 1461-1468.

Stokes, D., and May, G. 1997. Real-time control of reactive ion etching using neural networks, *Proc. American Control Conference*, pp. 1575-1578.

Woo, H.G., and Cho, H.S. 1988. Estimation of hardening layer sizes in laser surface hardening processes with variations of coating thickness, *Surface and Coatings Technology*, vol. 102, pp. 205-217.

Zurada, J.M. 1992, *Introduction to Artificial Neural Systems*, West Publishing, St. Paul, MN.

13 Computational Intelligence in Microelectronics Manufacturing

Gary S. May
Georgia Institute of Technology

13.1 Introduction

New knowledge and tools are constantly expanding the range of applications for semiconductor devices, integrated circuits, and electronic packages. The solid-state computing, telecommunications, aerospace, automotive and consumer electronics industries all rely heavily on the quality of these methods and processes. In each of these industries, dramatic changes are underway. In addition to increased performance, next-generation computing is increasingly being performed by portable, hand-held computers. A similar trend exists in telecommunications, where the user will soon be employing high-performance, multifunctional, portable units. In the consumer industry, multimedia products capable of voice, image, video, text, and other functions are also expected to be commonplace within the next decade.

The common thread in each of these trends is low-cost electronics. This multi-billion-dollar electronics industry is fundamentally dependent on the manufacture of semiconductor integrated circuits (ICs). However, the fabrication of ICs is extremely expensive. In fact, the last couple of decades have seen semiconductor manufacturing become so capital-intensive that only a few very large companies can participate. A typical state-of-the-art, high-volume manufacturing facility today costs over a billion dollars [Dax, 1996]. As shown in Figure 13.1, this represents a factor of over 1000 increase over the cost of a comparable facility 20 years ago. If this trend continues at its present rate, facility costs will exceed the total annual revenue of any of the four leading U.S. semiconductor companies at the turn of the century [May, 1994].

Because of rising costs, the challenge before semiconductor manufacturers is to offset capital investment with a greater amount of automation and technological innovation in the fabrication process. In other words, the objective is to use the latest developments in computer technology to enhance the manufacturing methods that have become so expensive. In effect, this effort in *computer-integrated*

0-8493-0592-6/01/$0.00+$.50

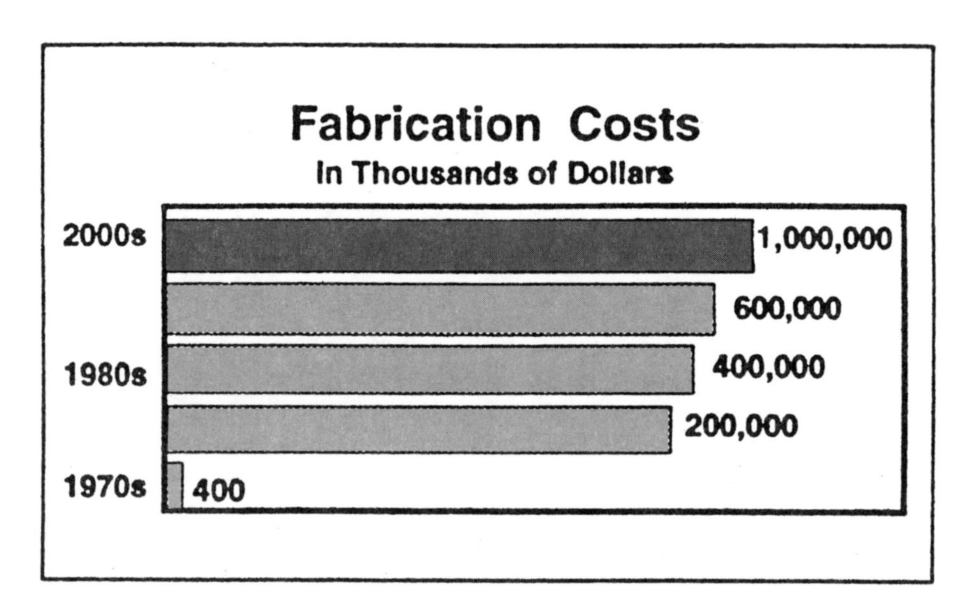

FIGURE 13.1 Graph of rising integrated circuit fabrication costs in thousands of dollars over the last three decades. (*Source:* May, G., 1994. Manufacturing ICs the Neural Way, *IEEE Spectrum*, 31(9):47-51. With permission.)

manufacturing of integrated circuits (IC-CIM) is aimed at optimizing the cost-effectiveness of integrated circuit manufacturing as *computer-aided design* (CAD) has dramatically affected the economics of circuit design.

Under the overall heading of reducing manufacturing cost, several important subtasks have been identified. These include increasing chip fabrication yield, reducing product cycle time, maintaining consistent levels of product quality and performance, and improving the reliability of processing equipment. Unlike the manufacture of discrete parts such as electrical appliances, where relatively little rework is required and a yield greater than 95% on salable product is often realized, the manufacture of integrated circuits faces unique obstacles. Semiconductor fabrication processes consist of hundreds of sequential steps, and yield loss occurs at every step. Therefore, IC manufacturing processes have yields as low as 20 to 80%. The problem of low yield is particularly severe for new fabrication sequences. Effective IC-CIM systems, however, can alleviate such problems. Table 13.1 summarizes the results of a Toshiba 1986 study that analyzed the use of IC-CIM techniques in producing 256K dynamic RAM memory circuits [Hodges et al., 1989]. This study showed that CIM techniques improved the manufacturing process on each of the four productivity metrics investigated.

Because of the large number of steps involved, maintaining product quality in an IC manufacturing facility requires strict control of literally hundreds or even thousands of process variables. The interdependent issues of high yield, high quality, and low cycle time have been addressed in part by the ongoing development of several critical capabilities in state-of-the-art IC-CIM systems: *in situ* process monitoring, process/equipment modeling, real-time closed-loop process control, and equipment malfunction diagnosis. Each of these activities increases throughput and reduces yield loss by preventing potential misprocessing, but each presents significant engineering challenges in effective implementation and deployment.

13.2 The Role of Computational Intelligence

Recently, the use of computational intelligence in various manufacturing applications has dramatically increased, and semiconductor manufacturing is no exception to this trend. Artificial neural networks

TABLE 13.1 Results of 1986 Toshiba Study

Productivity Metric	No CIM	With CIM
Turnaround Time	1.0	0.58
Integrated Unit Output	1.0	1.50
Average Equipment Uptime	1.0	1.32
Direct Labor Hours	1.0	0.75

Source: Hodges, D., Rowe, L., and Spanos, C., 1989. *Computer-Integrated Manufacturing of VLSI, Proc. IEEE/CHMT Int. Elec. Manuf. Tech. Symp.*, 1-3. With permission.

[Dayhoff, 1990], genetic algorithms [Goldberg, 1989], expert systems [Parsaye and Chignell, 1988], and other techniques have emerged as powerful tools for assisting IC-CIM systems in performing various process monitoring, modeling, control, and diagnostic functions. The following is an introduction to various computational intelligence tools in preparation for a more detailed description of the manner in which these tools have been used in IC-CIM systems.

13.2.1 Neural Networks

Because of their inherent learning capability, adaptability, and robustness, artificial neural nets are used to solve problems that have heretofore resisted solutions by other more traditional methods. Although the name "neural network" stems from the fact that these systems crudely mimic the behavior of biological neurons, the neural networks used in microelectronics manufacturing applications actually have little to do with biology. However, they share some of the advantages that biological organisms have over standard computational systems. Neural networks are capable of performing highly complex mappings on noisy and/or nonlinear data, thereby inferring very subtle relationships between diverse sets of input and output parameters. Moreover, these networks can also generalize well enough to learn overall trends in functional relationships from limited training data.

There are several neural network architectures and training algorithms eligible for manufacturing applications. However, the *backpropagation* (BP) algorithm is the most generally applicable and most popular approach for microelectronics manufacturing. Feedforward neural networks trained by BP consist of several layers of simple processing elements called "neurons" (Figure 13.2). These rudimentary processors are interconnected so that information relevant to input–output mappings is stored in the weight of the connections between them. Each neuron contains the weighted sum of its inputs filtered by a sigmoid transfer function. The layers of neurons in BP networks receive, process, and transmit critical information about the relationships between the input parameters and corresponding responses. In addition to the input and output layers, these networks incorporate one or more "hidden" layers of neurons that do not interact with the outside world, but assist in performing nonlinear feature extraction tasks on information provided by the input and output layers.

In the BP learning algorithm, the network begins with a random set of weights. Then an input vector is presented and fed forward through the network, and the output is calculated by using this initial weight matrix. Next, the calculated output is compared to the measured output data, and the squared difference between these two vectors determines the system error. The accumulated error for all of the input–output pairs is defined as the Euclidean distance in the weight space that the network attempts to minimize. Minimization is accomplished via the *gradient descent* approach, in which the network weights are adjusted in the direction of decreasing error. It has been demonstrated that, if a sufficient number of hidden neurons are present, a three-layer BP network can encode any arbitrary input–output relationship [Irie and Miyake, 1988].

The structure of a typical BP network appears in Figure 13.3. Referring to this figure, let $w_{i,j,k}$ = weight between the jth neuron in layer $(k\text{–}1)$ and the ith neuron in layer k; $in_{i,k}$ = input to the ith neuron in the kth layer; and $out_{i,k}$ = output of the ith neuron in the kth layer. The input to a given neuron is given by

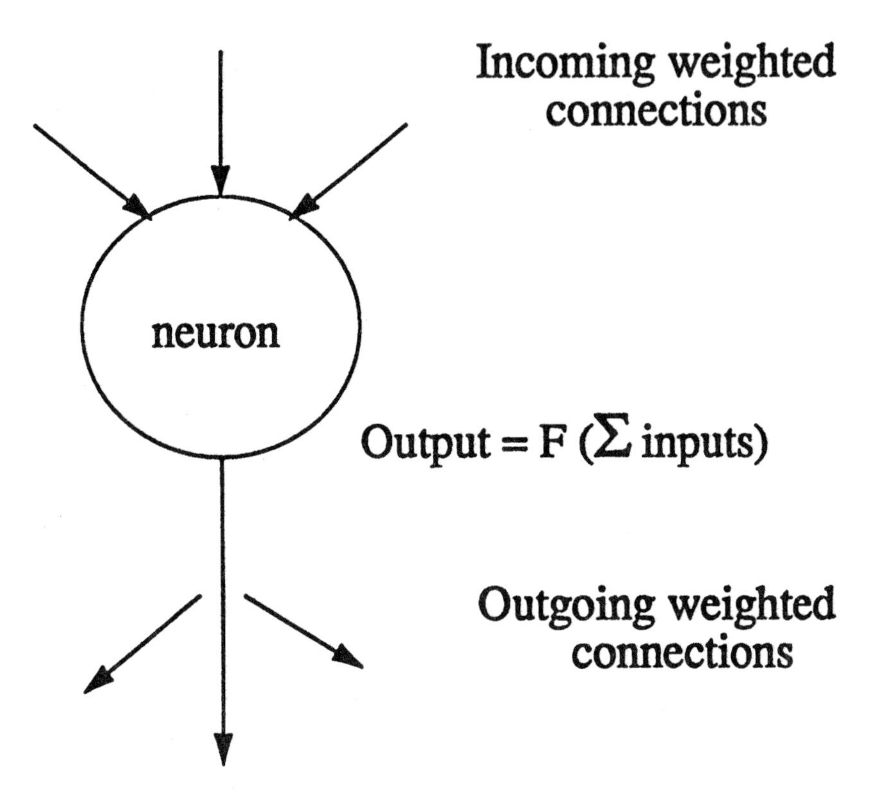

FIGURE 13.2 Schematic of a single neuron. The output of the neuron is a function of the weighted sum of its inputs, where F is a sigmoid function. Feedforward neural networks consist of several layers of interconnected neurons. (*Source:* Himmel, C. and May, G., 1993. Advantages of Plasma Etch Modeling Using Neural Networks over Statistical Techniques, *IEEE Trans. Semi. Manuf.*, 6(2):103-111. With permission.)

$$in_{i,k} = \sum_{j} \left[w_{i,j,k} \cdot out_{j,k-1} \right] \qquad \text{Equation (13.1)}$$

where the summation is taken over all the neurons in the previous layer. The output of a given neuron is a sigmoidal transfer function of the input, expressed as

$$out_{i,k} = \frac{1}{1+e^{-in_{i,k}}} \qquad \text{Equation (13.2)}$$

Error is calculated for each input–output pair as follows: Input neurons are assigned a value and computation occurs by a forward pass through each layer of the network. Then the computed value at the output is compared to its desired value, and the square of the difference between these two vectors provides a measure of the error (E) using

$$E = 0.5 \sum_{j=1}^{q} \left(d_j - out_{j,n} \right)^2 \qquad \text{Equation (13.3)}$$

where n is the number of layers in the network, q is the number of output neurons, d_j is the desired output of the jth neuron in the output layer, and $out_{j,n}$ is the calculated output of that same neuron.

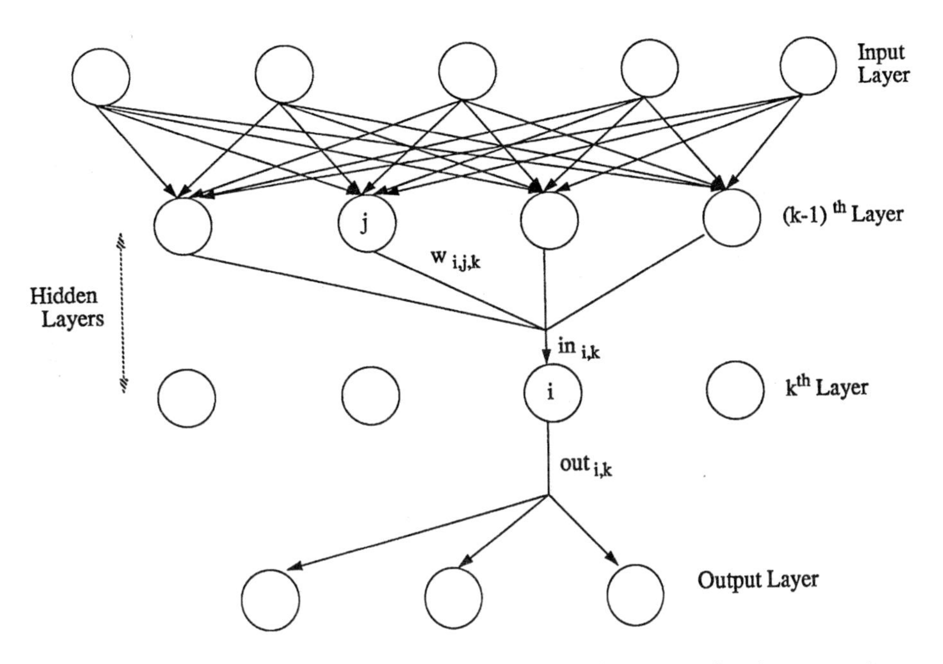

FIGURE 13.3 BP neural network showing input, output, and hidden layers, as well as interconnection strengths (weights), inputs and outputs of neurons in different layers. (*Source:* Himmel, C. and May, G., 1993. Advantages of Plasma Etch Modeling Using Neural Networks Over Statistical Techniques, *IEEE Trans. Semi. Manuf.*, 6(2):103-111. With permission.)

After a forward pass through the network, error is propagated backward from the output layer. Learning occurs by minimizing error through modification of the weights one layer at a time. The weights are modified by calculating the derivative of E and following the gradient that results in a minimum value. From Equations 13.1 and 13.2, the following partial derivatives are computed as

$$\frac{\partial in_{i,k}}{\partial w_{i,j,k}} = out_{j,k-1}$$
$$\frac{\partial out_{i,k}}{\partial in_{i,k}} = out_{j,k-1}\left(1 - out_{i,k}\right) \qquad \text{Equation (13.4)}$$

Now let

$$\frac{\partial E}{\partial in_{i,k}} = -\delta_{i,k}$$
$$\frac{\partial E}{\partial out_{i,k}} = -\phi_{i,k} \qquad \text{Equation (13.5)}$$

Using the chain rule, the gradient of error with respect to weights is given by

$$\frac{\partial E}{\partial w_{i,j,k}} = \left(\frac{\partial E}{\partial in_{i,k}}\right)\left(\frac{\partial in_{i,k}}{\partial w_{i,j,k}}\right) = -\delta_{i,k} \cdot out_{j,k-1} \qquad \text{Equation (13.6)}$$

In the previous expression, the $out_{j,k-1}$ is available from the forward pass. The quantity $\delta_{i,k}$ is calculated by propagating the error backward through the network. Consider that for the output layer

$$-\delta_{i,n} = \frac{\partial E}{\partial in_{i,n}} = \left(\frac{\partial E}{\partial out_{i,n}}\right)\left(\frac{\partial out_{i,n}}{\partial in_{i,n}}\right) \quad \text{Equation (13.7)}$$

where the expressions in Equations 13.3 and 13.4 have been substituted. Likewise, the quantity $\phi_{i,n}$ is given by

$$-\phi_{i,n} = \left(d_i - out_{j,n}\right) \quad \text{Equation (13.8)}$$

Consequently, for the inner layers of the network

$$-\phi_{i,k} = \frac{\partial E}{\partial out_{i,k}} = \sum_j \left(\frac{\partial E}{\partial in_{j,k+1}}\right)\left(\frac{\partial in_{j,k+1}}{\partial out_{i,k}}\right) \quad \text{Equation (13.9)}$$

where the summation is taken over all neurons in the $(k + 1)^{th}$ layer. This expression can be simplified using Equations 13.1 and 13.5 to yield

$$\phi_{i,k} = \sum_j \left[\delta_{j,k+1} \cdot w_{i,j,k+1}\right] \quad \text{Equation (13.10)}$$

Then $\delta_{i,k}$ is determined from Equation 13.7 as

$$\begin{aligned}\delta_{i,k} &= \phi_{i,k}\left(out_{i,k}\right)\left(1 - out_{i,k}\right) \\ &= out_{i,k}\left(1 - out_{i,k}\right)\sum_j \left[\delta_{j,k+1} \cdot w_{i,j,k+1}\right]\end{aligned} \quad \text{Equation (13.11)}$$

Note that $\phi_{i,k}$ depends only on the δ in the $(k + 1)^{th}$ layer. Thus, ϕ for all neurons in a given layer can be computed in parallel. The gradient of the error with respect to the weights is calculated for one pair of input–output patterns at a time. After each computation, a step is taken in the opposite direction of the error gradient. This procedure is iterated until convergence is achieved.

13.2.2 Genetic Algorithms

Neural networks are an extremely useful tool for defining the often complex relationships between controllable process conditions and measurable responses in electronics manufacturing processes. However, in addition to the need to predict the output behavior of a given process given a set of input conditions, one would also like to be able to use such models "in reverse." In other words, given a target response or set of response characteristics, it is often desirable to derive an optimum set of process conditions (or process "recipe") to achieve these targets. Genetic algorithms (GAs) are a method to optimize a given process and define this reverse mapping.

In the 1970s, John Holland introduced GAs as an optimization procedure [Holland, 1975]. Genetic algorithms are guided stochastic search techniques based on the principles of genetics. They use three operations found in natural evolution to guide their trek through the search space: *selection, crossover,* and *mutation.* Using these operations, GAs search through large, irregularly shaped spaces quickly,

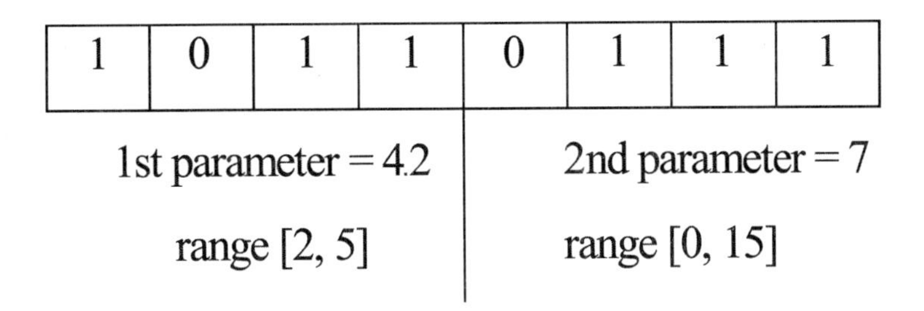

FIGURE 13.4 Example of multiparameter binary coding. Two parameters are coded into binary strings with different ranges and varying precision (π). (*Source:* Han, S. and May, G., 1997. Using Neural Network Process Models to Perform PECVD Silicon Dioxide Recipe Synthesis via Genetic Algorithms, *IEEE Trans. Semi. Manuf.*, 10(2):279-287. With permission.)

requiring only objective function values (detailing the quality of possible solutions) to guide the search. Furthermore, GAs take a more global view of the search space than many methods currently encountered in engineering optimization. Theoretical analyses suggest that GAs quickly locate high-performance regions in extremely large and complex search spaces and possess some natural insensitivity to noise. These qualities make GAs attractive for optimizing neural network based process models.

In computing terms, a genetic algorithm maps a problem onto a set of binary strings. Each string represents a potential solution. Then the GA manipulates the most promising strings in searching for improved solutions. A GA operates typically through a simple cycle of four stages: (i) creation of a population of strings; (ii) evaluation of each string; (iii) selection of "best" strings; and (iv) genetic manipulation to create the new population of strings. During each computational cycle, a new generation of possible solutions for a given problem is produced. At the first stage, an initial population of potential solutions is created as a starting point for the search process. Each element of the population is encoded into a string (the "chromosome"), to be manipulated by the genetic operators. In the next stage, the performance (or *fitness*) of each individual of the population is evaluated. Based on each individual string's fitness, a selection mechanism chooses "mates" for the genetic manipulation process. The selection policy is responsible for assuring survival of the most fit individuals.

A common method of coding multiparameter optimization problems is concatenated, multiparameter, mapped, fixed-point coding. Using this procedure, if an unsigned integer x is the decoded parameter of interest, then x is mapped linearly from $[0, 2^l]$ to a specified interval $[U_{min}, U_{max}]$ (where l is the length of the binary string). In this way, both the range and precision of the decision variables are controlled. To construct a multiparameter coding, as many single parameter strings as required are simply concatenated. Each coding has its own sub-length. Figure 13.4 shows an example of a two-parameter coding with four bits in each parameter. The ranges of the first and second parameter are 2-5 and 0-15, respectively.

The string manipulation process employs genetic operators to produce a new population of individuals ("offspring") by manipulating the genetic "code" possessed by members ("parents") of the current population. It consists of selection, crossover, and mutation operations. Selection is the process by which strings with high fitness values (i.e., good solutions to the optimization problem under consideration) receive larger numbers of copies in the new population. In one popular method of selection called elitist roulette wheel selection, strings with fitness value F_i are assigned a proportionate probability of survival into the next generation. This probability distribution is determined according to

$$P_i = \frac{F_i}{\sum F} \qquad \text{Equation (13.12)}$$

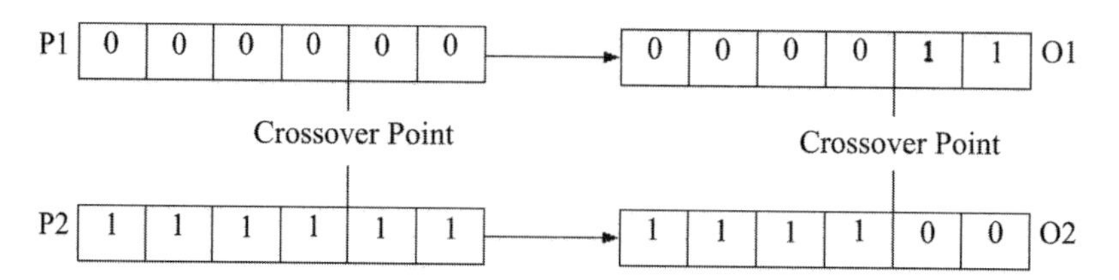

FIGURE 13.5 The crossover operation. Two parent strings exchange binary information at a randomly determined crossover point to produce two offspring. (*Source:* Han, S. and May, G., 1997. Using Neural Network Process Models to Perform PECVD Silicon Dioxide Recipe Synthesis via Genetic Algorithms, *IEEE Trans. Semi. Manuf.*, 10(2):279-287. With permission.)

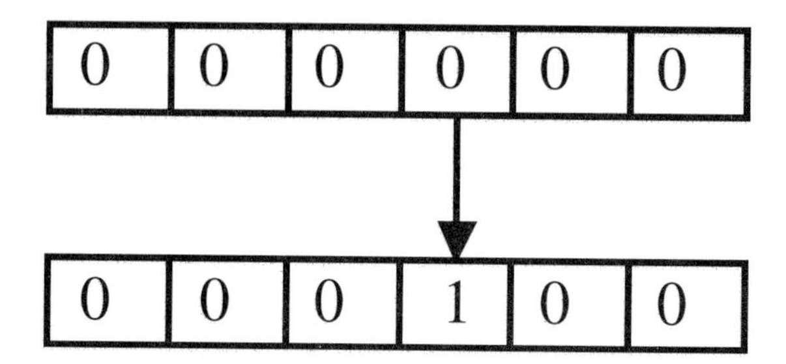

FIGURE 13.6 The mutation operation. A randomly selected bit in a given binary string is changed according to a given probability. (*Source:* Han, S. and May, G., 1997. Using Neural Network Process Models to Perform PECVD Silicon Dioxide Recipe Synthesis via Genetic Algorithms, *IEEE Trans. Semi. Manuf.*, 10(2):279-287. With permission.)

Thus, an individual string whose fitness is n times better than another's will produce n times the number of offspring in the subsequent generation. Once the strings have reproduced, they are stored in a "mating pool" awaiting the actions of the crossover and mutation operators.

The crossover operator takes two chromosomes and interchanges part of their genetic information to produce two new chromosomes (see Figure 13.5). After the crossover point is randomly chosen, portions of the parent strings (P1 and P2) are swapped to produce the new offspring (O1 and O2) based on a specified crossover probability. Mutation is motivated by the possibility that the initially defined population might not contain all of the information necessary to solve the problem. This operation is implemented by randomly changing a fixed number of bits in every generation according to a specified mutation probability (see Figure 13.6). Typical values for the probabilities of crossover and bit mutation range from 0.6 to 0.95 and 0.001 to 0.01, respectively. Higher rates disrupt good string building blocks more often, and for smaller populations, sampling errors tend to wash out the predictions. For this reason, the greater the mutation and crossover rates and the smaller the population size, the less frequently predicted solutions are confirmed.

13.2.3 Expert Systems

Computational intelligence has also been introduced into electronics manufacturing in the areas of automated process and equipment diagnosis. When unreliable equipment performance causes operating conditions to vary beyond an acceptable level, overall product quality is jeopardized. Thus, timely and accurate diagnosis is a key to the success of the manufacturing process. Diagnosis involves determining the assignable causes for the equipment malfunctions and correcting them quickly to prevent the subsequent occurrence of expensive misprocessing.

Neural networks have recently emerged as an effective tool for fault diagnosis. Diagnostic problem solving using neural networks requires the association of input patterns representing quantitative and qualitative process behavior to fault identification. Robustness to noisy sensor data and high-speed parallel computation makes neural networks an attractive alternative for real-time diagnosis. However, the pattern-recognition-based neural network approach suffers from some limitations. First, a complete set of fault signatures is hard to obtain, and representational inadequacy of a limited number of data sets can induce network overtraining, thus increasing the misclassification or "false alarm" rate. Also, approaches such as this, in which diagnostic actions take place following a sequence of several processing steps, are not appropriate, since evidence pertaining to potential equipment malfunctions accumulates at irregular intervals throughout the process sequence. At the end of process sequence, significant mis-processing and yield loss may have already taken place, making this approach economically undesirable.

Hybrid schemes involving neural networks and traditional expert systems have been employed to circumvent these inadequacies. Hybrid techniques offset the weaknesses of each individual method used by itself. Traditional expert systems excel at reasoning from previously viewed data, whereas neural networks extrapolate analyses and perform generalized classification for new scenarios. One approach to defining a hybrid scheme involves combining neural networks with an inference system based on the Dempster–Shafer theory of evidential reasoning [Shafer, 1976]. This technique allows the combination of various pieces of uncertain evidence obtained at irregular intervals, and its implementation results in time-varying, nonmonotonic belief functions that reflect the current status of diagnostic conclusions at any given point in time.

One of the basic concepts in Dempster–Shafer theory is the *frame of discernment* (symbolized by Θ), defined as an exhaustive set of mutually exclusive propositions. For the purposes of diagnosis, the frame of discernment is the union of all possible fault hypotheses. Each piece of collected evidence can be mapped to a fault or group of faults within Θ. The likelihood of a fault proposition A is expressed as a bounded interval $[s(A), p(A)]$ which lies in $\{0,1\}$. The parameter $s(A)$ represents the *support* for A, which measures the weight of evidence in support of A. The other parameter $p(A)$, called the *plausibility* of A, is defined as the degree to which contradictory evidence is lacking. Plausibility measures the maximum amount of belief that can possibly be assigned to A. The quantity $u(A)$ is the uncertainty of A, which is the difference between the evidential plausibility and support. For example, an evidence interval of [0.3, 0.7] for proposition A indicates that the probability of A is between 0.3 and 0.7, with an uncertainty of 0.4.

In terms of diagnosis, proposition A represents a given fault hypothesis. An evidential interval for fault is determined from a basic probability mass distribution (BPMD). The BPM $m\langle A\rangle$ indicates the portion of the total belief in evidence assigned exactly to a particular fault hypothesis set. Any residual belief in the frame of discernment that cannot be attributed to any subset of Θ is assigned directly to Θ itself, which introduces uncertainty into the diagnosis. Using the framework, the support and plausibility of proposition A are given by:

$$s(A)=\sum m\langle A_i\rangle \qquad \text{Equation (13.13)}$$

$$p(A)=1-\sum m\langle B_i\rangle \qquad \text{Equation (13.14)}$$

where $A_i \subseteq A$ and $B_i \subseteq \bar{A}$ and the summation is taken over all propositions in a given BPM. Thus the total belief in A is the sum of support ascribed to A and all subsets thereof.

Dempster's rules for evidence combination provide a deterministic and unambiguous method of combining BPMDs from separate and distinct sources of evidence contributing varying degrees of belief to several propositions under a common frame of discernment. The rule for combining the observed BPMs of two arbitrary and independent knowledge sources m_1 and m_2 into a third m_3 is

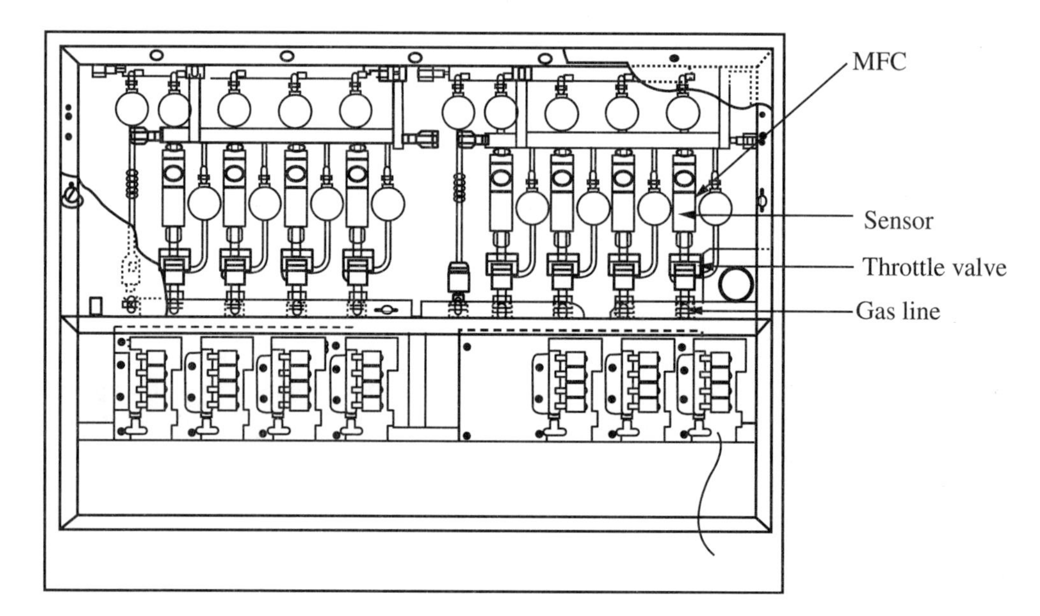

FIGURE 13.7 Partial schematic of RIE gas delivery system. (*Source:* Kim, B. and May, G., 1997. Real-Time Diagnosis of Semiconductor Manufacturing Equipment Using Neural Networks, *IEEE Trans. Comp. Pack. Manuf. Tech. C*, 20(1):39-47. With permission.)

$$m_3\langle Z\rangle = \frac{\sum m_1\langle X_i\rangle \bullet m_2\langle Y_j\rangle}{1-k} \qquad \text{Equation (13.15)}$$

where $Z = X_i \cap Y_j$ and

$$k = \sum m_1\langle X_i\rangle \bullet m_2\langle Y_j\rangle \qquad \text{Equation (13.16)}$$

where $X_i \cap Y_j = \varnothing$. Here X_i and Y_j represent various propositions which consist of fault hypotheses and disjunctions thereof. Thus, the BPM of the intersection of X_i and Y_j is the product of the individual BPMs of X_i and Y_j. The factor $(1 - k)$ is a normalization constant that prevents the total belief from exceeding unity due to attributing portions of belief to the empty set.

To illustrate, consider the combination of m_1 and m_2 when each contain different evidence concerning the diagnosis of a malfunction in a plasma etcher [Manos and Flamm, 1989]. Such evidence could result from two different sensor readings. In particular, suppose that the sensors have observed that the flow of one of the etch gases into the process chamber is too low. Let the frame of discernment $\Theta = \{A, B, C, D\}$, where $A, \ldots, D$ symbolically represent the following mutually exclusive equipment faults:

A = mass flow controller miscalibration
B = gas line leak
C = throttle valve malfunction
D = incorrect sensor signal

These components are illustrated graphically in the etcher gas flow system shown in Figure 13.7.

Suppose that belief in this frame of discernment is distributed according to the BPMDs:

$$m_1\langle A\cup C, B\cup D, \Theta\rangle = \langle 0.4, 0.3, 0.3\rangle$$
$$m_2\langle A\cup B, C, D, \Theta\rangle = \langle 0.5, 0.1, 0.2, 0.2\rangle$$

TABLE 13.2 Illustration of BPMD Combination

m_1				
$A \cup C$ 0.4	A 0.20	C 0.04	ϕ 0.08	$A \cup C$ 0.08
$B \cup D$ 0.3	B 0.15	ϕ 0.03	D 0.06	$B \cup D$ 0.6
θ 0.3	$A \cup B$ 0.15	C 0.03	D 0.06	θ 0.06
	$A \cup B$ 0.50	C 0.10	D 0.20	θ 0.20
		m_2		

Source: Kim, B. and May, G., 1997. Real-Time Diagnosis of Semiconductor Manufacturing Equipment Using Neural Networks, *IEEE Trans. Comp. Pack. Manuf. Tech. C,* 20(1):39-47. With permission.

The calculation of the combined BPMD (m_3) is shown in Table 13.2. Each cell of the table contains the intersection of the corresponding propositions from m_1 and m_2 along with the product of their individual beliefs. Note that the intersection of any proposition with Θ is the original proposition. The BPM attributed to the empty set, k, which originates from the presence of various propositions in m_1 and m_2 whose intersection is empty, is 0.11. By applying Equation 13.16, BPMs for the remaining propositions result in:

$$m_3\left\langle A, A \cup C, A \cup B, B, B \cup D, C, D, \Theta \right\rangle =$$
$$\left\langle 0.225, 0.089, 0.169, 0.169, 0.067, 0.079, 0.135, 0.067 \right\rangle$$

The plausibilities for propositions in the combined BPM are calculated by applying Equation 13.15. The individual evidential intervals implied by m_3 are $A[0.225, 0.550]$, $B[0.169, 0.472]$, $C[0.079, 0.235]$, $D[0.135, 0.269]$. Combining the evidence available from knowledge sources m_1 and m_2 thus leads to the conclusion that the most likely cause of the insufficient gas flow malfunction is a miscalibration of the mass flow controller (proposition A).

13.3 Process Modeling

The ability of neural networks to learn input–output relationships from limited data is beneficial in electronics manufacturing, where a plethora of nonlinear fabrication processes exist, and experimental data are expensive to obtain. Several researchers have reported noteworthy successes in using neural networks to model the behavior of a few key fabrication processes. In so doing, the basic strategy is usually to perform a series of statistically designed characterization experiments, and then to train BP neural nets to model the experimental data. The process characterization experiments typically consist of a factorial exploration of the input parameter space, which may be subsequently augmented by a more advanced experimental design. Each set of input conditions in the design corresponds to a particular set of measured process responses. This input–output mapping is what the neural network learns.

13.3.1 Modeling Using Backpropagation Neural Networks

As an example of the neural-network-based process modeling procedure, Himmel and May [1993] used BP neural nets to model plasma etching. Plasma etching removes patterned layers of material using reactive gases in an AC discharge (Figure 13.8). Because this process is popular, considerable effort has been expended developing reliable models that relate the response of process outputs (such as etch rate or etch uniformity) to variations in input parameters (such as pressure, radio-frequency power, or gas composition). These models are required to predict etch behavior under an exhaustive set of operating conditions with a very high degree of precision. However, plasma processing involves complex and dynamic interactions between reactive particles in an electric field. As a result of this inherent complexity, approaches to plasma etch modeling that preceded the advent of neural networks met with limited success.

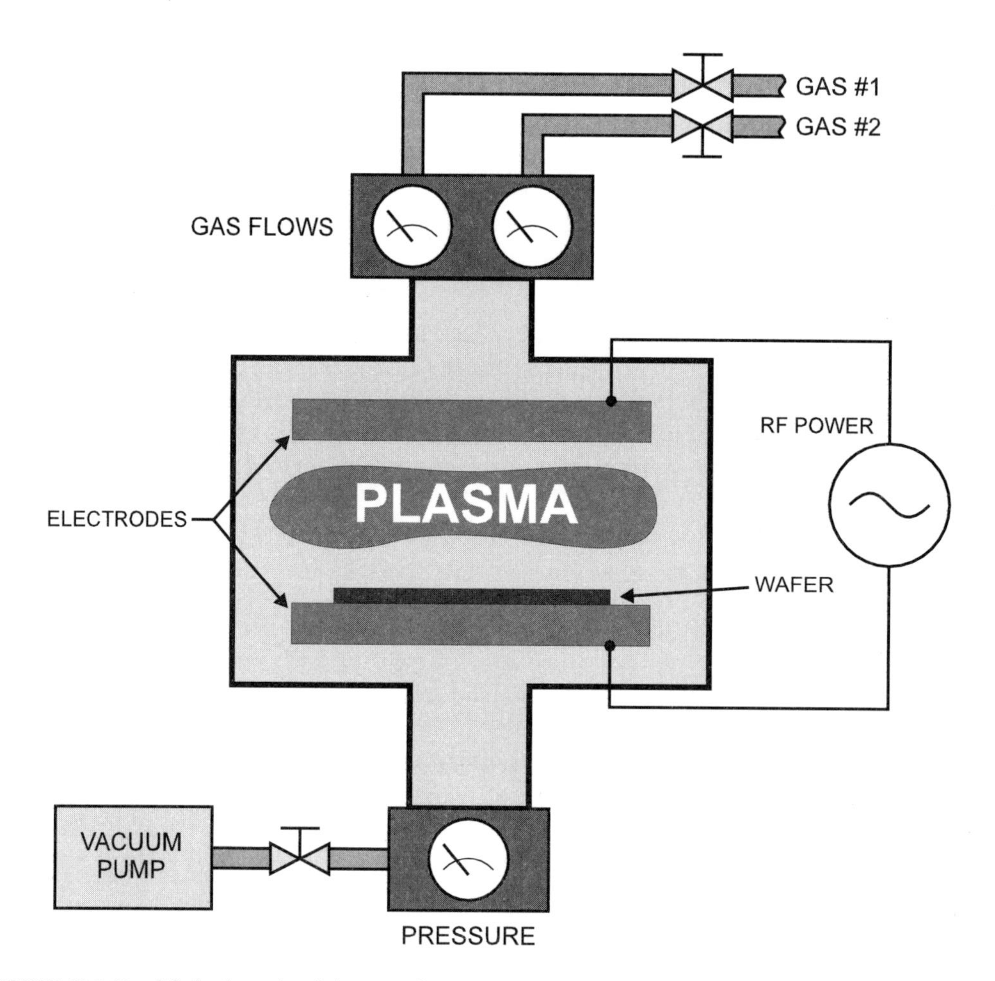

FIGURE 13.8 Simplified schematic of plasma etching system.

Plasma process modeling efforts have previously focused on statistical *response surface methods* (RSM) [Box and Draper, 1987]. RSM models can predict etch behavior under a wide range of operating conditions, but they are most efficient when the number of process variables is small (i.e., six or fewer). The large number of experiments required to adequately characterize the many significant variables in processes like plasma etching is costly and usually prohibitive, forcing experimenters to manipulate a reduced set of variables. Because plasma etching is a highly nonlinear process, this simplification reduces the accuracy of the RSM models.

Himmel and May compared RSM to BP neural networks for modeling the etching of polysilicon films in a carbon tetrachloride (CCl_4) plasma. To do so, they characterized the process by varying RF power, chamber pressure, electrode spacing, and gas composition in a partial factorial design, and trained the neural nets to model the effect of each combination of these inputs on etch rate, uniformity, and selectivity. Afterward, they found that the neural network models exhibited 40 to 70% better accuracy (as measured by *root-mean-square* error) than RSM models and required fewer training experiments. Furthermore, the results of this study also indicated that the generalizing capabilities of neural network models were superior to their conventional statistical counterparts. This fact was verified by using both the RSM and "neural" process models to predict previously unobserved experimental data (or test data). Neural networks showed the ability to generalize with an RMS error 40% lower than the statistical models even when built with less training data.

Investigators at DuPont, Bell Laboratories, the University of Texas at Austin, Michigan State University, and Texas Instruments have likewise reported positive results using neural nets for modeling plasma etching. Mocella et al. [1991] also modeled polysilicon etching, and found that BP neural nets consistently produced models exhibiting better fit than second- and third-order polynomial RSM models. Rietman and Lory [1993] modeled tantalum silicide/polysilicon etching of the gate of metal-oxide-semiconductor (MOS) transistors. They successfully used data from an actual production machine to train neural nets to predict the amount of silicon dioxide remaining in the source and drain regions of the devices after etching. Subsequently, they used their neural etch models to analyze the sensitivity of this etch response to several input parameters, which provided much useful information for process designers.

Huang et al. [1994] used neural networks to model the etching of silicon dioxide in a carbon tetrafluoride (CF_4)/oxygen plasma. This group found that neural nets consistently outperform RSM models, and they also showed that developing satisfactory models is possible from even fewer experimental data than coefficients in the neural network. Salam et al. [1997] modeled plasma etching in an electron cyclotron resonance (ECR) plasma. This group focused on novel variations of the BP learning algorithm that employed error functions different from the quadratic function described by Equation 13.3. They were able to successfully model ECR plasma responses using neural nets trained with a polynomial error function derived from the statistical properties of the error signal itself.

Other manufacturing processes have also benefited from the neural network approach. Specifically, chemical vapor deposition (CVD) processes, which are also nonlinear, have been modeled effectively. Nadi et al. [1991] combined BP neural nets and influence diagrams for both the modeling and recipe synthesis of low pressure CVD (LPCVD) of polysilicon. Bose and Lord [1993] demonstrated that neural networks provide appreciably better generalization than regression based models of silicon CVD. Similarly, Han et al. [1994] developed neural process models for the plasma-enhanced CVD (PECVD) of silicon dioxide films used as interlayer dielectric material in multichip modules.

13.3.2 Modifications to Standard Backpropagation in Process Modeling

In each of the previous examples, standard implementations of the BP algorithm have been employed to perform process modeling tasks. However, innovative modifications of standard BP have also been developed for certain other applications. In one case, BP has been combined with *simulated annealing* to enhance model accuracy. In addition, a second adjustment has been developed that incorporates knowledge of process chemistry and physics into a semi-empirical or hybrid model, with advantages over the purely empirical "black-box" approach previously described. These two variations of BP are described below.

13.3.2.1 Neural Networks and Simulated Annealing in Plasma Etch Modeling

Kim and May [1996] used neural networks to model etch rate, etch anisotropy, etch uniformity, and etch selectivity in a low-pressure form of plasma etching called *reactive ion etching* (RIE). The RIE process consisted of the removal of silicon dioxide films by a trifluoromethane (CHF_3) and oxygen plasma in a Plasma Therm 700 series dual chamber RIE system operating at 13.56 MHz. The process was initially characterized via a 2^4 factorial experiment with three center-point replications augmented by a central composite design. The factors varied included pressure, RF power, and the two gas flow rates.

Data from this experiment were used to train modified BP neural networks, which resulted in improved prediction accuracy. The new technique modified the rule used to update network weights. The new rule combined a memory-based weight update scheme with the simulated annealing procedure used in combinatorial optimization. Neural network training rules adjust synapse strengths to satisfy the constraints given to the network. In the standard BP algorithm, the weight update mechanism at the $(n + 1)^{th}$ iteration is given by

$$w_{ijk}(n + 1) = w_{ijk}(n) + \eta \Delta w_{ijk}(n) \qquad \text{Equation (13.17)}$$

where w_{ijk} is the connection strength between the jth *neuron* in layer $(k - 1)$ and the ith neuron in layer k, Δw_{ijk} is the calculated change in that weight that reduces the error function of the network, and η is the learning rate. Equation 13.17 is called the *generalized delta rule*. Kim and May's new *K-step prediction rule*, modified the generalized delta rule by using portions of previously stored weights in predicting the next set of weights. The new update scheme is expressed as

$$w_{ijk}(n+1) = w_{ijk}(n) + \eta \Delta w_{ijk}(n) + \gamma_K w_{ijk}(n-K) \qquad \text{Equation (13.18)}$$

The last term in this expression provides the network with long-term memory. The integer K determines the number of sets of previous weights stored and the γ_K factor allows the system to place varying degrees of emphasis on weight sets from different training epochs. Typically, larger values of γ_K are assigned to more recent weight sets.

This memory-based weight update scheme was combined with a variation of simulated annealing. In thermodynamics, annealing is the slow cooling procedure that enables nature to find the minimum energy state. In neural network training, this is analogous to using the following function in place of the usual sigmoidal transfer function:

$$\frac{1}{1+\exp\left[-\left(\frac{net_{ik}+\beta_{ik}}{\lambda T_0}\right)\right]} \qquad \text{Equation (13.19)}$$

where net_{ik} is the weighted sum of neural inputs and β_{ik} is the neural threshold. Network "temperature" gradually decreases from an initial value T_0 according to a decay factor λ(where $\lambda < 1$), effectively resulting in a time-varying gain for the network transfer function (Figure 13.9). Annealing the network at high temperature early leads to rapid location of the general vicinity of the global minimum of the error surface. The training algorithm remains within the attractive basin of the global minimum as the temperature decreases, preventing any significant uphill excursion. When used in conjunction with the K-step weight prediction scheme outlined previously, this approach is termed *annealed K-step prediction*.

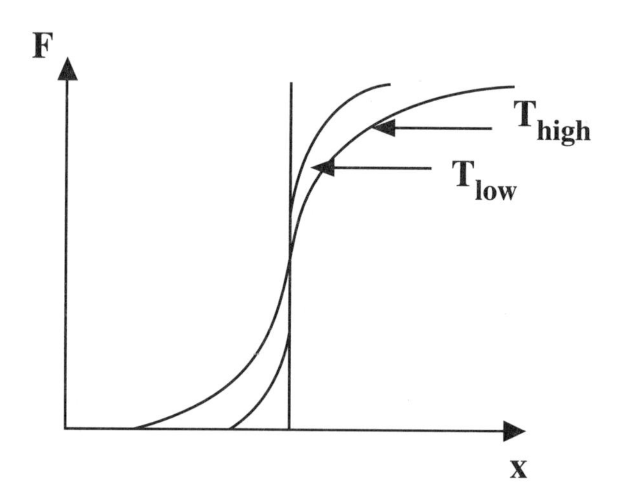

FIGURE 13.9 Plot of simulated annealing-based transfer function as temperature is decreased. (*Source:* Kim, B. and May, G., 1996. Reactive Ion Etch Modeling Using Neural Networks and Simulated Annealing, *IEEE Trans. Comp. Pack. Manuf. Tech. C*, 19(1): 3-8. With permission.)

BP neural networks were trained using this procedure with data from the 2^4 factorial array plus the three center-point replications. The remaining axial trials from the central composite characterization experiment were used as test data for the models. The annealed K-step training rule and the generalized delta rule were also compared. The RMS prediction errors are shown in Table 13.3, in which "% Improvement" refers to the improvement obtained with the annealed *K*-step training rule. Best results were achieved for $K = 2$, $\gamma_1 = 0.9$, $\gamma_2 = 0.08$, $T_0 = 100$, and $\lambda = 0.99$. It is clear that annealed *K*-step prediction improves network predictive ability.

13.3.2.2 Semi-Empirical Process Modeling

Though neural process models offer advantages in accuracy and robustness over statistical models, they offer little insight into the physical understanding of processes being modeled. This can be alleviated by neural process models that incorporate partial knowledge of the first-principles relationships inherent in the process. Two different approaches to accomplishing this include so-called *hybrid* neural networks and model transfer techniques.

13.3.2.2.1 The Hybrid Neural Network Approach

Nami et al. [1997] developed a semi-empirical model of the metal-organic CVD (MOCVD) process based on hybrid neural networks. Their model was constructed by characterizing the MOCVD of titanium dioxide (TiO_2) films by measuring the deposition rate over a range of deposition conditions. This was accomplished by varying susceptor and source temperature, flow rate of the argon carrier gas for the precursor (titanium tetra-iso-propoxide, or TTIP), and chamber pressure. Following characterization, a modified BP (hybrid) neural network was trained to determine the value of three adjustable fitting parameters in an analytical expression for the TiO_2 deposition rate.

The first step in this hybrid modeling technique involves developing an analytical model. For TiO_2 deposition via MOCVD, this was accomplished by applying the continuity equation to reactant concentration as the reactant of interest is transported from the bulk gas and incorporated into the growing film. Under these conditions and several key assumptions, the average deposition rate R for TiO_2 is given by

$$R = \frac{1200}{T_{inlet}} \frac{K_D}{1 + \left(K_D \frac{\delta}{D} \right)} P \frac{P_e}{P_0 - P_e} \frac{v}{Q} \qquad \text{Equation (13.20)}$$

where R is expressed in micrometers per hour, T_{inlet} is the inlet gas temperature in degrees Kelvin, P is the chamber pressure (mtorr), P_e is the equilibrium vapor pressure of the precursor (mtorr), P_0 is the total bubbler pressure (mtorr), v is the carrier gas flow rate (in standard cm^3/min), Q is the total flow rate (in standard cm^3/min), D is the diffusion coefficient of the reactant gas, δ is the boundary layer thickness, and K_D is the mass transfer coefficient given by $K_D = Ae^{-\Delta E/kT}$, where A is a pre-exponential factor related to the molecular "attempt rate" of the growth process, ΔE is the activation energy (cal/mol), k is Boltzmann's constant, and T is the susceptor temperature in degrees Kelvin. To predict R, the three unknown parameters that must be estimated are D, A, and ΔE. Estimating these parameters with hybrid neural networks is explained as follows.

In standard BP learning, gradient descent minimizes the network error E by adjusting the weights by an amount proportional to the derivative of the error with respect to previous weights. The weight update expression is the generalized delta rule given by Equation 13.17, where

$$\Delta w_{ijk}(n) = \frac{\partial E}{\partial w_{ijk}} \qquad \text{Equation (13.21)}$$

TABLE 13.3 Network Prediction Errors

Etch Response	Error (K-Step)	% Improvement
Etch Rate	8.0 Å/min	57.3
Uniformity	0.3 [%]	53.8
Anisotropy	3.9 [%]	51.1
Selectivity	0.12	59.8

Source: Kim B. and May, G., 1996. Reactive Ion Etch Modeling Using Neural Networks and Simulated Annealing, *IEEE Trans. Comp. Pack. Manuf. Tech. C*, 19(1):3-8. With permission.

The gradient of the error with respect to the weights is calculated for one pair of input–output patterns at a time. After each computation, a step is taken in the direction opposite to the error gradient, and the procedure is iterated until convergence is achieved.

In the hybrid approach, the network structure corresponding to the deposition of TiO_2 by MOCVD has inputs of temperature, total flow rate, chamber pressure, source pressure, precursor flow rate, and the actual (measured) deposition rate R_a. The outputs are D, A, and ΔE. These are fed into Equation 13.5, the predicted deposition rate, R_p is computed, and the result is compared with the actual (measured) deposition rate (see Figure 13.10). In this case, the error signal is defined as $E = 0.5(R_p - R_a)^2$. Because the expression for predicted deposition rate is differentiable, the new error gradient is computed by the chain rule as

$$\frac{\partial E}{\partial w_{ijk}} = \left(\frac{\partial E}{\partial R_p}\right)\left(\frac{\partial R_p}{\partial out_{ik}}\right)\left(\frac{\partial out_{ik}}{\partial w_{ijk}}\right) \qquad \text{Equation (13.22)}$$

where out_{ik} is the calculated output of the jth neuron in the kth layer. The first partial derivative in Equation 13.22 is $(R_p - R_a)$, and the third is the same as that of standard BP. The second partial derivative is computed individually for each unknown parameter to be estimated. Referring to Equation 13.20, the partial derivative of R_p with respect to activation energy is

$$\frac{\partial R_p}{\partial \Delta E} = \frac{1200}{T_{inlet}}\left(\frac{-1}{kT}\right)\frac{K_D}{\left[1+\left(K_D\frac{\delta}{D}\right)\right]^2}P\frac{P_e}{P_0 - P_e}\frac{v}{Q} \qquad \text{Equation (13.23)}$$

The partial derivatives for the other two parameters are computed similarly, and after error minimization, values of the three parameters for the TiO_2 MOCVD process are known explicitly.

Because hybrid neural networks rely on network training to predict only portions of a physical model, they require less training data. The hybrid network developed by Nami et al. [1997] was trained using only 11 training experiments. A three-layer neural network with six inputs, eight hidden neurons, and three outputs was the best network architecture for this case. After error minimization, the values of the diffusion coefficient, pre-exponential constant, and activation energy were 2.5×10^{-6} m²/s, 1.04 m/s, and 5622 cal/mol, respectively. Once trained, the hybrid neural network was subsequently used to predict the deposition rate for five additional MOCVD runs, which constituted a test data set not part of the original experiment. The RMS error of the deposition rate model predictions using the estimated parameters for the five test vectors was only 0.086 μm/h. The hybrid neural network approach, therefore, represents a general-purpose methodology for deriving semi-empirical neural process models that take into account underlying process physics.

13.3.2.2.2 The Model Transfer Approach

Model transfer techniques attempt to modify physically based neural network process models to reflect specific pieces of processing equipment. Marwah and Mahajan [1999] proposed model transfer

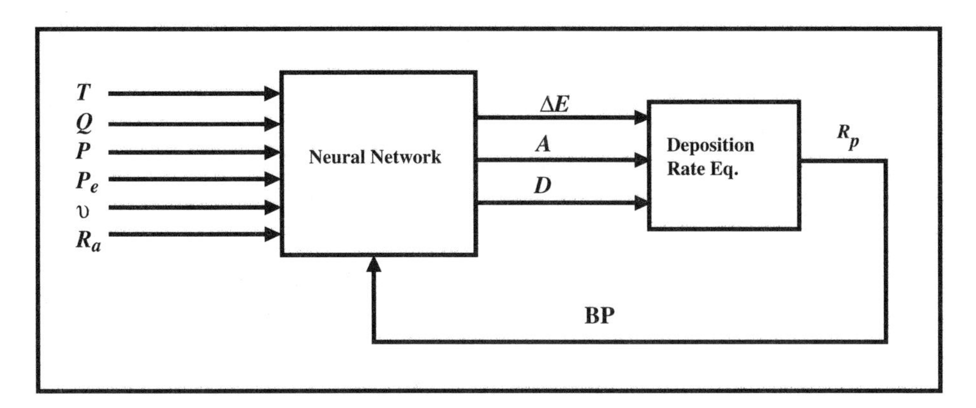

FIGURE 13.10 Illustration of the hybrid neural network process modeling architecture. A BP neural network is trained to model three adjustable parameters (D, A, and ΔE) from an analytical expression for predicted deposition rate (R_p). (*Source:* Nami, Z., Misman, O., Erbil, A., and May, G., 1997. Semi-Empirical Neural Network Modeling of Metal-Organic Chemical Vapor Deposition, *IEEE Trans. Semi. Manuf.*, 10(2):288-294. With permission.)

approaches for modeling a horizontal CVD reactor used in the epitaxial growth of silicon. The goal was to develop an equipment model that incorporated process physics, but was economical to build. The techniques investigated included (i) the *difference method,* in which a neural network was trained on the difference between the existing physical model (or "source" model) and equipment data; (ii) the *source weights method,* in which the final weights of the source model were used as initial weights of the modified model; and (iii) the *source input method,* in which the source model output was used as an additional input to the modified network.

The starting point for model transfer was the development of a physical neural network (PNM) model trained on 98 data points generated from a process simulator utilizing first principles. Training data was obtained by running the simulator for various combinations of input parameters (i.e., inlet silane concentration, inlet velocity, susceptor temperature, and downstream position) using a statistically designed experiment. The numerical data were then split into 73 training vectors and 25 test vectors, and the physical neural network source model was trained using BP to predict silicon growth rate and tested against the validation points for the desired accuracy. The average relative training and testing error obtained were 1.55% and 1.65%, respectively. The source model was then modified by training a new neural network with 25 extra experimentally derived data points obtained from central composite experiment.

In the difference method, the modified neural network model was trained on the difference between the source and equipment data (see Figure 13.11(a)). The inherent expectation was that if this difference was a simpler function of the inputs as compared to the pure equipment data, then fewer equipment data points would be required to build an accurate model. In the source weights method, the source model was retrained using the equipment data as test data. The final weights of the source model were then used as the initial weights of the modified model. The rationale for this approach was that training the source network with the experimental data as test data captures the common features of the source and final modified models. For the source input method, the source model is used as an additional input to the modified network (Figure 13.11(b)). Since the source model should be close to the final modified model, the source output should be some internal representation of the input data, which should be useful to the modified network. The expectation once again was that the additional input makes the learning task simpler for the modified network, thereby reducing the number of experimental data points required.

These investigators found that the source input method yielded the most accurate results (an average relative error of only 2.58%, as compared to 14.62% for the difference method and 14.59% for the source weights method), and the amount of training data required to develop the model modified using this

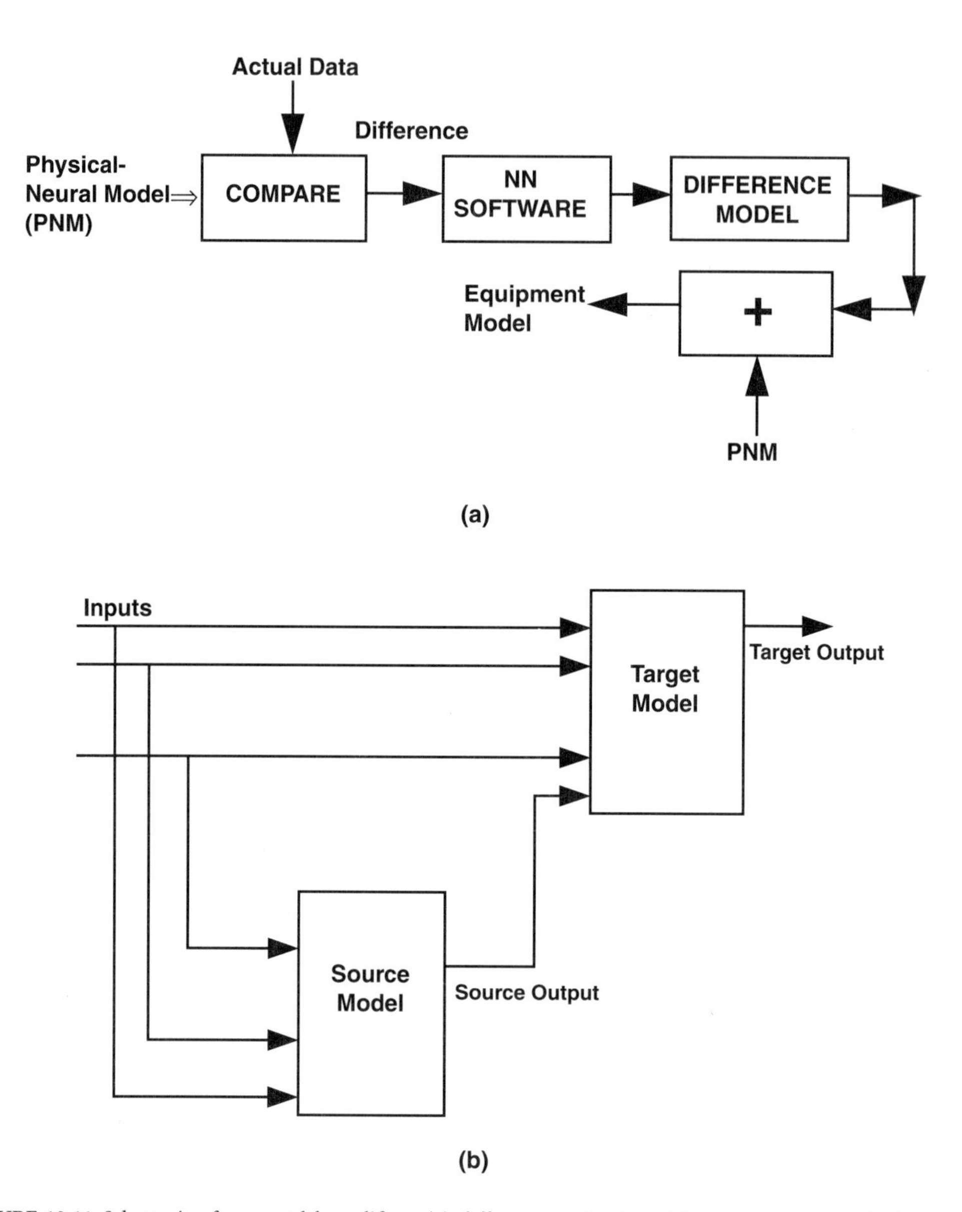

FIGURE 13.11 Schematic of two model modifiers: (a) difference method; and (b) source input method. (*Source:* Marwah, M. and Mahajan, R., 1999. Building Equipment Models Using Neural Network Models and Model Transfer Techniques, *IEEE Trans. Semi. Manuf.*, 12(3):377-380. With permission.)

technique was approximately 25% of that required to develop a complete equipment model from scratch. Furthermore, the source model can be reused for developing additional models of other similar equipment.

13.3.2.3 Process Modeling Using Modular Neural Networks

Natale et al. [1999] applied *modular neural networks* to develop a model of atmospheric pressure CVD (APCVD) of doped silicon dioxide films, a critical step in dynamic random access memory (DRAM) chip fabrication at the Texas Instruments fabrication facility in Avezzano, Italy. Modular neural networks consist of a group of subnetworks, or *modules*, competing to learn different aspects of a problem. As shown in Figure 12(a), "gating" network is applied to control the competition by assigning different regions of the input data space to different local modules. The gating network has as many outputs as

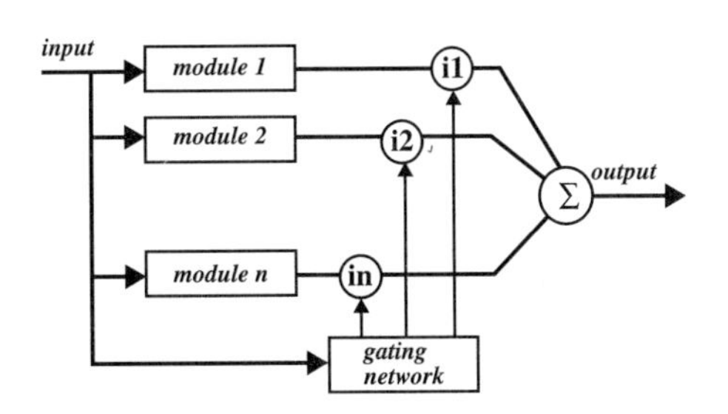

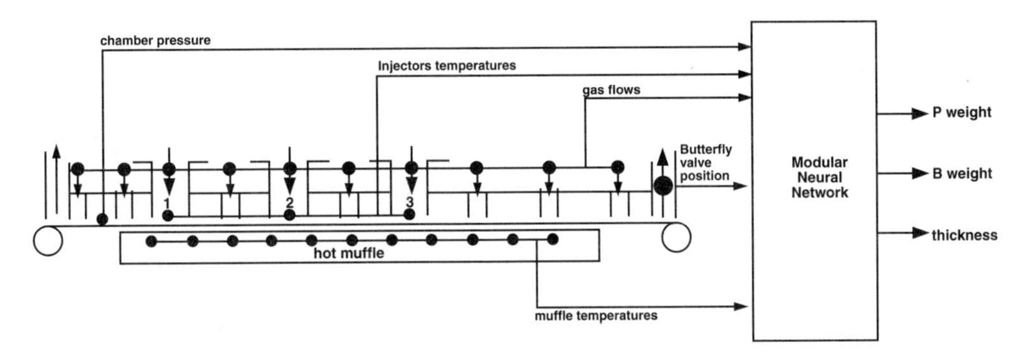

FIGURE 13.12 (a) Block diagram of a modular neural network. (b) Schematic of the location of the sensors inside the APCVD equipment. (*Source:* Natale, C. et al., 1999. Modeling of APCV-Doped Silicon Dioxide Deposition Process by a Modular Neural Network, *IEEE Trans. Semi. Manuf.*, 12(1):109-115. With permission.)

the number of modules. Both the modules and the gating network are trained by BP. The modular approach allows multiple networks to cooperate in solving the problem, as each module specializes in learning different regions of the input space. The outputs of each module are weighted by the gating network, thereby selecting a "winner" module whose output is closest to the target.

The deposition of both phosphosilicate glass (PSG) and boron-doped PSG (BPSG) were modeled using this approach. The inputs included 9 gas flows (three injectors each for silane, diborane, and phosphine gas), 3 injector temperatures, 6 nitrogen curtain flow rates, 12 thermocouple temperature readings, the chamber pressure, and a butterfly purge valve position reading. The outputs were the weight percentage of the boron and phosphorus dopants in the grown film, as well as the film thickness. An overall input/output schematic is shown in Figure 13.12(b). Since the data set was not homogeneous, but instead was formed by two classes representing both PSG and BPSG deposition, the modular approach was appropriate for this case. The final modular network developed in this investigation exhibited an excellent average relative error of approximately 1% in predicting the concentration of dopants and the thickness of the oxide.

13.4 Optimization

In electronics manufacturing, neural network-based optimization has been undertaken from two different viewpoints. The first uses statistical methods to optimize the neural process models themselves, with the goal of determining the network structure and set of learning parameters to minimize network training error, prediction error, and training time. The second approach focuses on using neural process models to optimize a given semiconductor fabrication process or to determine specific process recipes for a desired response.

13.4.1 Network Optimization

The problem of optimizing network structure and learning parameters has been addressed by Kim and May [1994] for plasma etch modeling and Han and May [1996] in modeling plasma-enhanced CVD. The former study performed a statistically designed experiment in which network structure and learning parameters are varied systematically, and used the results of this experiment to derive the optimal neural process model using the simplex search method. The latter study improved this technique by using genetic algorithms to search for the best combination of learning parameters.

13.4.1.1 Network Optimization Using Statistical Experimental Design and Simplex Search

Although they offer advantages over other methods, neural process models contain adjustable learning parameters whose proper values are unknown before model development. In addition, the structure of the network can be modified by adjusting the number of layers and the number of neurons per layer. As a result, the optimal network structure and values of network parameters for a given modeling application are not always clear. Systematically selecting an optimal set of parameters and network structure is an essential requirement for increasing the benefits of neural process modeling. Among the most critical optimality issues for neural process models are learning capability, predictive (or generalization) capability, and convergence speed.

Neural network architecture is determined by the number of layers and number of neurons per layer. Usually, the number of input-layer and output-layer neurons is determined by the number of process inputs and responses in the modeling application. However, specifying the number of hidden-layer neurons is less obvious. It is generally understood that an excessively large number of hidden neurons significantly increases training time and gives poorer predictions for unfamiliar facts. Aside from network architecture, several other parameters affect the BP algorithm, including learning rate, initial weight range, momentum, and training tolerance.

A number of efforts to obtain the optimal network structure have been described [Kim and May, 1994]. Other efforts have focused on the effect of variations in learning parameters on network performance. The consideration of interactions between parameters, however, has been lacking. Furthermore, much of the existing effort in this area has focused on improving networks designed to perform classification and pattern recognition. The optimization of networks that model continuous nonlinear processes (such as those in semiconductor manufacturing) has not been addressed as thoroughly. Kim and May, however, presented an experiment designed to comprehensively evaluate all relevant learning and structural network parameters. The goal was to design an optimal neural network for a specific semiconductor manufacturing problem, modeling the etch rate of polysilicon in a CCl_4 plasma.

To develop the optimal neural process model, these researchers designed a *D-optimal* experiment [Galil and Kiefer, 1980] to investigate the effect of six factors: the number of hidden layers, the number of neurons per hidden layer, training tolerance, initial weight range, learning rate, and momentum. This experiment determined how the structural and learning factors affect network performance and provided an optimal set of parameters for a given set of performance metrics. The network responses optimized were learning capability, predictive capability, and training time. The experiment consisted of two stages. In the first stage, statistical experimental design was employed to fully characterize the behavior of the etch process [May et al., 1991]. Etch rate data from these trials were used to train neural process models. Once trained, the models were used to predict the etch rate for 12 test wafers. Prediction error for these wafers was also computed, and these two measures of network performance, along with training time, were used as experimental responses to optimize the neural etch rate model as the structural and learning parameters were varied in the second stage (which consisted of the *D-optimal* design).

13.4.1.1.1 Individual Network Parameter Optimization

Independent optimization of each performance characteristic was then performed with the objective of minimizing training error, prediction error, and training time. A constrained multicriteria optimization technique based on the Nelder–Mead simplex search algorithm was implemented to do so. The optimal parameter set was first found for each criterion individually, irrespective of the optimal set for the other two. The results of the independent optimization are summarized in Table 13.4.

Several interesting interactions and trade-offs between the various parameters emerged in this study. One such trade-off can be visualized in two-dimensional contour plots such as those in Figures 13.13 and 13.14. Figure 13.13 plots training error against training tolerance and initial weight range with all other parameters set at their optimal values. Learning capability improves with decreased tolerance and wider weight distribution. Intuitively, the first result can be attributed to the increased precision required by a tight tolerance. Figure 13.14 plots network prediction error vs. the same variables as in Figure 13.13. As expected, optimum prediction is observed at high tolerance and narrow initial weight distribution. The latter result implies that the interaction between neurons within the restricted weight space during training is a primary stimulus for improving prediction. Thus, although learning degrades with a wider weight range, generalization is improved.

TABLE 13.4 Independently Optimized Network Inputs

Parameter	Training Error	Prediction Error	Training Time
Hidden Layer	1	1	1
Neurons/Hidden Layer	6	9	3
Training Tolerance	0.08	0.13	0.09
Initial Weight Range	+/– 2.00	+/– 1.04	+/– 1.00
Learning Rate	2.78	2.80	0.81
Momentum	0.35	0.35	0.95
Optimal Value	239 Å/*min*	162 Å/*min*	37.3 *s*

Source: Kim, B. and May, G., 1994. An Optimal Neural Network Process Model for Plasma Etching, *IEEE Trans. Semi. Manuf.*, 7(1):12-21. With permission.

13.4.1.1.2 Collective Network Parameter Optimization

The parameter sets in Table 13.4 are useful for obtaining optimized performance for a single criterion, but can provide unacceptable results for the others. For example, the parameter set that minimizes training time yields high training and prediction errors. Because it is undesirable to train three different networks corresponding to each performance metric for a given neural process model, it is necessary to optimize all network inputs simultaneously. This is accomplished by implementing a suitable cost function such as

$$Cost = K_1\sigma_t^2 + K_2\sigma_p^2 + K_3T^2 \qquad \text{Equation (13.24)}$$

where σ_t is the network training error, σ_p is the prediction error, and T is training time. The constants K_1, K_2, and K_3 represent the relative importance of each performance measure.

Prediction error is the most important quality characteristic. For modeling applications, a network need not be trained frequently, so training time is not a critical consideration. To optimize this cost function, the values chosen by Kim and May were $K_1 = 10$, $K_2 = 100$, and $K_3 = 1$. Optimization was performed on the overall cost function. The results of this collective optimization appear in Table 13.5. The parameter values in this table yield the minimum cost according to Equation 13.24. This combination resulted in a training error of 412 Å/min, a prediction error of 340 Å/min, and a training time of 292 s. Although this represents only marginal performance, these values may be further tuned by adjusting the cost function constants K_i and the optimization constraints until suitable performance is achieved.

13.4.1.2 Network Optimization Using Genetic Algorithms

Although Kim and May had success with designed experiments and simplex search to optimize BP neural network learning, the effectiveness of the simplex method depends on its initial search point. With an improper starting point, performance degrades, and the algorithm is likely to be trapped in local optima. Theoretical analyses suggest that genetic algorithms quickly locate high-performance regions in extremely

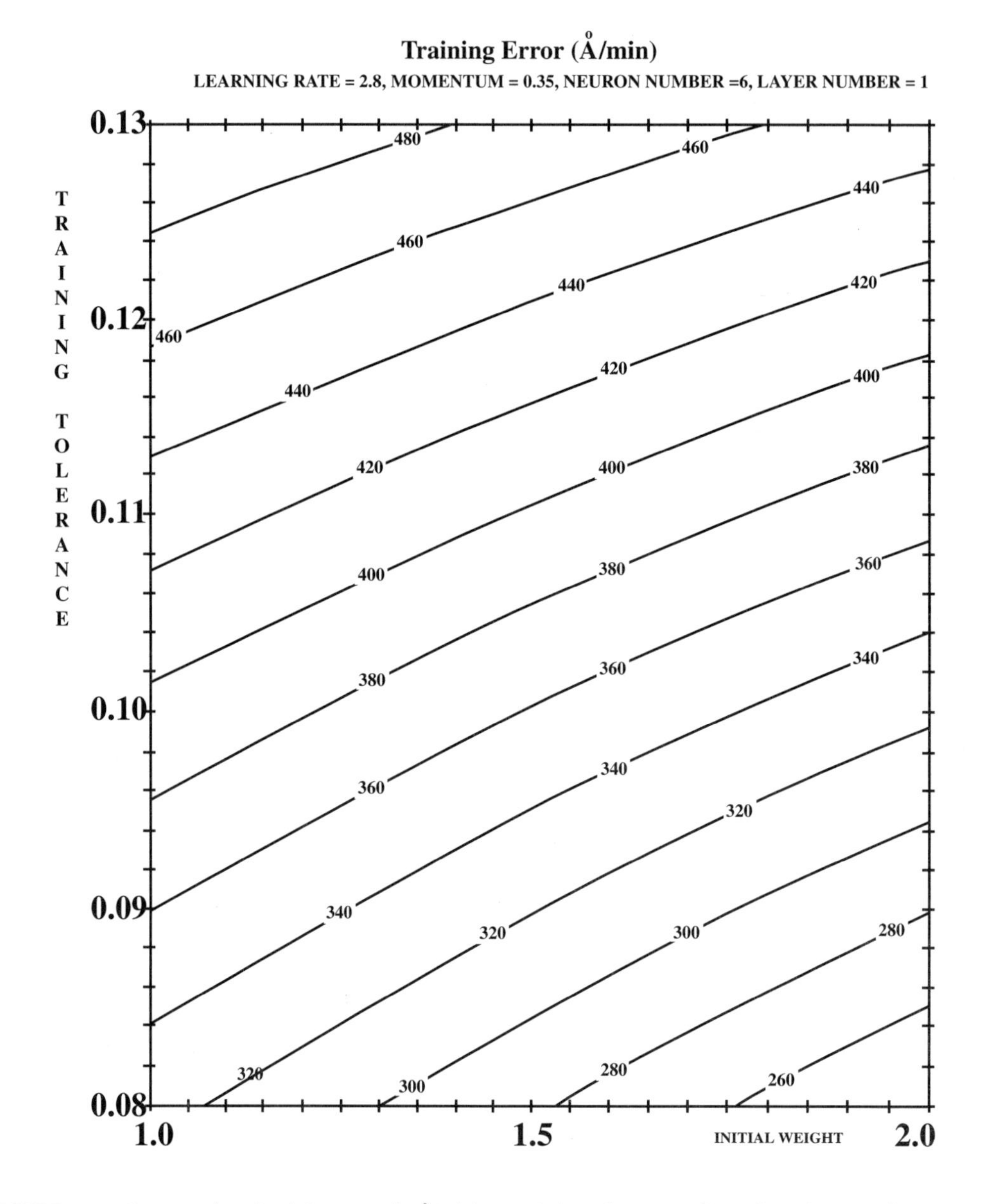

FIGURE 13.13 Contour plot of training error (in Å/min) vs. training tolerance and initial weight range (learning rate = 2.8, momentum = 0.35, number of hidden neurons = 6, number of hidden layers = 1). Learning capability is shown to improve with decreased tolerance and wider weight distribution. (*Source:* Kim, B. and May, G., 1994. An Optimal Neural Network Process Model for Plasma Etching, *IEEE Trans. Semi. Manuf.*, 7(1):12-21. With permission.)

large and complex search spaces and possess some natural insensitivity to noise, which makes GAs potentially attractive for determining optimal neural network structure and learning parameters.

Han and May [1996] applied GAs to obtain the optimal neural network structure and learning parameters for modeling PECVD. The goal was to design an optimal model for the PECVD of silicon dioxide as a function of gas flow rates, temperature, pressure, and RF power. The responses included film permittivity, refractive index, residual stress, uniformity, and impurity concentration. To obtain training data for developing the model, an experiment was performed to investigate the effect of the number of hidden-layer neurons, training tolerance, learning rate, and momentum. The network

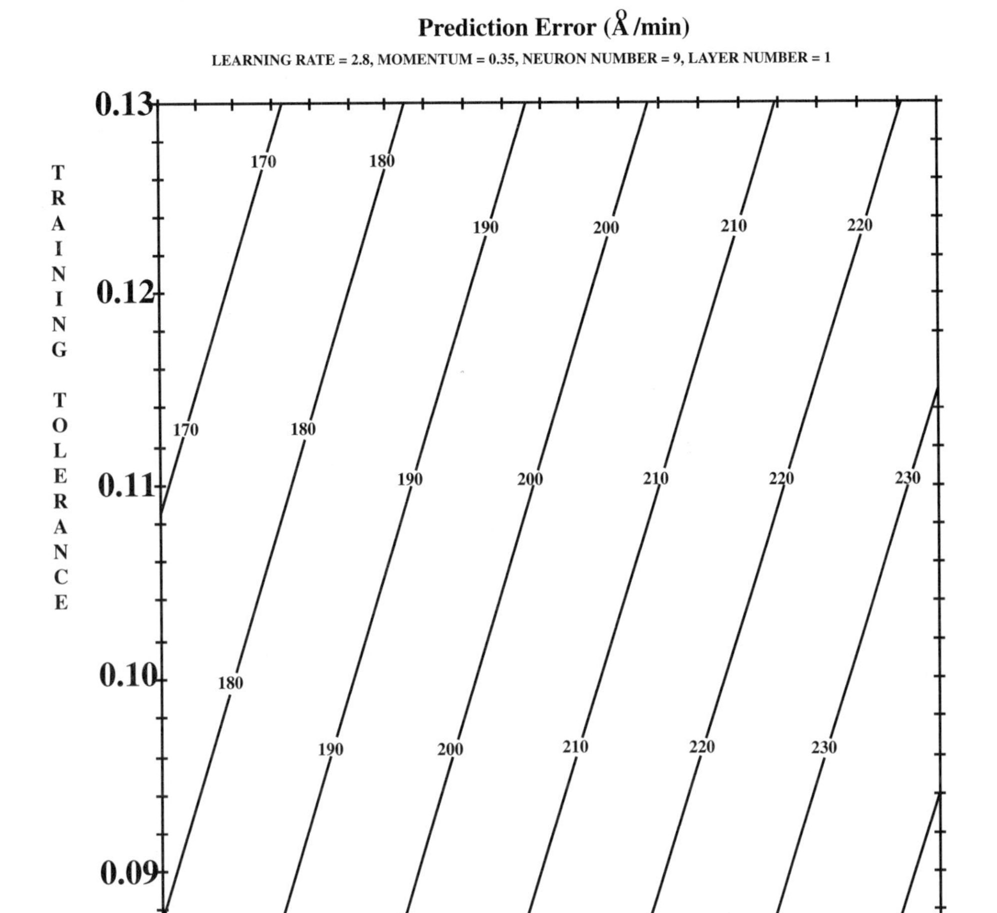

FIGURE 13.14 Contour plot of prediction error (in Å/min) vs. training tolerance and initial weight range (learning rate = 2.8, momentum = 0.35, number of hidden neurons = 6, number of hidden layers = 1). Optimum prediction occurs at high tolerance and narrow initial weight distribution. (*Source:* Kim, B. and May, G., 1994. An Optimal Neural Network Process Model for Plasma Etching, *IEEE Trans. Semi. Manuf.*, 7(1):12-21. With permission.)

responses were learning and predictive capability. Optimal parameter sets that minimized learning and prediction error were determined by genetic search, and this technique was compared with the simplex method.

Figure 13.15 shows the neural network optimization scheme. GAs generated possible candidates for neural parameters using an initial population of 50 potential solutions. Each element of the population was encoded into a 10-bit string to be manipulated by the genetic operators. Because four parameters were to be optimized (the number of hidden layer neurons, momentum, learning rate, and training tolerance), the concatenated total string length was 40 bits. The probabilities of crossover and mutation were set to 0.6 and 0.01, respectively.

TABLE 13.5 Collectively Optimized Network Inputs

Parameter	Optimized Value
Hidden Layers	1
Neurons/Hidden Layer	3
Training Tolerance	0.095
Initial Weight Range	+/– 1.50
Learning Rate	2.80
Momentum	0.35

Source: Kim, B. and May, G., 1994. An Optimal Neural Network Process Model for Plasma Etching, *IEEE Trans. Semi. Manuf.*, 7(1):12-21. With permission.

The performance of each individual of the population was evaluated with respect to the constraints imposed by the problem based on the evaluation of a fitness function. To search for parameter values that minimized both network training error and prediction error, the following performance index (PI) was implemented:

$$PI = K_1\sigma_t^2 + K_2\sigma_p^2 \qquad \text{Equation (13.25)}$$

where σ_t is the RMS training error, σ_p is the RMS prediction error, and K_1 and K_2 represent the relative importance of each performance measure. The values chosen for these constants were $K_1 = 1$ and $K_2 = 10$. The desired output was reflected by the following fitness function:

$$F = \frac{1}{1+PI} \qquad \text{Equation (13.26)}$$

Maximization of F continued until a final solution was selected after 100 generations. If the optimal solution was not found, the solution with the best fitness value was selected.

13.4.1.2.1 Optimization of Individual Responses

Individual response neural network models were trained to predict PECVD silicon dioxide permittivity, refractive index, residual stress, and nonuniformity, and impurity (H_2O and SiOH) concentration. The result of genetically optimizing these neural process models is shown in Table 13.6. Analogous results for network optimization by the simplex method are given in Table 13.7. Examination of Tables 13.6 and 13.7 shows that the most significant differences between the two optimization algorithms occur in the number of hidden neurons and learning rates predicted to be optimal.

Tables 13.8 and 13.9 compare σ_t and σ_p for the two search methods. (In each table, the "% Improvement" column refers to the improvement obtained by using genetic search). Although in two cases involving training error minimization the simplex method proved superior, the genetically optimized networks exhibited vastly improved performance in nearly every category for prediction error minimization. The overall average improvement observed in using genetic optimization was 1.6% for network training error and 60.4% for prediction error.

13.4.1.2.2 Optimization for Multiple PECVD Responses

The parameter sets called for in Tables 13.6 and 13.7 are useful for obtaining optimal performance for a single PECVD response, but provide suboptimal results for the remaining responses. For example, Table 13.6 indicates that seven hidden neurons are optimal for permittivity, refractive index, and stress, but only four hidden neurons are necessary for the nonuniformity and impurity concentration models. It is desirable to optimize network parameters for all responses simultaneously. Therefore, a multiple output neural process model (which includes permittivity, stress, nonuniformity, H_2O, and SiOH) was trained with that objective in mind.

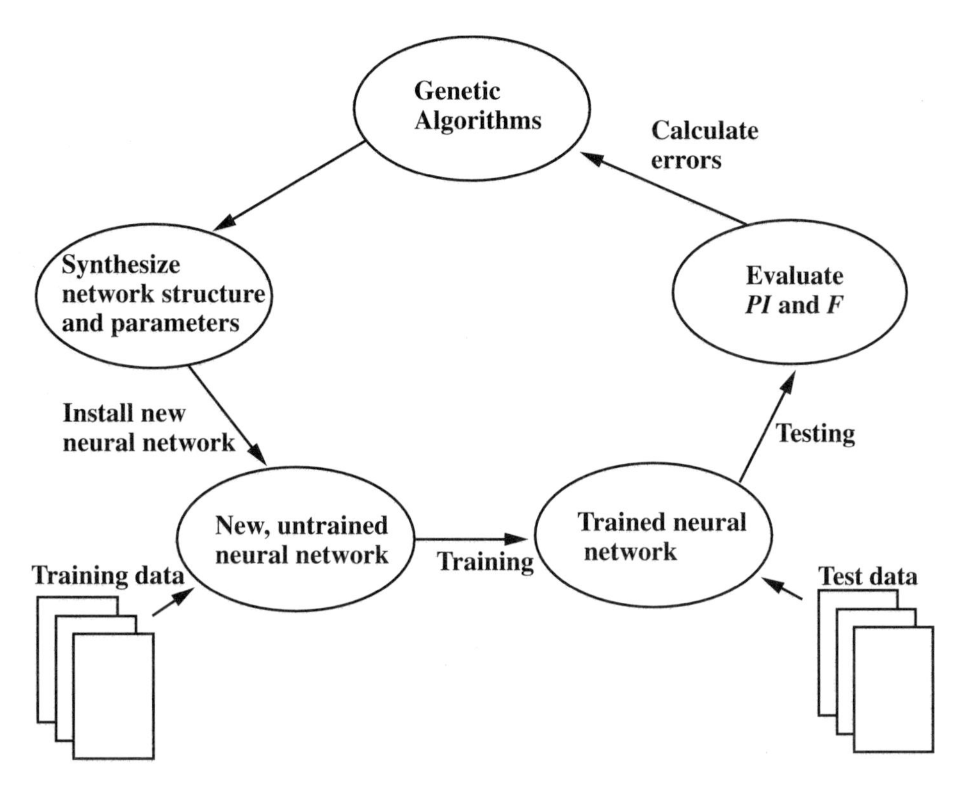

FIGURE 13.15 Block diagram of genetic optimization. Genetic algorithms are used to generate populations of neural networks with varying hidden layer neurons, momentum, learning rates, and training tolerances. The best network is selected after several generations according to a performance index. (*Source:* Han, S. and May, G., 1996. Optimization of Neural Network Structure and Learning Parameters Using Genetic Algorithms, *Proc. IEEE Int. Conf. AI Tools*, 8:200-206. With permission.)

TABLE 13.6 Network Parameters Optimized by Genetic Algorithms

PECVD Response	Hidden Neurons	Momentum	Learning Rate	Training Tolerance
Permittivity	7	0.41	0.19	0.01
Refractive Index	7	0.40	0.37	0.08
Residual Stress	7	0.39	0.07	0.06
Nonuniformity	4	0.43	0.06	0.11
Impurity Concentration	4	0.37	0.08	0.07

Source: Han, S. and May, G., 1996. Optimization of Neural Network Structure and Learning Parameters Using Genetic Algorithms, *Proc. IEEE Int. Conf. AI Tools*, 8:200-206. With permission.

TABLE 13.7 Network Parameters Optimized by Simplex Search

PECVD Response	Hidden Neurons	Momentum	Learning Rate	Training Tolerance
Permittivity	9	0.40	0.50	0.01
Refractive Index	9	0.40	0.50	0.01
Residual Stress	6	0.40	0.05	0.13
Nonuniformity	6	0.40	0.50	0.13
Impurity Concentration	6	0.40	0.05	0.07

Source: Han, S. and May, G., 1996. Optimization of Neural Network Structure and Learning Parameters Using Genetic Algorithms, *Proc. IEEE Int. Conf. AI Tools*, 8:200-206. With permission.

TABLE 13.8 Training Error Comparison of GA and Simplex Network Optimization

PECVD Response	Simplex	GAs	% Improvement
Permittivity	0.0578	0.0110	80.94
Refractive Index	0.0232	0.0822	–71.76
Residual Stress	0.0500	0.0571	–12.47
Nonuniformity	0.1146	0.1099	4.09
Impurity Concentration	0.0951	0.0841	7.35

Source: Han, S. and May, G., 1996. Optimization of Neural Network Structure and Learning Parameters Using Genetic Algorithms, *Proc. IEEE Int. Conf. AI Tools*, 8:200-206. With permission.

TABLE 13.9 Prediction Error Comparison of GA and Simplex Network Optimization

PECVD Response	Simplex	GAs	% Improvement
Permittivity	0.1788	0.0363	79.68
Refractive Index	0.2158	0.0591	72.61
Residual Stress	0.9659	0.4815	50.16
Nonuniformity	0.1361	0.0246	81.92
Impurity Concentration	0.0964	0.0795	17.54

Source: Han, S. and May, G., 1996. Optimization of Neural Network Structure and Learning Parameters Using Genetic Algorithms, *Proc. IEEE Int. Conf. AI Tools*, 8:200-206. With permission.

Table 13.10 shows optimized parameters for the multiple response PECVD model. Here, the optimal parameters derived by genetic and simplex search differ only slightly. However, they differ noticeably from the parameter sets optimized for individual responses. This is especially true for the number of hidden neurons and the learning rates. Further, slight differences in optimal parameters lead to significant differences in performance. This is indicated in Table 13.11, which shows training and prediction errors for the neural network models trained with the parameter sets in Table 13.10. If the improvements for the multiple response model are factored in, GAs provide an average benefit of 10.0% in training accuracy and 65.6% in prediction accuracy.

13.4.2 Process Optimization

A natural extension of process modeling is using models to optimize the processes (as opposed to the networks) or to generate specific process recipes. To illustrate the importance of process optimization, consider the PECVD of silicon dioxide films [Han et al., 1994]. In this process, one would like to grow a film with the lowest dielectric constant, best uniformity, minimal stress, and lowest impurity concentration possible (Figure 13.16). However, achieving these goals usually requires a series of trade-offs in growth conditions. Optimized neural process models can help a process engineer navigate the complex response surface and provide the necessary combination of process conditions (temperature, pressure, gas composition, etc.) or find the best compromise among potentially conflicting objectives to produce the desired results.

TABLE 13.10 Optimal Network Parameters for Multiple Response PECVD Model

Network Parameters	Simplex	GAs
Hidden Neurons	5	5
Momentum	0.35	0.37
Learning Rate	0.05	0.08
Training Tolerance	0.13	0.06

Source: Han, S. and May, G., 1996. Optimization of Neural Network Structure and Learning Parameters Using Genetic Algorithms, *Proc. IEEE Int. Conf. AI Tools*, 8:200-206. With permission.

TABLE 13.11 Optimal Network Parameters for Multiple Response PECVD Model

Error	Simplex	GAs	% Improvement
σ_t	0.2891	0.1391	51.88
σ_p	1.2979	0.1104	91.50

Source: Han, S. and May, G., 1996. Optimization of Neural Network Structure and Learning Parameters Using Genetic Algorithms, *Proc. IEEE Int. Conf. AI Tools*, 8:200-206. With permission.

Han and May used neural process models for the PECVD process to synthesize other novel process recipes. To characterize the PECVD of silicon dioxide (SiO_2) films, they first performed a 2^{5-1} fractional factorial experiment with three center-point replications [Box et al., 1978]. Data from these experiments were used to develop neural process models for SiO_2 deposition rate, refractive index, permittivity, film stress, wet etch rate, uniformity, silanol concentration, and water concentration. Then the recipe synthesis procedure was performed to generate the necessary deposition conditions to obtain specific film qualities, including zero stress, 100% uniformity, low permittivity, and minimal impurity concentration. This synthesis procedure compared GAs to other search procedures for determining optimal process recipes.

GAs offer the advantage of global search, but they are slow to converge. More traditional approaches to optimization include calculus-based "hill-climbing" methods, where it is critical to find the best direction for searching to optimize the response surface. One method in this category is Powell's algorithm, which generates successive quadratic approximations of the space to be optimized. This method involves determining a set of n linearly independent, mutually conjugate directions (where n is the dimensionality of the search space). Successive line minimizations put the algorithm at the minimum of the quadratic approximation. For functions that are not exactly quadratic, the algorithm does not find the exact minimum, but repeated cycles of n line minimizations converge in due course to the minimum. Another widely used searching technique is Nelder and Mead's simplex method. A regular simplex is defined as a set of $(n + 1)$ mutually equidistant points in n-dimensional space. The main idea of the simplex method is to compare the values of the function to be optimized at the $(n + 1)$ vertices of the simplex and move the simplex iteratively toward the optimal point.

In both Powell's algorithm and the simplex method, the initial starting search point profoundly affects overall performance. With an improper starting point, both algorithms are more likely to be trapped in local optima. However, if the proper initial point is given, the search is very fast. On the other hand, genetic algorithms search out the overall optimal area very fast, but converge slowly to a global optimum. Therefore, hybrid combinations of genetic algorithms with the other two algorithms (Powell's and simplex) sometimes offer improved results in both speed and accuracy. Hybrid algorithms start with genetic algorithms to initially sample the hypersurface and find the global optimum area. After some number of generations, the best point found by the GA is handed over to other algorithms as a starting point. With this initial point, both Powell's algorithm and simplex method quickly locate the optimum.

Han and May compared five optimization methods to synthesize PECVD recipes: (i) genetic algorithms; (ii) Powell's method; (iii) Nelder and Mead's simplex algorithm; (iv) a hybrid combination of genetic algorithms and Powell's method; and (v) a hybrid combination of genetic algorithms and the simplex algorithm. The desired output characteristics of the PECVD SiO_2 film to be produced are reflected by the following fitness function:

$$F = \frac{1}{1 + \sum_r \left| K_r \left(y_d - y \right) \right|} \qquad \text{Equation (13.27)}$$

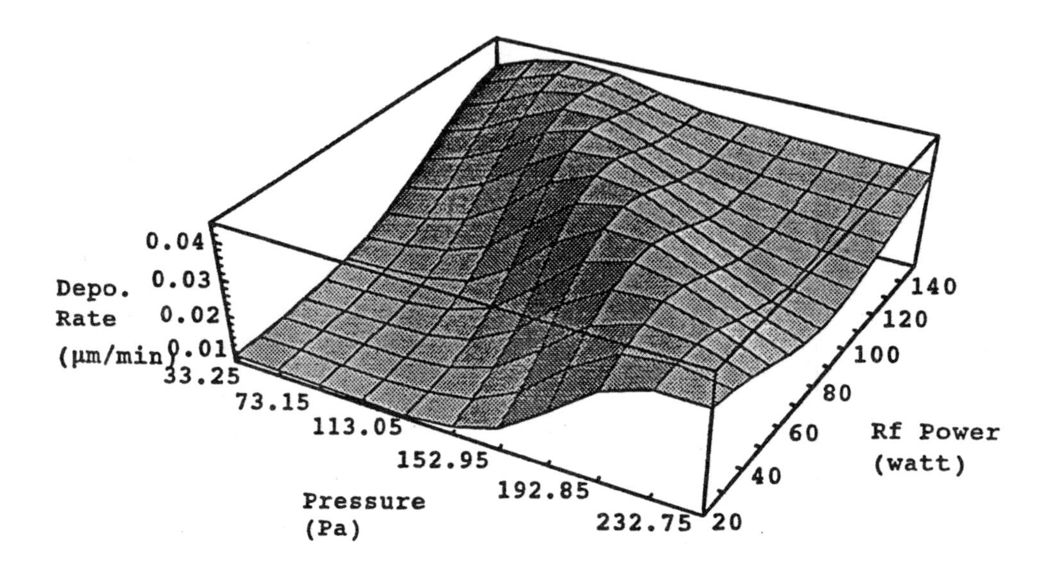

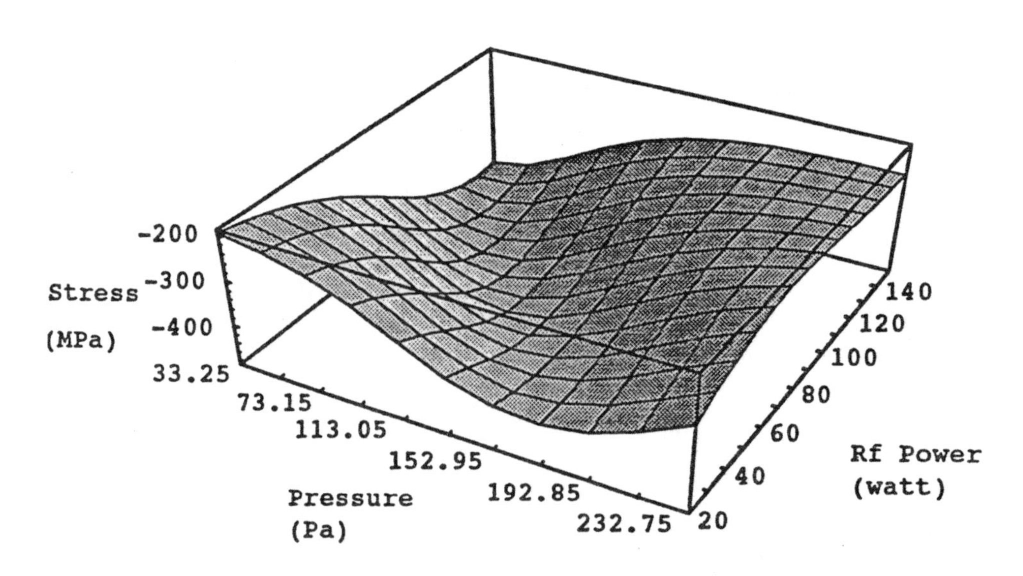

FIGURE 13.16 Examples of response surfaces generated by the application of neural process modeling to PECVD silicon dioxide films: (a) deposition rate vs. chamber pressure and RF power; (b) film stress vs. chamber pressure and RF power. These surfaces are typical of those that must be effectively searched to locate the optimal growth conditions. (*Source:* Han, S. et al., 1994. Modeling the Properties of PECVD Silicon Dioxide Films Using Optimized Back-Propagation Neural Networks, *IEEE Trans. Comp. Packag. Manuf. Technol.*, 17(2):174-182. With permission.)

where r is the number of process responses, y_d are the desired process responses, y are the process outputs dictated by the current choice of input parameters, and K_r is a constant which represents the relative importance of the rth process response.

For GAs, the probabilities of crossover and mutation were set to 0.6 and 0.01, and a population size of 100 was used in each generation. Each input parameter was coded as a 40-bit string, resulting in a total length of 200 bits. Maximization of F continued until a final solution was selected after 500 generations. Search in the simplex method is achieved by applying three basic operations: reflection, expansion, and contraction. Nelder and Mead recommend the coefficients of reflection α, expansion γ, and contraction β as 1, 2, and 0.5, respectively.

It was expected that hybrid methods would readily improve accuracy compared with genetic search alone if initiated immediately after the 500 GA generations. Such a large number of generations, however, severely affects the computational load of these techniques. Therefore, to reduce this computational burden in both hybrid methods, the GA portion of the search was limited to 100 generations. Then the resulting GA solution was handed over as the initial starting point for the simplex or Powell algorithms.

13.4.2.1 Results for Single Output Synthesis

The first objective of recipe synthesis was to find the optimal deposition recipes to achieve 100% thickness uniformity, low permittivity, zero residual stress, and low impurity concentration in the silicon dioxide film. The five previously mentioned recipe synthesis procedures were applied to determine the required deposition conditions. Afterward, films were grown using the synthesized recipes. Tables 13.12 through 13.16 show the recipes synthesized by each of these five methods, along with simulation results predicted by the neural process model using these recipes as inputs and actual measured values of the responses for the grown film.

Figure 13.17 ranks performance of each recipe synthesis method in simulation. A ranking of "1" represents the best performance, and "5" the worst. The hybrid method combining genetic algorithms and the simplex method gives the best overall simulation results. This indicates that even though the GAs can find the general optimal area quickly, convergence to a global optimum is not as fast or accurate as the simplex method. Once a good starting search point is provided by the GA, the simplex method can find an optimal point more efficiently than GAs alone. Figure 13.18 summarizes the rankings for each recipe synthesis method based measuring grown films. Genetic algorithms alone performed better than any other method overall. It is somewhat surprising that the hybrid methods did not rank best on the measured data. However, this indicates that more than 100 GA generations are required in the hybrid methods for them to exhibit the improved performance expected.

Powell's algorithm ranks the lowest in simulation, and the simplex method ranks worst in the measured data. These observations verify that the search spaces provided by the neural PECVD models are generally multimodal, rather than unimodal. Both the simplex method and Powell's algorithm become trapped in local optima in multimodal search spaces. The only major exception appears to be in the impurity concentration, where the recipe generated by Powell's algorithm produced the film with the lowest impurity level.

13.4.2.2 Results for Multiple Output Synthesis

The recipes generated above are useful for obtaining films optimized for a single criterion, but they can lead to unacceptable results for the other film characteristics. The recipe that provides optimal uniformity does not necessarily provide low residual stress. In fact, the recipe that yields the film with the greatest measured uniformity produces a film with a residual stress of –521.5 MPa, a value more than five times greater than the optimal measured stress of –102.6 MPa. The remaining film characteristics show that similar suboptimal results were obtained for the other recipes optimized for a single response. Clearly, it is desirable to grow films with the best combination of all the desired qualities. This involves processing trade-offs, and the challenge, therefore, is to devise a means for designating the importance of a given response variable in determining the optimal recipe.

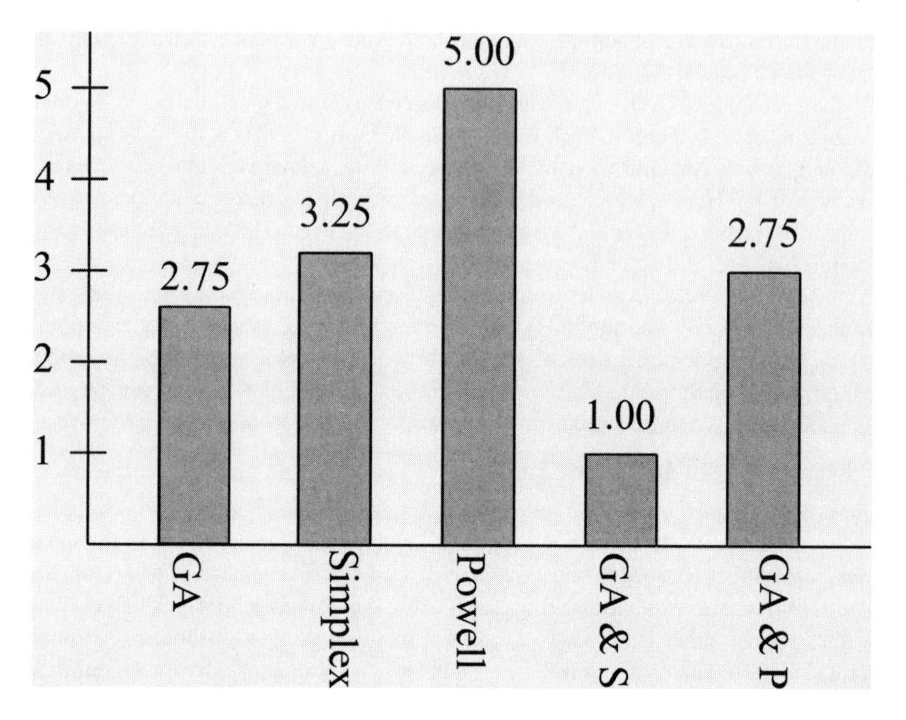

FIGURE 13.17 Average simulation performance of recipe synthesis methods. The hybrid method combining genetic algorithms and the simplex method gives the best average ranking. (*Source:* Han, S. and May, G., 1997. Using Neural Network Process Models to Perform PECVD Silicon Dioxide Recipe Synthesis via Genetic Algorithms, *IEEE Trans. Semi. Manuf.*, 10(2):279-287. With permission.)

TABLE 13.12 Recipes Synthesized Using Genetic Algorithms

	% Nonuniformity	Permittivity	Residual Stress	Impurity Level (H_2O/SiOH)
SiH_4 (sccm)	335	368	369	257
N_2O (sccm)	613	673	411	662
Temperature (°C)	352	376	289	214
Pressure (torr)	1.68	1.19	0.89	1.67
RF Power (Watts)	102	145	143	137
Simulation Result	0%	3.04	–0.001 MPa	0.37%/0.03%
Measured Result	8.67%	4.02	–102.6 MPa	1.64%/4.31%

Source: Han, S. and May, G., 1997. Using Neural Network Process Models to Perform PECVD Silicon Dioxide Recipe Synthesis via Genetic Algorithms, *IEEE Trans. Semi. Manuf.*, 10(2):279-287. With permission.

The only difference between this objective and single-output recipe synthesis is the number of desired characteristics. To illustrate, multiple-output synthesis objectives were accomplished by applying the fitness function in Equation 13.27 with a specific set of K_r weight coefficients chosen for growing films optimized for electronics packaging applications. Because one of the most important qualities in SiO_2 films for this application is permittivity, Han and May set the weight of permittivity to 100. Weights for both water and silanol concentration were set to 50, and the weights for both uniformity and stress were set to 1. Table 13.17 shows the results of the five different synthesis procedures for multiple outputs. The genetic algorithm and the hybrid GA/Powell algorithm provided the best compromise among the multiple objectives.

TABLE 13.13 Recipes Synthesized Using the Simplex Method

	% Nonuniformity	Permittivity	Residual Stress	Impurity Level (H_2O/SiOH)
SiH_4 (sccm)	299	400	344	261
N_2O (sccm)	650	441	597	436
Temperature (°C)	300	332	232	268
Pressure (torr)	1.38	1.43	0.25	1.80
RF Power (Watts)	85	111	150	150
Simulation Result	0%	3.19	–34.9 MPa	0.25%/1.14%
Measured Result	0.73%	4.35	–156.6 MPa	2.39%/5.22%

Source: Han, S. and May, G., 1997. Using Neural Network Process Models to Perform PECVD Silicon Dioxide Recipe Synthesis via Genetic Algorithms, *IEEE Trans. Semi. Manuf.*, 10(2):279-287. With permission.

TABLE 13.14 Recipes Synthesized Using Powell's Method

	% Nonuniformity	Permittivity	Residual Stress	Impurity Level (H_2O/SiOH)
SiH_4 (sccm)	300	325	398	300
N_2O (sccm)	650	705	861	650
Temperature (°C)	300	325	398	300
Pressure (torr)	1.31	1.13	1.36	1.28
RF Power (Watts)	85	92	113	85
Simulation Result	0%	3.67	–301.0 MPa	1.74%/2.31%
Measured Result	0.46%	4.40	–124.7 MPa	1.09%/2.38%

Source: Han, S. and May, G., 1997. Using Neural Network Process Models to Perform PECVD Silicon Dioxide Recipe Synthesis via Genetic Algorithms, *IEEE Trans. Semi. Manuf.*, 10(2):279-287. With permission.

TABLE 13.15 Recipes Synthesized Using Hybrid Method (GAs and Simplex)

	% Nonuniformity	Permittivity	Residual Stress	Impurity Level (H_2O/SiOH)
SiH_4 (sccm)	271	358	375	248
N_2O (sccm)	726	684	405	699
Temperature (°C)	266	388	225	225
Pressure (torr)	1.35	1.28	0.28	1.58
RF Power (Watts)	80	150	101	136
Simulation Result	0%	3.03	–0.00 MPa	0.00%/0.00%
Measured Result	0.67%	4.11	–152.8 MPa	2.57%/4.78%

Source: Han, S. and May, G., 1997. Using Neural Network Process Models to Perform PECVD Silicon Dioxide Recipe Synthesis via Genetic Algorithms, *IEEE Trans. Semi. Manuf.*, 10(2):279-287. With permission.

TABLE 13.16 Recipes Synthesized Using Hybrid Method (GAs and Powell)

	% Nonuniformity	Permittivity	Residual Stress	Impurity Level (H_2O/SiOH)
SiH_4 (sccm)	213	356	396	216
N_2O (sccm)	684	695	485	673
Temperature (°C)	337	389	245	208
Pressure (torr)	1.60	1.18	0.39	1.56
RF Power (Watts)	62	147	104	134
Simulation Result	0%	3.03	–0.003 MPa	0.90%/0.44%
Measured Result	0.19%	4.13	–185.2 MPa	1.99%/4.27%

Source: Han, S. and May, G., 1997. Using Neural Network Process Models to Perform PECVD Silicon Dioxide Recipe Synthesis via Genetic Algorithms, *IEEE Trans. Semi. Manuf.*, 10(2):279-287. With permission.

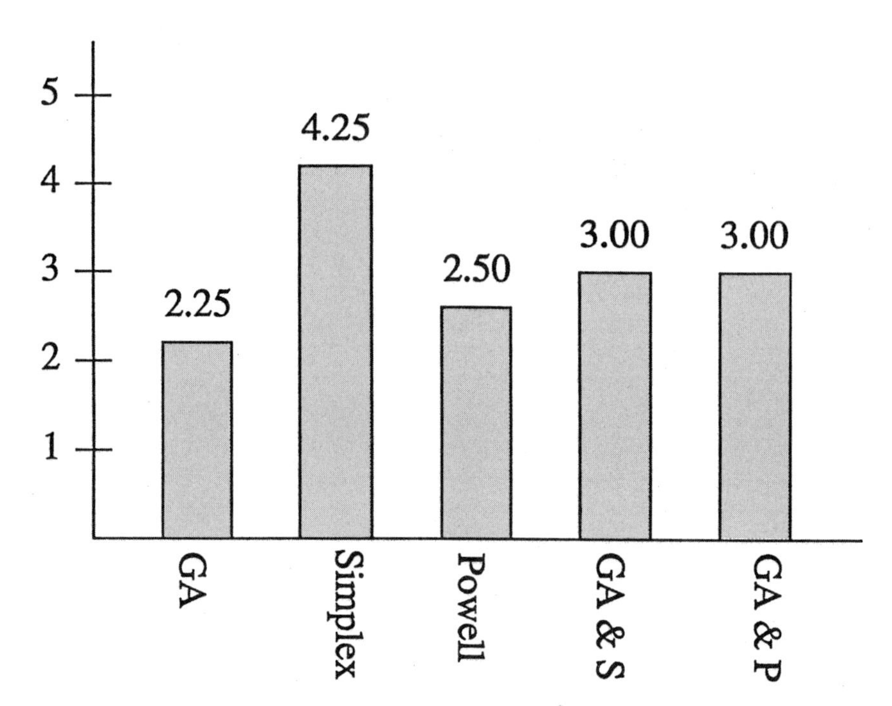

FIGURE 13.18 Average measured performance of recipe synthesis methods. Genetic algorithms ranked higher than any other method. (*Source:* Han, S. and May, G., 1997. Using Neural Network Process Models to Perform PECVD Silicon Dioxide Recipe Synthesis via Genetic Algorithms, *IEEE Trans. Semi. Manuf.*, 10(2):279-287. With permission.)

TABLE 13.17 Recipes Synthesized Using Hybrid Method (GAs and Powell)

Method	% Nonuniformity	Permittivity	Stress (MPa)	H_2O (Wt%)	SiOH (Wt%)
GAs	3.13	4.28	–209.3	1.99	5.08
Simplex	0.66	4.37	–173.4	1.28	3.17
Powell	0.26	4.26	–233.6	1.78	2.66
GA/Simplex	1.64	4.38	–216.3	1.25	4.51
GA/Powell	5.05	4.11	–264.3	1.19	4.01

Source: Han, S. and May, G., 1997. Using Neural Network Process Models to Perform PECVD Silicon Dioxide Recipe Synthesis via Genetic Algorithms, *IEEE Trans. Semi. Manuf.*, 10(2):279-287. With permission.

It is clear that the multiple-output recipe synthesis procedure yields permittivity comparable to the single response recipe synthesis procedures. The lowest permittivity value among the single-output synthesis procedure is 4.02 (generated by GAs), and the lowest permittivity for the multiple-output synthesis procedure is 4.11 (generated by the hybrid GA/Powell approach). Thus, the trade-offs for optimizing the other responses do not have an extremely adverse effect on permittivity. Permittivity is closely related to impurity concentration. Lowering the impurity concentration lowers the permittivity. Water and silanol concentration in the film deposited using the hybrid GA/Powell generated recipe are lowest in multiple output synthesis, thereby confirming the relationship between impurity concentration and permittivity.

13.5 Process Monitoring and Control

Due to demands to produce ICs with increased density and complexity, process control is an issue of growing importance. Robust process control techniques require accurately monitoring ambient process conditions. Historically, statistical process control (SPC) has been used to achieve the necessary control.

SPC minimizes misprocessing by applying control charts to monitor process fluctuations [Montgomery, 1991]. Although SPC detects undesirable process shifts, it is usually applied off-line. This delay results in fabricating devices that do not conform to required specifications. Neural networks have been used to address these issues from two different perspectives: (i) monitoring process conditions for real-time SPC; and (ii) using closed-loop control schemes that use *in situ* process sensors for on-line adjustments in process set points.

13.5.1 Process Monitoring and Statistical Process Control

The objective of real-time SPC is to take advantage of available on-line sensor data from fabrication equipment to identify process shifts and out-of-control states and generate real-time malfunction alarms. This offers on-line process monitoring for generating alarms at the very onset of a shift. The application of real-time SPC is complicated by the correlated nature of the sensor data. Traditional SPC is based on assuming that the data to be monitored in controlling a process are identically independent and normally distributed (IIND). This assumption is not valid when applied to real-time data. These data are often nonstationary (subject to mean and variance shifts), autocorrelated (dependent on data from previous time points), and cross-correlated (dependent on the values of other concurrently measured parameters).

Baker et al. [1995] addressed these difficulties by employing neural networks to develop time series models that filter cross- and autocorrelation from real-time sensor data. In applying this methodology to semiconductor manufacturing, Baker et al. developed a real-time equipment monitoring system that transfers data from an RIE system to a remote workstation. The processes monitored were the etching of silicon dioxide by trifloromethane (CHF_3) and aluminum by boron trichloride (BCl_3). The parameters monitored included gas flow rates, RF power, temperature, pressure, and DC bias. Data sampled at 50 Hz were used to train backpropagation neural networks. Then the trained networks were used both to forecast the time series data and to generate a malfunction alarm when the sampled data did not conform to its specification within a designated tolerance.

13.5.1.1 Time Series Modeling

Conventional SPC is based on the assumption of IIND data, which is not valid for data acquired in real time, because real-time data are nonstationary, autocorrelated, and cross-correlated. Time series modeling accounts for correlation in real-time data. The purpose of a time series model is to describe the chronological dependence among sequential samples of a given variable. Passing raw data through time series filters results in residual forecasting error that is IIND. One of the most basic time series models is the univariate Box–Jenkins autoregressive moving average (ARMA) model [Box and Jenkins, 1976]. Data collected from modern semiconductor manufacturing equipment can also be represented by means of time series models, and Baker et al. showed that neural networks may be used to generalize the behavior of a time series. They referred to this new genre of time series model as the *neural time series* (NTS) model. Like ARMA models, once an NTS model is developed, the forecast data can be used on conventional control charts. However, unlike ARMA models, the NTS model simultaneously filters both auto- and cross-correlated data. In other words, the NTS model accounts for correlation among several variables being monitored simultaneously.

The neural network used to model the RIE process was trained off-line on data acquired when the process was under control. The parameter of interest was BCl_3, but the same methodology could be extended to any other process variable. The NTS network was trained to model BCl_3 flow by a unique sampling technique that involved training the network to forecast the next BCl_3 value from the behavior of ten past values. The network was trained on a subset of the total autocorrelated data that consisted of the first 11 out of every 100 samples. The trained network was then tested on 11 midrange samples out of every 100.

Autocorrelation among consecutive measurements was accounted for by simultaneously training the network on the present value of the BCl_3 and ten past values. Cross-correlation among the BCl_3 and the other six parameters was modeled by including as inputs to the NTS network the present values of the

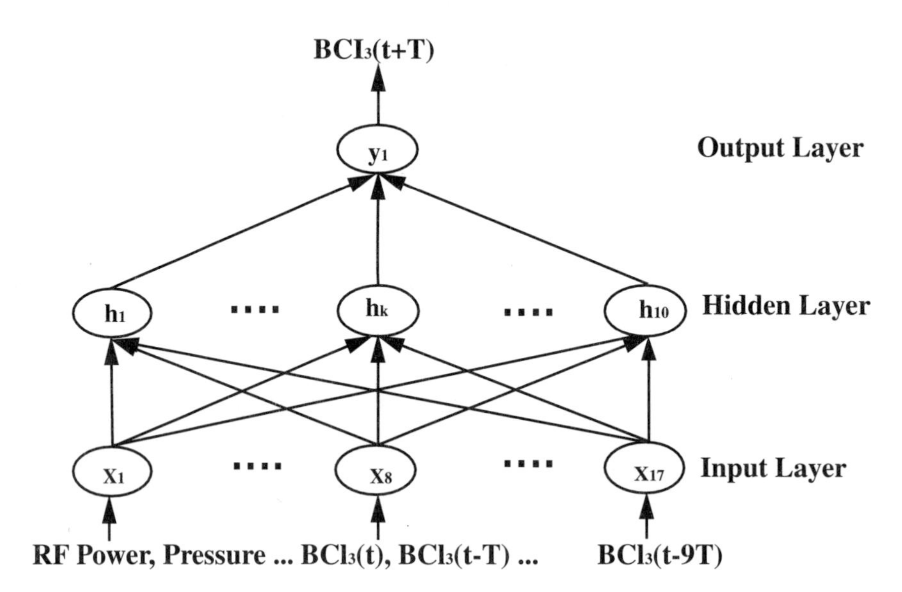

FIGURE 13.19 NTS network structure. Inputs to the network include historical values of the parameter being forecast (to account for autocorrelation) and concurrent values of other state variables (to account for cross-correlation). (*Source:* Baker, M., Himmel, C., and May, G., 1995. Time Series Modeling of Reactive Ion Etching Using Neural Networks, *IEEE Trans. Semi. Manuf.,* 8(1):62-71. With permission.)

temperature, pressure, incident and reflected RF power, chlorine, and BCl_3 itself. The resulting network topology had 17 input neurons, 10 hidden neurons, and a single output neuron (Figure 13.19). The future value of the BCl_3 at time $(t + T)$ was forecast at the network output (where T is the sampling period). Figure 13.20 shows the measured and NTS model predictions of the BCl_3 data. Each point on the graph represents 1 out of every 100 samples. The NTS model very closely approximates the actual value. Even when there are drastic changes in the BCl_3, the NTS network quickly adapted. This technique yielded an excellent RMS error of 1.40 standard cm^3/min.

13.5.1.2 Malfunction Detection

The NTS model generates a real-time alarm signal when sampled process data does not conform to their previously established pattern, indicating a possible equipment malfunction. This capability was demonstrated in the case of an aluminum etch in a CHF_3 and Cl_2 gas mixture. The malfunction consisted of an unstable feed condition in the CHF_3 mass flow controller. Figure 13.21 is a plot of the gas flows during the period leading up to the malfunction. Although the Cl_2 flow appears to fall out of compliance at the 200th sample, this was not the cause of the malfunction. The true cause may be observed in the behavior of the CHF_3 several samples earlier and comparing the instability of its flow to the more stable and consistent readings exhibited by the Cl_2 during the same time span. A careful study of this situation reveals that the CHF_3 mass flow controller was not able to regulate the gas flow correctly, and consequently the RIE control circuitry aborted the process, thus causing the Cl_2 to shut off.

An NTS model was used to generate an alarm signal warning of the impending out-of-control condition of CHF_3 flow even before the RIE aborted itself. Recall that the NTS model acts as a filter to remove autocorrelation and cross-correlation from the raw process data. Thus, the residuals that result from computing the difference between NTS model predictions and the measured values of the CHF_3 flow are IIND random variables. As a result, these residuals can be plotted on a standard control chart to identify process shifts. In this case, alarm generation was based on the well-known Western Electric Rules [Montgomery, 1991]:

1. One data point plots outside of the 3-σ control limits.
2. Two out of three consecutive points plot beyond the 2-σ warning limits.

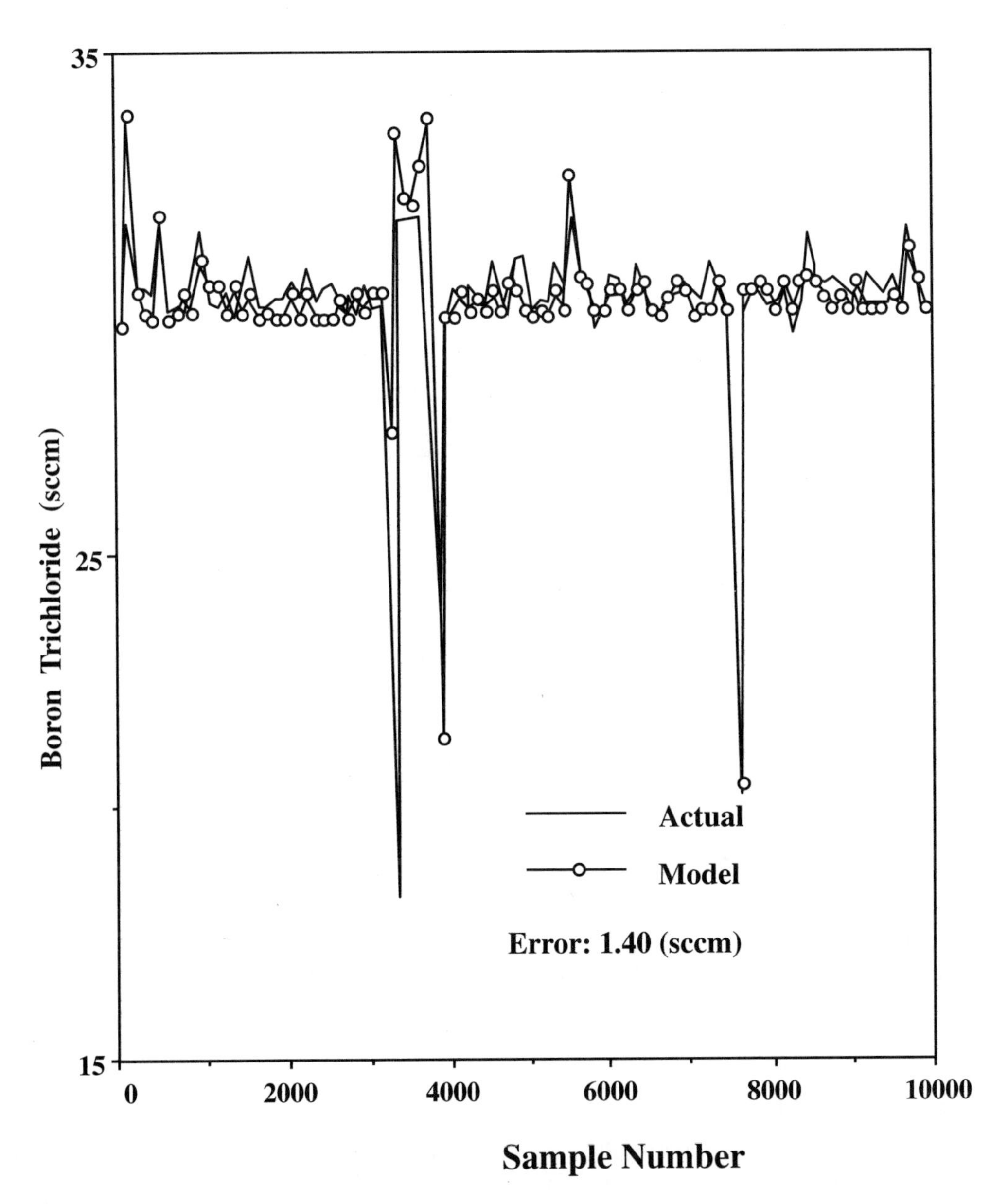

FIGURE 13.20 Measured BCl_3 flow and NTS model predictions. The NTS model very closely approximates the actual value, yielding an RMS error of 1.40 standard cm^3/min. (*Source:* Baker, M., Himmel, C., and May, G., 1995. Time Series Modeling of Reactive Ion Etching Using Neural Networks, *IEEE Trans. Semi. Manuf.*, 8(1):62-71. With permission.)

3. Four out of five consecutive points plot 1-σ or beyond from the center line.
4. Eight consecutive points plot on one side of the center line.

The violation of Rule 4 was invoked to generate the malfunction alarm. The data from the RIE malfunction was fed into the NTS network with CHF_3 as the forecast parameter. Figure 13.22 demonstrates

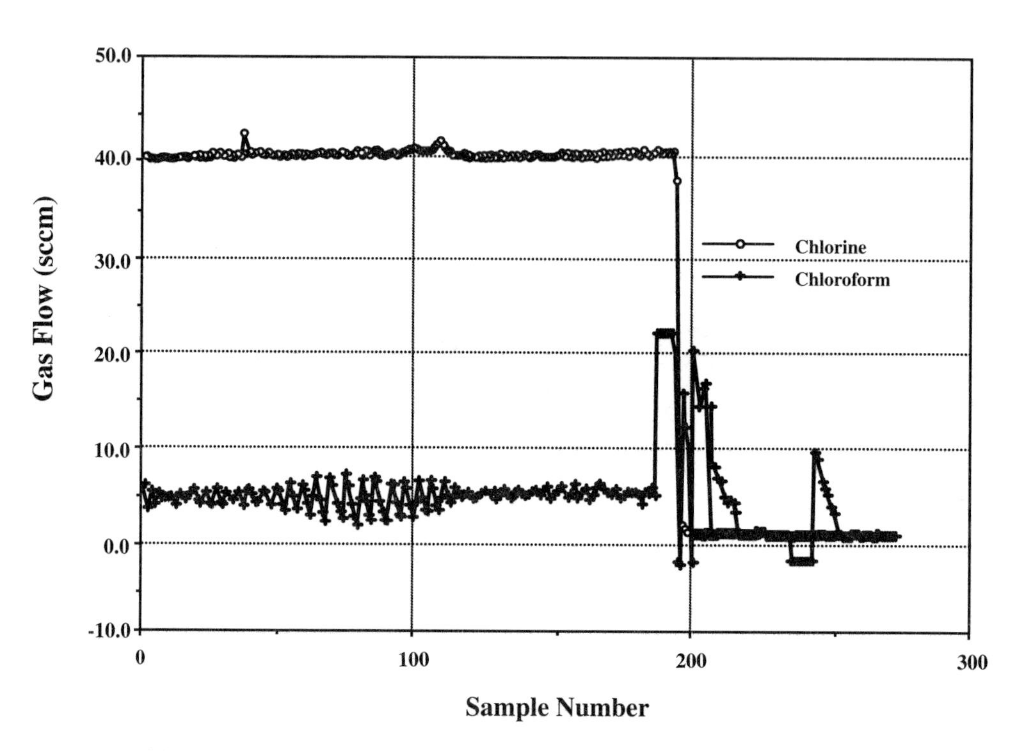

FIGURE 13.21 Chlorine and CHF_3 flow rates for an aluminum etch step just before an equipment malfunction. Although the Cl_2 flow appears to fall out of compliance at the 200th sample, the true cause of the malfunction was the earlier instability observed in the CHF_3 signal. (*Source:* Baker, M., Himmel, C., and May, G., 1995. Time Series Modeling of Reactive Ion Etching Using Neural Networks, *IEEE Trans. Semi. Manuf.*, 8(1):62-71. With permission.)

that the NTS model closely resembled the actual data sequence until the malfunction occurred, at which point the CHF_3 instability became too great and the NTS model predictions diverged. Figure 13.23 shows the measurement residuals from the difference between NTS model predictions and actual sensor data. When eight consecutive points plotted on one side the center line (which occurred at the 18th sample), the NTS network immediately responded by signaling an alarm. For the same malfunction, the internal RIE process control circuitry did not respond until significantly later. The rapid NTS response time can be instrumental in identifying incipient equipment faults and preventing subsequent misprocessing.

13.5.2 Closed-Loop Process Control

Neural networks have also been successfully applied to closed-loop control of a diverse array of processes. There are two basic approaches to so-called "neurocontrol" in semiconductor manufacturing: run-by-run and real-time (or *in situ*) control.

13.5.2.1 Run-by-Run Neurocontrol

The objective in run-by-run control is to adjust fabrication process conditions on a wafer-by-wafer basis. These adjustments are made by comparing measured wafer characteristics and a predictive model of these characteristics. Smith and Boning [1996] integrated neural networks into the run-by-run control of chemical–mechanical polishing (CMP), a process in which semiconductor wafers are planarized using a slurry of abrasive material in an acidic solution. Smith and Boning trained a neural network to map

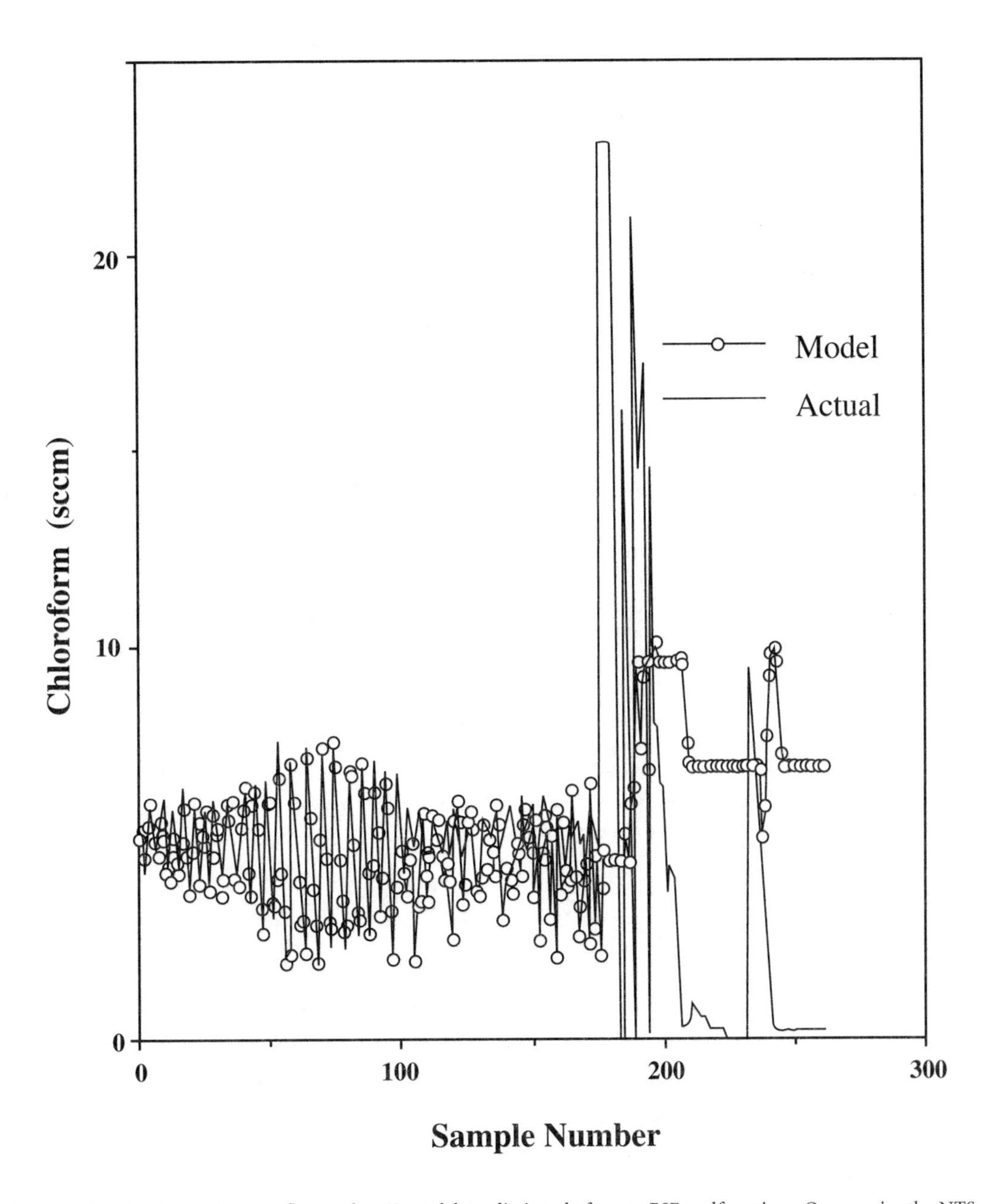

FIGURE 13.22 Measured CHF_3 flow and NTS model predictions before an RIE malfunction. Once again, the NTS model closely resembles the actual data sequence (until the occurrence of the malfunction). (*Source:* Baker, M., Himmel, C., and May, G., 1995. Time Series Modeling of Reactive Ion Etching Using Neural Networks, *IEEE Trans. Semi. Manuf.*, 8(1):62-71. With permission.)

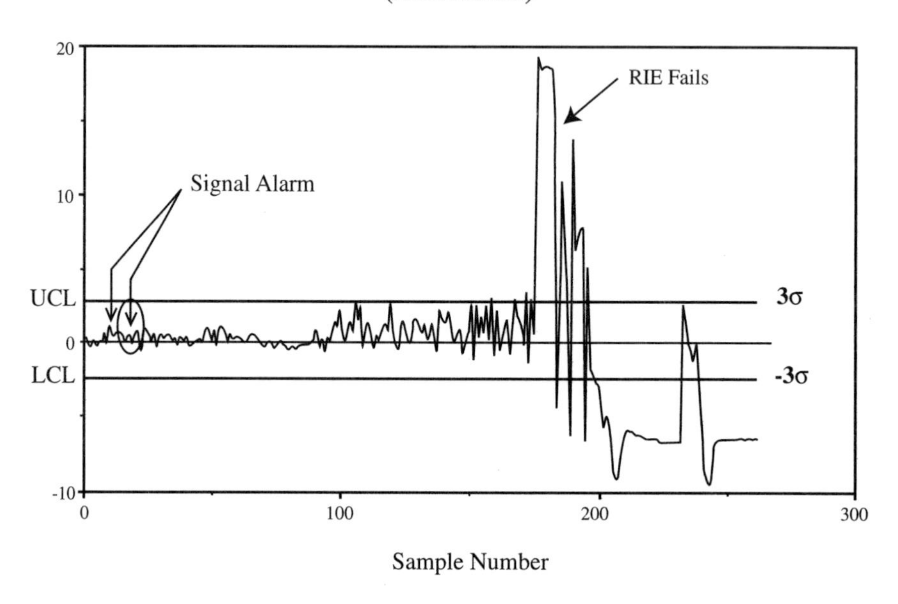

FIGURE 13.23 Measurement residuals from NTS model before an RIE malfunction plotted on a 3-σ control chart. The first arrow indicates the beginning of an eight-point sequence of data which plots above the center line, which is a violation of the Western Electric Rules. The second arrow (circled) indicates where an alarm is generated by the NTS model. (*Source:* Baker, M., Himmel, C., and May, G., 1995. Time Series Modeling of Reactive Ion Etching Using Neural Networks, *IEEE Trans. Semi. Manuf.,* 8(1):62-71. With permission.)

CMP process disturbances to optimal values for the coefficients in an exponentially weighted moving average (EWMA) controller. Statistical experimental design was used to generate a linearized multivariate model of the form $y_t = Ax_t + c_t$, where t is the run number, y_t is a vector of process responses, A is a constant gain matrix, x_t is vector of process inputs, and c_t is an offset vector, which is calculated recursively by an EWMA controller from the following relationship:

$$c_t = \alpha(y_t - Ax_t) + (1 - \alpha)c_{t-1} \qquad \text{Equation (13.28)}$$

The coefficient α is dynamically estimated from the neural network mapping according to the algorithm outlined in Figure 13.24. In designing this system, these researchers developed a self-tuning EWMA controller that dynamically updates its parameters by estimating the disturbance using the neural network mapping.

13.5.2.2 Real-Time Neurocontrol

The next evolutionary step in neurocontrol involves using neural nets to continuously correct process conditions. This real-time control approach has been pursued by Rietman et al. [1993] who have designed a neural network to compute in real time the overetch time for a plasma gate etch step. This time computation was based on a neural network mapping of the mean values of fluctuations about control variable set points and an *in situ* optical emission monitor. By monitoring a single optical emission wavelength during etching, these researchers inferred information about etch rate, etch uniformity, pattern density, and cleanliness of the reaction chamber. In neural network training, vectors representing process "signatures" inherent in the emission trace and set points were mapped to the etch time for a desired oxide thickness. This training procedure is illustrated in Figure 13.25. The backpropagation

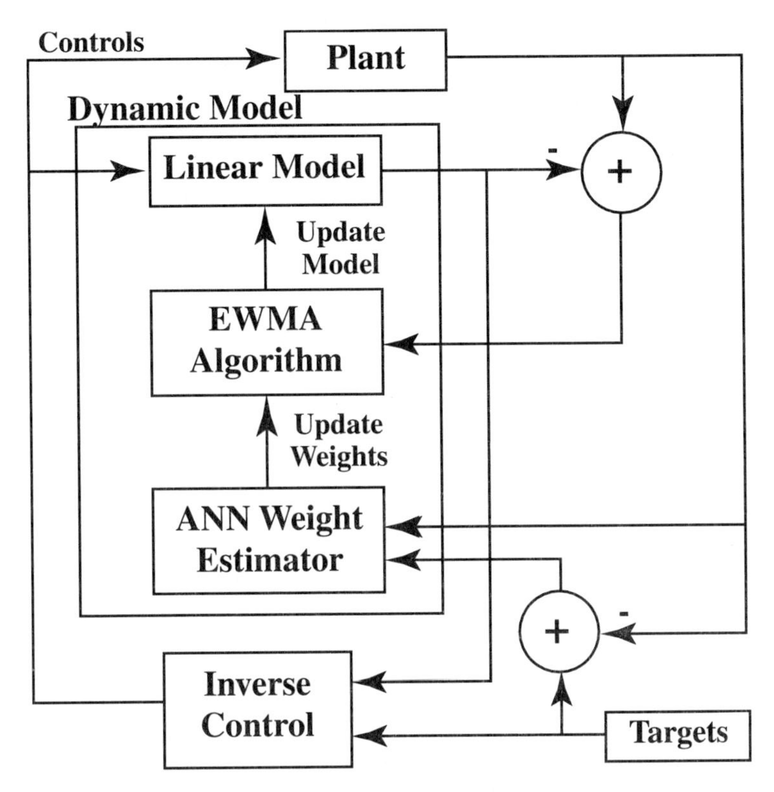

FIGURE 13.24 Block diagram of EWMA controller with a neural network weight estimator. The EWMA coefficient is dynamically estimated from the neural network mapping according to the algorithm shown. (*Source:* Smith, T. and Boning, D., 1996. A Self-Tuning EWMA Controller Utilizing Artifical Neural Network Function Approximation Techniques, *Proc. 1996 Int. Elec. Manuf. Tech. Symp.*, 18:355-361. With permission.)

network for the control operation consisted of 36 input nodes, 5 hidden neurons, and 1 output. This system has been learning on-line since 1993. During this time, the network trained on many thousands of wafers.

Stokes and May [1997] are pursuing a control scheme with two BP neural networks operating in unison, one trained to emulate the process in question and another trained to learn the inverse dynamics of the process and control the operation. In this arrangement, neural process models must predict the process outputs based on a set of operating conditions (forward models) and determine the appropriate set of operating conditions based on a desired output response (reverse models). If these conditions are met, then the process control technique is implemented, as shown in Figure 13.26. Here, the measured process output is compared with that of a forward neural process model (the process emulator). This control architecture allows simultaneous training of the emulator and controller networks continuously on-line.

In this indirect adaptive control (IAC) structure, the plant emulator is trained off-line with experimental data, whereas the controller is trained on-line with feedback from the plant emulator. Stokes and May have applied this system to simulate the real-time control of RIE. Their modified control scheme assumes that the plant emulator (PE) accurately models the RIE plant (i.e., $y = y_e$). The neural controller (NC) adjusts to the PE's inputs in real time to optimally match the output of the PE (y_e) to the control target (y^*). To train the neural controller, the control target is first fed through the neural controller to the plant emulator to obtain the process output $y_e(t)$. Second, using the generalized delta rule, the error between the control target and PE output ($e_2 = y^*(t) - y_e(t)$) is backpropagated through the plant emulator to calculate the weight adjustments for each layer of the PE. Next, the computed changes in the plant

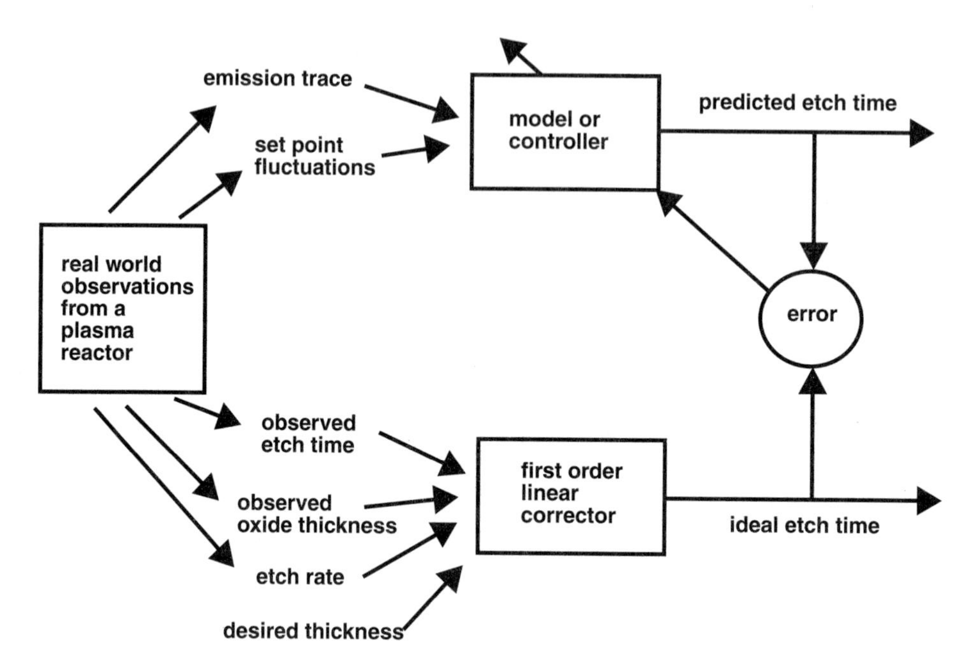

FIGURE 13.25 Illustration of training method for wafer-to-wafer neural network control of a plasma gate etch. Vectors representing etch process signatures inherent in the emission trace and set points are mapped to the ideal etch time for a desired oxide thickness. (*Source:* Rietman, E., Patel, S., and Lory, E., 1993. Neural Network Control of a Plasma Gate Etch: Early Steps in Wafer-to-Wafer Process Control, *Proc. Int. Elec. Manuf. Tech. Symp.*, 15:454-457. With permission.)

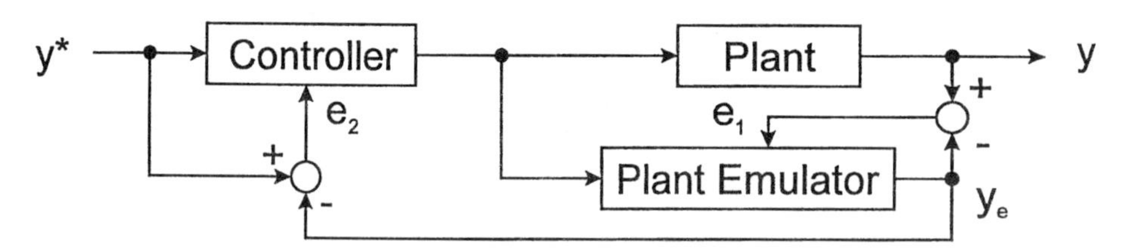

FIGURE 13.26 Illustration of adaptive process control scheme using two backpropagation neural networks: a plant emulator (PE) and a neural controller (NC). This control architecture allows simultaneous training of the emulator and controller networks continuously on-line. (*Source:* Stokes, C. and May, G., 1997. Real-Time Control of Reactive Ion Etching Using Neural Networks, *Proc. 1997 American Control Conf.*, III:1575-1579. With permission.)

emulator's inputs are used to estimate the output error of the neural controller. The neurocontroller's weights are then updated using the BP. This cycle is repeated for each control target ($y^*(t+1)$, $y^*(t+2)$, etc.).

To test this scheme, Stokes and May performed real-time simulations for RIE with data from a previous characterization experiment. The test materials were silicon dioxide and photoresist films on silicon wafers. The SiO_2 films were etched by a CHF_3 and oxygen plasma, with RF power, pressure, and the two gas flows as control variables. After training the plant emulator, the modified IAC scheme was used to train the controller on-line. The controller showed the best results when using a three-layer construction with one input neuron, seven hidden neurons, and four output neurons. An example of the performance of the control scheme for SiO_2 etch rate is shown in Figure 13.27. In this example, the control target was changed every 30 seconds. The neural controller network was trained for ten iterations to match each adjusted output of the plant emulator to the target value. After each change in the control target, the controller made adjustments in the recipe so that the output of the plant emulator matched the target value within 5 seconds or less.

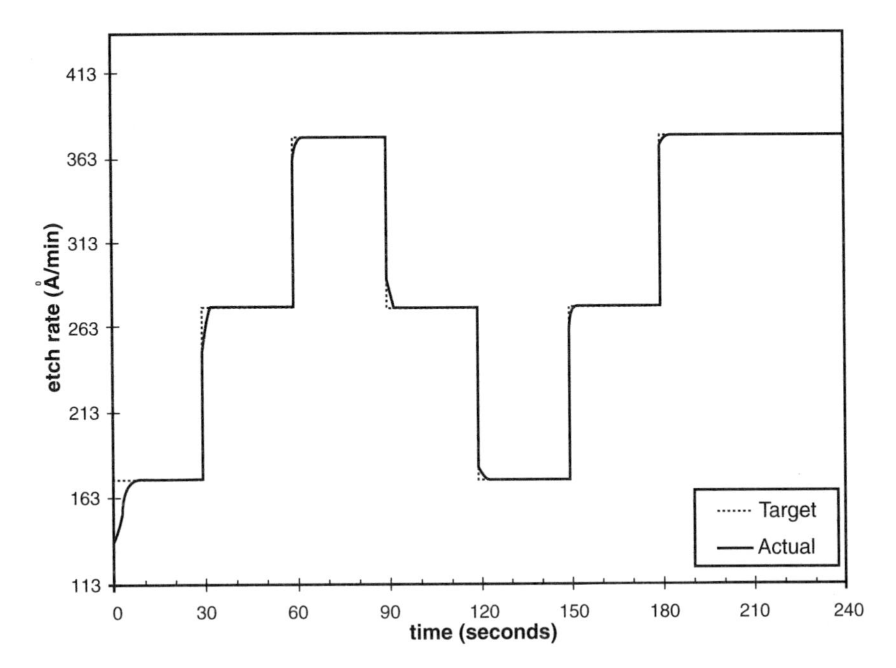

FIGURE 13.27 Real-time control of SiO_2 etch rate. The control target was changed every 30 seconds, and after each change in the control target, the neural controller made adjustments so that the etch rate predicted by the plant emulator matched the target value within 5 seconds or less. (*Source:* Stokes, C. and May, G., 1997. Real-Time Control of Reactive Ion Etching Using Neural Networks, *Proc. 1997 American Control Conf.*, III:1575-1579. With permission.)

13.6 Process Diagnosis

Neural networks have been widely used in process monitoring and diagnosis. Recently, neural nets have also begun to find use in electronics systems diagnosis. Using neural nets for process diagnosis in semiconductor manufacturing has also started to gain attention. The approaches undertaken by researchers in this area include diagnosis at three distinct levels of the manufacturing process: (i) the equipment level, (ii) the circuit level, and (iii) the wafer level.

13.6.1 Equipment-Level Diagnosis

13.6.1.1 Hybrid Expert System Approach

Kim and May employed a hybrid scheme that uses neural networks and traditional expert systems for real-time, automated malfunction diagnosis of IC fabrication equipment. Traditional expert systems excel at reasoning from previously viewed data, whereas neural networks extrapolate analyses and perform generalized classification for new scenarios. Kim and May's system has been implemented on a Plasma Therm 700 series RIE to outline a general diagnostic strategy applicable to other rapid single-wafer processes. Diagnostic systems that rely on post-process measurements and electrical test data alone cannot rapidly detect process shifts and also identify process faults. Because unreliable equipment jeopardizes product quality, it is essential to diagnose the root causes for the malfunctions quickly and accurately. May and Spanos [1993] have previously developed a real-time diagnostic system that integrates evidence from various sources using the Dempster-Shafer rules of evidential reasoning (see Section 13.2.3 above).

Extending this work, Kim and May [1997] integrated neural networks into this knowledge-based expert system. Diagnosis is conducted by this system in three chronological phases: the maintenance phase, the on-line phase, and the in-line phase. Neural networks were used in the maintenance phase to approximate the functional form of the failure history distribution of each component in the RIE system. Predicted failure rates were subsequently converted to belief levels. For on-line diagnosis of previously encountered faults, hypothesis testing on the statistical mean and variance of the sensor data was performed to search for similar data patterns and assign belief levels. Finally, neural process models of RIE figures of merit (such as etch or uniformity) were used to analyze the in-line measurements and identify the most suitable candidate among potentially faulty input parameters (i.e., pressure, gas flow, etc.) to explain process shifts.

13.6.1.1.1 Maintenance Diagnosis

During maintenance diagnosis, the objective is to derive evidence of potential component failures based on historical performance. The available data consists of only the number of failures a given component has experienced and the component age. To derive evidential support for potential malfunctions from this information, a neural-network-based reliability modeling technique was developed.

The failure probability and the instantaneous failure rate (or "hazard" rate) for each component may be estimated from a neural network trained on failure history. This neural reliability model may be used to generate evidential support and plausibility for each potentially faulty component in the frame of discernment. To illustrate, consider reliability modeling based on the *Weibull distribution.* The Weibull distribution has been used extensively as a model of time to failure in electrical and mechanical components and systems. When a system is composed of a number of components and failure is due to the most serious of a large number of possible faults, the Weibull distribution is a particularly accurate model.

The cumulative distribution function (which represents the failure probability of a component at time t) for the two-parameter Weibull distribution is given by

$$F_t = 1 - \exp\left[-\left(\frac{t}{\alpha} \right)^{\beta} \right] \qquad \text{Equation (13.29)}$$

where α and β are called scale and shape parameters. The hazard rate is given by

$$\lambda(t) = \frac{\beta t^{\beta-1}}{\alpha^{\beta}} \qquad \text{Equation (13.30)}$$

The failure rate may be computed by plotting the number of failures of each component vs. time and finding the slope of this curve at each time point. A scheme to extract the shape and scale parameters using neural networks was presented by Kim and May [1995]. After parameter estimation, the evidential support for each component is obtained from the Weibull distribution function in Equation 13.29. The corresponding plausibility is the *confidence level* (C) associated with this probability estimate, which is

$$C(t) = 1 - \left[1 - F(t)\right]^{n} \qquad \text{Equation (13.31)}$$

where n denotes the total number of component failures that have been observed at time t. Applying this methodology to the Plasma Therm RIE yields a ranked list of components faults similar to that shown in Table 13.18.

13.6.1.1.2 On-Line Diagnosis

In diagnosing previously encountered faults, neural time series (NTS) models are used to describe data indicating specific fault patterns (see Section 13.5.1.1 above). The similarity between stored NTS fault models and the current sampled pattern is measured to ascertain their likelihood of resemblance. An underlying assumption is that malfunctions are triggered by inadvertent shifts in process settings. This shift is assumed to be larger than the variability inherent in the processing equipment. To ascribe evidential support and plausibility to such a shift, statistical hypothesis tests are applied to sample means

TABLE 13.18 Fault Ranking After Maintenance Diagnosis

Component	Support	Plausibility
Capacitance manometer	0.353	0.508
Pressure switch	0.353	0.507
Electrode assembly	0.113	0.267
Exhaust valve controller	0.005	0.160
Throttle valve	0.003	0.159
Communication link	0.003	0.157
DC circuitry	0.003	0.157
Pressure transducer	0.003	0.157
Turbo pump	0.003	0.157
Gas cylinder	0.002	0.157

and variances of the time series data. This requires the assumption that the notion of statistical *confidence* is analogous to the Dempster–Shafer concept of *plausibility* [May and Spanos, 1993].

To compare two data patterns, it is assumed that if the two patterns are similar, then their means and variances are similar. Further, it is assumed that an equipment malfunction may cause either a shift in the mean or variance of a signal. The comparison begins by testing the hypothesis that the mean value of the current fault pattern ($\overline{X}_o$) equals the mean of previously stored fault patterns ($\overline{X}_i$). Letting s_o^2 and s_i^2 be the sample variances of current pattern and stored pattern, the appropriate test statistic is:

$$t_0 = \frac{\overline{X}_0 - \overline{X}_i}{\sqrt{\frac{s_0^2}{n_0} + \frac{s_i^2}{n_i}}} \qquad \text{Equation (13.32)}$$

where n_o and n_i are the sample sizes for the current and stored pattern, respectively. The statistical significance level for this hypothesis test (α_1) satisfies the relationship: $t_o = t_{\alpha 1}, v$ where v is the number of degrees of freedom. A neural network that takes the role of a t-distribution "learner" can be used to predict α_1 based on the values of t_o and v. After the significance level has been computed, the probability that the mean values of the two data patterns are equal (β_1) is equal to $1 - \alpha_1$.

Next, the hypothesis that the variance of the current fault pattern (σ_0^2) equals the variance of each stored pattern (σ_i^2) is tested. The appropriate test statistic is

$$F_0 = s_0^2 / s_i^2 \qquad \text{Equation (13.33)}$$

The statistical significance for this hypothesis test (α_2) satisfies the relationship $F_0 = F_{\alpha 2}, v_0, v_i$, where v_0 and v_i are the degrees of freedom for s_o^2 and s_i^2. A neural network trained on the F-distribution is used to predict α_2 using v_0, v_i, and F_o as inputs. The resultant probability of equal variances is $\beta_2 = 1 - \alpha_2$. After completing the hypothesis tests for equal mean and variance, the support and plausibility that the current pattern is similar to a previously stored pattern are defined as

$$\text{Support} = \text{Min}\ (\beta_1, \beta_2) \qquad \text{Equation (13.34)}$$
$$\text{Plausibility} = \text{Max}\ (\beta_1, \beta_2)$$

Using the rules of evidence combination, the support and plausibility generated at each time point are continuously integrated with their prior values.

To demonstrate, data corresponding to the faulty CHF_3 flow in Figure 13.28 was used to derive an NTS model. The training set for the NTS model consisted of one out of every ten data samples. The NTS fault model is stored in a database, from which it is compared to other patterns collected by sensors in real time so that the similarity of the sensor data to this stored pattern can be evaluated. In this example, the pattern of CHF_3 flow under consideration as a potential match to the stored fault pattern was sampled once for every 15 sensor data points. After evaluating the data, the evidential support and plausibility for pattern similarity are shown in Figure 13.29.

To identify malfunctions that have not been previously encountered, May and Spanos established a technique based on the *CUSUM* control chart [Montgomery, 1991]. The approach allows the detection of very small process shifts, which is critical for fabrication steps such as RIE, where slight equipment miscalibrations may only have sufficient time to manifest themselves as small shifts when the total processing time is on the order of minutes. CUSUM charts monitor such shifts by comparing the cumulative sums of the deviations of the sample values from their targets. This is accomplished by means of the moving "V-mask."

Using this method to generate support requires the cumulative sums

$$S_H(i) = \max\left[0, \bar{x} - (\mu_0 + b) + S_H(i-1)\right] \quad \text{Equation (13.35)}$$

$$S_L(i) = \max\left[0, (\mu_0 - b) - \bar{x} + S_L(i-1)\right] \quad \text{Equation (13.36)}$$

where S_H is the sum used to detect positive process shifts, S_L is used to detect negative shifts, $\bar{x}$ is the mean value of the current sample, and μ_0 is the target value. The parameter b is given by

$$b = \tan(2\theta\sigma_x) \quad \text{Equation (13.37)}$$

where σ_x is the standard deviation of the sampled variable and θ is the aspect angle of the V-mask, which has been selected to detect one-sigma process shifts with 95% probability. The chart has an average run length of 50 wafers between alarms when the process is in control.

When either S_H or S_L exceeds the decision interval (h), this signals that the process has shifted out of statistical control. The decision interval is

$$h = 2d\sigma_x \tan(\theta) \quad \text{Equation (13.38)}$$

where d is the V-mask lead distance. The decision interval may be used as the process tolerance limit and the sums S_H and S_L are to be treated as measurement residuals. Support is derived from the CUSUM chart using

$$s(S_{H/L}) = \frac{1-u}{1+\exp\left[-\left(\frac{S_{H/L}}{h} - 1\right)\right]} \quad \text{Equation (13.39)}$$

where the uncertainty u is dictated by the measurement error of the sensor. As S_H or S_L become large compared to h, this function generates correspondingly larger support values.

To illustrate this technique, the faulty CHF_3 data pattern in Figure 13.28 is used again, this time under the assumption that no similar pattern exists in the database. The two parameters b and h vary continuously as the standard deviation of the monitored sensor data is changing. Equation 13.35 was used to calculate the accumulated deviations of CHF_3 flow. Each accumulated shift was then fed into the sigmoidal belief function in Equation 13.38 to generate evidential support value. Figure 13.30 shows the incremental changes in the support values, clearly indicating the initial fault occurrence and the trend of process shifts.

13.6.1.1.3 In-Line Diagnosis

For in-line diagnosis, measurements performed on processed wafers are used in conjunction with inverse neural process models. Inverse models are used to predict etch recipe values (RF power, pressure, etc.)

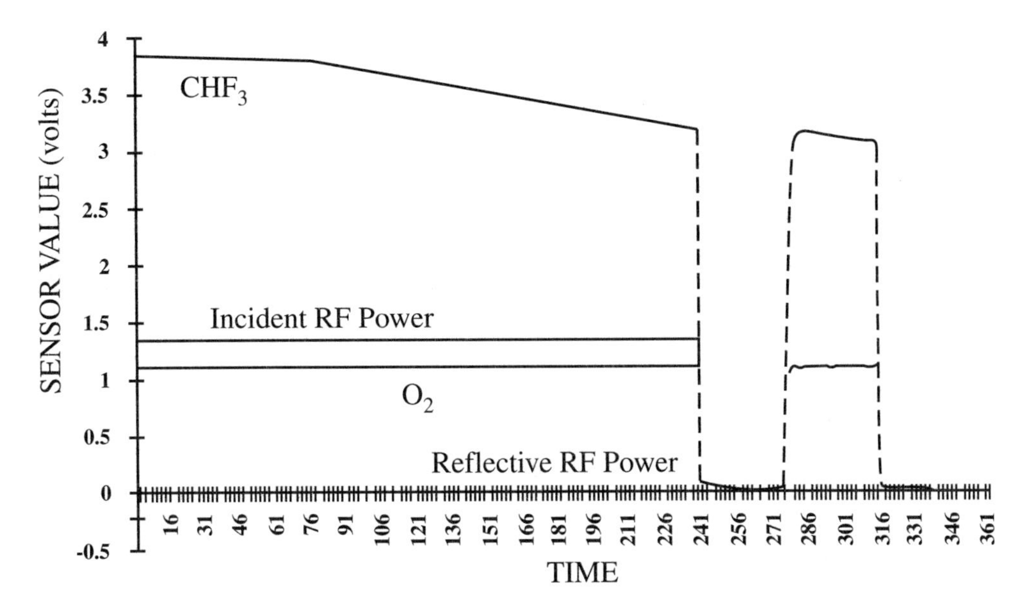

FIGURE 13.28 Data signatures for a malfunctioning chloroform mass flow controller. (*Source:* Kim, B. and May, G., 1997. Real-Time Diagnosis of Semiconductor Manufacturing Equipment Using Neural Networks, *IEEE Trans. Comp. Pack. Manuf. Tech. C*, 20(1):39-47. With permission.)

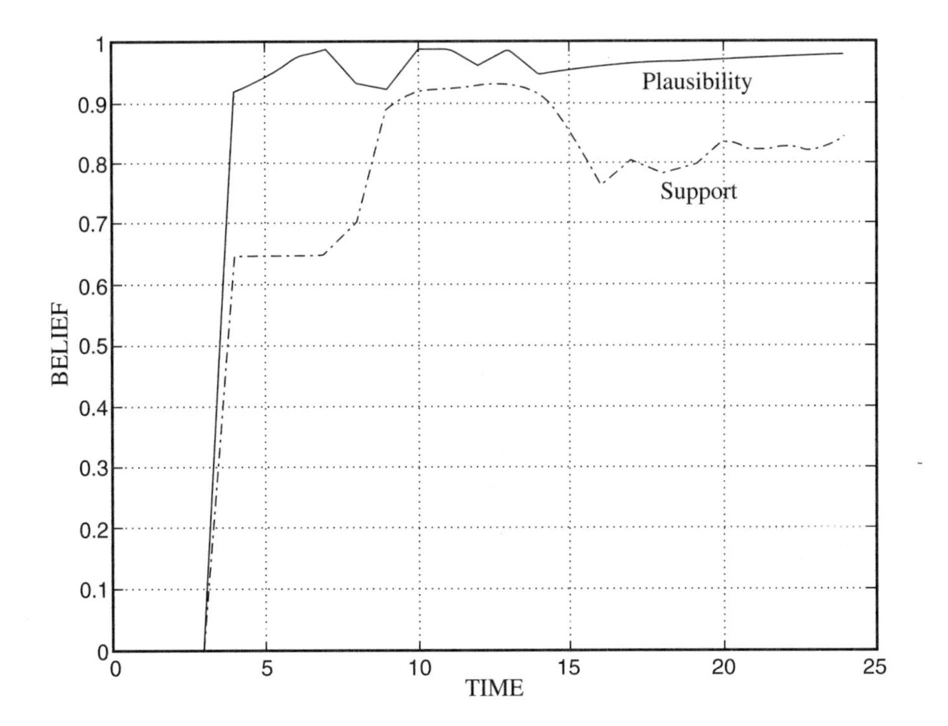

FIGURE 13.29 Plot of real-time support and plausibility for a recognized gas flow fault. (*Source:* Kim, B. and May, G., 1997. Real-Time Diagnosis of Semiconductor Manufacturing Equipment Using Neural Networks, *IEEE Trans. Comp. Pack. Manuf. Tech. C*, 20(1):39-47. With permission.)

which reduce deviations in the measured etch responses. Since the set point recipes are different from those predicted by the inverse model, the vector of differences between them (called ΔX_0) can be used in a hypothesis test to determine the statistical significance of the deviations. That statistical significance can be calculated by testing the hypothesis that $\Delta X_o = 0$.

Hotelling's T^2 statistic is employed to obtain confidence intervals on the incremental changes in the input parameters. The value of the T^2 statistic is

$$T^2 = n\Delta X^T{}_0 S^{-1} \Delta X_0 \qquad \text{Equation (13.40)}$$

where n and S are the sample size and covariance matrix of the q process input parameters. The T^2 distribution is related to the well-known F-distribution as follows:

$$T^2_{\alpha,q,n-q} = \frac{q(n-1)}{n-q} F_{\alpha,q,n-q} \qquad \text{Equation (13.41)}$$

Plausibility values calculated for each input parameter are equal to $1 - \alpha$.

To illustrate, consider a fault scenario in which increased RF power was supplied to an RIE during silicon dioxide etching due to an RF generator problem. The set points for this process were RF power = 300 W, pressure = 45 mtorr, O_2 = 11 sccm, CHF_3 = 45 sccm. The malfunction was simulated by increasing the power to 310 W and 315 W. In other words, due to the malfunction, the actual RF power being transmitted to the wafer is 310 or 315 W when it is thought to be 300 W. Forward neural models were used to predict etch responses for the process input recipes corresponding to the two different faulty values of RF power. A total of eight predictions (presumed to be the actual measurements) were obtained, and were then fed into the inverse neural etch models to produce estimates of their corresponding process input recipes. The T^2 value is calculated under the assumption that only one input parameter is the cause for any abnormality in the measurements. This leads to the different T^2 values for each process input. The resultant values of T^2 and $1 - \alpha$ are shown in Table 13.19. As expected, RF power was the most significant input parameter since it has the highest plausibility value.

Hybrid neural expert systems offer the advantage of easier knowledge acquisition and maintenance and extracting implicit knowledge (through neural network learning) with the assistance of explicit expert rules. The only disadvantage in neural expert systems is that, unlike other rule-based systems, the somewhat nonintuitive nature of neural networks makes it difficult to provide the user with explanations about how diagnostic conclusions are reached. However, these barriers are lessening as more and more successful systems are demonstrated and become available. It is anticipated that the coming decade will see neural networks integrated firmly into diagnostic software in newly created fabrication facilities.

13.6.1.2 Time Series Modeling Approach

Rietman and Beachy [1998] have used several variations of the time series modeling approach to show that neural networks can be used to detect precursors to failure in a plasma etch reactor. These authors showed that neural nets can detect subtle changes in process signals, and in some cases, these subtle changes were early warnings that a failure was imminent. The reactor used in this study was a Drytek Quad Reactor with four process chambers (although only a single chamber was considered). The process under investigation was a three-step etch used to define the location of transistors on silicon wafers.

During processing, several tool signatures were monitored (at 5-second intervals), including four gas flow rates, DC bias voltage, and forward and reflected RF power. Data was collected over approximately a 3.5-year period, which translated to over 140,000 processing steps on about 46,000 wafers. Models were built from the complete time streams, as well as from data consisting of time series summary statistics (mean and standard deviation values) for each process signature for each wafer. Samples that deviated by more than four standard deviations from the mean for a given response variable were classified as failure events. Based on this classification scheme, a failure occurred approximately every 9000 wafers.

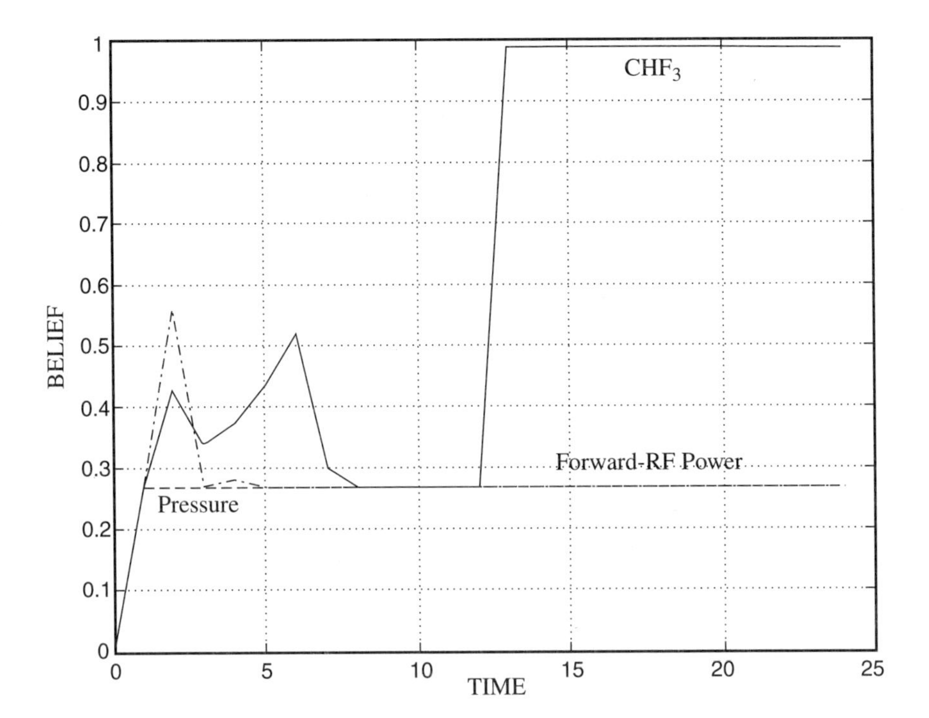

FIGURE 13.30 Support variations using CUSUM technique. (*Source:* Kim, B. and May, G., 1997. Real-Time Diagnosis of Semiconductor Manufacturing Equipment Using Neural Networks, *IEEE Trans. Comp. Pack. Manuf. Tech. C*, 20(1):39-47. With permission.)

TABLE 13.19 T^2 and Plausibility Values

Parameter	T_2	$1 - \alpha$
CHF_3	0.053	0.272
O_2	2.84	0.278
Pressure	2.89	0.280
RF Power	22.52	0.694

Source: Kim, B. and May, G., 1997. Real-Time Diagnosis of Semiconductor Manufacturing Equipment Using Neural Networks, *IEEE Trans. Comp. Pack. Manuf. Tech. C*, 20(1):39-47.

Rietman and Beachy focused on pressure for response modeling. The models constructed for summary statistical data had the advantage that the mean and standard deviation of the time series could be expected to exhibit less noise than the raw data. For example, a model was derived from process signatures for 3000 wafers processed in sequence. The means and standard deviations for each step of the three-step process served as additional sources of data. A network with 21 inputs (etch end time, total etch time, step number, mean and standard deviations for four gases, RF applied and reflected, pressure, and DC bias), 5 hidden units, and a single output was used to predict pressure. The results of this prediction for 1, 12, and 24 wafers in advance is shown in Figure 13.31.

To demonstrate malfunction prediction, these authors again elected to examine summary data, this time in the form of the standard deviation time streams. Their assumption was that fluctuations in these signatures would be more indicative of precursors to equipment failure. For this part of the investigation, a neural time series model was constructed with inputs consisting of five delay units, one current time unit, one recurrent time unit from the network output, and one bias unit. This network had five hidden units and a single output. Figure 13.32(a) shows the mean value of pressure at each of the three processing steps. This was the time stream to be modeled. A failure was observed at wafer 5770. Figure 13.32(b)

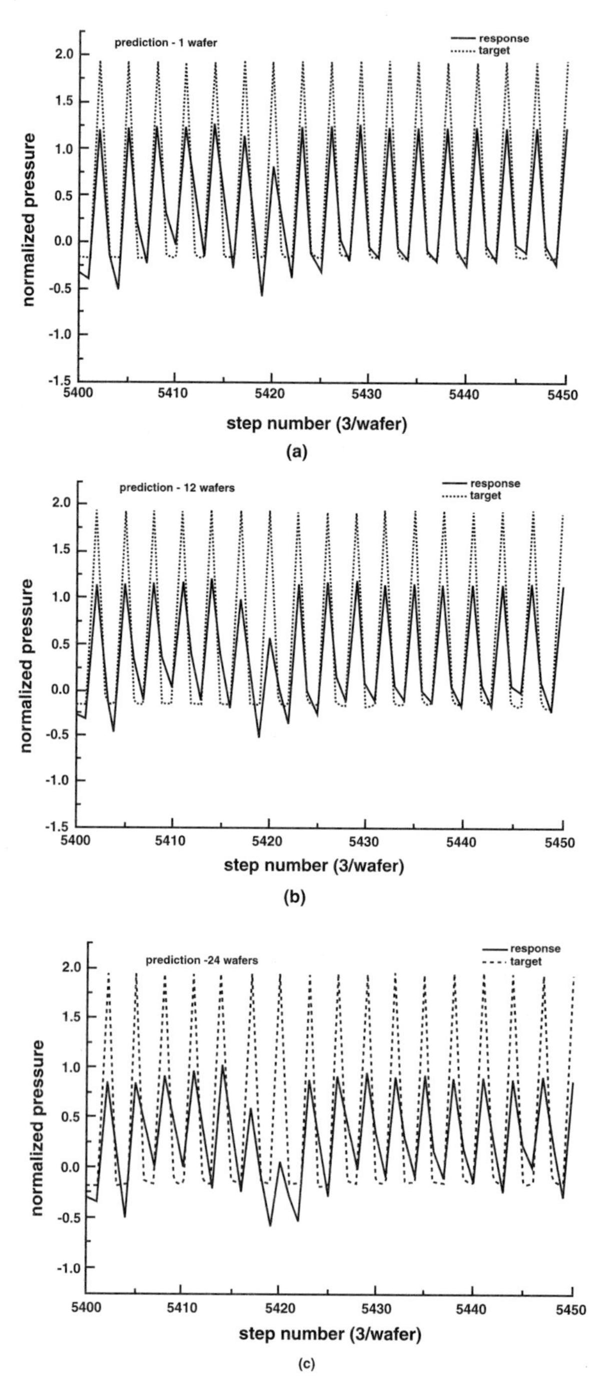

FIGURE 13.31 (a) Pressure prediction one wafer in the future; (b) pressure prediction 12 wafers in the future; and (c) pressure prediction 24 wafers in the future. (*Source:* Rietman, E. and Beachy, M., 1998. A Study on Failure Prediction in a Plasma Reactor, *IEEE Trans. Semi. Manuf.*, 11(4):670-680. With permission.)

shows the corresponding standard deviation time stream, with the failure at 5770 clearly observable, as well as precursors to failure beginning at 5710 to 5725. Figure 13.32(c) shows the RMS error of the network trained to predict the standard deviation signal as a function of the number of training iterations. Finally, Figure 13.32(d) compares the network response to the target values, clearly indicating that the network is able to detect the fluctuations in standard deviation that are indicative of the malfunction.

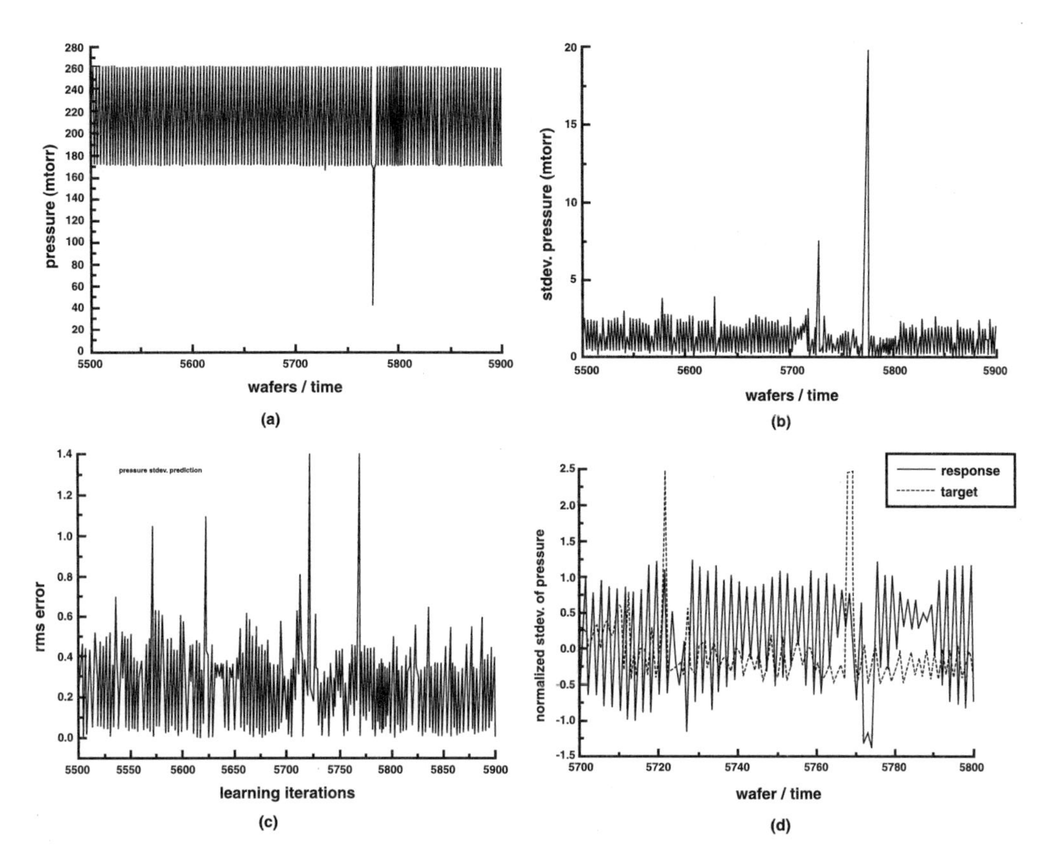

FIGURE 13.32 (a) Mean value of pressure between 5500 and 5900 samples. A failure can be seen, but no precursors to the failure are seen in the mean value data. (b) Standard deviation of pressure of the same time segment. Here, precursors are seen at about 5700 and the failure occurs at 5775. The precursors thus show up about 12 wafers prior to the actual failure. (c) Segment of neural network learning curve showing the detection of the precursors shown in (b). (d) Target and response curve for the same neural network predicting pressure. (*Source:* Rietman, E. and Beachy, M., 1998. A Study on Failure Prediction in a Plasma Reactor, *IEEE Trans. Semi. Manuf.*, 11(4):670-680. With permission.)

13.6.1.3 Pattern Recognition Approach

Bhatikar and Mahajan [1999] have used a neural-network-based pattern recognition approach to identify and diagnosis malfunctions in a CVD barrel reactor used in silicon epitaxy. Their strategy was based on modeling the spatial variation of deposition rate on a particular facet of the reactor. The hypothesis that motivated this work was that spatial variation, as quantified by a vector of variously measured standard deviations, encoded a pattern reflecting the state of the reactor. Thus, faults could be diagnosed by decoding this pattern using neural networks.

Figure 13.33 shows a schematic diagram of the CVD barrel reactor under consideration. In this reactor, silicon wafers are positioned in shallow pockets of a heated graphite susceptor. Reactive gases are introduced into the reactor through nozzles at the top of the chamber and exit from the outlet at the bottom. The six controllable reactor settings include flow velocity at the left and right nozzles, the settings of the nozzles in the horizontal and vertical planes, the main flow valve reading, and the rotational flow valve reading.

Bhatikar and Mahajan chose the uniformity of the deposition rate as the response variable to optimize. Each side of the susceptor held three wafers, and deposition rate measurements were performed on five sites on each wafer. Afterward, a polynomial regression model was developed that described the film thickness at each of the five measurement locations for each wafer as a function of the six reactor settings.

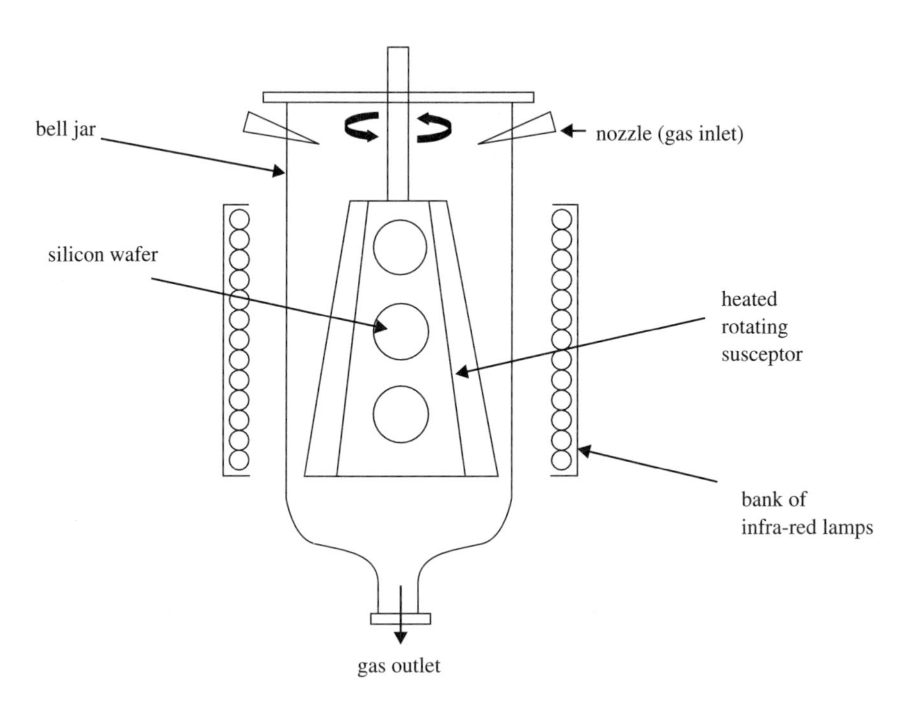

FIGURE 13.33 Vertical chemical vapor deposition barrel reactor. (*Source:* Bhatikar, S. and Mahajan, R., 1999. Neural Network Based Diagnosis of CVD Barrel Reactor, *Advances in Electronic Packaging*, 26-1:621-640. With permission.)

Next, backpropagation neural networks were trained as event classifiers to detect significant deviations from the target uniformity. Eight specific distributions of thickness measurements were computed; these are depicted in Figure 13.34. As a group, these eight standard deviations constituted a process signature. Patterns associated with normal and specific types of abnormal behavior were captured in these signatures.

Three disparate events were then simulated to represent deviations from normal equipment settings: (i) a mismatch between the left and right nozzles; (ii) a horizontal nozzle offset; and (iii) a vertical nozzle offset. The first event was simulated with a 5% mismatch, and offsets from 0% to 20% were simulated for both the vertical and horizontal directions. A neural network was then trained to match these events with their process signatures as quantified by the vector of eight standard deviations. The network had eight input neurons and three outputs (one for each event). The number of hidden layer neurons was varied from five to seven, with six providing the best performance. Each output was a binary response, with one or zero representing the presence or absence of a given event. The threshold for a binary "high" was set at 0.5. Training consisted of exposing the network to an equal number of representative signatures for each event. When tested on 12 signatures not seen during training (4 for each event), the network was able to discriminate between the three faults with 100% accuracy.

This scheme was then applied to a fault detection task (as opposed to fault classification only). This required the addition of a "non-event" representing normal equipment operation. Since there was only one signature corresponding to the non-event, this signature was replicated in the training data with the addition of white noise to the optimal equipment settings to simulate typical random process variation. The network used for detection had the same structure as that used for classification, with the exception of having seven hidden layer neurons rather than six. After an adjustment of the "high" threshold to a value of 0.78, 100% classification accuracy was again achieved.

13.6.2 Circuit-Level Diagnosis

At the integrated circuit level, Plummer [1993] has developed a process control neural network (PCNN) to identify faults in bipolar operational amplifiers (or op amps) based on electrical test data. The PCNN exploits the capability of neural nets to interpret multidimensional data and identify clusters of performance

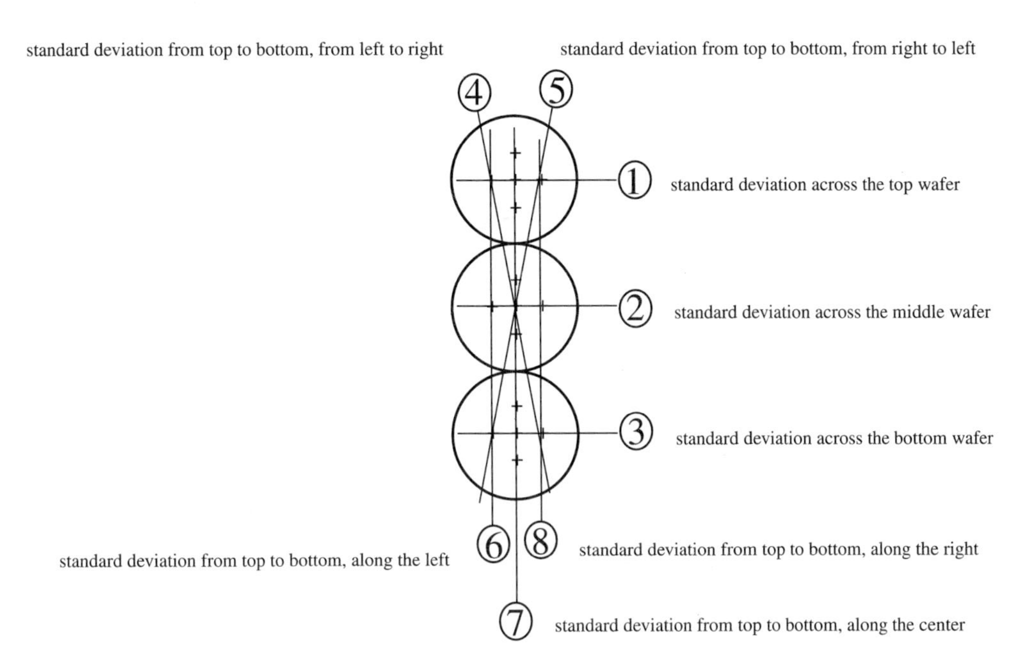

FIGURE 13.34 Extraction of vector to characterize spatial variation from thickness distribution. (*Source:* Bhatikar, S. and Mahajan, R., 1999. Neural Network Based Diagnosis of CVD Barrel Reactor, *Advances in Electronic Packaging*, 26-1:621-640. With permission.)

within such a data set. This provides enhanced sensitivity to sources of variation that are not distinguishable from observing traditional single-variable control charts. Given a vector of electrical test results as input, the PCNN can evaluate the probability of membership in each set of clusters, which represent different categories of circuit faults. The network can then report the various fault probabilities or select the most likely fault category.

Representing one of the few cases in semiconductor manufacturing in which backpropagation networks are not employed, the PCNN is formed by replacing the output layer of a probabilistic neural network with a Grossberg layer (Figure 13.35). In the probabilistic network, input data is fed to a set of pattern nodes. The pattern layer is trained using weights developed with a Kohonen self-organizing network. Each pattern node contains an exemplar vector of values corresponding to an input variable typical of the category it represents. If more than one exemplar represents a single category, the number of exemplars reflects the probability that a randomly selected pattern is included in that category. The proximity of each input vector to each pattern is computed, and the results are analyzed in the summation layer.

The Grossberg layer functions as a lookup table. Each node in this layer contains a weight corresponding to each category defined by the probabilistic network. These weights reflect the conditional probability of a cause belonging to the corresponding category. Then outputs from the Grossberg layer reflect the products of the conditional probabilities. Together, these probabilities constitute a Pareto distribution of possible causes for a given test result (which is represented in the PCNN input vector). The Grossberg layer is trained in a supervised manner, which requires that the cause for each instance of membership in a fault category be recorded beforehand.

Despite its somewhat misleading name, Plummer applied the PCNN in a diagnostic (as opposed to a control) application. The SPICE circuit simulator was used to generate two sets of highly correlated input–output operational amplifier test data, one representing an in-control process and the other a process grossly out of control. Even though the second data set represented faulty circuit behavior, its descriptive statistics alone gave no indication of suspicious electrical test data. Training the Kohonen

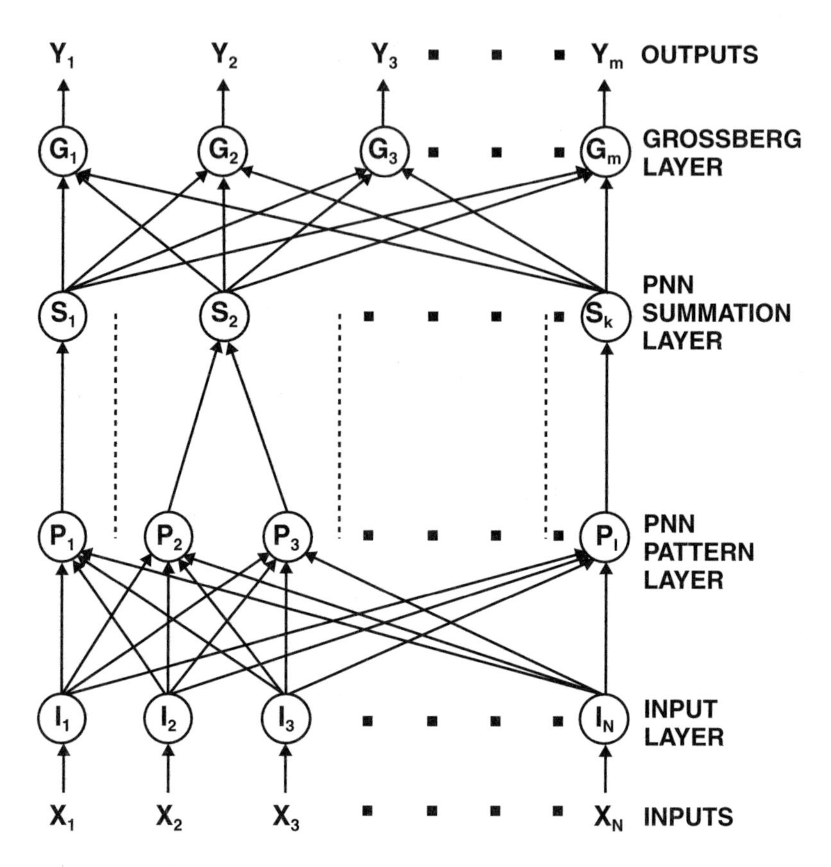

FIGURE 13.35 Process control neural network. This network is formed by replacing the output layer of a probabilistic neural network with a Grossberg layer whose outputs reflect probabilities which constitute a Pareto distribution of possible causes for a given input vector. (*Source:* Plummer, J., 1993. Tighter Process Control with Neural Networks, *AI Expert*, 10:49-55. With permission).

network with electrical test results from these data sets produced four distinct clusters (representing one acceptable and three faulty states).

With the Kohonen exemplars serving as weights in the pattern layer, the PCNN then was used to identify one of the three possible out-of-control conditions: (i) low npn β; (ii) high npn β and low resistor tolerance; or (iii) high npn β and high resistor tolerance. The summation layer of the PCNN reported the conditional probability of each of these conditions and the probability that the op amp measurements were acceptable for each input pattern of electrical test data. The PCNN was 93% accurate in overall diagnosis, and correctly sounded alarms for 86% of the out-of-control cases (no false alarms were generated). The PCNN was therefore an exceptional adaptive diagnostic tool.

13.7 Summary

In electronics manufacturing, process and equipment reliability directly influence cost, throughput, and yield. Over the next several years, significant process modeling and control efforts will be required to reach projected targets for future generations of microelectronic devices and integrated circuits. Computer-assisted methods will provide a strategic advantage in undertaking these tasks, and among such methods, neural networks have certainly proved to be a viable technique.

Thus far, the use of computational intelligence has not yet become routine in electronics manufacturing at the process engineering level. For example, the use of neural networks now is probably at a point in its evolution comparable to that of statistical experimental design or Taguchi methodology a decade or two ago, and now statistical methods such as these have become pervasive in the industry. The outlook for computational intelligence is therefore similarly promising. New applications are appearing and software is constantly being developed to meet the needs of these applications. The overall impact of computational intelligence techniques in this field depends primarily on awareness of their capabilities and limitations, coupled with a commitment to their implementation. With each new successful application, these techniques continue to gain acceptance, and thus their future is bright.

Defining Terms

Adaptive control: Advanced process control system capable of automatically adjusting or "adapting" itself to meet a desired output despite shifting control objectives and process conditions, or unmodeled uncertainties in process dynamics.

Backpropagation: Popular algorithm for training artificial neural networks; involves adjusting the weights of the network to minimize the squared error between the network output and a training signal.

CIM: Computer-integrated manufacturing; in electronics manufacturing, this refers to the effort to use the latest developments in computer hardware and software technology to enhance expensive manufacturing methods.

CVD: Chemical vapor deposition; semiconductor fabrication process in which material is deposited on a substrate using reactive chemicals in the vapor phase.

Dempster-Shafer theory: Set of techniques used to perform diagnostic inference that account for uncertainty in the system.

Expert systems: Experiential or algorithmic systems that attempt to encode and use human knowledge to perform inference procedures (such as fault diagnosis).

Factorial designs: Experimental designs in which multiple input variables are varied simultaneously at two or more discrete levels in every possible combination.

Fractional factorial designs: Factorial designs that reduce the number of experiments to be performed by exploring only a fraction (such as one half) of the input variable space in a systematic manner.

Genetic algorithms: Guided stochastic search techniques based on the principles of genetics.

Hybrid neural networks: Semi-empirical neural network process models that take into account known process physics.

IC-CIM: Computer-integrated manufacturing of integrated circuits.

LPCVD: Low-pressure chemical vapor deposition; CVD performed at pressures well below atmospheric pressure.

Modular neural networks: Neural networks that consist of a group of subnetworks ("modules") competing to learn different aspects of a problem.

Objective function: Criterion used in optimization applications to evaluate the suitability or "cost" of various solutions to the optimization problem at hand; also known as a "fitness" function in the context of genetic algorithms.

Neural networks: Artificial models that crudely mimic the functionality of biological neurological systems.

PECVD: Plasma-enhanced chemical vapor deposition; CVD performed in the presence of a plasma discharge.

Powell's algorithm: A method of optimization that generates successive quadratic approximations of the space to be optimized; involves determining a set of n linearly independent, mutually conjugate directions (where n is the dimensionality of the search space).

RIE: Reactive ion etching; method of removing material by reactive gases at low pressures in an electric field; also known as "plasma etching."

RSM: Response surface methodology; statistical method in which data from designed experiments is used to construct polynomial response models whose coefficients are determined by regression techniques.

Simplex method: A method of optimization; a regular simplex is defined as a set of $(n + 1)$ mutually equidistant points in n-dimensional space. The main idea of the simplex method is to compare the values of the function to be optimized at the $(n + 1)$ vertices of the simplex and move the simplex iteratively towards the optimal point.

Simulated annealing: Artificial algorithm for finding the minimum error state in optimizing a complex system; analogous to the slow cooling procedure that enables nature to find the minimum energy state in thermodynamics.

SPC: Statistical process control; method of continuous hypothesis testing to ensure that a manufactured product meets its required specifications.

Time series modeling: Statistical method for modeling chronologically sequenced data.

References

Baker, M., Himmel, C., and May, G., 1995. Time Series Modeling of Reactive Ion Etching Using Neural Networks, *IEEE Trans. Semi. Manuf.,* 8(1):62-71.

Bhatikar, S. and Mahajan, R., 1999. Neural Network Based Diagnosis of CVD Barrel Reactor, *Advances in Electronic Packaging,* 26-1:621-640.

Bose, C. and Lord, H., 1993. Neural Network Models in Wafer Fabrication, *SPIE Proc. Applications of Artificial Neural Networks,* 1965:521-530.

Box, G. and Draper, N., 1987. *Empirical Model-Building and Response Surfaces,* Wiley, New York, NY.

Box, G. and Jenkins, G., 1976. *Time Series Analysis: Forecasting and Control,* Holden-Day, San Francisco.

Box, G., Hunter, W., and Hunter, J., 1978. *Statistics for Experimenters,* Wiley, New York.

Burke, L. and Rangwala, S., 1991. Tool Condition Monitoring in Metal Cutting: A Neural Network Approach, *J. Intelligent Manuf.,* 2(5).

Cardarelli, G., Palumbo, M., and Pelagagge, P., 1996. Photolithography Process Modeling Using Neural Networks, *Semicond. Int.,* 19(7):199-206.

Dax, M., 1996. Top Fabs of 1996, *Semicond. Int.,* 19(5):100-106.

Dayhoff, J., 1990. *Neural Network Architectures: An Introduction,* Van Nostrand Reinhold, New York.

Galil, Z. and Kiefer, J., 1980. Time- and Space-Saving Computer Methods, Related to Mitchell's DETMAX, for Finding D-Optimum Designs, *Technometrics,* 22(3):301-313.

Goldberg, D., 1989. *Genetic Algorithms in Search, Optimization and Machine Learning,* Addison Wesley, Reading, MA.

Han, S. and May, G., 1996. Optimization of Neural Network Structure and Learning Parameters Using Genetic Algorithms, *Proc. IEEE Int. Conf. AI Tools,* 8:200-206.

Han, S. and May, G., 1997. Using Neural Network Process Models to Perform PECVD Silicon Dioxide Recipe Synthesis via Genetic Algorithms, *IEEE Trans. Semi. Manuf.,* 10(2):279-287.

Han, S., Ceiler, M., Bidstrup, S., Kohl, P., and May, G., 1994. Modeling the Properties of PECVD Silicon Dioxide Films Using Optimized Back-Propagation Neural Networks, *IEEE Trans. Comp. Packag. Manuf. Technol.,* 17(2):174-182.

Himmel, C. and May, G., 1993. Advantages of Plasma Etch Modeling Using Neural Networks over Statistical Techniques, *IEEE Trans. Semi. Manuf.,* 6(2):103-111.

Hodges, D., Rowe, L., and Spanos, C., 1989. Computer-Integrated Manufacturing of VLSI, *Proc. IEEE/CHMT Int. Elec. Manuf. Tech. Symp.,* 1-3.

Holland, J., 1975. *Adaptation in Natural and Artificial Systems,* University of Michigan Press, Ann Arbor, MI.

Hopfield, J. and Tank, D., 1985. Neural Computation of Decisions in Optimization Problems, *Biological Cybernetics,* 52:141-152.

Huang, Y., Edgar, T., Himmelblau, D., and Trachtenberg, I., 1994. Constructing a Reliable Neural Network Model for a Plasma Etching Process Using Limited Experimental Data, *IEEE Trans. Semi. Manuf.*, 7(3):333-344.

Irie, B. and Miyake, S., 1988. Capabilities of Three-Layered Perceptrons, *Proc. IEEE Int. Conf. on Neural Networks*, 641-648.

Kim, B. and May, G., 1994. An Optimal Neural Network Process Model for Plasma Etching, *IEEE Trans. Semi. Manuf.*, 7(1):12-21.

Kim, B. and May, G., 1995. Estimation of Weibull Distribution Parameters Using Modified Back-Propagation Neural Networks, *World Congress on Neural Networks*, I:114-117.

Kim, B. and May, G., 1996. Reactive Ion Etch Modeling Using Neural Networks and Simulated Annealing, *IEEE Trans. Comp. Pack. Manuf. Tech. C*, 19(1):3-8.

Kim, B. and May, G., 1997. Real-Time Diagnosis of Semiconductor Manufacturing Equipment Using Neural Networks, *IEEE Trans. Comp. Pack. Manuf. Tech. C*, 20(1):39-47.

Manos, D. and Flamm, D., 1989. Plasma Etching: An Introduction, Academic Press, San Diego, CA.

Marwah, M. and Mahajan, R., 1999. Building Equipment Models Using Neural Network Models and Model Transfer Techniques, *IEEE Trans. Semi. Manufac.*, 12(3):377-380.

May, G., 1994. Manufacturing ICs the Neural Way, *IEEE Spectrum*, 31(9):47-51.

May, G. and Spanos, C., 1993. Automated Malfunction Diagnosis of Semiconductor Fabrication Equipment: A Plasma Etch Application, *IEEE Trans. Semi. Manuf.*, 6(1):28-40.

May, G., Huang, J., and Spanos, C., 1991. Statistical Experimental Design in Plasma Etch Modeling, *IEEE Trans. Semi. Manuf.*, 4(2):83-98.

Mocella, M., Bondur, J., and Turner, T., 1991. Etch Process Characterization Using Neural Network Methodology: A Case Study, *SPIE Proc. on Module Metrology, Control and Clustering*, 1594.

Montgomery, D., 1991. *Introduction to Statistical Quality Control*, Wiley, New York.

Mori, H. and Ogasawara, T., 1993. A Recurrent Neural Network Approach to Short-Term Load Forecasting in Electrical Power Systems, *Proc. 1993 World Congress on Neural Networks*, I:342-345.

Murphy, J. and Kagle, B., 1992. Neural Network Recognition of Electronic Malfunctions, *J. Intelligent Manuf.*, 3(4):205-216.

Nadi, F., Agogino, A., and Hodges, D., 1991. Use of Influence Diagrams and Neural Networks in Modeling Semiconductor Manufacturing Processes, *IEEE Trans. Semi. Manuf.*, 4(1):52-58.

Nami, Z., Misman, O., Erbil A., and May, G., 1997. Semi-Empirical Neural Network Modeling of Metal-Organic Chemical Vapor Deposition, *IEEE Trans. Semi. Manuf.*, 10(2):288-294.

Natale, C., Proietti, E., Diamanti, R., and D'Amico, A., 1999. Modeling of APCV-Doped Silicon Dioxide Deposition Process by a Modular Neural Network, *IEEE Trans. Semi. Manuf.*, 12(1):109-115.

Nelson, D., Ensley, D., and Rogers, S., 1992. Prediction of Chaotic Time Series Using Cascade Correlation: Effects of Number of Inputs and Training Set Size, *SPIE Conf. on Applications of Neural Networks*, 1709:823-829.

Pan, J. and Tenenbaum, J., 1986. PIES: An Engineer's "Do-it-yourself" Knowledge System for Interpretation of Parametric Test Data, *Proc. 5th Nat. Conf. AI.*

Parsaye, K. and Chignell, M., 1988. *Expert Systems for Experts,* Wiley, New York.

Plummer, J., 1993. Tighter Process Control with Neural Networks, AI Expert, 10:49-55.

Rao, S. and Pappu, R., 1993. Nonlinear Time Series Prediction Using Wavelet Networks, *Proc. 1993 World Congress on Neural Networks*, IV:613-616.

Rietman, E. and Beachy, M., 1998. A Study on Failure Prediction in a Plasma Reactor, *IEEE Trans. Semi. Manufac.*, 11(4):670-680.

Rietman, E. and Lory, E., 1993. Use of Neural Networks in Semiconductor Manufacturing Processes: An Example for Plasma Etch Modeling, *IEEE Trans. Semi. Manuf.*, 6(4):343-347.

Rietman, E., Patel, S., and Lory, E., 1993. Neural Network Control of a Plasma Gate Etch: Early Steps in Wafer-to-Wafer Process Control, *Proc. Int. Elec. Manuf. Tech. Symp.*, 15:454-457.

Salam, F., Piwek, C., Erten, G., Grotjohn, T., and Asmussen, J., 1997. Modeling of a Plasma Processing Machine for Semiconductor Wafer Etching Using Energy-Function-Based Neural Networks, *IEEE Trans. Cont. Sys.*

Shafer, G., 1976. *A Mathematical Theory of Evidence*. Princeton University Press, Princeton, NJ.

Smith, T. and Boning, D., 1996. A Self-Tuning EWMA Controller Utilizing Artifical Neural Network Function Approximation Techniques, *Proc. 1996 Int. Elec. Manuf. Tech. Symp.*, 18:355-361.

Spanos, C., 1986. HIPPOCRATES: A Methodology for IC Process Diagnosis, *Proc. ICCAD.*

Stokes, C. and May, G., 1997. Real-Time Control of Reactive Ion Etching Using Neural Networks, *Proc. 1997 American Control Conf.*, III:1575-1579.

Wang, X. and Mahajan, R., 1996. Artifical Neural Network Model-Based Run-to-Run Process Controller, *IEEE Trans. Comp. Pack. Manuf. Tech. C*, 19(1):19-26.

Wasserman, P., Unal, A., and Haddad, S., 1991. Neural Networks for On-Line Machine Condition Monitoring, in *Intelligent Engineering Systems Through Artificial Neural Networks*, Ed. C. Dagli, pp. 693-699, ASME Press, New York.

White, D., Boning, D., Butler, S., and Barna, G., Spatial Characterization of Wafer State Using Principal Component Analysis of Optical Emission Spectra in Plasma Etch, *IEEE Trans. Semi. Manuf.*, 10(1):52-61.

Further Information

"Neural nets for semiconductor manufacturing" is presented by Gary S. May in volume 14 of the *Wiley Encyclopedia of Electrical and Electronic Engineering*, pp. 298-323, 1999. "Applications of neural networks in semiconductor manufacturing processes" are described by Gary S. May in Chapter 18 of the *Fuzzy Logic and Neural Network Handbook*, McGraw-Hill, 1996. "Manufacturing ICs the neural way" is discussed by Gary S. May in *IEEE Spectrum*, vol. 31, no. 9, September, 1994, pp. 47-51. "Neural networks in manufacturing: a survey" is authored by Samuel H. Huang and Hong-Chao Zhang and appears in the proceedings of the *15th International Electronics Manufacturing Technology Symposium*, Santa Clara, CA, October 1993, pp. 177-186. "Neural networks at work" (June 1993, pp. 26-32) and "Working with neural networks" (July 1993, pp. 46-53) are both presented by Dan Hammerstrom in *IEEE Spectrum.* These articles provide a comprehensive overview of neural network architectures, training algorithms, and applications. "Progress in supervised neural networks — What's new since Lippman" is discussed by Don R. Hush and William D. Horne in the *IEEE Signal Processing Magazine*, January 1993, pp. 8-38. "An introduction to computing with neural networks" is presented by Richard Lippman, *IEEE Acoustics, Speech, and Signal Processing Magazine*, April, 1987, pp. 4-22. In general, *IEEE Transactions on Neural Networks* and the *IEEE International Conference on Neural Networks* are widely acknowledged as the definitive IEEE publication and conference on neural network activities.

14

Monitoring and Diagnosing Manufacturing Processes Using Fuzzy Set Theory

R. Du*
University of Miami

Yangsheng Xu
Chinese University of Hong Kong

Abstract

Monitoring and diagnosis play an important role in modern manufacturing engineering. They help to detect product defects and process/system malfunctions early, and hence, eliminate costly consequences. They also help to diagnose the root causes of the problems in design and production and hence minimize production loss and at the same time improve product quality. In the past decades, many monitoring and diagnosis methods have been developed, among which the fuzzy set theory has demonstrated its effectiveness. This chapter describes how to use the fuzzy set theory for engineering monitoring and diagnosis. It introduces various methods such as fuzzy linear equation method, fuzzy C-mean method, fuzzy decision tree method, and a newly developed method, fuzzy transition probability method. By using good examples, it demonstrates step by step how the theory and the computation work. Two practical examples are also included to show the effectiveness of the fuzzy set theory.

14.1 Introduction

According to *Webster's New World Dictionary of the American Language,* "monitoring," among several other meanings, means checking or regulating the performance of a machine, a process, or a system. "Diagnosis" means deciding the nature and the cause(s) of a diseased condition of a machine, a process, or a system by examining the performance or the symptoms. In other words, monitoring detects suspicious symptoms while diagnosis determines the cause of the symptoms. There are several words and/or

* This work was completed when Dr. Du visited The Chinese University of Hong Kong.

0-8493-0592-6/01/$0.00+$.50

phrases that have similar or slightly different meanings, such as fault detection, fault prediction, in-process verification, on-line inspection, identification, and estimation.

Monitoring and diagnosing play a very important role in modern manufacturing. This is because manufacturing processes are becoming increasingly complicated and machines are much more automated. Also, the processes and the machines are often correlated; and hence, even small malfunctions or defects may cause catastrophic consequences. Therefore, a great deal of research has been carried out in the past 20 years. Many papers and monographs have been published. Instead of giving a partial review here, the reader is referred to two books. One by Davies [1998] describes various monitoring and diagnosis technologies and instruments. The reader should also be aware that there are many commercial monitoring and diagnosis systems available. In general, monitoring and diagnosis methods can be divided into two categories: a model-based method and a feature-based method. The former is applicable where a dynamic model (linear or nonlinear, time-invariant or time-variant) can be established, and is commonly used in electrical and aerospace engineering. The book by Gertler [1988] describes the basics of model-based monitoring. The latter uses the features extracted from sensor signals (such as cutting forces in machining processes and pressures in pressured vessels) and can be used in various engineering areas. This chapter will focus on this type of method.

More specifically the objective of this chapter is to introduce the reader to the use of fuzzy set theory for engineering monitoring and diagnosis. The presented method is applicable to almost all engineering processes and systems, simple or complicated. There are of course many other methods available, such as pattern recognition, decision tree, artificial neural network, and expert systems. However, from the discussions that follow, the readers can see that fuzzy set theory is simple and effective method that is worth exploring.

This chapter contains five sections. Section 14.2 is a brief review of fuzzy set theory. Section 14.3 describes how to use fuzzy set theory for monitoring and diagnosing manufacturing processes. Section 14.4 presents several application examples. Finally, Section 14.5 contains the conclusions.

14.2 A Brief Description of Fuzzy Set Theory

14.2.1 The Basic Concept of Fuzzy Sets

Since fuzzy set theory was developed by Zadeh [1965], there have been many excellent papers and monographs on this subject, for example [Baldwin et al., 1995; Klir and Folger, 1988]. Hence, this chapter only gives a brief description of fuzzy set theory for readers who are familiar with the concept but are unfamiliar with the calculations. The readers who would like to know more are referred to the above-mentioned references.

It is known that a crisp (or deterministic) set represents an exclusive event. Suppose A is a crisp set in a space $\mathbf{X}$ (i.e., $A \subset \mathbf{X}$), then given any element in $\mathbf{X}$, say x, there will be either $x \in A$ or $x \notin A$. Mathematically, this crisp relationship can be represented by a membership function, $\mu(A)$, as shown in Figure 14.1, where $x \notin$ (b,c). Note that $\mu(A) = \{0, 1\}$. In comparison, for a fuzzy event, A', its membership function, $\mu(A')$, varies between 0 and 1, that is $\mu(A) = [0, 1]$. In other words, there are cases in which the instance of the event $x \in A'$ can only be determined with some degree of certainty. This degree of certainty is referred to as *fuzzy degree* and is denoted as $\mu_{A'}(x \in A')$. Furthermore, the fuzzy set is denoted as $x/\mu_{A'}(x)$, $\forall\ x \in A'$, and $\mu_{A'}(x)$ is called the *fuzzy membership function* or the *possibility distribution.*

It should be noted that the fuzzy degree has a clear meaning: $\mu(x) = 0$ means x is impossible while $\mu(x) = 1$ implies x is certainly true. In addition, the fuzzy membership function may take various forms such as a discrete tablet,

$$\begin{array}{llllll} x: & x_1 & x_2 & \ldots & x_n & \\ \mu(x): & \mu(x_1) & \mu(x_2) & \ldots & \mu(x_n) & \text{Equation (14.1)} \end{array}$$

or a continuous step-wise function,

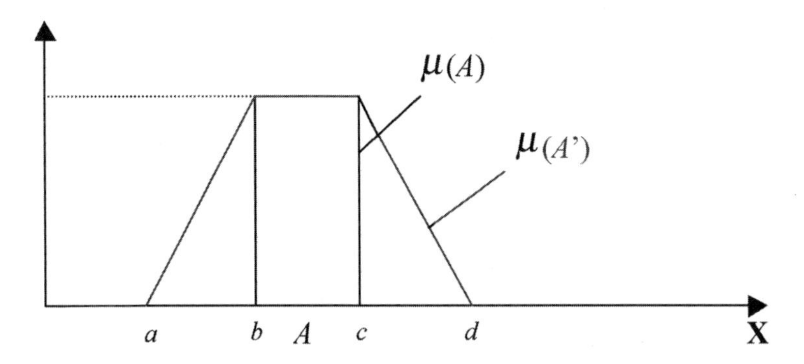

FIGURE 14.1 Illustration of crisp and fuzzy concept.

$$\mu(x)=\begin{cases} 0 & x \le a \\ \dfrac{x-a}{b-a} & a < x \le b \\ 1 & b < x \le c \\ \dfrac{d-x}{d-c} & c < x \le d \\ 0 & d < x \end{cases} \qquad \text{Equation (14.2)}$$

where a, b, c, and d are constants that determines the shape of $\mu(x)$. This is shown in Figure 14.1.

With the help of the membership functions, various fuzzy operations can be carried out. For example, given $A, B \subseteq \mathbf{X}$, we have

(a) union:

$$\mu(A \cup B) = \max\{\mu(A), \mu(B)\}, \forall\ x \in A, B \qquad \text{Equation (14.3)}$$

(b) intersection:

$$\mu(A \cap B) = \min\{\mu(A), \mu(B)\}, \forall\ x \in A, B \qquad \text{Equation (14.4)}$$

(c) contradiction:

$$\mu(\bar{A}) = 1 - \mu(A), \forall x \in A \qquad \text{Equation (14.5)}$$

To demonstrate these operations, a simple example is given below.

EXAMPLE 1: Given a discrete space $X = \{a, b, c, d\}$ and fuzzy events,

$$f = a\ /\ 1 + b\ /\ 0.7 + c\ /\ 0.5 + d\ /\ 0.1$$

$$g = a\ /\ 1 + b\ /\ 0.6 + c\ /\ 0.3 + d\ /\ 0.2$$

find $f \cup g$, $f \cap g$ and $\bar{f}$.

Solution: Using Equations 14.2 through 14.4, it is easy to see

$$f \cup g = a\ /\ 1 + b\ /\ 0.7 + c\ /\ 0.5 + d\ /\ 0.2$$

$$f \cap g = a\ /\ 1 + b\ /\ 0.6 + c\ /\ 0.3 + d\ /\ 0.1$$

$$\bar{f} = a\ /\ 0 + b\ /\ 0.3 + c\ /\ 0.5 + d\ /\ 0.9$$

14.2.2 Fuzzy Sets and Probability Distribution

There is often confusion about the difference between fuzzy degree and probability. The difference can be demonstrated by the following simple example: "the probability that a NBA player is 6 feet tall is 0.7" implies that there is an 70% chance of a randomly picked NBA player being 6 feet tall, though he may be just 5 feet 5. On the other hand, "the fuzzy degree that an NBA player is 6 feet tall is 0.7" implies that a randomly picked NBA player is most likely 6 feet tall (70%). In other words, the probability of an event describes the possibility of occurrence of the event while the fuzzy degree describes the uncertainty of appearance of the event.

It is interesting to know, however, that although the fuzzy degree and probability are different, they are actually correlated [Baldwin et al., 1995]. This correlation is through the probability mass function. To show this, let us consider a simple example below.

EXAMPLE 2: Given a discrete space $\mathbf{X} = \{a, b, c, d\}$ and a fuzzy event $f \subseteq \mathbf{X}$,

$$f = a \,/\, 1 + b \,/\, 0.7 + c \,/\, 0.5 + d \,/\, 0.1,$$

find the probability mass function of $Y = f$.

Solution: First, the possibility function of f is:

$$\mu(a) = 1, \mu(b) = 0.7, \mu(c) = 0.5, \mu(d) = 0.1$$

This is equivalent to:

$$\mu(\{a, b, c, d\}) = 1, \mu(\{b, c, d\}) = 0.7, \mu(\{c, d\}) = 0.5, \mu(\{d\}) = 0.1$$

Assuming $P(f) \leq \mu(f)$, and

$$P(a) = p_a, P(b) = p_b, P(c) = p_c, P(d) = p_d$$

it follows that

$$\begin{aligned}
&p_a + p_b + p_c + p_d = 1 \\
&p_b + p_c + p_d \leq 0.7 \\
&p_c + p_d \leq 0.5 \\
&p_d \leq 0.1 \\
&p_i \geq 0, i = a, b, c, d
\end{aligned}$$

Solving this set of equations, we have:

$$\begin{aligned}
&0.3 \leq p_a \leq 1 \\
&0 \leq p_b \leq 0.7 \\
&0 \leq p_c \leq 0.5 \\
&0 \leq p_d \leq 0.1
\end{aligned}$$

Therefore, the probability mass function of f is

$$m(a){:}\ [0.3, 1],\ m(b){:}\ [0, 0.7],\ m(c){:}\ [0, 0.5],\ m(d){:}\ [0, 0.1]$$

or

$$m = \{a\}{:}\ 0.3,\ \{a, b\}{:}\ 0.2,\ \{a, b, c\}{:}\ 0.4,\ \{a, b, c, d\}{:}\ 0.1$$

In general, suppose that $A \subseteq \mathbf{X}$ is a discrete fuzzy event, namely

$$A = x_1 / \mu(x_1) + x_2 / \mu(x_2) + \dots + x_n / \mu(x_n) \qquad \text{Equation (14.6)}$$

Then, the fuzzy set A induces a possibility distribution over **X**:

$$\Pi(x_i) = \mu(x_i)$$

Furthermore, assume (a) $\mu(x_1) = 1$, and (b) $\mu(x_i) \geq \mu(x_j)$ if $i < j$, then:

$$\Pi(\{x_i, x_{i+1}, \dots, x_n\}) = \mu(x_i) \qquad \text{Equation (14.7)}$$

If $P(A) \leq \Pi(A)$, $\forall A \in 2^{\mathbf{X}}$, then we have

$$\sum_{k=1}^{n} P(x_k) \leq \mu(x_i), \text{ for } i = 2, \dots, n \qquad \text{Equation (14.8a)}$$

$$\sum_{k=1}^{n} P(x_k) = 1 \qquad \text{Equation (14.8b)}$$

Solving Equation 14.8 results in

$$1 - \mu(x_2) \leq P(x_1) \leq 1 \qquad \text{Equation (14.9a)}$$

$$0 \leq P(x_i) \leq \mu(x_i), \text{ for } i = 2, \dots, n \qquad \text{Equation (14.9b)}$$

Finally, from the probability functions, the probability mass function can be found:

$$m_f = \{ \{x_1, \dots, x_i\}: \mu(x_i) - \mu(x_{i+1}), i = 1, \dots, n \} \text{ with } \mu(x_{n+1}) = 0$$

Equation (14.10)

It should be noted that, as shown in Equation 14.9, a fuzzy event corresponds to a family of probability distributions. Hence, it is necessary to apply a restriction to form a specific probability distribution. The restriction is to distribute the mass function with a focal element. For example, given a mass function $m = \{a, b, c\}$: 0.3, then there are three focal elements $\{a, b, c\}$ and its value is 0.3. Hence, applying the restriction, we have $m = (3, 0.3)$. In general, under the restriction a mass function can be denoted as $m = (L, M)$, where L corresponds to the size of the focal elements and M represents the value. In the example above, $L = 3$ and $M = 0.3$.

Also, it shall be noted that the mass function assignment may be incomplete. For example, if $f = a / 0.8 + b / 0.6 + d / 0.2$, $\mathbf{X} = \{a, b, c, d\}$, then the mass assignment would be

$$m_f = a: 0.2, \{a, b\}: 0.4, \{a, b, c\}: 0.2; \emptyset: 0.2$$

In this case, we need to normalize the mass assignment by using the formula:

$$\mu(x_i) = \mu(x_i) / \mu(x_1), i = 2, 3, .., n \qquad \text{Equation (14.11)}$$

and then do the mass assignment. For the above example, the normalization results in $f^* = a / (0.8/0.8) + b / (0.6/0.8) + d / (0.2/0.8) = a / 1 + b / 0.75 + d / 0.25$, and the corresponding mass assignment is

$$m_{f^*} = a: 0.25, \{a, b\}: 0.5, \{a, b, c\}: 0.25$$

It can be shown that the normalized mass assignment conforms the Dempster–Shafer properties [Baldwin et al., 1995]:

(a) $m(A) \geq 0$,
(b) $m(\varnothing) = 0$,
(c) $$\sum_{A \in F(X)} m(A) = 1$$

14.2.3 Conditional Fuzzy Distribution

Similar to condition probability, we can define the conditional fuzzy degrees (conditional possibility distribution). There are several ways to deal with the conditional fuzzy distribution. First, let g and g' be two fuzzy sets defined on $\boldsymbol{X}$, the mass function associated with the truth set of g given g', denoted by $m_{(g\,/\,g')}$, is another mass function defined over $\{t, f, u\}$ (t represents true, f represents false, and u stands for uncertain). Let $m_g = \{L_i\text{: } l_i\}$ and $m_{g'} = \{M_i\text{: } m_i\}$ and form a matrix

$$M = \left\{ T\left(L_i / M_j\right) : l_i . m_j \right\}, \text{ where } T\left(L_i / M_j\right) = \begin{Bmatrix} t & \text{if } M_j \subseteq L_i \\ f & \text{if } M_j \cap L_i = O \\ u & \text{otherwise} \end{Bmatrix} \quad \text{Equation (14.12)}$$

Then, the truth mass function $m_{(g\,/\,g')}$ is given below:

$$M_{(g/g')} = \begin{Bmatrix} t\text{:} & \sum_{i,j,T(L_i/M_j)=t} l_i . m_j \\ f\text{:} & \sum_{i,j,T(L_i/M_j)=f} l_i . m_j \\ u\text{:} & \sum_{i,j,T(L_i/M_j)=u} l_i . m_j \end{Bmatrix} \quad \text{Equation (14.13)}$$

where, $l_i.m_j$ denotes the element multiplication. The following example illustrates how a conditional mass function is obtained.

EXAMPLE 3: Let

$$g = a/1 + b/0.7 + c/0.2$$
$$g' = a/0.2 + b/1 + c/0.7 + d/0.1$$

be fuzzy sets defined on $\boldsymbol{X} = \{a, b, c, d\}$. Find the truth possibility distribution, $m_{(g\,/\,g')}$.

Solution: First, using Equation 14.10, it can be shown that

$$m_g = \{a\}\text{: } 0.3,\ \{a, b\}\text{: } 0.5,\ \{a, b, c\}\text{: } 0.2$$
$$m_{g'} = \{b\}\text{: } 0.3,\ \{b, c\}\text{: } 0.5,\ \{a, b, c\}\text{: } 0.1,\ \{a, b, c, d\}\text{: } 0.1$$

Hence, a matrix is formed (enclosed by the single line):

	{b} 0.3	{b,c} 0.5	{a,b,c} 0.1	{a,b,c,d} 0.1
{a} 0.3	f 0.09	f 0.15	u 0.03	u 0.03
{a,b} 0.5	t 0.15	u 0.25	u 0.05	u 0.05
{a,b,c} 0.2	t 0.06	t 0.1	t 0.02	u 0.02

The element of the matrix (enclosed by the bold line) may take three different values: t, f, and u, as defined by Equation 14.11. Take, for instance, the element in the first row and first column, since $\{a\} \cap \{b\} = 0$, it shall take a value f. For the element in the second row and first column, since $\{b\} \subseteq \{a, b\}$, it shall take a value of t. Also, for the element in second row and second column, since neither $\{a, b\} \cap \{b, c\}$ nor $\{a, b\} \subseteq \{b, c\}$, it shall take a value u. Finally, using Equation 14.12, it follows that

$$
\begin{aligned}
m = t&: (0.3)(0.3) + (0.2)(0.3) + (0.5)(0.2) + (0.1)(0.2) \\
&= 0.15 + 0.06 + 0.1 + 0.02 \\
&= 0.33 \\
f&: 0.09 + 0.15 = 0.24 \\
u&: 0.25 + 0.03 + 0.05 + 0.03 + 0.05 + 0.02 = 0.43
\end{aligned}
$$

If we are concerned only about the point value for the truth of g/g', there is a simple formula. Use the notations above to form the matrix

$$\mathbf{M} = \{m_{ij}\} = \left\{\frac{\text{card}(L_i \cap M_j)}{\text{card}(M_j)}\right\} l_i m_j \qquad \text{Equation (14.14)}$$

where, "card" stands for cardinality*. Then, the probability $P(g/g')$ is given below:

$$P(g/g') = \sum_{i,j} m_{ij} \qquad \text{Equation (14.15)}$$

EXAMPLE 4: Following Example 3, find the probability for the truth of g/g'.

Solution: From Example 3, it is known that

$$m_g = \{a\}: 0.3,\ \{a, b\}: 0.5,\ \{a, b, c\}: 0.2$$

$$m_{g'} = \{b\}: 0.3,\ \{b, c\}: 0.5,\ \{a, b, c\}: 0.1,\ \{a, b, c, d\}: 0.1$$

The following matrix can be formed:

	$\{b\}$ 0.3	$\{b,c\}$ 0.5	$\{a,b,c\}$ 0.1	$\{a,b,c,d\}$ 0.1
$\{a\}$ 0.3	0	0	0.01	0.00075
$\{a,b\}$ 0.5	0.15	0.125	0.0333	0.025
$\{a,b,c\}$ 0.2	0.06	0.1	0.02	0.015

*The cardinality of a set is its size. For example, given a set $A = [a, b, c]$, card$(A) = 3$.

Note that the matrix is found element by element. For example, for the element in the first row and first column, since $\{a\} \cap \{b\} = 0$, card$(L_1 \cap M_1) = 0$, thus $m_{11} = 0$. For the element in the second row and second column, since $\{a, b\} \cap \{b, c\} = \{b\}$, card$(L_2 \cap M_2)$ = card$(\{b\}) = 1$, card(M_2) = card$(\{b, c\})$ = 2, $m_{22} = (1/2)(0.5)(0.5) = 0.125$. The other components can be determined in the same way. Based on the matrix, it is easy to find $P(g/g') = 0 + 0 + 0.01 + \ldots + 0.015 = 0.53980$.

We can also determine the fuzzy degree of g given g'. It is a pair: the possibility of g/g' is defined as

$$\Pi(g/g') = \max(g \cap g') \qquad \text{Equation (14.16)}$$

and the necessity of g/g' is defined as

$$\pi(g/g') = 1 - \Pi(\bar{g}/g') \qquad \text{Equation (14.17)}$$

This is analogous to the probability support pair and provides the upper and lower bounds of the conditional fuzzy set.

EXAMPLE 5: Following Example 3, find its possibility support pair.

Solution: Since

$$g = a/1 + b/0.7 + c/0.2$$

$$g' = a/0.2 + b/1 + c/0.7 + d/0.1$$

it is easy to see

$$g \cap g' = a/0.2 + b/0.7 + c/0.2$$

$$\Pi(g \cap g') = 0.7$$

Furthermore,

$$\bar{g} = b/0.3 + c/0.8 + d/1$$

$$\bar{g} \cap g' = b/0.3 + c/0.7 + d/0.1$$

$$\pi(g \cap g') = 1 - \Pi(\bar{g} \cap g') = 0.3$$

Hence, the conditional fuzzy degree of g/g' is [0.3, 0.7].

14.3 Monitoring and Diagnosing Manufacturing Processes Using Fuzzy Sets

14.3.1 Using Fuzzy Systems to Describe the State of a Manufacturing Process

For monitoring and diagnosing manufacturing processes, two types of uncertainties are often encountered: the uncertainty of occurrence and the uncertainty of appearance. A typical example is tool condition monitoring in machining processes. Owing to the nature of metal cutting, tools will wear out. Through years of study, it is commonly accepted that tool wear can be determined by Taylor's equation:

$$VT^n = C \qquad \text{Equation (14.18)}$$

where V is the cutting speed (m/min), T is the tool life (min), n is a constant determined by the tool material (e.g., $n = 0.2$ for carbide tools), and C is a constant representing the cutting speed at which the

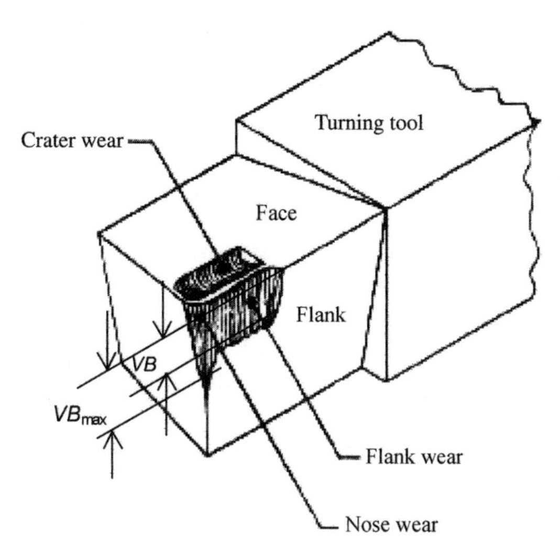

FIGURE 14.2 Illustration of tool wear.

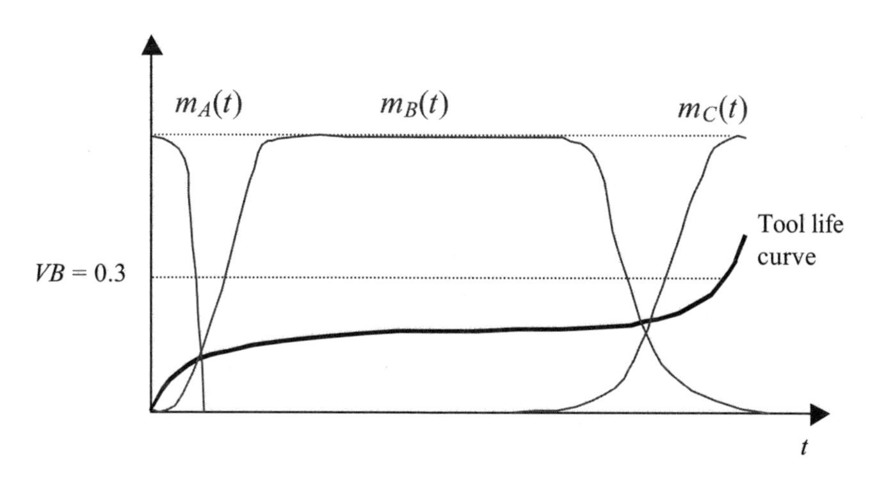

FIGURE 14.3 Illustration of the tool wear states and corresponding fuzzy sets.

tool life is 1 minute (it is dependent on the work material). Figure 14.2 shows a typical example of tool wear development, and the end of tool life is determined at VB = 0.3 mm for carbide tools (VB is the average flank wear), or VB_{max} = 0.5 mm (VB_{max} is the maximum average flank wear). However, it is also found that the tool may wear out much earlier or later depending on various factors such as the feed, the tool geometry, the coolant, just to name a few. In other words, there is an uncertainty of occurrence. Such an uncertainty can be described by the probability mass function shown in Figure 14.3. As shown in the figure, the states of tool wear can be divided into three categories: initial wear (denoted as A), normal tool (denoted as B), and accelerated wear (denoted as C). Their occurrences are a function of time.

On the other hand, it is noted that the state of tool wear may be manifested in various shapes depending on various factors, such as the depth of cut, the coating of the cutter, the coolant, etc. Consequently, even though the state of tool wear is the same, the monitoring signals may appear differently. In order words, there is an uncertainty of appearance. Therefore, in tool condition monitoring, the question to be answered is not only how likely the tool is worn, but also how worn is the tool. To answer this type of problem, it is best to use the fuzzy set theory.

14.3.2 A Unified Model for Monitoring and Diagnosing Manufacturing Processes

Although manufacturing processes are all different, it seems that the task of monitoring and diagnosing always takes a similar procedure, as shown in Figure 14.4. In Figure 14.4, the input to a manufacturing process is its process operating condition (e.g., the speed, feed, and depth of cut in a machining process). The manufacturing process itself is characterized by its process condition, $y \in \mathbf{Y} = \{y_i, i = 1, 2, \ldots, m\}$ (e.g., the state of tool wear in the machining process). Usually, the process operating conditions are controllable while the process conditions may be neither controllable nor directly observable. It is interesting to know that the process conditions are usually artificially defined. For example, as discussed earlier, in monitoring tool condition the end of tool life is defined as the flank wear, *VB*, exceeding 0.3 mm. In practice, however, tool wear can be manifested in various forms. Therefore, it is desirable to use fuzzy set theory to describe the state of the tool wear.

Sensing opens a window to the process through which the changes of the process condition can be seen. Note that both the process and the sensing may be disturbed by noises (an inherited problem in engineering practice). Consequently, signal processing is usually necessary to capture the process condition. Effective sensing and signal processing is very important to monitoring and diagnosing. However, it will not be discussed in this chapter. Instead, the reader is referred to [Du, 1998].

The result of signal process is a set of signal features, also referred to as *indices* or *attributes*, which can be represented by a vector $\mathbf{x} = [x_1, x_2, \ldots, x_n]$. Note that although the numeric values are most common, the attributes may also be integers, sets, or logic values. Owing to the complexity of the process and the cost, it is not unusual that the attributes do not directly reveal the process conditions. Consequently, decision-making must be carried out. There have been many decision-making methods; the fuzzy set theory is one of them and has been proved to be effective.

Mathematically, the unified model shown in Figure 14.4, as represented by the bold lines, can be described by the following relationship:

$$\mathbf{y} \bullet \mathbf{R} = \mathbf{x} \qquad \text{Equation (14.19)}$$

where **R** is the relationship function, which represents the combined effect of the process, sensing, and signal processing. Note that **R** may take different forms such as a dynamic system (described by a set of differential equations), patterns (described by a cluster center), neural network, and fuzzy logic. Finally, it should be noted that the operator "•" should not be viewed as simple multiplication. Instead, it corresponds to the form of the relationship.

The process of monitoring and diagnosing manufacturing processes consists of two phases. The first phase is learning. Its objective is to find the relationship **R** based on available information (learning from samples) and knowledge (learning from instruction). Since the users must provide information and instruction, the learning is a supervised learning. To facilitate the discussions, the available learning samples are organized as shown in Table 14.1.

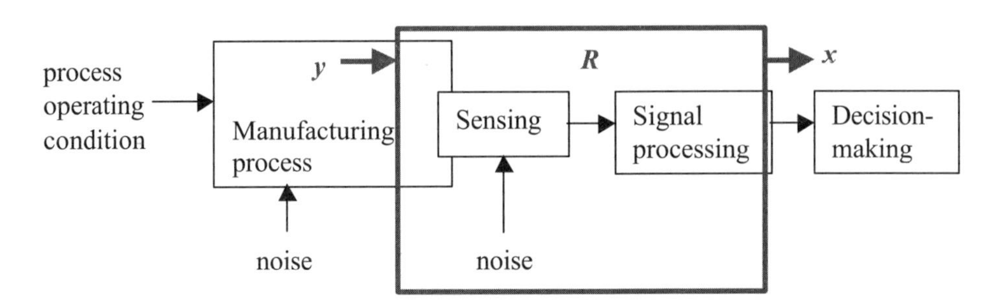

FIGURE 14.4 A unified model for monitoring and diagnosing manufacturing processes.

TABLE 14.1 Organization of the Available Learning Samples

Sample No.	x_1	x_2	...	x_n	$y(x)$
$\boldsymbol{x}_1$	x_{11}	x_{12}	...	x_{1n}	$y(\boldsymbol{x}_1)$
$\boldsymbol{x}_2$	x_{21}	x_{22}	...	x_{2n}	$y(\boldsymbol{x}_2)$
...	...	...	...	...	...
$\boldsymbol{x}_N$	x_{N1}	x_{N2}	...	x_{Nn}	$y(\boldsymbol{x}_N)$

Note: where, $y(\boldsymbol{x}_j) \in \mathbf{Y} = \{y_i, i = 1, 2, \ldots, m\}, j = 1, 2, \ldots, N$, represent the process condition and it must be known in order to conduct learning.

The second phase is classification. Given a new sample **x**, and the relationship **R**, the corresponding process condition of the sample $y(\boldsymbol{x})$ can be determined as follows:

$$\mathbf{y} = \mathbf{x} \bullet \mathbf{R}^{-1} \qquad \text{Equation (14.20)}$$

Here again, the operator "•" should not be viewed as simple multiplication. Instead, it may mean pattern matching, neural network searching, and fuzzy logic operations depending on the inverse of the relationship.

In the following subsection, we will show how to use fuzzy set theory to establish a fuzzy relationship function (Equation 14.19) and how to resolve it to identify the process condition of a new sample (Equation 14.20).

14.3.3 Linear Fuzzy Classification

One of the simplest fuzzy relationship functions is the linear equation defined below [Du et al., 1992]:

$$\mathbf{x} = \mathbf{Q} \bullet \mathbf{y} \qquad \text{Equation (14.21)}$$

where **Q** represents the linear fuzzy correlation between the classes and the attributes (signal features). Assuming that there are m different classes and n different attributes, then **y** is an m-dimensional vector and **x** is a n-dimensional vector. The fuzzy linear correlation function between the classes and the attributes may take various forms such as a tablet form (Equation 14.1) or a stepwise function (Equation 14.2). For simplicity, let us use the tablet form. First, each attribute is divided into K intervals. Note that just like the histogram in statistics, different definitions of intervals may lead to different results. As a rule of thumb, the number of intervals should be about one tenth of the total number of samples, that is,

$$K = \frac{N}{10} \qquad \text{Equation (14.22)}$$

and intervals should be evenly distributed. For the jth attribute, let

$$x_{j,\max} = \max\{x_{kj}\} \qquad \text{Equation (14.23a)}$$

$$x_{j,\min} = \min\{x_{kj}\} \qquad \text{Equation (14.23b)}$$

where, $k = 1, 2, \ldots, N$ correspond to the learning samples. The width of the interval will be

$$\Delta x_j = \frac{1}{K}\left(x_{j,\max} - x_{j,\min}\right) \qquad \text{Equation (14.24)}$$

The intervals would be

$$I_{j1} = [0, \Delta x_j] \qquad \text{Equation (14.25a)}$$

$$I_{jk} = I_{jk\text{-}1} + \Delta x_j,\ k = 2, 3, \ldots, K \qquad \text{Equation (14.25b)}$$

Note that attributes may be discrete numbers or sets. In these cases, the intervals will be reduced to discrete sets.

The fuzzy relationship function **Q** can be described as a matrix,

$$\mathbf{Q} = \begin{bmatrix} q_{11} & q_{12} & \cdots & q_{1m} \\ q_{21} & q_{22} & \cdots & q_{2m} \\ \vdots & \vdots & \ddots & \vdots \\ q_{n1} & q_{n2} & \cdots & q_{nm} \end{bmatrix} \qquad \text{Equation (14.26)}$$

where

$$q_{ij} = I_{j1} / \mu\left(I_{ij1}\right) + I_{j2} / \mu\left(I_{ij2}\right) + \ldots + I_K / \mu\left(I_{ijK}\right) \qquad \text{Equation (14.27)}$$

The fuzzy set q_{ij} represents the fuzzy correlation between the ith class and j^{th} attributes and the fuzzy degree $\mu(I_{ijk})$ can be obtained from the available training samples:

$$\mu\left(I_{ijk}\right) = \frac{N_{ijk}}{M_{ij}} \qquad \text{Equation (14.28)}$$

where N_{ijk} is the number of samples of the ith class in the kth interval of the jth attribute. M_{ij} is the number of samples of the ith class in the jth attribute. From a physical point of view, it represents the distribution of the training samples about the classes. We may also use a similar formula,

$$\mu\left(I_{ijk}\right) = \frac{N_{ijk}}{M_{jk}} \qquad \text{Equation (14.29)}$$

where M_{jk} is the number of samples in the kth interval of the jth attribute. From a physical point of view, it represents the distribution of the training samples about the attributes. Also, we can use the combination of both:

$$\mu\left(I_{ijk}\right) = \alpha \frac{N_{ijk}}{M_{ij}} + \left(1 - \alpha\right) \frac{N_{ijk}}{M_{jk}} \qquad \text{Equation (14.30)}$$

where $0 \leq \alpha \leq 1$ is a weighting factor. As an example, Figure 14.5 illustrates a fuzzy membership function, in which the attributes X_j is decomposed into ten intervals, and the fuzzy membership functions of two process conditions are overlapped.

When a new sample, $\boldsymbol{x}$, is provided, its corresponding process condition can be estimated by classification. The classification phase starts at checking the fuzzy degree of the new sample. Suppose the jth attribute of the new sample falls into the kth interval I_{jk}, then

$$q_{ij}(\boldsymbol{x}) = I_{jk} / \mu(I_{ijk}) \qquad \text{Equation (14.31)}$$

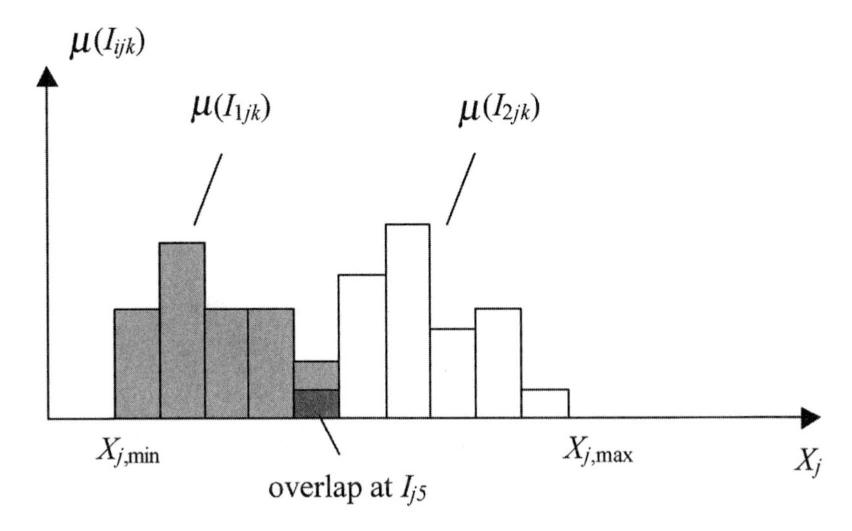

FIGURE 14.5 Illustration of a fuzzy membership function.

In other words, the fuzzy degree that the sample belong to j^{th} class is $\mu(I_{ijk})$. Next, using the max–min classification rule, the corresponding process condition of the sample can be estimated:

$$i^* = \arg\max_i \left\{ \min_j \left\{ \mu\left(I_{ijk}\right) \right\} \right\} \qquad \text{Equation (14.32)}$$

The other useful and often better performed classification rule is the max-average rule defined below:

$$i^* = \arg\max_i \left\{ \frac{1}{n} \sum_{j=1}^{n} \mu\left(I_{ijk}\right) \right\} \qquad \text{Equation (14.33)}$$

These operations are demonstrated by the example below.

EXAMPLE 6: Given the following discrete training samples:

	X_1	X_2	$y(x)$
x_1	a	c	A
x_2	a	d	A
x_3	b	d	B
x_4	b	c	B
x_5	a	d	C
x_6	b	d	C

find the linear fuzzy relationship function. Furthermore, suppose a new sample $\boldsymbol{x} = [a, d]$ is given, estimate its class.

Solution: There are two (discrete) attributes and three classes (A, B, C), and hence the fuzzy relationship function is

$$\mathbf{Q} = \begin{bmatrix} q_{A1} & q_{A2} \\ q_{B1} & q_{B2} \\ q_{C1} & q_{C2} \end{bmatrix}$$

Note that the attributes are discrete sets, and hence there is no need to use intervals. Based on the training samples, the elements of the fuzzy relationship function can be found. For example, for the first element, q_{A1}, since $N_{A1a} = 2$, $M_{A1} = 2$, and $N_{A1b} = 0$, using Equation 14.28, we have

$$q_{A1} = a \,/\, (2/2) + b \,/\, (0/2) = a \,/\, 1 + b \,/\, 0$$

Similarly, we can find

$$\begin{aligned}
q_{A2} &= c \,/\, (1/2) + d \,/\, (1/2) = c \,/\, 0.5 + d \,/\, 0.5 \\
q_{B1} &= a \,/\, (0/2) + b \,/\, (2/2) = a \,/\, 0 + b \,/\, 1 \\
q_{B2} &= c \,/\, (1/2) + d \,/\, (1/2) = c \,/\, 0.5 + d \,/\, 0.5 \\
q_{C1} &= a \,/\, (1/2) + b \,/\, (1/2) = a \,/\, 0.5 + b \,/\, 0.5 \\
q_{C2} &= c \,/\, (0/2) + d \,/\, (2/2) = c \,/\, 0 + d \,/\, 1
\end{aligned}$$

One can use Equation 14.29 or Equation 14.30 to get the fuzzy relationship function as well. Now, suppose the new sample, $\mathbf{x} = [a, d]$ is given, then

$$\begin{aligned}
q_{A1} &= a \,/\, 1 \\
q_{A2} &= d \,/\, 0.5 \\
q_{B1} &= a \,/\, 0 \\
q_{B2} &= d \,/\, 0.5 \\
q_{C1} &= a \,/\, 0.5 \\
q_{C2} &= d \,/\, 1
\end{aligned}$$

Using the max–min rule,

$$\begin{aligned}
i^* &= \operatorname{argmax}\{\min\{1, 0.5\}, \min\{0, 0.5\}, \min\{0.5, 1\}\} \\
&= \operatorname{argmax}\{0.5, 0, 0.5\} \\
&= A \text{ or } C.
\end{aligned}$$

Using the max–average rule,

$$\begin{aligned}
i^* &= \operatorname{argmax}\{(1+0.5)/2, (0+0.5)/2, (0.5+1)/2\} \\
&= \operatorname{argmax}\{0.75, 0.25, 0.75\} \\
&= A \text{ or } C.
\end{aligned}$$

14.3.4 Nonlinear Fuzzy Classification

The linear fuzzy classification method presented above assumes a linear correlation between the process condition and the attributes. In addition, as shown in Figure 14.5, the fuzzy membership functions are approximated by a bar chart. There are a number of methods to relax these assumptions leading to the nonlinear fuzzy classification methods. For example, we can assume that the fuzzy membership functions are trapezoid, as shown in Figure 14.1. In this case, however, the parameters of the trapezoid functions must be estimated through an optimization procedure. In fact, this is the idea of the fuzzy C-mean method.

The fuzzy C-mean method was first proposed by Bezdek [1981]. It uses the cost function defined below:

$$J(\mathbf{U},\mathbf{V},\mathbf{X}) = \sum_{k=1}^{N}\sum_{j=1}^{n}\sum_{i=1}^{m} u^r(k,j)\left\|x(k,i) - v(j,i)\right\|^r \qquad \text{Equation (14.34)}$$

where $\mathbf{X} = \{x(k, j)\}$ is the attributes of the samples, $k = 1, 2, \ldots, N$

$\mathbf{V} = \{v(i, j)\}$ is the fuzzy cluster center, $v(i, j)$ corresponds to the ith class and the jth attributes

$\mathbf{U} = \{u(k, i)\}$, $u(k, i) \in [0,1]$ is the fuzzy degree of the samples belonging to i^{th} class

r is a positive number that controls the shape of the membership function.

In the learning phase, the fuzzy cluster center, **V**, together with the membership function, **U**, can be found by the optimization

$$\min_{(\mathbf{U},\mathbf{V})\in\mathbf{M}}\left\{J(\mathbf{U},\mathbf{V},\mathbf{X})\right\} \qquad \text{Equation (14.35)}$$

where

$$M=\left\{u(k,i),v(i,j)\Big/\sum_{i=1}^{m}u(k,i)=1,\forall k=1,2,\ldots,N\right\}$$

It has been shown [Bezdek, 1981] that the necessary condition for a (**U, V**) to be an optimal solution of Equation 14.35 is

$$u(k,i)=\frac{1}{\sum_{j=1}^{n}\sum_{\alpha=1}^{n}\left(\frac{\left\|x(k,j)-v(i,j)\right\|}{\left\|x(k,j)-v(\alpha,j)\right\|}\right)^{1/r-1}} \qquad \text{Equation (14.36)}$$

$$v(i,j)=\frac{\sum_{k=1}^{N}u^{r}(k,i)x(k,i)}{\sum_{k=1}^{N}u^{r}(k,i)} \qquad \text{Equation (14.37)}$$

Equations 14.36 and 14.37 can be solved using an iteration procedure.

In the classification phase, for a given sample $\mathbf{x}_s = [x(s, 1), x(s, 2), \ldots, x(s, m)]$, Equation 14.19 provides a set of fuzzy membership functions $u(s, i)$, $i = 1, 2, \ldots, m$. Then, the estimated process condition is determined by the following equation:

$$i^* = \text{argmax}\{u(s, i)\} \qquad \text{Equation (14.38)}$$

The fuzzy C-mean method uses nonlinear fuzzy membership functions but assumes a linear correlation between the signal features and the process conditions as shown in Equation 14.29. A more flexible method is the fuzzy decision tree method [Du et al., 1995], in which the correlation between the signal features and the process conditions is represented as a tree:

$$c_1\colon d_1 \text{ is } \mu_1$$

$$c_2\colon d_2 \text{ is } \mu_2$$

......

$$c_p\colon d_p \text{ is } \mu_p \qquad \text{Equation (14.39)}$$

......

$$c_M\colon d_M \text{ is } \mu_M$$

where, c_p's are condition statements (e.g., "if $x_j > t_j$" where t_j is a threshold) and d_p's may be either a leaf of the tree that indicates a conclusion (e.g., "process condition = tool wear") or a sub-tree, and μ_p is the fuzzy membership function. Figure 14.6 shows such a fuzzy decision tree. In the decision tree, each node represents a partition that decomposes the problem space, **X**, into smaller subspaces. For example, the first node decomposes **X** into two subspaces $\mathbf{X} = \mathbf{X}_1 + \mathbf{X}_2$. Then the second node decomposes $\mathbf{X}_1$ into two sub-subspaces $\mathbf{X}_1 = \mathbf{X}_{11} + \mathbf{X}_{12}$, and so on. The end of tree are the leaves, which indicate the estimated process condition.

The decision tree can be constructed using the ID3 method developed by Quinlan [1986, 1987]. ID3 was originally developed for the problems in which the attributes are discrete sets. In order to accommodate the numeric attributes, the fuzzy membership function is therefore introduced. For each node of the decision tree, its fuzzy membership can be determined using the fuzzy C-mean method. In this way, the decision making will be more effective.

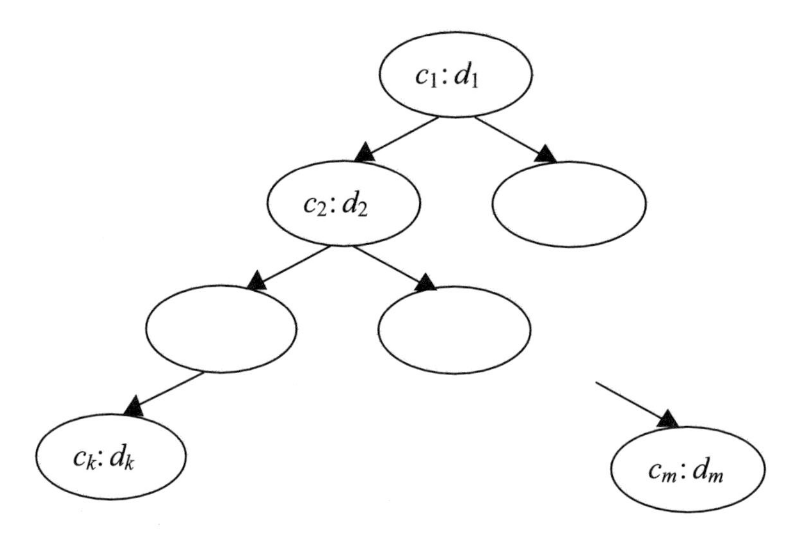

FIGURE 14.6 Illustration of a decision tree.

14.3.5 Fuzzy Transition Probability

In many manufacturing processes, some failures, such as wear and fatigue, are developed gradually. To monitor this type of failure, we can use the fuzzy transition matrix.

The fuzzy transition is analogous to the Markov process [Klyele and de Korvin, 1998]. For a system that transfers from one state to another state (e.g., from normal to worn out), the transition is Markov if the transition probability depends only on the current state of the system and not on any previous states the system may have experienced. In comparison, the fuzzy transition describes the phenomenon that the states of the system are not clearly defined, and hence, the probability of the transition depends on the current state and the corresponding fuzzy degrees, though the previous process states are not concerned.

As an example, let us consider the process of tool wear in machining processes. It has three states, $\mathbf{Y} = \{A, B, C\}$; state A represents the new tool, state B represents the normal wear, and state C represents the accelerated wear as shown in Figure 14.3. Note that, as pointed out earlier, since the tool wear may be manifested in various forms the definitions of the three states are somewhat fuzzy. Figure 14.7 shows the possible transitions of the process. There are six self- and/or forward transitions: $A \to A$, $A \to B$, $A \to C$, $B \to B$, $B \to C$, $C \to C$. Because the process of wear or fatigue is irrecoverable, there is no backward transition and $P(C \to C) = 1$. We wish to find the probability of the fuzzy transitions are as follows: $P(A \to A)$, $P(A \to B)$, $P(A \to C)$, $P(B \to B)$, $P(B \to C)$.

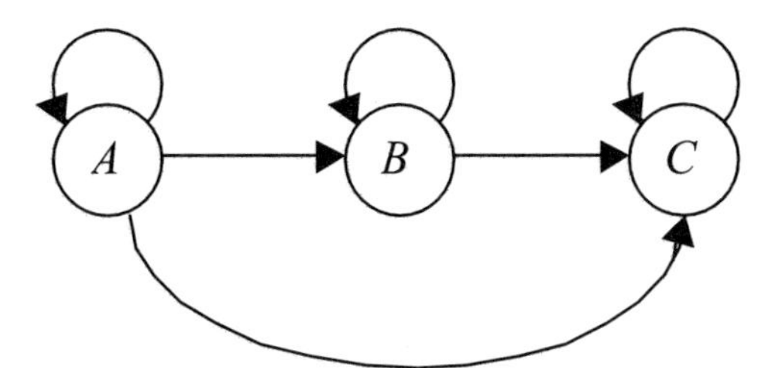

FIGURE 14.7 Illustration of transition of states in a simple process.

In general, the transitional probability of going from state A_i to state A_j can be defined as follows:

$$P(A_i \to A_j) = \sum_{k:A_i \cap A_k = A_i} m(A_k) \qquad \text{Equation (14.40)}$$

where the A_k' are the focal elements of $m(\bullet)$, and the probability mass function $m(A_k)$ describes how the system migrates from state A_i to state A_j. The way to calculate the transition probability is as follows: Let $\mathbf{Y} = \{A_1, A_2, \ldots, A\}$ denote the state space and for any state A_i, $A_i \in \mathbf{Y}$, let $m(\bullet / A_i)$ denote conditional mass function. Furthermore, suppose

$$S(A_{j1}, A_{j2}, \ldots, A_{jp}) = \{A_{j1}, A_{j2}, \ldots, A_{jp}\},\ 1 \le j_1 < j_2 < \ldots < j_p \le n,\ 1 \le p \le n. \qquad \text{Equation (14.41)}$$

is the intersection of the focal elements. Then the one-step fuzzy transition probability is

$$P(A_i \to A_j) = \sum_{j \in (j1, j2, \ldots jp)} m\left(S(A_{j1}, A_{j2}, \ldots, A_{jp}) / A_i\right) \div \left|S(A_{j1}, A_{j2}, \ldots, A_{jp})\right| \qquad \text{Equation (14.42)}$$

We can find the limiting transition probability as well [Klyele and de Korvin, 1998]. It can be shown that if the intersection of the focal elements, denoted as A_{jt}, $jt = 1, 2, \ldots, p$, is non-empty, then $m_\infty(A_{jt}) = 1$ and $m_\infty(B) = 0$ for all $B \ne A_{jt}$, which gives

$$P_\infty(A_i \to A_{jt}) = \frac{1}{p}, \forall A_{jt} \in S(A_{j1}, A_{j2}, \ldots, A_{jp}) \qquad \text{Equation (14.43)}$$

When the intersection of the focal elements is empty, it is necessary to examine the thresholds to limiting process and to determine which focal element has the highest transition probabilities. If there is a unique threshold state A such that

$$P(A \to A) = \max_{A_i \in T} P(A_i \to A_i) \qquad \text{Equation (14.44)}$$

where T denotes the set of all threshold states, and $P(A_i \to A_i)$ denotes the one-step transition probability of set state A_i into A_i, then we can still use Equation 14.43. However, if N threshold states are tied for the maximum transition probability, then

$$P_{\infty}\left(A_i \rightarrow A_{jt}\right)=\sum_{k=1}^{N}\left(\frac{1}{N} \div \left|A_k\right|\right), t=1,2,\ldots,p \qquad \text{Equation (14.45)}$$

where $A_1, A_2, \ldots, A_N$ denote the N tied threshold states. A demonstration example is shown below.

EXAMPLE 7: Consider a four-state process; suppose that for state A, the mass functions are

$$m(S(A, B, C, D) \,/\, A) = 0.2,$$
$$m(S(A, B, C) \,/\, A) = 0.3,$$
$$m(S(A, B, D) \,/\, A) = 0.3,$$
$$m(S(A, C, D) \,/\, A) = 0.2,$$

Find the one-step transition probability $P(A \rightarrow A)$, $P(A \rightarrow B)$, $P(A \rightarrow C)$, and $P(A \rightarrow D)$, as well as the limit transition probability $P_{\infty}(A \rightarrow A)$, $P_{\infty}(A \rightarrow B)$, $P_{\infty}(A \rightarrow C)$, and $P_{\infty}(A \rightarrow D)$.

Solution: Using Equation 14.42,

$$P(A \rightarrow A) = 0.2 \div 4 + 0.3 \div 3 + 0.3 \div 3 + 0.2 \div 3 = 0.3167$$
$$P(A \rightarrow B) = 0.2 \div 4 + 0.3 \div 3 + 0.3 \div 3 = 0.2500$$
$$P(A \rightarrow C) = 0.2167$$
$$P(A \rightarrow D) = 0.2167$$

Next, let us calculate the asymptotic (limiting) transition probability. Since the intersection of the focal elements for $m(\bullet \,/\, A)$ is $S(A) = \{A\}$, it follows that $m(\bullet \,/\, A) = 1$ and the limiting mass for all other subsets of the state space S equals zero. Consequently, the application of Equation 14.43 is trivial:

$$P_{\infty}(A \rightarrow A) = 1,\; P_{\infty}(A \rightarrow j) = 0,\; j = B \text{ and } C$$

In practice, the mass functions can be found by learning. Let us consider the tool condition monitoring example again. Since tool wear develops step by step, the time information is important. Accordingly, the learning samples are organized as shown in Table 14.2. As pointed out earlier, tool life follows the Taylor's equation. For each set of training samples, the tool life can be estimated as follows:

$$T_i = \left(\frac{C_i}{V_i}\right)^{\frac{1}{n_i}} \qquad \text{Equation (14.46)}$$

where $i = 1, 2, \ldots, M$ denotes the learning sample sets. In order to accommodate this information, a normalized time index is formed as follows:

$$s_{ij} = \frac{t_{ij}}{T_i} \qquad \text{Equation (14.47)}$$

where $j = 1, 2, \ldots, N$ indexes the time. Based on the time index, we can reorganize the learning samples. Let

$$s_{\max} = \max_{i,j}\left\{s_{ij}\right\} \qquad \text{Equation (14.48)}$$

TABLE 14.2 Organization of the Learning Samples with Timely Information

Learning Set 1			Learning Set 2			...
Time	Features	Class	Time	Features	Class	
t_{11}	$\mathbf{x}_{11}$	$c(\mathbf{x}_{11})$	t_{21}	$\mathbf{x}_{21}$	$c(\mathbf{x}_{21})$	...
t_{12}	$\mathbf{x}_{12}$	$c(\mathbf{x}_{12})$	t_{22}	$\mathbf{x}_{22}$	$c(\mathbf{x}_{22})$	...
t_{13}	$\mathbf{x}_{13}$	$c(\mathbf{x}_{13})$	t_{23}	$\mathbf{x}_{23}$	$c(\mathbf{x}_{23})$	...
...	...			...		...
t_{1i}	$\mathbf{x}_{1i}$	$c(\mathbf{x}_{1i})$	t_{2i}	$\mathbf{x}_{2i}$	$c(\mathbf{x}_{2i})$	...
...	...			...		...
t_{1n}	$\mathbf{x}_{1n}$	$c(\mathbf{x}_{1n})$	t_{2n}	$\mathbf{x}_{2n}$	$c(\mathbf{x}_{2n})$	...

Note: $\mathbf{x}_{ij}$ are the vectors representing the features of the learning samples, and $c(\mathbf{x}_{ij}) = A$, B, or C represents the corresponding class.

TABLE 14.3 Organization of the Learning Samples Using Normalized Time Index

	Learning Set 1		Learning Set 2			Learning Set M	
Time	Features	Class	Features	Class		Features	Class
ts_1	$\mathbf{x}_{11}$	$c(\mathbf{x}_{11})$	$\mathbf{x}_{21}$	$c(\mathbf{x}_{21})$	...	$\mathbf{x}_{M1}$	$c(\mathbf{x}_{M1})$
	$\mathbf{x}_{12}$	$c(\mathbf{x}_{12})$	$\mathbf{x}_{22}$	$c(\mathbf{x}_{22})$		$\mathbf{x}_{M2}$	$c(\mathbf{x}_{M2})$
	$\mathbf{x}_{13}$	$c(\mathbf{x}_{13})$				$\mathbf{x}_{M3}$	$c(\mathbf{x}_{M3})$
ts_2	$\mathbf{x}_{14}$	$c(\mathbf{x}_{14})$	$\mathbf{x}_{23}$	$c(\mathbf{x}_{23})$	...	$\mathbf{x}_{M4}$	$c(\mathbf{x}_{M4})$
	$\mathbf{x}_{15}$	$c(\mathbf{x}_{15})$	$\mathbf{x}_{24}$	$c(\mathbf{x}_{24})$		$\mathbf{x}_{M5}$	$c(\mathbf{x}_{M5})$
						$\mathbf{x}_{M6}$	$c(\mathbf{x}_{M6})$
	...	...	...	...	...	...	...
ts_K	$\mathbf{x}_{1i}$	$c(\mathbf{x}_{1i})$			...	$\mathbf{x}_{Mn}$	$c(\mathbf{x}_{Mn})$
	$\mathbf{x}_{1(i+1)}$	$c(\mathbf{x}_{1(i+1)})$					
	...	...					
	$\mathbf{x}_{1n}$	$c(\mathbf{x}_{1n})$					

$$s_{\min} = \min_{i,j}\{s_{ij}\} \qquad \text{Equation (14.49)}$$

Furthermore, set K transition steps ts_k, $k = 1, 2, \ldots, K$, will be evenly distributed within $s_{\max}$ and $s_{\min}$. Then, the learning samples can be organized as shown in Table 14.3. Note that each transition step may contain several samples or no sample from a certain learning set.

For each transition step, all available learning samples are used to form a fuzzy membership function. Again, various methods can be used to form the fuzzy linear equation method described in Section 14.3.3. The resulting fuzzy sets are

$$ts_1\text{: } \mathbf{x} \,/\, \mu_A(\mathbf{x}),\ \mathbf{x} \,/\, \mu_B(\mathbf{x}),\ \mathbf{x} \,/\, \mu_C(\mathbf{x})$$

$$ts_2\text{: } \mathbf{x} \,/\, \mu_A(\mathbf{x}),\ \mathbf{x} \,/\, \mu_B(\mathbf{x}),\ \mathbf{x} \,/\, \mu_C(\mathbf{x})$$

......

$$ts_K\text{: } \mathbf{x} \,/\, \mu_A(\mathbf{x}),\ \mathbf{x} \,/\, \mu_B(\mathbf{x}),\ \mathbf{x} \,/\, \mu_C(\mathbf{x})$$

Next, we can calculate the mass functions and conditional mass functions using Equations 14.10 and 14.14. Finally, the fuzzy transition probability and limiting fuzzy transition probability can be calculated using Equations 14.42 and 14.43, respectively.

In summary, the procedure of forming the fuzzy transition probabilities consists of six steps as shown below.

Step 1: Sort the learning sample sets according to predefined transition steps, ts_1, ts_2, …, ts_K.

Step 2: For each transition step, calculate fuzzy membership functions, $\mu_A(x)$, $\mu_B(x)$, $\mu_C(x)$. Note that various methods can be used, such as the fuzzy linear equation method described in Section 14.3.3.

Step 3: For each transition step, calculate the probability mass function, $m_A(x)$, $m_B(x)$, and $m_C(x)$ using the method described in Section 14.2.2.

Step 4: For each transition step, calculate the condition probabilities: $P(A/A)$, $P(B/A)$, $P(C/A)$, $P(B/B)$, and $P(C/B)$.

Step 5: For each transition step, calculate the conditional mass functions: $m(\{A, B\}/A)$, $m(\{A, C\}/A)$, $m(\{A, C\}/A)$, $m(\{A, B, C\}/A)$, and $m(\{B, C\}/B)$ using the method described in Section 14.2.3.

Step 6: For each transition step, calculate the probability of fuzzy transition $P(A \rightarrow A)$, $P(A \rightarrow B)$, $P(A \rightarrow C)$, $P(B \rightarrow B)$, and $P(B \rightarrow C)$.

One may calculate the limiting transition probabilities. However, we found that the limiting transition probabilities tend to be in favor of one or two states, and hence make biased decisions. A demonstration example is given below.

EXAMPLE 8: Continuing Example 6, suppose the available training samples are

Time	Learning Set		
	X_1	X_2	$y(x)$
t_1	a	c	A
t_2	a	d	A
t_3	b	d	B
t_4	b	c	B
t_5	a	d	C
t_6	b	d	C

Suppose we are interested in two transition steps: $ts_1 = \{t_1, t_2, t_3\}$ and $ts_2 = \{t_4, t_5, t_6\}$. Find the fuzzy transition probability $P(A \rightarrow A)$, $P(A \rightarrow B)$, and $P(A \rightarrow C)$ and the limiting fuzzy transition probability.

Solution: First, reorganize the training samples as follows:

Time	Learning Set		
	X_1	X_2	$y(x)$
st_1	a	c	A
	a	d	A
	b	d	B
st_1	b	c	B
	a	d	C
	b	d	C

Similar to Example 6, it can be shown that for the st_1, the fuzzy sets are

$$q_A = a / 1 + b / 0 + c / 0.5 + d / 0.5$$
$$q_B = a / 0 + b / 1 + c / 0 + d / 1$$
$$q_C = a / 0 + b / 0 + c / 0 + d / 0$$

For st_2, the fuzzy decision rule is

$$q_A = a / 0 + b / 0 + c / 0 + d / 0$$
$$q_B = a / 0 + b / 1 + c / 1 + d / 0$$
$$q_C = a / 0.5 + b / 0.5 + c / 0 + d / 1$$

Hence, the mass functions for transition step st_1 are

$$m_A: \{\{a\}: 0.5, \{a, c\}: 0, \{a, c, d\}: 0.5, \{a, c, d, b\}: 0\}$$
$$m_B: \{\{b\}: 0, \{b, c\}: 1, \{b, c, a\}: 0, \{b, c, a, d\}: 0\}$$
$$m_C: \{\{d\}: 0, \{d, a\}: 0, \{d, a, b\}: 0, \{d, a, b, c\}: 0\}$$

The mass functions for transition step st_2 are

$$m_A: \{\{a\}: 0, \{a, c\}: 0, \{a, c, d\}: 0, \{a, c, d, b\}: 0\}$$
$$m_B: \{\{b\}: 0, \{b, c\}: 1, \{b, c, d\}: 0, \{b, c, d, a\}: 0\}$$
$$m_C: \{\{d\}: 0.5, \{d, a\}: 0, \{d, a, b\}: 0.5, \{d, a, b, c\}: 0\}$$

Then, using Equation 14.14, the conditional probability mass function can be determined. For example, for st_1, we have

	$\{a\}$ 0.5	$\{a,c\}$ 0	$\{a,c,d\}$ 0.5	$\{a,b,c,d\}$ 0
$\{a\}$ 0.5	0.25	0	0.083	0
$\{a,b\}$ 0	0	0	0	0
$\{a,b,c\}$ 0.5	0.25	0	0.167	0
$\{a,b,c,d\}$ 0	0	0.1	0	0

Hence, $P(A/A) = 0.25 + 0.25 + 0.083 + 0.167 = 0.75$. Similarly, it can be shown that $P(B/A) = 0.167$, $P(C/A) = 0$. By normalization, it follows that $P(A/A) = 0.818$. Similarly, it can be shown that $P(B/A) = 0.182$, $P(C/A) = 0$. Hence, the conditional mass functions are

$$m^*_{/A} = \{\{A\}: 0.636, \{A, B\}: 0.182, \{A, B, C\}: 0\}$$

Taking the normalization again

$$m^*_{/A} = \{\{A\}: 0.778, \{A, B\}: 0.222, \{A, B, C\}: 0\}$$

Therefore,

$$P(A \to A) = 0.778 + 0.222 \div 2 + 0 \div 3 = 0.889$$
$$P(A \to B) = 0.222 \div 2 + 0 \div 3 = 0.111$$
$$P(A \to C) = 0$$

For st_2, no data are available to describe the transition from *A* to *B* or to *C*. However, it can be shown that the transition probabilities from *B* to *B* or *C* are

$$P(B \to B) = 0.667$$
$$P(B \to C) = 0.333$$

Using the fuzzy transition matrix for decision making can be done based on the *a posteriori* probability. Note that transition probabilities, $P(A \to A)$, $P(A \to B)$, and $P(A \to C)$, are now the *a priori* probability. Suppose the current transition step is ts_i and a new sample $\boldsymbol{x}$ is obtained, then the procedure for decision making is as follows:

Step 1: Calculate the conditional probability: $P(A/\mathbf{x})$, $P(B/\mathbf{x})$, $P(C/\mathbf{x})$,
Step 2: Calculate the total probability:

$$G(\mathbf{x}) = P(A/\mathbf{x})P(A \to A) + P(B/\mathbf{x})P(A \to B) + P(C/\mathbf{x})P(A \to C)$$

Step 3: Calculate the *a posteriori* probability:

$$P(A \to A/\mathbf{x}) = \frac{P(A/\mathbf{x})}{G(\mathbf{x})}$$

$$P(A \to B/\mathbf{x}) = \frac{P(B/\mathbf{x})}{G(\mathbf{x})}$$

$$P(A \to C/\mathbf{x}) = \frac{P(C/\mathbf{x})}{G(\mathbf{x})}$$

Step 4: Decision making based on the maximum *a posteriori* probability:

$$j^* = \text{argmax}\{P(A \to j / \boldsymbol{x})\}$$

where $j = A$, B, and C.

EXAMPLE 9: Following Example 8, suppose the current transition step is in ts_1 and a new sample $\mathbf{x} = [a\ d]$ is acquired, and estimate its corresponding state.

Solution: From Example 8, we have found that

$$q_A = a / 1 + b / 0 + c / 0.5 + d / 0.5$$
$$q_B = a / 0 + b / 1 + c / 0 + d / 1$$
$$q_C = a / 0 + b / 0 + c / 0 + d / 0$$

and the corresponding mass functions are

$$m_A: \{\{a\}: 0.5, \{a, c\}: 0, \{a, c, d\}: 0.5, \{a, c, d, b\}: 0\}$$
$$m_B: \{\{b\}: 0, \{b, c\}: 1, \{b, c, a\}: 0, \{b, c, a, d\}: 0\}$$
$$m_C: \{\{d\}: 0, \{d, a\}: 0, \{d, a, b\}: 0, \{d, a, b, c\}: 0\}$$

On the other hand, the new sample can be represented as

$$q_{\mathbf{x}} = a / 1 + b / 0 + c / 0 + d / 1$$

and the corresponding mass function is:

$$m_{\mathbf{x}}: \{\{a\}: 0, \{a, d\}: 1, \{a, d, b\}: 0, \{a, d, b, c\}: 0\}$$

Using Equation 14.14, it can be shown that $P(A/\mathbf{x}) = 0.5$, $P(B/\mathbf{x}) = 0.5$, and $P(C/\mathbf{x}) = 0$. From Example 8, it is known that $P(A \to A) = 0.8998$, $P(A \to B) = 0.1001$, and $P(A \to C) = 0$. Hence,

$$G(\boldsymbol{x}) = (0.5)(0.8998) + (0.5)(0.1001) + (0)(0) = 0.5$$
$$P(A \to A / \boldsymbol{x}) = (0.5)(0.8998) / (0.5) = 0.8998$$
$$P(A \to B / \boldsymbol{x}) = (0.5)(0.1002) / (0.5) = 0.1002$$
$$P(A \to B / \boldsymbol{x}) = (0)(0) / (0.5) = 0$$

Therefore, we will say that the current state is A. Recalling that in Example 6, we cannot reach a firm decision (the process state is either A or C in the same). Now, with the help of transition probability, we are able to make a certain decision.

14.4 Application Examples

14.4.1 Tool Condition Monitoring in Turning

This example were first used in [Du et al., 1992].

The cutting tests were performed on a 10 HP Standard Modern Lathe (model NC/17) and the schematic experiment setup is shown in Figure 14.8.

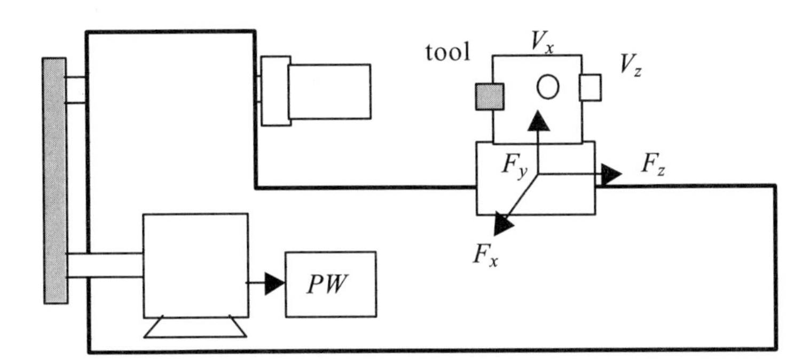

FIGURE 14.8 The schematic experiment setup.

As shown in the figure, six monitoring signals were used, including the three cutting force signals (F_x, F_y, F_z) measured using a dynamometer mounted under the tool holder; two vibration signals (V_x, V_z) measured using accelerometers mounted on the tool holder; and the cutting power (*PW*) measured from the armature current of the spindle motor. The monitoring signals were sampled simultaneously using an IBM PC compatible computer at 2 KHz. The cutting tests were conducted at 52 different cutting conditions. The variations of cutting conditions include the depth of cut (1.5 to 3.5 mm), the feed rate (32 to 126 mm/min), the cutting speed (92 to 322.2 m/min), the cutter inserts (K21, K6, and KC850), and the workpiece material (AISI 1040 steel and AISI 7065-T6 aluminum alloy). A total of eight process conditions were studied, as described in Table 14.4. From the six monitoring signals, 11 attributes were calculated, as defined in Table 14.5. These attributes were selected based on the pertinent literature and practical experience.

A total of 396 cutting experiments were conducted under various cutting and tool conditions as defined in Tables 14.4 and 14.5. Then, the fuzzy linear equation method was used to classify the results in five different tests, as described in Table 14.6. These tests have different meanings. Test 1 can be considered as self-classification, where all samples are used for training and classification. In Tests 2 and 3, part of the samples were randomly picked as learning samples, while the remaining samples were used as classification samples. Hence, the trained fuzzy linear equation were *estimates*. In Tests 4 and 5, the samples obtained under certain cutting conditions were used as training samples, while the remaining samples (obtained from different cutting conditions) were used as classification samples. Hence, the trained fuzzy linear equation were *predictions*. This situation is very common in many engineering applications and results in the most difficult tests, as one may expect. The testing results are summarized in Table 14.7.

From Table 14.7, it is seen that the overall success rate of the fuzzy linear equation method is about 75%. Besides, its performance in identifying slight tool wear and severe tool wear is inferior. A careful examination of the testing results reveals that many slight tool wear samples are classified as normal and many severe tool wear samples are classified as either medium tool wear or tool breakage. As pointed

TABLE 14.4 The Definition of Process Conditions

Class	Process Condition	Tool Identification	Work Identification
A	Normal	Wear < 0.1 mm	—
B	Air cut	—	—
C	Transient cut	—	A slot on workpiece
D	Tool breakage	Chipping > 0.04 mm	—
E	Slight wear	0.1 < VB < 0.15 mm	—
F	Medium wear	0.15 < VB < 0.3 mm	—
G	Severe wear	0.3 < VB < 0.6 mm	—
H	Chatter	—	Chatter marks

TABLE 14.5 The Definition of Attributes

Attribute	Definition
1	The mean of the resultant force $\bar{F} = \sqrt{\bar{F}_x + \bar{F}_y + \bar{F}_z}$
2	Crest factor of the feed force $CF = \dfrac{\max(F_z) - \min(F_z)}{\bar{F}_z}$
3	The ratio of the cutting forces $RF = \dfrac{\bar{F}_z}{\bar{F}_r} = \dfrac{\bar{F}_z}{\sqrt{\bar{F}_x^2 + \bar{F}_y^2}}$
4	The mean crossing rate of the feed force F_z
5	The energy of the feed force F_z in the frequency band [64–128] Hz
6	The energy of the feed force F_z in the frequency band [129–264] Hz
7	The energy of the feed force F_z in the frequency band [265–512] Hz
8	The energy of the vibration V_z in the frequency band [64–128] Hz
9	The energy of the vibration V_z in the frequency band [129–264] Hz
10	The energy of the vibration V_z in the frequency band [265–512] Hz
11	The mean of the cutting power *PW*

TABLE 14.6 The Design of Training and Classification Tests

Test 1	Training Samples	Classification Samples
1	All samples	All samples
2	Randomly pick 1/2 of the samples	The other 1/2 of the samples
3	Randomly pick 2/3 of the samples	The other 1/3 of the samples
4	Sequentially pick 1/2 of the samples	The other 1/2 of the samples
5	Sequentially pick 2/3 of the samples	The other 1/3 of the samples

TABLE 14.7 Class-by-Class Testing Results Using the Fuzzy Linear Equation (in %)

Process Condition	Test 1	Test 2	Test 3	Test 4	Test 5
Normal	90.2	88.6	97.6	92.4	94.5
Air cutting	100	100	100	70	100
Transient cutting	100	100	90	92.4	90
Tool breakage	82.7	72.5	81.3	79.9	91.3
Slight tool wear	16.7	25	18.9	17.8	25.2
Medium tool wear	81.8	67.2	72.4	67.2	77
Severe tool wear	7.7	10	32	7.9	12
Chatter	88.7	83.4	84.1	89.9	92.1
Overall	77.2	77.1	78.3	63.4	69

out earlier, the tool condition definitions are somewhat "fuzzy," and hence, the classification results are affected. In order to improve the success rate, it is recommended to use the "second-most-likely rule." That is, in addition to using Equation 14.32, the second-most-likely process condition is also taken into account:

$$i^{2nd} = \arg\max_{i}\left\{\min_{j}\left\{\mu\left(I_{ijk}\right)\right\}, i \neq i^{*}\right. \qquad \text{Equation (14.50)}$$

With the second-most likely rule, the success rate of the classification is improved, as shown in Table 14.8. It is seen that the success rate has been improved significantly, reaching 85% in average.

TABLE 14.8 Class-by-Class Classification with the Second-Most-Likely Rule (in %)

Process Condition	Test 1	Test 2	Test 3	Test 4	Test 5
Normal	100	98.6	97.6	100	84.8
Air cutting	100	100	100	100	100
Transient cutting	100	100	100	62.5	0
Tool breakage	90.4	82.5	91.3	59.3	100
Slight tool wear	63.3	75.0	68.9	6.7	44.4
Medium tool wear	80.0	67.2	72.4	51.7	64.7
Severe tool wear	69.2	80.0	82.0	7.1	28.5
Chatter	92.3	93.4	94.1	70.4	81.3
Overall	89.9	87.1	88.3	83.4	75.4

14.4.2 Tool Condition Monitoring in Boring

This example was first used in [Li, 1996]. The experiments were conducted on a CNC milling machining center with the experiment setup shown in Figure 14.9. The work material was mild steel and the cutter material was HSS. During the cutting experiments, the main spindle motor current and feed spindle motor current were recorded together with the tool usage time and cutting conditions (width of cut, feed and cutting speed). The tool was taken out to measure the wear once every 3 minutes or so using a tool maker's microscope. The tool wear is classified into three states: new tool ($VB_{max} < 0.1$ mm), normal tool ($0.1 < VB_{max} < 0.5$ mm), and worn tool ($VB_{max} > 0.5$ mm).

It is interesting to note that among various tool wear sensors, current sensor is not the most effective one. This is due to the fact that all the force and vibration information have been mixed up (so we cannot differentiate the directional forces and vibrations, the effect of plowing, built up edge, and so on), and filtered (the spindle and the motor can be considered as a low pass filter). However, the current sensor is arguably the cheapest and most reliable sensor, especially for the applications on the shop floor. Hence, it is worthwhile to study.

As pointed out earlier, tool life follows the Taylor's tool life equation. However, the exact state of the tool wear is somewhat fuzzy, and hence the fuzzy transition method is used. For each tool, three transition steps are used corresponding to the three states of tool wear (A: $ts_1 = [0, 0.3]$, B: $ts_2 = [0.3, 1.1]$, C: $ts_3 = [1.1, 3]$). The monitoring signal feature is the averaged spindle motor current and the feed motor current. The fuzzy membership functions are first determined using the fuzzy linear equation method with two intervals for each feature.

A total of 20 sets of experiments were conducted. Depending on the tool life, a set may contain more than 16 data points or just a few. First, all the data sets are used for training (to obtain the fuzzy transition probabilities), and then for classification. The one-step transition probabilities are summarized in Table 14.9. The average success rate is 100% (100% in ts_1, 100% in ts_2, and 100% in ts_3).

Second, the first 13 sets are used for training and the other 7 sets are used for testing. Table 14.10 shows the transition probabilities, and the success rate is also 100% (100% in ts_1, 100% in ts_2, and 100%

TABLE 14.9 The Probabilities of One-Step Fuzzy Transition

	A	B	C
(a) In Time Interval ts_1			
A	0.8554	0.1446	0
B	0	1	0
C	0	0	0
(b) In Time Interval ts_2			
A	0.4844	0.4844	0.0312
B	0	0.8783	0.1217
C	0	0	1
(c) In Time Interval ts_3			
A	0	0	0
B	0	0.8089	0.1911
C	0	0	1

TABLE 14.10 The Probabilities of One-Step Fuzzy Transition

	A	B	C
(a) In Time Interval ts_1			
A	0.7843	0.2157	0
B	0	0.7926	0.2074
C	0	0	1
(b) In Time Interval ts_2			
A	0.4841	0.4841	0.0317
B	0	0.8231	0.1769
C	0	0	1
(c) In Time Interval ts_3			
A	0	0	0
B	0	0.8231	0.1769
C	0	0	1

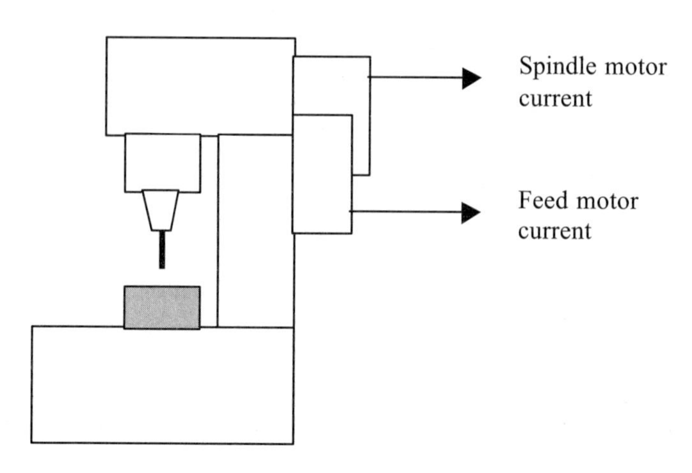

FIGURE 14.9 Schematic experiment setup of the tool condition monitoring in boring.

in ts_3). In comparison, using the fuzzy linear equation method (the machining time and cutting speed were used as features as well) alone results in a success rate of 61%.

Finally, tests were carried out without using the time intervals. For the self-classification test (all the data are for learning and testing), the success rate is still 100%. For the prediction test (first 13 sets of data are used for learning and the other 7 sets of data are for testing), the success rate is again 100%. Their fuzzy transition probabilities are shown in Tables 14.11 and 14.12. These tables indicate that the fuzzy transition method is indeed very effective.

TABLE 14.11 The Probabilities of One-Step Fuzzy Transition Without Time Intervals in Self-Classification Test

	A	*B*	*C*
A	0.4845	0.4890	0.0265
B	0	0.8851	0.1149
C	0	0	1

TABLE 14.12 The Probabilities of One-Step Fuzzy Transition Without Time Intervals in Prediction Test

	A	*B*	*C*
A	0.4929	0.4786	0.0265
B	0	0.8936	0.1064
C	0	0	1

14.5 Conclusions

In this chapter, we briefly described how to use fuzzy set theory for monitoring and diagnosing manufacturing engineering processes. This usually consists of two steps: (i) establish a fuzzy relationship between the sensor signal features and the process condition classes, and (ii) classify the new sensor signal using the fuzzy relationship. Note that the definitions of process condition classes are often imprecise (e.g., tool wear may be manifested into various forms), hence, the use of a fuzzy set is meaningful and effective.

Various methods can be used to establish the fuzzy relationship, and three of them are discussed in detail, including the fuzzy linear equation method, fuzzy C-mean classification method, and the fuzzy transition method. For the faults such as wear and fatigue, the fuzzy transition method takes the effect of time into consideration, and hence, provides more effective and reliable results. This method is new and may outperform other methods as it makes use of additional information.

The chapter uses several simple examples to demonstrate the computation procedures. In addition, two comprehensive examples are included to demonstrate how to solve practical problems. From these examples, it is seen the fuzzy set theory is effective. The reader should be aware that, like other monitoring and diagnosis methods such as neural networks and expert systems, fuzzy set theory is not a universal solution that can plug-and-play. Instead, successful applications rely on a complete understanding of the process, the sensors, and the sensor signals. Use of effective sensors is often the most effective and cost-effective way for monitoring and diagnosis. Also, signal processing is very important to capture the fault characteristics. Only through continuing investigation and improvement can success be assured.

Acknowledgments

First, the authors would like to thank Miss Ying Liu (University of Miami), who wrote the programs to test the fuzzy transition probability method. Second, the authors would like to thank Dr. Xiaoli Li (City University of Hong Kong), who provided the data for the boring tool condition monitoring example.

Many colleagues have contributed to this article, in particular, Dr. Jun Wang (Chinese University of Hong Kong), the co-editor of this handbook.

References

Baldwin, J. F., Martin, T. P., and Pilsworth, B. W., 1995, *Firl — Fuzzy and Evidential Reasoning in Artificial Intelligence*, Research Studies Press, England.

Bezdek, J. C., 1981, *Pattern Recognition with Fuzzy Objective Function Algorithm*, Plenum Press, New York.

Davies, A. (Ed.), 1998, *Handbook of Condition Monitoring, Techniques and Methodology*, Chapman & Hall, London.

Du, R., Elbestawi, M. A. and Li, S., 1992, Tool Condition Monitoring in Turning Using Fuzzy Set Theory, *Int. J. Mach. Tools Manuf.*, vol. 32, no. 6, pp. 781-796.

Du, R., Elbestawi, M. A. and Wu, S. M., 1995, Computer Automated Monitoring of Manufacturing Processes: Part 1, Monitoring Decision-Making Methods, *Trans. ASME, J. Eng. Industry*, vol. 117, no. 2, pp. 121-132.

Gertler, J., 1988, *Fault Detection and Diagnosis in Engineering Systems*, Marcel Dekker, New York.

Klir, G. J. and Folger, T. A., 1988, *Fuzzy Sets, Uncertainty, and Information*, Prentice-Hall, Englewood Cliffs, New Jersey.

Klyele, R. M. and de Korvin, A., 1998, Constructing One-Step and Limiting Fuzzy Transition Probabilities for Finite Markov Chains, *J. Intelligent Fuzzy Systems*, no. 6, pp. 223-235.

Quinlan, J. R., 1986, Induction of Decision Trees, *Machine Learning*, vol. 1, pp. 81-106.

Quinlan, J. R., 1987, Simplifying Decision Trees, *Int. J. Man-Machine Studies*, vol. 27, pp. 221-234.

Zadeh, L. A., 1965, Fuzzy Sets, *Information and Control*, vol. 8, pp. 338-353.

15

Fuzzy Neural Network and Wavelet for Tool Condition Monitoring

Xiaoli Li
Harbin Institute of Technology

15.1 Introduction

To reduce operating costs and improve product quality are two objectives for the modern manufacturing industries, so most manufacturing systems are fast converting to fully automated environments such as computer integrated manufacturing (CIM) and flexible manufacturing systems (FMS). However, many manufacturing processes involve some aspects of metal cutting operations. The most crucial and determining factor to successful maximization of the manufacturing processes in any typical metal cutting process is tool condition. It would seem be logical to propose that tool condition monitoring (TCM) will inevitably become an automated feature of such manufacturing environments. Due to failure, cutting tools adversely affect the surface finish of the workpiece and damage machine tools; serious failure of cutting tools may possibly endanger the operator's safety. Therefore, it is very necessary to develop tool condition monitoring systems that would alert the operator to the states of cutting tools, thereby avoiding undesirable consequences [1].

Initial TCM systems focused mainly on the development of mathematical models of the cutting process, which were dependent upon large amounts of experimental data. Due to the complexity of the metal cutting process, an accurate model for wear and breakage prediction of cutting tools cannot be obtained, so that many researchers resort to sensor integration methods for replacing model methods. These results in a series of problems such as signal processing, feature extraction, and pattern recognition. To overcome the difficulty of these problems, computational intelligence (fuzzy systems, neural networks, wavelet transforms, genetic algorithms, etc.) has been applied in some TCM systems in recent years. The TCM systems based on computational intelligence, such as wavelet transforms [2], fuzzy inference [3–5], fuzzy neural networks [6–9], etc., have been established, in which all forms of tool condition can be monitored.

Fuzzy systems and neural networks are complementary technologies in the design of intelligent systems. Neural networks are essentially low-level computational structures and algorithms that offer good performance in dealing with sensory data, while fuzzy systems often deal with issues such as reasoning on

0-8493-0592-6/01/$0.00+$.50

a higher lever than neural networks. However, since fuzzy systems do not have much learning capability, it is difficult for a human operator to tune the fuzzy rules and membership functions from the training data set. Also, because the internal layers of neural networks are always opaque to the user, the mapping rules in the network are not visible so that it is difficult to understand; furthermore, the convergence (learning time) is usually very slow or not guaranteed. Thus, it is very necessary to reap the benefits of both fuzzy systems and neural networks by combining them in a new integrated system, called a fuzzy neural network (FNN). FNN had been widely used in the TCM [10–12].

Spectral analysis and time series analysis are the most common signal processing methods in TCM. These methods have a good solution in the frequency domain but a very bad solution in the time domain, so that they lose some useful information during signal processing. In general, they are recommended only for processing stability stochastic signals. Recently, wavelet transforms (WT) have been proposed as a significant new tool in signal analysis and processing [13, 14]. They have been used to analyze some signals for tool breakage monitoring [15, 16]. WT has a good solution in the time–frequency domain so that it can extract more information in the time domain at different frequency bands from any signals [17].

Tool condition monitoring can be divided into the two types: tool breakage and tool wear. This chapter addresses how to apply the fuzzy neural network and wavelet transforms to TCM. First, the fuzzy neural network and the wavelet transforms are respectively introduced. Second, the continuous wavelet transforms (CWT) and discrete wavelet transforms (DWT) are used to decompose the spindle AC servomotor current signal and the feed AC servomotor current signal in the time–frequency domain, respectively. Real-time tool breakage detection of small-diameter drills is presented by using motor current decomposed. Third, analyzing the effects of tool wear as well as cutting parameters on the current signals, the models of the relationship between the current signals and the cutting parameters are established, and the fuzzy classification method is effectively used to detect tool wear states. Finally, wavelet packet transforms are applied to decompose AE signals into different frequency bands in the time domain; the root means square (RMS) values extracted from the decomposed signals of each frequency band are referred to as the features of tool wear. The fuzzy neural network is presented to describe the relationship between the tool wear conditions and the monitoring features.

15.2 Fuzzy Neural Network

15.2.1 Combination of Fuzzy System and Neural Network

Fuzzy system (FS) and neural networks (NN) are powerful tools for controlling the complex systems operating under a known or unknown environment. Fuzzy systems can easily be used to express approximate knowledge and to quickly implement a reaction, but have difficulty in executing learning processes [18]. Neural networks have strong learning abilities but are weak at expressing rule-based knowledge. Although the fuzzy system and neural networks possess remarkable properties when they are employed individually, there are great advantages to using them synergistically, resulting in what are generally referred to as *neuro-fuzzy approaches* [19].

Neural networks are organized in layers, each consisting of neurons or processing elements that are interconnected. The neurons or perceptions compute a weight sum of their inputs, generating an output. The connections between the neurons have weighted numerical inputs associated with them. There are a number of learning methods to train neural nets, but the backpropagation (BP) paradigm has emerged as the most popular training mechanism. The BP method works by measuring the difference between the system output and the observed output value. The values being calculated at the output layer are propagated to the previous layers and used for adjusting the connection weights. But there are potential drawbacks: (i) no clear guidelines on how to design neural nets; (ii) accuracy of results relies heavily on the size of the training set; (iii) the logic behind the estimate is hard to convey to the user; (iv) long learning time; (v) local convergence. In order to overcome its drawbacks, some hybrid models of neural network and fuzzy system are presented. There are many possible combinations of the two systems, but the four combinations shown in Figure 15.1 have been widely applied to actual systems [20].

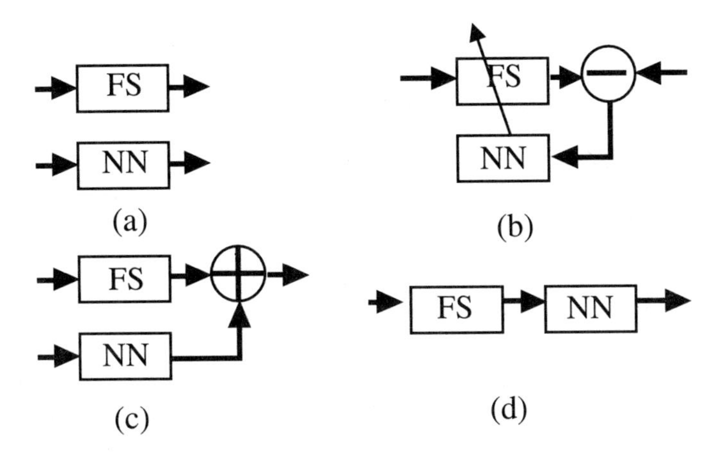

FIGURE 15.1 Combination type of neural network and fuzzy system. (Reprinted with permission of Springer-Verlag London, Ltd. From "Hybrid Learning for Tool Wear Monitoring," *Int. J. Adv. Manuf. Technol.*, 2000, 16, 303–307.)

Figure 15.1(a) shows the case where one piece of equipment uses the two systems for different purposes without mutual cooperation. The model in Figure 15.1(b) shows NN used to optimize the parameters of FS by minimizing the error between the output of FS and the given specification. Figure 15.1(c) shows a model where the output of FS is corrected by the output of NN to increase the precision of the final system output. Figure 15.1(d) shows a cascade combination of FS and NN where the output of the FS or NN becomes the input of another NN or FS. The models in Figures 15.1(b) and 15.1(c) are referred to as a combination model with net learning and a combination model with equal structure, respectively. These are shwon in greater detail in Figure 15.2. Figure 15.2(a) shows that the total system is controlled by means of fuzzy system, but the membership of the fuzzy system is produced and adjusted by the learning power of the neural network. The model in Figure 15.2(b) shows that the fuzzy system can be controlled by the neural network; the inference processing of the fuzzy system is responded to by the neural network.

15.2.2 Fuzzy Neural Network

In this chapter, a new neural network with fuzzy inference is presented. Let X and Y be two sets in [0,1] with the training input data $(x_1, x_2, \ldots, x_n)$ and the desired output value $(y_1, y_2, \ldots, y_m)$, respectively. The set of the corresponding elements of the weight matrix is $(w_{11}, w_{12}, \ldots, w_{nm})$. Based on the fuzzy inference, the definition is given as follows:

$$Y = X \circ W \qquad \text{Equation (15.1)}$$

and

$$y_j = \max(\min(x_i, w_{ij}))\ (i = 1, 2, \ldots, n;\ j = 1, 2, \ldots, m) \qquad \text{Equation (15.2)}$$

The fuzzy neural network topology is shown in Figure 15.3. Basically, the idea of backpropagation (BP) is used to find the errors of node outputs in each layer. Without any loss of generality, the detailed learning processes of a single layer for clarity are derived as follows. The derivation can easily be extended to the multiple-output case. The goal of the proposed learning algorithm is to minimize a least-squares error function:

$$E = \left(T_j - O_j\right)^2 / 2 \qquad \text{Equation (15.3)}$$

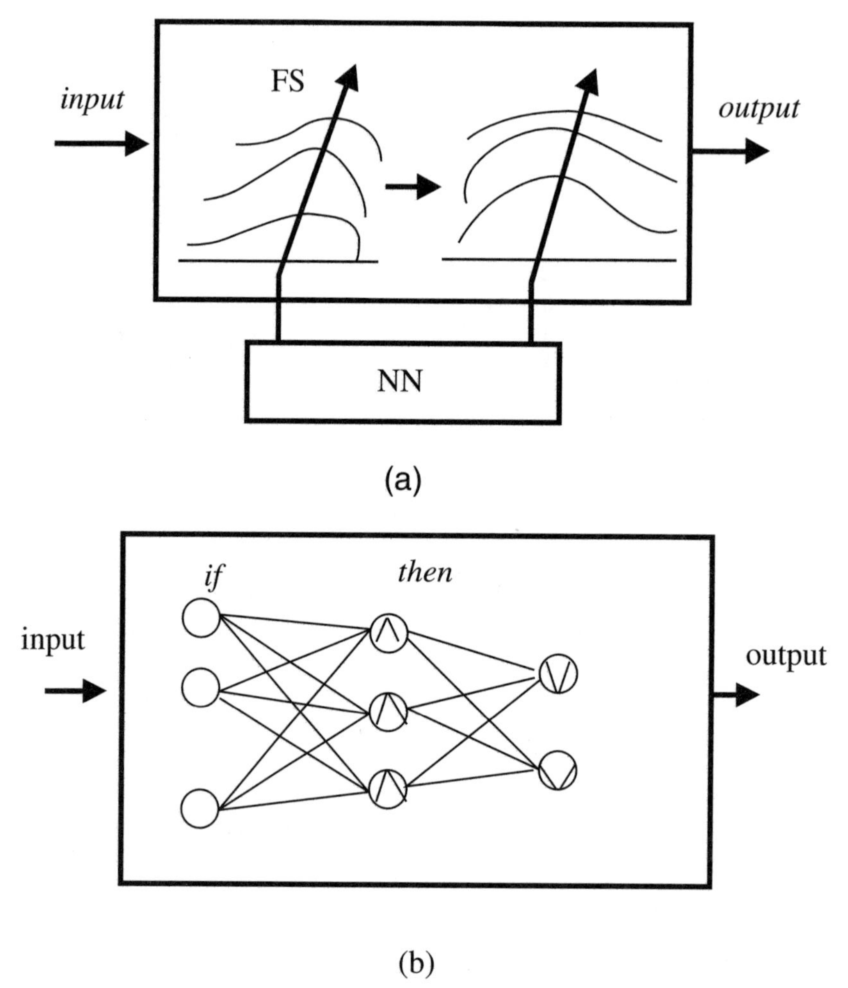

FIGURE 15.2 Combination model with (a) net learning, and (b) equal structure. (Reprinted with permission of Springer-Verlag London, Ltd. From "Hybrid Learning for Tool Wear Monitoring," *Int. J. Adv. Manuf. Technol.*, 2000, 16, 303–307.)

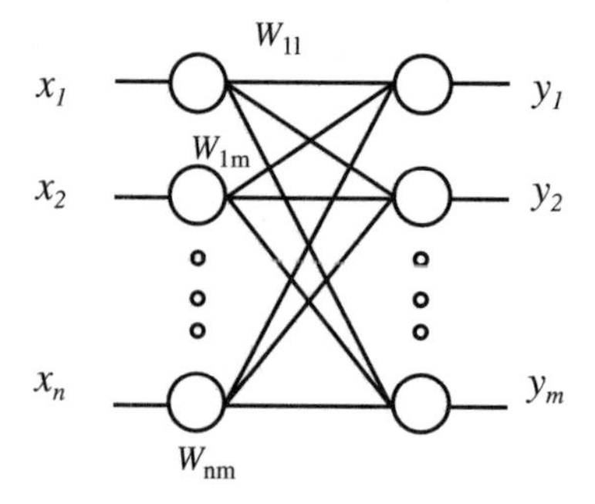

FIGURE 15.3 FNN net topology. (Reprinted with permission of Chapman & Hall, Ltd. From "On-line Tool Condition Monitoring System with Wavelet Fuzzy Neural Network," *Journal of Intelligent Manufacturing*, 1997, 8, 271–276.)

where $O_j = \max(\min(x_i, w_{ij}))$, T_j is desired output values, O_j is the actual values, the least-squares error between them is E. The general parameter learning rule used is as follows:

$$\frac{\partial E}{\partial w_{ij}} = \frac{\partial E}{\partial O_j} \cdot \frac{\partial O_j}{\partial w_{ij}} \qquad \text{Equation (15.4)}$$

where

$$\frac{\partial O_j}{\partial w_{ij}} = \frac{\partial \vee\left(\wedge\left(x_i, w_{ij}\right)\right)}{\partial \wedge\left(x_s, w_{sj}\right)} \frac{\partial \wedge\left(x_s, w_{sj}\right)}{\partial w_{sj}} \qquad \text{Equation (15.5)}$$

Set

$$\begin{aligned} a_1 &= \frac{\partial \vee\left(\wedge\left(x_i, w_{ij}\right)\right)}{\partial \wedge\left(x_s, w_{sj}\right)} = \frac{\partial \vee\left(\wedge\left(x_s, w_{sj}\right) \underset{i \neq s}{\vee}\left(\wedge\left(x_i, w_{ij}\right)\right)\right)}{\partial \wedge\left(x_s, w_{sj}\right)} \\ a_2 &= \frac{\partial \wedge\left(x_s, w_{sj}\right)}{\partial w_{sj}} \end{aligned} \qquad \text{Equation (15.6)}$$

Define

$$\text{when } \wedge\left(x_s, w_{sj}\right) \geq \underset{i \neq s}{\vee}\left(\wedge\left(x_i, w_{ij}\right)\right), a_1 = 1,$$

$$\text{otherwise } a_1 = \wedge\left(x_s, w_{sj}\right);$$

$$\text{when } x_s \geq w_{sj}, a_2 = 1, \text{ otherwise } a_2 = x_s$$

Assuming

$$\frac{\partial O_j}{\partial w_{sj}} = \Delta \qquad \text{Equation (15.7)}$$

According to fuzzy min–max and smooth derivative ideas, a fuzzy ruler is constructed as follows:

$$\begin{aligned} &\text{if } x_s < w_{sj} \text{ and } x_s \geq \underset{i \neq s}{\vee}\left(\wedge\left(x_i, w_{ij}\right)\right) \text{ then } \Delta = x_s \\ &\text{if } x_s < w_{sj} \text{ and } x_s < \underset{i \neq s}{\vee}\left(\wedge\left(x_i, w_{ij}\right)\right) \text{ then } \Delta = x_s^2 \\ &\text{if } x_s \geq w_{sj} \text{ and } w_{sj} \geq \underset{i \neq s}{\vee}\left(\wedge\left(x_i, w_{ij}\right)\right) \text{ then } \Delta = 1 \\ &\text{if } x_s \geq w_{sj} \text{ and } w_{sj} < \underset{i \neq s}{\vee}\left(\wedge\left(x_i, w_{ij}\right)\right) \text{ then } \Delta = w_s \end{aligned} \qquad \text{Equation (15.8)}$$

and

$$\frac{\partial E}{\partial O_i} = -\left(T_j - O_j\right) \quad \text{Equation (15.9)}$$

Set

$$\delta_j = \frac{\partial E}{\partial O_j} \quad \text{Equation (15.10)}$$

Then

$$\frac{\partial E}{\partial w_{ij}} = \delta_j \Delta \quad \text{Equation (15.11)}$$

the changes for the weight will be obtained from a δ-rule with expression

$$\Delta w_{ij} = \mu \delta_j \Delta \quad \text{Equation (15.12)}$$

where μ is learning rates , $\mu \in [0,1]$.

To test the fuzzy neural network (FNN), it is compared with the BP neural networks (BPNN) [22]. Under the same conditions (training sample, networks structure (5×5), learning rate (0.8), convergence error (0.0001)), the training iteration of FNN is 7, but that of BPNN is 427. Figure 15.4 shows each training process.

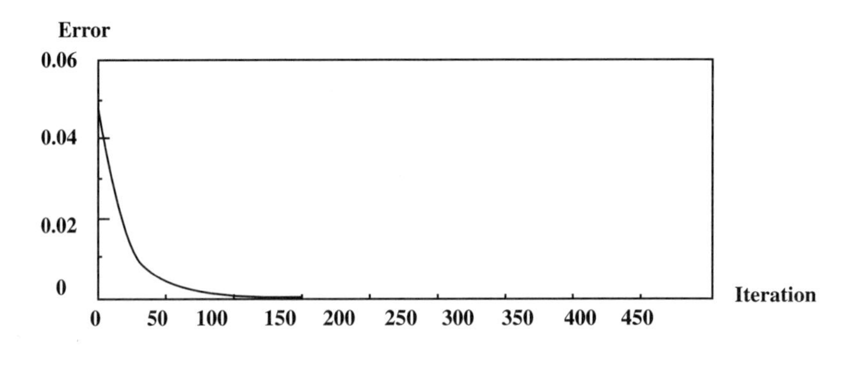

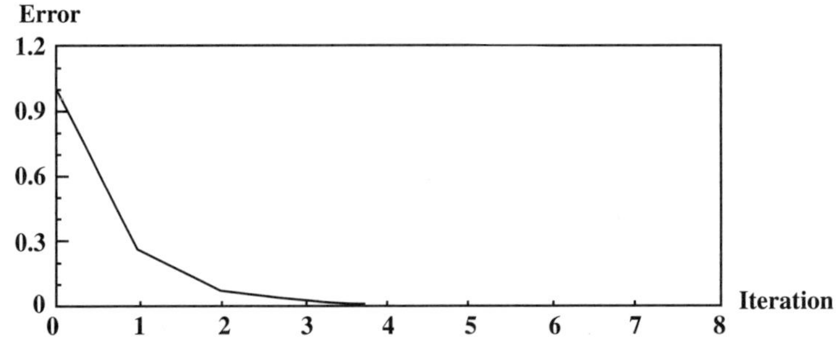

FIGURE 15.4 (Top): Training process BPNN and (Bottom): FNN. (Reprinted with permission of Chapman & Hall, Ltd. From "On-line Tool Condition Monitoring System with Wavelet Fuzzy Neural Network," *Journal of Intelligent Manufacturing*, 1997, 8, 271–276.)

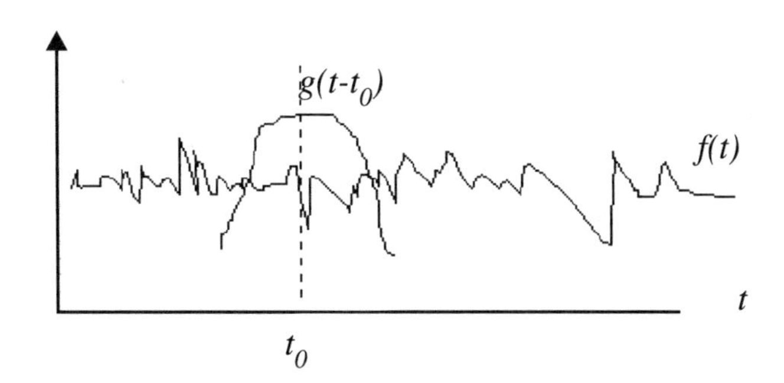

FIGURE 15.5 An illustration of the STFT. (Reprinted with permission of Elsevier Science, Ltd. From "Tool Wear Monitoring with Wavelet Packet Transform-Fuzzy Clustering Method," *Wear,* 1998, 219(2), 145–154.)

15.3 Wavelet Transforms

15.3.1 Wavelet Transforms (WT)

An energy limited signal $f(t)$ can be decomposed by its Fourier transforms $F(w)$, namely

$$f(t) = \frac{1}{2\pi}\int_{-\infty}^{+\infty} F(w)e^{iwt}dt \qquad \text{Equation (15.13)}$$

where

$$F(w) = \int_{-\infty}^{+\infty} f(t)e^{-iwt}dt \qquad \text{Equation (15.14)}$$

$f(t)$ and $F(w)$ are called a pair of Fourier transforms. Equation 15.13 implies that $f(t)$ signal can be decomposed into a family in which harmonics e^{iwt} and the weighting coefficient $F(w)$ represent the amplitudes of the harmonics in $f(t)$. $F(w)$ is independent of time; it represents the frequency composition of a random process that is assumed to be stationary so that its statistics do not change with time. However, many random processes are essentially nonstationary signals such as vibration, acoustic emission, sound, and so on. If we calculate the frequency composition of nonstationay signals in the usual way, the results are the frequency composition averaged over the duration of the signal, which can't adequately describe the characteristics of the transient signals in the lower frequency.

In general, a short-time Fourier transform (STFT) method is used to deal with nonstationary signals. STFT has a short data window centered at time (see Figure 15.5).

Spectral coefficients are calculated for this short length of data, and the window is moved to a new position and repeatedly calculated. Assuming an energy limited signal, $f(t)$ can be decomposed by STFT, namely

$$G(w, t_0) = \int_R f(t)g(t - t_0)e^{-iwt}dt \qquad \text{Equation (15.15)}$$

where $g(t{-}t_0)$ is called *window function.* If the length of the window is represented by time duration T, its frequency bandwidth is approximately $1/T$. Use of a short data window means that the bandwidth of each spectral coefficient is on the order $1/T$, namely its frequency band is wide. A feature of the STFT is that all spectral estimates have the same bandwidth. Clearly, STFT cannot obtain a high resolution in both the time and the frequency domains.

Wavelet transforms involve a fundamentally different approach. Instead of seeking to break down a signal into its harmonics, which are global functions that go on forever, the signals are broken down into a series of local basis functions called *wavelets*. Each wavelet is located at a different position on the time axis and is local in the sense that it decays to zero when sufficiently far from its center. At the finest scale, wavelets may be very long. Any particular local features of signals can be identified from the scale and position of the wavelets. The structure of nonstationary signals can be analyzed in this way, with local features represented by a close-packet wavelet of short length.

Given a time varying signal $f(t)$, wavelet transforms (WT) consist of computing a coefficient that is the inner product of the signal and a family of wavelets. In the continuous wavelet transforms (CWT), the wavelet corresponding to scale a and time location b is

$$\psi_{a,b} = \frac{1}{\sqrt{|a|}}\psi\left(\frac{1-b}{a}\right) \quad a,b \in R, a \neq 0 \qquad \text{Equation (15.16)}$$

where a and b are the dilation and translation parameters, respectively.

The continuous wavelet transform is defined as follows:

$$w_f(a,b) = \int x(t)\psi^*_{a,b}(t)dt \qquad \text{Equation (15.17)}$$

where "*" denotes the complex conjugation.

With respect to $w_f(a,b)$, the signals $f(t)$ can be decomposed into

$$f(t) = \frac{1}{c_\psi}\int_{-\infty}^{+\infty}\int_0^{+\infty} w_f(a,b)\frac{1}{\sqrt{|a|}}\psi\left(\frac{1-b}{a}\right)dadb \qquad \text{Equation (15.18)}$$

where c_ψ is a constant depending on the base function. Similar to the Fourier transforms, $w_f(a,b)$ and $f(t)$ constitute a pair of wavelet transforms. Equation 15.17 implies that WT can be considered as $f(t)$ signal decomposition. Compared with the STFT, the WT has a time-frequency function that describes the information of $f(t)$ in various time windows and frequency bands.

When $a = 2^j$, $b = k2^j$, $j, k \in Z$, the wavelet is in this case

$$\psi_{j,k} = 2^{-\frac{d}{2}}\psi\left({}^{-j}t - k\right) \qquad \text{Equation (15.19)}$$

The discrete wavelet transform (DWT) is defined as follows:

$$c_{j,k} = \int f(t)\psi_{j,k}(t) \qquad \text{Equation (15.20)}$$

where $c_{j,k}$ is defined as the wavelet coefficient, it may be considered as a time–frequency map of the original signal $f(t)$. Multi-resolution analysis is used in discrete scaling function:

$$\phi_{j,k} = 2^{-\frac{d}{2}}\phi\left(\frac{t-2^d k}{2^d}\right) \qquad \text{Equation (15.21)}$$

Set

$$d_{j,k} = \int f(t)\phi^*_{j,k}(t)dt \qquad \text{Equation (15.22)}$$

where $d_{j,k}$ is called the scaling coefficient, and is the sampled version of original signals. When $j = 0$, it is the sampled version of the original signals. Wavelet coefficients $c_{j,k}$ ($j = 1, 2, \ldots, J$) and scaling coefficients $d_{j,k}$ are given by

$$c_{j,k} = \sum_n x[n] h_j[n - 2^j k] \quad \text{Equation (15.23)}$$

and

$$d_{j,k} = \sum_n x[n] g_j[n - 2^j k] \quad \text{Equation (15.24)}$$

where $x[n]$ are discrete-time signals, $h_j[n - 2^j k]$ is the analysis discrete wavelets, and the discrete equivalents to $2^{-j/2}\psi(2^{-j}(t - 2^j k))$, $g_j[n - 2^j k]$ are called *scaling sequence.* At each resolution $j > 0$, the scaling coefficients and the wavelet coefficients can be written as follows:

$$c_{j+1,k} = \sum_n g[n - 2k] d_{j,k} \quad \text{Equation (15.25)}$$

$$d_{j+1,k} = \sum_n h[n - 2k] d_{j,k} \quad \text{Equation (15.26)}$$

In fact, the structure of computations in DWT is exactly an octave-band filter [23]. The terms g and h can be considered as high-pass and low-pass filters derived from the analysis wavelet $\psi(t)$ and the scaling function $\phi(t)$, respectively.

15.3.2 Wavelet Packet Transforms

Wavelet packets are particular linear combinations of wavelets. They form bases that retain many of the orthogonality, smoothness, and location properties of their parent wavelets. The coefficients in the linear combinations are computed by a factored or recursive algorithm, with the result that expansions in wavelet packet bases have low computational complexity.

The discrete wavelet transforms can be rewritten as follows:

$$\begin{aligned} c_j[f(t)] &= h(t) * c_{j-1}[f(t)] \\ d_j[f(t)] &= g(t) * c_{j-1}[f(t)] \\ c_0[f(t)] &= f(t) \end{aligned} \quad \text{Equation (15.27)}$$

Set

$$\begin{aligned} H\{\cdot\} &= \sum_k h(k - 2t) \\ G\{\cdot\} &= \sum_k g(k - 2t) \end{aligned} \quad \text{Equation (15.28)}$$

Then Equation 15.27 can be written as follows:

$$c_j[f(t)] = H\{c_{j-1}[f(t)]\}$$
$$d_j[f(t)] = G\{c_{j-1}[f(t)]\}$$
Equation (15.29)

Clearly, DWT is only the approximation $c_{j-1}[f(t)]$ but not the detail signals $d_{j-1}[f(t)]$; wavelet packet transforms don't omit the detail signals. Therefore, wavelet packet transforms is expressed as follows:

$$c_j[f(t)] = H\{c_{j-1}[f(t)]\} + G\{d_{j-1}[f(t)]\}$$
$$d_j[f(t)] = G\{c_{j-1}[f(t)]\} + H\{d_{j-1}[f(t)]\}$$
Equation (15.30)

Let $Q_j^i(t)$ be the ith packet on j^{th} resolution, then the wavelet packet transforms can also be computed by the recursive algorithm, as follows:

$$Q_0^1(t) = f(t)$$
$$Q_j^{2i-1}(t) = HQ_{j-1}^i(t)$$
$$Q_j^{2i}(t) = GQ_{j-1}^i(t)$$
Equation (15.31)

where $t = 1, 2, \ldots, 2^{J-i}$, $i = 1, 2, \ldots, 2^j$, $j = 1, 2, \ldots, J$, $J = \log_2 N$, N is data length. The wavelet packet transforms are represented by Figure 15.6.

15.4 Tool Breakage Monitoring with Wavelet Transforms

Tool breakage monitoring plays an important role in the cutting process. These monitoring systems have been developed over many years. A fair amount of techniques have been developed to detect tool breakage during the cutting process; the most common techniques reported in the industrial machining environment include force, acoustic emission (AE), and current. It is known that force measurement is the best method for detecting tool breakage. However, its main disadvantage is that each tool is required to a fitted sensor system, resulting in a very high cost; in addition, the installation of a force-measuring sensor system is difficult for the present machine tools. In recent years, the AE sensing technique has also been considered one of the most effective methods for tool breakage monitoring. One of the main obstacles is how to detect the AE signals from a rotating tool such as in boring, drilling, and milling. Moreover, AE signals analysis is a very difficult problem, for example, how to extract the features from AE signals that are related to tool condition. In Section 15.6, a method is described for using AE signals to monitor tool wear condition.

During the cutting processes, motor current is related to the tool conditions. Less power is consumed when a tool is broken, and this variance can be exploited for on-line tool breakage monitoring. The motor current of machine tools can be measured through a current transformer (such as a Hall Current Sensor); when the measured signals are processed, the result will be found to drop instantaneously and soon recover to a level prior to the drop when tool breakage occurs [24]. The current measurement system is relatively simple, and its mounting will not affect the machining operations [25]. But it is less sensitive for small tool breakage when compared to force sensing and AE sensing; the system of current measurement is only reliable in monitoring tool breakage at medium and heavy cuts [26].

This section presents on-line tool breakage detection of small diameter drills by sensing the AC servomotor current [27]. The continuous wavelet transforms (CWT) were used to decompose the spindle

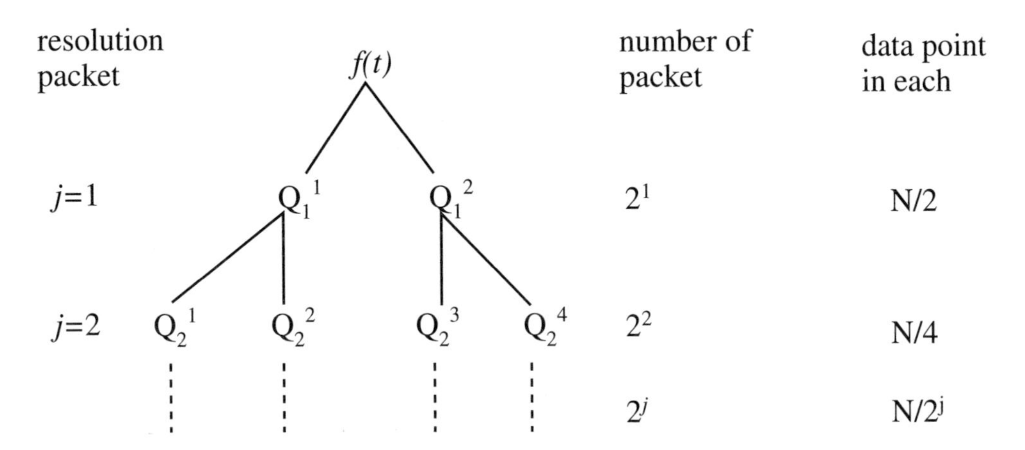

FIGURE 15.6 Tree structure of the wavelet packet transforms. (Reprinted with permission of Elsevier Science, Ltd. From "Tool Wear Monitoring with Wavelet Packet Transform-Fuzzy Clustering Method," *Wear,* 1998, 219(2), 145–154.)

AC servomotor current signals and the discrete wavelet transforms (DWT) were used to decompose the feed AC servomotor current signals in the time–frequency domain. The features of tool breakage were extracted from the decomposed signals. Experimental results showed that the proposed monitoring system could work in real time; in addition, it had a low sensitivity to changes in the cutting conditions and a high detection rate for the breakage of small diameter drills [28].

15.4.1 Experimental Setup

The schematic diagram of the experimental setup is shown in Figure 15.7. Cutting tests were performed on a Machine Center Makino-FNC74-A20. The four axles (spindle, X, Y, and Z) of the machine have recalculating ball screw drives and are directly driven by permanent magnet synchronous AC servomotors. The AC servomotor current signals of the Machine Center were measured through Hall Current Sensor. The signals were first passed though low-pass filters (cut-off frequency: 500 HZ) and sent via an A/D converter to a personal computer.

A successful tool breakage detecting method must be sensitive to tool change in tool condition, but insensitive to the variations of cutting conditions. Hence, cutting tests were conducted at different conditions to evaluate the performance of the proposed method. Table 15.1 shows the tool parameters and cutting conditions.

15.4.2 Wavelet Analysis of Tool Breakage Signals

Figures 15.8(a)and 15.8(b) show the spindle current signal and feed current signals of an AC servomotor, respectively. Figures 15.9(a) and 15.9(b) show the results of the spindle current CWT and the results of the feed current DWT at resolution $j = 2$, respectively.

To detect the tool breakage efficiently in drilling process monitoring, the method must fit the different kinds of cutting conditions. Figures 15.10, 15.12, and 15.14 show the spindle current signals and feed current signals under the different cutting conditions, respectively. Figures 15.11, 15.13, and 15.15 are the results of the above signal processing, respectively. Clearly, small differences between the normal wear and tool breakage can be observed by processed signals. Using a simple methodology, tool breakage can be reliably detected.

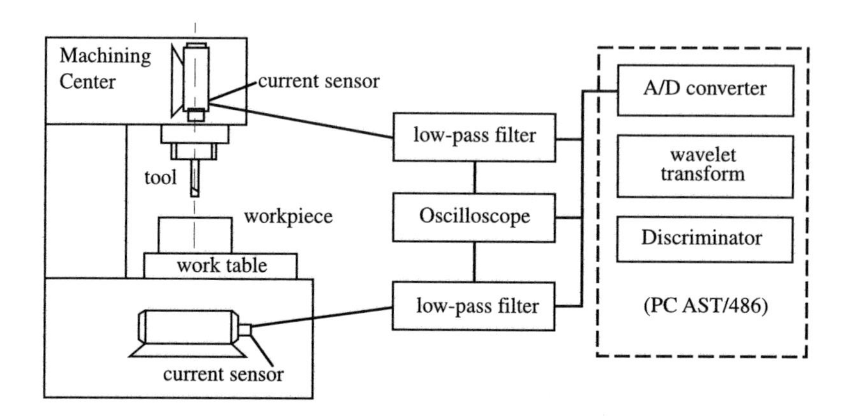

FIGURE 15.7 Schematic diagram of the experimental setup. (Reprinted with permission of Elsevier Science, Ltd. From "On-Line Detection of the Breakage of Small Diameter Drills Using Current Signature Wavelet Transform," *International Journal Machine Tools and Manufacture*, 1999, 39, 157–164.)

TABLE 15.1 Experimental Conditions

Tool	HSS-drill Diameter 3mm Tool material high-speed steel
Cutting conditions	Spindle speed 450 r/min Feed rate 30 mm/min Without coolant
Workpiece	45# quench steel

15.5 Identification of Tool Wear States Using Fuzzy Methods

In recent years, indirect methods that rely on the relationship between tool condition and measured signals (such as force, acoustic emission, vibration, current, etc.) for detecting the tool wear condition have been extensively studied. Among the methods used for detecting tool wear condition, motor current sensing constitutes one of major methods. The feasibility of motor power and current sensing for adaptive control and tool condition monitoring has been described [29]. Mannan and colleagues recommend using the spindle and feed current measured to estimate the static torque and thrust for monitoring tool wear condition [30]. The major advantages of using the measured motor current to detect malfunction in the cutting process are that the measuring apparatus does not disturb the machining process, and it can be applied in the manufacturing environment at almost no extra cost.

In the chapter, the spindle and feed currents measured are used to estimate the tool wear condition in boring. It is known that current signals depend on the cutting variable, namely, cutting speed v, feed speed f, the depth of cut d, as well as on the tool wear w. Moreover, tool wear itself also depends on the cutting variables. So, the measured currents are affected by the tool wear directly and the cutting variables indirectly. This section presents a new method to estimate tool wear condition by the current measured. The models with regression technology are first presented under a wide range of cutting conditions; then the method is used to classify the tool wear states by fuzzy classification. The key of the method is to model the relationship between the measured current and the tool wear states under different cutting conditions. Depending on the relationship, tool wear states can be estimated by known cutting parameters and current. Finally, a fuzzy inference method is presented to fuse the classification result of spindle and feed current signals. Experimental results showed that the method could be effectively employed in practical industry [31].

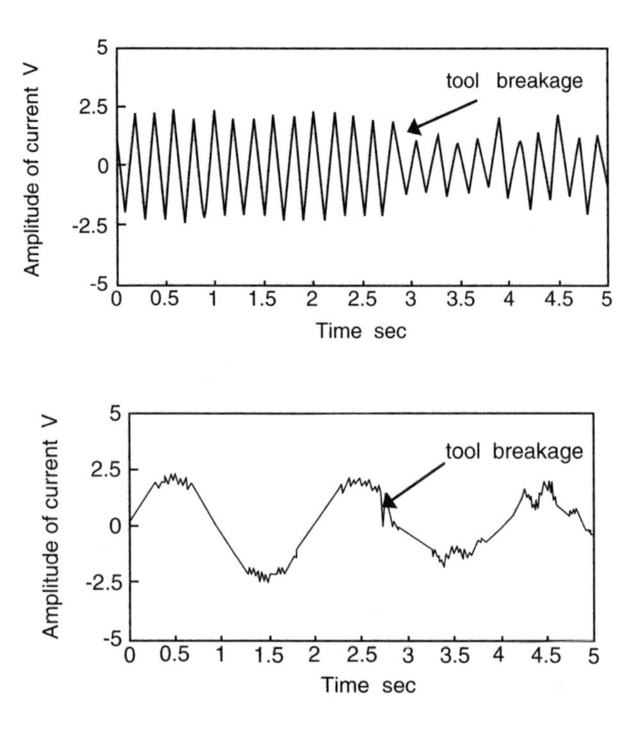

FIGURE 15.8 (Top): Live tool breakage spindle current signals, cutting speed 250 r/min, feed speed 30 mm/min, drill diameter 2 mm and (Bottom): Live tool breakage feed current signals, cutting speed 250 r/min, feed speed 30 mm/min, drill diameter 2 mm. (Reprinted with permission of Springer-Verlag London, Ltd. From "Real-Time Detection of the Breakage of Small Diameter Drills with Wavelet Transform," *Int. Adv. Manuf. Technol.*, 1999, 14, 539–543.)

15.5.1 Experimental Setup and Results

Figure 15.16 shows a schematic diagram of the experimental setup. Cutting tests were performed on a Machine Center Makino-FNC74-A20. The AC servomotor current signals of the Machine Center were measured through Hall Current Sensor. The signals were first passed though low-pass filters (cut-off frequency: 500 HZ), and then sent to a personal computer via an A/D converter. Table 15.2 shows the experimental conditions.

During the experiments, both spindle and feed current amplitude changed because of the change of tool wear, spindle speed, feed speed, and the depth of cut. The main conclusions are as follows:

1. Both spindle and feed current increase as tool wear increases; this is due to the increase of friction between tool and workpiece. Moreover, current increases almost linearly as tool wear. In addition, we found that tool wear had a more significant effect on feed current than spindle current.
2. Both spindle and feed current increase as the depth of cut increases. Moreover, feed current increases almost linearly as the depth of cut increases, while spindle current increase is proportional to the square of the depth of cut.
3. The current signal increases overall as the spindle speed increases, but current fluctuates at the range of 20 to 30 m/min, see Figure 15.17. The reason for the change of current signals is complex; the main influence factor is temperature, and the effect of temperature is small at the low speed, but increases as spindle speed increases.
4. The current signal increases overall as the feed speed increases, and current fluctuates, see Figure 15.18. The reason for the change of current signal is complex; see the discussion in [32].

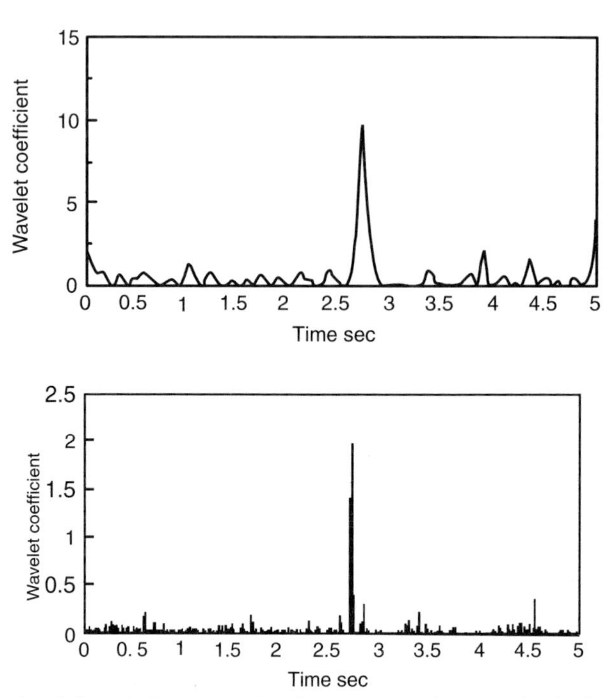

FIGURE 15.9 (Top): CWT of the spindle current signal, cutting speed 250 r/min, feed speed 30 mm/min, drill diameter 2 mm and (Bottom): DWT of feed current, resolution $j = 2$, cutting speed 250 r/min, feed speed 30 mm/min, drill diameter 2 mm. (Reprinted with permission of Springer-Verlag London, Ltd. From "Real-Time Detection of the Breakage of Small Diameter Drills with Wavelet Transform," *Int. J. Adv. Manuf. Technol.*, 1999, 14, 539–543.)

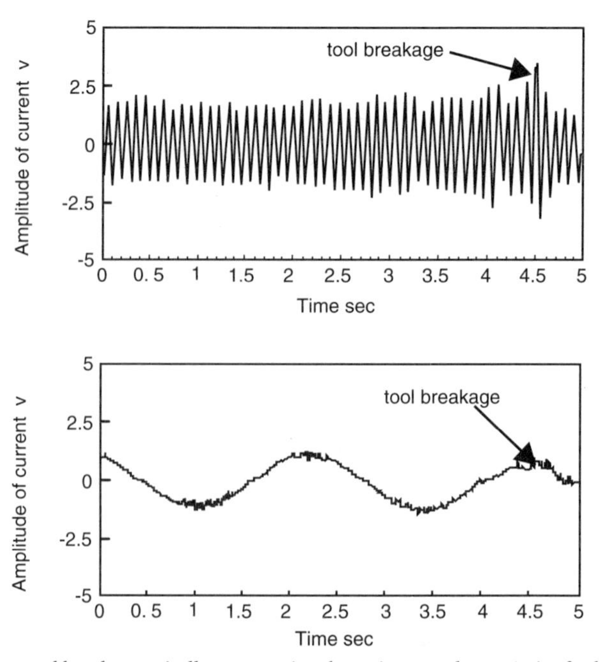

FIGURE 15.10 (Top): Live tool breakage spindle current signal, cutting speed 300 r/min, feed speed 25 mm/min, drill diameter 3 mm and (Bottom): Live tool breakage feed current signals, cutting speed 300 r/min, feed speed 25 mm/min, drill diameter 3 mm. (Reprinted with permission of Springer-Verlag London, Ltd. From "Real-Time Detection of the Breakage of Small Diameter Drills with Wavelet Transform," *Int. J. Adv. Manuf. Technol.*, 1999, 14, 539–543.)

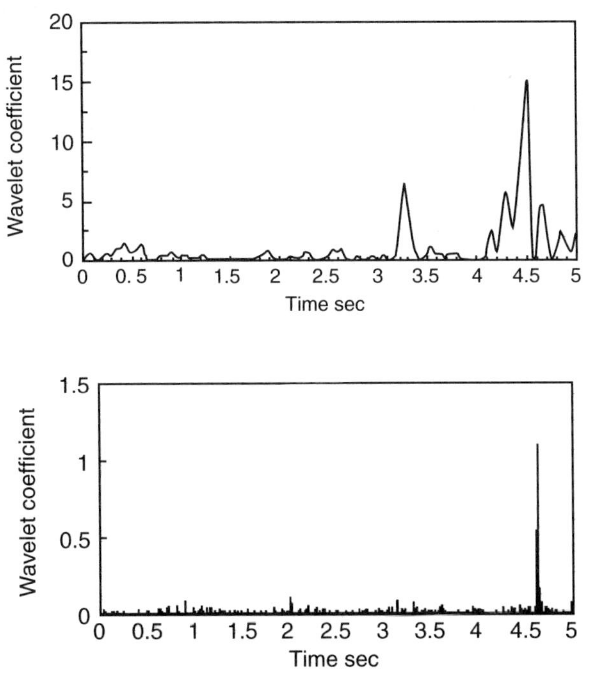

FIGURE 15.11 (Top): CWT of the spindle current signal, resolution $j = 2$, cutting speed 300 r/mim, feed speed 25 mm/min, drill diameter 3 mm and (Bottom): DWT of feed current resolution $j = 2$, cutting speed 300 r/min, feed speed 25 mm/min, drill diameter 3 mm. (Reprinted with permission of Springer-Verlag London, Ltd. From "Real-Time Detection of the Breakage of Small Diameter Drills with Wavelet Transform," *Int. J. Adv. Manuf. Technol.*, 1999, 14, 539–543.)

In brief, the effects on the current of tool wear, spindle speed, feed speed, and the depth of cut is significant. When establishing the model of current signal, the above parameters should be included; the fact that the spindle current and feed current can be selected as the features related to tool wear states is verified.

15.5.2 The Model and Fuzzy Classification

15.5.2.1 The Model

Based on the above studies, it is suggested that the effects of tool wear, spindle speed, feed speed, and the depth of cut should be taken into account when modeling current signals. In this paper, the tool wear state is divided into *A, B, C, D, E, F* classifications based on practical requirements in boring. The classification of tool wear states is shown in Table 15.3. The model is of spindle and feed current as functions of spindle speed v (m/min), feed speed f (mm/min), and the depth of cut d (mm) under different tool wear classifications, respectively.

The effect of the cutting variables v, f, and d on the spindle and feed current under each tool wear classification can be expressed as follows:

$$I_s = K_s v^{a_1} f^{a_2} d^{a_3}$$
$$I_F = K_F v^{b_1} f^{b_2} d^{b_3} \qquad \text{Equation (15.32)}$$

where I_S and I_F are spindle and feed current amplitude, respectively; K_S and K_F are constants with the tool geometry and workpiece material; parameter a_i, b_i $(i = 1, 2, 3)$ are the exponents of cutting variables.

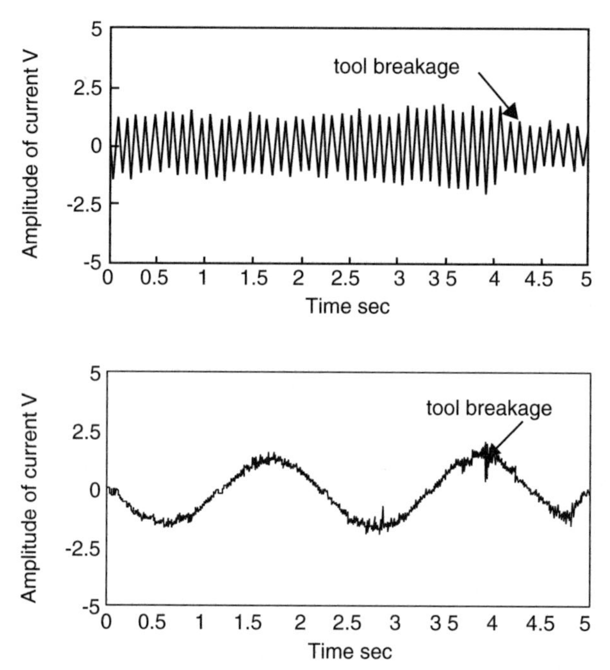

FIGURE 15.12 (Top): Live tool breakage spindle current signals, cutting speed 300 r/min, feed speed 30 mm/min, drill diameter 1 mm and (Bottom): Live tool breakage feed current signals, cutting speed 300 r/min, feed speed 30 mm/min, drill diameter 1 mm. (Reprinted with permission of Springer-Verlag London, Ltd. From "Real-Time Detection of the Breakage of Small Diameter Drills with Wavelet Transform," *Int. J. Adv. Manuf. Technol.*, 1999, 14, 539–543.)

When Equation 15.32 is taken by logarithm, the results under *A, B, C, D, E, F* classification are put in order as follows:

$$\begin{bmatrix} S_1 \\ S_2 \\ S_3 \\ S_4 \\ S_5 \\ S_6 \end{bmatrix} = \begin{bmatrix} a_{10} & a_{11} & a_{12} & a_{13} \\ a_{20} & a_{21} & a_{22} & a_{23} \\ a_{30} & a_{31} & a_{32} & a_{33} \\ a_{40} & a_{41} & a_{42} & a_{43} \\ a_{50} & a_{51} & a_{52} & a_{53} \\ a_{60} & a_{61} & a_{62} & a_{63} \end{bmatrix} \begin{bmatrix} 1 \\ \ln v \\ \ln f \\ \ln d \end{bmatrix} \, and \, \begin{bmatrix} F_1 \\ F_2 \\ F_3 \\ F_4 \\ F_5 \\ F_6 \end{bmatrix} = \begin{bmatrix} b_{10} & b_{11} & b_{12} & b_{13} \\ b_{20} & b_{21} & b_{22} & b_{23} \\ b_{30} & b_{31} & b_{32} & b_{33} \\ b_{40} & b_{41} & b_{42} & b_{43} \\ b_{50} & b_{51} & b_{52} & b_{53} \\ b_{60} & b_{61} & b_{62} & b_{63} \end{bmatrix} \begin{bmatrix} 1 \\ \ln v \\ \ln f \\ \ln d \end{bmatrix} \quad \text{Equation (15.33)}$$

where S_i, F_i $(i = 1, 2, \ldots, 6)$ is the logarithm value of the spindle current I_s, I_F, respectively (ln represents the usual logarithm).

15.5.2.2 Fuzzy Classification

Spindle and feed current models at the different wear states are established, respectively. The models can then be used to estimate tool wear states by known spindle current signal, feed current signal, and cutting parameters.

Measured currents S_0, F_0 are defined as real feature values. Estimated current values S_i, F_i $(i = 1, 2, \ldots, 6)$ are defined as the cluster centers of different tool wear classifications. Real feature values S_0, F_0 are compared with the estimated features S_i, F_i by fuzzy classification method. The membership degree of different tool wear classifications is calculated by

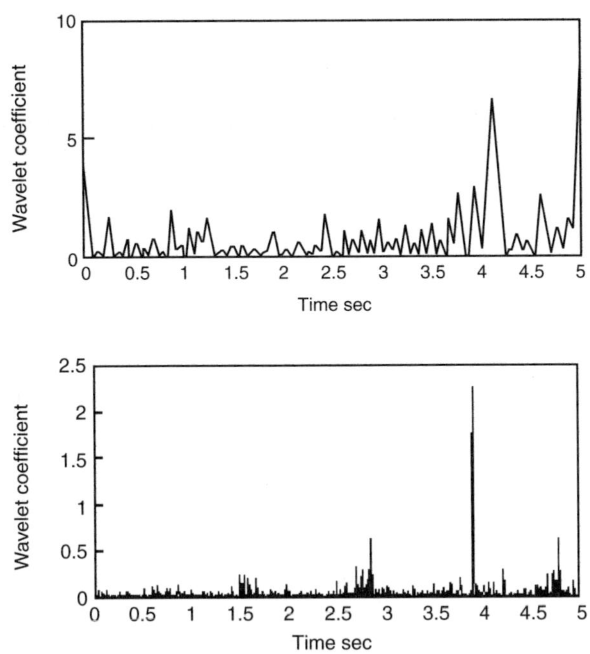

FIGURE 15.13 (Top): CWT of the spindle current signal, cutting speed 300 r/mim, feed speed 30 mm/min, drill diameter 1 mm and (Bottom): DWT of feed current signals, resolution $j = 2$, cutting speed 300 r/min, feed speed 30 mm/min, drill diameter 1 mm. (Reprinted with permission of Springer-Verlag London, Ltd. From "Real-Time Detection of the Breakage of Small Diameter Drills with Wavelet Transform," *Int. J. Adv. Manuf. Technol.*, 1999, 14, 539–543.)

$$\text{(1) if } S_0<S_1 \text{ then } \begin{cases} \mu_A(w)=1 \\ \mu_B(w)=\mu_C(w)=\mu_D(w)=\mu_E(w)=\mu_F(w)=0 \end{cases}$$

$$\text{(2) if } S_1<S_0<S_2 \text{ then } \begin{cases} \mu_A(w)=(S_2-S_0)/(S_2-S_1) \\ \mu_B(w)=(S_0-S_1)/(S_2-S_1) \\ \mu_C(w)=\mu_D(w)=\mu_E(w)=\mu_F(w)=0 \end{cases}$$

$$\text{(3) if } S_2<S_0<S_3 \text{ then } \begin{cases} \mu_A(w)=\mu_D(w)=\mu_E(w)=\mu_F(w)=0 \\ \mu_B(w)=(S_3-S_0)/(S_3-S_2) \\ \mu_C(w)=(S_0-S_2)/(S_3-S_2) \end{cases}$$

$$\text{(4) if } S_3<S_0<S_4 \text{ then } \begin{cases} \mu_A(w)=\mu_B(w)=\mu_E(w)=\mu_F(w)=0 \\ \mu_C(w)=(S_4-S_0)/(S_4-S_3) \\ \mu_D(w)=(S_0-S_3)/(S_4-S_3) \end{cases}$$

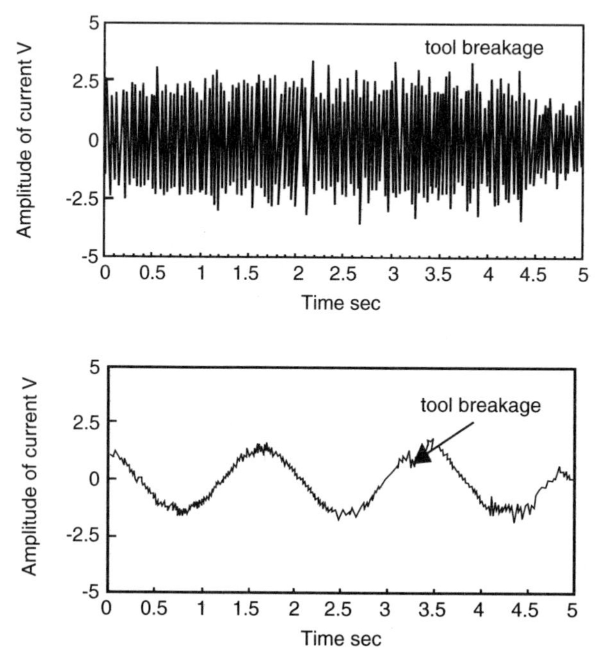

FIGURE 15.14 (Top): Live tool breakage spindle current signals, cutting speed 450 r/min, feed speed 45 mm/min, drill diameter 4.5 mm and (Bottom): Live tool breakage feed current signals, cutting speed 450 r/min, feed speed 45 mm/min, drill diameter 4.5 mm. (Reprinted with permission of Elsevier Science, Ltd. From "On-Line Detection of the Breakage of Small Diameter Drills Using Current Signature Wavelet Transform," *International Journal Machine Tools and Manufacture,* 1999, 39, 157–164.)

$$(5)\ \text{if } S_4 < S_0 < S_5 \text{ then } \begin{cases} \mu_A(w) = \mu_B(w) = \mu_C(w) = \mu_F(w) = 0 \\ \mu_D(w) = (S_5 - S_0)/(S_5 - S_4) \\ \mu_E(w) = (S_0 - S_4)/(S_5 - S_4) \end{cases}$$

$$(6)\ \text{if } S_5 < S_0 < S_6 \text{ then } \begin{cases} \mu_A(w) = \mu_B(w) = \mu_C(w) = \mu_D(w) = 0 \\ \mu_E(w) = (S_6 - S_0)/(S_6 - S_5) \\ \mu_F(w) = (S_0 - S_5)/(S_6 - S_5) \end{cases}$$

$$(7)\ \text{if } S_6 < S_0 \text{ then } \begin{cases} \mu_A(w) = \mu_B(w) = \mu_C(w) = \mu_D(w) = \mu_E(w) = 0 \\ \mu_F(w) = 1 \end{cases}$$

where $\mu_i(w)(i = A, B, \ldots, F)$ is the membership degree of current S_0 under each tool wear condition. The same method is fitted to feed current F_0.

15.5.3 Multi-Parameter Fusion with Fuzzy Inference

The membership degree of tool wear states, *i.e.*, $\mu_j^S(w)$ and $\mu_i^F(w)$ $(i = A, B, \ldots, F)$, with spindle current and feed current had been calculated using fuzzy classification. The above two parameters can be fused by fuzzy inference to obtain tool wear value accurately.

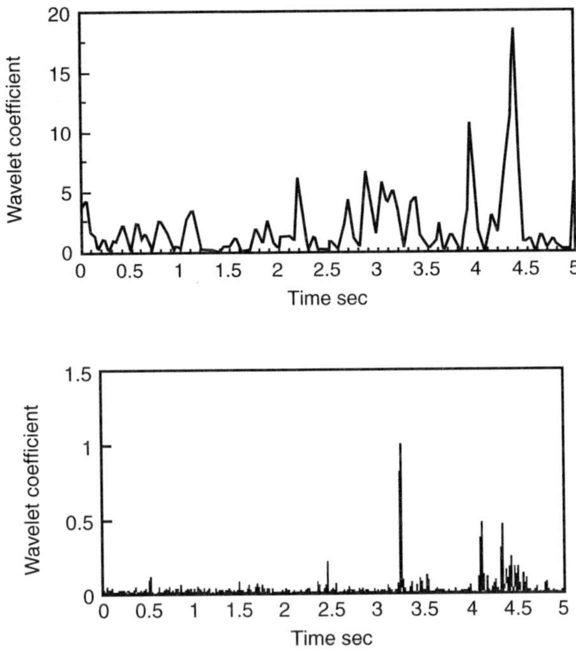

FIGURE 15.15 (Top): CWT of the spindle current signal, cutting speed 450 r/min, feed speed 45 mm/min, drill diameter 4.5 mm and (Bottom): DWT of feed current, resolution $j = 2$, cutting speed 450 r/min, feed speed 45 mm/min, drill diameter 4.5 mm. (Reprinted with permission of Elsevier Science, Ltd. From "On-Line Detection of the Breakage of Small Diameter Drills Using Current Signature Wavelet Transform," *International Journal Machine Tools and Manufacture,* 1999, 39, 157–164.)

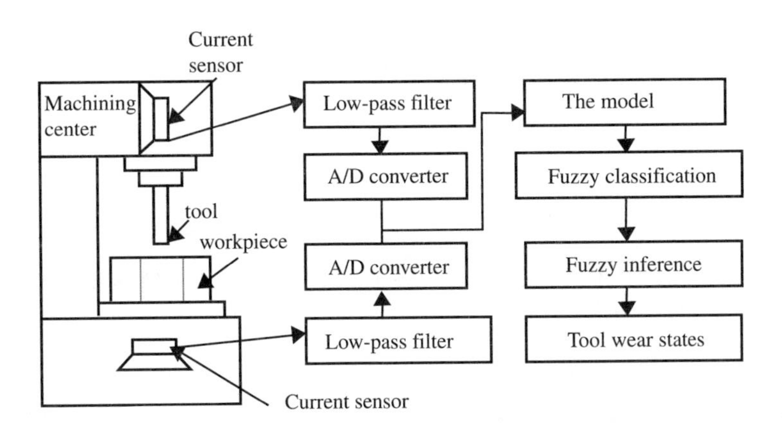

FIGURE 15.16 A schematic diagram of the experimental set-up. (Reprinted with permission of Taylor & Francis, Ltd. From "Identification of Tool Wear States with Fuzzy," *International Journal Computer Integrated Manufacturing,* 1999, 12, 503–509.)

15.5.3.1 Fusion

The relationship between input and output variables of a fuzzy system is defined by a set of linguistic statements, which are called *fuzzy rules*. There are two input variables and one output variable, which are all classified into six fuzzy sets. Based on the experimental results, 26 rules, as shown in Table 15.4, have been developed for tool wear states fusion. These rules are classified into six groups corresponding to six tool wear states.

TABLE 15.2 Experimental Conditions

Tool	Tool material: high-speed steel
Cutting conditions	Spindle speed: 15–35 mm/min Feed speed: 10–30 mm/min Depth of cut: 0.35, 0.45, 0.55, 0.8, 0.9, 1.15 mm Without coolant
Workpiece	45# quench steel

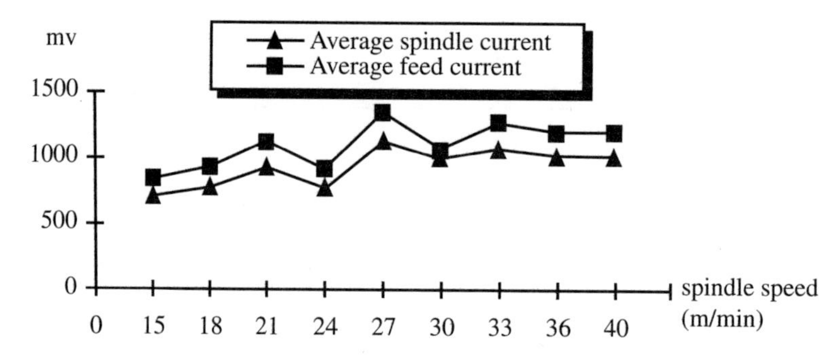

FIGURE 15.17 The effect of spindle speed on current signal. (Reprinted with permission of Taylor & Francis, Ltd. From "Identification of Tool Wear States with Fuzzy," *International Journal Computer Integrated Manufacturing*, 1999, 12, 503–509.)

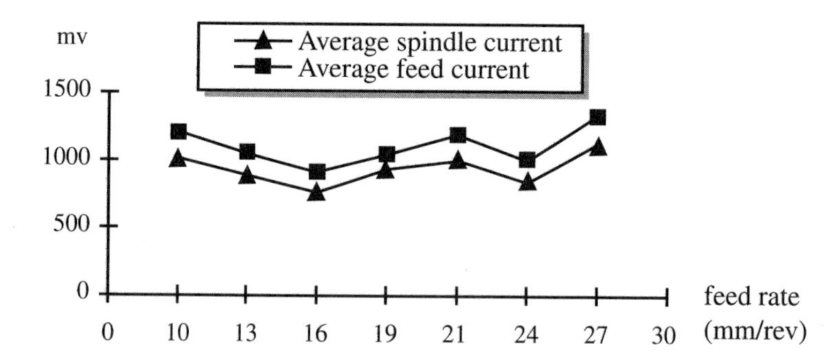

FIGURE 15.18 The effects of feed speed on current signal. (Reprinted with permission of Taylor & Francis, Ltd. From "Identification of Tool Wear States with Fuzzy," *International Journal Computer Integrated Manufacturing*, 1999, 12, 503–509.)

TABLE 15.3 Classifications of Tool Wear in Boring

Classification	A	B	C	D	E	F
Tool wear value (mm)	0–0.2	0.1–0.3	0.2–0.4	0.3–0.5	0.4–0.6	0.5–1

Through the use of fuzzy min–max algorithm, i.e., fuzzy intersection (AND) and fuzzy union (OR), the following equation can be generated to calculate the fuzzy membership values of tool wear states:

$$\mu_i(w) = \bigcup_{j=1}^{K} \left\{ \mu_i^S(w) \cap \mu_i^F(w) \right\} \qquad \text{Equation (15.34)}$$

TABLE 15.4 Fuzzy Rules for Tool Wear States Fusion

IF Rules	1 W_S W_F	2 W_S W_F	3 W_S W_F	4 W_S W_F	5 W_S W_F	6 W_S W_F
A	A A	B B	C C	D D	E E	F F
B	A B	B A	C B	D C	E D	F E
C	B A	A B	B C	C D	D E	E F
D		B C	C D	D E	E F	
E		C B	D C	E D	F E	
Then wear states	A	B	C	D	E	F

where $\mu_i(w)$ $(i = A, B, \ldots, F)$ is the fuzzy membership of tool wear states under the *A, B, C, D, E, F* classification and $j = 1, 2, \ldots, K$ represents the number of rules fired for the corresponding tool wear states.

15.5.3.2 Obtaining (Fuzzy) Tool Wear Value

The key to the fusion of tool wear states is the selection of appropriate shapes of fuzzy membership for process variables based on experimental results. Shown in Figure 15.19 is a membership function of tool wear states. The reason for choosing a trapezoid shape for tool wear states is that it is difficult to quantify an exact wear value. Using a wider range avoids defining an exact wear value for a certain level of linguistic variable of tool wear. This will also allow easy knowledge acquisition when developing a set of fuzzy rules for fuzzy inference [33]. Based on the classification of tool wear states, the trapezoid function is defined as follows:

$$\mu(w) = aw + b \quad k < w < l \qquad \text{Equation (15.35)}$$

where $\mu(w)$ is the fuzzy membership value for tool wear states, and *a*, *b*, *k*, and *l* are constants for different fuzzy sets, as shown in Table 15.5

15.5.3.3 Defuzzification of Tool Wear

The outputs of inference processes are still fuzzy values, and those need to be defuzzified. Basically, defuzzification maps from a space of fuzzy values into that of a non-fuzzy universe. At present, there are several strategies that can be used to perform a defuzzification process. The most commonly used strategy is the centered defuzzy method [34], which produces the center of area of the possibility distribution of inference output. Therefore, the defuzzified tool wear states can be obtained by using the formula

$$wear = \frac{\int_W \mu(w) w dw}{\int_W \mu(w) dw} \qquad \text{Equation (15.36)}$$

where *wear* represents the numerical value of tool wear states and $\mu(w)$ is the fuzzy membership degree fused by fuzzy inference..

15.5.4 Results and Discussion

A total of 77 tool wear cutting tests were collected under various cutting conditions. Of these, 50 samples were randomly picked as learning samples; 27 samples were used as the test samples in the classification phase. According to the classification of the tool wear states, 50 learning samples were divided into six groups. The parameter a_{ij}, b_{ij} values of Equation 15.2 were calculated by the least-squares method. The results are as follows:

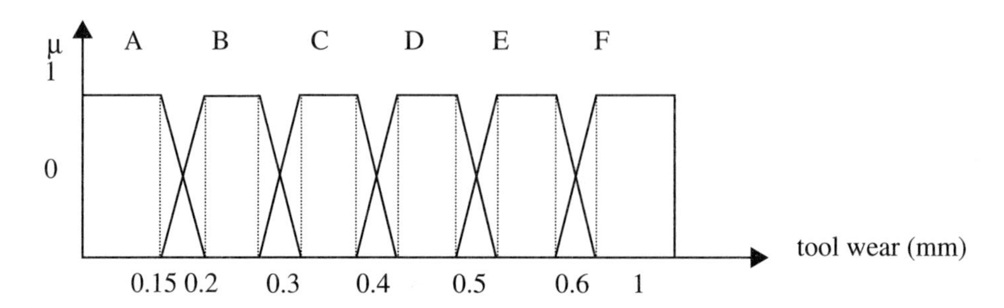

FIGURE 15.19 Fuzzy membership of tool wear state. (Reprinted with permission of Taylor & Francis, Ltd. From "Identification of Tool Wear States with Fuzzy," *International Journal Computer Integrated Manufacturing*, 1999, 12, 503–514.)

TABLE 15.5 Constants of Fuzzy Membership Functions for Tool Wear Condition

Tool Wear Classification	Constants of Fuzzy Membership Function A	B	k	l
A	0	1	0	0.15
	−20	4	0.15	0.20
B	20	−3	0.15	0.20
	0	1	0.20	0.25
	−20	6	0.25	0.30
C	20	−5	0.25	0.30
	0	1	0.30	0.35
	20	8	0.35	0.40
D	20	−7	0.35	0.40
	0	1	0.40	0.45
	−20	10	0.45	0.50
E	20	−9	0.45	0.50
	0	1	0.50	0.55
	−20	12	0.55	0.6
F	20	−11	0.55	0.6
	0	1	0.6	0.7

$$\begin{pmatrix} 5.2623 & 0.3206 & 0.0835 & 0.0928 \\ 5.0250 & 0.3477 & 0.1512 & 0.0366 \\ 5.3582 & 0.2532 & 0.2228 & 0.1801 \\ 5.5319 & 0.1993 & 0.2723 & 0.2382 \\ 6.3176 & 0.0138 & 0.2733 & 0.3425 \\ 6.4744 & -0.0249 & 0.3205 & 0.4285 \end{pmatrix} \text{and} \begin{pmatrix} 5.6904 & 0.2480 & 0.0916 & 0.0688 \\ 5.4995 & 0.2770 & 0.1385 & 0.0163 \\ 5.7568 & 0.2060 & 0.1945 & 0.1295 \\ 5.9368 & 0.1553 & 0.2315 & 0.1900 \\ 7.4501 & -0.1591 & 0.1151 & 0.4199 \\ 6.7410 & -0.0318 & 0.2736 & 0.3820 \end{pmatrix}$$

The correlation coefficients corroding to the weight value of each group are 0.9026, 0.8240, 0.7938, 0.7923, 0.9169, 0.9746, 0.9062, 0.8089, 0.7805, 0.7727, 0.9179, 0.9680. It is obvious that the correlation coefficients are very close to unity. It is indicated that the relationship between the current signals and the cutting parameters is well represented by the proposed models. In addition, 27 tests were conducted to examine the feasibility of using the above models to estimate tool wear states.

The above method is used to estimate tool wear value. First, the logarithm of the present cutting parameter v, f, d as well as 1 are inputted into Equation 15.33, and the estimated value of the spindle and feed current, namely, S_i and F_i (i = 1, 2, . . . , 6) are outputted. Second, the spindle and feed current detected are put into the logarithm, and are compared with the above estimated value of spindle and feed

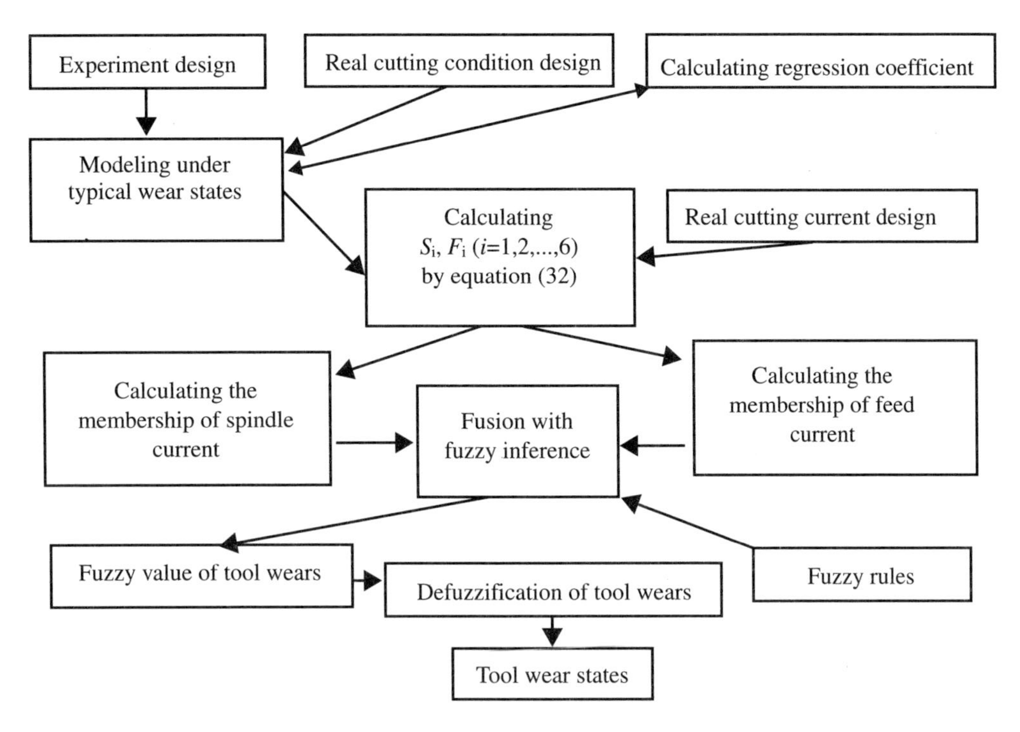

FIGURE 15.20 A flowchart of the tool wear states recognization method. (Reprinted with permission of Taylor & Francis, Ltd. From "Identification of Tool Wear States with Fuzzy," *International Journal Computer Integrated Manufacturing*, 1999, 12, 503–514.)

current S_i and F_i, respectively. The membership degrees of present tool wear states under different tool wear classifications are calculated based on Equation 15.34. Finally, the membership degree of tool states are fused by fuzzy inference, and the accurate tool wear value is detected using the centered defuzzy method. The above processing can be expressed in brief in Figure 15.20. In order to make clear the reliability of the above method, the comparison of actual tool wear values with those estimated is shown in Figure 15.21. The results showed that the above method had a more accurate estimation of tool wear states.

15.6 Tool Wear Monitoring with Wavelet Transforms and Fuzzy Neural Network

Flexible manufacturing systems (FMS) that employ automated machine tools for cutting operations require reliable process monitoring systems to oversee the machining operations. Among machine process variables monitored, tool wear plays a critical role in dictating the dimensional accuracy of the workpiece and guaranteeing the automatic cutting process. It is therefore essential to develop simple, reliable, and cost-effective on-line tool wear condition monitoring strategies in this vitally important area. Due to the complexity of the metal cutting mechanism, a reliable commercial tool wear monitoring system has yet to be developed.

Various methods for tool wear monitoring have been proposed in the past, although none of these methods have been universally successful due to the complex nature of the machining processes. These methods have been classified into direct (optical, radioactive, and electrical resistance, etc.) and indirect (acoustic emission (AE), spindle motor current, cutting force, vibration, etc.) sensing methods according to the sensors used. Recent attempts have concentrated on developing methods that monitor the cutting process indirectly. Among indirect methods, AE is the most effective means of sensing tool wear. The

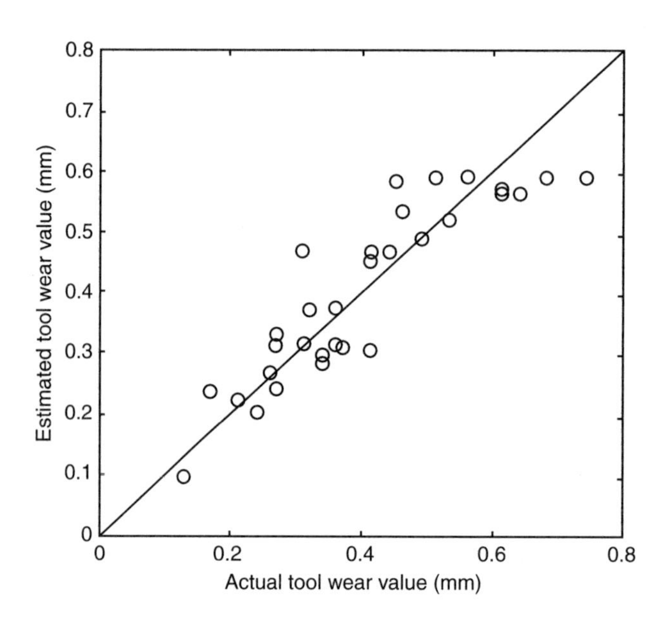

FIGURE 15.21 Tool wear comparison of actual with estimated. (Reprinted with permission of Taylor & Francis, Ltd. From "Identification of Tool Wear States with Fuzzy," *International Journal Computer Integrated Manufacturing*, 1999, 12, 503–514.)

major advantages of using AE to monitor the tool condition is that frequency range of the AE signal is much higher than that of the machine vibrations and environmental noises and does not interfere with the cutting operation. However, AE signals often have to be treated with additional signal processing schemes to extract the most useful information [35, 36, 27]. In the metal cutting process, AE is attributable to many courses, such as elastic and plastic deformations, tool wear, tool breakage, friction, etc. If the AE signal can effectively be analyzed, tool wear can be detected using the AE signal. The AE signal is usually detected by transducers, then amplified and transmitted to counter, RMS voltmeter, spectrum analysis, etc. Among various approaches taken to analyze AE signals, spectral analysis has been found to be the most informative for monitoring tool wear [37]. Spectral analysis, such as fast Fourier transforms (FFT), is the most common signal processing method in tool wear monitoring. A disadvantage of the FFT method is that it has a good resolution only in the frequency domain and a very bad resolution in the time domain, so that it loses some signal information in the time domain. FFT is only fitted to deal with stochastic stable signals.

Recently, wavelet transforms have been found to be a significant new tool in signal analysis and processing. They have been used to analyze tool failure monitoring signals [38]. Wavelet transforms have a good resolution in the frequency and time domains synchronously, and they can extract more information in the time domain at different frequency bands. Wavelet packets are particular linear combinations of wavelets. They form bases that retain many of the orthogonal, smooth, and locate properties of their parent wavelets. The wavelet packet transforms have been used for on-line monitoring machining process. They can capture important features of the sensor signal that are sensitive to the change of process condition (such as tool wear) but are insensitive to the variation of process working condition and various noises [39]. The wavelet packet transforms can decompose a sensor signal into different components in different time–frequency windows; the components, hence, can be considered as the features of the original signal [40].

This section presents a method of tool wear monitoring, consisting of a wavelet packet transforms preprocessor for generating features from acoustic emission (AE) signal, followed by a fuzzy neural network (FNN) for associating the preprocessor outputs with the appropriate decisions.

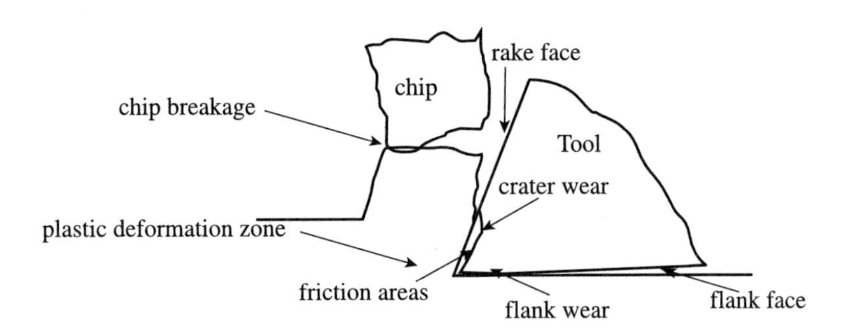

FIGURE 15.22 Acoustic emission sources in boring. (Reprinted with permission of Elsevier Science, Ltd. From "Tool Wear Monitoring with Wavelet Packet Transform-Fuzzy Clustering Method," *Wear*, 1998, 219(2), 145–154.)

15.6.1 Acoustic Emission Signals and Tool Wear

Research has shown that acoustic emission (AE), which refers to stress waves generated by the sudden release of energy in deforming materials, has been successfully used in laboratory tests to detect tool wear and fracture in single point turning operations [41]. Dornfeld [42] pointed out the possible sources of the AE in metal cutting:

1. Plastic deformation during the cutting process in the workpiece
2. Plastic deformation in the chip
3. Friction contact between the tool flank face and the workpiece resulting in flank wear
4. Friction contact between the tool rank face and the chip resulting in crater wear
5. Collisions between chip and tool
6. Chip breakage
7. Tool edge chipping

Acoustic emission sources in boring are shown in Figure 15.22.

Research results have shown that friction and plastic deformation have comparable importance with regard to the generation of the continuous AE. This is because the amplitude of the signals from the workpiece is reduced during wave transfer from workpiece to tool, possibly by reflection at the interface, so that the friction between workpiece and tool can be regarded as the most important source of the continuous AE [43]. In the present investigation, we verified the above results; therefore, we can consider the friction between workpiece and tool as the essential source of the AE.

The relationship between the RMS of continuous AE and the cutting parameters and tool wear can be established by experimental method. Results have shown that RMS is proportional to v_c a_p, tool flank wear VB, respectively, but it is independent of feed rate. The results are shown in Figures 15.23 through 15.26.

According to the above results, the RMS of AE can be calculated from the machining and tool wear parameters:

$$\mathrm{RMS} = \mathrm{K}\ v_c\ a_p\ \mathrm{VB} \qquad \text{Equation (15.37)}$$

where K is the area density of contact points, v_c the cutting speed, a_p the depth of cut, VB the wear land. K depends on the structure of the surface, which remains nearly constant with increasing wear.

During the experiment, the friction between workpiece and tool generated a continuous AE signal, giving information on tool wear. However, the experimental results show that sometimes burst-signals with high peak amplitudes interfered with the continuous AE signal. In fact, these burst signals relate to

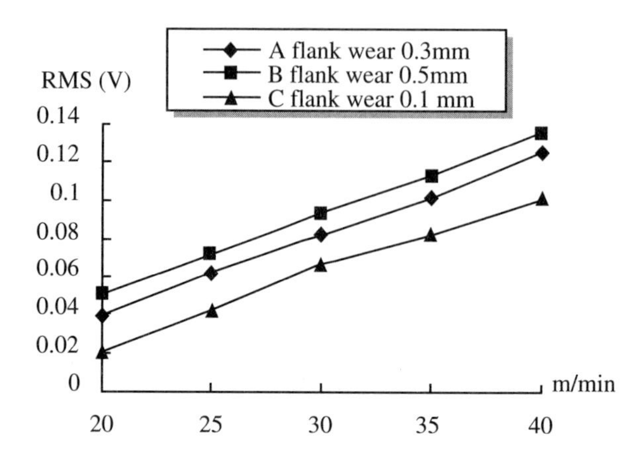

FIGURE 15.23 The relationship between the RMS of AE and cutting speed; feed rate 0.2 mm/rev, depth of cut 0.5 mm. (Reprinted with permission of Elsevier Science, Ltd. From "Tool Wear Monitoring with Wavelet Packet Transform-Fuzzy Clustering Method," *Wear*, 1998, 219(2), 145–154.)

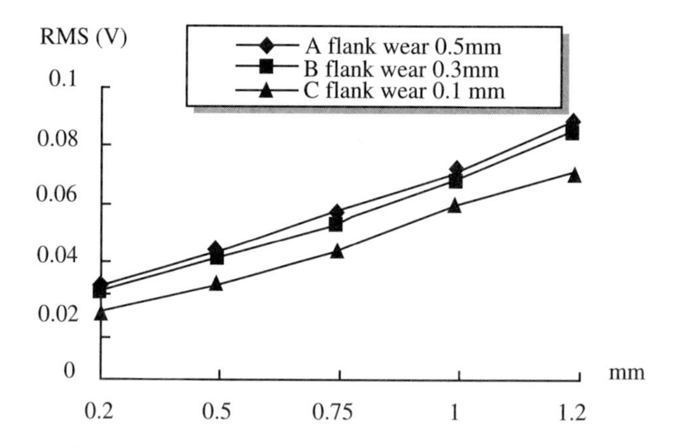

FIGURE 15.24 The relationship between the RMS of AE and the depth of cut; cutting feed 23 m/min, feed rate 0.2 mm/rev. (Reprinted with permission of Elsevier Science, Ltd. From "Tool Wear Monitoring with Wavelet Packet Transform-Fuzzy Clustering Method," *Wear*, 1998, 219(2), 145–154.)

the chip breakage, give information on the chip behavior, but not on tool wear. Therefore, it is essential to filter out these bursts from the continuous AE signal for reliable tool wear monitoring before further analysis is performed. The floating threshold value is defined, which is higher than the mean signal level. The constituents due to chip impact and breakage over this threshold are not considered as the determination of the mean signal level, and are filtered out from the AE signals. The signal constituents below the threshold represent the continuous AE that will be analyzed by the following signal processing method.

15.6.2 Signal Analysis and Features Extraction

In monitoring of tool wear, AE signals monitored contain complicated information on the cutting processing. To ensure the reliability of a tool monitoring system, it is important to extract the features of the signals that can respond to tool wear condition. From a mathematical point of view, the feature extraction can be considered as signal compression. Wavelet packet transforms can be represented as a compressed signal method. Therefore, it is ideal to use the wavelet packet transforms as a method for extracting features from AE signals [44].

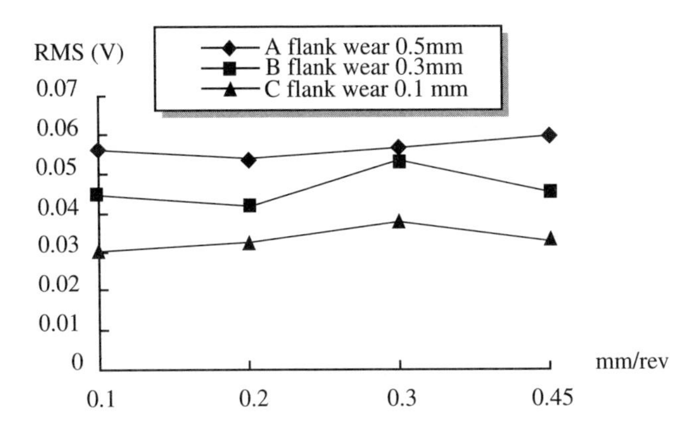

FIGURE 15.25 The relationship between the RMS of AE and feed rate; cutting feed 25 m/min, depth of cut 0.75 mm. (Reprinted with permission of Elsevier Science, Ltd. From "Tool Wear Monitoring with Wavelet Packet Transform-Fuzzy Clustering Method," *Wear*, 1998, 219(2), 145–154.)

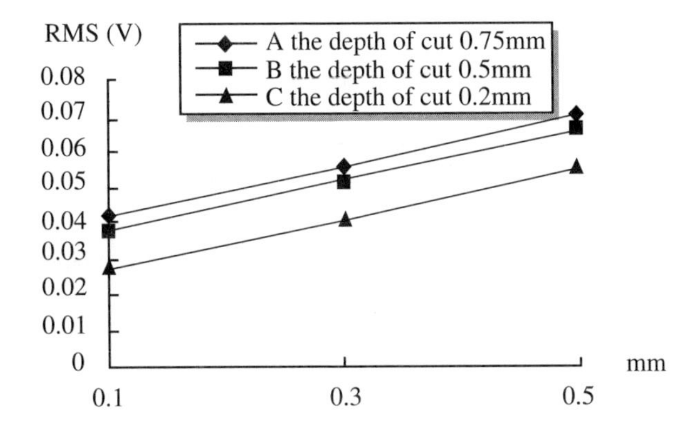

FIGURE 15.26 The relationship between the RMS of AE and tool flank wear; cutting feed 25 m/min, feed rate 0.2 mm/rev. (Reprinted with permission of Elsevier Science, Ltd. From "Tool Wear Monitoring with Wavelet Packet Transform-Fuzzy Clustering Method," *Wear*, 1998, 219(2), 145–154.)

Figure 15.27 shows an AE signal under a typical cutting condition for boring. The AE signals appear in the time domain. At the beginning of the cutting process, the magnitude of the AE signal is small because the tool is fresh, meaning that the cutting process is stable. As the tool wear increases, the magnitudes of the AE signal increase.

FFT is used to process the above AE signals; the frequency components of AE signals are shown in Figure 15.28. It is found that the magnitude of AE signals in frequency domain is sensitive to the change of tool states.

Figure 15.29 shows the decomposed results of AE signal through the wavelet packet transforms. Figure 15.30 shows the constituent parts of the AE signal at frequency band [0, 62.5], [62.5, 125], . . . , [937.5, 1000] KHz, respectively. Obviously, these decomposed results of AE signal not only keep the same typical features, but also provide more information such as the time domain constituent part of the AE signal at the frequency band. The mean values of the constituent parts of the AE signal of each frequency band can be represented by the energy level of the AE signal in the frequency band.

In FNN applications, the feature selection and feature number are very important. The selected features must be independent and the number of features must be large enough. For tool wear monitoring, the cutting conditions (cutting speed, feed rate, and cutting depth) are also the features related to wear. When

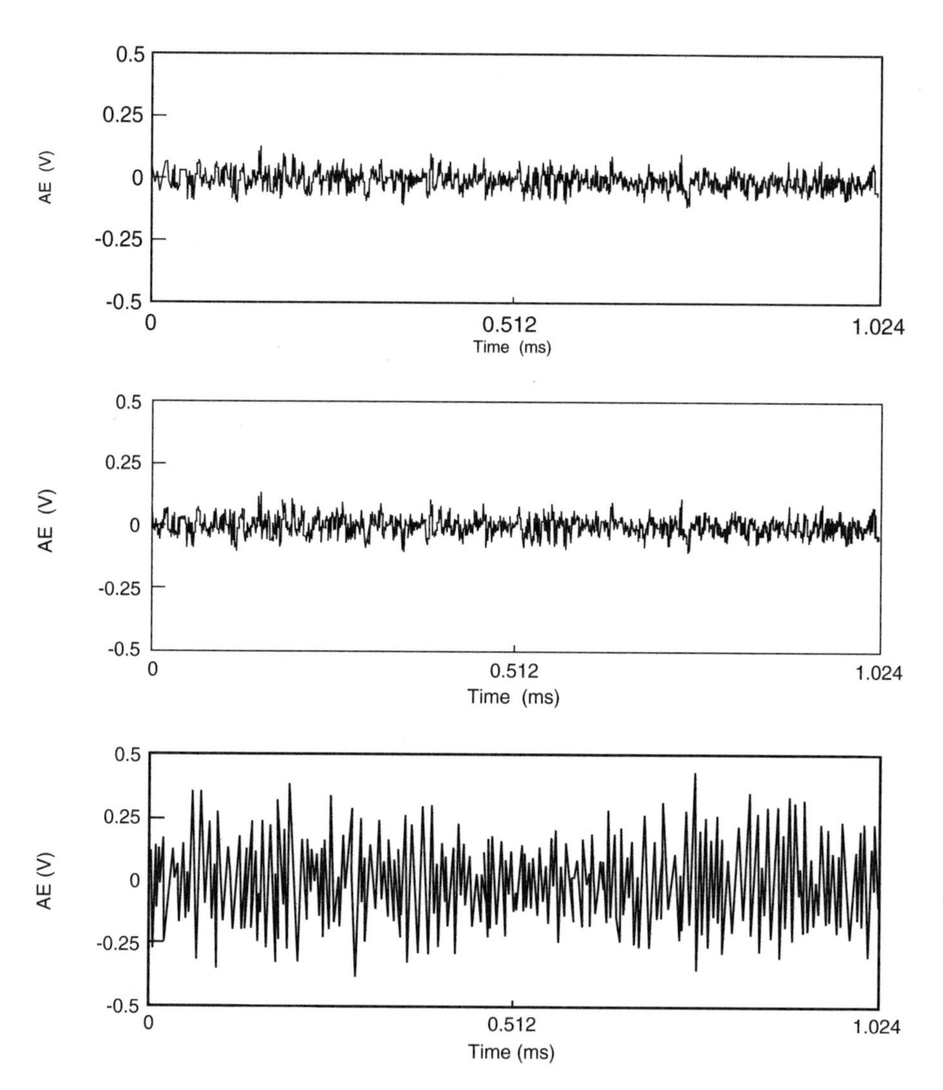

FIGURE 15.27 The AE signal in a typical tool wear cutting process, cutting speed 30 m/min, feed rate 0.2 mm/rev, depth of cut 0.5 mm; work material 40Cr steel, tool material is high-speed steel, without coolant. (Top): VB = 0.06 mm; (Middle): VB = 0.26 mm; and (Bottom): VB = 0.62 mm. (Reprinted with permission of Elsevier Science, Ltd. From "Tool Wear Monitoring with Wavelet Packet Transform-Fuzzy Clustering Method," *Wear*, 1998, 219(2), 145–154.)

signal features extracted from AE signal correspond to different cutting conditions, these cutting conditions are also represented by the features. In practice, the cutting conditions were not dependent on the features. Therefore, we hope that the selected features are not sensitive to changes in the cutting conditions, namely, tool wear monitoring system could be suitable for a wide range of machining conditions.

According to the results discussed above, the AE signal's RMS in each frequency band was used to describe the features of different tool wear conditions in boring. The selected features were summarized as follows [45]:

n_1 = RMS of wavelet coefficient in the frequency band [0, 62.5]KHz
n_2 = RMS of wavelet coefficient in the frequency band [62.5, 125]KHz

$\vdots \qquad \vdots \qquad \vdots$

n_{16} = RMS of wavelet coefficient in the frequency band [937.5, 1000]KHz

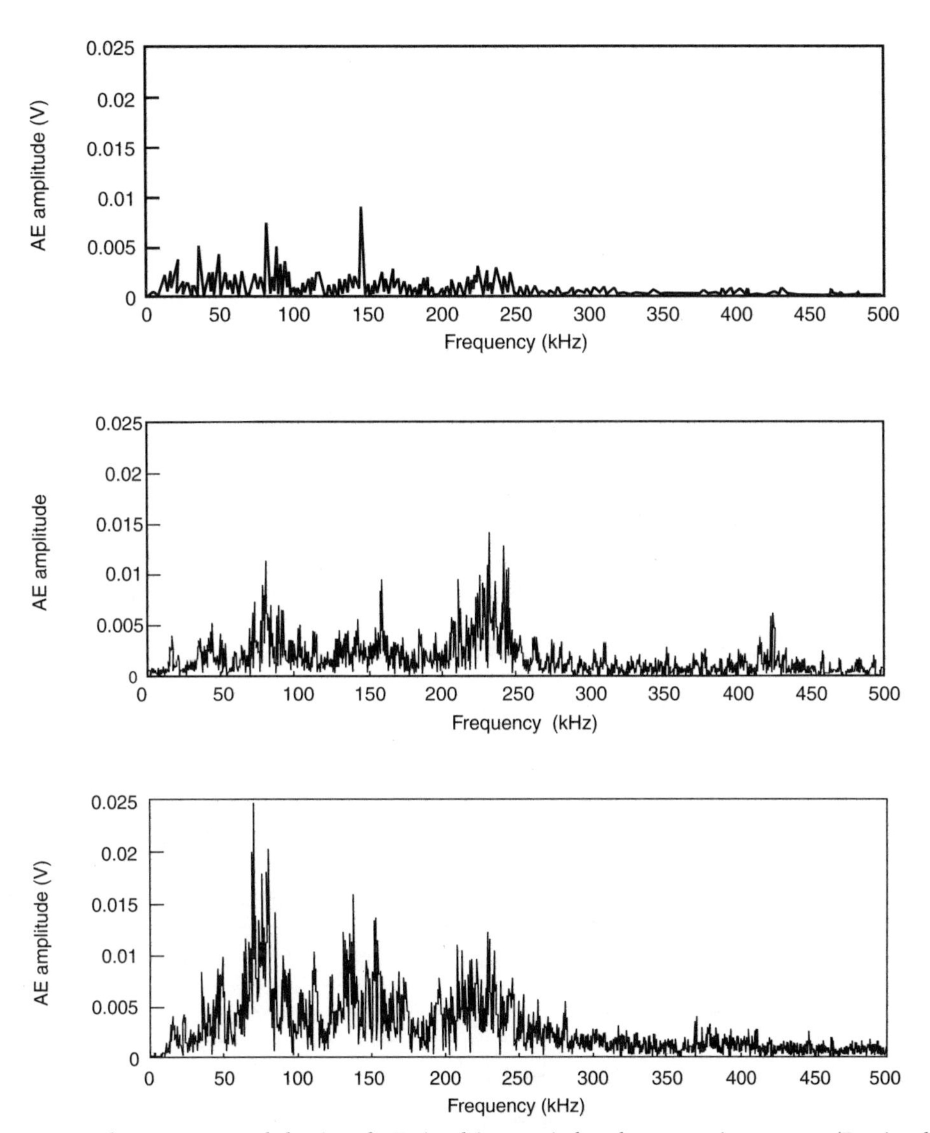

FIGURE 15.28 The power spectral density of AE signal in a typical tool wear cutting process. (Reprinted with permission of Elsevier Science, Ltd. From "Tool Wear Monitoring with Wavelet Packet Transform-Fuzzy Clustering Method," *Wear*, 1998, 219(2), 145–154.)

But all of the above features are insensitive to tool wear. According to large amounts of data analysis, we found that n_3, n_4, n_5, and n_7 are sensitive to tool wear, and they are shown in Figures 15.30 and 15.31. The above features were replaced by q_1, q_2, q_3, and q_4, respectively, which will be used to classify tool wear states.

According to Equation 15.37, the RMS of continuous AE signal is proportional to v_c a_p, tool flank wear VB, but it is independent of feed rate. For the purpose of eliminating the effects of cutting conditions on features, divide $v_c a_p$ into q_i ($i = 1, 2, \ldots, 4$) and get a new q_i value. The new q_i value is taken as the final monitoring features.

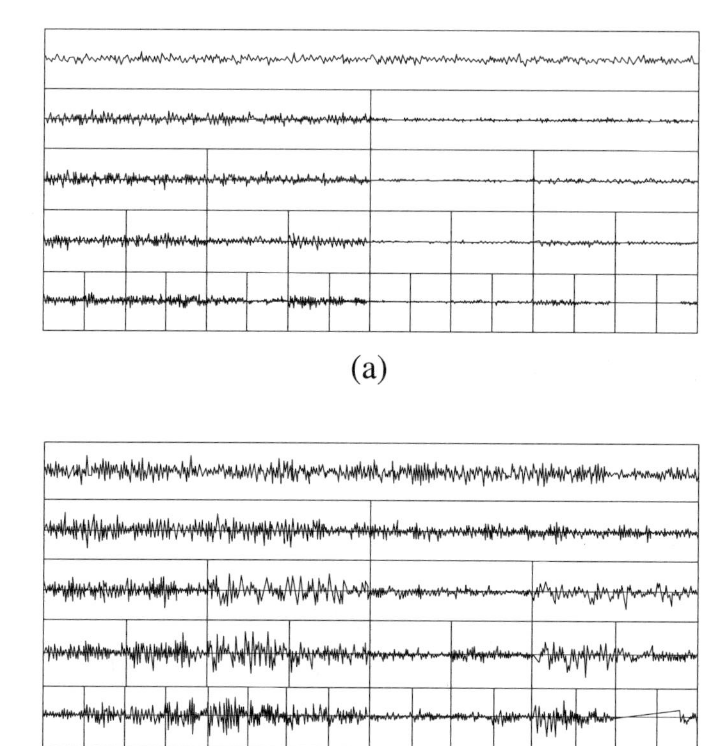

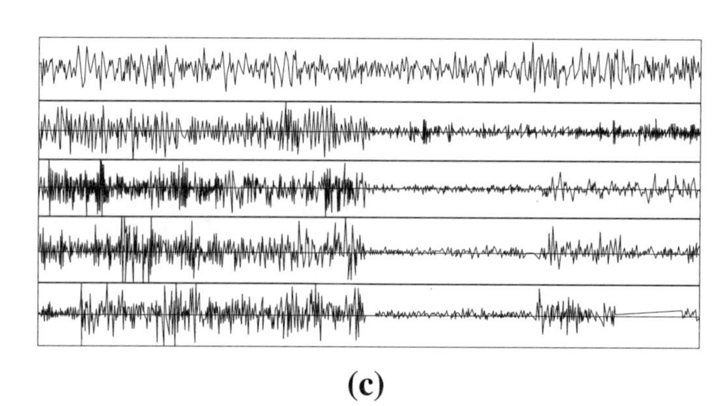

FIGURE 15.29 The composing results of AE by wavelet packet transformation. (Reprinted with permission of Elsevier Science, Ltd. From "Tool Wear Monitoring with Wavelet Packet Transform-Fuzzy Clustering Method," *Wear*, 1998, 219(2), 145–154.)

15.6.3 Experiments and Results

The schematic diagram of the experimental setup is shown in Figure 15.32. Cutting tests were performed on Machining Center Makino-FNC74-A20. In the experiments, a commercial piezoelectric AE transducer was mounted on the spindle. AE signals were transduced by magnetic fluid between spindle and tool. During the experiments, the monitored AE signals were amplified through a high-pass filter at 50 KHz and low-pass filter at 1 MHz and then were sent via an A/D converter to a personal computer (AST/486).

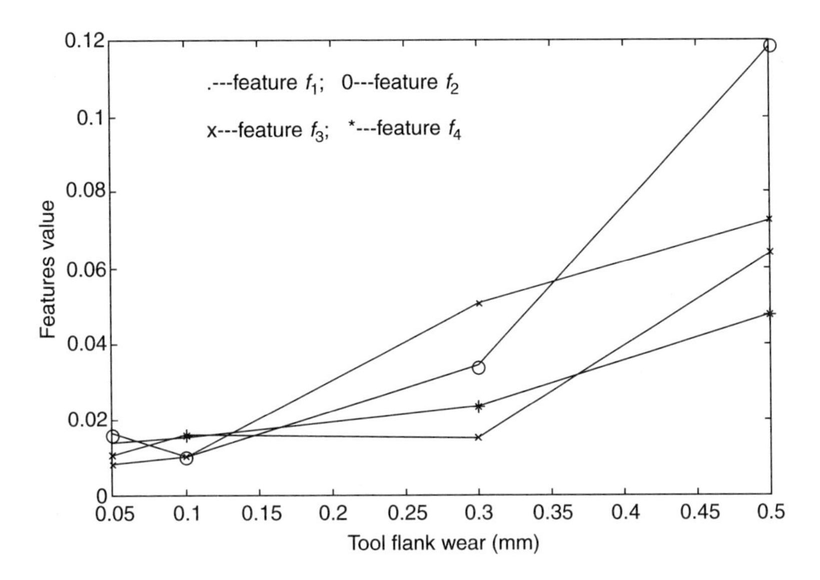

FIGURE 15.30 The relationship between features extracted and tool wear. Cutting speed 30 m/min, feed rate 0.2 mm/rev, depth of cut 0.5 mm; work material is 40Cr steel, tool material is high-speed steel, without coolant.

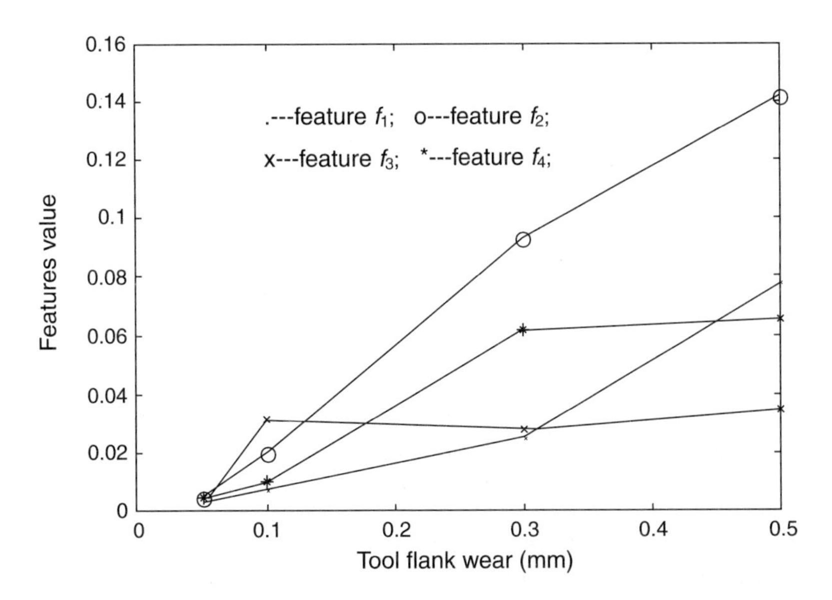

FIGURE 15.31 The relationship between features extracted and tool wear. Cutting speed 40 m/min, feed rate 0.3 mm/rev, depth of cut 1 mm; work material is 40Cr steel, tool material is high-speed steel, without coolant.

A successful tool wear detecting method must be sensitive to tool wear condition and insensitive to the variation of cutting conditions. Hence, cutting tests were conducted at different conditions to evaluate the performance of the proposed method. The tool parameters and cutting conditions are listed in Table 15.6.

Tool wear condition was divided into five states, including initial wear, normal wear, acceptable wear, severe wear, and failure. Based on flank wear of the tool, these conditions are summarized in Table 15.7.

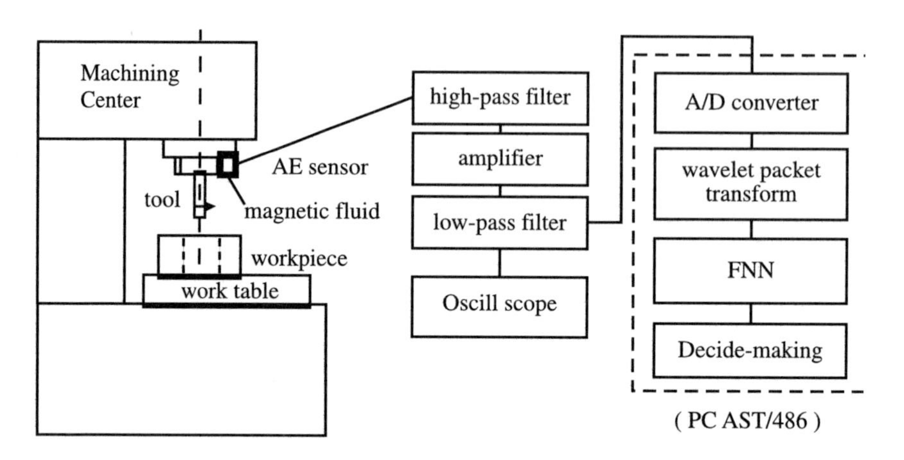

FIGURE 15.32 Schematic diagram of the experimental setup.

TABLE 15.6 Experimental Conditions for the Boring Example

Tool:	Bore tool material: high-speed steel
	Tool geometry: $\gamma = 10°$, $\alpha = 8°$, $\lambda = -2°$, $\chi = 90°$, $\kappa = 12°$, and $r = 0.3$ mm
Cutting conditions:	Cutting speed: 20–40 m/min
	Feed rate: 0.1, 0.2, 0.3 mm/min
	Depth of cutting: 0.1, 0.2, 0.5, 0.75, 1.0, 1.25 mm
	Without coolant
Workpiece:	45# quenching-and-tempering steel

TABLE 15.7 Tool Condition Classification

Tool Condition	Flank Wear
Initial wear	0 < wear < 0.1 mm
Normal wear	0.05 < wear < 0.3 mm
Acceptable wear	0.25 < wear < 0.5 mm
Severe wear	0.45 < wear < 0.6 mm
Failure	—

The above four features are taken as the input of FNN. Fuzzy membership functions of tool wear are set as the output indices of the FNN; the fuzzy membership function of tool wear condition is shown in Figure 15.19. The reason for choosing a trapezoid shape is that it is difficult to quantify what exact percentage of tool wear condition corresponds to a certain level of the linguistic variables. In order to improve the training speed of FNN, the tool wear conditions are coded as the following: initial (1,0,0,0,0), normal (0,1,0,0,0), acceptable (0,0,1,0,0), severe (0,0,0,1,0) and failure (0,0,0,0,1). Namely, if the tool condition is normal, the output values of the FNN are (0,1,0,0,0).

A total of 50 cutting tests corresponding to variable cutting states were collected. Thirty samples were randomly picked as learning samples; the remaining samples were used as the test samples in the classification phase. The final tool condition decision is made according to

$$J = \max(y_i) \ (i = 1,2,3,4,5) \qquad \text{Equation (15.38)}$$

where y_i is the output value of trained FNN. The maximum value of the y_i, namely J, is converted to 1, the others are converted to 0. For instance, if $J = y_2 = 0.8$, the output of FNN is (0,1,0,0,0), and the tool belongs to normal wear condition. The classification results are listed in Table 15.8. From Table 15.8, the

TABLE 15.8 Test Results

Tool Condition	Recognition Rate
Air cutting	100%
Initial	87%
Normal	85%
Acceptable	90%
Severe	95%
Failure	100%

detecting rate of tool wear is over 80%, therefore, the monitoring system based on an AE signal has a high success rate for detecting tool wear condition in boring.

References

1. Dimla, D. E., Jr., Lister, P. M. and Leighton, N. J., 1997, Neural Network Solutions to the Tool Condition Monitoring Problem in Metal Cutting — A Critical Review of Methods, *International Journal of Machine Tools and Manufacture*, vol. 37, no. 9, pp. 1219-1241.
2. Li, X., Yao, Y. and Yuan, Z., 1997, On-Line Tool Condition Monitoring System with Wavelet Fuzzy Neural Network, *Journal of Intelligent Manufacturing,* vol. 8, no. 4, pp. 271-276.
3. Lou, K.-N. and Lin, C.-J., 1997, Intelligent Sensor Fusion System for Tool Monitoring on a Machining Center, *International Journal of Advanced Manufacturing Technology*, vol. 13, no. 8, pp. 556-565.
4. Du, R. X., Elbestawi, M. A. and Li, S., 1992, Tool Condition Monitoring in Turning Using Fuzzy Set Theory, *International Journal of Machine Tools and Manufacture*, vol. 32, no. 6, pp. 781-796.
5. Li, S. and Elbestawi, M. A., 1996, Fuzzy Clustering for Automated Tool Condition Monitoring in Machining, *Mechanical Systems and Signal Processing,* vol. 10, no. 5, pp. 533-550.
6. Javed, M. A. and Hope, A. D., 1996, On-Line Multi-Sensor System Monitors Machine Tool Wear, *Noise and Vibration Worldwide*, vol. 27, no. 5, pp. 17-18.
7. Umeda, A., Sugimura, J. and Yamamoto, Y., 1998, Characterization of Wear Particles and Their Relations with Sliding Conditions, *Wear*, vol. 216, no. 2, pp. 220-228.
8. Mesina, O. S. and Langari, R., 1994, Neuro-fuzzy System for Tool Condition Monitoring in Metal Cutting, in *Dynamic Systems and Control* (vol. 2) *American Society of Mechanical Engineers, Dynamic Systems and Control Division*, vol. 55, no. 2, pp. 931-938.
9. Martin, K. F., 1994, Review by Discussion of Condition Monitoring and Fault Diagnosis in Machine Tools, *International Journal of Machine Tools and Manufacture*, vol. 34, no. 4, pp. 527-551.
10. Li, X., Yao, Y. and Yuan, Z., 1997, On-Line Tool Condition Monitoring with Improved Fuzzy Neural Network, *High Technology Letters.* vol. 3, no. 1, pp. 30-33.
11. Li, S. and Elbestawi, M A., 1995, Knowledge Updating for Automated Tool Condition Monitoring in Machining, *ASME Dynamic Systems and Control Division, American Society of Mechanical Engineers, Dynamic Systems and Control Division, (Publication) DSC*, vol. 57, no. 2, pp. 1063-1071.
12. Li, S. and Elbestawi, M. A., 1994, Tool Condition Monitoring in Machining by Fuzzy Neural Networks, in *Dynamic Systems and Control,* (vol. 2) *American Society of Mechanical Engineers, Dynamic Systems and Control Division*, vol. 55, no. 2, pp. 1019-1034.
13. Daubechies, I., 1990, The Wavelet Transforms, Time-Frequency Localization and Signal Analysis, *IEEE Transactions on Information Theory*, vol. 36, no. 5, pp. 961-1005.
14. Daubechies, I., 1988, Orthogonal Bases of Compactly Supported Wavelets, *Communications on Pure and Applied Mathematics*, vol. 41, pp. 909-996.
15. Tansel, I. N., Mekdeci, C. and McLaughlin, C. 1995, Detection of Tool Failure in End Milling with Wavelet Transformations and Neural Networks (WT-NN), *International Journal of Machine Tools and Manufacture*, vol. 35, no. 8, pp. 1137-1147.

16. Tansel, I. N., Mekdeci, C., Rodriguez, O. and Uragun, B., 1995, Monitoring Drill Conditions with Wavelet Based Encoding and Neural Network, *Int. J. Mach. Tools Manuf.*, vol. 33, no. 4, pp. 559-575.
17. Kasashima, N. M., Herrera R. K., and Taniguchi, G. N., 1995, Online Failure Detection in Face Milling Using Discrete Wavelet Transforms, *CIRP of Annals*, vol. 44, no. 1, pp. 483-487.
18. Wang, G. and Li, X., 1997, *Fuzzy Theory and Its Applications in Manufacturing*, China Yunnan Sciences Technology Press.
19. Halgamuge, S. K. and Glesner, M., 1994, Neural Networks in Designing Fuzzy Systems for Real World Applications, *Fuzzy Sets and Systems*, vol. 65, pp. 1-12.
20. Li, X. and Dong, S., 1999, *Intelligent Monitoring Technology in Advanced Manufacturing*, Science Press of China.
21. Li, X., Dong, S. and Venuvinod, P. K., 1999, Hybrid Learning for Tool Wear Monitoring, *International Journal of Advanced Manufacturing Technology* (in press).
22. Li, X., Yao, Y. and Yuan, Z., 1998, Study on Tool Monitoring Using Wavelet Fuzzy Neural Network, *Chinese Journal of Mechanical Engineering*, vol. 35, no. 1, pp. 67-72.
23. Cody, M. A., 1992, The Fast Wavelet Transforms, *Dr. Dobb's Journal*, April, pp. 16-28.
24. Li, D. and Mathew, J., 1990, Tool Wear and Failure Monitoring Techniques for Turning — A Review, *International Journal of Machine Tools and Manufacture*, vol. 30, no. 4, pp. 579-598.
25. Marti, K. F., Brandon, J. A., Grosvenor, B. I. and Dwen, B. I., 1986, A Comparison of In-Process Tool Wear Measurement Methods in Turning, in *Proc. 26th International Machine Tool Design and Research Conference*, pp. 289-296.
26. Novak, A. and Ossbahr, G., 1986, Reliability of the Cutting Force Monitoring in FMS-Installations, in *Proc. 26th International Machine Tool Design and Research Conference*, pp. 325-329.
27. Li, X., 1998, On-Line Detection of the Breakage of Small Diameter Drills Using Current Signature Wavelet Transforms, *International Journal Machine Tools and Manufacture*, vol. 39, no. 1, pp. 157-164.
28. Li, X., 1998, Real-Time Detection of the Breakage of Small Diameter Drills with Wavelet Transforms, *International Journal of Advanced Manufacturing Technology*, vol. 14, no. 8, pp. 539-543.
29. Mannan, M. A., Broms, S. and Lindustrom, B., 1989, Monitoring and Adaptive Control of Cutting Process by Means of Motor Power and Current Measurements, *Annals of the CIRP*, vol. 38, no. 1, pp. 347-350.
30. Mannan, M. A. and Nilsson, T., 1997, The Behavior of Static Torque and Thrust Due to Tool Wear in Boring, *Technical Papers of the North American Manufacturing Research Institution of SME*, pp. 75-80.
31. Li, X., Guan, X. and Wang, H., 1999, Identification of Tool Wear States with Fuzzy Classification, *International Journal Computer Integrated Manufacturing*, vol. 12, pp. 503-514.
32. Shaw, M. C., 1984, *Metal Cutting Principle*, Oxford University Press, Oxford, U.K.
33. Li, X. and Tso, S. K., 1999, Drill Wear Monitoring with Current Signal, *Wear*, vol. 231, pp. 172-178.
34. Lee, C. C., Fuzzy Logic in Control Systems Fuzzy Logic Controller — Part II, *IEEE Trans. System, Man Cybernetics*, vol. 20, pp. 419-435.
35. Souquet, P., Gsib, N., Deschamps, M., Roget, J., and Tanguy, J. C., 1987, Tool Monitoring with Acoustic Emission — Industrial Results and Future Prospects, *Annals of the CIRP*, vol. 36, no. 1, pp. 57-60.
36. Liang, S. and Dornfeld, D. A., 1989, Tool Wear Detection Using Time Series Analysis of Acoustic Emission, *Trans. ASME J. Eng. Ind.*, vol. 111, no. 2, pp. 199-204.
37. Emel, E. and Kannatey-Asibu, E., Jr., 1988, Tool Failure Monitoring in Turning by Pattern Recognition Analysis of AE Signal, *ASME J. Eng. Ind.*, vol. 110, pp. 137-145.
38. Tansel, I. N., Mekdeci, C., Rodriguez, O., and Uragun, B., 1993, Monitoring Drill Conditions with Wavelet Based Encoding and Neural Network, *Int. J. Mach. Tools and Manuf.*, vol. 33, no. 4, pp. 559-575.
39. Wu, Y. and Du, R. 1996, Feature Extraction and Assessment Using Wavelet Packets for Monitoring of Machining Process, *Mechanical Systems and Signal Processing*, vol. 10, no. 1, pp. 29-53.

40. Li, X., Yao, Y. and Yuan, Z., 1997, Experimental Investigate of Acoustic Emission Signal Transmitted Properties, *High Technology Letter*, vol. 3, no. 2, pp. 14-19.
41. Lan, M. S. and Dornfeld, D. A., 1982, Experimental Studies of Flank Wear via Acoustic Emission Analysis, in *10th NAMRC Proc.*, pp. 305-311.
42. Dornfeld, D. A., 1981, Tool Wear Sensing via Acoustic Emission Analysis, in *Proc. 8th NSF Grantees, Conference on Production, Research and Technology*, Stanford Univ., pp. 1-8.
43. Uehara, K., 1984, Identification of Chip Formation Mechanism through Acoustic Emission Measurement, *Annals of the CIRP*, vol. 33, no. 1. pp. 71-74.
44. Yao, Y. X., Li, X. and Yuan, Z., 1999, Tool Wear Detection with Fuzzy Classification and Wavelet Fuzzy Neural Network, *International Journal Machine Tools and Manufacture*, vol. 39, pp. 1525-1538.
45. Li, X. and Yuan, Z., 1998, Tool Wear Monitoring with Wavelet Packet Transform — Fuzzy Clustering Method, *Wear*, vol. 219, pp. 145-154.

V

Quality Assurance and Fault Diagnosis

16

Neural Networks and Neural-Fuzzy Approaches in an In-Process Surface Roughness Recognition System for End Milling Operations

Joseph C. Chen
Iowa State University

16.1 Introduction

Different machining processes produce different products with varying qualities. When evaluating the quality of a finished piece, surface roughness is the most important result of the machining process to consider, because many product attributes can be determined by how well the surface finish is produced. The quality of the surface finish, or surface roughness, affects several functional attributes of parts, such as surface friction, wear, reflectivity, heat transmission, porosity, coating adherence, and fatigue resistance. The desired surface roughness value is usually specified for individual parts, and a particular process is selected in order to achieve the specified roughness.

Typically, surface roughness measurement has been carried out by manually inspecting machined surfaces at fixed intervals. A surface profilometer containing a contact stylus is used in the manual inspection procedure. This procedure is both time-consuming and labor-intensive. In addition, a number of defective parts could be produced during the time needed to complete an off-line surface inspection, thereby creating additional production costs. Another disadvantage of using surface profilometers is that they register the serious interference of extraneous vibration generated in the surrounding environment. This extraneous vibration might significantly influence the accuracy of surface measurements. For these reasons, researchers are seeking solutions to model the surface roughness in an on-line or in-process fashion.

0-8493-0592-6/01/$0.00+$.50

The studies of Martellotti [1941, 1945] are among the earliest that represent a major contribution to the understanding of kinematics and the mechanism of surface generation in milling processes. Martellotti developed parametric equations to describe the trochoidal path that the tool follows. These studies also provide approximate analytical expressions for the ideal peak-to-valley roughness generated in up- and down-slab milling, and face milling.

Numerous other studies have explored the topography of milled surfaces. Many of these focused on predicting the two- or three-dimensional shape of a milled surface under ideal and non-ideal conditions. Kline et. al. [1982] demonstrated the effects of cutter runout on surface errors, and surface errors or dimensional inaccuracies were predicted using the cantilever beam theory for cutter runout. Another study by Babin et al. [1985] applied the cantilever beam theory to predict the topography of wall surfaces produced by end milling. Armarego and Deshpande [1989] presented one more milling process geometry model that incorporates cutter runout to predict cutting forces.

Sutherland and Babin [1985] demonstrated a two-dimensional worst-case analysis of the slot floor surface. However, the model for the slot floor surface significantly underpredicted surface roughness values. Research by Kolarits and DeVries [1989] extended the previous model to account for varying cut geometries and feed rates. This extended floor surface generation model improved prediction capabilities considerably. However, the roughness parameter predictions for some of the tests were found to deviate greatly from measured values.

You and Ehmann [1989] developed a comprehensive model to predict the three-dimensional surface texture generated by ball end mills. They also presented an algorithm for three-dimensional representations of the machined surface; however, the effect of flexibility of the cutter-workpiece system was not considered in this model. Montgomery and Altintas [1991] presented the effects of the cutter-workpiece system flexibility in their force and surface prediction model in order to analyze the surface generation mechanism in peripheral milling under dynamic cutting conditions.

All models previously discussed represent only deterministic cutting models, but most machined surfaces exhibit interrelated characteristics of both random and deterministic components. Zhang and Kapoor [1991] demonstrated the effect of random vibrations on surface roughness in the turning process. These vibrations were shown to occur due to random variations in the microhardness of the workpiece material. Ismail and others presented a surface generation model in milling that included both cutter vibrations and the effects of tool wear [Ismail et al., 1993]. Melkote and Thangaraj [1994] presented another enhanced end milling surface texture model including the effects of radial rake and primary relief angles. These three models, limited to laboratory usage or based on theoretical analysis, could not be implemented as an in-process monitoring system.

The findings of this literature review, in addition to communication with leading private industrial research and development laboratories in the state of Iowa (including Winnebago Co. in Forest City; Delavan Inc. in Des Moines; Sauer-Sundstrand Inc. in Ames), point to the feasibility of in-process surface roughness recognition (ISRR) systems for implementation in the newer generation of milling machines. The successful implementation of this surface roughness recognition system will enable metal cutting industries to reduce manufacturing costs by eliminating the relatively inefficient off-line quality control aspect of surface roughness inspection. Therefore, reductions in manufacturing costs will increase competitiveness in worldwide markets. This implication supports the development of an effective and inexpensive ISRR system. The development of this system will enable implementation of adaptive control in modern manufacturing environments.

16.2 Methodologies

In order to provide an adaptive control mechanism, ISRR systems require two major components: (i) sensors, which receive the dynamics signal from the machining cutting processes; and (ii) an intelligent technique able to learn the dynamics of the machining system while allowing for control features to be built in. The research described in this chapter employed an accelerometer to detect the dynamics mechanism of the tool and material interface. This study also used two major intelligent learning methodologies to incorporate data about the machining process through actual cuts. These methodologies were also employed to construct a control system that predicts surface roughness during the execution of the machining process. These two learning methodologies are artificial neural networks (ANN) and fuzzy neural (FN) systems. An overview of these two approaches follows in the next section.

16.2.1 Neural Networks Model

Several learning methods have been developed for ANNs. Many of these learning methods are closely connected with a certain network topology, with the main categorization method distinguished by supervised vs. unsupervised learning. Backpropagation was chosen from among various learning methods already existing in this field. This approach was adopted into this research for two reasons: primarily, it is the most representative and commonly used algorithm, in addition to being relatively easy to apply; additionally, it has been consistently successful when used in practical applications [Das et al., 1996; Huang and Chiou, 1996].

The backpropagation algorithm can be divided into two main processes, the process of *learning* and the process of *recalling*.

16.2.1.1 The Learning Process

Step 1. Given network parameters:
  Set all the necessary parameters, such as the number of input neurons (i), the number of hidden layers and the number of neurons included in each hidden layer (h), the number of output neurons (j), etc.

Step 2: Initialize the beginning weights and biases:
  Set all the initial weights and biases values randomly.

Step 3: Load the input vector X and the target output vector T of a training example.

Step 4: Calculate and infer the actual output vector Y.
  (a) Calculate the output vector H of hidden layers.

$$net_h = \sum_i W\_xh_{ih} \bullet X_i - \theta\_h_h \qquad \text{Equation (16.1)}$$

$$H_h = f\left(net_h\right) = \frac{1}{1+\exp^{-net_h}} \qquad \text{Equation (16.2)}$$

  (b) Infer the actual output vector Y.

$$net_j = {}_h W\_hy_{hj} \langle H_h - \theta\_y_j \qquad \text{Equation (16.3)}$$

$$Y_j = f\left(net_j\right) = \frac{1}{1+\exp^{-net_j}} \qquad \text{Equation (16.4)}$$

Step 5: Calculate the error term.
  (a) The error term of the output layer: $\delta_j = Y_j\left(1-Y_j\right)\left(T_j - Y_j\right)$ Equation (16.5)

  (b) The error term of the hidden layer: $\delta_h = H_h\left(1-H_h\right) \sum_j W\_hy_{hi} \bullet \delta_j$ Equation (16.6)

Step 6: Calculate the revised weight of the weight matrix and the revised bias of the bias vector.
(a) For the output layer: $\Delta W\_hy_{hj} = \eta\delta_j H_h,\ \Delta\theta\_y_j = -\eta\delta_j$ Equation (16.7)

(b) For the hidden layer: $\Delta W\_xh_{ih} = \eta\delta_h X_i,\ \Delta\theta\_h_h = -\eta\delta_h$ Equation (16.8)

Step 7: Adjust and renew the weight matrix and the bias vector.
(a) For the output layer:

$$W\_hy_{hj} = W\_hy_{hj} + \Delta W\_hy_{hj},\ \theta\_y_j = \theta\_y_j + \Delta\theta\_y_j \quad \text{Equation (16.9)}$$

(b) For the hidden layer:

$$W\_xh_{ih} = W\_xh_{ih} + \Delta W\_xh_{ih},\ \theta\_h_h = \theta\_h_h + \Delta\theta\_h_h \quad \text{Equation (16.10)}$$

Step 8: Repeat steps 3 through 7, until the energy function has converged or the specified learning cycles are completely executed.

16.2.1.2 The Recalling Process

Step 1: Set all the network parameters.
Step 2: Read the weight matrix W_xh and W_hy, and the bias vector θ_h and θ_y.
Step 3: Load the input vector X of a testing example.
Step 4: Calculate and infer the actual output Y.
(a) Calculate the output vector H of hidden layers.

$$net_h = \sum_i W\_xh_{ih} \bullet X_i - \theta\_h_h \quad \text{Equation (16.11)}$$

$$H_h = f(net_h) = \frac{1}{1+\exp^{-net_h}} \quad \text{Equation (16.12)}$$

(b) Infer the actual output vector Y.

$$net_j = \sum_h W\_hy_{hj} \bullet H_h - \theta\_y_j \quad \text{Equation (16.13)}$$

$$Y_j = f(net_j) = \frac{1}{1+\exp^{-net_j}} \quad \text{Equation (16.14)}$$

16.2.2 Fuzzy-Nets Modeling

The proposed fuzzy-nets system was developed by fuzzy rules generated from sampled input–output pairs. This model is built in five steps.

16.2.2.1 Step 1: Divide the Input and Output Spaces into Fuzzy Regions

Assume that the domain intervals of input variable x_i are $\left[x_i^-, x_i^+\right]$, and that the domain intervals of output variable y are $[y^-, y^+]$. Each domain interval can be divided into $2N + 1$ regions. The value of N is dynamic for different variables, and the lengths of each region can be equal or unequal. Each region is denoted by

SN (Small N), S(N-1) (Small N-1), …, MD (Medium), … , LN (Large N), Equation (16.15)

and then assigned a fuzzy membership function. The divisions of the input and output spaces are shown in Figure 16.1, where N is 2 for x_1, and 3 for x_2 and y. The width for each variable is the same.

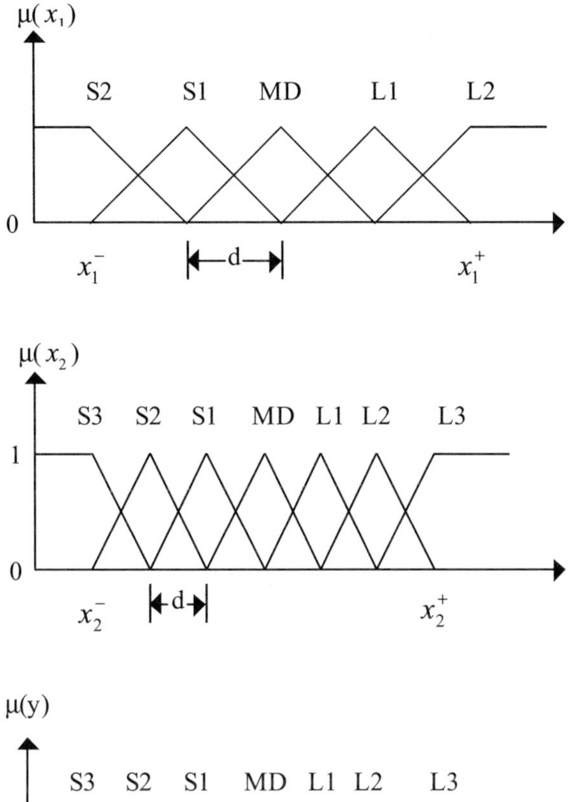

FIGURE 16.1 The domain intervals of the input–output variables and triangular membership function.

In this study, the input variables are spindle speed (S), feed rate (F), depth of cut (D), and vibration average per revolution (V). The output variable is the surface roughness average value, R_a. A triangular membership function specified by three parameters $\{a, b, c\}$ is employed as follows:

$$triangle(x;a,b,c)=\begin{cases} 0 & x \le a \\ \dfrac{x-a}{b-a} & a \le x \le b \\ \dfrac{c-x}{c-b} & b \le x \le c \\ 0 & c \le x \end{cases} \quad \text{Equation (16.16)}$$

The spread of the input feature shown in Figure 16.1 is defined as

$$d=\frac{x_i^+ - x_i^-}{2N} \quad (i = 1,2, ..., k), \qquad \text{Equation (16.17)}$$

where x_i^- and x_i^+ are the domain intervals of variable x_i, $x_i \in X_i$. There are $2N + 1$ fuzzy regions quantifying the universe of discourse X_i.

The center points of each linguistic variable are

$$\left(x_i^-, x_i^- + d, \ldots, x_i^- + \left(2N-1\right)d, x_i^+\right). \qquad \text{Equation (16.18)}$$

Equations 16.17 and 16.18 are also used for the output variable y.

16.2.2.2 Step 2. Generate Fuzzy Rules from Given Data Pairs through Experimentation

Three steps are used for generating fuzzy rules:

1. Determine the degree of input–output data obtained from the successful experiment.
2. Assign the input–output pairs to the region with the maximum degree.
3. Obtain one rule from one pair of designated input–output data.

In this study, the experimental input–output pairs were

$$\left[S^i, F^i, D^i, V^i, R_a^i\right], \qquad \text{Equation (16.19)}$$

where i denotes the number of input–output pairs.

1. A human expert examined these rules to ensure their usefulness and correctness.
2. The degrees of each data pair were determined by the function

$$\mu\left(x_i\right)=\begin{cases} 1-\dfrac{\left|x_i - x_c\right|}{x_s}, x_i \in \left[x_c, x_c + x_s\right] \\ 1-\dfrac{\left|x_c - x_i\right|}{x_s}, x_i \in \left[x_c - x_s, x_c\right]. \\ 0, \qquad \text{otherwise} \end{cases} \qquad \text{Equation (16.20)}$$

   where x_c is the center of the linguistic level x, and x_s is the width of the linguistic level x, equal to d.
3. After all of the input and output elements were determined, each element was assigned to the region with the maximum degree.
4. One rule from one pair of the desired input–output pair $[S^1,F^1,D^1,V^1,R_a^1]$, was assigned. For example, the degree of one input–output pair: was determined by Equation 16.20 as

$$\mu\left(S^1\right)=\left\{0.7\in \mathrm{MD},0.3\in \mathrm{S1}\right\}$$
$$\mu\left(F^1\right)=\left\{0.9\in \mathrm{L3},0.1\in \mathrm{L2}\right\}$$
$$\mu\left(D^1\right)=\left\{0.8\in \mathrm{L2},0.2\in \mathrm{L1}\right\} \qquad \text{Equation (16.21)}$$
$$\mu\left(V^1\right)=\left\{0.2\in \mathrm{MD},0.8\in \mathrm{S1}\right\}$$
$$\mu\left(R_a^1\right)=\left\{0.3\in \mathrm{S2},0.7\in \mathrm{S1}\right\}.$$

The region of each datum with a maximum degree was assigned as follows:

$$S^1\in MD,F^1\in L3,D^1\in L2,V^1\in S1,R_a^1\in S1. \qquad \text{Equation (16.22)}$$

Rule one was obtained by

IF S^1 is MD and F^1 is $L3$ and D^1 is $L2$ and V^1 is $S1$ THEN ${R_a}^1$ is $S1$, Equation (16.23)

where AND indicates that the conditions of the IF statement must all be met simultaneously in order for the result of the THEN statement to be true.

16.2.2.3 Step 3: Assign a Degree to Each Rule and Resolve the Conflicting Rules

If two or more rules generated in step 2 have the same IF command but a different THEN command, then the rules conflict. To resolve conflicts between the two data sets, a degree must be assigned to each rule, generated from the data pairs as

$$d\left(R_i\right)=\mu\left(S^i\right)\mu\left(F^i\right)\mu\left(D^i\right)\mu\left(V^i\right)\mu\left(R_a^i\right)\mu\left(E^i\right) \qquad \text{Equation (16.24)}$$

where $\mu(S^i)$ = the degree of the spindle speed variable,
$\mu(F^i)$ = the degree of the feed rate variable,
$\mu(D^i)$ = the degree of the depth of cut variable,
$\mu(V^i)$ = the degree of the vibration variable,
$\mu({R_a}^i)$ = the degree of the surface roughness variable,
$\mu(E^i)$ = the degree assigned by the human expert to determine the importance of this rule.

The following function resolved the conflict between rules:

$$\left|d\left(R_k\right)-d\left(R_l\right)\right|>\varepsilon \qquad \text{Equation (16.25)}$$

where R_k and R_l are two conflicting rules, $d(R_k)$ and $d(R_l)$ are the degree of rules, R_k and R_l, and ε is the user-defined parameter $0 < \varepsilon < 0.05$. Next, the rule with the maximum degree is selected. If the above function cannot resolve the conflict, ε may be decreased, or two more regions to one feature of the input vector may be increased and the input–output data pairs retrained. If these rules still conflict, the region number of the next input feature is extended to two more regions and then retrained. These procedures are repeated until all of the conflicting problems are resolved.

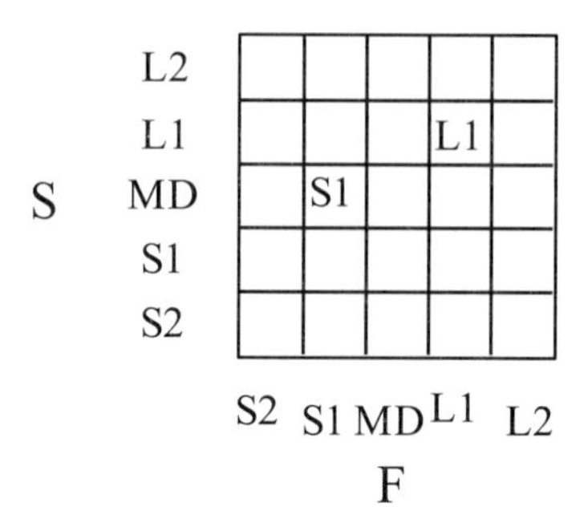

FIGURE 16.2 Illustration of a combined fuzzy rule base.

16.2.2.4 Step 4. Create a Combined Rule Base

A combined rule base consists of two kinds of rules: rules generated from numerical data by means of steps 1 through 3, and linguistic rules determined by a human expert. As shown in Figure 16.2, the combined rule base includes two fuzzy rules:

IF S is L1 and F is L1 THEN R_a is L1;

IF S is MD and F is S1 THEN R_a is S1. Equation (16.26)

16.2.2.5 Step 5. Determine a Mapping Based on the Combined Fuzzy Rule Base

A defuzzification strategy is used to determine the output control y for any given input datum. There are many methods for defuzzification. In this study, the following centroid of area method was applied:

$$y = \frac{\sum \mu_0^i y^i}{\sum \mu_0^i} \qquad \text{Equation (16.27)}$$

where $\mu_o^i = \min\left\{\mu\left(S^i\right), \mu\left(F^i\right), \mu\left(D^i\right), \mu\left(V^i\right)\right\}$, y^i = the center value of the region, and y = the output for a given input datum. This is the most widely adopted defuzzification strategy today, and it is reminiscent of the calculation of expected values of probability distributions.

16.3 Experimental Setup and Design

Figure 16.3 shows the complete experimental setup in this research. A computer numerical control (CNC) program was written to perform the end milling cutting processes. The electromagnetic proximity sensor was fixed at a close distance to the spindle, as shown in Figure 16.4, and the accelerometer sensor was mounted on the vise beneath the workpiece (Figure 16.5).

Rotation and vibration data were collected simultaneously by the proximity sensor and the accelerometer sensor, respectively, when the cutter had cut the workpiece at a distance of 0.35 in. The main concern was to avoid the impact of initially unstable or significant vibration. Figure 16.6 illustrates the two types of signals (i.e., rotation data and vibration data). These two types of data were connected to an analog-to-digital (A/D) board (the vibration data from the accelerometer sensor had to be amplified by the PCB battery power unit beforehand) and then transmitted to a 486 personal computer for further data recording, processing, and analysis.

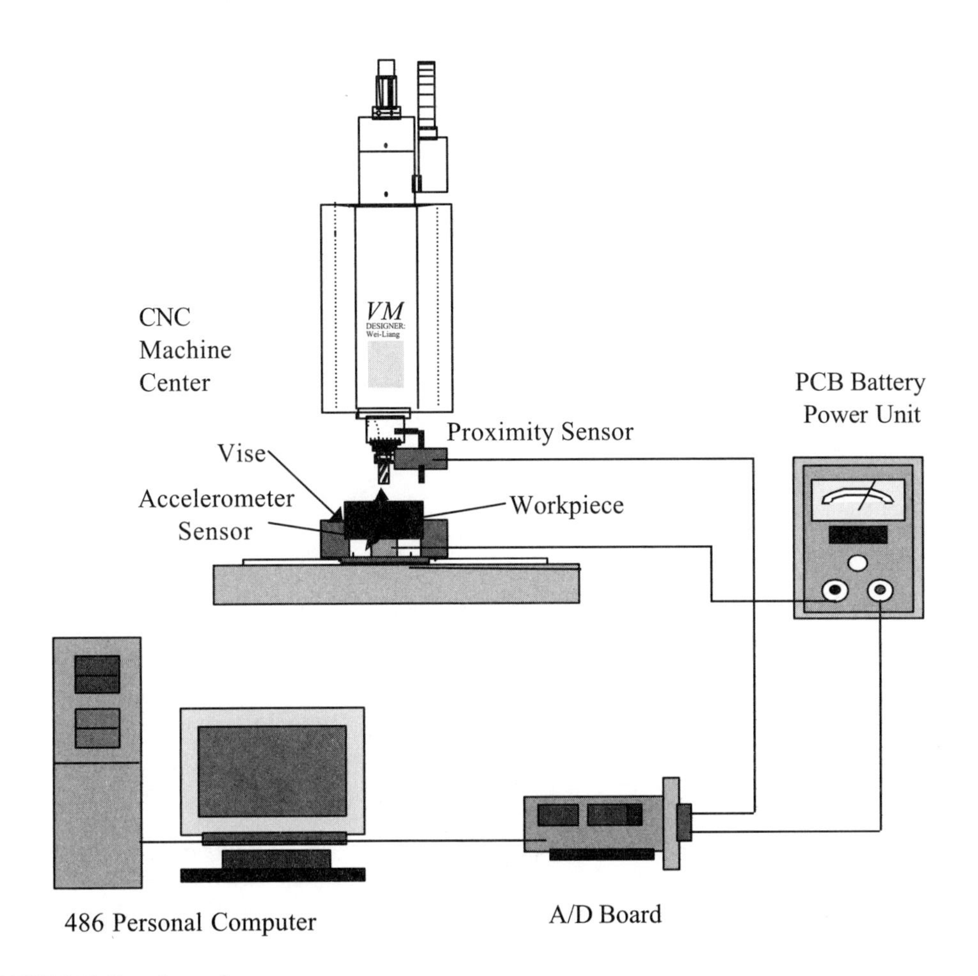

FIGURE 16.3 Experimental setup.

A computer program was written in C language that allowed the collection of the two kinds of data transformed from analog to digital signals by an A/D converter. The collection time for each run was about 0.54 s, and the runs contained 6000 rotation or vibration data from the proximity sensors or accelerometer, respectively.

The cutting parameters (spindle speed, feed rate, and depth of cut) were changed manually according to different cutting conditions for each run. Also, after each specimen was cut, the cutting tool was cleaned to avoid chip formation or a built-up edge (BUE), which would affect the surface roughness of the following cut. The tool condition was also checked to ensure that it was free from defects.

All specimens in this experiment were conducted under dry cutting conditions without coolants. Coolants are not generally used in order to reduce costs and prevent tool breakage due to thermal shock. Moreover, the decision to use dry cutting conditions was based on the need to isolate the correlation between cutting vibrations and surface roughness in end milling.

With the experimental setup complete, the next step was to develop the ISRR-ANN and -FN models. In this study, identifying the parameters of the training and testing data sets was a very important factor in establishing the experimental runs. These runs could not exceed the suggested cutting parameters, which were based on machine capabilities and the nature of the material composition of both the workpiece and end mill.

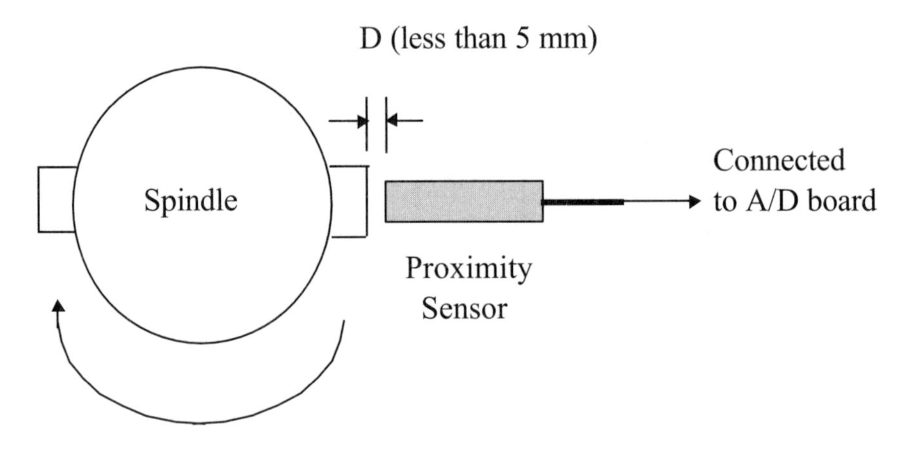

FIGURE 16.4 Top view of the proximity sensor and spindle.

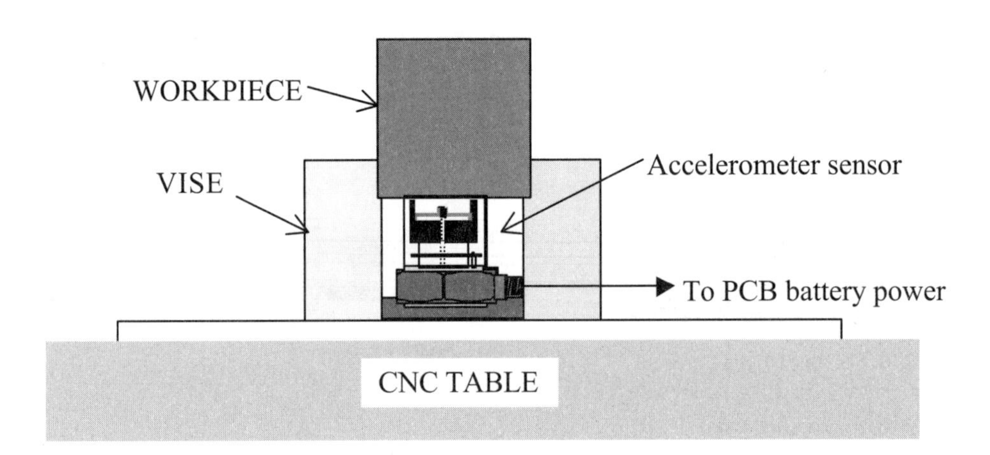

FIGURE 16.5 The accelerometer setup.

16.3.1 Training and Testing Experiments

A total of 48 specimens were cut based on the following combination of cutting parameters: four levels of spindle speed (S = 750, 1000, 1250, and 1500 rpm); four levels of feed rate (F = 6, 12, 18, and 24 in./min); and three levels of depth of cut (D = 0.01, 0.03, and 0.05 in.).

After these cuts were made, all specimens were measured off-line with a stylus profilometer to obtain their roughness average R_a values. The average R_a value shown in Table 16.1 is the average of three R_a measurements of each specimen.

During the machining process, the accelerometer sensor produces vibration data. Each cut produces 13 or more revolutions of vibration data. The statistical average of the vibration voltages (V), transformed from analog to digital data, is based on one revolution of the spindle. This vibration average served as the fourth independent variable for the ISRR systems. Three statistical averages of vibration voltage (V) data were collected per specimen to provide a better empirical representation of the training data set. Therefore, a total of 492 data sets were available for training.

In this research, 92 data sets were selected randomly from the 492 training data. In addition, 36 experimental cuts under different cutting conditions than those in the training data set were used for a

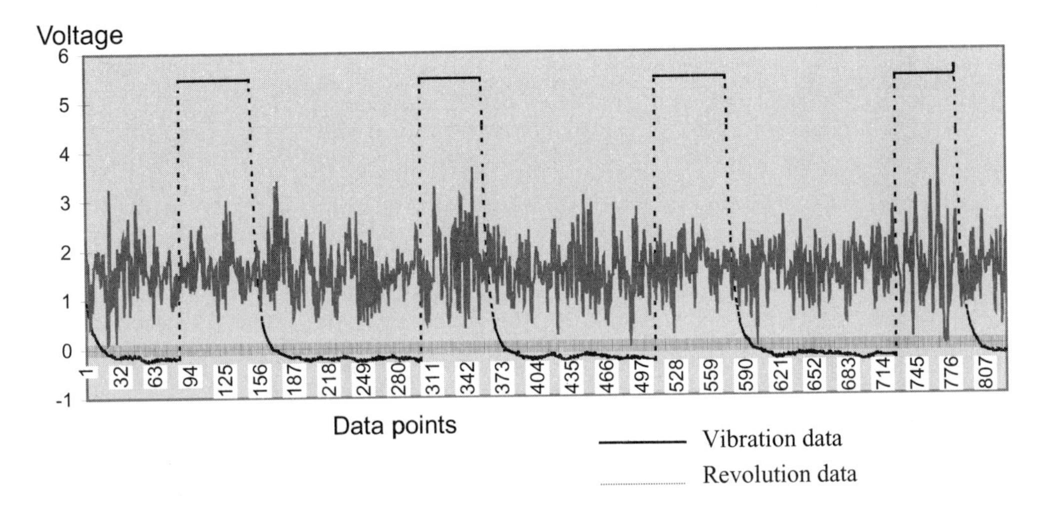

FIGURE 16.6 Vibration and rotation signals.

flexible testing data set (Table 16.1). Therefore, a total of 128 pieces of data were used to evaluate the accuracy of ISRR systems. The evaluation criteria are summarized in the next section.

16.3.2 Test Criteria

Criteria used in this experiment to judge the predictive capabilities of the ISRR–ANN and FN systems included the percentage deviation (ϕ_i) of each testing sample, given as:

$$\phi_i = \frac{\left| Ra_i' - Ra_{i,M}^{\sim} \right|}{Ra_i'} \times 100\%, \qquad \text{Equation (16.28)}$$

where ϕ_i = percentage deviation of single sample data

Ra_i' = actual R_a measured by a profilometer

$Ra_{i,M}^{\sim}$ = predicted R_a generated by the ISRR systems; M indicates the predicted value using ISRR-ANN or ISRR-FN models, respectively.

Additionally, the overall average percentage deviation ($\overline{\phi}$) of all 128 samples is given as

$$\overline{\phi_i} = \frac{\sum_{i=1}^{m} \phi_i}{m} \qquad \text{Equation (16.29)}$$

where $\overline{\phi}$ = average percentage deviation of all sample data

m = the size of testing samples; in this study $m = 128$.

16.4 The In-Process Surface Roughness Recognition Systems

In this research, two ISRR systems were developed and tested. Their training and testing processes are presented in the following sections.

TABLE 16.1 Experimental Results

Sample No.	Spindle Speed (rpm)	Feed Rate (ipm)	Depth of Cut (in.)	Roughness (R_a: μin.)
1	750	6	0.01	65.40
2	750	6	0.03	62.75
3	750	6	0.05	72.40
4	750	12	0.01	143.85
5	750	12	0.03	101.90
6	750	12	0.05	94.05
7	750	18	0.01	184.80
8	750	18	0.03	146.60
9	750	18	0.05	121.05
10	750	24	0.01	186.55
11	750	24	0.03	170.40
12	750	24	0.05	172.40
13	1000	6	0.01	58.40
14	1000	6	0.03	78.30
15	1000	6	0.05	62.20
16	1000	12	0.01	129.90
17	1000	12	0.03	83.60
18	1000	12	0.05	92.05
19	1000	18	0.01	137.50
20	1000	18	0.03	124.15
21	1000	18	0.05	85.75
22	1000	24	0.01	163.15
23	1000	24	0.03	153.30
24	1000	24	0.05	142.30
25	1250	6	0.01	49.95
26	1250	6	0.03	63.30
27	1250	6	0.05	70.85
28	1250	12	0.01	101.30
29	1250	12	0.03	98.75
30	1250	12	0.05	84.95
31	1250	18	0.01	115.00
32	1250	18	0.03	92.25
33	1250	18	0.05	94.70
34	1250	24	0.01	155.45
35	1250	24	0.03	108.85
36	1250	24	0.05	120.65
37	1500	6	0.01	36.55
38	1500	6	0.03	55.70
39	1500	6	0.05	55.65
40	1500	12	0.01	87.55
41	1500	12	0.03	81.65
42	1500	12	0.05	94.05
43	1500	18	0.01	119.45
44	1500	18	0.03	86.50
45	1500	18	0.05	104.25
46	1500	24	0.01	119.20
47	1500	24	0.03	103.30
48	1500	24	0.05	109.40

16.4.1 ISRR-ANN Model

Figure 16.7 shows the structure of the ISRR-ANN system. The input variables were spindle speed (S), feed rate (F), depth of cut (D), and the statistical average of the vibration voltages (V) based on one revolution. In their original form, the four independent variables (S, F, D, and V) for both data sets (training and testing) could not be trained by neural networks due to the wide range of values distributed among them.

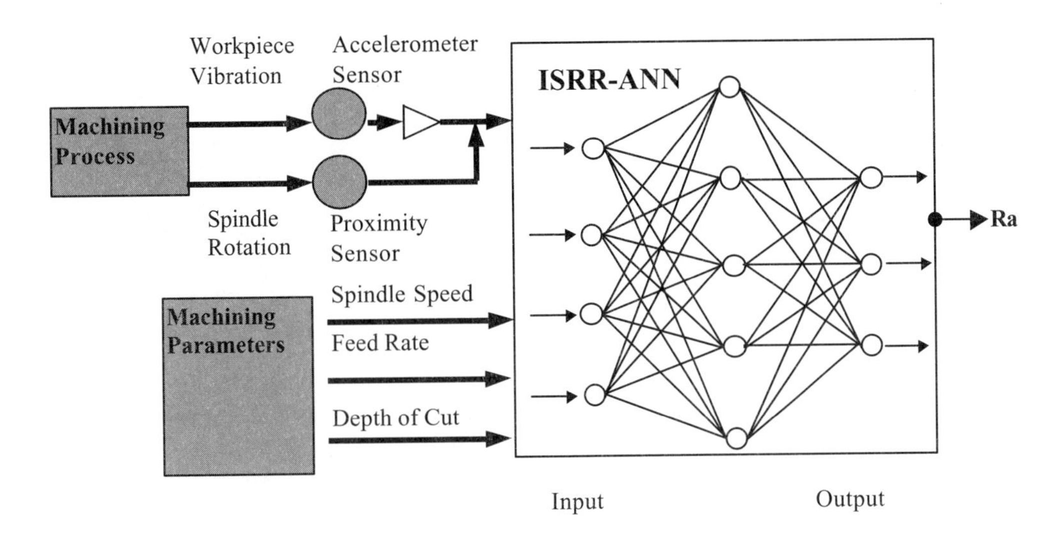

FIGURE 16.7 Structure of the ISRR-ANN.

In order to make them feasible as input neurons, all values in the first four columns (i.e., input neurons) were preprocessed by normalizing, and then transformed into a value between zero and one.

Based on the RMS error of the training examples and testing examples, it is clear that the 4-5-1 structure had the lowest RMS-error among all the structures with one hidden layer, and 4-7-7-1 was better than the other structures with two hidden layers. In the final step of development for this ANN model, two programs were written in C language to predict the R_a values for one or two hidden layers by retrieving the weighted files that resulted from the training and testing.

16.4.2 ISRR-FN System

The structure of the ISRR-FN, as shown in Figure 16.8, consisted of the sensing system, machining parameters, and ISRR-FN. In this sensing system, an accelerometer sensor was used to measure the real-time vibration of the workpiece, a proximity sensor was used to measure the real-time rotation of the spindle of the CNC machine center, and an A/D board and interface program were applied for analog-to-digital conversion with 12-bit resolution. Machining parameters, such as spindle speed, feed rate, and depth of cut, were transmitted to the ISRR-FN before or during machining.

The primary objective was to train the fuzzy system by generating fuzzy rules from input–output pairs, and combining these generated and linguistic rules into a common fuzzy rule base. After input vectors were fuzzified by the fuzzification interface, the fuzzy inference engine generated output values by inferencing these fuzzified input values based on the fuzzy rule bank. Finally, the defuzzification interface determined the final prediction of R_a values based on input variables. Through training, the ISRR-FN learned to detect different conditions for individual machines, build the fuzzy rule base, and infer the surface roughness, R_a. All of these processes were based on the experimental data.

Before the fuzzy-nets training takes up the experimental data, some parameters need to be preset based on the limitations of the machine and the machining processes. They are summarized as follows:

1. The domain intervals of the input–output data pairs were assigned as follows:
   - Spindle speed: [500, 2000] rpm
   - Feed rate: [6, 42] inches per minute
   - Depth of cut: [0.01, 0.07] inches
   - Vibration: [780, 2460] microvolts
   - R_a: [38, 168] micro inches

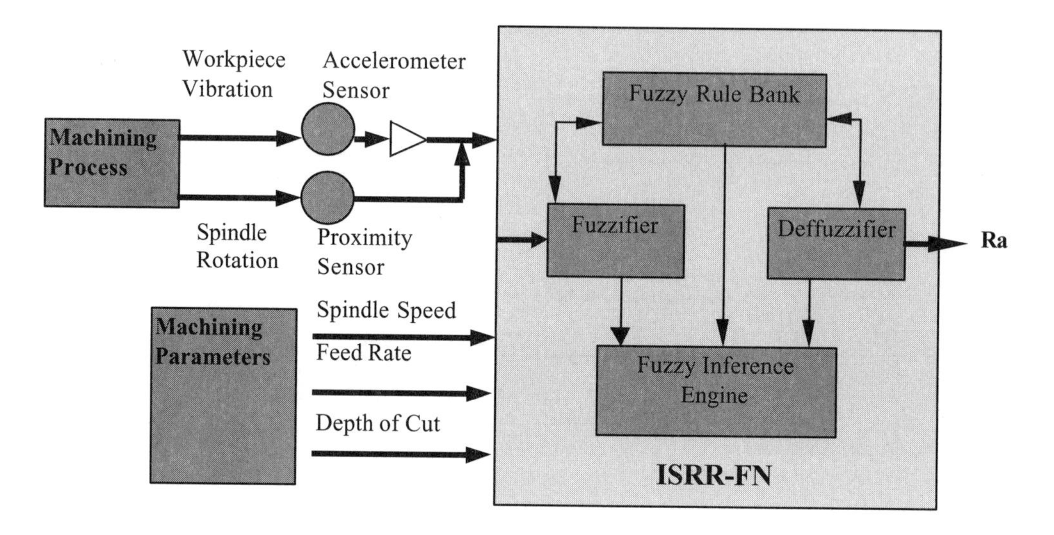

FIGURE 16.8 Structure of the ISRR-FN.

2. Each domain interval was divided into $2N + 1$ regions. To increase the accuracy of prediction and properly decrease the number of fuzzy rules, N was selected as follows:
   - $N = 3$ for input variables: spindle speed, feed rate, depth of cut, and vibration
   - $N = 5$ for the output variable R_a
3. The ε, as shown in Equation 14.10, was assigned the value of 0.01. The degree assigned by a human expert, $\mu(E')$is 1, since all the experiments assume the tool to be in good condition, free from chatter or tool-wear conditions during data collection.

16.5 Testing Results and Conclusions

16.5.1 Testing Results

After the experiment was conducted and both the ANN and FN systems were developed, testing experiments were designed to test the predictive ability of this model. The two sets of testing data mentioned in Section 3.1 were used for testing the accuracy of these two systems.

Using the same four variables (spindle speed, feed rate, depth of cut, and the VAPR) as independent variables or input neurons, predicted roughness R_a values (response variable or output neuron) were generated for either the ANN or FN model. Tables 16.2 and 16.3 contain the measured R_a values and their predicted R_a values, as well as the percentage deviation of both ISRR systems with the 92-item testing set and the 36-item flexible testing set, respectively. Considering the percentage deviation, both models performed well in their ability to predict the roughness of machined surfaces. In summary, the prediction accuracy of ISRR-FN, ISRR-ANN 4-5-1, and ISRR-ANN 4-7-7-1 models are 95.78%, 95.87%, and 99.27%, respectively.

16.5.2 Conclusions

The fuzzy logic and neural-networks-based ISRR models demonstrated that learning and reasoning capabilities could be used for an in-process surface roughness recognition system. With better than 95%

TABLE 16.2 Testing Data Set—92 Samples

Spindle Speed	Feed Rate	Depth of cut	Vibration	R_a	R_a ANN 4-5-1	R_a ANN 4-7-7-1
750	24	1	0.1123	187	190	186
750	18	3	0.1488	147	155	147
1500	12	3	0.1162	82	84	82
1000	12	3	0.1330	84	91	84
750	24	3	0.1659	171	177	171
1250	6	5	0.0977	71	67	67
1250	24	5	0.1764	121	121	121
1000	18	5	0.1674	86	105	86
1500	12	5	0.1225	94	91	94
750	12	3	0.1509	102	104	100
1500	12	1	0.1249	88	83	87
1000	24	5	0.1859	142	146	142
1000	6	3	0.1013	78	73	76
1250	6	1	0.8855	50	44	49
1250	12	1	0.1418	100	104	101
750	6	3	.01045	63	68	64
1500	18	5	0.1509	104	102	104
1000	18	5	0.1647	86	105	86
1500	12	5	0.1180	94	92	94
1250	6	5	0.0974	71	67	71
1500	12	3	0.1232	82	85	82
1000	18	3	0.1691	124	108	124
750	18	5	0.1656	121	128	121
1250	18	1	0.1252	115	116	116
1500	12	1	0.1203	88	84	84
1250	18	1	0.1266	115	117	116
1500	6	3	0.1119	56	57	55
1000	12	3	0.1391	84	91	84
1250	18	5	0.2122	95	102	94
1500	24	3	0.2164	103	106	103
1000	6	5	0.0989	62	69	63
1250	12	5	0.1253	85	93	85
1250	18	3	0.1700	92	100	92
1500	24	1	0.1372	120	125	118
1000	18	3	0.1592	124	109	124
1500	12	1	0.1151	88	85	84
750	18	5	0.1699	121	126	121
750	6	1	0.0899	66	63	64
750	6	1	0.1003	66	62	63
1250	24	5	0.1740	121	121	121
1250	12	3	0.1575	99	91	98
750	24	5	0.1705	172	168	172
1000	24	1	0.1202	163	166	162
1500	24	5	0.1900	110	107	109
1500	12	3	0.1053	82	83	82
1000	18	5	0.1610	86	105	86
1500	18	5	0.1410	104	102	104
1250	18	5	0.2160	95	102	94
1250	24	3	0.2196	109	118	108
1000	24	5	0.1932	142	144	143
1500	6	5	0.0895	56	62	56
1250	24	1	0.1249	156	155	155
1250	12	5	0.1331	85	93	85
1500	12	5	0.1155	94	92	94
1500	18	1	0.1674	120	129	119

TABLE 16.2 (continued) Testing Data Set—92 Samples

Spindle Speed	Feed Rate	Depth of cut	Vibration	R_a	R_a ANN 4-5-1	R_a ANN 4-7-7-1
1500	12	3	0.1101	82	84	82
1500	18	5	0.1500	104	102	105
1250	12	1	0.1288	100	106	101
1500	6	3	0.1084	56	57	55
1250	12	5	0.1253	85	93	85
1500	6	1	0.0647	37	39	36
1500	18	3	0.1359	87	96	87
1000	24	1	0.1187	163	166	163
1500	12	3	0.1151	82	84	82
750	6	3	0.1028	63	68	65
1500	24	1	0.1394	120	125	118
1500	12	3	0.1233	82	85	82
1250	24	1	0.1316	156	153	155
1000	12	5	0.1832	92	92	92
1500	24	1	0.1339	120	126	118
1500	24	1	0.1374	120	125	118
1000	18	1	0.1011	138	139	137
750	18	3	0.1734	147	145	147
1250	6	1	0.0759	50	46	49
1000	24	3	0.1723	153	157	153
1250	24	3	0.2145	109	120	108
750	12	5	0.1331	94	101	96
1500	18	1	0.1300	120	118	119
1500	18	5	0.1513	104	102	105
1250	24	5	0.1857	121	119	121
750	6	5	0.0916	72	69	70
1000	24	5	0.2035	142	141	143
1000	18	1	0.0979	138	140	137
750	18	3	0.1446	147	155	147
1250	6	3	0.1060	63	67	64
1500	18	5	0.1516	104	102	105
1250	24	1	0.1216	156	155	155
1000	12	5	0.1599	92	92	91
1500	18	3	0.1536	87	98	87
750	18	5	0.1622	121	129	129
1000	18	1	0.0979	138	140	137
1500	12	5	0.1183	94	92	94

accuracy of prediction, this system could be implemented as an in-process surface roughness prediction system; however, some directions for further research could make future implementation more effective:

1. The cutting tool used in the study was a four-flute, high-speed steel cutter. Use of the model with more diverse cutter materials or tools with a different number of flutes will benefit from further investigation.
2. The workpiece material used in the study was Aluminum 6061 T6. Different workpiece materials widely used in industry, such as carbon steel, aluminum alloys 380 and 390, or alloy steel, merit further exploration in order to build an overall ISRR system capable of extensive application in actual manufacturing environments.
3. Feedback control is necessary for automated production systems. The ISRR should be a closed-loop system, ensuring that output of the ISRR feeds back to the CNC machine center, causing the machine to adapt to the predicted real-time surface roughness value by adjusting the feed rate of the machine table. Thus, a study of interface techniques between the ISRR system and the CNC machining center is also necessary.

TABLE 16.3 Flexible Testing Data — 36 Samples

Sample No.	Spindle Speed (rpm)	Feed Rate (ipm)	Depth of Cut (in.)	VAPR	Measured R_a (μin.) ISRR-FN	R_a ISRR-FN
1	1500	9	0.01	0.0883	53	53
2	1500	9	0.03	0.1110	74	74
3	1500	9	0.05	0.1056	70	70
4	1500	15	0.01	0.1464	110	110
5	1500	15	0.03	0.1256	84	84
6	1500	15	0.05	0.1638	99	99
7	1500	21	0.01	0.1473	119	119
8	1500	21	0.03	0.1787	102	102
9	1500	21	0.05	0.1980	113	113
10	1250	9	0.01	0.1197	80	80
11	1250	9	0.03	0.1381	82	82
12	1250	9	0.05	0.1202	92	92
13	1250	15	0.01	0.1338	107	107
14	1250	15	0.03	0.1521	97	97
15	1250	15	0.05	0.1444	87	87
16	1250	21	0.01	0.1300	129	129
17	1250	21	0.03	0.1726	99	98
18	1250	21	0.05	0.1846	105	105
19	1000	9	0.01	0.0911	95	92
20	1000	9	0.03	0.1226	97	96
21	1000	9	0.05	0.1426	102	102
22	1000	15	0.01	0.1001	129	129
23	1000	15	0.03	0.1487	108	108
24	1000	15	0.05	0.1598	95	92
25	1000	21	0.01	0.1034	149	149
26	1000	21	0.03	0.1680	145	145
27	1000	21	0.05	0.1687	112	112
28	750	9	0.01	0.0931	109	109
29	750	9	0.03	0.1255	99	99
30	750	9	0.05	0.1171	95	95
31	750	15	0.01	0.0950	126	125
32	750	15	0.03	0.1514	122	122
33	750	15	0.05	0.1530	104	104
34	750	21	0.01	0.1135	178	178
35	750	21	0.03	0.1624	163	163
36	750	21	0.05	0.1659	150	150

4. The ISRR systems used in this study were controlled by a personal computer system. To increase operation speed, minimize size, reduce costs, and enhance efficiency, the ISRR could be created in an electronic model operated by a microprocessor, fuzzy chip, memory chip, and other related circuits. These kinds of techniques are important and may be required for further development of ISRR technology for the next century.

References

Armarego, E. J. A. and Deshpande, N. P., 1989, Computerized predictive cutting models for forces in end-milling including eccentricity effects, *Annals of the CIRP*, 38(1), pp. 45-49.

Babin, T. S., Lee, J. M., Sutherland, J. M., and Kapoor, S. G., 1985, A model for end milling surface topography, *Proc. of 13th North American Metalworking Research Conference*, pp. 362-368.

Das, S., Roy, R., and Chaptopadhyay, A. B., 1996. Evaluation of wear of turning carbide inserts using neural networks, *Int. J. Mach. Tools Manuf.*, 36(7), pp. 789-797.

Huang, S. J. and Chiou, K. C., 1996. The application of neural networks in self-tuning constant force control, *Int. J. Mach. Tools Manuf.*, 36, pp. 17-31.

Ismail, F., Elbestawi, M. A., Du, R., and Urbasik, K., 1993, Generation of milled surfaces including tool dynamics and wear, *ASME J. Eng. Ind.*, 115, pp. 245-252.

Kline, W. A., DeVor, R. E., and Shareef, I. A., 1982, The prediction of surface accuracy in end milling, *ASME J. Eng. Ind.*, 104, pp. 272-278.

Kolarits, F. M. and DeVries, W., 1989, A model of the geometry of the surface generated in end milling with variable process inputs, in *Mechanics of Deburring and Surface Finishing Processes,* J. R. Stango, and P. R. Fitzpatrick, Eds., ASME, PED, vol. 38, pp. 63-78.

Martellotti, M. E., 1941, An analysis of the milling process, *Trans. ASME.* 12:677-700.

Martellotti, M. E., 1945, An analysis of the milling process, part II — Down milling, *Trans. ASME*, 67, pp. 233-251.

Melkote, S. N. and Thangaraj, A. R., 1994, An enhanced end milling surface texture model including the effects of radial rake and primary relief angles, *ASME J. Eng. Ind.*, 116, pp. 166-174.

Montgomery, D. and Altintas, Y., 1991, Machanism of cutting force and surface generation in dynamic milling, *ASME J. Eng. Ind.*, 113 (1), pp. 160-168.

Sutherland and Babin 1985,

You, S. J. and Ehmann, K. F., 1989, Scallop removal in die milling by tertiary cutter motion, *ASME J. Eng. Ind.*, 111, pp. 213-219.

Zhang, G. M. and Kapoor, S. G., 1991, Dynamic generation of machined surface, part 1: Description of a random excitation system, *ASME J. Eng. Ind.*, 113(3), pp. 137-144.

Zhang, G. M. and Kapoor, S. G., 1991b, Dynamic generation of machined surface, part 2: Construction of surface topography, *ASME J. Eng. Ind.*, 113(3), pp. 145-153.

17

Intelligent Quality Controllers for On-Line Parameter Design

Ratna Babu Chinnam
Wayne State University

17.1 Introduction

Besides aggressively innovating and incorporating new materials and technologies into practical, effective, and timely commercial products, in recent years, many industries have begun to examine new directions that they must cultivate to improve their competitive position in the long term. One thing that has become clearly evident is the need to push the quality issue farther and farther upstream so that it becomes an integral part of every aspect of the product/process life cycle. In particular, many have begun to recognize that it is through engineering design that we have the greatest opportunity to influence the ultimate delivery of products and processes that far exceed the customer needs and expectations.

For the last two decades, classical experimental design techniques have been widely used for setting critical product/process parameters or targets during design. But recently their potential is being questioned, for they tend to focus primarily on the mean response characteristics. One particular design approach that has gained a lot of attention in the last decade is the *robust parameter design approach* that borrows heavily from the principles promoted by Genichi Taguchi [1986, 1987]. Taguchi views the design process as evolving in three distinct phases:

1. *System Design Phase* — Involves application of specialized field knowledge to develop basic design alternatives.
2. *Parameter Design Phase* — Involves selection of "best" nominal values for the important design parameters. Here "best" values are defined as those that "minimize the transmitted variability resulting from the noise factors."
3. *Tolerance Design Phase* — Involves setting of tolerances on the nominal values of critical design parameters. Tolerance design is considered to be an economic issue, and the loss function model promoted by Taguchi can be used as a basis.

0-8493-0592-6/01/$0.00+$.50

Besides the basic parameter design method, Taguchi strongly emphasized the need to perform *robust* parameter design. Here "robustness" refers to the insensitive behavior of the product/process performance to changes in environmental conditions and noise factors. Achieving this insensitivity at the design stage through the use of designed experiments is a corner stone of the Taguchi methodology.

Over the years, many distinct approaches have been developed to implement Taguchi's parameter design concept; these can be broadly classified into the following three categories:

1. Purely analytical approaches
2. Simulation approaches
3. Physical experimentation approaches.

Due to the lack of precise mechanistic models (models derived from fundamental physics principles) that explain product/process performance characteristics (in terms of the different controllable and uncontrollable variables), the most predominant approach to implementing parameter design involves physical experimentation. Two distinct approaches to physical experimentation for parameter design include (i) orthogonal array approaches, and (ii) traditional factorial and fractional factorial design approaches.

The orthogonal array approaches are promoted extensively by Taguchi and his followers, and the traditional factorial and fractional factorial design approaches are normally favored by the statistical community. Over the years, numerous papers have been authored comparing the advantages and disadvantages of these approaches. Some of the criticisms for the orthogonal array approach include the following [Box, 1985]: (i) the method does not exploit a sequential nature of investigation, (ii) the designs advocated are rather limited and fail to deal adequately with interactions, and (iii) more efficient and simpler methods of analysis are available.

In addition to the different approaches to "generating" data on product/process performance, there exist two distinct approaches to "measuring" performance:

1. Signal-to-Noise (S/N) Ratios — Tend to combine the location and dispersion characteristics of performance into a one-dimensional metric; the higher the S/N ratio, the better the performance.
2. Separate Treatment — The location and dispersion characteristics of performance are evaluated separately.

Once again, numerous papers have been authored questioning the universal use of the S/N ratios suggested by Taguchi and many others. The argument is that the Taguchi parameter design philosophy should be blended with an analysis strategy in which the mean and variance of the product/process response characteristics are modeled to a considerably greater degree than practiced by Taguchi. Numerous papers authored in recent years have established that one can achieve the primary goal of the Taguchi philosophy, i.e., to obtain a target condition on the mean while minimizing the variance, within a response surface methodology framework. Essentially, the framework views both the mean and the variances as responses of interest. In such a perspective, the dual response approach developed by Myers and Carter [1973] provides an alternate method for achieving a target for the mean while also achieving a target for the variance. For an in-depth discussion on response surface methodology and its variants, see Myers and Montgomery [1995]. For a panel discussion on the topic of parameter design, see Nair [1992].

17.1.1 Classification of Parameters

A block diagram representation of a product/process is shown in Figure 17.1. A number of parameters can influence the product/process response characteristics, and can be broadly classified as *controllable* parameters and *uncontrollable* parameters (note that the word *parameter* is equivalent to the word *factor* or *variable* normally used in parameter design literature).

1. **Controllable Parameters:** These are parameters that can be specified freely by the product/process designer and or the user/operator of the product/process to express the intended value for the

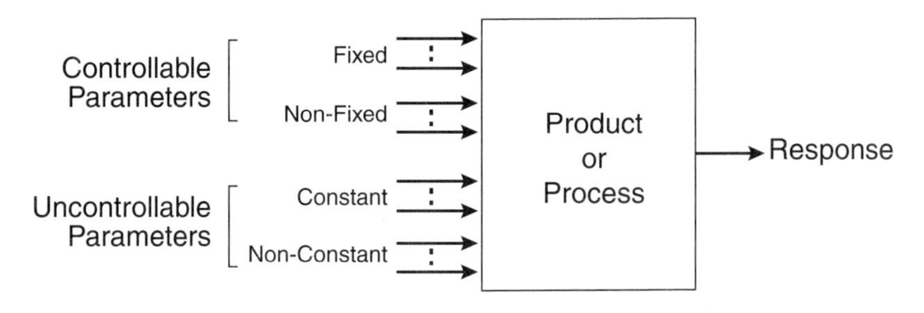

FIGURE 17.1 Block diagram of a product/process.

response. These parameters can be classified into two further groups: *fixed* controllable parameters and *non-fixed* controllable parameters.

a. *Fixed Controllable Parameters:* These are parameters that are normally optimized by the product/process designer at the design stage. The parameters may take multiple values, called *levels*, and it is the responsibility of the designer to determine the best levels for these parameters. Changes in the levels of certain fixed controllable parameters may not have any bearing on manufacturing or operation costs; however, when the levels of certain others are changed, the manufacturing and/or operation costs might change (these factors that influence the manufacturing cost are also referred to as *tolerance* factors in the parameter design literature). Once optimized, these parameters remain fixed for the life of the product/process. For example, parameters that influence the geometry of a machine tool used for a machining process, and/or its material/technology makeup fall under this category.
b. *Non-Fixed Controllable Parameters:* These are controllable parameters that can be freely changed before or during the operation of the product/process (these factors are also referred to as *signal* factors in the parameter design literature). For example, the cutting parameters such as speed, feed, and depth of cut on a machining process can be labeled non-fixed controllable parameters.

2. **Uncontrollable Parameters:** These are parameters that cannot be freely controlled by the process/process designer. Parameters whose settings are difficult to control or whose levels are expensive to control can also be categorized as uncontrollable parameters. These parameters are also referred to as *noise factors* in the parameter design literature. These parameters can be classified into two further groups: *constant* uncontrollable parameters and *non-constant* uncontrollable parameters.

a. *Constant Uncontrollable Parameters:* These are parameters that tend to remain constant during the life of the product or process but are not easily controllable by the product/process designer. Certainly, the parameters representing variation in components that make up the product/process fall under this category. This variation is inevitable in almost all manufacturing processes that produce any type of a component and is attributed to *common causes* (representing natural variation or true process capability) and *assignable/special causes* (representing problems with the process rendering it out of control). For example, the nominal resistance of a resistor to be used in a voltage regulator may be specified at 100 KΩ. However, the resistance of the individual resistors will deviate from the nominal value affecting the performance of the individual regulators. Please note that the parameter (i.e., resistance) is to some degree uncontrollable; however, the level/amplitude of the uncontrollable parameter for any given individual regulator remains more or less constant for the life of that voltage regulator.
b. *Non-Constant Uncontrollable Parameters:* These parameters normally represent the environment in which the product/process operates, the loads to which they are subjected, and their deterioration. For example, in machining processes, some examples of non-constant uncontrollable variables include room temperature, humidity, power supply voltage and current, and amplitude of vibration of the shop floor.

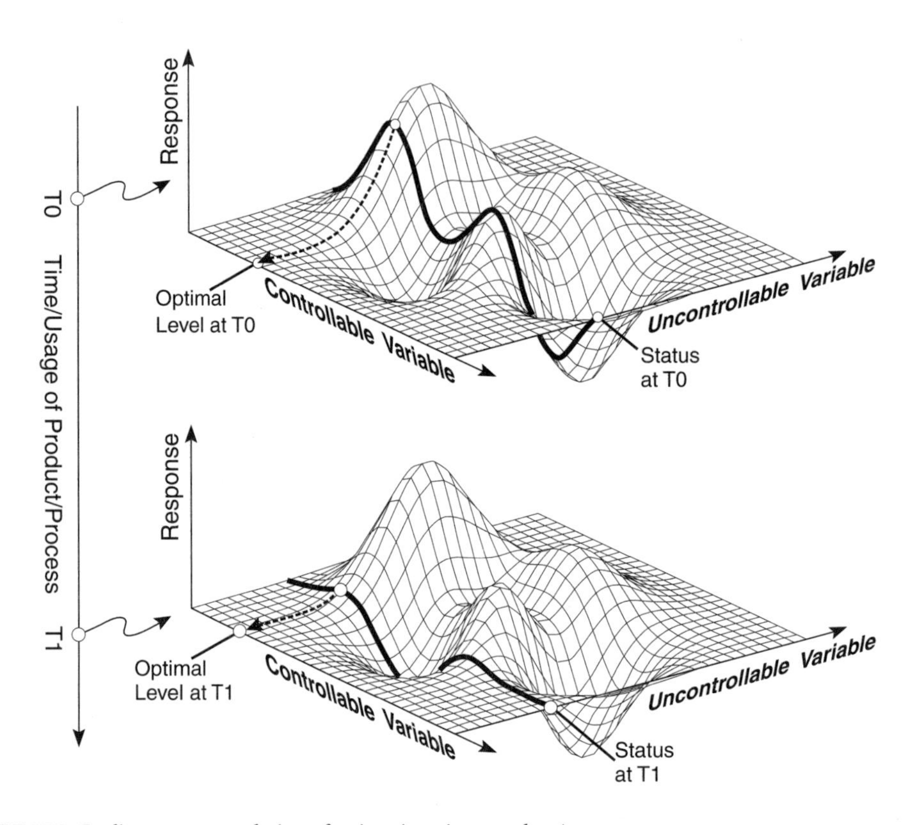

FIGURE 17.2 On-line parameter design of a time-invariant product/process.

17.1.2 Limitations of Existing Off-Line Parameter Design Techniques

Whatever the method of design, in general, parameter design methods do not take into account the common occurrence that some of the uncontrollable variables are observable during production [Pledger, 1996] and part usage. This extra information regarding the levels of non-constant uncontrollable factors enhances our choice of values for the non-fixed controllable factors, and, in some cases, determines the viability of the production process and or the product. This process is hypothetically illustrated for a time-invariant product/process in Figure 17.2. Here T_0 and T_1 denote two different time/usage instants during the life of the product/process. Given the level of the uncontrollable variable at any instant, the thick line represents the response as a function of the level of the controllable variable. Given the response model, the task here is to optimize the controllable variable as a function of the level of the uncontrollable variable. In depicting the optimal levels for the controllable variable in Figure 17.2, the assumption made is that it is best to maximize the product/process response (i.e., larger the response, the better it is). The same argument can be extended to cases where the product/process has multiple controllable/uncontrollable variables and multiple outputs. In the same manner, it is possible to extend the argument to smaller-is-better and nominal-is-best response cases, and combinations thereof.

Given the rapid decline in instrumentation costs over the last decade, the development of methods that utilize this additional information will facilitate optimal utilization of the capability of products/processes. Pledger [1996] described an approach that explicitly introduces uncontrollable factors into a designed experiment. The method involves splitting uncontrollable factors into two sets, *observable* and *unobservable*. In the first set there may be factors like temperature and humidity, while in the second there may be factors such as chemical purity and material homogeneity that may be unmeasurable due to time, physical, and economic constraints. The aim is to find a relationship between the controllable factors and the observable uncontrollable factors while simultaneously minimizing the variance of the response and keeping the mean response on target. Given the levels of the observable uncontrollable variables, appropriate values for the controllable factors are generated on-line that meet the stated objectives.

As is also pointed out by Pledger [1996], if an observable factor changes value in wild swings, it would not be sensible to make continuous invasive adjustments to the product or process (unless there is minimal cost associated with such adjustments). Rather, it would make sense to implement formal control over such factors. Pledger derived a closed-form expression, using Lagrangian minimization, that facilitates minimization of product or process variance while keeping the mean on target, when the model that relates the quality response variable to the controllable and uncontrollable variables is linear in parameters and involves no higher order terms. However, as Pledger pointed out, if the model involves quadratic terms or other higher order interactions, there can be no closed-form solution.

17.1.3 Overview of Proposed Framework for On-Line Parameter Design

Here, we develop some general ideas that facilitate on-line parameter design. The specific objective is to not impose any constraint on the nature of the relationship between the different controllable and uncontrollable variables and the quality response characteristics, and allow multiple quality response characteristics. In particular, we recommend feedforward neural networks (FFNs) for modeling the quality response characteristics. Some of the reasons for making this recommendation are as follows:

1. *Universal Approximation.* FFNs can approximate any continuous function $f \in (R^N, R^M)$ over a compact subset of R^N to arbitrary precision [Hornik et al., 1989]. Previous research has also shown that neural networks offer advantages in both accuracy and robustness over statistical methods for modeling processes (for example, Nadi et al. [1991]; Himmel and May [1993]; Kim and May [1994]). However, there is some controversy surrounding this issue.
2. *Adaptivity.* Most training algorithms for FFNs are incremental learning algorithms and exhibit a built-in capability to adapt the network to changes in the operating environment [Haykin, 1999]. Given that most product and processes tend to be time-variant (nonstationary) in the sense that the response characteristics change with time, this property will play an important role in achieving on-line parameter design of time-variant systems.

Besides proposing nonparametric neural network models for "modeling" quality response characteristics of manufacturing processes, we recommend a gradient descent search technique and a stochastic search technique for "optimizing" the levels of the controllable variables on-line. In particular, we consider a neural network iterative inversion scheme and a stochastic search method that utilizes genetic algorithms for optimization of controllable variables. The overall framework that facilitates these two on-line tasks, i.e., modeling and optimization, constitutes a *quality controller.* Here, we focus on development of quality controllers for manufacturing processes whose quality response characteristics are static and time-invariant. Future research can concentrate on extending the proposed controllers to deal with dynamic and time-variant systems. In addition, future research can also concentrate on modeling the signatures of the uncontrollable variables to facilitate feedforward parameter design.

17.1.4 Chapter Organization

The chapter is organized as follows: Section 17.2 provides an overview of feedforward neural networks and genetic algorithms utilized for process modeling and optimization; Section 17.3 describes an approach to designing intelligent quality controllers and discusses the relevant issues; Section 17.4 presents some results from the application of the proposed methods to a plasma etching semiconductor manufacturing process; and Section 17.5 provides a summary and gives directions for future work.

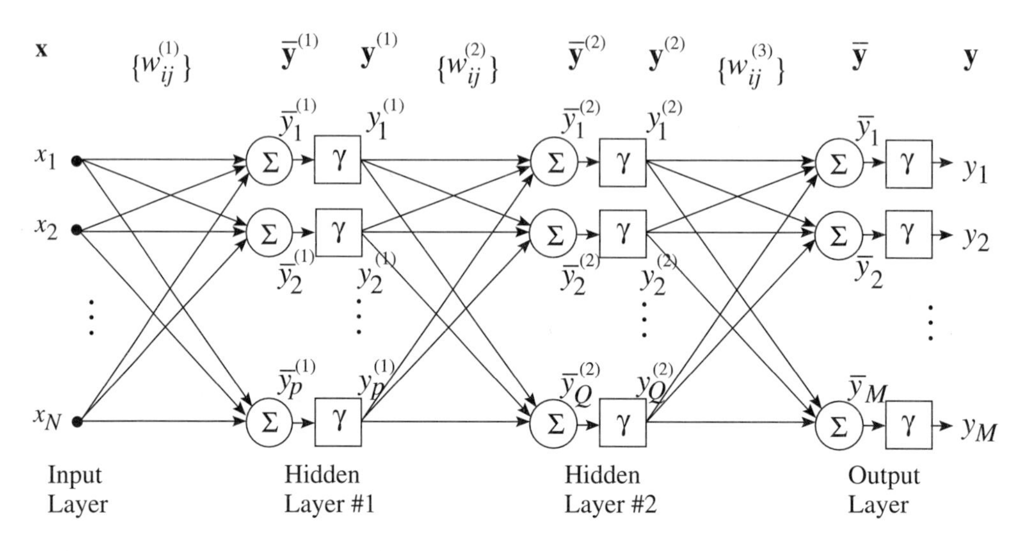

FIGURE 17.3 A three-layer neural network.

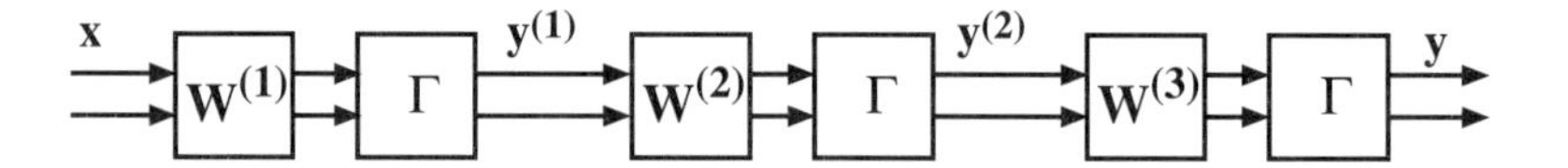

FIGURE 17.4 A block diagram representation of a three-layer network.

17.2 An Overview of Certain Emerging Technologies Relevant to On-Line Parameter Design

17.2.1 Feedforward Neural Networks

In general, feedforward artificial neural networks (ANNs) are composed of many nonlinear computational elements, called *nodes*, operating in parallel, and arranged in patterns reminiscent of biological neural nets [Lippman, 1987]. These processing elements are connected by weight values, responsible for modifying signals propagating along connections and used for the training process. The number of nodes plus the connectivity define the topology of the network, and range from totally connected to a topology where each node is just connected to its neighbors. The following subsections discuss the characteristics of a class of feedforward neural networks.

17.2.1.1 Multilayer Perceptron Networks

A typical multilayer perceptron (MLP) neural network with an input layer, an output layer, and two hidden layers is shown in Figure 17.3 (referred to as a three-layer network; normally, the input layer is not counted). For convenience, the same network is denoted in block diagram form as shown in Figure 17.4 with three weight matrices $\mathbf{W}^{(1)}$, $\mathbf{W}^{(2)}$, and $\mathbf{W}^{(3)}$ and a diagonal nonlinear operator Γ with identical sigmoidal elements γ following each of the weight matrices. The most popular nonlinear nodal function for multilayer perceptron networks is the sigmoid [*unipolar* $\rightarrow \gamma(x) = 1/(1 + e^{-x})$ where $0 \leq \gamma(x) \leq 1$ for $-\infty < x < \infty$ and *bipolar* $\rightarrow \gamma(x) = (1 - e^{-x})/(1 + e^{-x})$ where $-1 \leq \gamma(x) \leq 1$ for $-\infty < x < \infty$]. It is necessary to either scale the output data to fall within the range of the sigmoid function or use a linear nodal function in the outermost layer of the network. It is also common practice to include an externally applied *threshold* or *bias* that has the effect of lowering or increasing the net input to the nodal function. Each layer of the network can then be represented by the operator

$$\mathbf{N}_l\left[\mathbf{x}\right]=\boldsymbol{\Gamma}\left[\mathbf{W}^{(l)}\mathbf{x}\right], \qquad \text{Equation (17.1)}$$

and the input–output mapping of the MLP network can be represented by

$$\mathbf{y}=\mathbf{N}\left[\mathbf{x}\right]=\boldsymbol{\Gamma}\left[\mathbf{W}^{(3)}\boldsymbol{\Gamma}\left[\mathbf{W}^{(2)}\boldsymbol{\Gamma}\left[\mathbf{W}^{(1)}\boldsymbol{\Gamma}\right]\right]\right]=\mathbf{N}_3\mathbf{N}_2\mathbf{N}_1\left[\mathbf{x}\right]. \qquad \text{Equation (17.2)}$$

The weights of the network $\mathbf{W}^{(1)}$, $\mathbf{W}^{(2)}$, and $\mathbf{W}^{(3)}$ are adjusted (as described in Section 17.2.1.2) to minimize a suitable function of error $\mathbf{e}$ between the predicted output $\mathbf{y}$ of the network and a desired output $\mathbf{y}_d$ (error-correction learning), resulting in a mapping function $\mathbf{N}[\mathbf{x}]$. From a systems theoretic point of view, multilayer perceptron networks can be considered as versatile nonlinear maps with the elements of the weight matrices as parameters.

It has been shown in Hornik et al., [1989], using the Stone–Weierstrass theorem, that even an MLP network with just one hidden layer and an arbitrarily large number of nodes can approximate any continuous function $f \in C(\Re^N, \Re^M)$ over a compact subset of $\Re^N$ to arbitrary precision (universal approximation). This provides the motivation to use MLP networks in modeling/identification of any manufacturing process' response characteristics.

17.2.1.2 Training MLP Networks Using Backpropagation Algorithm

If MLP networks are used to solve the identification problems treated here, the objective is to determine an adaptive algorithm or rule that adjusts the weights of the network based on a given set of input–output pairs. An error-correction learning algorithm will be discussed here, and readers can see Zurada [1992] and Haykin [1999] for information regarding other training algorithms. If the weights of the networks are considered as elements of a parameter vector θ, the error-correction learning process involves the determination of the vector θ^*, which optimizes a performance function J based on the output error. In error-correction learning, the gradient of the performance function with respect to θ is computed and adjusted along the negative gradient as follows:

$$\theta(s+1)=\theta(s)-\eta\frac{\partial J(s)}{\partial\theta(s)} \qquad \text{Equation (17.3)}$$

where η is a positive constant that determines the rate of learning (step size) and s denotes the iteration step.

In the three-layered network shown in Figure 17.3, $\mathbf{x} = (x_1, \ldots, x_N)^T$ denotes the input pattern vector while $\mathbf{y} = (y_1, \ldots, y_M)^T$ is the output vector. The vectors $\mathbf{y}^{(1)} = (y_1^{(1)}, \ldots, y_P^{(1)})$ and $\mathbf{y}^{(2)} = (y_1^{(2)}, \ldots, y_Q^{(2)})^T$ are the outputs at the first and the second hidden layers, respectively. The matrices $\left\{w_{ij}^{(1)}\right\}_{P\times N}$, $\left\{w_{ij}^{(2)}\right\}_{Q\times P}$, and $\left\{w_{ij}^{(3)}\right\}_{M\times Q}$ are the weight matrices associated with the three layers as shown in Figure 17.3. Note that the first subscript in weight matrices denotes the neuron in the next layer and the second subscript denotes the neuron in the current layer. The vectors $\bar{\mathbf{y}}^{(1)} \in \Re^P$, $\bar{\mathbf{y}}^{(2)} \in \Re^Q$, and $\bar{\mathbf{y}} \in \Re^M$ are as shown in Figure 17.3 with, $\gamma\left(\bar{y}_i^{(1)}\right)=y_i^{(1)}$, $\gamma\left(\bar{y}_i^{(2)}\right)=y_i^{(2)}$, and

$\gamma(\bar{y}_i) = y_i$ where $\bar{y}_i^{(1)}$, $\bar{y}_i^{(2)}$, and $\bar{y}_i$ are elements of $\bar{\mathbf{y}}^{(1)}$, $\bar{\mathbf{y}}^{(2)}$, and $\bar{\mathbf{y}}$ respectively. If $\mathbf{y}_d = (y_{d1}, \ldots, y_{dM})^T$ is the desired output vector, the output error of a given input pattern $\mathbf{x}$ is defined as $\mathbf{e} = \mathbf{y} - \mathbf{y}_d$.

Typically, the performance function J is defined as

$$J = \frac{1}{2} \sum_{S} \|\mathbf{e}\|^2, \qquad \text{Equation (17.4)}$$

where the summation is carried out over all patterns in a given training data set S. The factor 1/2 is used in Equation 17.4 to simplify subsequent derivations resulting from minimization of J with respect to free parameters of the network.

While strictly speaking, the adjustment of the parameters (i.e., weights) should be carried out by determining the gradient of J in parameter space, the procedure commonly followed is to adjust it at every instant based on the error at that instant. A single presentation of every pattern in the data set to the network is referred to as an *epoch*. In the literature, a well-known method for determining this gradient for MLP networks is the backpropagation method. The analytical method of deriving the gradient is well known in the literature and will not be repeated here. It can be shown that the back-propagation method leads to the following gradients for any MLP network with L layers:

$$\frac{\partial J(s)}{\partial w_{ij}^{(l)}(s)} = -\delta_i^{(l)}(s) y_j^{(l-1)}(s) \qquad \text{Equation (17.5)}$$

$$\delta_i^{(L)}(s) = e_i \gamma_i'\left(\bar{y}_i^{(L)}(s)\right) \text{ for neuron } i \text{ in output layer } L \qquad \text{Equation (17.5a)}$$

$$\delta_i^{(l)}(s) = \gamma_i'\left(\bar{y}_i^{(l)}(s)\right) \sum_k \delta_k^{(l+1)}(s) w_{ki}^{(l+1)}(s) \text{ for neuron } i \text{ in hidden layer } l \qquad \text{Equation(17.5b)}$$

Here, $\delta_i^{(l)}(s)$ denotes the local gradient defined for neuron i in layer l and the use of prime in $\gamma_i'(\bar{y}_i^{(L)}(s))$ signifies differentiation with respect to the argument. It can be shown that for a *unipolar* sigmoid function, $g'(x) = x(1 - x)$ and for a *bipolar* function, $g'(x) = 2x(1 - x)$. One starts with local gradient calculations for the outermost layer and proceeds backwards until one reaches the first hidden layer (hence the name backpropagation). For more information on MLP networks, see Haykin [1999].

17.2.1.3 Iterative Inversion of Neural Networks

In error backpropagation training of neural networks, the output error is "propagated backward" through the network. Linden and Kindermann [1989] have shown that the same mechanism of weight learning can be used to iteratively invert a neural network model. This approach is used here for on-line parameter design and hence the discussion. In this approach, errors in the network output are ascribed to errors in the network input signal, rather than to errors in the weights. Thus, iterative inversion of neural networks proceeds by a gradient descent search of the network input space, while error backpropagation training proceeds through a search in the synaptic weight space.

Through iterative inversion of the network, one can generate the input vector, $\mathbf{x}$, that gives an output as close as possible to the desired output, $\mathbf{y}_d$. By taking advantage of the duality between the synaptic weights and the input activation values in minimizing the performance criterion, the iterative gradient descent algorithm can again be applied to obtain the desired input vector:

$$\mathbf{x}(s+1)=\mathbf{x}(s)-\eta\cdot\frac{\partial J(s)}{\partial \mathbf{x}(s)} \qquad \text{Equation (17.6)}$$

where η is a positive constant that determines the rate of iterative inversion and the superscript refers to the iteration step. For further information, see Linden and Kindermann [1989].

17.2.2 Genetic Algorithms

Genetic algorithms (GAs) are a class of stochastic optimization procedures that are based on natural selection and genetics. Originally developed by John H. Holland [1975], the genetic algorithm works on a population of solutions, also called individuals, represented by fixed bit strings. Although there are many possible variants of the basic GA, the fundamental underlying mechanism operates on a population of individuals, is relatively standard, and consists of three operations [Liepins and Hilliard, 1989]: (i) evaluation of individual fitness, (ii) formation of a gene pool, and (iii) recombination and mutation, as illustrated in Figure 17.5(a). The individuals resulting from these three operations form the next generation's population. The process is iterated until the system ceases to improve. Individuals contribute to the gene pool in proportion to their relative fitness (evaluation on the function being optimized); that is, well performing individuals contribute multiple copies, and poorly performing individuals contribute few copies, as illustrated in Figure 17.5(b). The recombination operation is the crossover operator: the simplest variant selects two parents at random from the gene pool as well as a crossover position. The parents exchange "tails" to generate the two offspring, as illustrated in Figure 17.5(c). The subsequent population consists of the offspring so generated. The mutation operator illustrated in Figure 17.5(d) helps assure population diversity, and is not the primary genetic search operator. A thorough introduction to GAs is provided in Goldberg [1989].

Due to their global convergence behavior, GAs are especially suited for the field of continuous parameter optimization [Solomon, 1995]. Traditional optimization methods such as steepest-descent, quadratic approximation, Newton method, etc., fail if the objective function contains local optimal solutions. Many papers suggest (see, for example, Goldberg [1989] and Mühlenbein and Schlierkamp-Voosen [1994]) that the presence of local optimal solutions does not cause any problems to a GA, because a GA is a multipoint search strategy, as opposed to point-to-point search performed in classical methods.

17.3 Design of Quality Controllers for On-Line Parameter Design

The proposed framework for performing on-line parameter design is illustrated in Figure 17.6. In contrast to the classical control theory approaches, this structure includes two distinct control loops. The process control loop "maintains" the controllable variables at the optimal levels, and will involve schemes such as feedback control, feedforward control, and adaptive control. It is the quality controller in the quality control loop that "determines" these optimal levels, i.e., performs parameter design. The quality controller includes both a model of the product/process quality response characteristics and an optimization routine to find the optimal levels of the controllable variables. As was stated earlier, the focus here is on time-invariant products and processes, and hence, the model building process can be carried out off-line. In time-variant systems the quality response characteristics have to be identified and constantly tracked on-line, and call for an experiment planner that facilitates constant and optimal investigation of the product/process behavior.

In solving this on-line parameter design problem, the following assumptions are made:

1. *Quality response characteristics of interest can be expressed as static nonlinear maps in the input space (the vector space defined by controllable and uncontrollable variables).* This assumption implies that there exits no significant memory or inertia within the system, and that the process response state is strictly a function of the "current" state of the controllable and uncontrollable variables. In other

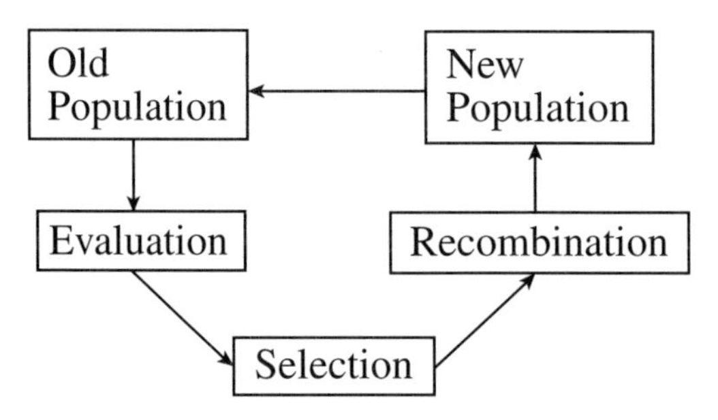

a) Basic genetic algorithm cycle.

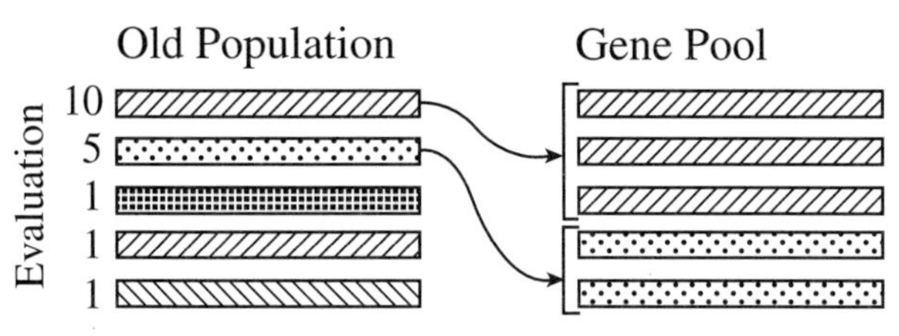

b) Evaluation and contribution to the gene pool.

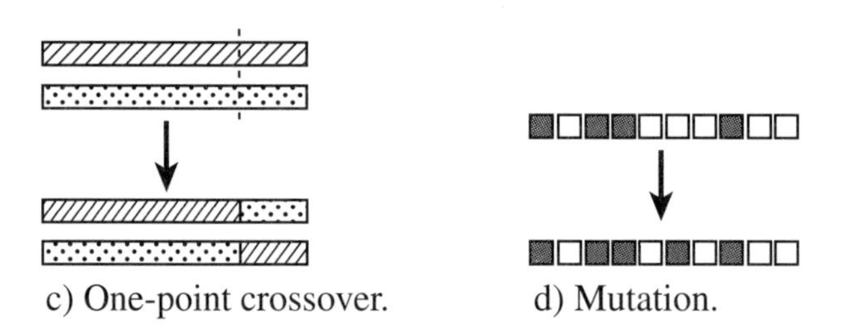

c) One-point crossover. d) Mutation.

FIGURE 17.5 The fundamental cycle and operations of a basic genetic algorithm.

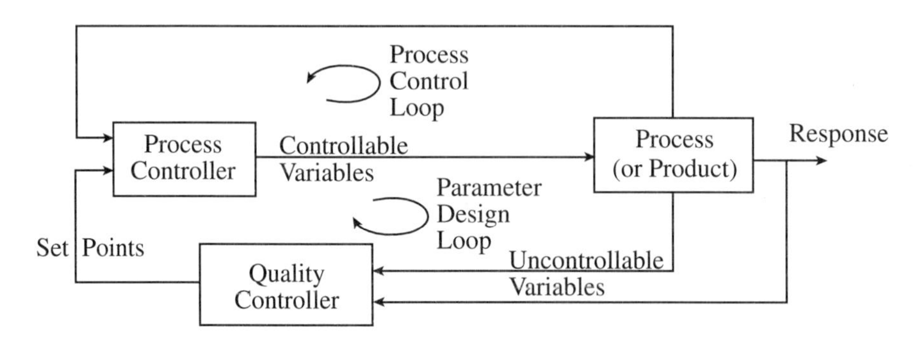

FIGURE 17.6 Proposed framework for performing on-line parameter design.

words, the response is neither dependent on the history of the levels of the controllable/uncontrollable variables nor dependent on past process response history.

2. *The process time constant is relatively large in comparison with the rate of change of uncontrollable variables.* The assumption implies that there exists enough time to respond to changes in the levels of the uncontrollable variables (i.e., perform parameter design). If the rate of change is too high, one ends up constantly chasing the uncontrollable variables and may not be able to enhance

product/process quality. In fact, if the rate of change of uncontrollable variables is relatively high in comparison with the process time constant, attempting to perform on-line parameter design might even deteriorate product/process quality in the long term.

3. *Uncontrollable variables are observable during production and unit operation.* The need for this assumption is rather obvious. If the vast majority of significant uncontrollable variables are unobservable during product/process operation, one has to resort to off-line robust parameter design strategies.
4. *Uncontrollable variables are autocorrelated and change smoothly over time.* This assumption is critical if one cannot or does not desire to significantly change the levels of the non-fixed controllable variables in a more or less random fashion within different ranges. For example, in a pultrusion process that produces reinforced composite material, normally one cannot quickly change the temperature of the pultrusion die (given the large mass and specific and thermal conductivity properties of most die materials). However, given "adequate" time, one can control the temperature of the die to follow a relatively smooth desired trajectory.
5. *Scales for the controllable variables and response variables are assumed to be continuous.* This constraint is necessary if one desires to work with gradient search techniques in the non-fixed controllable variable space in performing parameter design. The assumption can be relaxed by using traditional mixed-integer programming methods and their variants to perform parameter design. Please note that it is certainly possible to perform product/process identification (model building) using neural networks even in the event that certain controllable variables and response variables are noncontinuous. Also, genetic algorithms discussed in Section 17.2.2 are extremely popular for their ability to solve mixed-integer programming type problems.

Throughout the rest of this chapter, the focus will be on on-line parameter design of time-invariant static products and processes using artificial neural networks for product/process modeling. Performing on-line parameter design of dynamic systems is not as challenging as it appears originally; however, the task of dealing with time-variant products/processes is truly daunting. Future research efforts will focus on extending the proposed on-line parameter design methods to time-variant dynamic systems.

Given the above discussion, the role of a quality controller can be broken into two distinct tasks: product/process identification and product/process parameter design.

17.3.1 Identification Mode

Let $\mathbf{x} = (x_1, \ldots, x_K, x_{K+1}, \ldots, x_N)^T$ be a column vector of K controllable variables, x_1 through x_K, and N-K uncontrollable variables, x_{K+1} through x_N, where $K \leq N$. Let $\mathbf{y} = (y_1, \ldots, y_M)^T$ be a vector of M quality response characteristics of interest. The quality vector $\mathbf{y}$ is a function of x_1 through x_N, and hence can be written (for time-invariant systems) as

$$\mathbf{y} = \mathbf{f}(\mathbf{x}) \qquad \text{Equation (17.7)}$$

Here, $\mathbf{f}(\mathbf{x}) = (f_1(\mathbf{x}), \ldots, f_M(\mathbf{x}))^T$ denotes a column vector of functions, where $y_i = f_i(\mathbf{x})$ for $i = 1, \ldots, M$. In most cases, due to economic, time, and knowledge constraints, there exists no accurate mechanistic model for $\mathbf{f}$, and it has to be estimated in an empirical fashion. We recommend MLPs for modeling $\mathbf{f}$, given their universal approximation properties [Hornik et al., 1989] and extreme success discussed in the literature with regard to accurate approximation of complex nonlinear functions [Haykin, 1999].

In contrast to some pattern recognition problems and other function approximation problems, in general, off-line planning, design, and execution of experiments for modeling product/process response characteristics can be very time-consuming and expensive. At the initial stage, it is not uncommon to see fractional factorial designs being utilized for screening significant controllable and uncontrollable variables. Even second phase experiments tend to use some form of a central composite design, typically used for empirical modeling of response surfaces. The point here is that the size of the data set normally available for product/process identification is very limited. This makes division of the data set between training and testing more difficult, but does not prevent it. As the name implies, a training data set will be used for training the MLP network to approximate **f** from Equation 17.7 as follows:

$$\tilde{\mathbf{f}}(\mathbf{x}) \approx \mathbf{f}(\mathbf{x}), \quad \text{Equation (17.8)}$$

such that

$$J_I = \left\| \tilde{\mathbf{f}}(\mathbf{x}) - \mathbf{f}(\mathbf{x}) \right\|_{Identification} \leq \varepsilon \quad \text{Equation(17.9)}$$

for some specified constant $\varepsilon \geq 0$ and a suitably defined norm (denoted by $\|\cdot\|_{Identification}$). The testing data set will facilitate evaluation of the generalization characteristics of the network, i.e., the ability of the network to perform interpolations and make unbiased predictions. We use an S-fold cross-validation method [Weiss and Kulikowski, 1991] for designing (i.e., determining the architecture in terms of number of hidden layers, nodes per different hidden layers, and connectivity) and building MLP models for approximating quality response characteristics. This involves dividing the data set into S mutually exclusive subsets, using $S - 1$ subsets for training the network (as discussed in Section 17.2.1.2) and the remaining subset for testing, repeating the training process S times, holding out a different subset for each run, and totaling the resulting testing errors. The performance of the different network configurations under consideration will be compared using the S-fold cross-validation error, and the network with the least error will be used for product/process identification. Once the optimal configuration has been identified, the complete data set can be used for training the final network. Several other guidelines are discussed in the literature regarding selection of potential network configurations and their training, and are not repeated here [Haykin, 1999; Weigand et al., 1992; Solla, 1989; Baum and Haussler, 1989].

17.3.2 On-Line Parameter Design Mode

Once the product/process identification is completed, parameter design can be performed on-line using the MLP model, $\tilde{\mathbf{f}}(\mathbf{x})$. Let $\mathbf{y}_d = (y_{d1}, \ldots, y_{dM})^T$ denote the vector of M desired/target quality response characteristics of interest. The objective is to determine the optimal levels for the K controllable variables, x_1 through x_K, to

$$\text{minimize } J_{PD} = \left\| \mathbf{y} - \mathbf{y}_d \right\|_{ParameterDesign} = \left\| \mathbf{f}(\mathbf{x}) - \mathbf{y}_d \right\|_{ParameterDesign}, \quad \text{Equation (17.10)}$$

for a suitably defined norm (denoted by $\|\cdot\|_{ParameterDesign}$) on the output space. In Equation 17.10, $\mathbf{f}(\mathbf{x})$ denotes the output of the product/process and hence $\mathbf{f}(\mathbf{x}) - \mathbf{y}_d \equiv e_d$ is the difference between the product/process output and the desired output $\mathbf{y}_d$. In the absence of any knowledge about $\mathbf{f}(\mathbf{x})$, the objective is to minimize the performance criterion

$$J_{PD} = \left\| \mathbf{y} - \mathbf{y}_d \right\|_{ParameterDesign} \cong \left\| \tilde{\mathbf{f}}(\mathbf{x}) - \mathbf{y}_d \right\|_{ParameterDesign}, \quad \text{Equation (17.11)}$$

The constraints would be those restricting the levels of the controllable variables to an acceptable domain.

17.3.2.1 Iterative Inversion Method

This section introduces an iterative inversion method to determine the optimal controllable variable levels. The approach utilizes the MLP product/process model from the identification phase.

As discussed in Section 17.2.1.3, through iterative inversion of the network, one can generate the optimal controllable variable input vector, $[x_1, \ldots, x_K]$, that gives an output as close as possible to y_d. By taking advantage of the duality between the synaptic weights and the input activation values in minimizing the performance criterion, J_{PD}, the iterative gradient descent algorithm can be applied to obtain the desired input vector:

$$x_j(s+1) = x_j(s) - \eta \cdot \frac{\partial J_{PD}(s)}{\partial x_j(s)} + \alpha\big(x_j(s) - x_j(s-1)\big) \text{ for } 1 \le j \le .K.$$

Equation (17.12)

Here s denotes iteration step, η and α are the rates for inversion and momentum, respectively, in the gradient descent approach. If the least-mean-square criterion was used as the performance criterion, for any MLP network, a derivation that parallels backpropagation algorithm will lead to the following gradient:

$$\frac{\partial J_{PD}(s)}{\partial x_j(s)} = -\sum_{i}^{all} \delta_i^{(1)}(s) w_{ij}^{(1)}(s) \qquad \text{Equation (17.13)}$$

The iterative inversion is performed until the controllable variables converge:

$$\frac{\partial J_{PD}(s)}{\partial x_j(s)} \approx 0; \quad x_j(s+1) \approx x_j(s). \qquad \text{Equation (17.14)}$$

If J_{PD} meets all the criteria for a strictly convex function in the input domain, gradient descent techniques lead to a global optimal solution. However, if J_{PD} is not a convex function, the iterative inversion method, being a gradient descent technique by definition, leads to a local optimal solution. Hence, it is necessary that the quality controller search through all the basins (multiple basins might exist in the case of nonconvex energy functions) to locate the global optimal levels for the controllable variables. Under the assumption that the step sizes taken along the negative gradient (a function of η and α as shown in Equation 17.12) are not large enough to move into a different basin, the quality controller converges to the local minimum in the basin holding the starting point. However, it is not difficult to incorporate a simulated annealing module (or other enhanced optimization techniques) to converge toward a global optimal solution.

17.3.2.2 Stochastic Search Method

Here we utilize the genetic-algorithm-driven search method discussed in Section 17.2.2 to determine the optimal controllable variable levels. Once again, the approach utilizes the MLP product/process model from the identification phase.

As discussed in Section 17.2.2, utilizing a genetic algorithm for searching the controllable variable space, one can generate the optimal controllable variable input vector, $[x_1, \ldots, x_K]$, that gives an output as close as possible to $\mathbf{y}_d$. In essence, (i) J_{PD} from Equation 17.12 will play the role of the fitness evaluation function, and (ii) individual solutions at any given generation are represented by chromosomes or floating point strings of length K (equal to the number of controllable variables, x_1 through x_K). Factors that need to be determined include the population size for any single generation, number of generations involved in the search, and the nature of recombination and mutation operators. The factor selection process significantly impacts the quality of parameter design and the associated computational complexity.

17.4 Case Study: Plasma Etching Process Modeling and On-Line Parameter Design

To facilitate evaluation of the proposed on-line intelligent quality controllers (IQCs), we chose to work with a semiconductor fabrication process, in particular, an ion-assisted plasma etching process (used for removing layers of materials through AC discharge). The process is inherently complex and is popular in semiconductor manufacturing industry. The raw data for this case study comes from May et al., [1991] and Himmel and May [1993]. The primary focus of their papers is on efficient and effective modeling of plasma etching processes using statistical response surface models and artificial neural networks, with the intent of utilizing the models for recipe generation, process control, and equipment malfunction diagnosis. They report that plasma modeling from a fundamental physical standpoint has had limited success [Himmel and May, 1993]. They state that most physics-based models attempt to derive self-consistent solutions to first-principle equations involving continuity, momentum balance, and energy balance inside a high-frequency, high-intensity electric field (normally accomplished through expensive numerical simulation methods that are subject to many simplifying assumptions and tend to be unacceptably slow). They also state that the complexity of practical plasma processes at the equipment level is presently ahead of theoretical comprehension, and hence, most practical efforts have focused on empirical approaches to plasma modeling involving response surface models [May et al., 1991; Riley and Hanson, 1989; Jenkins et al., 1986] and neural network models [Himmel and May, 1993; Kim and May, 1994; Rietman and Lory, 1993]. The next two sections briefly discuss the experimental technique (Section 17.4.1) and the experimental design (Section 17.4.2) used by May, Huang, and Spanos [1991] in collecting the data. The sections that follow 17.4.1 and 17.4.2 discuss the implementation of on-line parameter design methods proposed in Section 17.3.

17.4.1 Experimental Technique

The study focuses on the etch characteristics of n^+-doped polysilicon using carbon tetrachloride as the etchant. May, Huang, and Spanos [1991] performed the experiment on a test structure designed to facilitate the simultaneous measurement of etch rates of polysilicon, SiO_2, and photoresist. Test patterns were fabricated on 4-in. diameter silicon wafers. Approximately 1.2 μm of phosphorous-doped polysilicon was deposited over 0.5 μm of thermal SiO_2 by low-pressure chemical vapor deposition. The thick layer of oxide was grown to prevent etching through the oxide by the less selective experimental recipes. Poly resistivity was measured at 86.0 Ω-cm. Oxide was grown in a steam ambient at 1000°C. One micron of Kodak 820 photoresist was spun on and baked for 60 s at 120°C.

The etching apparatus used by May and colleagues [1991] consisted of a Lam Research Corporation Autotech 490 single-wafer parallel-plate system operating at 13.56 MHz. Film thickness measurements were performed on five points per wafer using a Nanometrics Nanospec AFT system and an Alphastep 200 Automatic Step Profiler. Vertical etch rates were calculated by dividing the difference between the pre- and post-etch thickness by the etch time. Expressions for the selectivity of etching poly with respect to oxide (S_{ox}) and with respect to resist (S_{ph}) are percent nonuniformity (U), respectively, are given below:

$$S_{ox} = \frac{R_p}{R_{ox}} \qquad \text{Equation (17.15)}$$

$$S_{ph} = \frac{R_p}{R_{ph}} \qquad \text{Equation (17.16)}$$

TABLE 17.1 Ranges of Input Factors

Parameter	Range	Units
Pressure (P)	200–300	W
RF Power (Rf)	300–400	mtorr
Electrode Gap (G)	1.2–1.8	cm
CCl_4 flow	100–150	sccm
He flow	50–200	sccm
O_2 flow	10–20	sccm

$$U = \frac{\left|R_{pc} - R_{pe}\right|}{R_{pc}} * 100 \qquad \text{Equation (17.17)}$$

where R_p is the mean vertical poly etch rate over the five points, R_{ox} is the mean oxide etch rate, R_{ph} is the mean resist etch rate, R_{pc} is the poly etch rate at the center of the wafer, and R_{pe} is the mean poly etch rate of the four points located about 1 in. from the edge. The overall objectives are to achieve high vertical poly etch rate, high selectivities, and low nonuniformity. For a detailed discussion of the process, see May et al., [1991].

17.4.2 Experimental Design

Of the nearly dozen different factors that have been shown to influence plasma etch behavior in the literature, the study by May, Huang, and Spanos [1991] focused on the following parameters, regarded as the most critical: chamber pressure (*P*), RF power (*Rf*), electrode spacing (*G*), and the gas flow rate of CCl_4. The primary etchant gas is CCl_4, but He and O_2 are added to the mixture to enhance uniformity and reduce polymer deposition in the process chamber, respectively. The six input factors and their respective ranges of variation are shown in Table 17.1.

The experiments were conducted in two phases at the Berkeley Microfabrication Laboratory. In the first phase (screening experiment), a 2^{6-1} fractional factorial design requiring 32 runs was performed to reduce the experimental budget. Experimental runs were performed in two blocks of 16 trials, each in such a way that no main effects or first-order interactions were confounded. Three center points were also added to check the model for nonlinearity. Analysis of the first stage of the experiment revealed significant nonlinearity, and showed that all six factors are significant [May et al., 1991]. In order to obtain higher order models, the original experiment is augmented with a second experiment, which employed a central composite circumscribed (CCC) Box–Wilson design [Box et al., 1978]. In this design, the two-level factorial box was enhanced by further replicated experiments at the center as well as symmetrically located star points. In order to reduce the size of the experiment and combine it with results from the screening phase, a half replicate design was again employed. The entire second phase required 18 additional runs. In total, there were 53 data points.

17.4.3 Process Modeling Using Multilayer Perceptron Networks

The task here is to design and train a neural network to recognize the interrelationships between the process input variables and outputs, using the 53 input–output data pairs provided by May et al. [1991]. Experimental investigation has revealed that an MLP network with a single hidden layer can adequately model the input–output relationships of the process. An MLP network with 6 input nodes (matching the 6 process input factors), 12 nodes in the hidden layer (using a bipolar sigmoid nodal function in the hidden layer), and 4 nodes in the output layer (matching the 4 process outputs and carrying a linear nodal function), trained using the standard backpropagation algorithm, proved to be optimal (optimal based on a full-factorial design considering multiple hidden nodes per layer, multiple learning rates, and multiple nodal functions), and yielded very good prediction accuracy. More information regarding the neural network configuration and training scheme is available in Tables 17.2 and 17.3, respectively.

TABLE 17.2 MLP Neural Network Configuration

Layer	Nodes Per Layer	Nodal Function	Data Scaling[a]
Input	6	Not relevant	Yes (−1 to +1)
Hidden	12	Bipolar sigmoid	Not relevant
Output	4	Linear	Yes (−1 to +1)

[a] Before presenting the data as inputs and desired outputs to the neural network, it is scaled to a level easily managed by the network. In general, this facilitates rapid learning, and more importantly, gives equal importance to all the outputs in the network during the learning process (eliminating the undue influence of differing amplitudes and ranges of the outputs on the training process).

TABLE 17.3 MLP Neural Network Training Scheme

Training algorithm	Backpropagation
Starting learning rate (η)	0.1
Learning adaptation rate	–10% (reduction)
Minimum learning rate	0.00001
Starting momentum (α)	0.001
Momentum adaptation rate	+15% (growth)
Maximum momentum	0.95
Parameter adaptation frequency	250 epochs
Maximum training epochs[a]	20,000

[a] The training phase is also terminated if the percentage change of error with respect to training time/epochs is too small (<0.01% over 500 epochs) or if the training error is consistently increasing (>1% over 50 epochs).

TABLE 17.4 Performance Comparison of Neural Network Model vs. RSM Model

Output	Sqrt(MSE_{RSM})	Sqrt(MSE_{NN})	% Improvement
Rp	306.45 Å/min	114.76 Å/min	62.6
U	6.60 [%]	2.63 [%]	60.2
S_{ox}	0.90	0.50	44.4
S_{ph}	0.26	0.09	65.4

Table 17.4 compares the performance of the neural network model with quadratic response surface method (RSM) models reported by May et al. [1991] built using the same 53 data points, in terms of square root of the residual mean square error (MSE) for each response. MSE is calculated as follows for any given response y_i:

$$MSE_i = \frac{1}{D}\sum_{d=1}^{D}\left(y_i(d) - \hat{y}_i(d)\right)^2 \qquad \text{Equation (17.18)}$$

where D is the number of experiments (data points), $y_i(d)$ is the measured value, and (d) is the corresponding model prediction, for data point d. Figure 17.7 illustrates the neural network learning curve, plotting the training epoch number against the total residual mean squared error (totaled over all the four network outputs). As is normally observed, learning is very rapid initially, where network parameters, i.e., weights, change rapidly from the starting random values toward approximate "optimal" final values. After this phase, the network weights go through a fine-tuning phase for a prolonged period to accurately model the input–output relationships present in the data. Figure 17.8 provides "goodness-of-fit" plots, which depict the neural network predictions vs. actual measurements. In these plots, perfect model predictions lie on the diagonal line, whereas scatter in the data is indicative of experimental error and bias in the model.

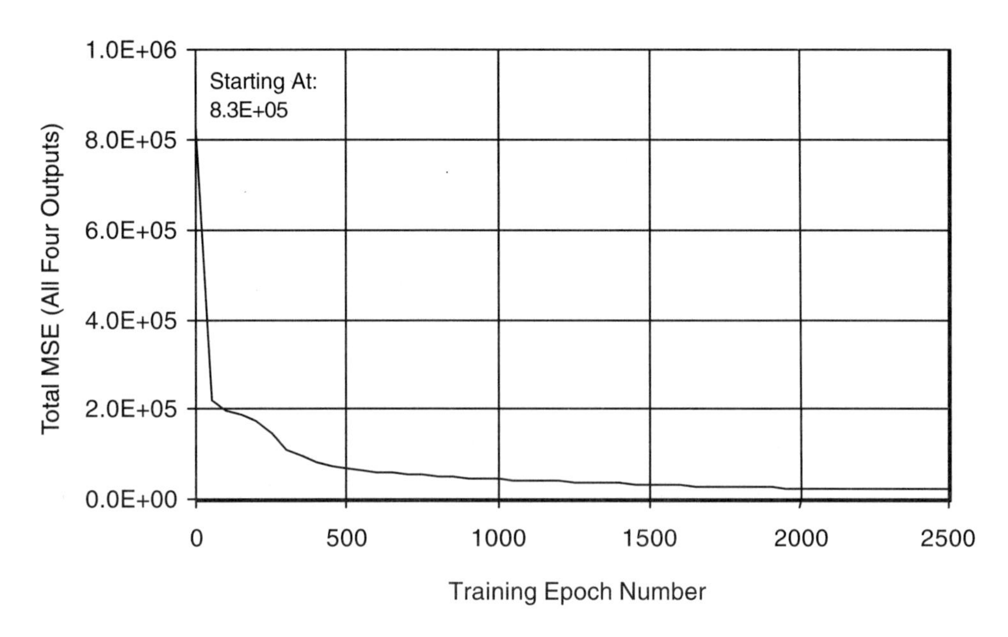

FIGURE 17.7 Neural network learning curve.

17.4.4 On-Line Process Parameter Design Using Neural Network Model

To illustrate and evaluate the performance of the proposed IQCs, we simulate the manufacturing process, using the established MLP neural network model, with varying degrees of fluctuation in the uncontrollable variables. Here the three process gas flow rates (that of CCl_4, He, and O_2) are treated as uncontrollable variables, with different degrees of uncontrollability during different simulations. The strategy involves evaluating the performance of the IQCs in the form of average deviation from desired target process outputs in standard deviations. Here, the standard deviations for the four different process outputs are calculated from the 53 point data set. To facilitate comparison of the performance of the IQCs with traditional off-line parameter design approaches, we work with a method labeled pSeudo Parameter Design (SPD) approach. The idea behind the SPD approach is to determine the "best" combination of controllable variable settings that in the long run lead to the least deviation from desired target process outputs, in light of the variation in the uncontrollable variables. In other words, the focus is to determine the levels for the controllable variables that are robust to variation in uncontrollable variables and minimize the "expected" deviations from desired target process outputs. Here, we determine the best settings for the controllable variables through process simulation using some sort of an experimental design in the controllable variable space.

17.4.4.1 Establishing Target Process Outputs

For the plasma etching process at hand, as was stated earlier, the overall objectives are to achieve high vertical poly etch rate (R_p), low nonuniformity (U), high oxide selectivity (S_{ox}), and high resist selectivity (S_{ph}). The optimum etch recipe that will lead to best etch responses was determined using the iterative inversion scheme and allowing all the six process input factors to be controllable. We utilized the iterative inversion method for locating the optimal process parameters. As was mentioned earlier, since the neural network is trained using scaled outputs, the iterative inversion process gives equal importance to all the process outputs during optimization (eliminating the undue influence of differing amplitudes and ranges of the outputs). A comparison between the standard recipe (normally used for plasma etching) and the optimized recipe appears in Table 17.5. Estimated etch responses for the standard and optimal recipes were determined using the neural network process model. Notably, significant improvement was to be

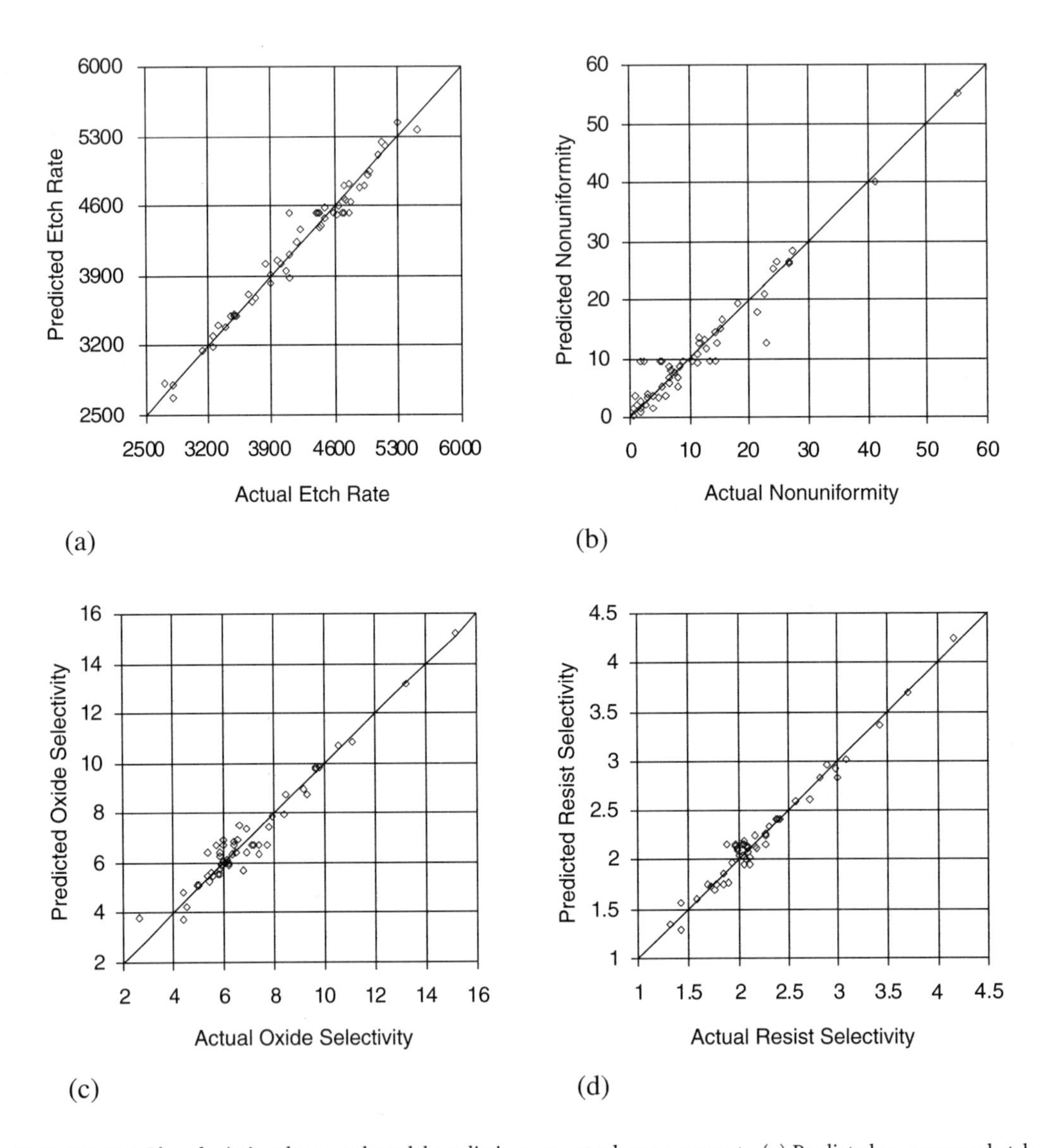

FIGURE 17.8 Plots depicting the neural model predictions vs. actual measurements. (a) Predicted vs. measured etch rate. (b) Predicted vs. measured nonuniformity. (c) Predicted vs. measured oxide selectivity. (d) Predicted vs. measured resist selectivity.

TABLE 17.5 Standard and Optimized Etch Recipe

Parameter	Standard Recipe	Optimized Recipe
Pressure (mtorr)	300	300
RF power (W)	280	300
Electrode spacing (cm)	1.5	1.43
CCl_4 flow (sccm)	130	150
He flow (sccm)	130	50
O_2 flow (sccm)	15	10

TABLE 17.6 Estimated Standard and Optimized Responses

Response	Standard Recipe	Optimized Recipe	% Change
Rp	4100.00 Å/min	4663.33 Å/min	13.74
U	12.17 [%]	9.11 [%]	–25.14
S_{ox}	9.26	15.38	66.09
S_{ph}	3.10	4.90	58.06

obtained in all the four process responses. After the optimum recipe was determined, an experiment was undertaken to confirm the improvement of the etch responses. In this experiment, six wafers were identically prepared and divided into two equal groups and were approximately subjected to standard and optimized treatments. The results were consistent with the estimations. Hence, during the evaluation of the performance of IQCs in comparison with SPD, it would be best to attempt to constantly achieve the optimized response (i.e., the optimized response values shown in Table 17.6 will be used as targets) in spite of variation in the uncontrollable variables (i.e., the three gas flow rates).

17.4.4.2 Simulation of Uncontrollable Variables

In general, the process gas flow rates tend to exhibit strong autocorrelation with respect to time. Here, we simulate the gas flow rates as an autocorrelated process known as an AR(1) process, that carries the following model:

$$x(t) = \phi x(t-1) + \varepsilon_t \qquad \text{Equation (17.19)}$$

where t denotes time, and the ε_t's are *iid* normal with zero mean and variance of σ_ε^2. The value of ϕ has to be restricted within the open interval of (–1, 1) for the AR(1) process to be stationary. In fact, additional simulations treating the gas flow rates as a random walk and other ARMA models led to results similar to those reported here.

Here, the simulations were conducted by setting ϕ at 0.9. As was stated earlier, for evaluation of the performance of the proposed IQC, we need to simulate the manufacturing process with varying degrees of fluctuation in the uncontrollable variables. The strategy here is to generate the AR(1) process data by setting σ_ε equal to one, and then linearly scaling the generated data to the desired range of variation. The degree of variation is allowed to be between zero and one, where zero denotes no change in the level of the uncontrollable variable and one denotes the case where the range of the generated data spans the tolerated range for the particular variable, shown in Table 17.1. It is important to note that when the degree of variation is set to zero, the final levels generated should coincide with the optimal recipe levels, and any increase in the degree of variation will oscillate the levels of the uncontrollable variables in and around the optimal recipe levels. Figure 17.9 illustrates the data generated for CCl_4 gas flow rate at different degrees of variation. Note that when the degree of variation is set to zero, the values match with the optimized recipe level (150 sccm) for the variable (CCl_4). All the simulations discussed here involve 10,000 discrete time instants, and the starting random seeds are different for the three AR(1) processes for the uncontrollable variables. This ensures that the patterns are not the same for all the three uncontrollable variables even if the degree of variation is set the same for any given simulation. The degrees of variation chosen for evaluation of the proposed IQC are as follows: 0.001, 0.005, 0.01, 0.02, 0.03, 0.04, 0.05, 0.1, 0.25, 0.5, 0.75, and 1.0.

17.4.4.3 Comparison of Performance of IQCs and SPD

For the different simulations, the iterative inversion for the IQC, referred to from now on as IQC_{II}, is performed using an iterative inversion rate of 0.05 allowing a maximum of 5000 iterative inversions at any discrete time instant. Additional information regarding the iterative inversion scheme used by the IQC_{II} is available in Table 17.7. The GA search for the IQC, referred to from now on as IQC_{GA}, is performed by allowing 50 individuals per generation and 250 generations for the complete search.

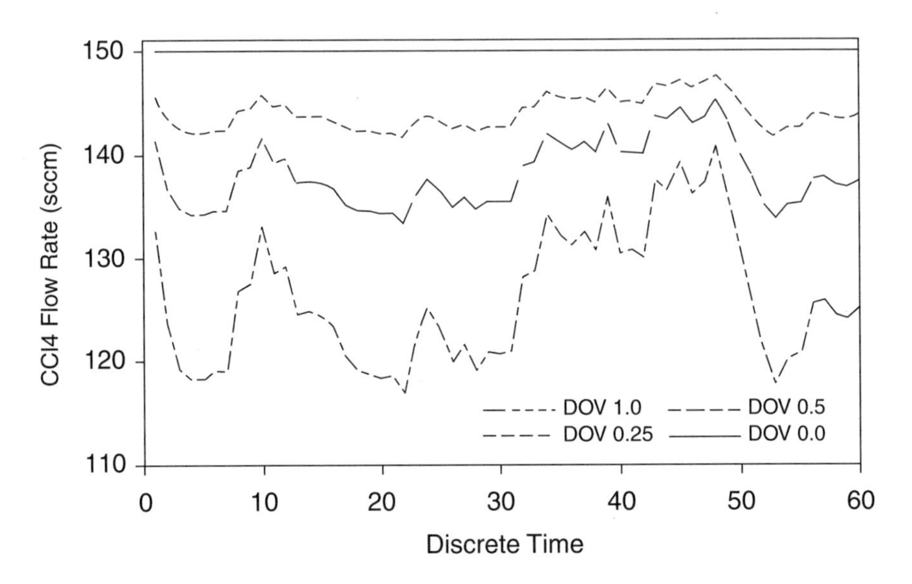

FIGURE 17.9 Simulation of CCl_4 flow rate with different degrees of variation (DOV).

TABLE 17.7 IQC, Iterative Inversion Scheme Parameters

Starting iterative inversion rate	0.05
Inversion adaptation rate	–10% (reduction)
Minimum inversion rate	0.00001
Iterative inversion momentum	0.0
Parameter adaptation frequency	25 iterations
Maximum inversion iterations[a]	5,000
Number of iterative inversion starting points	9[b]

[a] The iterative inversion is also terminated if the percentage change or error with respect to time is too small (<0.1% over 10 iterations) or if the error is consistently increasing (>1% over 50 iterations).

[b] Two levels per controllable variable (dividing the range into three equal parts), leading to 2^3 combinations for the three controllable variables. An additional starting point is the center of the overall search space.

TABLE 17.8 IQC, Genetic Algorithm Scheme Parameters

Individual per generation	50
Maximum number of generations	250
Mutation scale factor	30%

Note: These parameters are optimized by conducting a full factorial design. In these experiments, the number of individuals per generation was allowed to vary from 25 to 75 in steps of 25, the maximum number of generations was allowed to vary between 100 and 500, and the mutation operator was allowed to have an impact between 1% and 50%.

Additional information regarding the GA scheme used by the IQC_{GA} is available in Table 17.8. For SPD, the potential combinations for the levels of the controllable inputs were generated using a full factorial design with two different resolutions per variable, five and eight, resulting in 5^3 and 8^3 combinations. All the combinations are evaluated with the 10,000 point data set generated for each degree of variation, and the combination that leads to best overall performance is picked to represent the performance of SPD. What constitutes "best" overall performance is measured in terms of an "average deviation from target" metric defined as follows:

$$\frac{1}{M}\sum_{i=1}^{M}\sum_{t=0}^{T} w\left[i\right]\cdot\left|y_{di}\left(t\right)-\hat{y}_{oi}\left(t\right)\right| \qquad \text{Equation (17.20)}$$

where $y_{di}(t)$ denotes the desired target output for process output i at time instant t, $\hat{y}_{oi}(t)$ denotes the corresponding optimal output level achieved using iterative inversion scheme, $w[i]$ denotes the weight assigned to process output i, T denotes the length of the simulation run in terms of discrete time instants, and M denotes the number of process outputs. The weights for calculating the average deviation are defined as follows:

$$w\left[i\right]=\frac{1}{\sigma_{y_i}} \qquad \text{Equation (17.21)}$$

where σ_{yi}, denotes the standard deviation of the ith process output determined from the complete experimental data set. Such a weight definition facilitates a relatively fair calculation of the combined average deviation in standard deviations.

The performance comparisons of the IQCs and SPD schemes in terms of average deviation from target for different degrees of variation in the uncontrollable variables, shown in Figures 17.10 and 17.11, clearly illustrate the ability of IQCs to significantly reduce the average deviation from target, when feasible. In addition, as expected, the IQCs (both the IQC_{II} and IQC_{GA}) always outperform the SPD approach. Obviously, the improvement in general will depend on the particular process at hand and its sensitivity to deviations in uncontrollable variables. For illustrative purposes, Figure 17.12 illustrates the performance of the IQC_{II} when the degree of variation in the uncontrollable variables is set at 0.1. Note that RF power and pressure remained the same throughout the first 251 discrete time instants shown in the figure (of the 10,000 time instant simulation). This is attributed to the fact that the iterative inversion procedure was suggesting that RF power be increased beyond 300 W and pressure beyond 300 mtorr; however, these levels already represent the boundary of the acceptable range for these variables (see Table 17.1). With respect to computational complexity, for the simulations discussed above, on the average the processing time at any time instant was on the order of 0.01 s on a Pentium II 300 MHz processor for iterative inversion (this includes the 10 to 100 iterations necessary on the average to converge toward the optimized controllable variables for each of the iterative inversion starting points) and 1 s for GA search. Even though the iterative inversion method (with multiple starting points) and the GA method can lead to solutions of equal quality (as evidenced for the plasma etching case in Figure 17.10), due its superiority with regard to computational complexity, we recommend the iterative inversion method for products/processes that exhibit relatively smooth quality cost surfaces.

17.5 Conclusion

Off-line parameter design and robust design techniques have gained a lot of popularity over the last decade. The methods introduced here allow us to extend these off-line techniques to be performed on-line. In particular, the methods proposed account for extra information available about observable uncontrollable factors in products and processes. All the methods introduced are compatible with traditional statistical modeling approaches (such as response surface models) for modeling product/process behavior. However, we recommend feedforward neural networks for modeling the quality response characteristics due to their nonparametric nature, strong universal approximation properties, and compatibility with adaptive systems. An iterative inversion scheme and a genetic algorithm scheme are proposed for on-line optimization of controllable variables. Once again, note that the proposed methods are compatible with traditional linear and nonlinear optimization methods popular in the domain of operations research.

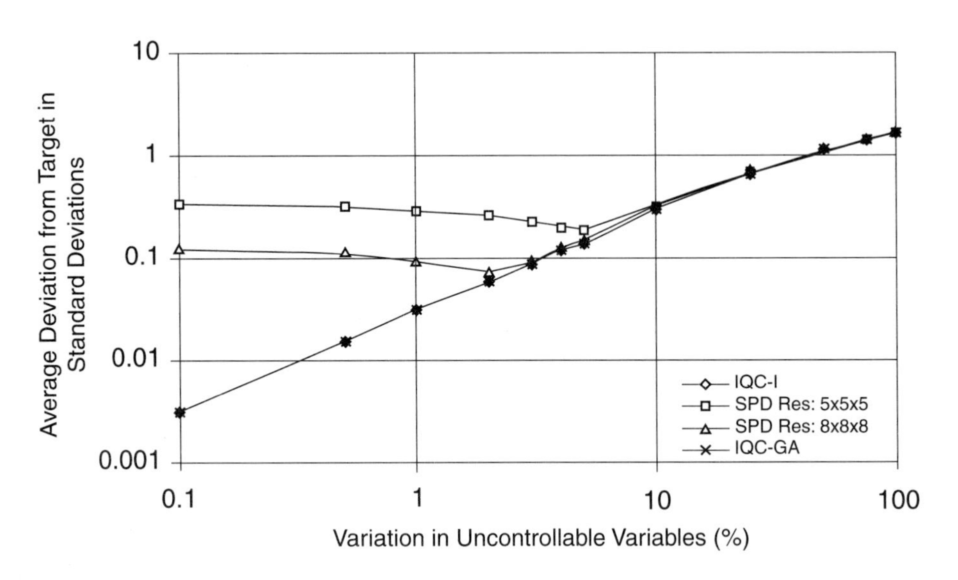

FIGURE 17.10 Performance comparison of IQCs and SPD.

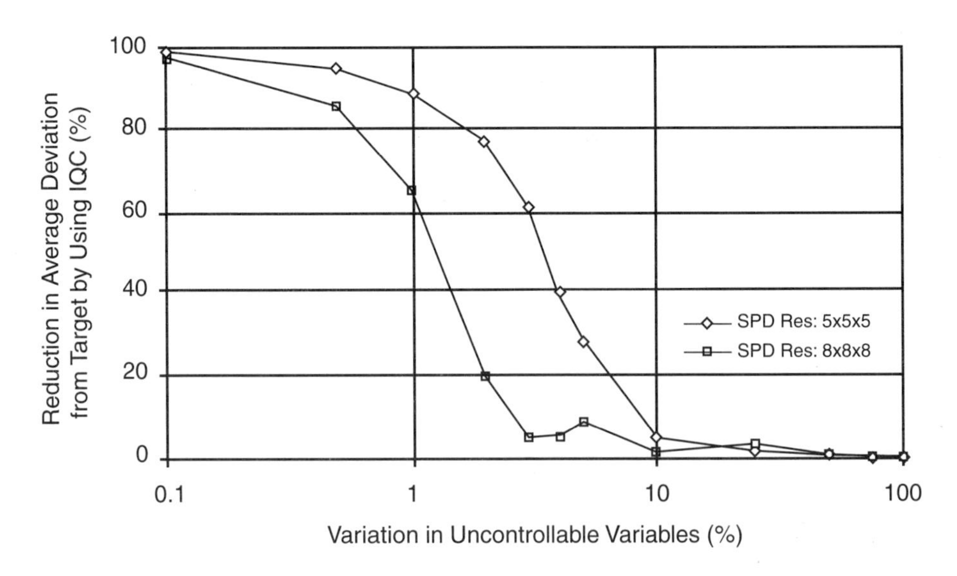

FIGURE 17.11 Reduction in average deviation from target by using IQC over SPD.

Deployment of the proposed on-line parameter design methods on an reactive ion plasma etching semiconductor manufacturing process revealed the ability of the method to significantly improve product/process quality beyond contemporary off-line parameter design approaches. However, more research is necessary to extend the proposed methods to dynamic time-variant systems. In addition, future research can also concentrate on modeling the signatures of the uncontrollable variables to facilitate feedforward parameter design.

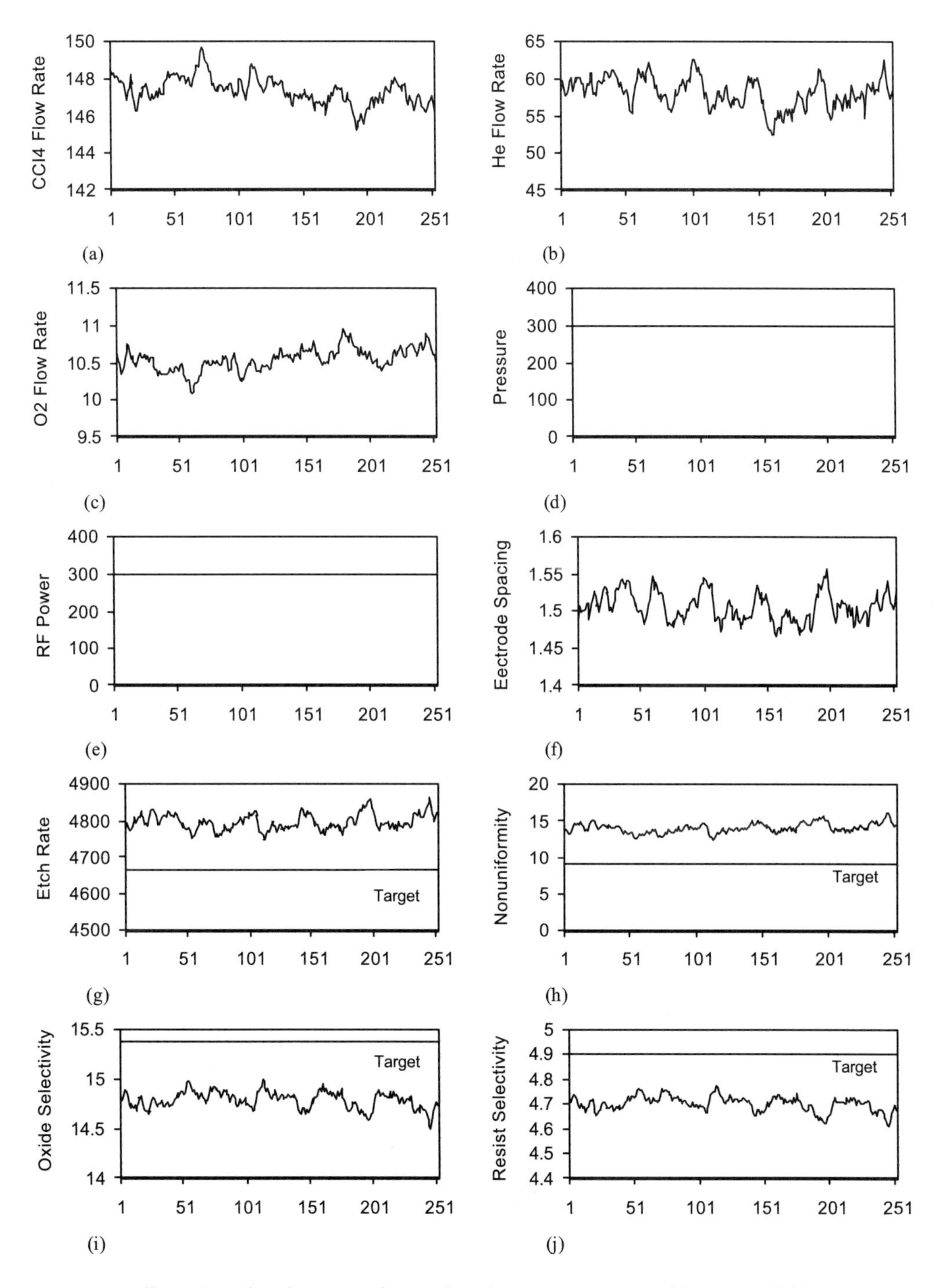

FIGURE 17.12 Illustration of performance of IQC when the DOV is set at 0.1. (a) Uncontrollable but observable CCl_4 flow rate (sccm). (b) Uncontrollable but observable He flow rate (sccm). (c) Uncontrollable but observable O_2 flow rate (sccm). (d) On-line optimized pressure (mtorr). (e) On-line optimized RF power (W). (f) On-line optimized electrode gap (cm). (g) Process etch rate (Å/min). (h) Etch nonuniformity (%). (i) Oxide selectivity. (j) Photoresist selectivity.

References

Baum, E. B. and Haussler, D., 1989, What Size Net Gives Valid Generalization? *Neural Computation*, vol. 1, pp. 151-160.

Box, G. E. P., 1985, Discussion of "Off-Line Quality Control, Parameter Design, and the Taguchi Method" by R. N. Kacker, *Journal of Quality Technology*, vol. 17, pp. 189-190.

Box, G. E. P., Hunter, W., and Hunter, J., 1978, *Statistics for Experimenters*, Wiley, New York.

Goldberg, D. E., 1989, *Genetic Algorithms in Search, Optimization, and Machine Learning*, Addison-Wesley, Reading, MA.

Haykin, S., 1999, *Neural Networks: A Comprehensive Foundation*, 2nd ed., Prentice-Hall, Upper Saddle River, NJ.

Himmel, C. D. and May, G. S., 1993, Advantages of Plasma Etch Modeling Using Neural Networks over Statistical Techniques, *IEEE Trans. Semiconductor Manuf.*, vol. 6, pp. 103-111.

Holland J. H., 1975, *Adaptation in Natural and Artificial Systems*, University of Michigan Press, Ann Arbor, MI.

Hornik, K., Stinchcombe, M., and White, H., 1989, Multi-Layer Feedforward Networks are Universal Approximators, *Neural Networks*, vol. 2.

Jenkins, M. W., Mocella, M. T., Allen, K. D., and Sawin, H. H., 1986, The Modeling of Plasma Etching Processes Using Response Surface Methodology, *Sol. St. Tech.*, April.

Kim, B. and May, G. S., 1994, An Optimal Neural Network Process Model for Plasma Etching, *IEEE Trans. Semiconductor Manuf.*, vol. 7, pp. 12-21.

Liepins, G. E. and Hilliard, M. R., 1989, Genetic Algorithms: Foundations and Applications, *Annals of Operations Research*, vol. 21, pp. 31-58.

Linden, A. and Kindermann J., 1989, Inversion of Multi-Layer Nets, *Int. Joint Conf. Neural Networks (IJCNN)*, pp. 425-430, June.

Lippman, R. P., 1987, An Introduction to Computing with Neural Nets, *IEEE ASSP Magazine*, April, pp. 4-22.

May G. S., Huang J., and Spanos, C. J., 1991, Statistical Experimental Design in Plasma Etch Modeling, *IEEE Trans. Semiconductor Manuf.*, vol. 4, pp. 83-98.

Mühlenbein, H. and Schlierkamp-Voosen, D., 1994, The Science of Breeding and its Application to the Breeder Genetic Algorithm, in *Evolutionary Computation*, MIT Press, Boston.

Myers, R. H. and Carter, W. H., Jr., 1973, Response Surface Techniques for Dual Response Surfaces, *Technometrics*, vol. 15, pp. 301-317.

Myers, R. H. and Montgomery, D. C., 1995, *Response Surface Methodology: Process and Product Optimization Using Designed Experiments*, John Wiley, New York.

Nadi, F., Agogino, A., and Hodges, D., 1991, Use of Influence Diagrams and Neural Networks in Modeling Semiconductor Manufacturing Processes, *IEEE Trans. Semiconductor Manuf.*, vol. 4, pp. 52-58.

Nair, V. N., 1992, Taguchi's Parameter Design: A Panel Discussion, *Technometrics*, vol. 34, pp. 127-161.

Pledger M., 1996, Observable Uncontrollable Factors in Parameter Design, *Journal of Quality Technology*, vol. 28, pp. 153-162.

Rietman, E. A. and Lory E. R., 1993, Use of Neural Networks in Modeling Semiconductor Manufacturing Processes: An Example for Plasma Etch Modeling, *IEEE Trans. on Semiconductor Manufacturing*, vol. 6, no. 4, pp. 343-347, November.

Riley, P. E. and Hanson, D. A., 1989, Study of Etch Rate Characteristics of SF_6/He Plasmas by Response Surface Methodology: Effects of Inter-Electrode Spacing, *IEEE Trans. Semiconductor Manuf.*, vol. 2, no. 4, November.

Salomon, R., 1995, Genetic Algorithms and the $O(n \ln n)$ Complexity on Selected Test Functions, *Proc. Artificial Neural Networks Engineering Conference*, St. Louis, MO, November 12-15, pp. 325-330.

Solla, S. A., 1989, Learning and Generalization in Layered Neural Networks: The Contiguity Problem, in *Neural Networks from Models to Applications* (Eds., G. Dreyfus and L. Personnaz), I.D.S.E.T, Paris.

Taguchi, G., 1986, *Introduction of Quality Engineering*, UNIPUB/Kraus International, White Plains, NY.
Taguchi, G., 1987, *System of Experimental Design: Engineering Methods to Optimize Quality and Minimize Cost*, UNIPUB/Kraus International, White Plains, NY.
Weigand, A. S., Rumelhart, D. E., and Huberman, B. A., 1992, *Advances in Neural Information Processing Systems*, Morgan Kaufmann, San Mateo, CA.
Weiss, S. M. and Kulikowski, C. A., 1991, *Computer Systems That Learn*, Morgan Kaufmann, San Mateo, CA.
Zurada, J. M., 1992, *Introduction to Artificial Neural Systems*, West Publishing, St. Paul, MN.

For Further Information

For additional information on contemporary robust parameter design techniques and the issues associated with these techniques, see the following:

Books:

Myers, R. H. and Montgomery, D. C., 1995, *Response Surface Methodology: Process and Product Optimization Using Designed Experiments*, John Wiley, New York.
Taguchi, G. and Wu, Y., 1980, *Introduction to Off-Line Quality Control*, Central Japan Quality Control Association, Tokyo (available from American Supplier Institute, Dearborn, MI).

Selected Journal Articles:

Box, G. E. P., 1988, Signal-to-Noise Ratios, Performance Criteria, and Transformations (with discussion), *Technometrics*, vol. 30, pp. 1-40.
Kackar, R. N., 1985, Off-Line Quality Control, Parameter Design and the Taguchi Methods, *Journal of Quality Technology*, vol. 17, pp. 176-188.
Khattree, R., 1996, Robust Parameter Design: A Response Surface Approach, *Journal of Quality Technology*, vol. 28, pp. 187-198.
Myers, R. H., Khuri, A. I., and Vining, G., 1992, Response Surface Alternatives to the Taguchi's Robust Parameter Design Approach, *American Statistician*, vol. 46, pp. 131-139.
Nelder, J. A. and Lee, Y., 1991, Generalized Linear Models for the Analysis of Taguchi-Type Experiments, *Applied Stochastic Models and Data Analysis*, vol. 7, pp. 107-120.
Shoemaker, A. C., Tsui, K. L., and Wu, C. F. J., 1991, Economical Experimentation Methods for Robust Parameter Design, *Technometrics*, vol. 33, pp. 415-427.
Vining, G. G. and Myers, R. H., 1990, Combining Taguchi and Response Surface Philosophies: A Dual Response Approach, *Journal of Quality Technology*, vol. 22, pp. 38-45.
Welch, W. J., Yu, T., Kang, S. M., and Sacks, J., 1990, Computer Experiments for Quality Control by Parameter Design, *Journal of Quality Technology*, vol. 22, pp. 15-22.

To learn more about artificial neural networks, see the following:

Book:

Haykin, S., 1999, *Neural Networks: A Comprehensive Foundation*, 2nd ed., Prentice-Hall, Upper Saddle River, NJ.

Journals:

Neural Networks published by the International Neural Network Society. For more information see *http://cns-web.bu.edu/INNS/*. *IEEE Transactions on Neural Networks* published by IEEE Neural Networks Council. For more information see *http://www.ieee.org/*.

To learn more about genetic algorithms and evolutionary computation, see the following:

Book:

Goldberg, D. E., 1989, *Genetic Algorithms in Search, Optimization, and Machine Learning*, Addison-Wesley, Reading, MA.

Journals:

Evolutionary Computation by MIT Press, Boston, MA. For more information see *http://mitpress.mit.edu/*.

IEEE Transactions on Evolutionary Computation published by IEEE Neural Networks Council. For more information see *http://www.ieee.org/*.

18

A Hybrid Neural Fuzzy System for Statistical Process Control

Shing I Chang
Kansas State University

Abstract

A hybrid neural fuzzy system is proposed to monitor both process mean and variance shifts simultaneously. One of the major components of the proposed system is composed of several feedforward neural networks that are trained off-line via simulation data. Fuzzy sets are also used to provide decision-making capability on uncertain neural network output. The hybrid control chart provides an alternative to traditional statistical process control (SPC) methods. In addition, it is superior in that (1) it outperforms other SPC charts in most situations in terms of faster detection and more accurate diagnosis, and (2) it can be used in automatic production processes with minimal human intervention — a feature the other methods ignore. In this chapter, theoretical base, operations, user guidelines, chart properties, and examples are provided to assist those who seek an automatic SPC strategy.

18.1 Statistical Process Control

Statistical process control (SPC) is one of the most often applied quality improvement tools in today's manufacturing as well as service industries. Instead of inspecting end products or services, SPC focuses on processes that produce products and services. The philosophy of a successful SPC application is to identify sources of special causes of production variation as soon as possible during production rather than wait until the very end. Here "production" is defined as either a manufacturing or service activity. SPC provides savings over traditional inspection operations on end products or service because it eliminates accumulations of special causes of variation by monitoring key quality characteristics during production. Imagine how much waste is generated when a production mistake enters a stream of products during mid-day but inspection doesn't take place until the end of an 8-hour shift. SPC can alleviate this situation by frequently monitoring the production process via product quality characteristics.

0-8493-0592-6/01/$0.00+$.50

A quality characteristic (QC) is a measure of quality on a product or service. Examples of QC are weight of a juice can, length of a cylinder part, the number of errors made during payroll operations, etc. A QC can be mathematically defined as a random variable, which is a function that takes values from a population or distribution. Denote a QC as random variable $\boldsymbol{x}$. If a population Ω only contains discrete members, that is, $\Omega = \{x_1, x_2, \ldots, x_n\}$, then QC $\boldsymbol{x}$ is a discrete random variable. For example, if $\boldsymbol{x}$ is the number of errors made during payroll operations, then member x_1 is the value in January, x_2 is the value in February, and so on. In this case, attribute control charts can be used to monitor a QC with discrete distribution. A control chart for fraction nonconforming, also known as a P chart, based on binomial distribution, is the most frequently used chart (Montgomery, 1996). However, in this chapter, we will focus only on a more interesting class of control charts when QC $\boldsymbol{x}$ is a continuous random variable where $\boldsymbol{x}$ can take a value in a continuous range, i.e., $\boldsymbol{x} \in \Omega = \{\boldsymbol{x} \mid L \leq \boldsymbol{x} \leq U\}$. For example, $\boldsymbol{x}$ is the weight of a juice can with a target weight of 8 oz.

The central limit theorem (CLT) implies that the sample mean of a continuous random variable $\boldsymbol{x}$ is approximately normally distributed where the sample mean is calculated by $\boldsymbol{n}$ independently sampled observations of $\boldsymbol{x}$. The approximation improves when the size of $\boldsymbol{n}$ increases. In much of the quality control literature, $\boldsymbol{n}$ is chosen to be 5 to 10 when the approximation is considered good enough. Note that CLT does not impose any restriction on the original distribution on $\boldsymbol{x}$, which provides the foundation for control charts. Since the sample mean of $\boldsymbol{x}$, $\bar{\boldsymbol{x}}$, is approximately normal distributed, i.e., $N(\mu, \sigma^2/n)$ where μ and σ are the mean and standard deviation of $\boldsymbol{x}$, respectively, we can collect $\boldsymbol{n}$ observations of a QC, calculate its sample mean, and plot it against a control chart with three lines. If both μ and σ are known, the centerline is μ with lower control limit $\mu - 3\frac{\sigma}{\sqrt{n}}$ and upper control limit $\mu + 3\frac{\sigma}{\sqrt{n}}$. If CLT holds and the process defined by QC $\boldsymbol{x}$ remains in control, 99.73% of the sample population will fall within the two control limits. On the other hand, if either μ or σ shifts from its target, this will increase the probability that sample points plot outside the control limits, which indicates an out-of-control condition.

A pair of control charts are often used simultaneously to monitor QC $\boldsymbol{x}$ — one for the mean μ and the other for the standard deviation σ. The goal is to make sure the process characterized by QC $\boldsymbol{x}$ is under statistical control. In other words, SPC charts are used to verify that the distribution of $\boldsymbol{x}$ remains the same over time. Since a probability distribution is usually estimated by two major parameters, μ and σ, SPC charts monitor the distribution through these two parameters. Figure 18.1 (Montgomery, 1996) demonstrates two out-of-control scenarios. At time t_1, the mean μ_0 of x starts to shift to μ_1. One of the most often used control charts, $\overline{X}$ chart, can be used to detect this situation. On the other hand, at time t_2, the mean is on target but the standard deviation has increased from σ_0 to σ_1 where $\sigma_1 > \sigma_0$. In this case, a control chart for ranges (R chart) can be used to detect the variation change. Notice that SPC charts are designed to detect assignable causes of variation as indicated by mean or standard deviation shifts and at the same time tolerate the chance variation as shown by the bell-shaped distribution of $\boldsymbol{x}$. Such a chance variation is inevitable in any production process.

Statistical process control charts have been applied to a wide range of manufacturing and service industries since Shewhart first introduced the concept in the 1920s. There have been several improvements on the traditional control charts since then. Page (1954) first introduced cumulative sum (CUSUM) control charts to enhance the sensitivities of detecting small process shifts. Instead of depending solely on data collected in the most recent sample period for plotting in the traditional Shewhart-type control chart, the CUSUM chart's plotting statistic involves all data points previously collected and assigns an equal weight factor for every point. If a small shift occurs, CUSUM statistic can accumulate such a deviation in a short period of time and thus increase the sensitivity of an SPC chart. However, CUSUM charts cannot be plotted as easily as the Shewhart-type control charts. Roberts (1959) proposes an exponential weighted moving average (EWMA) control chart that weighs the most recent observations more heavily than remote data points. EWMA charts were developed to have the structure of the traditional Shewhart charts, yet match the CUSUM charts' capability of detecting small process shifts.

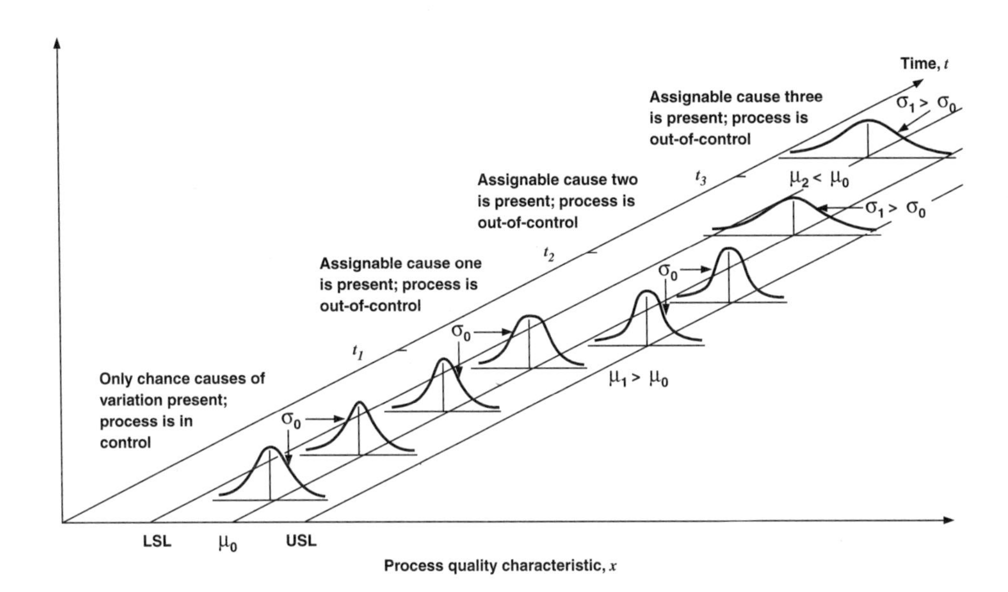

FIGURE 18.1 In-control and out-of-control scenarios in SPC. (From Montgomery, D.C., 1996, *Introduction to Statistical Quality Control*, 2nd ed. p. 131. Reproduced with the permission of John Wiley & Sons, Inc.)

Most control chart improvements over the years have been focused on detecting process mean shifts, with a few exceptions that are discussed in the following section.

Shewhart R, S, and S^2 charts are the first statistical control charts for monitoring process variance changes. Johnson and Leone (1962a, 1962b) and Page (1963) later proposed CUSUM charts based on sample variance and sample range. As an alternative, Crowder and Hamilton (1992) developed an exponential weighted moving average (EWMA) scheme based on the log transformation of the sample variance $ln(S^2)$. Their experimental results show that the EWMA chart outperforms the Shewhart S^2 chart and is comparable to the CUSUM chart for variation proposed by Page (1963). Using the concept of log transformation of sample variance, Chang and Gan (1995) suggest a CUSUM scheme based on $ln(S^2)$, which performs as well as the corresponding EWMA. Performances of Chang and Gan's (1995) CUSUM and Crowder and Hamilton's (1992) EWMA are not significantly better than Page's (1963) CUSUM; however, their development of design strategies and procedures are relatively easier for practitioners to use.

18.2 Neural Network Control Charts

In recent years, attempts to apply neural networks to process control have been investigated by several researchers. Guo and Dooley (1992) proposed network models that identify positive mean or variance changes using backpropagation training. Their best network performs 40% better on the average error rate than conventional control chart heuristic tests.

Pugh (1989, 1991) also successfully trained backpropagation networks for detecting process mean shifts with subgrouping size of five. He found his networks equal in average run length (ARL) performance to a 2-σ control chart in both type I and II errors.

Hwarng and Hubele (1991, 1993) trained a backpropagation pattern recognition classifier to detect six unnatural control chart patterns — trend, cycle, stratification, systematic, mixture, and sudden shift. Their results were promising in recognizing various special causes in out-of-control situations.

Smith (1994) and Smith and Yazici (1993) described a combined X-bar and R chart backpropagation model to investigate both mean and variance shifts. They found their networks performed 50% better in average error rate when compared to Shewhart control charts. However, the majority of the wrong classification is of type I error. That is, the network signals too many out-of-control false alarms when the process is actually in control.

Chang and Aw (1994) proposed a four-layer backpropagation network and a fuzzy inferencing system for detecting process mean shifts. Their network outperforms conventional Shewhart control charts in terms of both type I and type II errors, while Pugh's and Smith's charts have larger type I errors than that of the 3σ $\overline{X}$ chart. Further, Chang and Aw's scheme has the advantage of identifying the magnitude of shifts. None of the Shewhart-type charts, or the other neural network charts, offer this feature. Chang and Ho (1999) further introduced a two-stage neural network approach for detecting and classifying process variance shifts. The performance of the proposed method is comparable to that of the other control charts for detecting variance changes as well as being capable of estimating the magnitude of the variance change, which is not supported by the other control charts. Furthermore, Ho and Chang (1999) integrated both neural network control chart schemes and compared this with many other approaches for monitoring process mean and variance shifts. In this chapter, we will summarize the proposed hybrid neural fuzzy system for monitoring both process mean and variance shifts, provide guidelines and examples for using this system, and list the properties.

18.3 A Hybrid Neural Fuzzy Control Chart

As shown in Figure 18.2 (Ho and Chang, 1999), the proposed hybrid neural fuzzy control chart, called C-NN (C stands for "combined" and NN means "neural network"), is composed of several modules — data input, data processing, decision making, and data summary. The data input module takes observations from QC $\boldsymbol{x}$ and transforms them into appropriate types for both control charts for mean M-NN and for variance V-NN, which are the major components of the data processing module. The decision-making module is responsible for interpreting the neural network outputs from the previous module. There are four distinct possibilities: no process shift, process mean shift only, process variance shift only, and both process mean and variance shifts. Note that two different classifiers — fuzzy and neural network — are adopted for the process mean and variance components, respectively. Finally, the data summary module calculates estimated shift magnitudes according to appropriate diagnosis. Details of each module will be discussed in the following sections.

18.3.1 Data Input Module

The data input module takes samples or observations of QC $\boldsymbol{x}$ in two ways. Sample observations, x_1, x_2, ..., and x_n in the first input method are independent of each other. In the proposed system, $\boldsymbol{n}$ is chosen as five, that is, each plotting point consists of a sample of five observations. Traditional Shewhart-type control charts normally use this input method.

A moving window of five observations is used for the second method to select incoming observations. For example, the first sample point consists of observations $x_1, x_2, \ldots, x_5$ and the second sample point is composed of $x_2, x_3, \ldots, x_6$, and so on. This method is explored due to the fact that both CUSUM and EWMA charts for mean shifts are capable of taking individual observations. The proposed moving range method comes close to individual observation in terms of the number of observations used for decision making. Unlike the "true" individual observation input method, the moving range method must wait until the fifth observation to complete the first sample point to start using the proposed chart. After this point, it is on pace with the "true" individual observation input method in that it uses the most recent and four immediately passed observations. The reason for maintaining a few observations in a sample point is due to the need to evaluate process variation. An individual observation does not provide such information.

Transformation is also a key component in the data input module. As we will discuss later, both neural networks were trained "off-line" from simulated observations. In order to make the proposed schemes work for various applications, data transformation is necessary to standardize the raw data into the value range that both neural network components can work with. Formulas for data transformation are as follows:

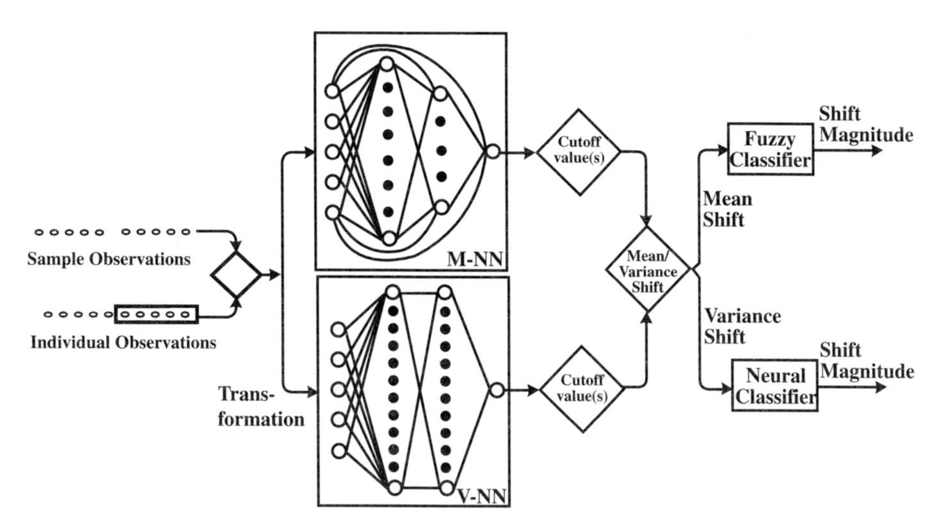

FIGURE 18.2 A schematic diagram of C-NN (combined neural network) control chart. (Adapted from Ho and Chang, 1999, Figure 3, p. 1891.)

18.3.1.1 Transformation for M-NN Input

$$z_{ti} = \frac{x_{ti} - \bar{x}}{s}, i = 1,2,3,\ldots,5 \qquad \text{Equation (18.1)}$$

where i is the index of observations in a sample or window; t is the index for the sample period, and $\bar{x}$ and s are estimates of process mean and standard deviation, respectively. In traditional control charts, it takes 100 to 125 observations, e.g., 25 samples of 4 or 5 observations each, to establish the control limits. However, in this case, 20 to 30 observations can provide reasonably good estimates.

18.3.1.2 Transformation for V-NN Input

Given the data standardization in Equation 18.1, the input for V-NN of variance detection needs to further process as

$$I_{ti} = \left| z_{ti} - \bar{z}_t \right|, I = 1,2,\ldots,5 \qquad \text{Equation (18.2)}$$

where t and i are the same as those defined in Equation 18.1, and $\bar{z}_t$ is the average of five transformed observations z_{ti} of the sample at time t.

18.3.2 Data Processing Module

The heart and soul of the proposed system is a module composed of two independently developed neural networks: M-NN and V-NN. M-NN, developed by Chang and Aw (1996), is a 5–8–5–1 four-layer neural network for detecting process mean shift. On the other hand, Chang and Ho's (1999) V-NN is a 5–12–12–1 neural network for detecting process variance shift. Data from transformation formulas (Equations 18.1 and 18.2) are fed into M-NN and V-NN, respectively. Both neural networks have single output nodes. M-NN's output values range from –1 to +1. A value that falls into a negative range indicates a decrease in process mean value, while a positive M-NN output value indicates a potential increase in process mean shift. On the other hand, V-NN's output ranges from 0 to 1 with larger values meaning larger shifts. Note that both neural networks were trained off-line using simulations. By incorporating the trained weight matrices, one can start using the proposed method. The only setup required is to estimate both process mean and variance for transformation. The central limit theorem guarantees that transformed data is

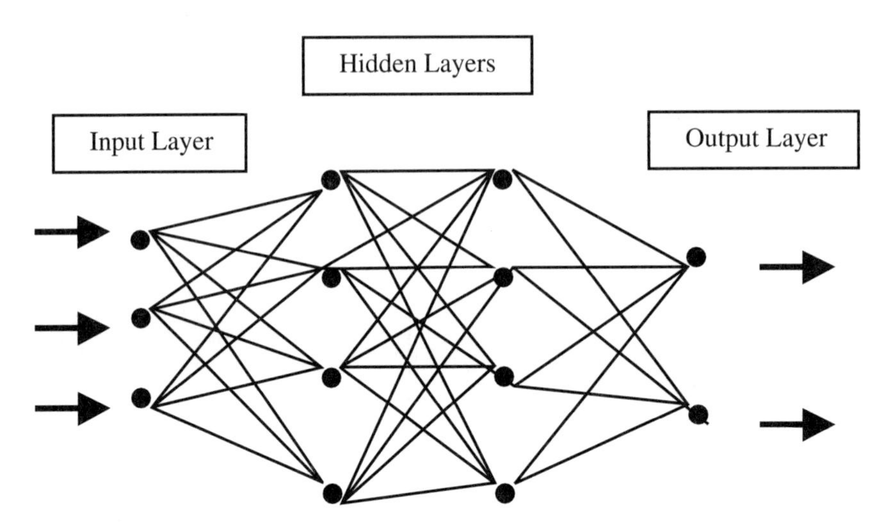

FIGURE 18.3 An example of a multilayer neural network.

similar to the simulated data used for training. Thus the proposed method can be applied to many applications with various data types as long as they can be defined as QC $\boldsymbol{x}$. Before M-NN and V-NN are introduced in detail, we first summarize calculation and training of any feedforward, multiple-layer neural networks as follows.

18.3.2.1 Computing in a Neural Network

The most commonly implemented neural network is the multilayer backpropagation network, which adapts weights according to the steepest gradient descent rule along a nonlinear transformation function. The reason for this popularity is due to the versatility of its paradigm in solving diverse problems, and its strong mathematical foundation. An example of a multilayer neural network is shown in Figure 18.3. In neural networks, information propagates from input nodes (or neurons) through the system's weight connections in the middle layers (or hidden layers) of nodes, finally passing out the last layer of nodes — the output nodes.

Each node, for example node j in the hidden and output layers, contains the input links with weights w_{ij}, an activation function (or transfer function) f, and output links to other nodes as shown in Figure 18.4. Assuming k input links are connected to node j, the output V_j of node j is processed by the activation function

$$V_j = f\left(I_j\right),$$

$$I_j = \sum_{i=1}^{k} w_{ij} V_{pi}, \qquad \text{Equation (18.3)}$$

where V_{pi} is the output of node i from the previous layer.

Many activation functions, e.g., sigmoidal and hyperbolic-tangent functions, are available. We choose to use the sigmoidal function

$$f\left(I\right) = \frac{1}{1+e^{-cI}} \qquad \text{Equation (18.4)}$$

where c is a coefficient that adjusts the abruptness of the function.

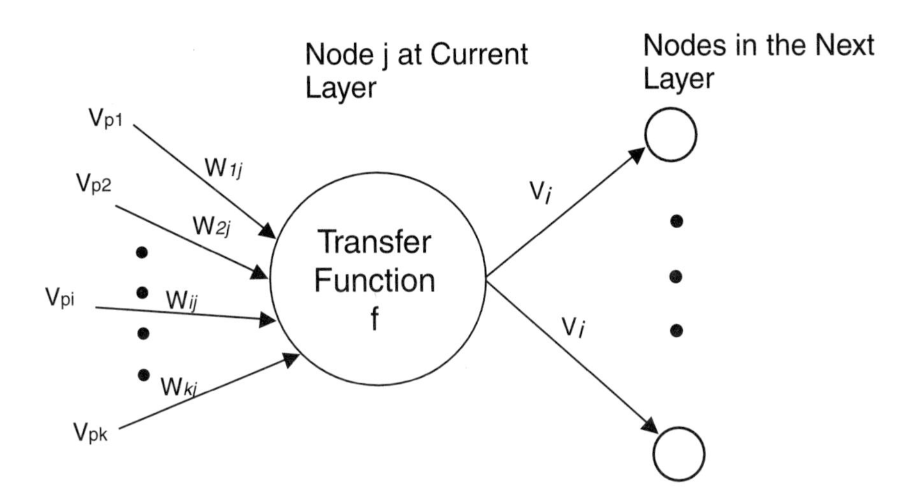

FIGURE 18.4 Node j and its input–output values in a multilayer neural network.

18.3.2.2 Training of a Neural Network

Backpropagation training is the most popular supervised neural network training algorithm. The training is designed to modify the thresholds and weights so that the overall error will be minimized. At each iteration, we first calculate error signals δ_o, o = 1, 2, . . . , n_o, for the output layer nodes as follows:

$$\delta_o = f'(I_o)(t - f(I_o)) = 0.5(1 + V_o)(1 - V_o)(t - V_o), \qquad \text{Equation (18.5)}$$

where $f'(I)$ is the first-order derivative of the activation function $f(I)$; t_o is the desired target value; and V_o is the actual output for output node o. We then update the weights connected between hidden layer nodes and output layer nodes:

$$w_{ho}(\text{new}) = w_{ho}(\text{old}) + \eta\delta_o V_h + \alpha[\Delta w_{ho}(\text{old})], \qquad \text{Equation (18.6)}$$

where η is a constant chosen by users for adjusting training rate; α is a momentum factor; δ_o is obtained from Equation 18.5; V_h is the output of node h in the last hidden layer; and Δw_{ho} is the previous weight change between node h and output node o. Subsequent steps include computing the error signals for a hidden layer(s) and propagating the errors backward toward the input layer. The error signals for node h in the current hidden layer are

$$\delta_h = f'(I_h)\sum_{i=1}^{n'} w_{ih}\delta'_i = 0.5(1 + V_h)(1 - V_h)\sum_{i=1}^{n'} w_{ih}\delta'_i, \qquad \text{Equation (18.7)}$$

where V_h is the output for node h in the current hidden layer under consideration; w_{ih} is the weight coefficient between node h in the current hidden layer and node i in the next hidden layer; δ'_i is the error signal for node i in the next hidden layer; and n' is the number of nodes in the next hidden layer. Given the error signals from Equation 18.7, the weight coefficient w_{jh} between node j in the lower hidden layer and node h in the current hidden layer can be updated as follows:

$$w_{jh}(\text{new}) = w_{jh}(\text{old}) + \eta\delta_h V_j + \alpha[\Delta w_{jh}(\text{old})], \qquad \text{Equation (18.8)}$$

where the indices η and α are defined in Equation 18.6 and V_j is the actual output from node j in the lower hidden layer. In summary, the procedure of backpropagation training is as follows:

Step 1. Initialize the weight coefficients.
Step 2. Randomly select a data entry from the training data set.
Step 3. Feed the input data of the data entry into the network under training.
Step 4. Calculate the network outputs.
Step 5. Calculate the error signals between the network outputs and desired targets using Equation 18.5.
Step 6. Adjust the weight coefficients between the output layer and closest hidden layer using Equation 18.6.
Step 7. Propagate the error signals and weight coefficients backward using Equations 18.7 and 18.8.
Step 8. Repeat steps 2 to 7 for each entry in the training set until the network error term drops to an acceptable level.

Note that calculations in steps 2 to 4 are done from the input layer toward the output layer, while weight updates in steps 5 and 7 are calculated in a backward manner. The term "backpropagation" comes from the way the network weight coefficients are updated.

18.3.2.3 Computing and Training of M-NN

The first neural network is a backpropagation type trained by Chang and Aw (1996). It is a 5–8–5–1 four-layer network, i.e., five input nodes with two hidden layers, each having eight and five neurons and one output node. This network has a unique feature in that the input layer is connected to all nodes in the other three layers, as shown in Figure 18.5. They trained M-NN by using 900 samples, each with five observations, simulated from $N(\mu_o \pm \delta\sigma_o, \sigma_o^2)$ where $\mu_o = 0$ and $\sigma_o = 1$ and $\delta = 0, \pm1, \pm2, \pm3$, and ±4. These observations were fed directly to the network and trained by a standard backpropagation algorithm to achieve a desired output between –1 and 1. The network was originally developed to detect both positive and negative mean shifts. Since we will analyze positive shifts only, our interest here is in positive output values between 0 and 1. A value close to zero indicates the process is in control while it triggers an out-of-control signal when the output value exceeds a set of critical cutoff points. The larger the output value, the larger the process mean shift.

18.3.2.4 Computing and Training of V-NN

Chang and Ho (1999) trained a neural network to detect process variation shifts henceforth called V-NN. V-NN is a standard backpropagation network with a 5-12-12-1 structure. The number of nodes for input and output were kept the same so that parallel use of both charts is possible.

In training V-NN, 600 exemplar samples were taken from simulated distributions $N(\mu_o, (\rho\sigma_o)^2)$ where $\mu_o = 0$ and $\sigma_o = 1$, and $\rho = 1, 2, 3, 4, 5$. They were then transformed into input values for the neural network by using Equation 18.2.

The desired output, which represents different shift magnitudes, has values between 0 and 1. The network was trained by a standard backpropagation algorithm with adaptive learning rates. A V-NN output value close to 0 means the process variation is likely to be in control, while larger values indicate that the process variation increases. The larger the V-NN output, the larger the magnitude of increase.

18.3.3 Decision-Making Module

The decision-making module is responsible for interpreting neural network outputs from both M-NN and V-NN. The fuzzy set theory is applied to justify human solutions to these problems. Before the decision rules for evaluating both M-NN and V-NN are given, fuzzy sets and fuzzy computing related to this module are briefly reviewed in the following sections.

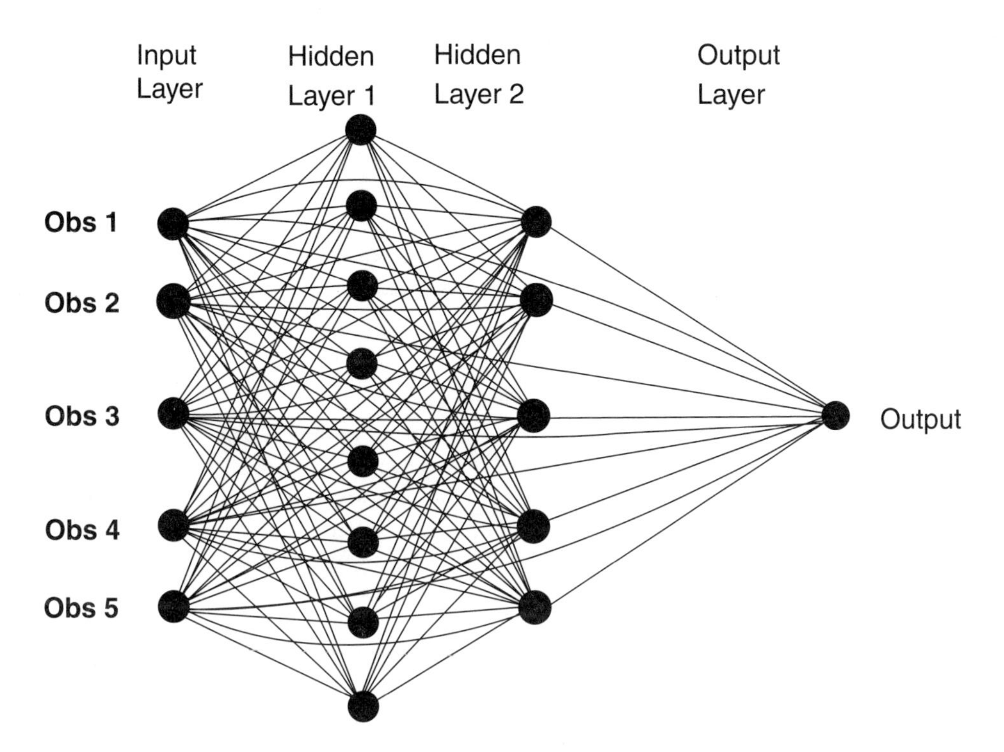

FIGURE 18.5 A proposed two-sided mean shift detection neural network model. (Adapted from Chang and Aw, 1996, Figure 1, p. 2266.)

18.3.3.1 Fuzzy Sets and Fuzzy Computing

Zadeh (1965) emphasized that applications based on fuzzy logic start with a human solution, which is distinctly different from a neural network solution. Motivated by solving complex systems, Zadeh observed that system models based on the first principle, such as physics, are not always able to solve the problem. Any attempt to enhance details of modeling of complex systems often leads to more uncertainties. On the other hand, a human being is able to offer a solution for such a system from his or her experiences. The fact is that human beings can handle uncertainties much better than a system model can.

18.3.3.1.1 Fuzzy Sets and Fuzzy Variables

Zadeh (1965) first introduced the concept of the fuzzy set. A member in a fuzzy set or subset has a membership value between [0, 1] to describe how likely it is that this member belongs to the fuzzy set. Let U be a collection of objects denoted generically by $\{x\}$, which could be discrete or continuous. U is called the universe of discourse and u represents a member of U (Yager and Filev, 1994). A fuzzy set F in a universe of discourse U is characterized by a membership function μ_F which takes values in the interval [0, 1], namely,

$$\mu_F: U \rightarrow [0, 1]. \qquad \text{Equation (18.9)}$$

That fuzzy set F can be represented as $F = \{(x, \mu_F(x)), x \in U\}$.

An ordinary set may be viewed as a special case of the fuzzy set whose membership function only takes two values, 0 or 1. An example of probability modeling is throwing a dice. Assuming a fair dice, outcomes can be modeled as a precise set $\boldsymbol{A} = \{1, 2, 3, 4, 5, 6\}$ with probabilities 1/6 for the occurrence of each member in the set. To model this same event in fuzzy sets, we need six fuzzy subsets ONE, TWO, THREE, FOUR, FIVE, and SIX that contain the outcomes of coin flipping. In this case, the universe of discourse U is same as set $\boldsymbol{A}$ and six membership functions $\mu_1, \mu_2, \ldots, \mu_6$ are for the members in fuzzy sets ONE, TWO, …, and SIX, respectively. We can now define fuzzy sets ONE = {(1, 1), (2, 0), (3, 0), (4, 0), (5, 0), (6, 0)}, TWO = {(1, 0), (2, 1), (3, 0), (4, 0), (5, 0), (6, 0)}, and so on. There is no ambiguity about this event — you receive a number from 1 to 6 when a dice is thrown. Conventional set theory is appropriate in this case.

In general, the grade of each member u in a fuzzy set is given by its membership function $\mu_F(u)$ whose value is between 0 and 1. Thus the vagueness nature of the real world can be modeled. In this paper, the output from M-NN, for example, can be modeled as a fuzzy variable in that we cannot precisely define the meaning of an M-NN output value. For example, a value 0.4 can mean either a positive small or a positive medium process mean shift because of the way we define the M-NN output target values. Target values 0.3 and 0.5 are for one-sigma and two-sigma positive mean shifts, respectively. One way to model this vagueness is to define the meaning of M-NN output by several linguistic fuzzy variables as discussed in the following section.

18.3.3.1.2 Membership Functions and Linguistic Fuzzy Variables

A fuzzy membership function, as shown in Equation 18.9, is a subjective description of how likely it is that an element belongs to a fuzzy set. We propose a fuzzy set of nine linguistic fuzzy variables to define M-NN outputs that take values within the range [–1, 1]. The universe of disclosure U is [–1,1] in this case. Nine linguistic fuzzy variables for the M-NN outputs are Extremely Negative Large Shift (XNL), Negative Large Shift (NL), Negative Medium Shift (NM), Negative Small Shift (NS), No Shift (NO), Positive Small Shift (PS), Positive Medium Shift (PM), Positive Large Shift (PL), and Extremely Large Shift (XPL). Each fuzzy set is responsible for one process status; e.g., NS means that the process experiences a negative small mean shift. Due to the nature of neural network output, some fuzzy sets overlap each other; that is, different fuzzy sets share the same members.

We use two of the most popular membership functions, triangular and trapezoidal functions, to define these linguistic fuzzy variables, as shown in Figure 18.6. Note that an M-NN output value 0.4 will generate two non-zero fuzzy membership values, i.e., $\mu_{PS}(x = 0.4) = 0.5$ and $\mu_{PM}(x = 0.4) = 1$. In other words, an M-NN output with value 0.4 most likely belongs to a positive mean shift, although there is a 50% possibility that it may be a small mean shift.

18.3.3.1.3 Fuzzy Operators

An α-cut set of a fuzzy set is defined as the collection of members whose membership values are equal to or larger than α where α is between 0 and 1. An α-cut set of F is defined as

$$F_\alpha = \{x \in U |\ \mu_F(x) \geq \alpha\}. \qquad \text{Equation (18.10)}$$

For example, the 0.5 cut set of PS (positive small mean shift) contains M-NN values between 0.05 and 0.4. Perhaps this concept can be best demonstrated by visualization. In Figure 18.6, this represents the x axis interval supporting the portion of the PS trapezoid above the horizontal 0.5 alpha-level line. For a NO fuzzy variable, its α-cut members $\{x \in U | -0.15 \leq x \leq 0.15\}$ support a triangular shape.

We can rewrite the α-cut of a fuzzy set F as an interval of confidence (IC)

$$F_\alpha = [f_1^{(\alpha)}, f_2^{(\alpha)}] \qquad \text{Equation (18.11)}$$

which is a monotonic decreasing function of α, that is,

$$(\alpha' > \alpha) \Rightarrow (F_{\alpha'} \subset F_\alpha) \qquad \text{Equation (18.12)}$$

where for every $\alpha, \alpha' \in [0,1]$. Note that the closer the value of α is to 1, the more the element u belongs to the fuzzy set. On the other hand, the closer the value of α to 0, the more uncertain we are of the set membership.

For a given α level, if a neural network output value falls into the IC of a fuzzy set, we can classify the NN output into this fuzzy set; that is, the process status can be identified. The ICs for the fuzzy decision

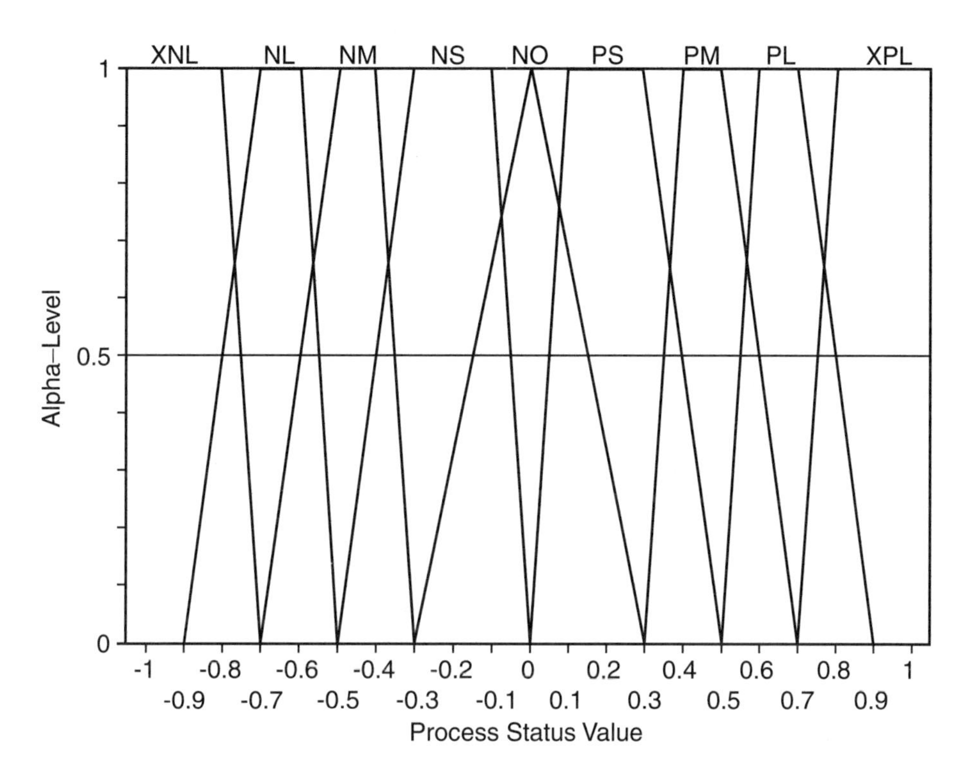

FIGURE 18.6 Fuzzy membership functions of the neural fuzzy control charts. (Adapted from Chang and Aw, 1996, Figure 2, p. 2270.)

sets are defined in the third column of Table 18.1. Following Equation 18.11, we define the IC of a fuzzy set X at α-level as

$$IC(X)_\alpha = [x_1, x_2] \qquad \text{Equation (18.13)}$$

where x_1 and x_2 are interceptions between the fuzzy membership function of the fuzzy set X and the horizontal line at α level, for example, the horizontal line at $\alpha = 0.5$ shown in Figure 18.6.

We notice that the higher the α level, the more certain the classification. However, there will be some gaps between adjacent ICs. On the other hand, the smaller the α level, the more likely there will be some overlapping between adjacent ICs. In order to obtain crisp classifications, we thus propose a fuzzy classifier (FC) that provides nonoverlapping and connecting ICs for classifications.

The connecting intervals of decision (ID) are listed in column four in Table 18.1. At a given α level, the ID of the fuzzy set NO remains the same as the IC. This choice will reduce type I errors of the NF charts. For fuzzy sets of negative mean shifts, i.e., XNL, NL, NM, and NS, the left boundary point of the IC will be that of the ID. In this case, we can define the ID for fuzzy set Y as

$$ID(Y)_\alpha = [y_1, y_2)$$

where y_1 is the same as the x_1 in $IC(Y)_\alpha$; y_2 is the same as the left-hand extreme point from the adjacent interval of decision; and $Y \in \{XNL, NL, NM, NS\}$. For example, y_2 of $ID(NS)_\alpha$ is defined by x_1 in $IC(NO)_\alpha$.

Similarly, we can also define the IDs for fuzzy sets of positive mean shifts, i.e., PS, PM, PL, and XPL as

$$ID(Y)_\alpha = (y_1, y_2]$$

TABLE 18.1 Intervals of Confidence and Decision (Adapted from Chang and Aw, 1996, Figure 1, p. 2271.)

Fuzzy Set	Shift Magnitude	Interval of Confidence $[\alpha_1, \alpha_2]$	Interbal of Decision $[(b_1, b_2)]$
XNL	–4σ shift	$\left[-1.0, \frac{-\alpha-7}{10}\right]$	$\left[-1.0, \frac{\alpha-4.5}{5}\right]$
NL	–3σ shift	$\left[\frac{\alpha-4.5}{5}, \frac{-\alpha-5}{10}\right]$	$\left[\frac{\alpha-4.5}{5}, \frac{\alpha-3.5}{5}\right]$
NM	–2σ shift	$\left[\frac{\alpha-3.5}{5}, \frac{-\alpha-3}{10}\right]$	$\left[\frac{\alpha-3.5}{5}, \frac{\alpha-2.5}{5}\right]$
NS	–1σ shift	$\left[\frac{\alpha-2.5}{5}, \frac{-\alpha}{10}\right]$	$\left[\frac{\alpha-2.5}{5}, \frac{3\alpha-3}{10}\right]$
NO	No shift	$\left[\frac{3\alpha-3}{10}, \frac{-3\alpha+3}{10}\right]$	$\left[\frac{3\alpha-3}{10}, \frac{-3\alpha+3}{10}\right]$
PS	1σ shift	$\left[\frac{\alpha}{10}, \frac{-\alpha+2.5}{5}\right]$	$\left[\frac{-3\alpha+3}{10}, \frac{-\alpha+2.5}{5}\right]$
PM	2σ shift	$\left[\frac{\alpha+3}{10}, \frac{-\alpha+3.5}{5}\right]$	$\left[\frac{-\alpha+2.5}{5}, \frac{-\alpha+3.5}{5}\right]$
PL	3σ shift	$\left[\frac{\alpha+5}{10}, \frac{-\alpha+4.5}{5}\right]$	$\left[\frac{-\alpha+3.5}{5}, \frac{-\alpha+4.5}{5}\right]$
XPL	4σ shift	$\left[\frac{\alpha+7}{10}, 1.0\right]$	$\left[\frac{-\alpha+4.5}{5}, 1.0\right]$

where y_2 is the same as the x_2 in $IC(Y)_\alpha$; y_1 is the same as the right-hand extreme point from the adjacent interval of decision; and $Y \in$ {PS, PM, PL, XPL}.

18.3.3.2 Decision Rules for M-NN

A two-in-a-row decision rule is used to detect a mean shift. We now restrict our discussion to positive shifts only. Since negative shift cases are symmetric to the positive ones, similar rules will apply to both situations. This two-in-a-row rule uses two values of cutoff points, C_{m1} and C_{m2}, where $C_{m1} < C_{m2}$. If an M-NN output is smaller than C_{m1}, we conclude that the process is in control. Otherwise, if it is greater than C_{m2}, the process is immediately declared out-of-control. But when the output falls between C_{m1} and C_{m2}, an additional sample will be drawn to obtain another M-NN output. This is when the two-in-a-row rule is used. If two consecutive outputs are greater than C_{m1}, the process is said to be out-of-control. Otherwise, the first sample is deemed as a false alarm. The advantage of the two-in-a-row rule is a decrease in type I errors.

C_{m1} and C_{m2} were chosen and justified by fuzzy computing from the previous sections. C_{m2} is chosen to be 0.3 because the right-hand boundary of NO (no shift) fuzzy variable ends at 0.3. If an M-NN output gives a value larger than this point, it is very likely there is a positive process mean shift. C_{m1} is defined by the α cut chosen, that is, $C_{m1} = -0.3\ \alpha + 0.3$. For $\alpha = 0.5$, C_{m1} is set at 0.15.

Suppose the first M-NN output gives a value 0.2. Because this value is between $C_{m1} = 0.15$ and $C_{m2} = 0.3$, a second sample observation is necessary. Assume the second M-NN output is 0.1. The process will be deemed in control. Had the second value been 0.25, one would have concluded an out-of-control process with the most possible shift in PS (positive mean shift). Any M-NN output that is greater than

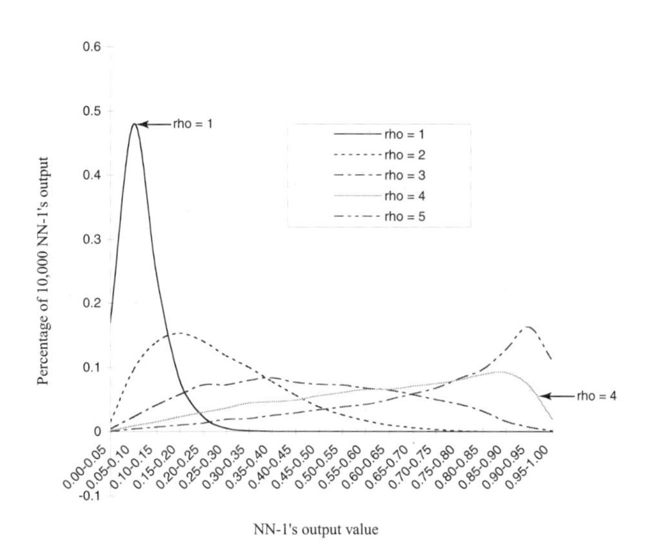

FIGURE 18.7 Patterns of process variation shifts. (Adapted from Chang and Ho, 1999, Figure 3, p. 1590.)

0.3 and smaller than –0.3 will generate an out-of-control classification in any one of the following categories: XNS, NL, NM, PS, NS, NM, NL, and XNL.

18.3.3.3 Decision Rules for V-NN

We name the first neural network in V-NN as NN-1. The major task of NN-1 is to detect whether there is a shift in process variance. As shown in Figure 18.7, the big spike belongs to the case where process has no shift, which is very different from the rest of the group. We can model V-NN's NN-1 output as two fuzzy variables — VS and NO. VS means variance shift and NO means no shift. Under this model, similar to M-NN, we can define the cutoff values for the one-point rule or two-in-a-row rule for V-NN's NN-1. The cutoff values for the two-in-a-row rule are $C_{v1} = 0.19$ and $C_{v2} = 0.28$. However, the parameter pairs (C_{v1}, C_{v2}) and (C_{m1}, C_{m2}) will be fine-tuned to obtain necessary type I and type II errors when both M-NN and V-NN are used simultaneously.

18.3.4 Data Summary Module

The data summary module consists of two classifiers to summarize the information given so far from previous modules and estimate the magnitudes of shift for each diagnostic scenario. For M-NN, the fuzzy classifier previously introduced is used. For V-NN, a bootstrapping sample and another neural network, NN-2, will be used together for this task. In the following subsection, we will discuss the use of the fuzzy classifier followed by the details of bootstrapping sampling and NN-2.

18.3.4.1 Fuzzy Classifier for M-NN

After M-NN gives an out-of-control signal, a fuzzy classifier is used to estimate the shift magnitude. This is achieved linguistically by a fuzzy variable expression $(y, \mu_F(y))$ where y is the neural network output value and $F \in \{XNL, NL, NM, NS, NO, PS, PM, PL, XPL\}$. For the example $y = 0.4$, as mentioned earlier, we obtain $(y, \mu_{PS}(y)) = (0.4, 0.5)$, $(y, \mu_{PM}(y)) = (0.4, 1)$, and the membership values for the other fuzzy variables are 0. The fuzzy classifier simply picks the fuzzy variable with the largest membership value. In this case, the fuzzy variable PM is chosen due to the possibility value of 1. In other words, the proposed fuzzy classifier concludes that this process is more likely to have a two-sigma process mean increase. However, this classifier also identifies a possibility of 0.5 that the shift magnitude is one-sigma.

18.3.4.2 Neural Network Classifier for V-NN

A closer examination of the M-NN output patterns as shown in Figure 18.7 shows that most of the output values from various shift magnitudes (defined by rho) overlap with each other. Had we applied fuzzy sets to NN-1 of V-NN using the principle similar to that used in M-NN, linguistic fuzzy variables XL, L, M, S, and N could have been defined for extremely large shift, large shift, medium shift, small shift, and no shift, respectively. However, many inferior results would have been obtained in terms of the classification rate. To overcome this problem, we propose a bootstrap method to provide more distinguishable patterns and apply a second neural network, NN-2, to interpret them.

18.3.4.2.1 A Bootstrap Resampling Scheme to Produce Input Data for a Neural Network Classifier

Once an output of NN-1 exceeds a specified cutoff point, the next step is to classify the change magnitude by another neural network (NN-2). In order to train NN-2, we must obtain the patterns associated with each variance-shift category as shown in Figure 18.7. When an out-of-control signal is detected, it is reasonable to assume that a few inspected sample observations are available. With these few observations, we need to generate many samples to feed NN-1 and obtain an output pattern with respect to an out-of-control signal. The pattern is formed by percentage values falling in ranges specified between zero and one. Bootstrap sampling is primarily used to find a distribution of a statistic when the distribution is not known and the number of sample data is limited (Efron and Tibshirani, 1993; Seppala et al., 1995). We thus adopt a bootstrap sampling scheme to generate many samples from limited observations by assuming that these samples could represent underlying process distribution. The implementation of the scheme is as follows.

When an NN-1's output exceeds the cutoff point, the current out-of-control sample of five measurements and the previous sample of five measurements are taken to form a finite population Ω. The limitation of this scheme is that there must exist a total of ten observations once current observations signal that the process is out of control. This assumption is reasonable, because in practice, a production engineer usually will not discover an abnormal process variance change until a few samples are taken. From the pool of ten observations $(x_1, x_2, \ldots, x_{10}) \subseteq \Omega$, five of them $(X_1, \ldots, X_5) \in \Omega$ are taken randomly one by one. The probability of each x_k to be selected at each draw is 1/10, i.e., the random samples are taken with replacements. This resampling procedure is repeated up to 2000 times. Each time a sample is fed into NN-1, a corresponding output is counted when it falls in a specified range. Ranges of the NN-1 outputs are specified into 20 ranges, [0.0, 0.05], [0.05, 0.10], … , and [0.95, 1.0], as shown in Table 18.2.

Since the sample population of ten observations consists of the current sample of five observations and five observations in the previous sample, it is possible some bootstrap observations may carry observations taken from in-control process conditions. Therefore, we resample 2000 times in order to include samples that reflect both out-of-control and in-control situations. These 2000 samples produce 2000 NN-1 outputs, which scatter in different specified ranges. Total numbers falling into each specified range are transformed into percentage values. The set of percentages, which consists of 20 values, is an input vector to NN-2 for classifying what change magnitude it represents. Since the resampling is conducted by computer, processing time is negligible. The advantage of this technique is that the "data mining" task does not require many measurements. Table 18.2 shows a sample result of NN-1 outputs in percentages falling into specified ranges, which is fed by samples of 2000 bootstrap samples for each process shift.

18.3.4.2.2 Neural Network NN-2 for Variance Change Magnitude Classification

NN-2 is a backpropagation neural network that classifies change magnitudes as soon as NN-1 signals an out-of-control situation. This second neural network for classification has a structure of 20 input nodes, 2 hidden layers, and 5 output nodes. Input values are the decimal values equivalent to the percentage values shown in Table 18.2 and are obtained by using the method described in the previous section. The number of training vectors and corresponding target vectors for each variance change category are listed in Tables 18.3 and 18.4. When a large number of training patterns are generated by simulation, we observe that some patterns in adjacent groups are very similar, especially for larger variance changes. This is

TABLE 18.2 Percentage of NN-1 Output that Falls into Specified Output Ranges Based on 10,000 Sets for Each Distribution (Adapted from Chang and Ho, 1999, Table 3, p. 1589.)

Output Range	Input Data ρ = 1 vs. Target = 0.05	Input Data ρ = 2 vs. Target = 0.275	Input Data ρ = 3 vs. Target = 0.50	Input Data ρ = 4 vs. Target = 0.725	Input Data ρ = 5 vs. Target = 0.95
0.00–0.05	17.00%	1.50%	0.40%	0.10%	0.00%
0.05–0.010	47.90%	9.70%	2.30%	0.90%	0.40%
0.10–0.15	24.30%	14.00%	4.10%	1.50%	0.70%
0.15–0.20	8.00%	15.30%	5.80%	2.30%	1.00%
0.20–0.25	2.30%	14.10%	7.30%	3.00%	1.30%
0.25–0.30	0.50%	11.80%	7.30%	3.60%	1.90%
0.30–0.35	0.10%	10.00%	8.00%	4.40%	2.00%
0.35–0.40	0.00%	7.80%	8.40%	4.60%	2.50%
0.40–0.45	0.00%	5.70%	7.70%	4.90%	2.90%
0.45–0.50	0.00%	3.80%	7.40%	5.60%	3.40%
0.50–0.55	0.00%	2.60%	7.30%	6.10%	3.80%
0.55–0.60	0.00%	1.70%	6.80%	6.60%	4.20%
0.60–0.65	0.00%	1.00%	6.50%	6.60%	5.00%
0.65–0.70	0.00%	0.60%	5.80%	7.10%	5.90%
0.70–0.75	0.00%	0.30%	5.00%	7.50%	6.80%
0.75–0.80	0.00%	0.10%	4.30%	8.10%	8.30%
0.80–0.85	0.00%	0.00%	3.20%	8.90%	9.70%
0.85–0.90	0.00%	0.00%	1.60%	9.20%	12.80%
0.90–0.95	0.00%	0.00%	0.70%	7.20%	16.30%
0.95–1.00	0.00%	0.00%	0.10%	1.90%	11.00%

TABLE 18.3 Input Training Patterns for NN-2 (Adapted from Chang and Ho, 1999, Table 5, p. 1592.)

Group	Distribution from Which Observations are Fed into NN-1	No. of Samples for NN-2 (obtained from bootstrap resampling after NN-1 signals out of control)[a]
A	$N(0, 1^2)$; ρ = 1	100
B	$N(0, 2^2)$; ρ = 2	100
C	$N(0, 3^2)$; ρ = 3	100
D	$N(0, 4^2)$; ρ = 4	100
E	$N(0, 5^2)$; ρ = 5	100

[a] Each sample has 20 percentage values.

TABLE 18.4 Target Vectors for Variance Classification NN-2 (Adapted from Chang and Ho, 1999, Table 6, p. 1592.)

	Variance Shifts Category				
	ρ = 1	ρ = 2	ρ = 3	ρ = 4	ρ = 5
Output node 1	1	0	0	0	0
Output node 2	0	1	0	0	0
Output node 3	0	0	1	0	0
Output node 4	0	0	0	1	0
Output node 5	0	0	0	0	1

expected, because the number of sample sizes considered is very limited and actual outputs from NN-1 for different change magnitudes overlap each other. Several patterns are first generated by procedures discussed in the previous section for all normal distributions under consideration. Among those, 100 patterns representing each change magnitude are selected. The selection is made such that patterns in a variance change magnitude are at least slightly different from patterns in other change magnitudes.

Finally, 20 patterns for each change magnitude are mixed in order, according to $\rho = 1, 2, 3, 4, 5$ for all 500 input data. The arranged data set is then fed to NN-2 for training the network.

NN-2 learns according to the same algorithm as does NN-1. Network structure of 20–27–27–5 is trained with two different initial random weight vectors, since different initial weights affect error convergence rates. The learning rate is adaptively changed when the minimum sum of absolute error stops decreasing. The 20–27–27–5 network gives minimum sum 11.7333 of cumulative error vectors for all training patterns, [0.0428, 1.1410, 5.5964, 4.5882, 0.3649]. The element in the error vector describes the sum of absolute differences between actual and target outputs for all training patterns regarding five different process variances. Note that NN-2 has learned quite well for $\rho = 1$ and $\rho = 5$ because their absolute error-sums are 0.0428 and 0.3649, respectively, near 0.0. The learning rate of training NN-2 starts at 0.12 and ends at 0.035.

The proposed neural network classifier NN-2 of V-NN is easy to use. Since there are five NN-2 output nodes corresponding to no shift and two to five times the original process standard deviations, respectively, we simply pick the output node with the largest value. The name of the chosen node provides the information for the shift magnitude.

18.4 Design, Operations, and Guidelines for Using the Proposed Hybrid Neural Fuzzy Control Chart

The proposed hybrid neural fuzzy control chart consists of two major components — M-NN and V-NN which are independently developed but now integrated together. In order to make sure the performance of the joined chart is satisfactory, we can adjust the parameters (C_{m1}, C_{m2}) of M-NN and (C_{v1}, C_{v2}) of V-NN for the two-in-a-row decision rule. The performance criterion used here is the average run length (ARL), defined as the average number of points plotted before a control chart scheme gives an out-of-control signal. For the hybrid neural fuzzy chart or any other joint control charts such as $\overline{X}$ and R charts, an out-of-control situation is either a mean shift, a variance shift, or both.

An in-control ARL value corresponds to a type I error, which means the process is in control but the control charts indicate that the process is out of control. In this case, we would prefer a large in-control ARL value or a small type I error (usually called false alarm). On the other hand, we would prefer a small ARL value or small type II error when a process has shifted. The rationale is that a quick detection of a shifted process is preferred. There are many out-of-control ARL values but only one in-control ARL value. Various shift magnitudes correspond to different out-of-control ARL values.

Designing a hybrid neural fuzzy control chart, we use ARL curve diagrams similar to the operation characteristic (OC) curves commonly used for $\overline{X}$ and R charts. Two ARL curve diagrams — one for mean shifts and the other for variance shifts — are shown in Figures 18.8 and 18.9. Two sets of curves are plotted in each ARL curve diagram. One set is for the first sampling method where an independent subgroup sample of five is considered. The second set is for the moving-window sampling method where each subgroup contains the current observation and four previous observations. Each subgroup is correlated with the others. This is called *individual observations* in Figures 18.8 and 18.9.

We must first decide which input method to use in order to select ARL curves for designing the proposed chart. Then we can decide which type I error (or false alarm) is acceptable. Two choices are available. A large in-control ARL provides the minimum chance of false alarm at the cost of losing sensitivity for catching small process mean or variance shifts. On the contrary, a smaller in-control ARL provides a larger chance of false alarm but is able to catch small process shifts faster. After choosing the in-control ARL value, performance of the hybrid control chart is fully defined. We can then analyze its capability. Examples of this will be given in the following subsections.

18.4.1 Example 1: Design a Hybrid Chart for Small Process Shifts

Suppose that catching small process shifts, either mean or variance, is critical, and frequent production stops are acceptable. For many mission-critical manufacturing parts, such as those for space shuttles, any

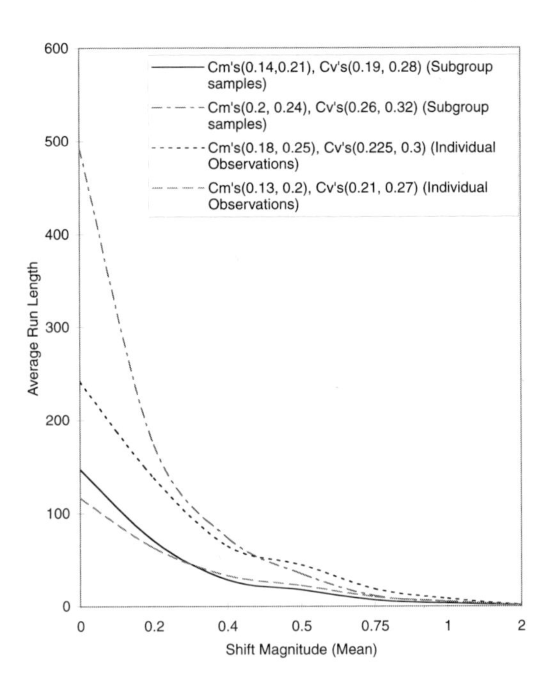

FIGURE 18.8 ARL curves of C-NN for mean shifts. (Adapted from Ho and Chang, 1999, Figure 4, p. 1893.)

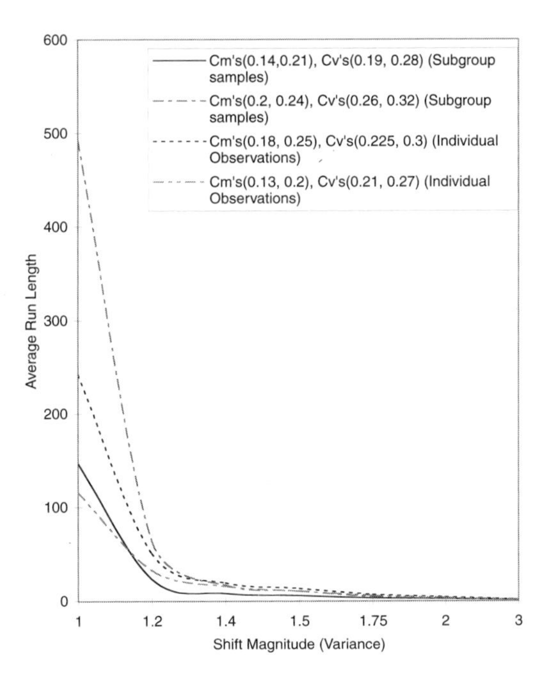

FIGURE 18.9 ARL curves of C-NN for variance shifts. (Adapted from Ho and Chang, 1999, Figure 5, p. 1893.)

slight deviation from target is not tolerable. The parameter sets with smaller in-control ARL are preferred. For the independent sampling method, we choose to use $(C_{m1}, C_{m2}) = (0.14, 0.21)$ and $(C_{v1}, C_{v2}) = (0.19, 0.28)$, which give an in-control ARL 150. This chart design will respond very quickly to any slight process shifts. For example, a 0.5 sigma shift in mean value can be detected within 20 samples on average. The other design for in-control ARL 500 will take twice as many samples to catch the same mean shift. The sensitivity of this design for variance shifts can also be defined as shown in Figure 18.9. For a 1.4 times standard deviation shift, the out-of-control ARL value is within 15 samples for the proposed design.

18.4.2 Example 2: Design a Hybrid Chart for Quality Assurance

Suppose we are interested in implementing a quality assurance plan on a very stable production process. The goal is to make sure that the process is operating smoothly. Fewer false alarms are preferred in this case. Again, for the independent sampling method, we would choose to use $(C_{m1}, C_{m2}) = (0.2, 0.24)$ and $(C_{v1}, C_{v2}) = (0.26, 0.32)$, which give an in-control ARL 500. Note the ARL values also dramatically increase when small shifts occur, e.g., 0.2 sigma mean shift and 1.2 times standard deviation shift.

So far we have only given examples about cases where independent sampling is used. In fact, the design philosophy is the same when the moving-window sample method is chosen, except that ARL values of the designs using the moving-window sample method are in general smaller than those using the independent method, considering the meaning of ARL. Specifically, if we consider the average time to signal ATS (Montgomery, 1996), then ATS = ARL * h, where h is the length of a sample period. It is easy to see that the h value of the subgroup or independent sample is five times as large as that of the individual or moving-window sample.

18.5 Properties of the Proposed Hybrid Neural Fuzzy Control Chart

In this section, we compare the proposed hybrid neural fuzzy control charts to other mean and variance chart combinations in terms of ARL values and correct classification rates. In general, we first fix the in-control ARL values for all methods and then compare out-of-control ARL values, in which case the smaller the better. By correct classification rate, we mean the percentage of correct diagnosis by a combined chart. For example, if the R chart indicates a variance shift but the real process shift is a mean shift, this will result in a misclassification. We will summarize only a few typical cases. For details, please refer to Ho and Chang (1999).

18.5.1 Performance Comparison for Subgroup Samples

Tables 18.5(a) and (b) show the ARL values of the proposed hybrid chart vs. $\overline{X}$ and R charts. In Table 18.5(a), only mean shifts are considered. The proposed C-NN charts consistently outperform $\overline{X}$ and R charts in terms of ARL values and correct classification percentages. On the other hand, similar conclusions can be drawn when only variance shifts are considered, as shown in Table 18.5(b).

18.5.2 Performance Comparison for Moving-Window Samples

For the moving-window sampling method, we compared the proposed charts to Acosta and Pignatiello's (1996) CUSUM scheme and EWMA charts, as shown in Tables 18.6(a) and (b), to Acosta and Pignatiello's (1996) EWMA chart which adopts Roberts' (1959) EWMA for mean and Wortham's and Heinrich's (1972) EWMA for variance.

To obtain a fair comparison, we adjusted cutoff values for C-NN to achieve in-control ARLs comparable to those of CUSUM and EWMA. The resultant values are $C_{m1} = 0.13$, $C_{m2} = 0.20$, $C_{v1} = 0.21$, and $C_{v2} = 0.27$. We ran 10,000 simulation trials for each shift. Since the purpose of CUSUM and EWMA is for small shifts, only these cases are studied. From Table 18.6(a), we observe that C-NN and EWMA are comparable and CUSUM exhibits the worst performance. While C-NN has better classification rates, especially for $\delta = 0.1$, the EWMA has smaller ARL values. From Table 18.6(b), where small variance shifts were studied, C-NN outperforms both EWMA and CUSUM in terms of both ARL and correct classification rate. Therefore, we conclude that the overall performance of the proposed C-NN is the best.

TABLE 18.5 Comparisons between C-NN and $\overline{X}$ and R Charts (Adapted from Ho and Chang, 1999, Table 1, p. 1894.)

(a) Comparisons between C-NN and $\overline{X}$ and R charts for mean shifts

δ[a]	ρ	C-NN (ARL)	$\overline{X}$ & R (ARL)	C-NN (Correct %)	$\overline{X}$ & R (Correct %)
0.0	1.0	147.1	142.0	—	—
0.2	1.0	70.2	100.6	72.00	55.81
0.4	1.0	28.0	45.7	88.29	79.98
0.5	1.0	17.6	29.6	93.40	87.39
0.75	1.0	6.4	10.2	97.87	95.72
1.0	1.0	3.0	4.4	98.90	98.22
2.0	1.0	1.0	1.1	99.59	99.59

(b) Comparisons between C-NN and $\overline{X}$ and R charts for variance shifts

δ	ρ[a]	C-NN (ARL)	$\overline{X}$ & R (ARL)	C-NN (Correct %)	$\overline{X}$ & R (Correct %)
0.0	1.0	147.1	142.0	—	—
0.0	1.2	23.3	24.0	75.61	70.36
0.0	1.4	7.9	8.4	81.61	75.56
0.0	1.5	5.6	5.7	84.03	77.65
0.0	1.75	3.0	3.1	87.00	80.86
0.0	2.0	2.1	2.1	89.71	83.01
0.0	3.0	1.2	1.2	94.74	91.25

[a] δ is the mean shift magnitude in σ as defined in Table 18.1 and ρ is the variance shift magnitude in σ as defined in Table 18.3.

TABLE 18.6 Comparisons between C-NN and Acosta's CUSUM and EWMA

(a) Comparisons between C-NN and Acosta's CUSUM and EWMA for mean shifts

δ[a]	ρ	C-NN (ARL)	CUSUM (ARL)	EWMA (ARL)	NN (Correct %)	CUSUM (Correct %)	EWMA (Correct %)
0.0	1.0	116.1	111.74	102.01	—	—	—
0.1	1.0	87.0	102.61	87.12	58.42	43.94	49.62
0.2	1.0	62.7	78.61	59.91	70.71	63.42	70.05
0.3	1.0	43.7	55.52	39.94	79.67	76.40	81.00

(b) Comparisons between C-NN and Acosta's CUSUM and EWMA for variance shifts

δ	ρ[a]	C-NN (ARL)	CUSUM (ARL)	EWMA (ARL)	NN (Correct %)	CUSUM (Correct %)	EWMA (Correct %)
0.0	1.0	116.1	111.74	102.01	—	—	—
0.0	1.1	57.3	64.33	63.95	63.48	46.79	39.04
0.0	1.2	32.7	37.4	36.91	68.08	59.86	53.32
0.0	1.3	21.3	24.8	24.13	71.65	65.78	59.64

[a] δ is the mean shift magnitude in σ as defined in Table 18.1 and ρ is the variance shift magnitude in σ as defined in Table 18.3.

18.6 Final Remarks

In this chapter, we introduced an alternative statistical process control method for monitoring process mean and variance. A hybrid neural fuzzy system consists of four modules for data input, data processing, decision making, and data summary. Major components of the proposed system are several fuzzy sets and neural networks combined to automatically detect process status without human involvement once the system is set up. Examples were given to demonstrate how to choose system parameters to design the hybrid neural fuzzy control charts. We also compared the proposed hybrid neural fuzzy charts to other combined control chart schemes. Average run length and correct classification rate were used to judge chart performance. In general, the proposed charts outperform other combined SPC charts in the current SPC literature.

Development of the proposed hybrid chart is far from complete. Many situations such as correlated data and short run, and low volume production have not been accounted for by the proposed method. Applying the neural fuzzy method in SPC makes the pursuit of automation in quality engineering a possibility. More research in this direction is needed as manufacturing and service industries move into a new millenium.

References

Acosta, C. A. and Pignatiello, J. J. (1996) Simultaneous monitoring for process location and dispersion without subgrouping, *5th Industrial Engineering Research Conference Proceedings*, 693-698.

Chang, S. I. and Aw, C. (1996) A neural fuzzy control chart for detecting and classifying process mean shifts, *International Journal of Production Research*, 34(8), 2265-2278.

Chang, T. C. and Gan, F. F. (1995) A cumulative sum control chart for monitoring process variance, *Journal of Quality Technology*, 27(2), 109-119.

Chang, S. I. and Ho, E. S. (1999) A neural network approach to variance shifts detection and classification, *International Journal of Production Research*, 37(7), 1581-1599.

Cheng, C. S. (1995) A multilayer neural network model for detecting changes in the process mean, *Computers and Industrial Engineering*, 28(1), 51-61.

Crowder, S. T. and Hamilton, M. D. (1992) An EWMA for monitoring a process standard deviation, *Journal of Quality Technology*, 24(1), 12-21.

Efron, B. and Tibshirani, R. J. (1993) *An Introduction to the Bootstrap*, Chapman & Hall, New York.

Fausett, L. (1994) *Fundamentals of Neural Networks, Architectures, Algorithms, and Applications*, Prentice-Hall, Englewood Cliffs, NJ.

Gan, F. F. (1989) Combined cumulative sum and Shewhart variance charts, *Journal of Statistical Computation and Simulation*, 32, 149-163.

Gan, F. F. (1995) Joint monitoring of process mean and variance using exponentially weighted moving average control charts, *Technometrics*, 37(4), 446-453.

Guo, Y. and Dooley, K. J. (1992) Identification of change structure in statistical process control, *International Journal of Production Research*, 30(7), 1655-1669.

Ho, E. S. and Chang, S. I. (1999) An integrated neural network approach for simultaneous monitoring of process mean and variance shifts — a comparative study, *International Journal of Production Research*, 37(8), 1881-1901.

Johnson, N. L. and Leone, F. C. (1962a) Cumulative sum control charts — Mathematical principles applied to their construction and use, Part I, *Industrial Quality Control*, 18, 15-21.

Johnson, N.L. and Leone, F.C. (1962b) Cumulative Sum Control Charts: Mathematical Principles Applied to Their Construction and Use, Part II, *Industrial Quality Control*, 18, 29-36.

Montgomery, D. C. (1996) *Introduction to Statistical Quality Control*, 2nd ed., John Wiley & Sons, New York.

Page, E. S. (1954) Continuous Inspection Schemes, *Biometrics*, Vol. 41.

Page, E. S. (1963) Controlling the Standard Deviation By Cusums and Warning Lines, *Technometrics*, Vol. 3.

Pugh, G. A. (1989) Synthetic neural networks for process control, *Computers and Industrial Engineering*, 17, 24-26.

Pugh, G. A., (1991) A comparison of neural networks to SPC charts, *Computers and Industrial Engineering*, 21, 253-255.

Roberts, S. W. (1959) Control chart tests based on geometric moving averages, *Technometrics*, 1, 239-250.

Seppala, T., Moskowitz, H., Plante, R. and Tang, J. (1995) Statistical process control via the subgroup bootstrap, *Journal of Quality Technology*, 27(2), 139-153.

Smith, A. E. (1994) X-bar and *R* control chart interpretation using neural computing, *International Journal of Production Research*, 32(2), 309-320.

Wortham, A. W. and Heinrich, G. F. (1972) Control charts using exponential smoothing techniques, *ASQC Technical Conference Transactions*, ASQC, Milwaukee, WI, 451-458.

Yazici, H. and Smith, A. E. (1993) Neural network control charts for location and variance process shifts, *Proceedings of the World Congress on Neural Networks*, 1993, I-265-268.

Yager, R. R. and Filev, D. P. (1994) *Essentials of Fuzzy Modeling and Control*, Wiley, New York.

Zadeh, L. A. (1965) Fuzzy sets, *Information and Control*, 8, 338-353.

19

RClass*: A Prototype Rough-Set and Genetic Algorithms Enhanced Multi-Concept Classification System for Manufacturing Diagnosis

Li-Pheng Khoo
Nanyang Technological University

Lian-Yin Zhai
Nanyang Technological University

19.1 Introduction

Inductive learning or classification of objects from large-scale empirical data sets is an important research area in artificial intelligence (AI). In recent years, many techniques have been developed to perform inductive learning. Among them, the decision tree learning technique is the most popular. Using such a technique, Quinlan [1992] has successfully developed the Inductive Dichotomizer 3 (ID3), and its later versions C4.5 and C5.0 (See 5.0) in 1986, 1992, and 1997, respectively. Essentially, decision support is based on human knowledge about a specific part of a real or abstract world. If the knowledge is gained by experience, decision rules can possibly be induced from the empirical training data obtained.

In reality, due to various reasons, empirical data often has the property of granularity and may be incomplete, imprecise, or even conflicting. For example, in diagnosing a manufacturing system, the opinions of two engineers can be different, or even contradictory. Some earlier inductive learning systems such as the once prevailing decision tree learning system, the ID3, are unable to deal with imprecise and inconsistent information present in empirical training data [Khoo et al., 1999]. Thus, the ability to handle imprecise and inconsistent information has become one of the most important requirements for a classification system.

0-8493-0592-6/01/$0.00+$.50

Many theories, techniques, and algorithms have been developed to deal with the analysis of imprecise or inconsistent data in recent years. The most successful ones are fuzzy set theory and Dempster–Shafer theory of evidence. On the other hand, **rough set** theory, which was introduced by Pawlak [1982] in the early 1980s, is a new mathematical tool that can be employed to handle uncertainty and vagueness. Basically, rough set handles inconsistent information using two approximations, namely the upper and lower approximations. Such a technique is different from fuzzy set theory or Dempster–Shafer theory of evidence. Furthermore, rough set theory focuses on the discovery of patterns in inconsistent data sets obtained from information sources [Slowinski and Stefanowski, 1989; Pawlak, 1996] and can be used as the basis to perform formal reasoning under uncertainty, machine learning, and rule discovery [Ziarko, 1994; Pawlak, 1984; Yao et al., 1997]. Compared to other approaches in handling uncertainty, rough set theory has its unique advantages [Pawlak, 1996, 1997]. It does not require any preliminary or additional information about the empirical training data such as probability distribution in statistics; the basic probability assignment in the Dempster–Shafer theory of evidence; or grades of membership in fuzzy set theory [Pawlak et al., 1995]. Besides, rough set theory is more justified in situations where the set of empirical or experimental data is too small to employ standard statistical method [Pawlak, 1991].

In less than two decades, rough set theory has rapidly established itself in many real-life applications such as medical diagnosis [Slowinski, 1992], control algorithm acquisition and process control [Mrozek, 1992], and structural engineering [Arciszewski and Ziarko, 1990]. However, most literature related to inductive learning or classification using rough set theory is limited to a binary concept, such as *yes* or *no* in decision making or *positive* or *negative* in classification of objects.

Genetic algorithms (GAs) are stochastic and evolutionary search techniques based on the principles of biological evolution, natural selection, and genetic recombination. GAs have received much attention from researchers working on optimization and machine learning [Goldberg, 1989]. Basically, GA-based learning techniques take advantage of the unique search engine of GAs to perform machine learning or to glean probable decision rules from its search space. This chapter describes the work that leads to the development of RClass*, a prototype multi-concept classification system for manufacturing diagnosis. RClass* is based on a hybrid technique that combines the strengths of rough set, genetic algorithms, and Boolean algebra. In the following sections, the basic notions of rough set theory and GAs are presented. Details of RClass*, its validation, and a case study using the prototype system are also described.

19.2 Basic Notions

19.2.1 Rough Set Theory

Large amounts of applications of rough set theory have proven its robustness in dealing with uncertainty and vagueness, and many researchers attempted to combine it with other inductive learning techniques to achieve better results. Yasdi [1995] combined rough set theory with neural network to deal with learning from imprecise training data. Khoo et al. [1999] developed RClass*, a prototype system based on rough sets and a decision-tree learning methodology, and the predecessor of RClass*, for inductive learning under noisy environment.

Approximation space and the lower and upper approximations of a set form two important notions of rough set theory. The *approximation space* of a rough set is the classification of the domain of interest into disjoint categories [Pawlak, 1991]. Such a classification refers to the ability to characterize all the classes in a domain. The upper and lower approximations represent the classes of indiscernible objects that possess sharp descriptions on concepts but with no sharp boundaries. The basic philosophy behind rough set theory is based on equivalence relations or indiscernibility in the classification of objects. Rough set theory employs a so-called **information table** to describe objects. The information about the objects are represented in a structure known as an **information system,** which can be viewed as a table with its rows and columns corresponding to objects and attributes, respectively (Table 19.1). For example, an information system (S) with 4-tuple can be expressed as follows:

$$S = \langle U, Q, V, \rho \rangle$$

TABLE 19.1 A Typical Information System Used by Rough Set Theory

Objects	Attributes		Decisions
U	q_1	q_2	d
x_1	1	0	0
x_2	1	1	1
x_3	1	2	1
x_4	0	0	0
x_5	0	1	0
x_6	0	2	1
x_7	0	1	1
x_8	0	2	0
x_9	1	0	0
x_{10}	0	0	0

where U is the *universe* which contains a finite set of objects,

Q is a finite set of attributes,
$V = \bigcup_{q \in Q} V_q$
V_q is a domain of the attribute q,
$\rho : U \times Q \rightarrow V$ is the information function such that $\rho(x, q) \in$ for every $q \in Q$ and $x \in U$ and $\exists(q, v)$, where $q \in Q$ and $v \in V_q$ is called a *descriptor* in S.

Table 19.1 shows a typical information system used for rough set analysis with x_i s ($i = 1, 2, \ldots 10$) representing objects of the set U to be classified; q_i s ($i = 1, 2$) denoting the *condition attributes*; and d representing the *decision attribute*. As a result, q_i s and d form the set of attributes, Q.

More specifically,

$$U = \{x_1, x_2 \ldots x_{10}\};$$
$$Q = \{q_1, q_2, d\}; \text{ and}$$
$$V = \{V_{q1}, V_{q2}, V_d\} = \{\{0,1\}, \{0,1,2\}, \{0,1\}\}.$$

A typical information function, $\rho(x_1, q_1)$, can be expressed as

$$\rho(x_1, q_1) = \{1\}$$

Any attribute-value pair such as $(q_1, 1)$ is called a descriptor in S.

Indiscernibility is one of the most important concepts in rough set theory. It is caused by imprecise information about the observed objects. The *indiscernibility relation* (R) is an equivalence relation on the set U and can be defined in the following manner:

If $x, y \in U$ and $P \in Q$, then x and y are *indiscernible* by the set of attributes P in S.

Mathematically, it can be expressed as follows

$$x\hat{P}y \quad \text{if } \rho(x, q) = \rho(y, q) \text{ for } \exists q \in P.$$

For example, using the information system given in Table 19.1, objects x_5 and x_7 are indiscernible by the set of attributes $P = \{q_1, q_2\}$. The relation can be expressed as $x_5 \hat{P} x_7$ because the information functions for the two objects are identical and are given by

$$\rho(x_5, q_1, q_2) = \rho(x_7, q_1, q_2) = \{1,0\}.$$

Hence, it is not possible to distinguish one from another using attributes set $\{q_1, q_2\}$.

The equivalence classes of relation, $\hat{P}$, are known as *P-elementary sets* in *S*. Particularly, when $P = Q$, these *Q*-elementary sets are known as the *atoms* in *S*. In an information system, *concepts* can be represented by the *decision*-elementary sets. For example, using the information system depicted in Table 19.1, the $\{q_1\}$*-elementary sets, atoms,* and *concepts* can be expressed as follows:

$\{q_1\}$*-elementary sets*

$E_1 = \{x_1, x_2, x_3, x_9\}$ for $\rho(x, q_1) = \{1\}$

$E_1 = \{x_4, x_5, x_6, x_7, x_8, x_{10}\}$ for $\rho(x, q_1) = \{0\}$

Atoms

$A_1 = \{x_1, x_9\}$ $A_2 = \{x_2\}$ $A_3 = \{x_3\}$ $A_4 = \{x_4, x_{10}\}$

$A_5 = \{x_5\}$ $A_6 = \{x_6\}$ $A_7 = \{x_7\}$ $A_8 = \{x_8\}$

Concepts

$C_1 = \{x_1, x_4, x_5, x_8, x_9, x_{10}\}$ $\Rightarrow$ Class $= 0$ $(d = 0)$

$C_2 = \{x_2, x_3, x_6, x_7\}$ $\Rightarrow$ Class $= 1$ $(d = 1)$

Table 19.1 shows that objects x_5 and x_7 are indiscernible by condition attributes q_1 and q_2. Furthermore, they possess different decision attributes. This implies that there exists a *conflict* (or *inconsistency*) between objects x_5 and x_7. Similarly, another conflict also exists between objects x_6 and x_8.

Rough set theory offers a means to deal with inconsistency in information systems. For a concept (C), the greatest definable set contained in the concept is known as the **lower approximation** of C ($\underline{R}(C)$). It represents the set of objects (Y) on U that can be *certainly* classified as belonging to concept C by the set of attributes, R, such that

$$\underline{R}(C) = \cup\{Y \in U / R : Y \subseteq C\}.$$

where U/R represents the set of all atoms in the approximation space (U, R). On the other hand, the least definable set containing concept C is called the **upper approximation** of C ($\overline{R}(C)$). It represents the set of objects (Y) on U that can be *possibly* classified as belonging to concept C by the set of attributes R such that

$$\overline{R}(C) = \cup\{Y \in U / R : Y \cap C \neq \varnothing\}$$

where U/R represents the set of all atoms in the approximation space (U, R). Elements belonging only to the upper approximation compose the *boundary region* (BN_R) or the *doubtful area.* Mathematically, a boundary region can be expressed as

$$BN_R(C) = \overline{R}(C) - \underline{R}(C).$$

A boundary region contains a set of objects that cannot be certainly classified as belonging to or not belonging to concept C by a set of attributes, R. Such a concept, C, is called a *rough set.* In other words, rough sets are sets having non-empty boundary regions.

Using the information system shown in Table 19.1 again, based on rough set theory, the upper and lower approximations, concepts C_1 for $d = 0$ and C_2 for $d = 1$, can be easily obtained. For example, the lower approximation of concept C_1 ($d = 0$) is given by

$$\underline{R}(C_1) = \{x_1, x_4, x_9, x_{10}\};$$

and its upper approximation is denoted as

$$\overline{R}(C_1) = \{x_1, x_4, x_5, x_6, x_7, x_8, x_9, x_{10}\}.$$

Thus, the boundary region of concept C_1 is given by

$$BN_R(C_1) = \overline{R}(C_1) - \underline{R}(C_1) = \{x_5, x_6, x_7, x_8\}.$$

As for concept C_2 ($d = 1$), the approximations can be similarly obtained as follows.

$$\underline{R}(C_2) = \{x_2, x_3\};$$
$$\overline{R}(C_2) = \{x_2, x_3, x_5, x_6, x_7, x_8\}; \text{ and}$$
$$BN_R(C_2) = \overline{R}(C_2) - \underline{R}(C_2) = \{x_5, x_6, x_7, x_8\}.$$

As already mentioned, rough set theory offers a powerful means to deal with inconsistency in an information system. The upper and lower approximations make it possible to mathematically describe classes of indiscernible objects that possess sharp descriptions on concepts but with no sharp boundaries. For example, universe U (Table 19.1) consists of ten objects and can be described using two concepts, namely "$d = 0$" and "$d = 1$." As already mentioned, two conflicts, namely objects x_5 and x_7, and objects x_6 and x_8, exist in the data set. These conflicts cause the objects to be indiscernible and constitute doubtful areas, which are denoted by $BN_R(0)$ or $BN_R(1)$, respectively (Figure 19.1). The lower approximation of concept "0" is given by object set $\{x_1,x_4,x_9,x_{10}\}$, which forms the **certain training data set** of concept "0." On the other hand, the upper approximation is represented by object set $\{x_1,x_4,x_5,x_6,x_7,x_8,x_9,x_{10}\}$, which contains the **possible training data set** of concept "0." Concept "1" can be similarly interpreted.

19.2.2 Genetic Algorithms

As already mentioned, GAs are stochastic and evolutionary search techniques based on the principles of biological evolution, natural selection, and genetic recombination. They simulate the principle of "survival of the fittest" in a population of potential solutions known as *chromosomes*. Each chromosome represents one possible solution to the problem or a rule in a classification. The population evolves over time through a process of competition whereby the fitness of each chromosome is evaluated using a fitness function. During each generation, a new population of chromosomes is formed in two steps. First, the chromosomes in the current population are selected to reproduce on the basis of their relative fitness. Second, the selected chromosomes are recombined using idealized genetic operators, namely crossover and mutation, to form a new set of chromosomes that are to be evaluated as the new solution of the problem. GAs are conceptually simple but computationally powerful. They are used to solve a wide variety of problems, particularly in the areas of optimization and machine learning [Grefenstette, 1994; Davis, 1991].

Figure 19.2 shows the flow of a typical GA program. It begins with a population of chromosomes either generated randomly or gleaned from some known domain knowledge. Subsequently, it proceeds to evaluate the fitness of all the chromosomes, select good chromosomes for reproduction, and produce

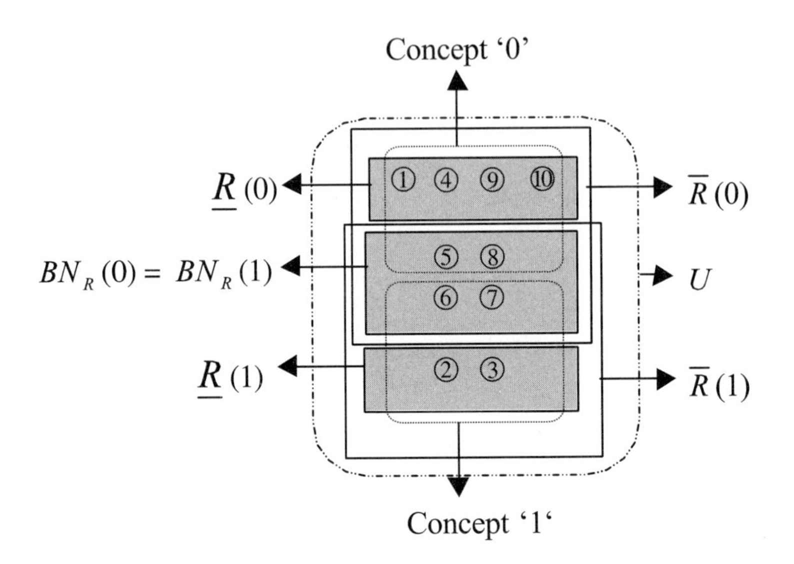

FIGURE 19.1 Basic notions of rough sets.

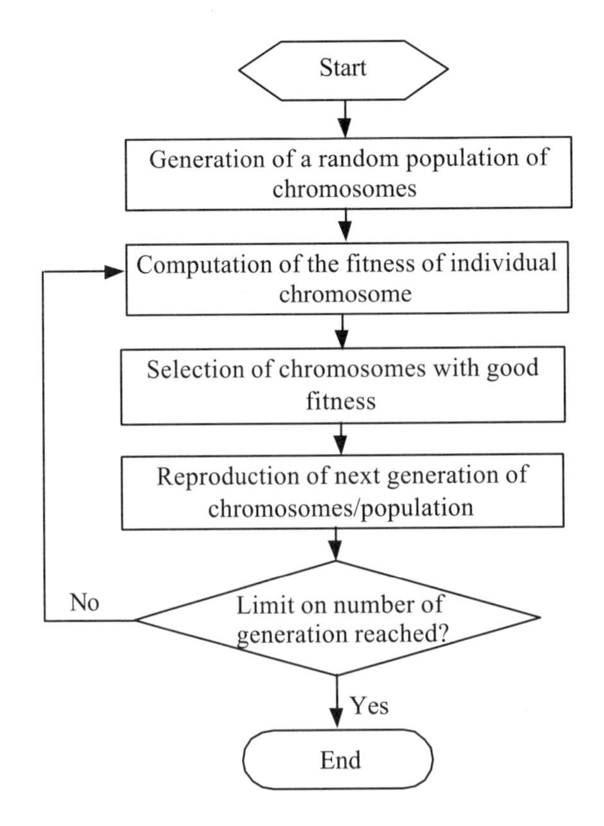

FIGURE 19.2 A typical GA program flow.

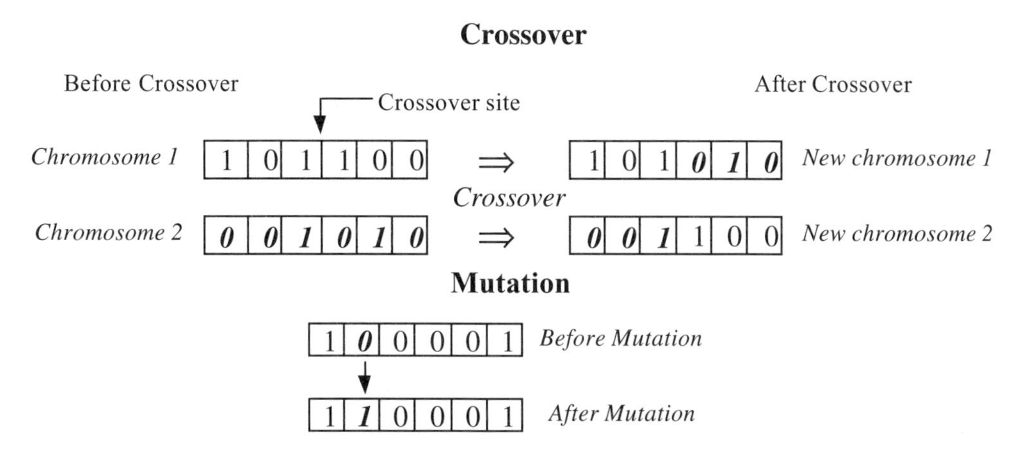

FIGURE 19.3 Genetic operators.

the next generation of chromosomes. More specifically, each chromosome is evaluated according to a given performance criterion or *fitness function*, and assigned a *fitness score*. Using the fitness value attained by each chromosome, good chromosomes are selected to undergo reproduction. Reproduction involves the creation of offspring using two operators namely crossover and mutation (Figure 19.3). By randomly selecting a common crossover site on two parent chromosomes, two new chromosomes are produced. During the process of reproduction, mutation may take place. For example, the binary value of bit 2 in Figure 19.3 has been changed from 0 to 1. The above process of fitness evaluation, chromosome selection, and reproduction of next generation of chromosomes continues for a predetermined number of generations or until an acceptable performance level is reached.

19.3 A Prototype Multi-Concept Classification System

19.3.1 Twin-Concept and Multi-Concept Classification

The basic principle of rough set theory is founded on a twin-concept classification [Pawlak, 1982]. For example, in the information system shown in Table 19.1, an object belongs either to "0" or "1." However, binary-concept classification, in reality, has limited application. This is because in most situations, objects can be classified into more than two classes. For example, in describing the vibration experienced by a rotary machinery such as a turbine in a power plant or a pump in a chemical refinery, it is common to use more than two states such as *normal, slight vibration, mild vibration,* and *abnormal*, rather than just *normal* or *abnormal* to describe the condition. As a result, the twin-concept classification of rough set theory needs to be generalized in order to handle multi-concept problems. Based on rough set theory, Grzymala-Busse [1992] developed an inductive learning system called LERS to deal with inconsistency in training data. Basically, LERS is able to perform multi-concept classification. However, as observed by Grzymala-Busse [1992], LERS becomes impractical when it encounters a large training data set. This can possibly be attributed to the complexity of its computational algorithm. Furthermore, the rules induced by LERS are relatively complex and difficult to interpret.

19.3.2 The Prototype System — RClass*

19.3.2.1 The Approach

RClass* adopts a hybrid approach that combines the basic notions of rough set theory, the unique searching engine of GAs, and Boolean algebraic operations to carry out multi-concept classification. It possesses the ability of

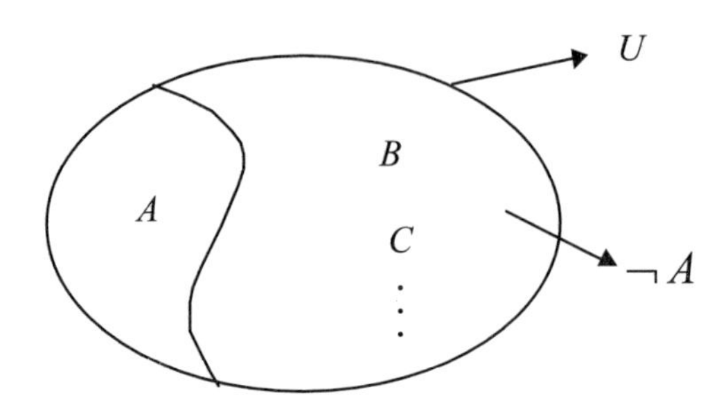

FIGURE 19.4 Partitioning of universe *U*.

1. Handling inconsistent information. This is treated by rough set principles.
2. Inducing probable decision rules for each concept. This is achieved by using a simple but effective GA-based search engine.
3. Simplifying the decision rules discovered by the GA-based search engine. This is realized using the Boolean algebraic operators to simplify the decision rules induced.

Multi-concept classification can be realized using the following procedure.

1. Treat all the concepts (classes) in a training data set as component sets (sets *A*, *B*, *C* . . .) of a universe, *U* (Figure 19.4).
2. Partition the universe, *U*, into two sets using one of the concepts such as *A* and '*not A*' ($\neg A$). This implies that the rough set's twin-concept classification can be used to treat concept *A* and its complement, $\neg A$.
3. Apply the twin-concept classification to determine the upper and lower approximations of concept *A* in accordance to rough set theory.
4. Use Steps 2 and 3 repeatedly to classify other concepts on universe *U*.

19.3.2.2 Framework of RClass*

The framework of RClass* is shown in Figure 19.5. It comprises four main modules, namely a preprocessor, a rough-set analyzer, a GA-based searching engine, and a rule pruner.

The raw knowledge or data gleaned from a process or experts is stored and subsequently forwarded to RClass* for classification and rule induction. The preprocessor module performs the following tasks:

1. Access input data.
2. Identify attributes and their value.
3. Perform redundancy check and reorganize the new data set with no superfluous observations for subsequent use.
4. Initialize all the necessary parameters for the GA-based search engine, such as the length of chromosome, population size, number of generation, and the probabilities of crossover and mutation.

The rough set analyzer carries out three subtasks, namely, consistency check, concept forming, and approximation. It scans the training data set obtained from the preprocessor module and checks its consistency. Once an inconsistency is spotted, it will activate the concept partitioner and the approximation operator to carry out analysis using rough set theory. The concept partitioner performs set operations for each concept (class) according to the approach outlined previously. The approximation operator employs the lower and upper approximators to calculate the lower and upper approximations, during which the training data set is split into *certain training data set* and *possible training data set*. Subsequently, these training sets are forwarded to the GA-based search engine for rule extraction.

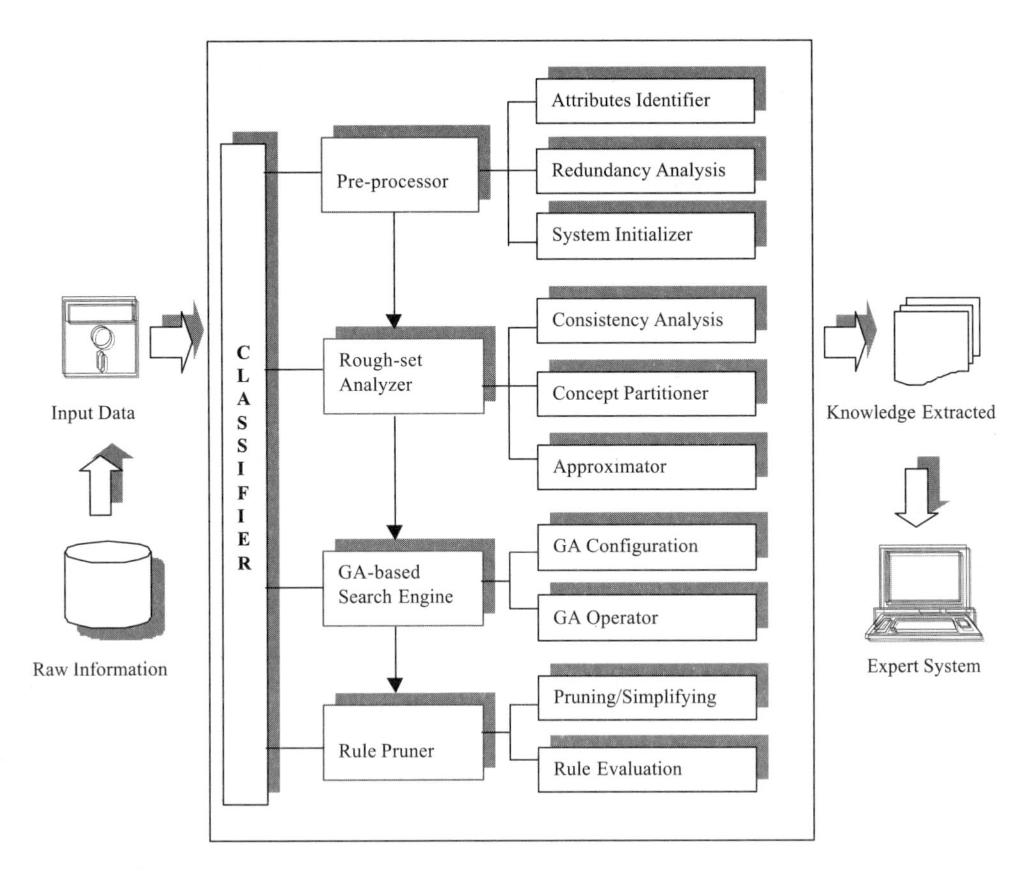

FIGURE 19.5 Framework of RClass*.

The GA-based search engine, once invoked, performs the bespoke genetic operations such as crossover, mutation, and reproduction to gather **certain rules** and **possible rules** from the certain training data set and possible training data set, respectively.

The rule pruner performs two tasks: pruning (or simplifying) and rule evaluation. It examines all the rules, both *certain* and *possible* rules, extracted by the GA-based search engine and employs Boolean algebraic operators such as union and intersection, to prune and simplify the rules. During the pruning operation, redundant rules are removed, whereas related rules are clustered and generalized during simplification. As *possible* rules are not definitely *certain*, the quality and reliability of these *possible* rules must therefore be assessed. For every *possible* rule, RClass* also estimates its reliability using the following index:

$$Reliability\ index = \frac{Observation\_Possible\_Rule}{Observation\_Possible\_Original\_Data}$$

where *Observation_Possible_Rule* is the number of observations that are correctly classified by a possible rule, and *Observation_Possible_Original_Data* is the number of observations with condition attributes covered by the same rule in the original data set.

This index can be viewed as the probability of classifying an inconsistent training data set correctly. For each *certain* rule extracted from the *certain* training data set, RClass* uses a so-called completeness index to indicate the number of observations in the original training data set that are related to the certain rule. Such an index is defined as follows:

$$Completeness\ index = \frac{Observation\_Certain\_Rule}{Observation\_Certain\_Original\_Data}$$

where *Observation_Certain_Rule* is the number of observations that are correctly classified by a certain rule, and *Observation_Certain_Original_Data* is the number of observations with condition attributes covered by the same rule in the original training data.

In other words, the completeness index represents the usefulness or the effectiveness of a *certain* rule. The reliability and completeness indices are included as part of RClass*'s output and are displayed in the parentheses following the rules induced.

19.4 Validation of RClass*

The training example on the classification of hypothermic post-anesthesia patients used by Gryzmala-Busse [1992] for the verification of LERS is adopted here to validate the prototype system RClass*. Briefly, the attributes (symptoms) used to describe the condition of patients are *body temperature, hemoglobin, blood pressure*, and *oxygen saturation*. Attributes *body temperature* and *blood pressure* can be represented by three discrete conditions — namely, low, normal, and high. Attributes *hemoglobin* and *oxygen saturation* can be expressed using linguistic terms such as poor, fair, or good. The level of *comfort* experienced by the patients may be clustered into three different classes or concepts — namely, very low, low, and medium. In this example, nine observations are recorded and summarized in Table 19.2.

The **linguistic description** of the condition of patients (symptoms) and the level of comfort experienced by them (decision) need to be transformed into real numbers. The transformation is achieved by using the following conversion scheme.

For attributes/symptoms

Low/Poor ⇒ 1; Normal/Fair ⇒ 2; High/Good ⇒ 3.

For decision/concept

Very low ⇒ 1; Low ⇒ 2; Medium ⇒ 3.

The results of the conversion are depicted in Table 19.3. It is clear that the comfort levels experienced by patients 3 and 4 contradicts one another.

As an inconsistency is detected in this information system, the rough set analyzer proceeds to perform concept forming and carry out approximation. Three concepts, namely C_1(Comfort = Very low), C_2(Comfort = Low), and C_3(Comfort = Medium) can be formed. The lower and upper approximations of these concepts are then calculated. At the same time, the *certain* and *possible* training data sets are identified. Upon completion, the GA-based search engine is invoked to look for classification rules from the *certain* and *possible* training data sets obtained from the rough set analyzer. It randomly generates 50 chromosomes to form an initial population of possible solutions (chromosomes). These chromosomes are coded using the scheme shown in Table 19.4.

For chromosome representation and genetic operations, RClass* adopts the traditional binary string representation, and its corresponding crossover and mutation operators. Using this scheme, each chromosome is expressed as a binary string comprising "0" and "1" genes. As a result, a classification rule can be represented by an 8-bit chromosome. For instance, the rule "*If (Body Temperature = low) and (Hemoglobin = fair) Then (Comfort = low)*" can be coded as 01100000. Such a representation is rather effective in performing crossover and mutation operations.

Other than choosing a good scheme for chromosome representation, it is important to define a reasonable fitness function that rewards the right kind of chromosomes. The objective of using GAs here is

TABLE 19.2 Training Data Set for the Validation of RClass*

	Attributes/Symptoms				Decision/Concept
Patient	Body Temperature	Hemoglobin	Blood Pressure	Oxygen Saturation	Comfort
1	Low	Fair	Low	Fair	Low
2	Low	Fair	Normal	Poor	Low
3	Normal	Good	Low	Good	Low
4	Normal	Good	Low	Good	Medium
5	Low	Good	Normal	Good	Medium
6	Low	Good	Normal	Fair	Medium
7	Normal	Fair	Normal	Good	Medium
8	Normal	Poor	High	Good	Very low
9	High	Good	High	Fair	Very low

TABLE 19.3 Results of Conversion

	Attributes/Symptoms				Decision/Concept
Patient	Body Temperature	Hemoglobin	Blood Pressure	Oxygen Saturation	Comfort
1	1	2	1	2	2
2	1	2	2	1	2
3	2	3	1	3	2
4	2	3	1	3	3
5	1	3	2	3	3
6	1	3	2	2	3
7	2	2	2	3	3
8	2	1	3	3	1
9	3	3	3	2	1

TABLE 19.4 Chromosome Coding Scheme

Bit Number	Interpretation
Bit 1–2: Attribute value of Body Temperature	00 = Ignore this attribute 01 = Low 10 = Normal 11 = High
Bit 3–4: Attribute value of Hemoglobin	00 = Ignore this attribute 01 = Poor 10 = Fair 11 = Good
Bit 5–6: Attribute value of Blood Pressure	00 = Ignore this attribute 01 = Low 10 = Normal 11 = High
Bit 7–8: Attribute value of Oxygen Saturation	00 = Ignore this attribute 01 = Poor 10 = Fair 11 = Good

to extract rules that can maximize the probability of classifying objects correctly. Thus, the fitness of a chromosome is calculated by testing the rules using existing training data set. Mathematically, it is given by

$$\textit{fitness of chromosome} = \left(\frac{\text{number of examples classified correctly by the rule}}{\text{number of examples related to the rule}} \right)^2 .$$

The above fitness function favors rules that classify examples correctly. It satisfies both completeness and consistency criteria. A rule is said to be consistent if it covers no negative samples and is complete if it covers all the positive samples [De Jong et al., 1993]. Chromosomes with above-average fitness values are selected for reproduction. In this case, the probability of crossover and mutation are fixed at 0.85 and 0.01, respectively.

The rule set induced by the GA-based search engine may contain rules with identical fitness values. Some of these rules can be combined to form a more general or concise rule using Boolean operations. As previously mentioned, the rule pruner is assigned to detect and solve the redundancy problem. For example, two of the rules extracted are found to have the same fitness values.

Rule 1: *If (Temperature = low) and (Hemoglobin = fair) Then (Comfort = low)*
Rule 2: *If (Temperature = low) Then (Comfort = low)*

It is obvious that Rule 1 $\subset$ Rule 2. The rule pruner proceeds to combine the rules and produce the following rule:

$$If\ (Temperature = low)\ Then\ (Comfort = low).$$

The fitness value attained by the rule remains the same.

The induction rules generated using the information system are depicted in Figure 19.6. As already mentioned, two kinds of rules namely certain and possible rules, are available. The value in the parenthesis following each of the rules represents the bespoke completeness or reliability indices. All the indices are represented in fraction form, with the numerator and denominator corresponding to the number of correctly classified observations and the number of observations whose condition attributes are covered by the rule, respectively.

The results show that RClass* is able to support multi-concept classification of objects. It has successfully integrated the basic notions of rough set theory with a GA-based search engine and Boolean algebraic operations to yield a new approach for inductive learning under uncertainty. RClass* has enhanced the performance of its predecessor, RClass [Khoo et al., 1999], and expands rough set's twin-concept to multi-concept classification.

Through this integration, RClass* has combined the strengths of rough set theory and the GA-based search mechanism. With the help of Boolean algebraic operations, the rules produced are simple and concise compared to those derived by LERS. The ability to induce simple and concise rules has the following advantages:

- Easy to understand
- Easy to interpret and analyse
- Easy to validate and cross-check.

19.5 Application of RClass* to Manufacturing Diagnosis

The diagnosis of critical equipment in a manufacturing system is an important issue. Frequently, it is also a difficult task as vast amount of experience or knowledge about the equipment is needed. A computerized system that can assist domain experts in extracting diagnostic knowledge from historical operation records of equipment becomes necessary. Figure 19.7 shows a rotary machinery that comprises a motor and a pump.

Three main types of mechanical faults — namely, *machine unbalance*, *misalignment*, and *mechanical loosening* — are considered here. All these mechanical faults will result in abnormal vibration. Among them, machine unbalance is the most common fault, contributing nearly 30% of abnormal vibration. Misalignment and mechanical loosening are mainly caused by improper installation of the machine. Figure 19.8 shows a typical vibration signature (presented in frequency domain) of the equipment. The frequency of the vibration signature can be broadly divided into five bands based on the number of

```
************************************************************************************
*-----------------------Rough Set - GA Enhanced Rule Induction under Uncertainty---------------------*
************************************************************************************
Last compiled on Dec 11 1998, 15:47:44.
Length of chromosome: 8 (bits)
Decoding sites: 1 3 5 7

Rules extracted from data file <lers.dat>:

Rule                                                                    Confidence level
1. Certain rules for concept 1 :
  1: IF(Blood Pressure=3) THEN Comfort=1                                  (2/2=100%)
  2: IF(Hemoglobin=1) THEN Comfort=1                                      (1/1=100%)
  3: IF(Temperature=3) THEN Comfort=1                                     (1/1=100%)

2. Certain rules for concept 2 :
  1: IF(Temperature=1)&(Hemoglobin=2) THEN Comfort=2                      (2/2=100%)
  2: IF(Blood Pressure=1)&(Oxygen_Saturation=2) THEN Comfort=2            (1/1=100%)
  3: IF(Temperature=1)&(Blood Pressure=1) THEN Comfort=2                  (1/1=100%)
  4: IF(Hemoglobin=2)&(Blood Pressure=1) THEN Comfort=2                   (1/1=100%)
  5: IF(Hemoglobin=2)&(Oxygen_Saturation=2) THEN Comfort=2                (1/1=100%)
  6: IF(Oxygen_Saturation=1) THEN Comfort=2                               (1/1=100%)

 Possible rules for concept 2 :
  7: IF(Blood Pressure=1) THEN Comfort=2                                  (2/3=67%)
  8: IF(Hemoglobin=2) THEN Comfort=2                                      (2/3=67%)

3. Certain rules for concept 3 :
  1: IF(Blood Pressure=2)&(Oxygen_Saturation=3) THEN Comfort=3            (2/2=100%)
  2: IF(Hemoglobin=3)&(Blood Pressure=2) THEN Comfort=3                   (2/2=100%)
  3: IF(Temperature=1)&(Hemoglobin=3) THEN Comfort=3                      (2/2=100%)
  4: IF(Temperature=1)&(Oxygen_Saturation=3) THEN Comfort=3               (1/1=100%)
  5: IF(Blood Pressure=2)&(Oxygen_Saturation=2) THEN Comfort=3            (1/1=100%)
  6: IF(Temperature=2)&(Hemoglobin=2) THEN Comfort=3                      (1/1=100%)
  7: IF(Hemoglobin=2)&(Oxygen_Saturation=3) THEN Comfort=3                (1/1=100%)
  8: IF(Temperature=2)&(Blood Pressure=2) THEN Comfort=3                  (1/1=100%)

 Possible rules for concept 3 :
  9: IF(Blood Pressure=2) THEN Comfort=3                                  (3/4=75%)
  10: IF(Hemoglobin=3)&(Oxygen_Saturation=3) THEN Comfort=3               (2/3=67%)
  11: IF(Temperature=1)&(Blood Pressure=2) THEN Comfort=3                 (2/3=67%)
  12: IF(Oxygen_Saturation=3) THEN Comfort=3                              (3/5=60%)
  13: IF(Hemoglobin=3) THEN Comfort=3                                     (3/5=60%)

                                 --------END---------
```

FIGURE 19.6 Rules extracted by RClass*.

revolution, X, of the equipment, namely, less than 1X (0 ~ 0.9X), about 1X (0.9 ~ 1.1X), 2X (1.9 ~ 2.1X), 3X (2.9 ~ 3.1X), and more than 4X. Within each of the bands, the largest amplitude is indicative of a fault symptom at a particular frequency.

Seven attributes are used to describe the condition of the targeted equipment. These attributes are defined as follows.

A0: the ratio of peak amplitudes in bands less than '1X' and '1X'
A1: the ratio of peak amplitudes in band '1X' and its initial record (in good condition)
A2: the ratio of peak amplitudes in bands '2X' and '1X'
A3: the ratio of peak amplitudes in bands '3X' and '1X'
A4: the ratio of peak amplitudes in bands more than '4X' and '1X'
A5: mode of vibration denoted by horizontal (H), vertical (V), or axial (A)
A6: the overall vibration level.

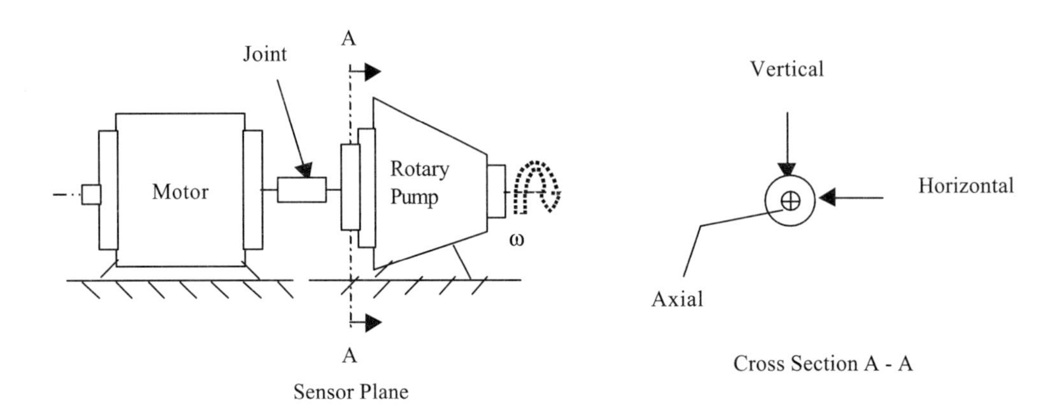

FIGURE 19.7 A motor and pump assembly.

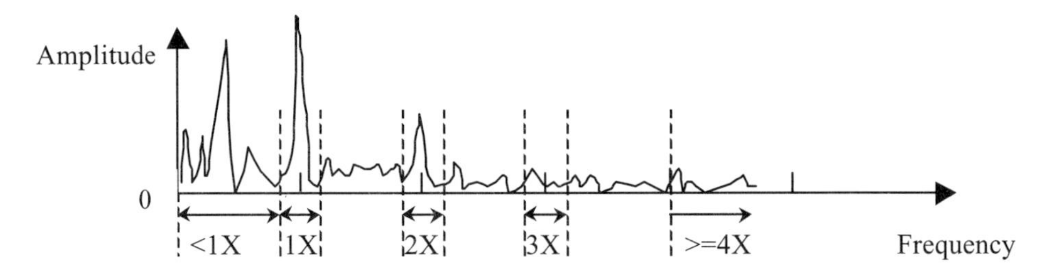

FIGURE 19.8 A typical vibration signature.

Attributes A0 – A4 are continuous variables. On the other hand, attributes A5 and A6 are discrete variables. Attribute A6 registers the overall vibration level using three states, namely normal vibration (N), moderately high vibration (M), and extremely high vibration (E). A sample set of data with 72 observations is depicted in Table 19.5.

The continuous attributes, A0 ~ A4, are first discretized into different intervals (Table 19.6) using classification codes 1, 2, 3, and 4. Details of the discretization will not be discussed here. As for attributes A5 and A6, and the decision attribute, they are transformed into integers using the following conversion scheme.

Discrete Attributes

Attribute A5: A (Axial) ⇒ 1; H (Horizontal) ⇒ 2; V (Vertical) ⇒ 3.

Attribute A6: N (Normal) ⇒ 1; M (Moderate) ⇒ 2; E (Extreme) ⇒ 3.

Decision Attributes

NC (Normal Condition) ⇒ 1; MA (Misalignment) ⇒ 2;

ML (Mechanical Loosening) ⇒ 3; UB (Unbalance) ⇒ 4.

Using the classification codes and conversion scheme given above, the original data set is transformed into an information system with elements denoted by integers which is the format required by RClass*. Fifty-one certain rules and forty-three possible rules are extracted from the sample data set (see Appendix for the rules generated).

TABLE 19.5 Sample Data

Observation	A0	A1	A2	Attributes A3	A4	A5	A6	Decision
1	0.02	1.60	1.45	0.17	0.10	A	E	MA
2	0.01	2.80	0.10	0.06	0.02	V	E	UB
3	0.30	1.90	0.42	0.28	0.15	V	E	ML
4	0.05	1.20	0.10	0.09	0.08	H	N	NC
5	0.03	2.34	0.09	0.07	0.05	A	M	UB
6	0.08	1.32	1.21	0.07	0.08	H	N	MA
7	0.02	2.50	0.15	0.04	0.09	H	E	UB
8	0.06	1.45	0.16	0.03	0.05	V	M	NC
9	0.02	2.75	0.18	0.05	0.07	V	E	UB
10	0.04	1.35	0.20	0.05	0.03	A	N	NC
11	0.04	1.58	1.36	0.08	0.11	V	M	MA
12	0.25	1.85	0.35	0.18	0.16	H	E	ML
⋮								
64	0.28	1.95	0.85	0.22	0.20	V	E	ML
65	0.02	1.55	1.35	0.10	0.11	A	M	MA
66	0.34	1.76	0.52	0.27	0.15	H	E	ML
67	0.03	2.60	0.12	0.06	0.04	H	E	UB
68	0.05	1.22	0.14	0.06	0.02	H	N	NC
69	0.16	2.20	0.34	0.31	0.23	A	M	ML
70	0.03	1.50	1.00	0.16	0.06	H	M	MA
71	0.05	1.30	0.15	0.04	0.04	A	N	NC
72	0.08	1.15	0.12	0.07	0.03	V	N	NC

Notes: MA denotes misalignment; UB stands for machine unbalanced; NC denotes normal condition; ML stands for mechanical loosening.

TABLE 19.6 Discretization of Continuous Attributes

Attribute	Classification Code 1	2	3	4
A0	0<A0<0.13	A0>0.13	/	/
A1	0<A1<1.50	1.50<A1<1.715	1.715<A1<2.26	A1>2.26
A2	0<A2<0.205	0.205<A2<0.97	A2>0.97	/
A3	0<A3<0.075	0.075<A3<0.105	0.105<A3<0.215	A3>0.215
A4	0<A4<0.085	0.085<A4<0.145	A4>0.145	/

From the rules extracted, the characteristics of the equipment can be summarized as follows.

Machine operating under normal condition. The peak amplitudes of all the frequency bands are relatively low and do not deviate much from the initial values. The vibration level remains relatively stable with respect to the rotating speed of the equipment.

Machine with misalignment problem. Large peak amplitude is expected at '2X'. It is normally larger than that in Band '1X'.

Mechanical loosening. The amplitude at higher frequency increases rapidly. The vibration level increases with the rotating speed of the equipment.

Unbalanced machine. The vibration level at the rotating frequency (1X) of the equipment increases significantly compared to historical record under normal condition. Moreover, it becomes more and more violent with the increase in rotating speed.

Some of the rules corresponding to the machine characteristics described above are depicted in Table 19.7.

TABLE 19.7 Sample Rules Corresponding to Machine Characteristics

Working States	Sample Rules Extracted
Machine operating under normal condition	IF(A1 = 1) & (A6 = 1) THEN Machine State = 1
Machine with misalignment problems	IF(A2 = 3) THEN Machine State = 2
Mechanical loosening	IF(A3 = 3) & (A4 = 3) & (A6 = 3) THEN Machine State = 3
Unbalanced machine	IF(A1 = 4) THEN Machine State = 4

Generally, the rules extracted are quite consistent with those experienced by domain experts. They are reasonable and logical. Furthermore, they are concise and easy to understand. With the rules extracted, a knowledge-based system can possibly be developed to assist engineers in diagnosing the equipment.

19.6 Conclusions

The work has successfully shown that the RClass* is able to combine the strengths of rough set theory and GA-based search algorithm to deal with rule induction under uncertainty. RClass* has incorporated a novel approach that extends rough set's twin-concept to perform multi-concept classification. This has made RClass* more practical in dealing with real-life problems compared to its predecessor, RClass. Using RClass*, two kinds of rules, certain rules and possible rules, can be induced from examples. RClass* was validated using an example gleaned from literature. Results show that the rules induced are concise, sensible, and complete. For all the rules extracted, RClass* is also able to provide an estimation of the expected reliability. This would assist users in ascertaining the appropriateness of the rules extracted. A case study was used to illustrate the possibility of using RClass* in performing machine diagnosis in a manufacturing environment. In this case, machine vibration is studied. Results show that the rules extracted are quite consistent with those experienced by domain experts. They are reasonable and logical. Using the rules extracted, it is envisaged that a knowledge-based system can be developed to assist engineers in diagnosing the equipment.

Defining Terms

Certain rules: Rules that can definitely classify some observations into a certain concept.

Certain training data set: The data set that all the observations contained can be definitely classified into a given concept.

Genetic algorithms: Genetic algorithms are a stochastic and evolutionary search technique based on the principles of biological evolution, natural selection, and genetic recombination.

Inductive learning: A procedure that learns general knowledge from a finite set of examples.

Information system: A set of objects whose properties can be described by a number of multi-valued attributes.

Information table: A table that describes a finite number of objects, represented by a structure with its rows and columns corresponding to objects and attributes, respectively.

Linguistic description: Using natural language to qualitatively describe the state(s) of the target observations.

Lower approximation: For a given concept, its lower approximation refers to the set of observations that can all be classified into this concept.

Possible rules: Rules that cannot definitely classify some observations into a certain concept.

Possible training data set: The data set that contains observations that cannot be definitely classified into a given concept.

Rough set: In an information system, a rough set refers to a concept (or class) that contains observations that cannot be definitely classified into this concept (or class).

Upper approximation: For a given concept, its upper approximation refers to the set of observations that can be possibly classified into this concept.

References

Arciszewski, T. and Ziarko, W. 1990. Inductive Learning in Civil Engineering: A Rough Sets Approach, *Microcomputers in Civil Engineering*, 5(1): 19-28.

Davis, L. (Ed.). 1991. *Handbook of Genetic Algorithms*. Van Nostrand Reinhold, New York.

De Jong, K. A., Spears, W. M., and Gordon, D. F. 1993. Using Genetic Algorithms for Concept Learning, *Machine Learning*, 12(13): 161-188.

Goldberg, D. E. 1989. *Genetic Algorithms in Search, Optimization and Machine Learning*. Addison-Wesley, Reading, MA.

Grefenstette, J. J. (Ed.). 1994. *Genetic Algorithms for Machine Learning*. Kluwer Academic Publishers, Dordrecht, The Netherlands.

Grzymala-Busse, J. W. 1992. LERS — A System for Learning from Examples Based on Rough Sets. In *Intelligent Decision Support — Handbook of Applications and Advances of the Rough Sets Theory*, Ed. R. Slowinski, pp. 3-8. Kluwer Academic Publishers, Dordrecht, Netherlands.

Khoo, L. P., Tor, S. B., and Zhai, L. Y. A Rough-Set Based Approach for Classification and Rule Induction, *Int. Journal of Advanced Manufacturing*, in press.

Mrozek, A. 1992. Rough Sets in Computer Implementation of Rule-Based Control of Industrial Process. In *Intelligent Decision Support — Handbook of Applications and Advances of the Rough Sets Theory*, Ed. R. Slowinski, pp. 19-32. Kluwer Academic Publishers, Dordrecht, The Netherlands.

Pawlak, Z. 1982. Rough Sets, *Int. Journal of Computer and Information Sciences*, 11(5): 341-356.

Pawlak, Z. 1984. Rough Classification, *Int. Journal of Man-Machine Studies*, 20(4): 469-483.

Pawlak, Z. 1991. *Rough Sets — Theoretical Aspects of Reasoning about Data*. Kluwer Academic Publishers, Dordrecht, The Netherlands.

Pawlak, Z. 1996. Why Rough Sets. In *1996 IEEE Int. Conference on Fuzzy Systems* (vol. 2), pp. 738-743. IEEE, Piscataway, NJ.

Pawlak, Z., Grzymala-Busse, J., Slowinski, R., and Ziarko, W. 1995. Rough Sets, *Communications of the ACM*, 38(11): 89-95.

Quinlan, J. R. 1992. *C4.5: Programs for Machine Learning*. Morgan Kaufmann, San Mateo, CA.

Slowinski, K. 1992. Rough Classification of HSV Patients. In *Intelligent Decision Support — Handbook of Applications and Advances of the Rough Sets Theory*, Ed. R. Slowinski, pp. 77-94. Kluwer Academic Publishers, Dordrecht, The Netherlands.

Slowinski, R. and Stefanowski, J. 1989. Rough Classification in Incomplete Information Systems, *Mathematical & Computer Modeling*, 12(10/11): 1347-1357.

Yao, Y. Y., Wong, S. K. M., and Lin, T. Y. 1997. A Review of Rough Sets Models. In *Rough Sets and Data Mining — Analysis for Imprecise Data*, Ed. T. Y. Lin and N. Cercone, pp. 47-76. Kluwer Academic Publishers, Boston, MA.

Yasdi, R. 1995. Combining Rough Sets Learning and Neural Learning Method to Deal with Uncertain and Imprecise Information, *Neurocomputing*, 7: 61-84.

Ziarko, W. 1994. Rough Sets and Knowledge Discovery: An Overview. In *Rough Sets, Fuzzy Sets and Knowledge Discovery — Proceedings of the Int. Workshop on Rough Sets and Knowledge Discovery (RSKD '93)*, Ed. W. Ziarko, pp. 11-15. Springer-Verlag, London.

Appendix

```
Diagnostic Knowledge Extracted from the Real Application

****************************************************************************************************************
*----------------------Rough Set - GA Enhanced Rule Induction under Uncertainty-------------------*
****************************************************************************************************************

Last compiled on Jan 21 1999, 11:41:39.
Length of chromosome: 16 (bits)
Decoding sites: 1 4 6 9 11 13 15

Rules extracted from data file <machine.dat>:

Rule                                                                                   Confidence level
1.  Certain rules for concept ë1í:

 1: IF( A2=1)&( A6=1) THEN  Machine State=1                                           (10/10=100%)
 2: IF( A1=1)&( A4=1) THEN  Machine State=1                                           (10/10=100%)
 3: IF( A2=2)&( A6=1) THEN  Machine State=1                                           (8/8=100%)
 4: IF( A1=1)&( A2=2) THEN  Machine State=1                                           (8/8=100%)
 5: IF( A3=2)&( A4=1)&( A6=1) THEN  Machine State=1                                   (7/7=100%)
 6: IF( A2=2)&( A3=2) THEN  Machine State=1                                           (6/6=100%)
 7: IF( A1=1)&( A5=2)&( A6=1) THEN  Machine State=1                                   (6/6=100%)
 8: IF( A5=1)&( A6=1) THEN  Machine State=1                                           (6/6=100%)
 9: IF( A0=2)&( A3=2)&( A6=1) THEN  Machine State=1                                   (4/4=100%)
10: IF( A3=2)&( A5=2)&( A6=1) THEN  Machine State=1                                   (4/4=100%)
11: IF( A0=2)&( A1=1)&( A6=1) THEN  Machine State=1                                   (4/4=100%)
12: IF( A1=1)&( A3=1)&( A6=1) THEN  Machine State=1                                   (4/4=100%)
13: IF( A4=1)&( A5=3)&( A6=1) THEN  Machine State=1                                   (4/4=100%)
14: IF( A0=1)&( A1=1)&( A5=2) THEN  Machine State=1                                   (4/4=100%)
15: IF( A0=2)&( A4=1)&( A6=1) THEN  Machine State=1                                   (3/3=100%)
16: IF( A0=2)&( A3=2)&( A4=1) THEN  Machine State=1                                   (3/3=100%)
17: IF( A1=1)&( A2=1)&( A5=2) THEN  Machine State=1                                   (3/3=100%)
18: IF( A1=1)&( A2=1)&( A3=2)&( A5=1) THEN  Machine State=1                           (3/3=100%)
19: IF( A0=2)&( A5=3)&( A6=1) THEN  Machine State=1                                   (2/2=100%)
20: IF( A0=1)&( A4=2)&( A5=2)&( A6=1) THEN  Machine State=1                           (2/2=100%)
21: IF( A2=2)&( A3=1)&( A4=2) THEN  Machine State=1                                   (1/1=100%)
22: IF( A0=2)&( A4=1)&( A5=2) THEN  Machine State=1                                   (1/1=100%)
23: IF( A3=1)&( A4=2)&( A6=1) THEN  Machine State=1                                   (1/1=100%)
24: IF( A2=1)&( A3=2)&( A5=2) THEN  Machine State=1                                   (1/1=100%)

   Possible rules for concept ë1í:

25: IF( A1=1)&( A6=1) THEN  Machine State=1                                           (18/20=90%)
26: IF( A3=2)&( A6=1) THEN  Machine State=1                                           (14/16=88%)
27: IF( A0=1)&( A1=1)&( A6=1) THEN  Machine State=1                                   (14/16=88%)
28: IF( A1=1)&( A5=2) THEN  Machine State=1                                           (6/7=86%)
29: IF( A4=1)&( A6=1) THEN  Machine State=1                                           (10/12=83%)
30: IF( A0=1)&( A3=2)&( A6=1) THEN  Machine State=1                                   (10/12=83%)
31: IF( A1=1)&( A4=2)&( A6=1) THEN  Machine State=1                                   (8/10=80%)
32: IF( A0=2)&( A6=1) THEN  Machine State=1                                           (4/5=80%)
33: IF( A1=1)&( A3=1) THEN  Machine State=1                                           (4/5=80%)
34: IF( A1=1)&( A3=2)&( A5=1) THEN  Machine State=1                                   (4/5=80%)
```

```
35: IF( A3=1)&( A6=1) THEN  Machine State=1                     (4/5=80%)
36: IF( A6=1) THEN  Machine State=1                             (18/23=78%)
37: IF( A3=2)&( A4=2)&( A6=1) THEN  Machine State=1             (7/9=78%)
38: IF( A0=1)&( A1=1)&( A4=2)&( A6=1) THEN  Machine State=1     (7/9=78%)
39: IF( A0=1)&( A4=1)&( A6=1) THEN  Machine State=1             (7/9=78%)
40: IF( A0=1)&( A1=1)&( A3=2) THEN  Machine State=1             (10/13=77%)
41: IF( A0=1)&( A3=2)&( A4=2)&( A6=1) THEN  Machine State=1     (6/8=75%)
42: IF( A3=2)&( A5=3)&( A6=1) THEN  Machine State=1             (6/8=75%)
43: IF( A3=1)&( A4=1)&( A6=1) THEN  Machine State=1             (3/4=75%)
44: IF( A1=1)&( A4=2)&( A5=2) THEN  Machine State=1             (3/4=75%)
45: IF( A4=2)&( A5=2)&( A6=1) THEN  Machine State=1             (3/4=75%)
46: IF( A4=2)&( A6=1) THEN  Machine State=1                     (8/11=73%)
47: IF( A1=1) THEN  Machine State=1                             (18/25=72%)
48: IF( A1=1)&( A3=2) THEN  Machine State=1                     (14/20=70%)
49: IF( A5=2)&( A6=1) THEN  Machine State=1                     (6/9=67%)
50: IF( A0=2)&( A3=2)&( A5=2) THEN  Machine State=1             (2/3=67%)
51: IF( A1=1)&( A3=2)&( A4=2)&( A5=1) THEN  Machine State=1     (2/3=67%)
52: IF( A0=2)&( A1=1)&( A5=2) THEN  Machine State=1             (2/3=67%)
53: IF( A4=1)&( A5=2)&( A6=1) THEN  Machine State=1             (3/5=60%)
54: IF( A3=2)&( A4=2)&( A5=2) THEN  Machine State=1             (3/5=60%)

2. Certain rules for concept ë2í:

 1: IF( A2=3) THEN  Machine State=2                                      (18/18=100%)
 2: IF( A1=2)&( A3=2)&( A4=2)&( A6=2) THEN  Machine State=2              (5/5=100%)
 3: IF( A1=2)&( A3=3)&( A4=2) THEN  Machine State=2                      (2/2=100%)
 4: IF( A3=3)&( A6=1) THEN  Machine State=2                              (2/2=100%)
 5: IF( A3=3)&( A4=2)&( A6=2) THEN  Machine State=2                      (1/1=100%)
 6: IF( A1=2)&( A3=1)&( A4=2) THEN  Machine State=2                      (1/1=100%)
 7: IF( A0=2)&( A3=3)&( A4=2) THEN  Machine State=2                      (1/1=100%)
 8: IF( A3=3)&( A4=2)&( A5=2) THEN  Machine State=2                      (1/1=100%)
 9: IF( A1=1)&( A3=1)&( A6=2) THEN  Machine State=2                      (1/1=100%)

3. Certain rules for concept ë3í:

 1: IF( A3=3)&( A4=3)&( A6=3) THEN  Machine State=3                      (4/4=100%)
 2: IF( A0=2)&( A2=2)&( A3=3)&( A5=2)&( A6=3) THEN  Machine State=3      (2/2=100%)
 3: IF( A0=2)&( A1=3)&( A2=2)&( A3=3)&( A6=3) THEN  Machine State=3      (2/2=100%)
 4: IF( A1=3)&( A3=3)&( A4=3)&( A5=2) THEN  Machine State=3              (2/2=100%)
 5: IF( A1=3)&( A2=2)&( A3=3)&( A4=3) THEN  Machine State=3              (2/2=100%)
 6: IF( A0=2)&( A1=3)&( A3=3)&( A5=2)&( A6=3) THEN  Machine State=3      (2/2=100%)

4. Certain rules for concept ë4í:

 1: IF( A1=4) THEN  Machine State=4                                      (18/18=100%)
 2: IF( A0=1)&( A2=1)&( A6=3) THEN  Machine State=4                      (10/10=100%)
 3: IF( A0=1)&( A3=1)&( A5=2)&( A6=3) THEN  Machine State=4              (6/6=100%)
 4: IF( A0=1)&( A2=2)&( A4=1)&( A6=3) THEN  Machine State=4              (4/4=100%)
 5: IF( A2=1)&( A3=1)&( A4=1)&( A6=3) THEN  Machine State=4              (4/4=100%)
 6: IF( A2=1)&( A3=2)&( A4=1)&( A6=3) THEN  Machine State=4              (4/4=100%)
 7: IF( A0=1)&( A2=2)&( A4=1)&( A5=2) THEN  Machine State=4              (2/2=100%)
 8: IF( A2=2)&( A3=1)&( A5=2) THEN  Machine State=4                      (2/2=100%)
 9: IF( A2=1)&( A3=1)&( A5=3) THEN  Machine State=4                      (2/2=100%)
10: IF( A2=2)&( A3=1)&( A5=3) THEN  Machine State=4                      (2/2=100%)
11: IF( A2=1)&( A3=1)&( A5=1) THEN  Machine State=4                      (2/2=100%)
```

```
12: IF( A2=1)&( A4=2)&( A5=1)&( A6=3) THEN  Machine State=4          (2/2=100%)

   Possible rules for concept ë4í:

13: IF( A2=2)&( A3=1)&( A4=1) THEN  Machine State=4                  (6/7=86%)
14: IF( A2=1)&( A3=1) THEN  Machine State=4                          (8/10=80%)
15: IF( A2=1)&( A6=3) THEN  Machine State=4                          (12/16=75%)
16: IF( A2=2)&( A3=1) THEN  Machine State=4                          (6/8=75%)
17: IF( A0=1)&( A2=1)&( A3=1) THEN  Machine State=4                  (6/8=75%)
18: IF( A3=1)&( A4=1) THEN  Machine State=4                          (10/14=71%)
19: IF( A3=1) THEN  Machine State=4                                  (14/21=67%)
20: IF( A0=1)&( A2=2)&( A4=1) THEN  Machine State=4                  (4/6=67%)
21: IF( A2=1)&( A5=1)&( A6=3) THEN  Machine State=4                  (4/6=67%)
22: IF( A0=1)&( A2=1)&( A4=1)&( A5=1) THEN  Machine State=4          (2/3=67%)
23: IF( A2=1)&( A4=2)&( A5=2) THEN  Machine State=4                  (2/3=67%)
24: IF( A0=1)&( A3=1) THEN  Machine State=4                          (10/16=63%)
25: IF( A2=2)&( A4=1) THEN  Machine State=4                          (6/10=60%)

                        --------END---------
```

Index

A

B

C

D

E

F

G

H

I

J

K

L

M

N

O

P

Q

R

S

T

V

W

X